Regions

"MIKE RAFFERTY"

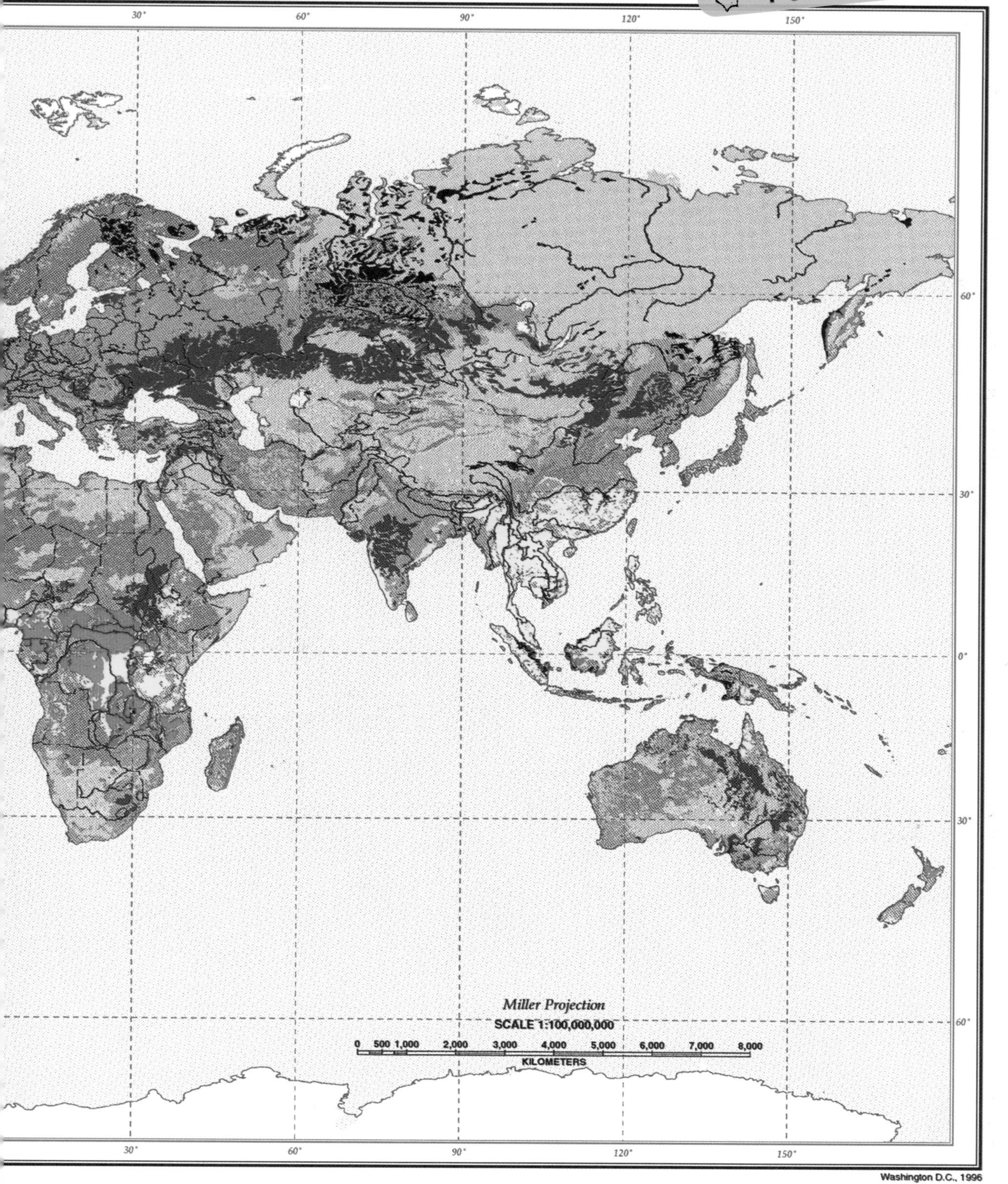

Washington D.C., 1996

In Our Environment

Eighth Edition

Raymond W. Miller

Department of Plants, Soils, and Biometeorology Emeritus
Utah State University

Duane T. Gardiner

Department of Agronomy and Resource Sciences
Texas A&M University-Kingsville

Joyce U. Miller

Editorial Assistant

PRENTICE HALL
UPPER SADDLE RIVER, NEW JERSEY 07458

Library of Congress Cataloging-in-Publication Data

Miller, Raymond W.
Soils in our environment / Raymond W. Miller, Duane T. Gardiner;
Joyce U. Miller, editorial assistant. — 8th ed.
p. cm.
Includes bibliographical references and index.
ISBN 0-13-610882-2
1. Soil science. 2. Crops and soils. 3. Soil management.
I. Gardiner, Duane T. II. Title.
S591.M733 1998
631.4—dc21

96-53481
CIP

Acquisitions Editor: **Charles Stewert**
Production Editor: **Kelli Rahlf/Carlisle Publishers Services**
Production Liaison: **Barbara Marttine Cappuccio**
Director of Manufacturing & Production: **Bruce Johnson**
Managing Editor: **Mary Carnis**
Manufacturing Buyer: **Marc Bove**
Marketing Manager: **Debbie Yarnell**
Assistant Editor: **Kate Linsner**
Art Director: **Marianne Frasco**
Cover Designer: **Paul Gourhan**
Cover photograph: **Uniphoto Picture Agency**
Formatting/page make-up: **Carlisle Communications, Ltd.**

Simon & Schuster/A Viacom Company
Upper Saddle River, New Jersey 07458

Printed in the United States of America

10 9 8 7 6 5 4 3 2 1

ISBN 0-13-610882-2

Prentice-Hall International (UK) Limited, *London*
Prentice-Hall of Australia Pty. Limited, *Sydney*
Prentice-Hall Canada Inc., *Toronto*
Prentice-Hall Hispanoamericana, S.A., *Mexico*
Prentice-Hall of India Private Limited, *New Delhi*
Prentice-Hall of Japan, Inc., *Tokyo*
Simon & Schuster Asia Pte. Ltd., *Singapore*
Editora Prentice-Hall do Brasil, Ltda., *Rio de Janeiro*

This book is a superficial composite of the research and field experiences of hundreds of researchers and producers. Their work was partly possible because of the foresight of men and women who promoted and funded agricultural research through the Hatch Act (Land Grant Institutions) and the United States Department of Agriculture. To all of those facilitators and collectors of knowledge about the soils of the World, we dedicate this book. We hope our generation leaves the soil in a better condition than when we began to use it.

Raymond W. Miller
Duane T. Gardiner

Brief Table of Contents

Contents

Preface

Soils do not usually change rapidly, but the knowledge we accumulate and the methods we use do change, often quickly. It is some of these new methods and data that characterize many of the changes in this new edition. The use of Geophysical Information Systems (GISs) and the Global Positioning System (GPS) has changed our accuracy and ease in locating ourselves on the land or the sea and has multiplied the information we can store, evaluate and use. Many disciplines may now interact in the activities of day-to-day agriculture and land use. It is important to be somewhat familiar with their systems. An overview of engineering soil mechanics is mentioned, particularly the involvement of particle-size soil classifications.

New methods are being refined to aid in the measurement of water, particularly to reduce the need to use the radioactive neutron probe, and its inconvenient monitoring and storage requirements. Methods to increase water efficiencies in irrigation are crucial as water supplies become more and more in demand for non-agricultural uses. Some of these, and variable-rate irrigation, are described. A short addition about wetlands has been added as an introduction to drainage.

The United States Soil Classification System is changed in some of its suborders, particularly in adding several new suborders to the order Aridisols. Many classifications or use systems exist in the World. Two of these, the World system of FAO (more a map legend than a classification system) and the limited system of Canada, are related to the U.S. system.

Precision agriculture, introduced to agriculture in the 1990s, involves spatial variability and eases into variable-rate technology. Remote sensing, plant health, on-the-go variable-rate planting, fertilizing, pesticide application, and yield monitoring is part of the farming *by the foot, not the field* technology.

The chapter on environmental pollution related to agriculture has been increased to two chapters. The first chapter illustrates the nature of the problem contaminants. The second chapter discusses some solutions for those contaminations. The solutions involve, among others, bioremediation, phytoremediation and Best Management Practices (BMPs). The road to a cleaner World is difficult and will be expensive.

Finally, the use of non-soil growth media and hydroponics is now common in producing plants. The relation of these techniques to fertilizers, lime, salt hazards, water control and general plant growth, even in space, is reviewed in a brief final chapter.

Two additional inclusions are new. The first is an extended listing of models used by researchers for depicting many systems, including erosion, hydrology, solute movement, weed control, and plant growth. The second addition is a limited but referenced list of Decision Cases, real examples of situations with their resolutions, which can be used as class discussion problems for students to use their reasoning, to debate, to evaluate data and to wrestle with regulatory mandates in the study of actual situations.

Raymond W. Miller
Duane T. Gardiner

Soil Composition and Importance

Our entire society rests upon—and is dependent upon—our water, our land, our forests, and our minerals. How we use these resources influences our health, security, economy, and well-being.

—John F. Kennedy

1:1 Preview and Important Facts

PREVIEW

Water, air, sunlight, and seemingly lifeless soil combine to produce life on earth. Soil, the outermost layer of the earth, is a product of geologic processes and human intervention. It is composed of minerals physically and chemically altered from original bedrock, organic chemicals and biomass, and pore spaces filled with air, water, and dissolved material.

We rely on soil for many purposes, especially for the production of food and fiber. The quality and scarcity of soil has always affected human civilization. With a burgeoning human population and dwindling resources for agricultural input, the wise use of soil will be ever more critical in the years to come.

IMPORTANT FACTS TO KNOW

1. The definition of soil
2. Materials from which soils form
3. Important early studies and observations contributing to modern soil science
4. Reasons soil is important
5. How soil facilitates plant life

1:2 What Is Soil?

We have a working knowledge of what is and what is not soil. We know it as the earthy material that we learned to call *dirt* when we played in it as children. We marvel at this seemingly lifeless material that gives life to plants. We understand that it is the most basic of building materials and the foundation on which we build structures. Yet, whatever this stuff is, it nearly defies formal definition.

According to the U.S. Department of Agriculture, **soil** is the collective term for "natural bodies, made up of mineral and organic materials, that cover much of the Earth's surface, contain living matter, and can support vegetation out of doors. Soils have in places been changed by human activity. The upper limit of soil is air or shallow water."[1] The lower limit of soil is more difficult to define, but it generally coincides with the common rooting depth of native perennial plants.

Soils do not cover all of the earth's land. **Nonsoil** land surfaces, which will not grow plants, include the *ice lands* of polar and high-elevation regions, recent *hard lava* flows, *salt flats, bare rock* mountain slopes and ridges, and areas of *moving dunes.* Engineers generally ignore the biological component of soil and consider soil to be material that can be excavated with a shovel or compacted into roadbeds or other support base. More formally, engineers may consider soil as "rock particles and minerals derived from preexisting rocks."[2]

Earth's Crust

Prevailing geologic theory states that the Earth's mass condensed from material in space about 4.7 billion years ago. Subsequent heating caused the most dense materials, especially iron, to migrate toward the Earth's center. The Earth differentiated into layers referred to as the **inner core, outer core, mantle,** and **crust.** Only the outermost 5 km to 30 km (3 mi to 19 mi) of the Earth's 6391 km (3971 mi) radius is crust. Nevertheless, the crust is of particular interest:

- It is the only part of the Earth most of us ever experience.
- It contains the Earth's least dense materials.
- It is highly dynamic, following a cycle of deposition of sediments, formation of sedimentary rock, conversion to metamorphic and igneous rock, and weathering and erosion to form sediments again.

Commonly, the unconsolidated sediments undergoing geologic cycling are **sands, silts,** and **clays** (i.e., materials that one would normally call *soil*). Miners excavating many meters into the Earth, and oil rigs drilling many kilometers into the Earth, encounter sedimentary materials that had been surface accumulations of soil in an earlier era (Figure 1-1).

Mineral Composition of Soils

A volume of soil typically contains about 50 percent solids and 50 percent pore space. The solids are mineral and organic particles; mineral particles usually make up 40 percent to 49 percent of the soil volume. The pores are nonsolid spaces filled with varying amounts of water, solutes (dissolved substances), and air.

The rocks and minerals of the Earth's crust provide the mineral constituents of soils. **Minerals** are inorganic (nonliving) substances that are homogeneous, have a definite composition, and have characteristic physical properties such as shape, color, melting temperature, and hardness. Minerals may be either **primary** (formed by the cooling of molten rock) or **secondary** (precipitated or recrystallized from solutions that contained elements from the dissolution of other minerals).

---

[1]Soil Survey Staff, *Keys to Soil Taxonomy. Sixth Ed.,* USDA—Soil Conservation Service, Washington, DC, 1994, 1.
[2]A. E. Kehew, *Geology for Engineers and Environmental Scientists,* Prentice-Hall, Englewood Cliffs, NJ, 1995, 137.

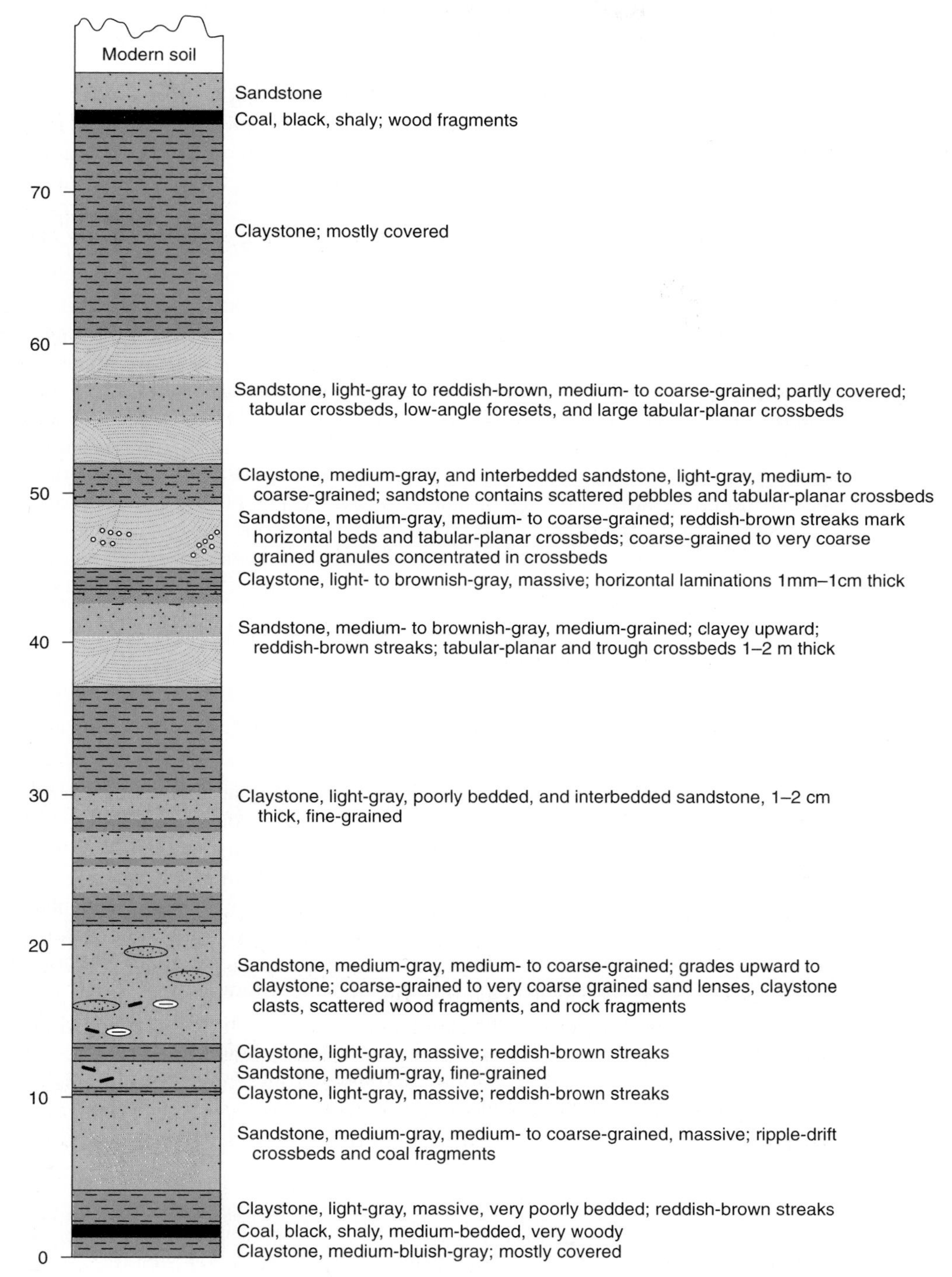

FIGURE 1-1 What is below the surface of the soil? It varies among locations. This stratigraphy (sequence of rock and sediment layers) of the upper 80 m of the Earth in South-Central Alaska has predominantly sandstone rocks, some thought to be more than 3 million years old. Vertical scale is in meters. (*Source:* Modified from U.S. Geological Survey.)

Rocks are mixtures of minerals. They are classified into three major divisions:

- **Igneous:** cooled molten rock (cooled magma)
- **Sedimentary:** sediments deposited in water and consolidated (hardened into a mass of rock)
- **Metamorphic:** igneous or sedimentary rocks changed by heat, pressure (hardened or changed mineral orientations), or chemical solution

When molten magma from under the Earth's crust is exposed on the surface or at different depths in the Earth, it forms **igneous rocks** as it cools. These are **volcanic** or **extrusive** igneous rocks. The expelled igneous rocks (from volcanic eruptions) cool rapidly; thus, they are glassy and amorphous (no crystals). Those minerals in material less rapidly cooled form small crystals in the rock mass. Magma near the surface, but not expelled into exposure to air, cools more slowly and forms **plutonic** or **intrusive** igneous rocks composed of large crystals. *Changes* in the cooling rate during solidification and crystal growth result in the formation of **porphyrys,** igneous rocks having both some large and some small crystals of their minerals. Figure 1-2 is a simplified classification of some common igneous rocks by crystal size and mineral composition. The relative vertical widths of the mineral portions in the headings represent the range of composition (e.g., granites have few dark minerals and vary in composition from small amounts of quartz to nearly equal portions of feldspars and quartz).

Sedimentary rocks are formed from what were at one time rock, mineral, and soil particles or soluble substances that became consolidated or cemented into hard masses. The cementing material is identified as part of the name for some sedimentary rocks, as follows:

- **Calcareous:** for carbonates (lime), as in *calcareous sandstone*
- **Ferruginous:** for iron oxides, the red and yellow coloring of many sandstones
- **Siliceous:** for silica (SiO_2)

Sedimentary **conglomerates** and **breccias** are made up of various-sized fragments of rocks cemented together. **Sandstones** are cemented sands (particles 2 mm to 0.05 mm in diameter, mostly quartz) and lesser quantities of particles smaller than 0.05 mm. **Shales** are consolidated clays and silts (particles 0.05 mm to 0.002 mm) with varying amounts of cementation. **Limestones** are clays, silts, and sands cemented in calcium carbonate or mixtures of calcium and magnesium carbonates; limestones have more than 50 percent of their mass as carbonates. **Dolomites (dolostones)** are similar to limestones but have more magnesium carbonate as part of the carbonate mixture. Sedimentary **quartzites** are silica-cemented sands in which the cement is as hard as the sands.

Metamorphic rocks may be as hard as, or harder than, the igneous, sedimentary, or other metamorphic rock from which they formed, but they weather to produce similar soils. Common metamorphic rocks include the following:

- **Gneiss** (pronounced *nice*): lighter-colored igneous rocks; minerals are segregated and oriented in light and dark bands (formed mostly from granites, rhyolites, andesites, and other similar minerals)
- **Schist:** fissile or foliated (appears flaky or layered); comprised of many rocks or minerals, especially micas (the latter are called *mica schists*)
- **Slate:** hardened shale or siltstone; very hard; used for early chalkboards and pool tables

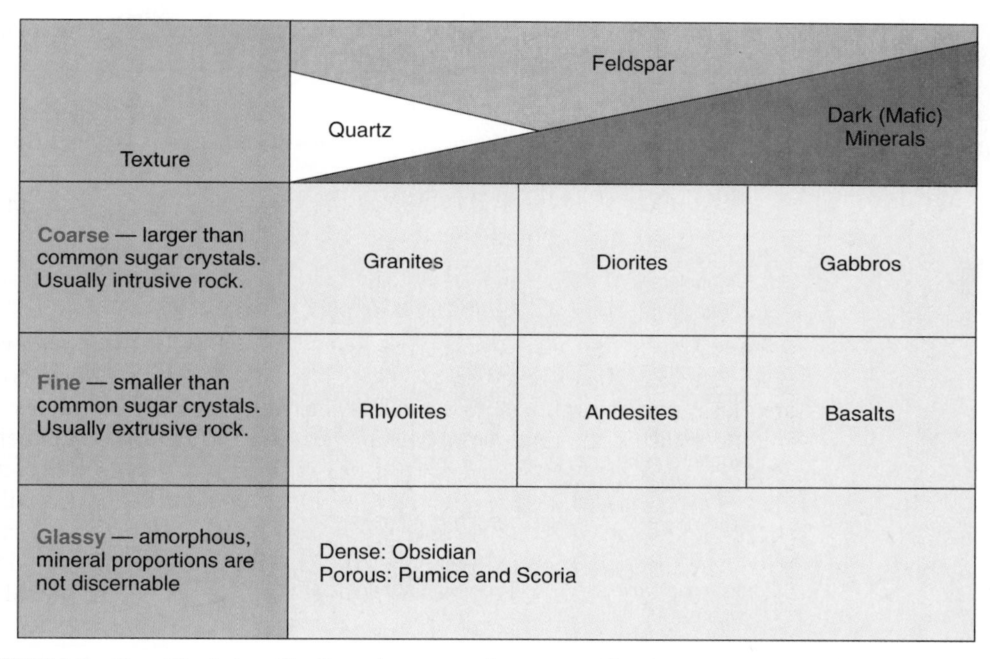

Feldspar
Quartz
Dark (Mafic) Minerals
Texture
Coarse — larger than common sugar crystals. Usually intrusive rock.
Granites
Diorites
Gabbros
Fine — smaller than common sugar crystals. Usually extrusive rock.
Rhyolites
Andesites
Basalts
Glassy — amorphous, mineral proportions are not discernable
Dense: Obsidian
Porous: Pumice and Scoria

FIGURE 1-2 Simplified classification of common igneous rocks. A complete classification would include many other named rocks. The range in proportions of the minerals in each rock group is indicated by the vertical widths of the mineral sections above that column (e.g., granites have few dark minerals and have high but varying amounts of quartz). The dark, mafic minerals have relatively high contents of base cations such as iron and magnesium, and relatively low silica and alumina contents.

- **Quartzite:** recrystallized quartzitic sandstone; formed by heat and pressure; very slow to weather; produces sandy, shallow soils
- **Marble:** limestone or dolomite hardened enough to polish (easily decomposed; acid-forming gases in the air dissolve in rainfall and slowly destroy marble statues by dissolving them)

The substances from which mineral soils develop are mixtures of literally hundreds of different minerals. Some of the more common of these minerals are shown in Table 1-1. Minerals in soils are mostly quartz, feldspars, and dark (mafic) minerals (biotite, hornblende, augite, epidote, and others) with various quantities of clay minerals and precipitated salts such as lime and gypsum. **Silica** (SiO_2) and **alumina** (Al_2O_3) groups predominate soil minerals, and there are lesser amounts of the common cations iron, calcium, sodium, potassium, and magnesium. The summation of these eight elements (Si, Al, Fe, Ca, Na, K, Mg, O) equals more than 90 percent of the mineral volume. Much of the mineral mass in soils is silicon and oxygen (See Detail 1-1).

Organic Materials in Soil

Most soils are **mineral soils;** only about 0.5 percent of U.S. soils and 0.9 percent of world soils are organic soils. **Organic soils** are formed from plant and animal residues in ponded or cold, wet areas where organic-matter decomposition is slow; they are referred to as **peat** or **muck** soils. Many of these soils have been drained in the past century.

Table 1-1 Some Important Minerals in Soils

Name	Formula	Comments
Primary Minerals		
Quartz	SiO_2	Hard; weathers slowly; major material of most sands (see Secondary Minerals, below)
Feldspars		Hard; weather slowly or moderately, but provide important nutrients and clay in the weathered products
Orthoclase	$(K,Na)AlSi_3O_8$[a]	
Plagioclase	$(Ca,Na)Al(Al,Si)Si_2O_8$	
Micas		"Glitter" in rocks or wet sands; important source of potassium and clay
Muscovite	$KAl_3Si_3O_{10}(OH)_2$	
Biotite	$KAl(Mg,Fe)Si_3O_{10}(OH)_2$[a]	
Dark minerals (augite,hornblende, biotite mica, others)	$Ca_2(Al,Fe)_4(Mg,Fe)_4Si_6O_{24}$[a]	Include several minerals that weather moderately fast; good clay formers
Apatite	$3Ca_3(PO_4)_2 \cdot CaF_2$[b]	Most common mineral supplying phosphorus
Calcite, dolomite, gypsum	(See Secondary Minerals, below)	Can be either primary or secondary
Secondary Minerals		
Calcite	$CaCO_3$	Slightly soluble materials in limestone or dolomite rock common in arid-region soils; calcium or magnesium source
Dolomite	$(Ca,Mg)(CO_3)_2$[a]	
Gypsum	$CaSO_4 \cdot 2H_2O$[b]	A soft, moderately soluble mineral found in arid-region soils
Iron oxides	$Fe_2O_3 \cdot x\,H_2O$[b,c]	A group of minerals with different amounts of water giving soils their yellow-to-red colors; iron source
Quartz	SiO_2	Reprecipitated forms such as opal, agate, and petrified wood (see Primary Minerals, above)
Clays Kaolinite, montmorillonite Vermiculite, illite	(Complex)	(See Chapter 5)

[a]Two elements in the same set of parentheses means that either or both may be part of the mineral in the crystal site.
[b]The dot within a formula indicates that the compounds on either side of the dot both exist as part of the formula.
[c]The *x* indicates a variable amount of absorbed water.

To be considered an **organic soil** by the U.S. Department of Agriculture, a soil must meet the following criteria:

1. Soil that is never saturated with water for more than a few days must contain more than 20 percent organic carbon.
2. If a soil is saturated for periods longer than a few days, it is organic if it contains the following:

Detail 1-1 The Importance of Silicon and Oxygen

The Earth's crust is almost three-fourths silicon (Si) and oxygen (O). The Earth's crust is composed of the following:

Oxygen (O)	46.6%	Calcium (Ca)	3.6%
Silicon (Si)	27.7%	Sodium (Na)	2.8%
Aluminum (Al)	8.1%	Potassium (K)	2.6%
Iron (Fe)	5.0%	Magnesium (Mg)	2.1%

All other elements make up the remainder, only about 1.5 percent.

Quartz (SiO_2) is the most common form of silicon and oxygen and is an important constituent of soils. Most sands and many silts (smaller than sands) consist of quartz particles. Quartz exists as **coarse crystalline, microcrystalline** (cryptocrystalline), and **amorphous** varieties. Some of these materials are described below.

Coarse crystalline (glassy, greasy, or vitreous luster; hardness 7)

Rock crystal: Colorless, hexagonal crystals

Amethyst: Hexagonal crystals; violet colors from ferric iron

Rose quartz: Without crystals; smoky from free silicon atoms from exposure of the rock to radioactive materials

Milky quartz: Without crystals; milky from minute *fluid* inclusions

Granular varieties: Without crystals; fibrous inclusions in quartz

Microcrystalline

Fibrous varieties: **Chalcedony, agate** (alternating layers of different colored chalcedony), **petrified wood,** and **onyx** (similar to agate but with parallel layers)

Granular varieties: **Flint, chert** (much alike; flint has dark-colored inclusions), and **jasper** (usually colored red by hematite)

Amorphous

Opal of many colors and quality, comprises some parts of petrified wood.

- at least 12 percent organic carbon if the soil has no clay,
- at least 18 percent organic carbon if the soil has 60 percent or more clay, or
- if it contains a proportional amount of organic carbon for intermediate amounts of clay.

The usage of the term *organic soil* in this text is not to be confused with the commercial term *organic,* which is used to imply the absence of pesticides and artificial fertilizers in the production of food materials, such as *organically grown carrots.*

Organic soils are derived from plants growing in environments that decompose dead residues slowly. Usually, stagnant waters (swamps, marshes, moors, bogs) in temperate climates allow good growth of adapted plants, which die or shed leaves into the water. The decomposition of plant material in stagnant water is mostly anaerobic (lacking O_2) and is very slow. Over centuries such organic accumulations may reach depths of several meters (Figure 1-3). Examples of organic-accumulating areas are California's Stockton Delta, Florida's Everglades, and the many ponded waters in depressions in northern Michigan, Minnesota, and Canada. Common plants from which organic soils develop include many mosses (such as sphagnum), pondweed, cattails, sedges, reeds, grasses, and various water-loving deciduous and coniferous shrubs and trees.

Although most soils are classified as mineral soils rather than organic soils, mineral soils nearly always contain some organic matter. This organic matter is very important for several soil properties, especially to supply plants with nitrogen as the organic matter is decomposed. The soil organic matter (called **humus**) consists primarily of plant residues in various stages of decomposition and the **biomass** (living cells) of soil microorganisms.

FIGURE 1-3 Fibric organic materials (peat) interbedded with silt in the Yukon region of Alaska, where preservation is helped by the cold climate. (*Source:* U.S. Geological Survey.)

1:3 Historical Perspectives

History of Soil Use

Historians place the beginnings of agriculture in Mesopotamia about 7000 years ago. In Mesopotamia the Sumarian irrigation infrastructure dates back to 3500 B.C. About 1000 B.C. Phoenicians were cutting the famous Lebanon cedar forests to send to King Solomon, leaving a heritage of denuded and eroded rocky barrens. Hammurabi of Babylon, in the eighteenth century B.C., left descriptions of brick-lined canals having asphalt mortar and of an irrigation system encompassing 25,900 km^2 (10,000 mi^2—the size of Vermont), which supported about 15 to 20 million people.[3] The same area, now known as Iraq, supports only one-fourth as many people as in the days of its Hanging Gardens. Silts plagued and plugged the canals that were eventually abandoned.

In ancient Egypt, civilization flourished along the Nile. Here silts were a blessing as the river replenished the land with a new addition of soil with each flood. Farming the river delta was a sustainable practice until the construction of dams on the river. W. C. Loudermilk, a scientist, philosopher, and historian working for the U.S. Department of Agriculture, noted that it was perhaps 6000 years ago along the Nile that a farmer "hitched an ox to a hoe and invented the plow, thus originating power-farming to disturb the social structure of those times much as the tractor disturbed the social structure of our country in recent years. By this means farmers became more efficient in growing food; a single farmer released several of his fellows from the vital task of growing food."[4] He speculates that the pyramids arose out of a need to keep a surplus work force occupied.

In China the once deep, fertile soils of the north continue to erode so severely that the Yellow River (known also as China's Sorrow) requires watchful vigil each flood season. Deforestation and the planting of annual crops has resulted in enough silt being lost from the land and deposited in the river to change the course of the river many times. The Yellow

---

[3] Vernon Gill Carter and Tom Dale, *Topsoil and Civilization,* University of Oklahoma Press, Norman, OK, 1974.
[4] W. C. Loudermilk, "Conquest of the Land Through Seven Thousand Years," *Agriculture Information Bulletin* 99, USDA—Soil Conservation Service, Washington, DC, 1975.

River, like many old rivers in populated areas, follows a bed elevated dangerously above the floodplain by construction of artificial levees and dikes. The present-day threat of flooding is but a continuation of China's age-old eroding–silting–flooding sequence documented by construction of the first dikes 4000 years ago.

The Western Hemisphere bears evidence of historical water control and soil conservation. The Incas of Peru left a few remnants of detailed water and soil distribution. Machu Picchu (believed to be the final retreat of the Incan leaders during the Spanish conquest of South America) was extensively terraced and contained waterways cut into rock. The mountaintop location and steep slopes made it necessary to terrace the slopes and fill them with soil carried from the valleys below to help produce the food they needed (Figure 1-4).

In North America European settlers encountered a more hostile environment than they had known in the Old World. In a remarkably short period of time, damage to plowed soils by violent rains and turbulent winds was extensive. Then came the drought and dust storms of the 1930s (Figure 1-5). On 12 May 1934 a single dust storm over parts of Texas, Oklahoma, Colorado, and Kansas carried 185 million Mg of soil (400 billion lb) to East coast cities and out to the sea.[5] Enormous soil losses were suffered. By some estimates half of the original topsoil in the American Midwest has eroded in the past 150 years. Citizens have responded by creating government agencies; farmers have responded by adopting soil conservation practices on a limited scale and with limited success.

History of Soil Science

The earliest investigators into the nature of soils could be called **edaphologists** (those who study soil as a habitat for organisms and, particularly, as the medium in which plants grow). **Edaphology** still accounts for many soil investigators. Many scientists who call themselves foresters, agronomists, ecologists, range scientists, or botanists are edaphologists. Those who study soil as a geologic entity—its origin, morphology, geography, and taxonomy—are **pedologists.** Developments in **pedology** are more recent. Today most scientists who study soil for primarily agricultural purposes refer to their profession by the term **soil science.**

History of Edaphology Xenophon, a Greek historian (430–355 B.C.), is credited with first recording the merits of soil-enriching crops when he wrote, "But then whatever weeds are upon the ground, being turned into the earth, enrich the soil as much as dung."

Expanding upon this concept, Cato (234–149 B.C.) wrote a practical handbook in which he recommended intensive cultivation, crop rotations, the use of legumes for soil improvements, and the use of manure in a system of livestock farming. Cato was also the first to classify land according to its relative value for specific crops.

By the time the barbarians of the north conquered Rome and ended Roman scientific efforts, Alexander had already overrun the eastern Mediterranean; the Mongol invasion in the thirteenth century burned and destroyed Babylonia and completed the cultural destruction of the area. The knowledge gained in ancient civilizations was lost or forgotten and scientific agriculture was arrested until nearly 1600.

Early in the seventeenth century, an experiment performed by Jan Baptiste Van Helmont (1577–1644) in Holland began a new era in agricultural research. He put a 2.3 kg (5 lb) willow tree in 90.8 kg (200 lb) of soil. The tree received only rainwater for five years. At the end of this period, the soil weighed only 57 g less than when the experiment began (he assumed the loss to be weighing errors), but the willow tree weighed 76.8 kg (169 lb 3 oz).

---

[5] F. R. Troeh, J. Arthur Hobbs, and Roy L. Donahue. Second Ed., *Soil and Water Conservation,* Prentice-Hall, Englewood Cliffs, NJ, 1991, 35.

FIGURE 1-4 Machu Picchu, the Lost City of the Incas, near Cuzco, Peru. Its hidden and remote setting, atop steep cliffs rising 760 m (about 2500 ft) above the valley floor, made it necessary for the inhabitants to become self-sufficient in food production by terracing the slopes and filling them with transported soil. The tall peak in (b), from which photo (a) was taken, also has terraces (note arrow). The terraces, 2 m to 3 m (about 6.5 ft–9.8 ft) wide, have stone retaining walls and extend down steep slopes as shown in (c). (Courtesy of Raymond W. Miller, Utah State University.)

FIGURE 1-5 A half-buried shed in the mid-1930s is evidence that too much soil is moved and deposited by wind. Many farmsteads became temporarily unusable, and the homesteaders went bankrupt and left the land. Only when more permanent cover, windbreaks, and other preventive measures were used did the land again become productive. (Courtesy of USDA—Soil Conservation Service.)

Because the tree was given only water, Van Helmont reasoned that water was the "principal of vegetation." The experiment was advanced for the time—but Van Helmont was only partially correct. Much of the weight gain in the willow tree would have been from water, and the hydrogen atoms in the cellulose were derived from water. However, he was also partially wrong for the following two reasons:

- The loss of 57 g of soil, which was considered to be experimental error, actually consisted of minerals such as calcium, phosphorus, and potassium that were absorbed by the tree.
- The willow's dry weight consisted mostly of carbon and oxygen from carbon dioxide.

By the time of the American Revolution, late in the eighteenth century, little was still known of plant needs. Joseph Priestley (1733–1804), who would later discover oxygen, recognized that air played a part in plant growth. He observed that animal respiration and rotting residues impaired the air; however, sprigs of *living* mint plants improved the air, making it wholesome. We now know that rotting and animal breathing reduce the concentration of O_2 in the air and increase CO_2 (and some odoriferous gases). Plants improve the air by using CO_2 and giving off O_2. This fact was later proposed by Theodore de Saussure in 1804, after oxygen was discovered, but his views were not then accepted. After Justus von Liebig (1803–1873), a German chemist, chastised plant physiologists for their blind adherence to the unfounded concept that carbon was derived from soluble soil forms rather than from the CO_2 in air, scientists finally came to accept, and later to prove, the concepts of de Saussure.

Liebig also postulated the **law of the minimum,** which states that the growth of plants is limited by the essential element present in the least *relative* amount. This principle is valid and worth remembering. Each essential element is needed by a plant in some minimum amount. At any one time whichever element is absent or available in an amount below its critical minimum is the one that retards plant growth. Liebig's list of essential plant nutrients included potassium, magnesium, calcium, phosphorus, sulfur, carbon, oxygen, nitrogen, silica, and hydrogen. In modern times silica is not included among essential elements (it is listed as a *beneficial* element), and seven additional elements have been added to the list. Actually, Liebig's law can be expanded to include all growth factors (the growth of the plant will be limited by the growth factor in *least relative* amount, whether nutrient, temperature, or water).

The first modern agricultural experiment station was established in 1843 by J. B. Lawes and J. H. Gilbert, at Rothamsted, England, near London. In 1862 the U.S. Congress passed the Hatch Act, creating colleges that included agriculture in their curriculum, and in 1887 created agricultural experiment stations in every state. Much of the agricultural advancement of the twentieth century sprang from the U.S. system and others patterned after it.

History of Pedology Early pedologists viewed soil as weathered rock from the earth's mantle, having characteristics inherited from the underlying rock. Beginning in 1870 the Russian school of soil science, under the leadership of V. V. Dokuchaiev and N. M. Sibertsev, was developing a new concept of soil. The Russians saw soils as independent, natural bodies, each with unique properties resulting from a unique combination of climate, living matter, parent material, relief, and time. They hypothesized that properties of each soil reflected the combined effects of the particular set of genetic factors responsible for the soil's formation. The Russian concept was revolutionary: Properties of soils were no longer based wholly on inferences from the nature of the rocks, climate, or other environmental factors, considered singly or collectively; rather, a soil's properties were determined directly from the soil itself and then explained on the basis of soil-forming factors.

In the 1920s C. F. Marbut adapted the Russian approach to the American system. In 1938 the U.S. Department of Agriculture Yearbook contained the first extensive Soil Classification System for the United States.[6] In 1941 Hans Jenny published a system of quantitative pedology.[7] Building upon these works the U.S. National Cooperative Soil Survey, under the leadership of Guy Smith, developed a system of soil classification in 1975.[8] This system, continually updated, is the system of soil classification in use in the United States today.

1:4 Soil—A Precious Resource

One cannot overstate the value of soil and the many purposes it serves for man and nature. This book highlights the role of soil in plant production but recognizes that other roles touch people in important ways as soils and their uses influence food prices, environmental quality, quiet havens where we can stop a while, and spaces where we build and live.

Many Roles

Like air and water, soil plays unique and essential roles in our ecosystems. Our hopes to preserve species diversity and to maintain a favorable carbon dioxide–oxygen balance in the atmosphere hinge on the quality of the soil. Indeed, each **biome** of the earth is an interwoven system in which the soil, climate, and living organisms affect each other in a complex way.

---

[6] M. Baldwin, Charles E. Kellogg, and J. Thorp, *Soil Classification,* in *Soils and Men,* USDA Yearbook of Agriculture, U.S. Government Printing Office, Washington, DC, 1938, 979–1001.

[7] Hans Jenny, *Factors of Soil Formation: A System of Quantitative Pedology,* McGraw Hill Book Company, New York, 1941, 281.

[8] Soil Survey Staff, *Soil Taxonomy, A Basic System of Soil Classification for Making and Interpreting Soil Surveys, Agriculture Handbook 436,* USDA—Soil Conservation Service, 1975, 754.

FIGURE 1-6 Garbage strewn along a rural road near El Paso, Texas. During heavy rains some of the trash washes onto the farmland below. (Courtesy of USDA—Soil Conservation Service.)

Just as air and water are susceptible to pollution, soil, too, is often our depository for human wastes such as sewage, garbage, and milling residues—sometimes deliberately, sometimes thoughtlessly (Figure 1-6). Depending upon the technology employed, application of these wastes can contribute to the quality of soils, or it can render them unfit for food crops. The most costly pollution cleanup sites in the United States tend to be sites with contaminated soil (Figure 1-7). Many such sites are military installations with organic fuels or solvents spilled on the ground, threatening surface or groundwaters. Other sites, such as the Hanford Nuclear Reservation in Washington and the Lawrence Livermore National Laboratory in California, are sites with soil contaminated by radioactive waste. The Hanford site alone will require at least $30 billion for remediation. When the Chernobyl reactor exploded in 1986, 70 percent of the fallout fell on the republic of Byelorussia; yet, because food shortages are so common to the region, crops and dairy products continue to be produced on the contaminated soils. Reports state, though, that farm managers, recognizing the seriousness of the problem, do not feed their own children foods from Byelorussian fields.[9]

To an engineer, soil is construction material for roadbeds and landfills or it is a foundation for buildings or dams. The engineer's interest in the soil centers on its ability to support a load or to withstand stress without deforming excessively. Soils differ tremendously in their suitability for engineering purposes. A major function of the Natural Resource Conservation Service is to determine the degree of limitation placed on each soil regarding its use for such purposes as roadbeds, campsites, dwellings, and septic tank absorption fields (Figure 1-8).

A soil's ability to produce crops or sustain development is reflected in land values. This indicates the intrinsic value of soils to people, and for good reason: Soils provide the settings in which human enterprises thrive or fail to thrive. Because this was also true in ancient times, soil is also the domain of archaeologists. A soil buried and protected by later deposits of

---

[9]Rebecca Lair, *in* Gary E. McCuen and Ronald P. Swanson (eds.) *Toxic Nightmare: Ecocide in the USSR and Eastern Europe,* Gary E. McCuen Publications, Hudson, WI, 1993, 34–35.

FIGURE 1-7 Oil-waste land in Jim Hogg County, Texas, is wasteland caused by excessive pollution by oil, saltwater, and oil-drilling muds left in the early days of oil well drilling. Modern drilling techniques control waste disposal and do relatively little damage. Many of these areas will improve gradually, but most will require the help of irrigation, tillage, and lots of time. (*Source:* USDA—Soil Conservation Service, Temple, Texas.)

sediment or glacial debris can be a rich archaeological find. At a site in the Sinai, archaeologists found evidence of soil that may have supported a hunter-gatherer civilization in the late Pleistocene period. Buried soils in the great plains of the United States show accumulations of clay and lime, indicating that dry conditions prevailed when the buried soils formed. Studying soil buried beneath a Roman amphitheater in England, archaeologists attempted to reconstruct conditions that prevailed during the Roman era. In Central Texas stone tools and lithic manufacturing debris were found in a soil that formed 8000 years ago and was then buried under sediments from successive flooding. Characteristics of this buried soil indicate a prolonged period of landscape stability and human occupation.[10] Combining soil science with archaeological investigations provides detailed environmental studies of the Holocene (Quaternary period from 10,000 to 12,000 years ago up to the present time). Such studies are important to understand the chronology, evidence of environmental change, and impact of people on landscapes during this time.[11]

Plants and Soil

Soils provide the basic growth factors to plants: **support, oxygen (O_2)** for roots, **water,** and **nutrients,** but some soils are better than others at providing optimum amounts of these growth factors.

---

[10] Vance T. Holliday (ed.), *Soils in Archaeology: Landscape Evolution and Human Development,* The Smithsonian Institution, Washington, DC, 1992.
[11] S. J. Scudder, J. E. Foss, and M. E. Collins, "Soil Science and Archaeology," *Advances in Agronomy* **57** (1996), 1–76.

FIGURE 1-8 Landslides are caused by steep slopes and unstable soil materials. Good highways are only as good as their foundations. This one in Muskingum County, Ohio, was not adequately protected from sliding. (Courtesy of USDA—NRCS and the Ohio Department of Natural Resources, *Soil Survey of Muskingum County, Ohio.*)

Soils provide increasing support to plants as their roots penetrate to greater depths. Forests on shallow mountainsides, which may allow only about 30 cm of tree roots, often have catastrophic windthrow of trees. A combination of rain—which softens and lubricates the soils—and high winds will overcome the roots' ability to hold the trees erect.

Oxygen is supplied to roots from air in the pore spaces of soil. Plant roots constantly respire during growth, which requires the availability of O_2 directly to the roots. Only a few plants (rice, for example) have mechanisms to transport air *internally* through the aerial part of the plant to its roots. Sandy soils with large pores are well aerated because the pores tend to contain air; clay soils with many small pores are typically not well aerated because the pores tend to be filled with water and because the small pores often do not allow sufficiently rapid gaseous exchange between the atmosphere and the soil. Without adequate O_2 replacement in deeper soil layers, roots will not grow very deep or very well. Roots may penetrate only 40 cm to 60 cm (about 16 in to 23 in) deep in some clays because of poor aeration.

The water needed by plants is usually retained in the soil reservoir. Most crops need from 300 mm to 800 mm (11.7 in to 31.2 in) of water during a four- or five-month growing season. Even the most productive soils normally hold only about 75 mm to 100 mm (3 in to 4 in) of usable water at one time. Thus, the soil must be wetted several times during the growing season by rainfall or irrigation. Some soils (shallow or sandy) are able to hold only enough water to supply plants for a week or so. A soil's water-holding capacity and the rainfall pattern in a given location are two factors that determine which plants will grow in a nonirrigated soil (Figure 1-9).

Scientists generally agree that plants require 16 chemical elements (Detail 1-2). Additional elements, such as nickel (see Detail 1-3), may be essential for some, but not all, plants. Still other elements may improve a plant's growth but are not considered essential because the plant can complete its life cycle without them. Of the 16 elements, carbon and oxygen are supplied from carbon dioxide in the air; hydrogen is derived from water. The remaining 13 nutrients are supplied by the soil. Nitrogen is a major component of soil organic materials (plant residues and animal residues). As these organic materials decompose, the nitrogen is released as soluble ions. These nitrogen ions (nitrate and ammonium) often persist in the soil for only a short time; therefore, *nitrogen is the nutrient most often deficient in soil.* The other essential elements are solubilized from weathering minerals, and several are held

FIGURE 1-9 A typical Hopi dryland cornfield in Jeddito loamy sand in Arizona. Where irrigation water is not available, the need to adapt has motivated people to develop unique systems. This area of 200 mm to 300 mm (8 in to 12 in) of rainfall requires widely spaced hills, according to the farmers' estimate of future rainfall. Other unique practices are involved, too, by the Hopi. (Courtesy of USDA—NRCS, the Bureau of Indian Affairs, and the Arizona Agricultural Experiment Station, *Soil Survey of Hopi Area, Arizona, Parts of Coconino and Navajo Counties.* Photo by Fred Kootswatewa, Hopi Tribe.)

on the surfaces of clay particles by electrostatic attraction. The fertility of a soil depends upon the amount of plant nutrients it contains and on the availability or solubility of those nutrients.

Crops and the Human Race

Soils produce the plants that produce most of the world's food and fiber. The mushrooming human population imposes increasing pressure on farmers to produce more food each year. The quality and quantity of food available for the human diet is determined by the area of **arable** (farmable) land available per person and by the productivity of the land (Figure 1-10).

A subsistence diet requires about 180 kg (400 lb) of grain per person per year; a typical affluent diet requires about four times as much grain:

- The affluent diet includes excessive direct consumption of grain products.
- The affluent diet is high in meat. Producing one pound of feedlot beef requires several pounds of feed.

Producing a subsistence diet of 180 kg grain requires about 0.085 hectare (0.21 acre). As shown in Table 1-2, by the year 2020 a worldwide average of 0.186 hectare per person will be available if all arable land is cultivated and growing grain. This is sufficient land for a subsistence diet but not enough for the entire world's population to enjoy an affluent diet. Considerable farmable land is presently used to feed animals, which people eat as meat. If all arable land were farmed, there would be none available for wildlife habitat, military reserves, parks, watersheds, or golf courses.

Food, however, is not the only issue. More people need more clothing, which requires more cotton, flax, wool, and petrochemicals. Also, as world petroleum reserves are used up in the next few decades, pressure will mount for alternative fuels such as ethanol (grain alcohol). With present technology about 6 hectares (15 acres) are required to produce the grain to make ethanol to run a car about 16,000 km (9920 mi) per year, if the car gets 13 km/L (30 mi/gal). Furthermore, in the near future competition for cropland will also mount as demand

Detail 1-2 Plant Nutrients and Other Important Elements in Soils

Element Name	*Chemical Symbol*	*Ionic Soil Forms*	*Comments Concerning Forms of the Element and the Element's Importance in Soils*
Aluminum	Al	Al^{3+} $Al(OH)_2^{+}$	Can be toxic to plants in strongly acid soils; occurs as various hydroxyl forms
Boron*	B	H_3BO_3	A water-soluble plant nutrient in small concentrations
Cadmium	Cd	Cd^{2+}	High atomic weight ("heavy metal"); retained in animals and people and is highly toxic
Calcium*	Ca	Ca^{2+}	Essential plant nutrient; the cation often most prevalent in nonacidic soils
Carbon*	C	HCO_3^{-} CO_3^{2-}	The basic element of organic substances (mostly made by living organisms); component of carbon dioxide (CO_2)
Chlorine*	Cl	Cl^{-}	Occurring in small amounts except when it is a part of soluble salts
Cobalt	Co	Co^{2+}	Needed in N_2 fixation; essential to animals
Copper*	Cu	Cu^{2+}	May be as Cu^{+} (cuprous) in poorly aerated soils
Hydrogen*	H	H^{+}	A small, active, strongly absorbed and chemically active ion
Iron*	Fe	Fe^{3+}	Low solubility in most soils; may be as Fe^{2+} in minerals and poorly aerated soil; as iron oxide (Fe_2O_3), it causes the reddish and yellowish coloring in soils
Lead	Pb	Pb^{2+}	Toxic heavy metal; also as PbO_2 in soil
Magnesium*	Mg	Mg^{2+}	Similar in properties and reactions to calcium
Manganese*	Mn	Mn^{2+}	Also as MnO_2 in soil
Mercury	Hg	Hg^{2+}	Toxic heavy metal; also as HgO in soil
Molybdenum*	Mo	MoO_4^{2-}	An essential plant nutrient required in very small amounts
Nickel	Ni	Ni^{2+}	Essential plant nutrient; toxic metal in large amounts
Nitrogen*	N	NO_3^{-} NH_4^{+}	Necessary for proteins; in complex organic forms; both ionic forms are usable by plants
Oxygen*	O	O^{2-} OH^{-}	As free gaseous form, O_2, it is essential to all respiration
Phosphorous*	P	HPO_4^{2-} $H_2PO_4^{-}$	Forms many low-solubility phosphates with Ca, Al, Fe, and other heavy metals
Potassium*	K	K^{+}	Soluble in soils, except mineral forms are very insoluble
Selenium	Se	SeO_4^{2-}	Toxic heavy metal; essential to animals
Silicon	Si	Si^{4+}	Common in minerals holding oxygens together; sands and quartz are mostly SiO_2, beneficial to some plants
Sodium	Na	Na^{+}	It may be essential to some plants; very soluble; part of "soluble salts"; causes sealing of soil
Sulfur*	S	SO_4^{2-}	Forms S^{2-} (sulfide) form or toxic hydrogen sulfide gas (H_2S) in poorly aerated soil
Zinc*	Zn	Zn^{2+}	Often deficient in calcareous and eroded or leveled soils

*Essential plant nutrient.

Detail 1-3 Nickel (Ni^{2+}), A New Seventeenth Essential Element?*

Some plant physiologists now believe nickel to be the seventeenth essential plant nutrient, but this discovery took some effort. Frequently, the nickel needed by the plant could be supplied by *the nickel already in the seed* (100 ng to 200 ng per gram of dry seed—1 g of nickel per million kg of dry seed)! To prove the need for nickel, the plants had to be grown for three or four generations in nonickel solutions to get seed sufficiently low in Ni to be Ni-deficient. For barley (*Hordeum vulgare*), seeds needed 90 ng of nickel per gram of dry seed weight. In plant tissues, Ni is usually 0.05 mg to 5 mg kg^{-1} of dry weight. Nickel deficiency depresses seedling vigor, causes chlorosis, and may cause necrotic lesions in leaves. Nickel is a component of at least two enzymes—urease and hydrogenase.

*D. A. Dalton, S. A. Russell, and H. J. Evans, "Nickel as a Micronutrient Element for Plants," *Biofactors* **1** (1988), 11–16; and William G. Hopkins, *Introduction to Plant Physiology,* John Wiley & Sons, Inc., New York, NY, 1995, 78–79.

increases for industrial crops such as those used for latex, pharmaceuticals, plastics, and paper.

If the population continues to increase at the current rate (10,000 more people per hour), one can predict that the world will experience more critical food shortages during the lifetime of the young people alive today. By 1996 the world's fisheries and rangelands were thought to be near their maximum sustainable production potential; meaning that only farmers—with little help from ranchers or fishermen—will bear the burden of feeding the increasing population. This comes at a time when crop production per area of farmland is *not* increasing much and crop production per person is declining. Less land is available each year as roads,

FIGURE 1-10 To supply themselves with food, people have moved into new areas, cut forests, and cleared rocks, where necessary. Rolling hills have been graded and windbreaks planted to make living comfortable. Fewer and fewer areas for expansion, such as this farm in Barnes County, North Dakota, exist. People ask, "Where is the unsettled 50 percent of uncultivated, arable land?" Much of it is dry, hard to work, erosive, or fraught with other problems. (Courtesy of USDA—Natural Resources Conservation Service. Photo by W. B. Sebens.)

Table 1-2 World Population and Arable Land					
	Population[a] (millions)			Arable Land[b] (ha/person)	
Region	1960	1990	2020	1990	2020
Africa	282	626	1351	0.296	0.137
Asia	1685	3131	4660	0.144	0.097
China	651	1137	1425	0.085	0.068
Latin America and Caribbean	218	443	652	0.406	0.276
North America	199	277	358	0.852	0.659
United States	181	250	323	0.760	0.588
Europe	425	501	530	0.279	0.264
Former Soviet Union	214	289	335	0.806	0.696
Oceania	16	27	37	1.815	1.324
World	3038	5293	7924	0.278	0.186

[a]Source: U.S. Department of Commerce. World Population Profile: 1994.
[b]Assuming arable land is constant and as reported in FAO Production Yearbook, 1988.

shopping centers, homes, and office buildings consume former farms and pastures. Also, many of the water sources we have exploited so recklessly in the past, such as the Ogallala Aquifer under the high plains, are drying up faster than they recharge. Even our expected recharge waters vary from year to year. When rains and snows are low, many people may lack the water they depend on (Figure 1-11). Meanwhile, larger urban populations require more water to drink, to bathe in, and to sprinkle on lawns. Obviously, water for irrigation is becoming less available as time goes by. Not surprisingly, world grain reserves in 1996 reached

FIGURE 1-11 Large reservoirs supply water for many people. When a few dry years string together, water can be in short supply. This reservoir in California in 1977 exhibits large, barren slopes (white) along the greatly lowered lake. More people require more water, and even one dry year can cause concern and shortages in some areas. (Courtesy of *California Agriculture Magazine,* University of California.)

FIGURE 1-12 Farmland is not the only land subject to erosion. This Douglas-fir stand near Corvallis, Oregon, is harvested by clear-cutting, leaving the soil easily erodible. (Courtesy of USDA—Natural Resources Conservation Service.)

an historic low of 49 days.[12] In short, the world is increasingly demanding more food, more fiber, and more industrial crops grown on less land using less water.

From the facts available, several deductions are evident:

- The apparent excess of food and arable land in some countries exists only because many people there and elsewhere live in poverty and eat less affluent diets. Current levels of agricultural productivity do not yield enough food to provide everyone in the world with an affluent diet.
- As more arable land is developed into industrial sites or public works, or used to produce nonfood products, the amount of land available for food production will decrease.
- Soil is a finite resource. As Anthony Bailey said, "The basic problem is that God stopped making land some time ago, but He still is making people." We must retain all the soil possible by using soil conservation practices and by eliminating soil erosion (Figure 1-12).
- To avoid intolerable strain on the food supply, two options are available: Produce more or consume less. Neither is easily accomplished.

1:5 Computers, Models, and Soil Systems

It has been said that even if you are on the right track, you must keep going or you'll get run over. The advent of computers and the interest in more accurately depicting nature has increased the attraction to, and use of, computer models.

Models serve at least two purposes: (1) They simulate (imitate) the real thing by using data, postulates, and sometimes mathematical descriptions; and (2) they can be a technique to organize knowledge about a subject into a system to predict desired results. Models help answer questions such as, "If I use less irrigation, how much will it affect yield?" and "What are the influences of temperature and rainfall on nitrogen cycling in hardwood forests?"

---

[12] Lester R. Brown, *State of the World,* W. W. Norton, New York, 1996.

There are many models—some good, some poor, some complex, some simple—all attempting to illustrate a bigger, more complete picture of phenomena in nature. For those interested, and to illustrate the expanding role of modeling in environmental and agricultural practices, a list of numerous models and source references is given in **Appendix A.**

My earliest lessons on environmental protection were about the prevention of soil erosion on our family farm.

—Al Gore

Questions

1. (a) What is soil? (b) Describe the composition of soils.
2. Name the eight most common chemical elements making up soil minerals.
3. Discuss the quantity of arable soil in the world as it relates to future food supplies.
4. (a) What is a mineral? (b) How does a mineral differ from a rock?
5. Differentiate between igneous, sedimentary, and metamorphic rocks.
6. Differentiate between primary and secondary minerals.
7. How is an organic soil different from a mineral soil?
8. (a) Why is soil important to man? (b) What roles does soil play in facilitating plant growth?
9. (a) What is an essential plant nutrient? (b) How many essential plant nutrients are known?
10. What is the projected world population for the year 2020?

2 Soil Formation and Morphology

Soils were not formed in past ages. They are being formed and are ever changing.

—R. L. Cook and B. G. Ellis

We all know what light is; but it is not easy to tell what it is.

—Samuel Johnson, 1776

2:1 Preview and Important Facts

PREVIEW

Everyone has seen, played on, and used soils, but what is this material called *soil* and how is it formed? Most soils are **mineral soils** formed by the weathering of hard (solid) rock masses to loose, unconsolidated, and often transported materials. They contain living and dead organic materials, perhaps coarse rocks or gravel, some sand, silt, and clays, and some soluble ions. **Organic soils** (peats and mucks) develop mostly from plant accumulations that have fallen into stagnant water, where the decomposition is much slower than the plant production, causing a buildup of organic materials in various stages of decomposition.

Soil formation is the production of unconsolidated material by weathering processes and soil profile development, which are the changes involved in the development of horizons. The first action, **weathering,** involves the changes from a consolidated mass (rock) *not capable of growing plants* to the development of an unconsolidated (loose) layer of material that can support plants if climate is suitable and water is available. The second action, **soil development,** involves the changes occurring within the loose material over time.

Soils usually develop various layers that look different from one another, called **horizons.** Most horizon differences involve (1) organic matter accumulation in the surface layers, (2) formation of hardpans deeper down, (3) leaching (washing) downward of soluble salts and very small particles (clays, organic particles, mineral oxides), and (4) accumulations of these leached materials when they lodge in small pores or cease to move when water flow stops.

Horizons tell much about the characteristics of a soil. They include information about hardpans, depth and amounts of organic matter accumulation, soil denseness from cementation and clay deposition, and the extent of leaching. The total accumulation of horizons in one soil with depth show the **soil profile**—the view of a vertical slice of soil.

Soils have extreme variations in horizons and profiles and range from blowing sand dunes to the deep, wet muds of river deltas and from rocky glacial deposits to the wind-blown plains of the American Midwest. Fence builders in the mountains of the Western United States remember painfully the shallow, rocky soils and carbonate-cemented hardpans that made digging post holes a blister-raising job. Farm youth soon learn not to plow clayey soils when the fields are wet because the soil becomes a sticky mire that fouls the plow—and tempers. Many Northeastern U.S. farmers surround their fields with rock fences built from glacier-strewn stones they have removed from the fields. Tourists in tropical areas can see deep, uniformly red or yellow soil profiles as deep as 7 or more m (23 ft) in exposed road cuts. The sands of the Sahara differ greatly from the soils of the jungles of the Congo or the Amazon; the sun-baked plains of India contrast sharply with the highly cultivated fields of the Netherlands. Much of the earth's soil cover is far from perfect for its many uses by people.

As soil differences are recognized, the question arises of how good a particular soil is. What is that soil's *health* or *quality*? An evaluation of soil quality and what improves or degrades it are of increasing concerns to those interested in maintaining the quality of soil.

Soils can change with every few steps we take in any direction. It will be important to be able to identify those changes and to accurately locate them for later activities. The enormous amount of data about the area can only be collected, stored, synthesized, and used with a computer set up with a manager of multiple sets of data. This manager and integrator of sets of geographic data is called a *geographic information system* (GIS). The equipment to physically identify where you are on the Earth is called a *global positioning system* (GPS). The use of both of these systems is revolutionizing data storage, integration, manipulation, and location.

IMPORTANT FACTS TO KNOW

1. The meaning of and processes of *soil formation;* compare with *soil development*
2. The five soil-forming factors and how each affects soil development
3. The meaning of the term *landform* and how each of these landforms affects a soil's development and properties
4. The definitions of, and differences between, *soil profile, soil horizons,* and *diagnostic horizons*
5. The *letter horizon* you would give each of the more common *diagnostic horizons* (There isn't always a sure match.)
6. One-sentence definitions of these diagnostic horizons: *mollic, argillic, albic, ochric, oxic, spodic, kandic, calcic, duripan,* and *fragipan*
7. The difference between a *soil* (pedon, polypedon) and a *mapping unit*
8. The definition of *geographical information systems* (GISs) and how they are used
9. The general usefulness of GISs and other database systems compared to that of hand-drafted maps
10. The concept of *spatial variation* and its relation to soils
11. The meaning of *soil quality* and what constitutes a good soil and a degraded soil

2:2 Weathering of Soil Minerals

Weathering is the disintegration of **primary** (original) minerals and the reformation of some of those dissolved materials into new, **secondary** minerals. Only when a solid rock mass is disintegrated to **unconsolidated** material is a soil formed from rock. Water, air, and roots may then move into this loosened material, at which time the material is referred to as *soil.*

When the original material is a hard rock, such as **granite,** the disintegration to form even a couple of centimeters (an inch or so) of soil can require many hundred to several thousand years. Soils form more rapidly from **sandstones** (cemented sands), **shales** (consolidated clays, silts, and sands), and **limestones** than from granites, but it still takes tens of thousands of years (Figure 2-1).

Processes of Weathering

What causes minerals to weather, and what are the products of that weathering? Weathering includes breaking or grinding particles to smaller sizes (**physical weathering**) and the dissolution or chemical alteration of minerals (**chemical weathering**). Particles suspended in flowing water or in blowing winds grind the minerals into smaller particles when the suspended moving particles hit them. Water in cracks in rocks helps split those rocks when it freezes and swells; temperature changes will weaken some internal bonding, thereby making both chemical and physical weathering more effective.

Chemical weathering is usually the most active and effective weathering process. Water in the soil or rock splits into $\mathbf{H^+}$ and $\mathbf{OH^-}$ and causes dissolution of minerals (**hydrolysis**). In some situations water also softens minerals that absorb the water (**hydration**) and develop swelling pressures internally. Carbon dioxide from the atmosphere and from the respiration of organisms dissolves in water, forming carbonic acid, which acidifies the water and increases its solvent action (**carbonation**). Oxygen in the air combines with iron in minerals and forms iron oxides (causing orange and reddish-yellow discoloration). These **oxidation-reduction** reactions increase the rates of disintegration of many dark minerals that have appreciable iron in them. All of these chemical processes will produce in the soil solution many **ions** dissolved from the weathering minerals.

Products of Weathering

What happens to the dissolved mineral substances? Some dissolved ions are retained in the soil; the more soluble ions are washed away wherever leaching (percolation) water flows,

FIGURE 2-1 Weathering takes a long time. We are concerned about the logging of our old and valuable forests, such as these enormous redwoods cut in the first quarter of this century. Soils take many times longer than the forests to form a few centimeters from hard rock. We should be even more concerned about the lost soil than about these prime forests. (Courtesy of *California Agriculture* magazine, University of California.)

even deeper into or below the soil profile. The least soluble materials (iron hydrous oxides and aluminum hydrous oxides) resolidify, and some precipitate and persist as part of the soil solids. Soil clays form in this way from these and other low-solubility materials during hundreds of years of weathering. By calculating the soluble mineral materials in river waters, one scientist estimated that the rivers of Europe and North America carried an average of 40 tons of soluble salts dissolved from each square kilometer of watershed. This quantity corresponds to the dissolving and removal of about 1 cm of soil depth per thousand years.

The expected products of weathering are indicated by a weathering scheme of *clay-sized particles* (Table 2-1). The minerals found in a soil indicate the extent to which it has weathered. For example, soils that still contain gypsum or crystals of sodium salts have been only slightly weathered in their present setting and contain the most easily weathered minerals. In their present location these soils have little loss of soluble materials from weathering. Arid regions with little rainfall for leaching have these kinds of soils. In contrast, soils consisting mostly of iron oxides (as much as 60 percent to 70 percent of the total material in some wet tropical soils) have been extensively weathered. Most of their primary minerals, including forms of silica, have been washed out by centuries of weathering and extensive leaching.

Two considerations may modify the patterns shown in Table 2-1. First, if particles are of silt and sand sizes, some quartz and feldspars will still remain in some highly weathered

Table 2-1 A Weathering Scheme for Clay-Sized Minerals in Soils

Typical Clay-Sized Mineral	*Other Expected Minerals*	*Comments*
Minerals Most Easily Weathered; Present Only in Slightly Weathered Soil		
1– Gypsum	Halite, other sodium salts	In arid-region soils; only in slightly leached soils
2– Calcite	Dolomite, apatite	In most limestones
3– Olivine-hornblende	Pyroxenes (dark minerals)	Easily weathered primary minerals
4– Biotite mica	Iron-magnesium chlorite	
5– Albite feldspar	Other K-feldspars	
6– Quartz	Cristobolite, other quartz	The most resistant of the primary minerals
Secondary (Reformed) Minerals Produced from Soluble Products of Weathering Primary Minerals		
7– Illite clay	Muscovite, other micas	Least resistant of clays to weathering breakdown
8– Vermiculite clay	Muscovite, other micas	Moderately resistant clay
9– Montmorillonite clay	Beidellite clay	Moderately resistant clay in arid climates
10– Kaolinite clay	Halloysite clay	Moderate-to-high resistance (used for fine china)
11– Gibbsite [Al $(OH)_3$]	Boehmite, allophane	Resists breakdown; in bauxite (aluminum ores)
12– Hematite [Fe_2O_3]	Goethite, limonite	Very resistant (in bog iron, sesquioxides)
13– Anatase [TiO_2]	Zircon, rutile, corundum	Most resistant minerals
Minerals Most Resistant to Weathering (Insoluble)		

soils, along with the iron and aluminum oxide clay materials. Second, soils formed somewhere else and eroded may have been deposited as *new alluvial deposits.* These new alluvial deposits start to form a soil at "time zero," yet the material contains *residues* of highly weathered minerals (*inherited minerals from prior weathering*).

Weathering is a slow process, and clay mineral suites in soils from alluvium and from sedimentary rocks can be **inherited.** Thus, it is usually more important to know the minerals a soil now contains rather than how it was formed. This *inherited* mineralogy in alluvium (or in rocks such as limestone and sandstone formed from alluvium) must be considered when characterizing any soil being developed from alluvium. Limestones, sandstones, shales, and conglomerates (formed from alluvium in ocean floors) are now uplifted land areas. Soils forming from them may contain clay minerals formed in weathered soils eons before and in a different climate, before the soils were eroded to the sea, where they formed rocks in an ocean bottom.

2:3 Soil Formation: Building a Matrix for Living Organisms

The term **soil formation** describes a mixture of minerals or hard rock as it changes into loosened material in which plants and other organisms will be able to live and into which air and water can move.

From Rock and Sediment to Soil

To become a soil many of the minerals or organic residues must be transformed into separate particles of sand, silt, clay, and humus. The soil is a mass of materials with pore space, air, and water that permit root growth. Plants, with a few exceptions, depend on decomposing soil organic materials to supply their nitrogen and portions of at least twelve other nutrients. To form soil the mineral mass weathers to form some clays and to accumulate humus. Soil has a depth of at least a few centimeters (an inch) and often is a meter (39 in) or more deep. In this soil microorganisms will be active and roots will grow. The change from minerals or rocks to soil involves changes over centuries of weathering time. The many changes can be summarized as additions, losses, transformations, and translocations.

Additions are made by water (rainfall, irrigation), nitrogen from bacterial fixation, energy as sunlight, sediment from wind and water, salts, organic residues, fertilizers, and other substances. **Losses** result from chemicals soluble in soil water carried away, eroded small-sized particles, nutrients removed in grazed and harvested plants, water losses, carbon losses as carbon dioxide, and denitrification loss of N_2. **Transformations** happen because of the many chemical and biological reactions that decompose organic matter, form insoluble materials from some of the soluble substances, and alter or dissolve some minerals. As water and organisms move within the soil, many substances are **translocated** (moved) to different depths. For example, clays are carried to deeper layers in percolating water; soluble salts are carried into the groundwater or back to the soil surface as soil water evaporates. Even *localized* sorting occurs, such as deposits of carbonates or iron oxides accumulating in the soil pores deeper in the profile.

Soil Profile

The vertical cross-section of a soil with depth is referred to as a **soil profile** (Figure 2-2). There are hundreds of soil profiles, all a bit different from each other. In arid regions there will be mostly a thin, light brown surface (A) over a thin parent material or horizon slightly higher in clay. There may be carbonate or salt accumulation. Figure 2-2 is for a wet forested soil, perhaps in the Midwest, where organic matter accumulates on the surface and the top layers of mineral soil are leached of dark-colored organic matter and some clays (E horizon),

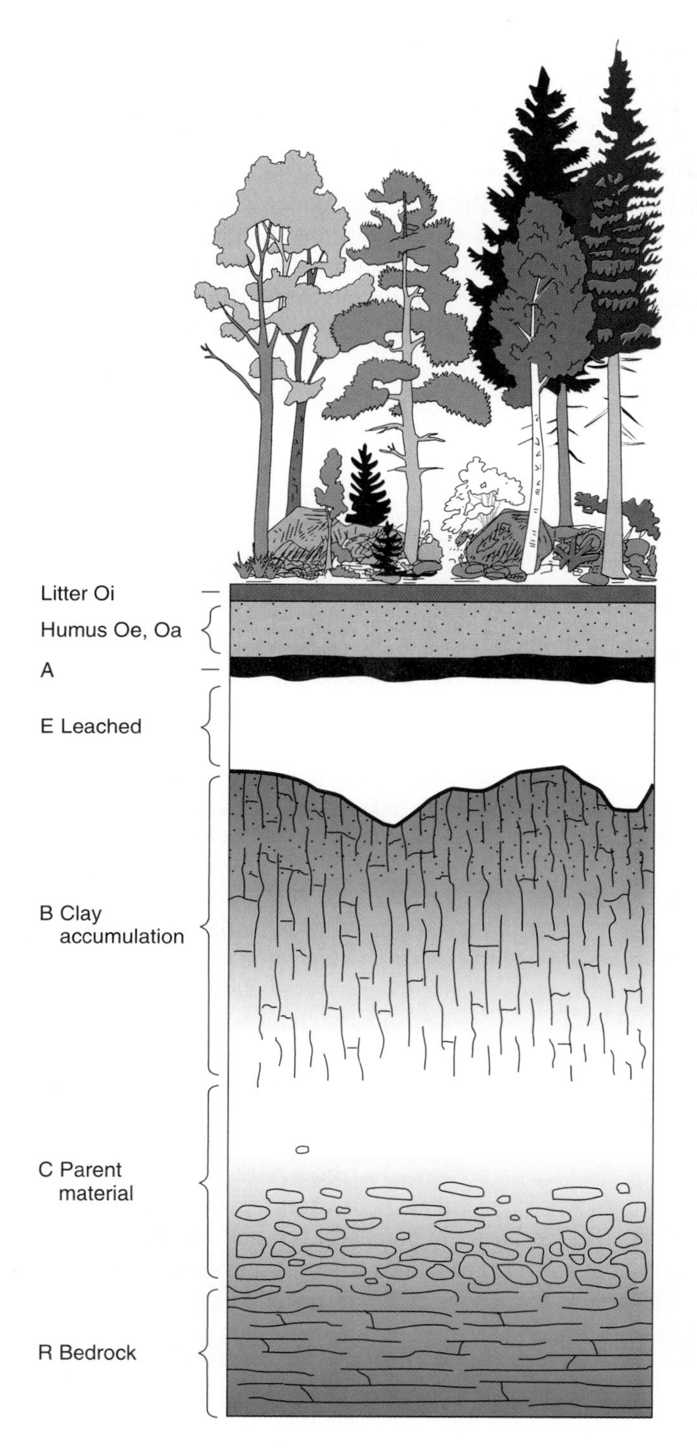

FIGURE 2-2 A soil profile is a view of the vertical slice of a soil. Depths of many developed soils are about 1 m to 2 m deep. As soils develop they produce layers that are different from each other, which are called *horizons.* The above soil under mostly conifer forests in a cool, wet climate would accumulate litter and some partially decomposed organic material **(O)**. Perhaps it would form a thin, dark mineral layer **(A)**, a layer leached by water that removes much of its organic matter and some clays **(E)**, a layer in which the leached materials accumulate **(B)**, below which the loose but otherwise slightly developed parent material **(C)** and, at some depth, the bedrock **(R)** occur. (Courtesy of Raymond W. Miller, Utah State University.)

which are moved to a horizon below the surface. Most soils have the parent material (C horizon) and perhaps a solid rock (R horizon) from which they formed.

2:4 Soil-Forming Factors

Throughout the world more than 15,000 different soils have been described. There are thousands of kinds of minerals that can be mixed in different proportions. Some soils have been weathering for hundreds of centuries. Some soils exist in hot, wet climates; others exist in arid, dry climates. All of these conditions cause differences in the final soil developed. Five **soil-forming factors** are primarily responsible for the character of the developed soil:

- **Parent material** (unconsolidated material or rock in which soil development occurs)
- **Climate** (temperature and precipitation, for the most part)
- **Biota** (living organisms, organic residues)
- **Topography** (slope, aspect, elevation)
- **Time**

Parent Materials and Soil Formation

Parent materials are those surface materials from which soils form. They include exposed bedrock, but the largest soil areas are formed from *unconsolidated materials* from eroded soils elsewhere laid down by water or wind, or are sediments from glacial depositions. The sediments from erosion may be rocky, clayey, sandy (sand dunes), or many mixtures of particle sizes. The parent material developing a new soil can also be sediment from the erosion of well-weathered soils or relatively new, slightly weathered soils. Many kinds of primary and secondary minerals can be present. An example of various kinds of parent materials is shown in Figure 2-3.

In slightly weathered soils the parent material may be dominant in determining many soil properties, and the parent materials of glacial till, loess, or bedrock beginning to weather will be little changed except to accumulate organic material, develop some structure, and perhaps translocate soluble salts and some carbonates. Even moderately- to well-weathered soils may not erase the presence of rocks in glacial moraines or greatly change the clays in deltas

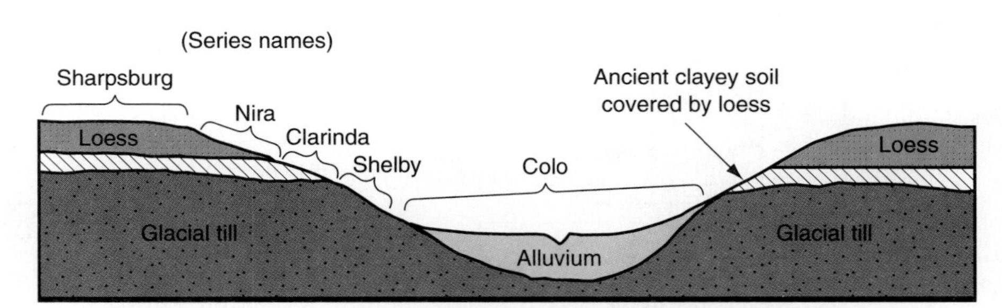

FIGURE 2-3 Cross-section of a landscape in Iowa, illustrating how different times of deposition can result in different soils due to different kinds of parent material, different development times in place, and varying topography. Soils on recent alluvium are the youngest (alluvium = sediment from streams). Eroded loess soils are next youngest (Nira soils). Ancient soil surface (a paleosol now exposed by erosion) is oldest (Clarinda soils). Glacial till now exposed by erosion (Shelby) is next oldest. Stable loess surfaces (Sharpsburg) are more weathered than eroding ones (Nira) (see Figure 2-7). (*Source:* Drawn from information in *Soil Survey of Adair County, Iowa,* Soil Conservation Service, 1980, by Raymond W. Miller, Utah State University.)

or the sands in river terraces (Figure 2-4). It is not surprising that many early soil scientists were also geologists and attributed to parent materials the *dominant role* in determining the properties of soils.

Climate and Soil Formation

Climate is increasingly a dominant factor in soil formation over time, chiefly because of the effects of precipitation and temperature. Some direct effects of climate on soil formation include the following:

- Lime (carbonates) can accumulate at shallow depths in areas having low rainfall because calcium bicarbonates (from dissolving carbon dioxide, minerals, and lime) are not leached when little water is present. During drying the carbonates precipitate. Such soils are usually alkaline.
- Acidic soils form in humid areas due to intense weathering and leaching out of basic cations (calcium, sodium, magnesium, potassium), which are replaced by H^+ and $Al(OH)_2^+$ or other aluminum species.
- Erosion of soils on sloping lands constantly removes developing soil layers.
- Deposition of soil materials downslope covers developing soil layers.
- Weathering, leaching, and erosion are more intense and of longer duration in warm and humid regions such as Hawaii, where the soil does not freeze. The converse is true in cold climates, such as Alaska; soil is too cold for rapid weathering.

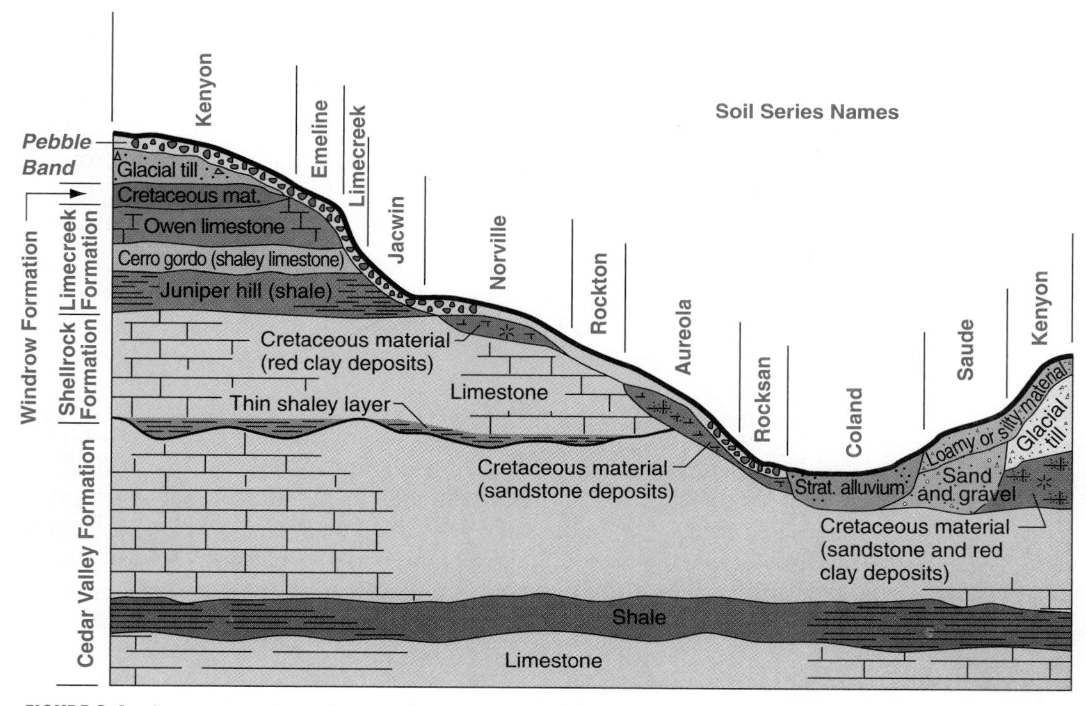

FIGURE 2-4 A cross-section showing how the underlying parent material has an important part in the formation of the eleven soils, in Floyd County, Iowa. Time and other soil-forming factors can alter parent material effects, but many soils are still greatly influenced by parent material. (Courtesy of USDA—NRCS and Iowa State University, *Soil Survey of Floyd County, Iowa.*)

Moisture is also important in soil formation. A soil is said to have some *development* when it has detectable layers (horizons), such as accumulated clay (**B**) horizons, organic matter-enriched (**A**) horizons, carbonates, or soluble salt accumulations that have been moved downward by water. The extent of colloid movement and the depth of their depositions are controlled partly by the amount and pattern of precipitation, which produces the leaching.

Climate also influences soil formation indirectly through its action on vegetation. Semiarid climates encourage only scattered shrubs and grasses.[1] Drier climates supply only enough moisture for sparse short grasses or shrubs which may not be dense enough to protect the soil against wind and water erosion. Many such arid soils exhibit only slight organic matter accumulation and small amounts of profile development.

Soils resulting from year-round hot, humid weathering—such as that which occurs in some parts of Hawaii on old, stable, and well-drained surfaces—are typically deep (often more than 3 m; 10 ft), reddish in color (due to the presence of oxidized iron), have well-decomposed organic matter, and are low in essential elements because of leaching by the high rainfall.

Biota and Soil Formation

Living plants and animals and their organic wastes and residues (the **biota**) have marked influences on soil development. Differences in soils that have resulted primarily from differences in vegetation are especially noticeable in the transition where trees and grasses meet. Minnesota, Illinois, Missouri, Oklahoma, Texas, and most Western U.S. states are places where such differences can be observed.

Some soils beneath humid forest vegetation may develop many horizons. These soils are leached (washed, eluviated) in the surface layers and have slowly decomposing organic-matter layers on the surface. In contrast, some grassland soils near the transition zone of forests are rich in well-decomposed organic matter, with dark organic matter accumulations to depths of 30 cm (1 ft) or more into the mineral soil (Figures 2-5 and 2-6).

Burrowing animals—such as moles, gophers, prairie dogs, earthworms, ants, and termites—are important in soil formation when they exist in large numbers. Soils that harbor many burrowing animals have fewer but deeper horizons because of the constant mixing within the profile, which nullifies the tendency for downward movement of organic colloids and clay.

Microorganisms help soil development by slowly decomposing organic matter and forming weak acids that dissolve minerals faster than does pure water. Some of the first plants to grow on weathering rocks are crust-like **lichens,** which are a beneficial (symbiotic) combination of algae and fungi.

Topography and Soil Formation

The Earth's surface contour is called its **topography** (sometimes called *relief*). Topography influences soil formation primarily through its modification of water and temperature. Soils on steep hillsides within the same general climatic area developing from similar parent material will typically have thin developing horizons because less water moves down through the profile as a result of rapid surface runoff. Also, the soil surface erodes quite rapidly. In

---

[1] J. C. F. Tedrow, J. V. Drew, D. E. Hill, and L. A. Douglas, "Major Genetic Soils of the Arctic Slope of Alaska," in *Selected Papers in Soil Formation and Classification,* Soil Science Society of America Special Publication Series 1, SSSA, Madison, WI, 1967, 164–176. *Note:* Although arctic Alaska, receiving about 13 cm to 26 cm (5 in to 10 in) of precipitation a year, is climatically a desert, ecologically it is a humid region because of the constantly high relative humidity that condenses as dew at night and thereby adds to the total moisture supply without being recorded as precipitation.

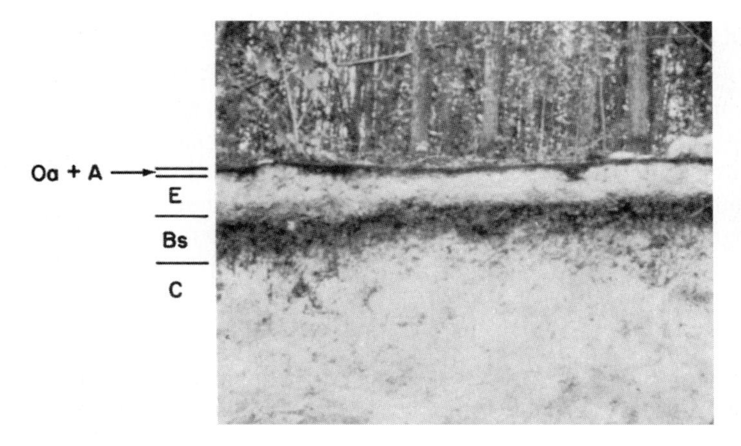

FIGURE 2-5 Under forest vegetation in humid regions, the humus accumulation horizon is more highly leached and there is a well-developed colloid accumulation horizon. Minnesota. (Compare with Figure 2-6.) (*Source:* USDA.)

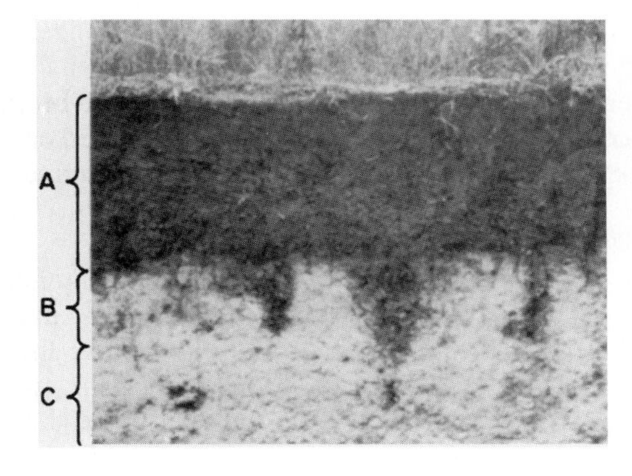

FIGURE 2-6 Under grass vegetation in humid regions, the humus accumulation **(A)** horizon is dark and deep and the colloid accumulation **(B)** is thin or absent. Minnesota. (Compare with Figure 2-5.) (*Source:* USDA.)

contrast, similar materials on gently sloping hillsides have more water passing vertically through them. The profile on gentle slopes is usually deeper, the vegetation more luxuriant, and the organic-matter content greater than in similar materials on steep topography.

Materials lying in landlocked depressions receive runoff waters from surrounding higher areas. Such conditions favor greater production of vegetation but slower decomposition of dead plant remains. The oxygen deficiency in waterlogged (saturated) soil results in accumulations of large amounts of organic materials. If the area above the soil surface is wet for many months of the year, organic (peat or muck) soils develop. If the accumulation waters dissolve salts from surrounding soils, the depression may become a salt marsh with unique and tolerant plants growing, or it may develop toxic salt conditions with no plants at all. When soils on the watershed are strongly acidic, iron may leach from them and be deposited in depressions to form bog iron (limonite). Alkaline soils on sloping topography may leach the soluble carbonates into the ponded depression, forming **marl** (a bog lime).

In the northern hemisphere, soils on south- and west-facing slopes receive more direct rays of sun and are, therefore, warmer and drier than north- and east-facing slopes. In arid climates these drier south- and west-facing slopes are often less productive than soils on north- and east-facing slopes. The opposite slopes are affected in the southern hemisphere. In cold, wet areas these warmer sites may be the most productive. A higher temperature on south and west slopes results in greater loss of water by evaporation; the net result in regions where water is limiting is often soils with thinner horizons and less vegetative cover than soils on north and east slopes.

Time and Soil Formation

Time is of the essence! For soil development, the length of time required to develop layers, called *genetic horizons,* depends upon many interrelated factors of climate, nature of the parent material, the organisms, and topography. Horizons tend to develop most rapidly in warm, humid, forested climates where there is adequate water to move **colloids** (subvisible-sized clays, humus, and other materials). Acid, sandy, medium-textured soils lying on gently sloping topography appear to be the soils most conducive to rapid soil-profile development.

Under ideal conditions a recognizable soil profile may develop within 200 years in loose sediment; under less favorable circumstances, such as from weathering rock, the time may be extended to thousands of years. **Soil development** proceeds at a rate determined by the effects of time plus the intensities of climate and biota, further modified by the effect of topography on which the soil is situated and the parent material from which it is developed. Even former soils, now buried, may be part of a new soil profile (Figure 2-7).

Different surfaces of the Earth's land have been exposed for different lengths of time. Some plateau soils have been exposed for hundreds of thousands of years. Glacial till surfaces are more recent but may still be a few hundred thousand years old. More recently, rivers have flooded and covered flood plains and valley bottoms with recent deposition; these land surfaces are only a few years or decades old, but soil development is beginning there. Under humid conditions in the central United States, an organic matter-darkened (**A**) horizon and genetic soil structure have developed in surface mine spoils in 5 years to 64 years.[2]

Interactions of Soil-Forming Factors

The formation of soil is a diverse and complex process, and effects from the five major factors work in combination. For instance, soil with good drainage (topography or parent material) and mild temperature and high rainfall (climate) will probably support ample plant life (biota) because favorable drainage provides an aerated material in which plant roots can grow. The plants, in turn, decompose, producing carbon dioxide (CO_2), which combines with water from rainfall to form carbonic acid (H^+ and HCO_3^-). Acidity increases the solubility of parent minerals; sodium, potassium, calcium, and magnesium are dissolved and some of them leach away in the draining soil water; the soil becomes ever more acidic. More rain moves some clays and small organic-matter particles deeper into soil layers where they accumulate.

The same climate (or clayey parent material) but different topography (a low-lying valley or depression instead of a gentle slope) might produce a waterlogged soil with poor drainage. Poor drainage results in stagnant water and nonaerated soil, often reducing both plant growth and the rates of organic matter decomposition. The accumulated drainage waters may also amass dissolved salts. The horizons of these adjacent soils, developed from common soil-forming factors except for the differing topography, will be very different.

Areas of high rainfall and good drainage typically develop acidic soils, horizons of organic matter accumulation in the upper layers (**A**), and movement of colloids out of upper horizons to accumulate in the deeper horizons (**B**). Soluble materials are moved deeper or completely below the rooting depth. If the soils have been developing a long time (dozens of centuries), they tend to have high clay content, horizon differentiation, acidity in wet climates, and salt accumulation in some soils of arid regions. Wet, cool forests can become strongly acidic, accumulate slowly decomposing organic matter on top of the soil, and have extensive colloid translocation. Wet, warm forests have faster organic-matter decomposition because of

---

[2]David Thomas and Ivan Jansen, "Soil Development in Coal Mine Spoils," *Journal of Soil and Water Conservation* **40,** (1985), No. 5, 439–442.

FIGURE 2-7 Road cut exposing the buried soil from preglacial times that is now part of the currently forming Clarinda soils (see also Figure 2-3). The ancient clayey subsoil (a paleosol) is very sticky and of low water permeability. Its upper boundary is indicated by the line between the arrows. (*Source:* USDA—Soil Conservation Service, State of Iowa.)

warmer temperatures and are extensively weathered and acidic. In drier climates grasslands have thicker layers of dark organic matter accumulation from grass roots and decomposed tops. In semiarid areas vegetation is sparse and carbonates often precipitate to form whitish carbonate zones called **lime zones;** accumulated soluble salts are common in the arid soils.

A land body may not be greatly altered during its exposure to the factors of soil development because of other conditions that retard soil profile development, such as the following:

- Low rainfall (slow weathering; little soluble material is moved within the soil)
- Low relative humidity (little growth of microorganisms such as algae, fungi, and lichens)
- High lime content of parent material (keeps clays less mobile)
- Parent materials that are mostly quartz sands and have few easily weathered minerals and little clay (slow weathering; few colloids to move)
- High clay content (causes poor aeration; slow water movement)
- Resistant parent rock materials, such as quartzite (slow weathering)
- Very steep slopes (erosion removes soil as fast as the top horizon develops; little water intake lessens leaching)
- High water tables (slight leaching; low weathering rate)
- Cold temperatures (all chemical processes and microbial activity slowed)
- Constant accumulations of soil material by deposition (continuously new material on which soil development must begin anew)

- Severe wind or water erosion of soil material (exposes new material to soil-forming processes) (See Figure 2-7.)
- Mixing by animals and humans (tillage, digging) minimizes net downward colloid movement
- Presence of substances toxic to plants, such as excess salts, heavy metals, or excess herbicides, that hinder plant growth and their effects on soil development

The opposite conditions favor more rapid rates of soil development. For example, material with nearly level topography and little or no erosion, in an area of high rainfall, developing from easily weathered parent materials in warm climates where plants are growing, would develop a differentiated soil profile rapidly.

Ancient Soils: Paleosols

Normal soils are defined as those soils developed or developing in the existing climatic environment. Some soils have been developing for hundreds of thousands of years, and increased development time changes some horizons and increases their degree of development. Climates of many areas have changed during the tens of thousands of years of soil development. In eons past, Earth climates, which are now temperate, changed as the glacial and interglacial periods alternated. These previously different climates caused distinctive soils to develop. It can be difficult to know whether the climate in which a soil exists has changed within the several thousand years needed for most soils to form.

At least three terms describe nonnormal soils that are classified as paleosols: *Paleosols, relict soils,* and *fossil soils.* **Paleosols** are *soils formed mostly in previously existing climates* and are now either buried or not. Most of these soils formed during the Quaternary Period (from the beginning of the Ice Age until recently). **Relict soils** are *exposed paleosols that were formed in a previous time under environmental conditions quite different from those currently existing.* **Fossil soils** are *paleosols buried in the geological section of the Earth's crust but deeper than present soil development processes.* These paleosols are essentially unchanged since burial.

A fourth soil category used by archaeologists is *anthrosol.*[3] **Anthrosols** are *people-made soils,* those soils that exhibit the deliberate actions of people. The *plaggen epipedon,* the long-term manuring of fields, is related to this. Anthrosols are as diverse as the cultures that produced them. They include simple *middens* (aboriginal garbage heaps) to complex layers of depleted or eroded soil resulting from repeated cycles of unfertilized cropping, poor irrigation, and overstocking. Soil scientists find value in the study of these anthrosols.

Information about paleosols is used for many purposes, such as to reconstruct ancient history related to climate, crops, and ground cover of the sites before burial. Such information has helped to verify and even explain conditions existing when catastrophic floods or earthflows buried cities.

Radiocarbon dating is a method that permits the dating of many sites, even the dates of sequential burials by several catastrophic events. It is the most useful *dating method* and is based on the fact that radioactive carbon (C^{14}) exists now and has existed in the past in the Earth's atmosphere in an exact and constant proportion to the total air carbon dioxide ($C^{14}O_2/CO_2$ = a constant ratio). Radioactive carbon in the air's CO_2 is produced by *cosmic rays* (protons, alpha particles, other, from space) bombarding atmospheric N_2 molecules forming C^{14} atoms ($_7N^{14}$ + neutron $\rightarrow$ $_6C^{14}$ + proton off). Any living organism using air CO_2 will incorporate a definite proportion of radioactive CO_2. As long as life continues, the car-

[3] S. J. Scudder, J. E. Foss, and M. E. Collins, "Soil Science and Archaeology," *Advances in Agronomy* **57** (1996), 1–76.

bon is continuously replaced by new carbon and the *proportion* of radioactive carbon stays constant. As soon as the organism dies, no further continual replacement of carbon occurs. The amount of radioactive carbon in the dead organism decreases as the organism's C^{14} undergoes radioactive decay. The decay reaction is $_6C^{14}$—decay → $_7N^{14}$ + beta ray off. Some properties of radioactive carbon include the following:

Half-life (time to lose half its initial activity): 5568 years
Proportion in atmosphere: 1×10^{-12} g of C^{14} per 1 g of total C
Usable time period for dating: 100 to 50,000 years
Sample size needed: Varies, 0.1 g to 10 g of carbon

To radiocarbon-date a site a sample of buried carbonaceous materials (wood, bone, coal, oil, carbonate shells, soil humus, charcoal, or other carbon) is required. Some examples of dates follow:

- Sediments in Catalina Basin, California (0 m to 0.05 m deep)

  Carbonate fraction — 2320 years ± 130 years
  Organic carbon fraction — 1970 years ± 150 years

- Sediments in Catalina Basin, California (4 m to 4.05 m deep)

  Carbonate fraction — 23,100 years ± 1000 years
  Organic carbon fraction — 18,400 years ± 600 years

- Linen wrapping enclosing Book of Isaiah from the Dead Sea Scrolls

  1917 years ± 200 years

- Wood and humic acid from a 58-cm-thick soil buried under 0.7 m of leached sand and then covered by glacial till

  Wood in buried soil — 29,000 years
  Humic acid in buried soil — 41,000 years

How would you distinguish between a paleosol and a normal soil? If the soil is buried because of some cataclysmic change (burial by heavy volcanic ash fall, glaciation, or flood alluvium), the soil was obviously formed before burial. The properties (horizons) of any paleosol are different from those of a soil that would develop at that site in the climate currently present or in the time that soil has been developing.

In order to justify calling any soil a paleosol, the scientist must first verify that the buried layer is a *formed soil,* not just deep, unconsolidated, earthen deposits. The evidence that the deposit is truly a soil involves many factors, some of which are the presence of (1) profile horizons (**O, A, E,** and/or **B**), (2) roots or root channels in place, (3) worm channels that indicate long-time activity of soil organisms, (4) translocated and deposited clays, (5) formed structural peds, particularly those cemented by organic substances, (6) accumulated concretions or translocated lime or gypsum, (7) accumulated plant silica **phytoliths** (plant silica that is formed in various shapes, usually peculiar to the kind of plant), and (8) numerous other features found only in soils as they develop. Some of these other features include horizon sequences, hardpan formation, and fossils of soil organisms or evidence of their extensive activity (earthworm casts). Even land animals, trapped in muds and later buried, are found in the sedimentary rocks or sediments (Figure 2-8). The phytoliths and other

FIGURE 2-8 Archaeologists use clues in buried soils to characterize the environment in which many ancient and extinct organisms, such as this dinosaur, lived. Display at Vernal, Utah. (Courtesy of Raymond W. Miller, Utah State University.)

buried evidence helps characterize the environment in which the organisms lived. The more easily changed soil properties include the diagnostic horizons mollic, gypsic, and cambic. Other features may be very slow to change. Plinthite, durinodes, clay layers, fragipans, and other hardpans have slow weathering and/or cementation. These horizons and features are defined later in this chapter.

2:5 Landforms and Soil Development

When a soil begins to undergo profile changes because of weathering and leaching, the original material in which the changes are happening is called the **parent material.**

There are numerous possible combinations of parent materials in the rocks, the mineral mixtures, flood-deposited sediments, glacial deposits, and wind deposits. To organize and describe the properties of these parent materials, they are grouped into the categories of surface rocks, sediments deposited from water, sediments deposited from wind, sediments deposited from melting ice, and unconsolidated masses accumulated because of gravitational forces.

The Earth's surface materials have distinct shapes and often have characteristic particle sizes. These characteristic mineral or organic masses are **landforms.** Some of the more familiar landforms are mesas, buttes, plateaus, plains, glacial moraines, terraces, and playas. For example, flood plains tend to be flat and fine-textured, but glacial till remnants are rocky and seldom very smooth-surfaced unless covered by other sediments. Figure 2-9 is an outline of parent materials.

Soils Formed from Rocks

Many soils form *in place* on hard rocks (residual soils). This is a very slow process. In subtropical Zimbabwe, Africa, weathering granite studied at two sites was estimated to form soil at the rate of 11.0 mm (0.43 in) and 4.1 mm (0.16 in) per 100 years.[4] Many of the soils forming on rock are shallow because even slight yearly erosion can exceed soil formation rates (Figures 2-10 and 2-11). Yet on stable slopes, soils in tropical areas can be several meters deep, particularly on porous rocks such as sandstones, limestones, and volcanic ash.

---

[4]L. B. Owens and J. P. Watson, "Rates of Weathering and Soil Formation on Granite in Rhodesia," *Soil Science Society of America Journal* **43** (1979), 160–166.

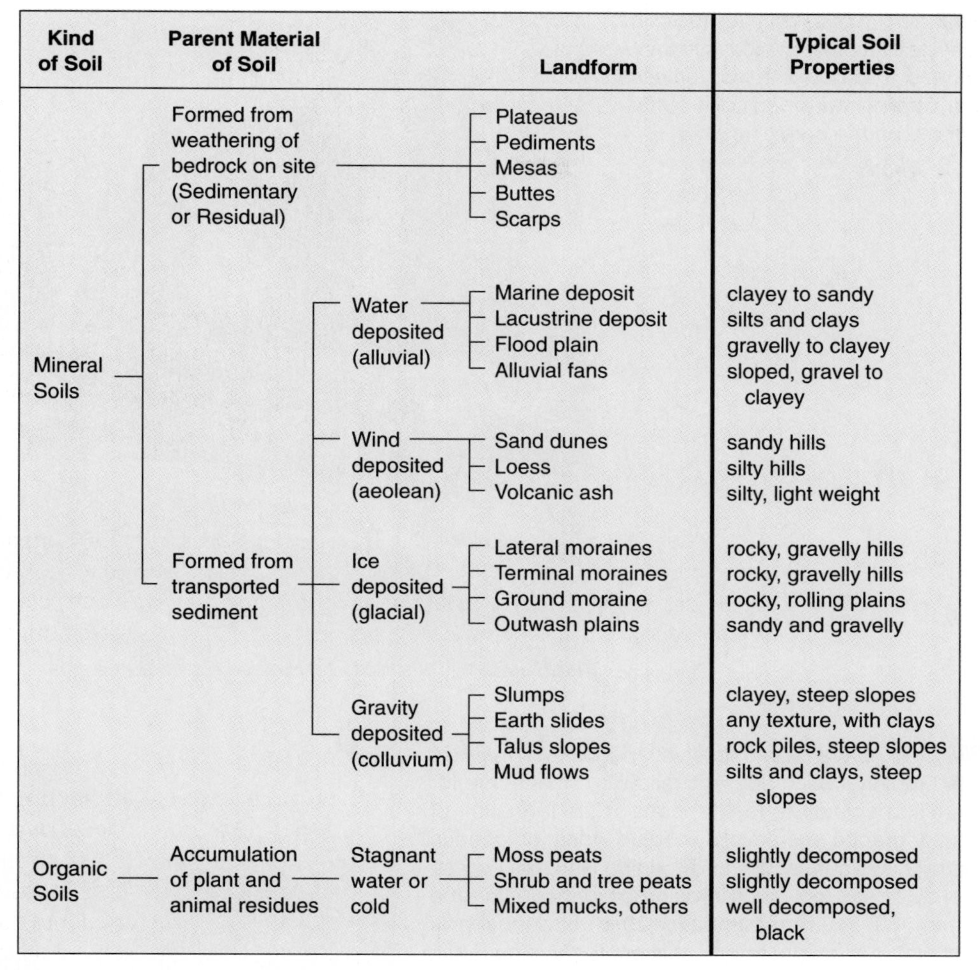

Kind of Soil
Parent Material of Soil
Landform
Typical Soil Properties
Formed from weathering of bedrock on site (Sedimentary or Residual)
Plateaus
Pediments
Mesas
Buttes
Scarps
Mineral Soils
Water deposited (alluvial)
Marine deposit
Lacustrine deposit
Flood plain
Alluvial fans
clayey to sandy
silts and clays
gravelly to clayey
sloped, gravel to clayey
Wind deposited (aeolean)
Sand dunes
Loess
Volcanic ash
sandy hills
silty hills
silty, light weight
Formed from transported sediment
Ice deposited (glacial)
Lateral moraines
Terminal moraines
Ground moraine
Outwash plains
rocky, gravelly hills
rocky, gravelly hills
rocky, rolling plains
sandy and gravelly
Gravity deposited (colluvium)
Slumps
Earth slides
Talus slopes
Mud flows
clayey, steep slopes
any texture, with clays
rock piles, steep slopes
silts and clays, steep slopes
Organic Soils
Accumulation of plant and animal residues
Stagnant water or cold
Moss peats
Shrub and tree peats
Mixed mucks, others
slightly decomposed
slightly decomposed
well decomposed, black

FIGURE 2-9 Outline of soil parent materials for all soils. Transported parent materials are by far the most extensive materials on the Earth's surface to several meters deep. (Courtesy of Raymond W. Miller, Utah State University.)

The landforms of soils forming directly from underlying bedrock are seldom unique. Geologic weathering and erosion tend to produce an equalizing effect in the long term, gradually resulting in bedrock slopes with a shallow, loose covering of soil. These erosion-formed slopes are called **pediments.** Water cutting gullies into gently sloping pediments or nearly level rock layers can leave large, flat **plateaus,** smaller **mesas,** or tiny **buttes** if the upper rock layers are more resistant to erosion than underlying layers. Erosion undercuts the top layer until it caves in. Nearly vertical **bluffs** and steep **scarps** slope down to the area below (Figure 2-12).

Materials Deposited from Water

Sediment deposited from flowing water is called **alluvium. Flood plains** are the landforms built by these deposits in the low areas where streams and rivers overflow. Older flood plains now at levels higher than river bottoms, as the waters cut deeper, are called **river terraces.**

FIGURE 2-10 This granite rock in Yosemite National Park will require centuries to produce a soil unless slope wash or winds deposit some sediment on the area. (Courtesy of H. W. Turner, U.S. Geological Survey.)

FIGURE 2-11 Lava is an extrusive igneous rock that has cooled rapidly, which does not allow sufficient time for minerals to crystallize (as happens in the formation of basalt or granite, more slowly cooled types, or igneous rocks). This is a lava flow in Coconino National Forest, Arizona, where too much erosion or inadequate time and the 380–510 mm (about 15–20 in) of annual precipitation have not yet been sufficient to develop much soil material. The lone, stunted ponderosa is typical in such environments. (Courtesy of Michigan State University.)

If soil-laden water is flowing out of mountain canyons, it deposits material as it reaches less steep slopes and flow slows down. These flooding waters spread out from the canyon mouth to deposit material in fan shapes called **alluvial fans.**

If the alluvial fans along the mountain extend to each other and coalesce into one continuous slope, the coalesced fans are called **bajadas.** Nearly level **peneplains** can be formed near stream level. If no outlet occurs, a lake in wet climates or a **playa** in more arid climates forms in the low bottoms. The particles of these alluvial deposits are gravelly (2 mm to 7.6 cm) where water flows rapidly and more clayey in more level bottoms (flood plains, playas).

Sedimentation in standing water (especially lakes) are **lacustrine deposits** (lake bottoms) (Figure 2-13). Extensive ancient lakes are now dried-up land areas. Large valley areas of lacustrine sediments from ancient glacial Lake Lahontan and Lake Bonneville are now

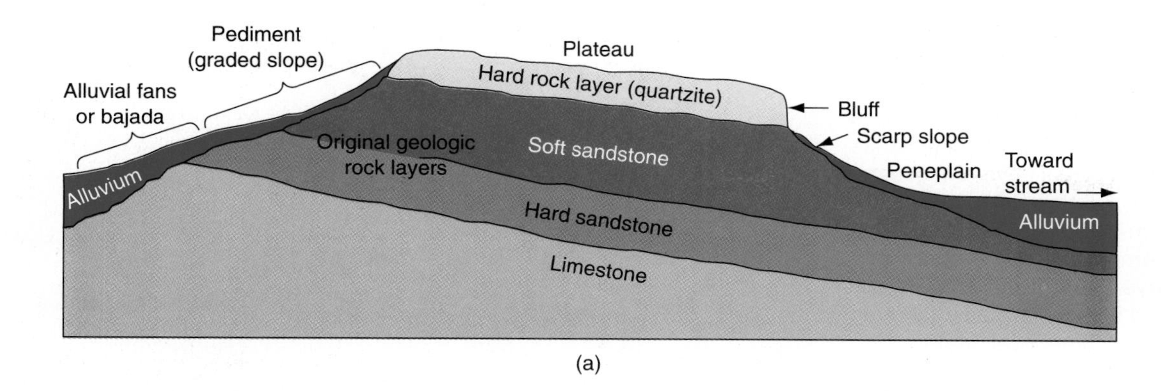

(a)

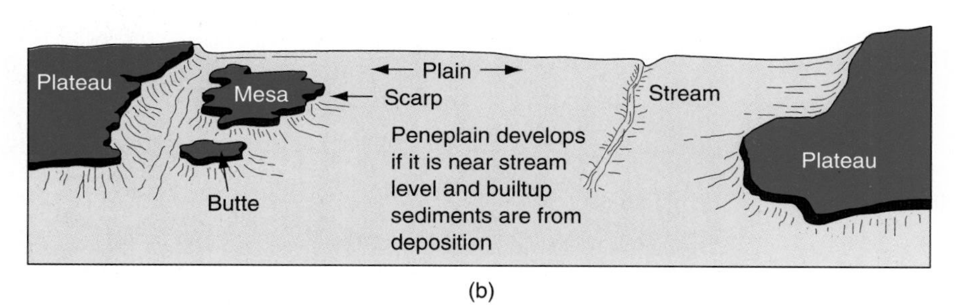

(b)

FIGURE 2-12 Some landforms of land areas where soil is forming from bedrock. These are common where the bedrock is sandstones or limestones, particularly horizontally oriented geologic layering. Such landforms typically occur in arid and semiarid regions of the world. (a) is a cross-section of parent materials that may result in the relief view shown in (b). (Courtesy of Raymond W. Miller, Utah State University.)

valleys and sloped farmland in Nevada and Utah. Many lacustrine deposits in the United States are found bordering the Great Lakes, in eastern North Dakota and northwestern Minnesota (Red River Valley), northwestern Nevada, and northeastern Washington.

Sediments deposited in oceans or reworked by oceans are called **marine sediments.** Many of today's land areas were once marine sediments (salty shales, some dolomites, limestones, and sandstones) that are now lifted up as large land areas and mountains. Parts of the Great Plains, the Colorado River delta, the Imperial Valley of California, and many coastal areas along the Gulf of Mexico and the Atlantic Ocean have marine sediments. **Beaches** are landforms of marine origin. **Deltas** are formed at the mouths of rivers from waterborne sediment. Where the deposits fall into calm waters (lakes, some bays, calm oceans), deltas form. Very large deltas exist at the mouths of the Mississippi and Nile Rivers.

Materials Deposited from Winds

Wind-transported materials form **eolian deposits.** Carried materials may be coarse and fine sands, silts (smaller than sands), and smaller dust-like particles (clays). Small-sized soil materials that were wind-deposited, many following the last glacial period, are known as **loess** (lewr-ess or luss) (Detail 2-1). Present-day deposits of dominantly silt-sized particles (0.05 mm to 0.002 mm diameter) are also called *loess.* Silts are smaller than sands and larger than clays. Large areas of volcanic ash deposits, most very ancient, are common.

FIGURE 2-13 These horizontal lacustrine beds of sands, silts, and clays were formed from previous soils that later were eroded and deposited during glacial times over a period of several thousand years in a freshwater lake in northeastern Washington state. As this area erodes the materials of different layers are mixed and are deposited somewhere else to begin soil development yet again.
(*Source:* U.S. Geological Survey.)

Loessial soil materials are 60 percent to 90 percent silt and, in the United States, occur mainly in the Mississippi Valley. There are large areas of loess in Kansas, Nebraska, Iowa, Missouri, Illinois, Indiana, Kentucky, Tennessee, and Mississippi. Extensive deposits also occur in Washington and Idaho. Loess deposits in Shaanxi province, China, are said to reach about 300 m (984 ft) deep. Other extensive deposits are in the plains of Germany, interior China, and the pampas of Argentina.

Landforms of eolian deposits are **dunes** (sand sizes), **loess hills** or **plains,** and ash deposits. Dunes are further categorized and given names according to their sizes and shapes.

Materials Deposited from Ice

From perhaps 1 million years ago to as recently as 10,000 years ago, continental ice sheets intermittently occupied the land that is now the northern border area of the conterminous United States. Some geologists claim that we are currently in another such interglacial period. Parts of Alaska, Greenland, Canada, Siberia, and Iceland, and the mountains of northern Europe, Switzerland, and Antarctica, are now occupied by a mass of ice similar to that which once covered parts of Canada and the northern United States.

The general name for glacial deposits is **glacial till.** Identifying the various landforms of glacial till helps better describe the parent material. When the ancient ice front melted about as fast as the ice moved, deposits of sediment built up along the melting boundary, resulting in a series of stony hills at the ice front known as **terminal moraines.** Stony ridges deposited along the outer edges (sides) of the ice mass are known as **lateral moraines** (Figure 2-14). When the ice front melted faster than it advanced, the glacier shrank and a larger and smoother deposition resulted, known as **ground moraines (till plains)** (Figure 2-15). Water gushing from a rapidly melting ice mass carried fairly coarse sand and gravel particles and deposited them in a somewhat gently sloping plain over ground moraine at the outer boundaries of the glacier. These are **outwash plains** (Figure 2-16).

Detail 2-1 Particle Sizes and Loess

Soil mineral particle sizes are defined as follows:

Dimensional Units	Size Fraction of the Fine Portion of Soils		
	Sands	Silt	Clay
Inches	0.08–0.002	0.002–0.00008	<0.00008
Millimeters	2–0.05	0.05–0.002	<0.002

When soil contains about 20 percent or less clay and nearly equal amounts of sand and silt in the rest, the soil mixture is called a **loam.** When the words *sandy, silty,* or *clayey* are added as adjectives to the *loam* classification, it indicates a change toward a soil with a higher percentage of sand, silt, or clay, respectively. **Loess,** a German word from *löss,* meaning *pour, dissolve, or loosen,* is the term for wind-carried deposits dominantly silt-sized (about 60–90 percent silt). Loess, although forming generally productive soils, is a little unusual in some properties, such as the following:

- It is quite open and porous.
- Vertical cuts are more stable than slopes. Grade recommendations are 1 unit distance vertical for each 1/4 unit distance horizontally for 9-m-high (about 30-ft) cuts.* Flatter slopes erode easily by *liquefaction*—supersaturated silt *soup* with water; also known as *mudflows.*
- When loess is used for earthen dams, problems occur, such as the formation of sinkholes, cavities, and subsurface channel cutting (*piping*).

*Highway Research Board, Highway Research Record 212, Publication 1557, HRB, Washington, DC, 1968.

FIGURE 2-14 Summer view of a glacier that is active during current winters in the province of Alberta, Canada, showing a fresh deposit of lateral moraine (as a ridge) in the left foreground as well as directly in front of the glacier on both sides of the U-shaped outlet. (*Source:* U.S. Geological Survey.)

FIGURE 2-15 Glacial ground moraine (till plain) materials are usually stony on the surface (above) as well as throughout the entire deposit (below). (Composite photo courtesy of Julian P. Donahue [above] and U.S. Geological Survey [below].)

FIGURE 2-16 Fresh deposits of coarse sand and gravel from water gushing forth from a glacier (this one in the Gulf region of Alaska) are known as outwash plains. (*Source:* U.S. Geological Survey.)

As the glacier bottom melted and tunnels were formed in the ice, sands were deposited along the flow channel, forming **eskers.** When the ice met hard rock hills, it scraped over them, leaving convex mounds (hills) called **drumlins.** Where blocks of ice were left and outwash sediment built up around it, the later melting of the ice left a **kettle** (hole). Unusually large boulders left occasionally on the landscape are called **erratics.**

It is not unusual to have quite impermeable rocky ground moraines severely compacted by the ice weight, later covered by outwash sediment, and finally with a top layer of locally

FIGURE 2-17 Colluvial material, known as a talus cone, is formed by rocks moving in response to gravity. Glacier National Park, Montana. (*Source:* U.S. Geological Survey.)

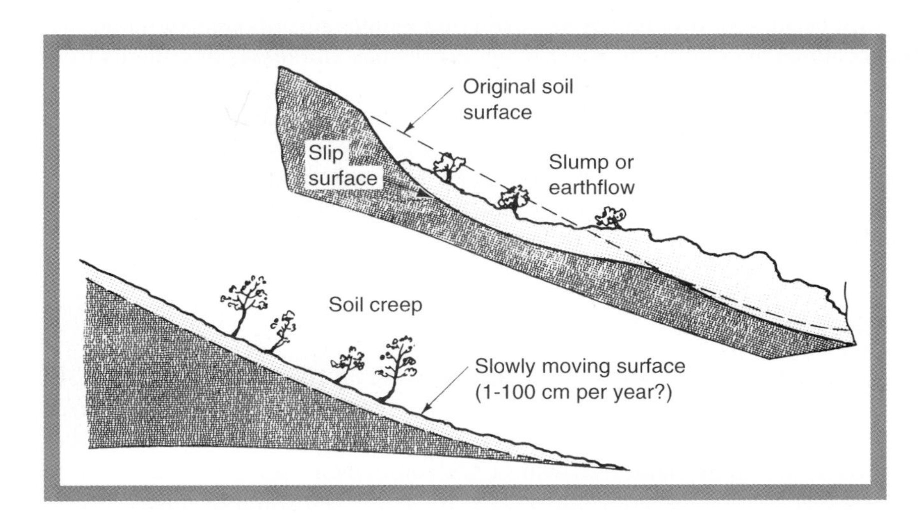

FIGURE 2-18 Effects of gravity on the flow of soil downslope. Rapid movement (a few weeks or less) would be an earthslip or flow, but some areas of soil creep may move only a centimeter or so (fraction of an inch) per month or year. (Courtesy of Raymond W. Miller, Utah State University.)

sorted sediment from wind or from surface flooding. The gravel and rock hinder tillage and the compacted till can cause poor drainage and restrict root growth. The thickness of outwash or finer covering over the till is crucial to the use of the land.

Sediments Moved by Gravity

The collective term for all downslope movements caused by the effect of gravity on weathered rock debris and sediments is **mass-wasting.** The material moved is called **colluvium.** Some colluvium moves in a relatively dry condition. Rocky accumulations along or at the base of steep slopes are called **talus** (tay′-lus) (*scree; talus cone,* if fan-shaped, Figure 2-17). Slowly moving slopes caused by expansion and contraction of soil are called **soil creep** (or rock creep). Wetting–drying, freezing–thawing, and other temperature changes may all aid movement. In lubricated conditions movement may be a *flow process.* Flow may be from a few centimeters per day (**solifluction**) to rapid **debris flow, mudflow,** or **earth flows** (*slumps,* Figure 2-18). Some classifications list all mass-wasting flow along discrete shear planes under the general category of **slides.** More rapid flow is termed **avalanche.** The landform for colluvium from these processes could be specified as **soil creep colluvium, earthflow colluvium,** or **slide colluvium.**

2:6 Development of Soil Horizons

What is a soil horizon and why do horizons form? **Horizons** are *soil layers that are approximately parallel with the soil's surface.* Each horizon is different from other horizons in the profile. Usually, but not always, the differences between adjacent horizons, above or below, are quite obvious to the eye (color, structure differences), or the difference can be felt by the fingers (variation in clay content). The **boundaries** between horizons in a profile range from *indistinct* to *abrupt and clear.* Some boundaries are relatively *smooth* (the horizon is the same thickness in that volume of soil); other boundaries have *tonguing* patterns (large vertical variations in the boundary), as is evident in the bottom of the **E** horizon of the Spodosol (color photo 8 in Table 7-2).

In any particular observation of a single profile, the horizons can only be seen in the very thin observable surface of the soil slice, which is called a **profile.** Horizons can be *thin, indistinct,* and *discontinuous* and, as such, may be seen in one profile observation but not be clearly visible in a second nearby examined profile.

Horizons form because of differences in weathering with depth, amounts of humus accumulated, translocations of colloids by water to deeper depths, and losses of colloids from the profile in percolating waters, to name only a few processes. Usually a horizon is separated from others because (1) the horizon has accumulated more humus and is dark-colored, (2) the horizon has had some of its clay and humus moved to greater depths and is a leached horizon, (3) the horizon has accumulated or produced more clay than is in other horizons, and (4) the horizon has accumulated some secondary minerals (calcite, gypsum, silica, or iron oxides), forming horizons such as lime zones or hardpans.

2:7 Letter Horizons

The first scientists to study soils put labels on the different soil layers, using the letters **A, B, C,** and **D.** Since that time, letter-horizon nomenclature has been modified and standardized by the Soil Survey Staff of the U.S. Department of Agriculture's Natural Resources Conservation Service. The latest major overhaul of horizon nomenclature was published in May 1981. Because much of the literature currently being published still uses older systems of letters, both the new and the previous system of letter horizons are presented in Table 2-2.

General rules for usage and a few examples follow:

- Capital letters are used to designate *master horizons.*
- Lowercase letters are used as suffixes to capital letters to indicate specific characteristics of the master horizon.
- Arabic numerals *following* horizon letters indicate vertical subdivisions within a horizon.
- Arabic numerals *in front of* master horizons indicate discontinuities (used in place of the Roman numerals previously used). *Example:* Loess **A** over limestone **B** would have the **B** written as **2B** to show parent material change (discontinuity).

Examples. An **EB** horizon is a transition horizon between the leached **E** (eluvial) and the accumulation (**B**) horizons. An accumulation of carbonates in a **B** also with clay is a **Btk.**

A parent material with strong gleying and of differing parent material than the horizon above it would be a **2Cg;** the **C** subhorizon below it might be a **2C2g.**

Letter horizons can be subdivided because of differences even within a letter horizon. Thus, a **B** could be divided into a **B1** and **B2** (read as *B one* and *B two*). The **C** horizon is

Table 2-2 Letter Horizons Most Commonly Encountered in Soils

New	*Old*	*Description*
		Usually Surface Horizons
Oi, Oe	**O1**	*Organic* horizon in which most leaves, needles, stems, and other *plant parts* are still *identifiable* (includes recent litter). Usually quite thin—a centimeter or so thick.
Oa, Oe	**O2**	*Organic* horizon so extensively altered that *identification of the parts of* plant materials *is not usually possible.* Can be many centimeters thick.
A	**A1**	*Mineral horizon darkened by organic matter accumulation.* Under **Oa** horizons it is usually thin; in a cultivated soil it is the surface horizon and may be labeled **Ap.** An **Ap** horizon may be a mixture of several thin horizons, even including part of a shallow **B.**
		Usually Subsurface Horizons
E	**A2**	A *mineral* horizon lighter colored than an **A** or **Oa** above it or **B** below it. Fine *clays* and minute organic substances have been *washed* (*eluviated, leached*) out of it by percolating waters. Usually common in high-rainfall areas, especially under forests [see Figure 2-19(A)].
AB or **EB**	**A3**	A transition horizon more like the **A** or **E** above it than like the **B** below it.
BA or **BE**	**B1**	A transition horizon more like the **B** below it than like the **A** or **E** above it.
B or **Bw**	**B2**	*Layer of illuvial colloids* (*accumulation*) or evidence of weathering below the **A** horizon(s). Small particles that have washed from the **O, A,** or **E** horizons have accumulated because of filtration (lodging) or lack of enough water to move them deeper. Early **B** horizon development stages of soils may have only redder (orange, yellow, brown) colors of weathering caused by the colored iron hydrous oxides. Often higher in clay than the **A;** *always* higher in clay than the **E.** The top of the **B** may start at a depth ranging from about 15–50 cm (6–20 in) below the soil surface.
BC or **CB**	**B3**	A transition horizon from **B** to **C** horizons.
C	**C**	*Unconsolidated material* (unless consolidated *during soil development* by carbonates, silica, gypsum, or other material) below **A** or **B** horizons. Little evidence of profile development.
R	**R**	Underlying *consolidated* (hard) *rock,* **Cr** for softer material.
		Horizon Subscripts
Subscripts are added to letter horizons for further detail, always as *lowercase letters.* See Glossary Tables G-2 and G-3 for complete list or more detail.		
k	**ca**	A depositional accumulation of calcium and magnesium *carbonates* (lime).
g	**g**	*Strong gleying,* which is a result of long-time poor aeration, usually because of excess water. Soil colors are grays to pastel blues and greens. *Example:* **Cg.**
h	**h**	*Deposited (illuvial) humus* from percolating water (**Bh**).
s	**ir**	*Deposited (illuvial) iron hydrous oxides* (**Bs** or **Bir**).
m	**m**	*Strong cementation* into hardpans (as by carbonates, gypsum, and silica).
p	**p**	*Plowed* or other farming disturbance, usually but not limited to **A** horizons.
t	**t**	*Deposited (illuvial) clay* from horizons above; usually labeled as **Bt.**
x	**x**	*Fragipan* (hard, silty texture, brittle hardpan).

Source: Adapted by Raymond W. Miller, Utah State University, from (1) USDA—Soil Conservation Service draft of the *Soil Survey Manual,* Chapter 4, 4–39 to 4–50, May 1981; (2) R. L. Guthrie and J. E. Witty, "New Designation for Soil Horizons and Layers and the New *Soil Survey Manual,*" *Soil Science Society of America Journal* **46** (1982), 443–444.

FIGURE 2-19 Profiles of soils typical of contrasting conditions of soil formation. The Adams soil (a) formed under forests in humid Vermont and the Dixie soil (b) formed in semiarid southern Utah. Scale is in feet (ft × 30.5 = cm). (*Sources:* (A) Vermont Agricultural Experiment Station; (B) USDA—Soil Conservation Service.)

divided into **C1, C2, C3,** and so on. A change in parent material (such as loess over limestone) is designated by the Arabic numerals **1, 2,** and so on in front of the horizon symbols. The **1** is understood as the top material and is omitted. However, if the parent material at a deeper horizon is different than the top horizon's parent material, a **2** is written in front of the horizon to indicate that change in parent material. Examples of typical profile sequences follow. Note the changes in parent material at **C2** (written **2C2**) of the grassland and changes in parent material in **B** and **R** in the conifer-forested soil (written **2Bs** and **3R**).

A natural grassland: **A1, A2, AB, B1, B2, BC, C1, 2C2, 2R**
An arid-area soil: **A, B1, B2, C, R**
Conifer-forested soil: **Oi, Oa, E, Bh, 2Bs, 2C, 3R**
Plowed prairie soil: **Ap1, Ap2, BA, Bk, C1, C2, R**

All examples use the new horizon terminology. See Figure 2-19 for two examples of profiles.

2:8 Diagnostic Horizons

Diagnostic horizons are used in the new Soil Taxonomy System (Chapter 7) to differentiate among soil orders, suborders, great groups, and subgroups. **Diagnostic horizons** that form at the soil surface are called *epipedons;* those forming below the surface, *endopedons.*[5] The most dominant and frequent horizons of U.S. soils are indicated by two asterisks; common

[5]*Epipedon,* the surface-developed horizon, is accepted terminology; *endopedon* (from *endo-* in, within), the subsurface horizon, is not established terminology but has been coined by the authors.

but less frequent ones are indicated by one asterisk. The brief definition of each horizon in italics is *not official;* it is given for quickly orienting the reader. Examples and soil names (Spodosols) referred to are in Chapter 7 in the color photos of the soil orders (see Table 7-2).

Epipedons and Other Horizons Formed at or Near the Soil Surface

***Albic horizon:** *A strongly leached E horizon.* Common as a white or light-colored layer, **E** horizon, near the surface of cold and wet soils (Spodosols, others). A surface or subsurface horizon that is light-colored (>4 in color value, moist; >5 in value, dry) caused by eluviation (leaching out) of coatings of clay and free iron oxides. The light shade is the color of the remaining sand and silt. The clay deposited below it may cause a perched water table.

Anthropic epipedon: *A people-made mollic horizon.* A surface horizon formed during use of soil by people for long periods of time as homesites or as sites for growing irrigated crops. Basic cation saturation is high; when not irrigated, the epipedon is dry for seven years out of ten.

Histic epipedon: *An organic surface horizon underlain by mineral soil.* A surface horizon that is saturated with water at some season unless artificially drained, generally between 20 cm and 30 cm (8 in and 12 in) thick and containing at least 20 percent to 30 percent organic matter if not plowed or at least 14 percent to 28 percent organic matter if plowed. In each case the limiting organic matter content depends on the amount of mineral portion that is clay. The lower percentage is used if the horizon has no clay, and the higher percentage is used if the horizon has 60 percent or more clay.

Melanic epipedon: *A thick, black, friable horizon formed in volcanic materials.* Usually more than 10 percent humus content, low bulk density, and at least 30 cm thick. Allophane clays are common, with adsorbed humus.

****Mollic epipedon:** *A thick, dark, friable surface horizon, not strongly acidic.* This horizon is found in grassland and broadleaf forested soils (Mollisols and some Vertisols). A surface horizon that is dark-colored, contains more than 1 percent organic matter, and is generally more than 25 cm (10 in) thick unless sandy or shallow to an impermeable layer. It has more than 50 percent basic cation saturation (50 percent of adsorption capacity is Ca + Mg + Na + K) and is not both hard and massive when dry. The mollic must have Munsell color *values* darker than 3.5 when moist. For sandy soils the mollic epipedon may be as shallow as 18 cm (7 in) or one-third of the depth to a hardpan or bottom of an argillic or natric horizon or to a lime zone, whichever is deeper.

***Ochric epipedon:** *A thin or light-colored surface.* This horizon is common in arid and slightly developed soils (Aridisols, Entisols, Inceptisols). It is a surface horizon that is too light in color (higher value of chroma than mollic epipedon), too low in organic matter, or too thin to be either a mollic or an umbric epipedon.

Plaggen epipedon: *A people-caused, high-humus horizon.* An anthropic surface layer of soil 50 cm (about 20 in) or more in thickness that has been produced by long-continued manuring.

***Umbric epipedon:** *An acidic dark-surface horizon.* This horizon is found in wet, warm, subtropical soils (Ultisols). It is a surface horizon similar to a mollic epipedon but has less than 50 percent basic cation saturation.

A diagrammatic illustration of relationships among surface-formed diagnostic horizons is shown in Figure 2-20.

Endopedon (B) Horizons

These horizons are commonly given letter symbols of **B,** indicating colloid accumulation or weathering changes in a subsurface horizon. See Chapters 5 and 9 for details on kinds of clays, cation exchange capacity (CEC), and sodium adsorption ratio (SAR).

Agric horizon: *A tillage-caused clay and humus accumulation horizon.* A subsurface horizon that has formed under a plowed layer by the movement of silt, clay, and humus into voids created by worms, shrink–swell cracks, and capillary pores.

****Argillic horizon:** *A clay accumulation horizon.* It is thick in wet, well-leached temperate soils (Alfisols and Ultisols). Argillic horizons are common in most grassland soils and in some dry area soils (Mollisols and Aridisols). It is a subsurface horizon into which clay has moved; the presence of clay films on ped surfaces and in soil pores is evidence of clay movement. The argillic horizon is 15 cm thick or at least one-tenth as thick as the sum of all overlying developed horizons. An argillic horizon has the following clay contents:

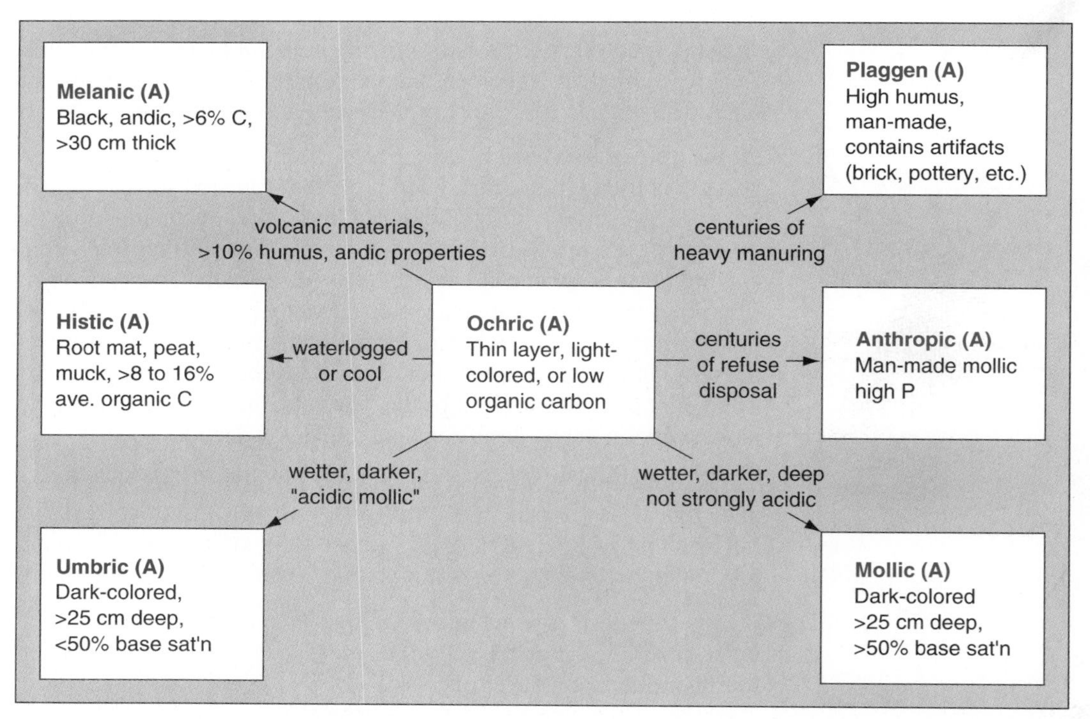

FIGURE 2-20 The relationship among epipedons (surface-formed horizons) using the least developed, thin or light-colored ochric as the central concept. (Courtesy of Raymond W. Miller, Utah State University.)

Any Part of Overlying (eluvial) Horizon Has:	*Argillic Horizon Must Have:*
Less than 15% clay	At least 3% more clay than the eluvial horizon
15%–40% clay	1.2 times more clay than the eluvial horizon
More than 40% clay	At least 8% more clay than the eluvial horizon

***Cambic horizon:** *A color or weakly developed* **B** *horizon.* This is common in weakly developed soils (Inceptisols). A subsurface horizon that has textures more clayey than a fine sandy soil (loamy fine sand) and in which materials have been altered or removed but not accumulated. Evidences of alteration include the changes caused by wetness, such as gray colors and mottling, and redistribution of carbonates; and yellower or redder colors than in the underlying horizons. A common horizon but weakly developed.

Kandic horizon: *An argillic horizon of kaolinite-like clays* (see Chapter 5 for definition of kaolinite and CEC). The presence of clay films is *not* required. [The layer has a low CEC of <16 $cmol_c/kg$ of clay and is at least 30 cm (about 1 ft) thick.]

***Natric horizon:** *Like an argillic but with a high exchangeable sodium content.* It is a subsurface horizon that is a special kind of argillic horizon with enough salt (sodium) to disperse the soil and make it less permeable. [Contains 15 percent or more of the exchangeable sodium, or SAR (see Chapter 9) if the saturation extract is 13 or higher within a 40 cm (15.6 in) depth.]

Oxic horizon: *A thoroughly weathered* **B** *horizon.* It is common in strongly weathered soils in wet, tropical areas (Oxisols). It is a subsurface horizon that is a mixture principally of kaolinite, hydrated iron and aluminum oxides, quartz, and other highly insoluble primary minerals and containing very little water-dispersible clay. It is at least 30 cm (1 ft) thick and has more than 15 percent clay of less than 16 $cmol_c/kg$ of clay for its CEC. (See Chapter 5 for definitions of clays and CEC.)

Sombric horizon: *An acidic, humus accumulation, tropical* **B** *horizon.* It is a subsurface horizon formed in well-drained mineral soils, consisting of illuvial (leached from above) humus. Basic cation saturation is low (less than 50 percent of adsorption sites are Ca, Mg, K, and Na). Formed only in tropical and subtropical climates.

***Spodic horizon:** *An acidic, cool area* **B** *horizon with an accumulation of humus and/or iron and aluminum hydrous oxides (sesquioxides).* It is common in cold, wet, sandy, evergreen forest areas where strongly acidic well-leached soils exist (Spodosols). A subsurface horizon in which amorphous materials consisting of organic matter plus compounds of aluminum, and usually iron, have accumulated. The spodic is either sandy-textured and has a >2.5-cm-thick (1-in) layer cemented by sesquioxide or humus or has an iron-plus-aluminum content divided by the clay content of 0.2.

Glossic horizon: *A degrading argillic, kandic, or natric horizon.* This has an *albic horizon* being developed in a previously *argillic, kandic,* or *natric horizon.* The glossic must be at least 5 cm thick and have 15 percent to 85 percent of it as albic materials. Usually the glossic will form between an overlying albic and an underlying argillic, kandic, or natric.

A diagrammatic illustration of relationships among diagnostic **B** (colloid accumulation) horizons is shown in Figure 2-21, but *glossic* horizons are not included.

Endopedons from Accumulations of Solubilized Substances

****Calcic horizon:** *A calcium carbonate accumulation horizon.* (Common in arid soils—Aridisols.) A subsurface horizon (or surface, if exposed by erosion) more than 15 cm (6 in) thick that has more than 15 percent calcium carbonate equivalent, at least 5 percent more carbonates than the **C** horizon, and evidence of carbonate translocation (lime-filled cracks, precipitated lime nodules, or stalagmites on undersides of rocks and stones).

Gypsic horizon: *A gypsum accumulation horizon.* A weakly cemented or noncemented subsurface horizon (or on the surface, when eroded severely) that contains a high concentration of gypsum, mostly $CaSO_4 \cdot 2H_2O$. Thickness in cm × gypsum percentage is equal to or greater than 150 cm-percent.

Salic horizon: *A soluble salt accumulation horizon.* A salic horizon, usually below the surface, at least 15 cm (6 in) thick, that contains at least 2 percent salt. Thickness in cm × salt percentage is equal to or greater than 60 cm-percent.

Sulfuric horizon: *A horizon high in sulfides.* A surface or subsurface horizon rich in sulfide minerals or high-sulfur organic matter that, when drained, oxidizes to

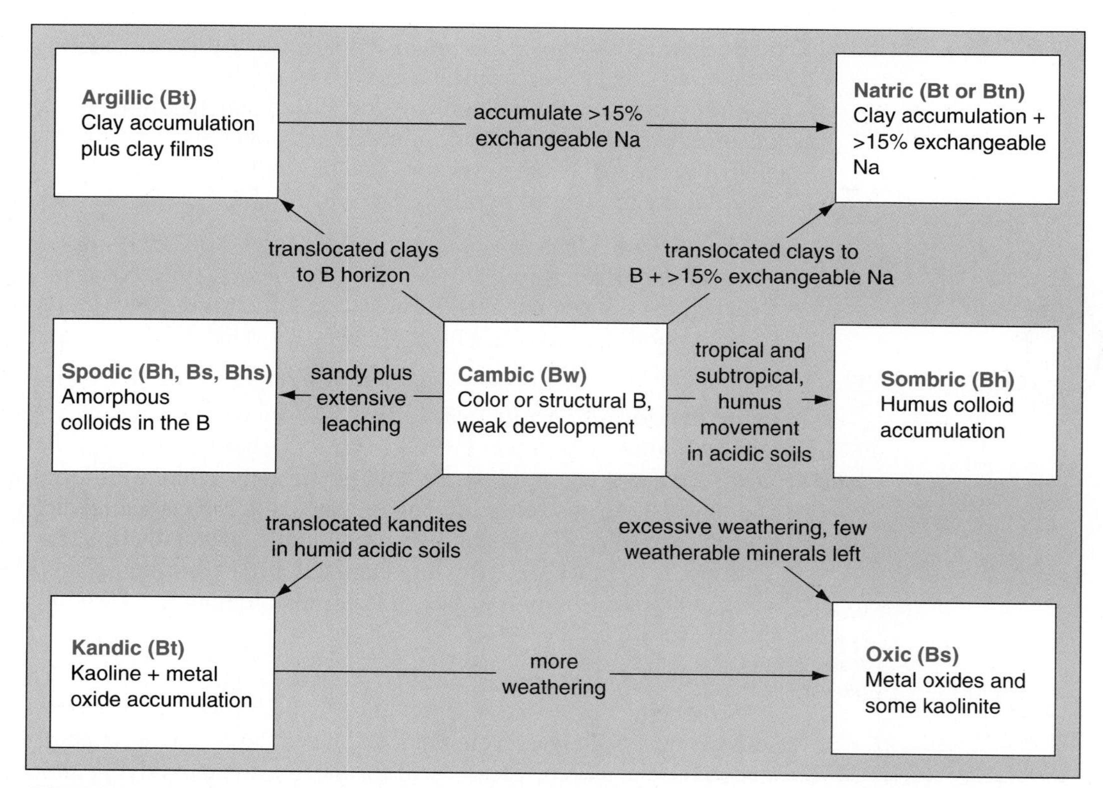

FIGURE 2-21 The **B** diagnostic horizons as they might form from the early-stage horizon, the weakly developed structural or color **B,** called a *cambic.* (Courtesy of Raymond W. Miller, Utah State University.)

sulfuric acid. Oxidized soil pH is less than 3.5 and is, therefore, toxic to most plants.

A diagrammatic illustration of relationships among diagnostic horizons formed by leaching or by precipitation of secondary minerals (soluble salts, gypsum, lime, silica, sulfides, and iron and aluminum oxides) is shown in Figure 2-22.

Hardpan Horizons

***Duripan:** *A silica-cemented hardpan usually having some carbonates.* A subsurface horizon that is cemented mostly by silica. Although carbonates may be present, duripans will not slake in water nor in 8 percent hydrochloric (HCl) acid, but they will disintegrate in hot, concentrated KOH solution or alternating acidic and basic solutions. Basic (alkaline) solutions dissolve silica.

Fragipan: *A dense, brittle, minimal-cemented hardpan.* A natural subsurface horizon with high bulk density relative to **A** and **B** horizons (the solum) above, seemingly cemented (small amounts of silica at contact points) when dry but showing a moderate-to-weak brittleness when moist. The layer is low in organic

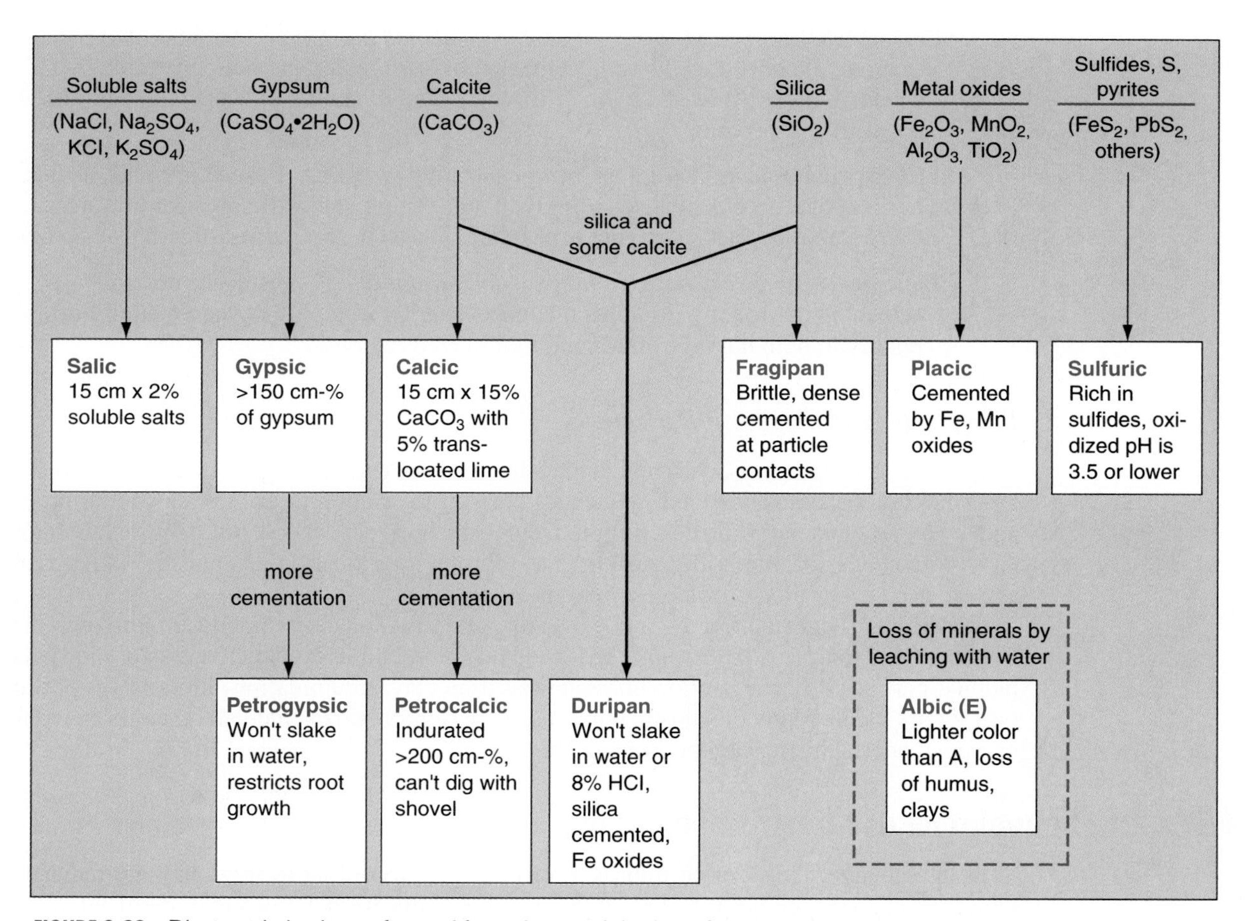

FIGURE 2-22 Diagnostic horizons formed from the precipitation of secondary minerals other than clays and by the loss of minerals by leaching. (Courtesy of Raymond W. Miller, Utah State University.)

FIGURE 2-23 Example of overlapping diagnostic horizons. Master letter horizons are mutually exclusive, but diagnostic horizons are not. As long as the defined criteria fit, a layer of soil (horizon or part of a horizon) may fit several diagnostic horizons. An example is shown in the diagram.

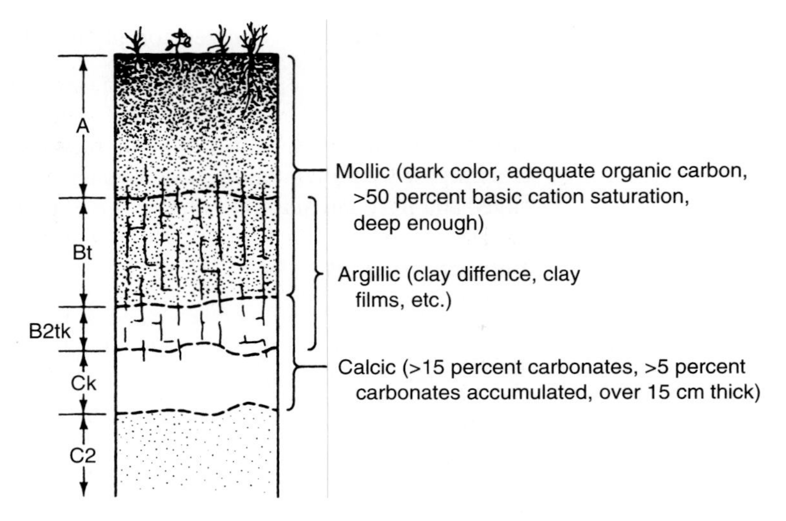

matter, mottled, slowly or very slowly permeable to water, and usually shows occasional or frequent bleached cracks, forming polygons. It may be found in profiles of either cultivated or virgin soils but not in calcareous material.

***Petrocalcic horizon:** *A hard, carbonate-cemented hardpan.* An indurated (hardened) subsurface horizon cemented by carbonates and not penetrable by spade or auger. At least 2.5 cm (1 in) thick, and thickness times percentage $CaCO_3$ is $\geq$ 200 cm-percent.

Petrogypsic horizon: *A hard, gypsum-cemented hardpan.* A surface or subsurface horizon that is cemented so strongly that dry fragments will not slake in water. Cementation restricts plant root penetration. Fits other requirements for gypsic horizon.

Placic horizon: *A very hard, iron-cemented hardpan.* A subsurface horizon cemented by iron, by iron plus manganese, or by iron plus organic matter. Forms readily in both humid tropical and humid cold regions.

Using Letter and Diagnostic Horizons

Both letter horizons (**A, E,** etc.) and diagnostic horizons (mollic, argillic, etc.) are used. The letter horizons characterize each sequential horizon in relation to each other in that profile and are a part of each profile description. Diagnostic horizons, in contrast, require laboratory data (percentages of lime, salt, gypsum, clay) for their verification. Diagnostic horizons are seldom used in detailed profile descriptions.

Although letter horizons cannot overlap, often two or more diagnostic horizons may overlap. For example, a *mollic* horizon (based on color, humus content, base saturation) may include part or all of the *argillic* horizon (based on clay accumulation) in its lower portion (see Figure 2-23). Many diagnostic horizons, by definition, do exclude all other horizons (for example, ochric, histic, cambic, spodic, fragipan).

2:9 Degradation and Destruction of Soils

Soils can change; they can be improved or damaged according to how they are managed. Often the changes are not planned by the user but result from actions that are of interest to the user in producing immediate yields.

Degradation of Soil

People refer to some soils as worn out or as degraded and unproductive. Even the term *destroyed* has been applied to soils buried under deep deposits of flood debris, volcanic ash falls, or wind-blown debris. Burned organic soils and eroded shallow soils certainly are destroyed.

In the sense that a **degraded soil** has become less good for its planned use, many soils have been degraded. The worn-out soils of parts of the southeastern United States in the middle 1800s probably suffered deficiencies of nitrogen, phosphorus, and potassium as the yearly crops of cotton and tobacco mined these nutrients from the soils faster than they were supplied by fertilizer (which was rarely applied, except as manures) and weathering. Resting the soil for a few years without harvesting crops allows some time for the buildup of soil organic materials (plant residues) to supply nitrogen and an increased accumulation of available phosphorus and potassium from mineral weathering and humus. Soils used for filtering wastewater are degraded when their permeability is reduced or their absorption or adsorption capacity for certain chemicals is exceeded.

Which Activities Degrade Soils? **Soil degradation** is *any added substance or unnatural loss or alteration that makes the soil less productive or usable than it was before the change.* This definition could include any changes in the soil—nutrient losses, compaction, or additions—that cause reduced plant growth or that make the soil less satisfactory for use. Using this definition, reduction in soil humus (the source of nitrogen and much phosphorus and sulfur) would degrade a soil. Cultivation of a virgin humid-area soil without deliberate application of organic materials usually decreases the soil humus over several decades of use to between one-third and one-half of its original amount. This loss of humus is a slow degradation of the soil, which can be partly reversed by good management (adding manure, using green manure crops, using reduced tillage, and using crop rotations). Natural salt accumulation would degrade soils.

Soil erosion degrades soils by (1) losing soil from areas eroded, (2) reducing permeability, and (3) sometimes depositing poor soil material on top of good soil. Soil erosion often removes the soil's surface, the most fertile portion of the soil. Deposition of sediment (clays, sands, or rocky materials) on a good soil may lower the quality of that soil (Figure 2-24).

FIGURE 2-24 Large plow used for special tillage needs, near Pullman, Washington. Plowing to depths of about 90 cm (about 35 in) was done to mix impermeable clay layers with coarser textures for better water and root penetration. Such a plow was also used in California to invert good soil back to the surface after it had been covered by alluvial deposits of sand. A plow of about 183 cm (about 6 ft) turned the soil about 120 cm (nearly 4 ft) deep. Such deep tillage is expensive. (*Source:* USDA—Soil Conservation Service; photo by E. E. Rowland.)

Soluble salts can hinder plant growth. Soils that were once productive can lose their structure because of high sodium contents and enough salts to limit water uptake. Nonsalty soils can become salty as irrigation water (with its soluble salt) is applied to them year after year. Some soils do not become salty because there is enough water leaching through them to keep the salts washed out. A soil is not completely destroyed if the salt and sodium can be washed out to help reclaim the soil.

Which Other Additions to Soils May Degrade Them? When detrimental materials are added to soils, plants may not grow well or may be poor food. Such chemicals as waste oils, gasoline, and herbicides reduce plant growth. Toxic levels of boron (often accompanying other soluble salts), selenium, heavy metals (from sewage sludge or ore-smelting waste tailings), radioactive fallout elements (strontium, cesium, iodine), and many other materials can degrade the soil. This degradation may cause a direct reduction in plant growth or leave an unacceptable residue (such as heavy metals or pesticides) in the plant when it is used as a food.

Complete Destruction of Soils

The most obvious destruction of soils occurs in shallow soils, some over rock or hardpans, that (1) are organic, such as peat, and are burned away; or (2) are on steep mountain slopes and are washed away or slip away. When these soils are gone, they have been destroyed.

Contaminated soils that can no longer grow plants are essentially destroyed if their reclamation is excessively costly or impractical. Some salty soils in low depressions are not easily drained to allow removal of the inhibiting salts. To install tile drains and pump water up out of a depression can be economically impractical. Heavy loadings of some toxic heavy metals (chromium, cadmium, nickel, selenium, lead, mercury) by large applications of sewage sludge or other materials may be impossible to remove because heavy metals do not easily leach from soils. Such contaminated soils are *effectively destroyed,* even though reclamation may be possible sometime in the future.

Some people are so accustomed to getting everything they need from supermarkets that they forget that foods originate on the land or in the sea. To survive we must respect those sources and resist destroying or degrading them. What constitutes a good soil? What needs to be done to retain or build better soils? These are questions about which everyone should have some knowledge or concept. As Winston Churchill said, "To build may have to be the slow and laborious task of years. To destroy can be the thoughtless act of a single day."

2:10 Assessing the Quality of Soils

The need for environmental and agricultural sustainability has dramatically redefined what is meant by *soil quality.* **Soil quality** is the suitability of the soil for the use planned. Some people refer to a soil's quality as the **soil's health.** A good-quality soil for making a roadbed (should have gravel, rock, not much clay) may be a poor-quality soil for growing crops. A good-quality soil for cleansing wastewater must be permeable yet able to absorb to its particles' surfaces many pollutants, such as heavy metals; soluble, toxic, organic chemicals, and so on.

For production of many common plants, soil is of high quality when it has relatively high soil organic matter, has obvious active soil organisms (many earthworms, bacteria, fungi, etc.), has slight soil erosion, is easy to till, has good drainage, and has suitable pH (soil acidity). These are only a few of the properties influencing a soil's quality. In a recent survey of farmers about what a healthy soil is, they described healthy soils as *"loose, soft, crumbly, flexible, mellow, dark-colored, and loamy.*"[6] Level land isn't necessarily the best soil; water needs a place to drain away. A variety of plants and crop rotations help improve a soil's health.

Soils of good health on which to recycle wastes need active microorganisms to decompose organic materials to make room for additional applications of the wastes. Some soils must have strong adsorption characteristics to become long-term storage and inactivation reservoirs of other pollutant wastes, such as heavy metals.

Soil quality needs to be defined in terms of the various functions that soils perform in the ecosystems.[7] Soil is a great *buffer.* Soils modify extremes of temperature, relative humidity, and concentrations of toxic and beneficial chemicals, allowing organisms to adapt to changes. Soils *store* water and nutrients. It is the Earth's soils that *partition energy* that drives circulation of air over land masses.

The challenge in evaluating soil quality is to devise a system that can be adapted to different values and uses without being biased toward values associated with any single use. Such evaluations require that observations and measurements be done on the soil to various depths. The soil and its quality is not determined by the land's surface, but rather by its *volume.* Soils have depths and can change in numerous properties with changes in depth. These layers that change with depth are the soil horizons—collectively called a soil profile—in a given soil location.

One list of basic soil properties to use as indicators for screening soil quality and health and to sustain the quality of the environment includes the following[8]:

Physical properties. (1) Soil texture; (2) depth of topsoil, soil profile and rooting; (3) infiltration and bulk density; and (4) water-holding capacity.

Chemical properties. (5) Soil organic matter, including total nitrogen; (6) soil pH; (7) soil salinity; and (8) extractable (available) N, P, and K.

Biological properties. (9) Microbial biomass C and N; (10) potentially mineralizable N; and (11) soil respiration, water content, and temperature.

Obviously, most of these eleven properties are slanted toward the growth of plants and maintenance of the ecosystem. Some are interrelated, such as texture and water content.

2:11 Soil Individuals and Mapping Units

A profile makes visible the soil horizons and their thickness, the profile depth, sandiness or clayey nature of horizons, and other features of the soil *at that slice.* However, a second slice of soil taken just a few feet away—another profile—would likely appear a bit different than the first profile in exact horizon colors, horizon thicknesses, or other properties.

Pedons and Polypedons

To minimize the frustration in encountering constantly changing soil properties with very short distances, a soil individual was defined, and called a *pedon.* A **pedon** is the smallest

[6] E. E. Romig, M. J. Garlynd, R. E. Harris, and K. McSweeny, "How Farmers Assess Soil Health and Quality," *Journal of Soil and Water Conservation* **50** (1995), No. 3, 229–236.

[7] B. P. Warkentin, "The Changing Concept of Soil Quality," *Journal of Soil and Water Conservation* **50** (1995), No. 3, 226–228.

[8] J. W. Doran, M. Sarrantonio, and M. A. Liebig, "Soil Health and Sustainability," *Advances in Agronomy* **56,** 1996, 1–54.

volume that can be called a soil individual. It has three dimensions. It extends downward to the depth of plant roots or to the lower limit of the genetic (naturally-formed) soil horizons. Its lateral cross-section is roughly hexagonal and ranges from 1 m to 10 m^2 in size, depending on the variability in the horizons with distance (faster changes with distance mean that only a smaller volume can be a soil individual).

Adjacent pedons may be slightly different but still be similar enough to fit within the most detailed description listed for a soil by the soil surveyor—a **soil series** (see Chapter 7). Two or more *contiguous pedons,* all of which are within the defined limits as having the same horizons of similar thicknesses, same pH, same humus contents, same color, same clay content, and from the same parent material, are called **polypedons** (many pedons). Sometimes parent materials, slopes, plant cover, and other features change so rapidly across a distance on land that many different polypedons exist in an area of the land's surface barely large enough to separate on a map. Thus, maps may separate areas in which each delineated area is a single polypedon; another area may contain several *different* polypedons (different soil series).

These gradual, even abrupt, changes in soils as one traverses an area is one concern of soil mappers. Compare the problem of separating a land area into different soils with the one of separating visible light wavelengths into colors. Visible light has continuously lengthening wavelengths from yellows to greens to blues to reds. Where does one put the dividing lines on the wavelengths to say that these wavelengths are yellow light, these are green, and those are blue? In this same manner soil mappers must make arbitrary separations between adjacent but slightly different soils, because usually the changes in soils occur gradually over a distance. The mappers must define the different mapping units to avoid undue confusion.

Changes Needed in Mapping

Traditionally, soil surveys were produced for a nontechnical audience. Environmental awareness now develops a need for more precise inventories and data for many persons, including numerous technical users. This means that broad generalized data is no longer adequate. Local specific data is needed. Previously, when a soil survey party finished an area and were relocated to a new survey, much of the details and knowledge of the area was in the surveyors' minds. As they left the area, much of the details of the area were, in essence, abandoned and lost for later users. The scale of the maps were usually 1:15,840 or 1:24,000. These scales did not allow detailed soil separations, and they restricted mappers' abilities to report some of the site-specific soil variability and other details they had observed.

In the 1960s the United States Department of Agriculture (USDA) divided the United States into **Land Resource Regions (LRRs)** and **Major Land Resource Areas (MLRAs).** There are 24 LRRs and 212 MLRAs in the United States. An **MLRA** *is a geographically associated area that has a common pattern of soils, climate, water resources, and land uses.* Some of the significant improvements claimed for recent MLRA soil surveys include the following[9,10]:

- Information is based on natural rather than political boundaries.
- The MLRA has *one* legend instead of a different one for each small survey area, which occurred in older soil surveys.

---

[9]R. L. McLeese, "Updating and Maintaining the Soil Survey in Illinois: A Major Land Resource Area Approach," *Illinois GIS & Mapnotes* **11** (1992), 36–38.

[10]R. L. McLeese, S. J. Indorante, D. R. Grantham, D. E. Calsyn, and D. J. Fehrenbacher, "Soil Survey Updates by Major Land Resource Areas or Illinois Soil Surveys: The Next Generation," *Agronomy Abstracts,* American Society of Agronomy, Madison, WI, 1991.

Detail 2-2 Proposed MLRA Soil Survey Organization*

"We advocate a way to provide soils information to customers that is current, readily accessible, more easily understood by users, and can be easily updated—in short the update and maintenance of digitized soil surveys by Major Land Resource Areas." Some of the details are as follows:

- Using the MLRAs as the survey study areas will effectively reduce the number of soil survey areas from more than 3000 to 212. The MLRA areas would range in size from 110 square miles (MLRA #165 is "Subhumid Intermediate Mountain Slopes") to 110,000 square miles (MLRA #133A is "Southern Coastal Plain"). Within each MLRA there would be smaller units as needed.
- There would be *permanent* MLRA field soil survey offices. The MLRA *update staff* would systematically acquire data necessary to refine and update soil surveys data already available. A new emphasis would be to (1) gather information, (2) interpret soil-landscape behavior, and (3) interact with users. The soil scientist would become an interpreter and teacher, refining and sharing personal knowledge with colleagues and users. The update office should become the soils information and technology center of the region. Scientists working at these offices would be permanent, not moving on as each survey area was finished.
- The information heart of the MLRA update office would be a *geographic information system* (**GIS**), which would include digital soil maps, other digital soil information, and additional natural resource information. Some of these other digital information models would include *digital elevation models* (**DEM**), *global position services* (**GPS**), **LANDSAT** and **SPOT** photographic imagery, as well as low-level aerial photography (infrared and false color).
- A 20-year program to get to an up-to-date condition is proposed. Users would be expected to share in the cost of this updating/digitizing cost. Cost sharing methods are yet to be developed.

*S. J. Indorante, R. L. McLeese, R. D. Hammer, B. W. Thompson, and D. L. Alexander, "Positioning Soil Survey for the 21st Century," *Journal of Soil and Water Conservation* **51** (1996), No. 1, 21–28; and K. J. McSweeney, R. L. Bell, S. J. Indorante, and C. Love, "Soil Survey by MLRA: Positioning the NCSS for the 21st Century," Report to the 1993 National Cooperative Soil Survey Conference, 1993.

- Map scale, mapping intensity, data recorded, and documentation are uniform.
- Statements about the expected reliability of maps and interpretations are more precise.
- A digital soils data layer is provided within a GIS with spatial and attribute data that meet National Map Accuracy Standards. The digital data will be available to all users.
- Digitized soil interpretations can be easily analyzed and updated by entries into the database.

In 1996 a proposal was made to structure these LRRs and MLRAs into an organization to modernize and keep current survey data and related details for the entire United States (see Detail 2-2).[11]

[11]J. C. Antenucci, K. Brown, P. L. Croswell, M. J. Kevany, and H. Archer, *Geographic Information Systems—A Guide to the Technology,* Van Nostrand Reinhold, New York, 1991.

Modernization: Spatial Variation and Databases

Modernization of mapping really means to increase the involvement of the computer and some of the innovative software being made available to store and manipulate data. Models are developed to predict results from the interactions of data values of many **attributes** (properties). Nearly all of these attributes change from location to location and, thus, involve *spatial variability.*

Spatial variability *is the change in a soil property with distance in a given direction.* Spatial variability will occur in most attributes in natural systems. For example, changes in *nutrient contents, clay contents, water permeability, erosion hazard, organic matter content,* and in almost anything else occur with distance from any point toward some direction. Sometimes the changes are small; thus, an area can be considered to be the same because changes are small. When maps are made, these all-the-same areas are separated out (**delineated**) from different areas that are not quite similar in attributes.

The preparation of maps by hand is arduous and very time consuming. Aerial photographs of the soil area are often used on which to draw the map to enable locating exactly the position for each point of the delineating line drawn to separate different mapping units. The work of penciling in areas and then drafting the map in ink is tedious and final. The finished map can be photocopied, but it is difficult and wearisome to modify the finished map very much. Overlays can be traced to make new delineated areas to show such features as salty areas, poorly drained areas, flood hazard areas, and so on (see Chapter 20).

The large collection of data to make these maps of different properties of the same area or large lists of items, such as addresses or customers, are referred to as **databases** of that area or group. A particular database may be for a field, valley, watershed, county, or even larger areas. But, the enormous amount of labor and work to tabulate and maintain these databases is costly and, at times, frustrating.

2:12 Geographical Information Systems (GISs)

Fortunately, an essential tool has been developed to help organize and store data—the personal computer. Even more recently, and involving the personal computer, **geographical information systems (GISs)** software allows enormous data storage, retrieval, and manipulation with remarkable ease and capacity. GIS is bringing about one of the greatest technological revolutions in the country, if not the world.

What Is a GIS?

What are GISs? Because of the complexity of the systems, one author listed eleven "selected definitions."[12] One of the simplest definitions of a GIS is "A computer system capable of holding and using data describing places on the earth's surface."[13] It is an information integrator. One user said, ". . . it should not be shackled by strict definitions. We are only beginning to see its potential; let's not limit its growth."[14]

---

[12] D. J. Maquire, "An Overview and Definition of GIS," Chapter 1, p. 9–20 *in* D. J. Maquire, M. F. Goodchild, and D. W. Rhind (eds.), "Geographical Information Systems: Principles and Applications," Longman Scientific & Technical (John Wiley & Sons, New York), 1991.

[13] Environmental Systems Research Institute, Inc. "Understanding GIS," John Wiley & Sons, New York, 3rd ed., 1995, 1–2.

[14] Chris Harlow, "What the Heck Is GIS, Anyway?" *Professional Surveyor,* (1993) Mar–Apr, 24, 25, 28–30.

Some of the kinds of data include *hydrology, topography, land use, soils, utilities, streets, vegetation, fertility, soluble salts,* and *rainfall averages.* A more elaborate definition would include any information management system which can do these things[15]:

- Collect, store, and retrieve information based on its spatial location
- Identify locations within a selected environment that meet specific criteria
- Explore relationships among data sets within that environment
- Analyze the related data spatially as an aid to making decisions about that environment
- Facilitate selecting and passing data to application-specific analytical models that can assess the impact of alternatives on the chosen environment
- Display the selected environment both graphically and numerically either before or after analysis

What a GIS Can Do

A GIS will make and display databases (maps, lists, etc.) from data entered into the program. Redelineating areas of a map for different properties becomes quite a bit easier to do, even in color, and has only the limitation of factual data and one's imagination in resorting the area under study. Visual displays and printouts are easily obtained. Collecting valid data, using valid calculations, and interpreting the maps and values obtained are still mostly done by hard work, intelligent thought, and careful evaluation of the process and its product.

Three views about a GIS might include the following: a GIS (1) can produce maps, (2) will include one or more databases, and (3) can allow spatial analysis.

The first view, that of the map view advocates, visualizes the GIS as a map (cartographic) processing or display system. It should be evident that no single map can show all attributes. Additional maps, easily done with GIS, may conceivably be required by a user (Detail 2-3).

The second view, that of GISs as databases, considers GISs as techniques to group data placed within a specific management system for a particular use, user, system, or program. All uses of the GIS are based on data entered into the system or already accumulated and entered (Detail 2-4). These data collections are called **databases.**

The third view emphasizes spatial analysis. **Spatial analysis** indicates that *there are changes with space or location within an area.* This concept can be related in four ways. First, a **point** (mountain peak, intersection, isolated building) exists in relation to other parts of the area. Second, a **line** (length of road, shortest route between points) is related to other parts of the area. Third, an **area (or polygon),** such as a lake, an aspen grove, or a landslide area, can be related to the larger total area. Fourth, a **three-dimensional surface** (slope of a hill, relief, volume of a lake) exists relative to other features in the area. One scientist stated that mapped and related statistical data form the greatest storehouse of knowledge about the conditions of the living space of mankind.

With the use of a GIS, questions such as the following can be asked of the data:

- What are the engineering information needs for planning, designing, and constructing roads, bridges, buildings, and waterways?
- What is the distance between points, the size of an area, or the volume of a lake?

---

[15]Maquire, 1991, ibid.

Detail 2-3 City of Long Beach Uses GIS*

Besides boasting 15 miles of ocean beaches, home of the ocean liner Queen Mary, and a Grand Prix automobile race, Long Beach has substantial oil properties and a maze of underground pipelines. It is a mapping nightmare. In the early 1980s there were more than 4000 maps of stormdrains, sanitary sewers, oil pipelines, subdivision records, street lighting, zoning, drainage, and house numbering, among others. When a serious pipeline rupture was followed by an explosion in the early 1980s, there was a major problem in identifying the pipeline and its owner to get the pipe shut off. The City reevaluated its situation.

A full GIS was begun in 1983. Digital maps were prepared of the roadways, driveways, sidewalks, building outlines, contours, railroads, trees, and spot heights, among other items. The detailed database includes 25 attributes of each pipeline section (emergency contacts, ownership, surrounding soil condition and depths, plus pipe material, diameter, thickness, and flow direction). The primary database had 78 layers by the early 1990s and is still growing.

*John C. Antenucci, Kay Brown, Peter L. Croswell, Michael J. Kevany, and Hugh Archer, *Geographic Information Systems*, Van Nostrand Reinhold, New York, 1991.

- How does one evaluate storm water flows, assess property values, and measure salt or pesticide movement in runoff water?
- How does one process polygons (areas), such as soil patterns, erosion patterns, and flood plain areas?
- How can one determine population distributions, densities, and patterns of people's ages within an area?
- How can one determine where to site certain structures (sewage plants, landfill), based upon certain soils, distances from the city, elevation, and other restrictions?

With questions such as, "What is at . . . ? (location)," "Where is it? (converse of 'What is at . . . ?')," "What has changed since. . . ? (trends)," "What spatial patterns exist? (patterns)," and "What if . . . happens? (modeling to predict changes)," you need a lot of data. A GIS is an aid to perform all these operations by its use of geography (space) and the *linking* of various sets of data. **Linkage** is the ability to connect to and draw from more than one set of data (databases). For example, suppose one has a database with soil survey data (delineations of soils, kinds of soils, topography) and wants to see if another database (nitrogen and phosphorus contents in the same area) has any relationship to the soil survey data. Linkage by the GIS allows a study of this relationship. Linking the various files is like using several transparent overlays, each with one kind of data (vegetation pattern, kinds of soils, nitrogen contents, etc.). It would be possible to look down through all of the transparent overlays and see the relationships of areas of one attribute to areas of the other attributes being considered (Figure 2-25).

GISs and Precision Farming

Precision farming involves managing smaller and smaller portions of a field differently. The use of a GIS is critical to know where the boundaries of each soil area are, how the areas are

Detail 2-4 A TIGER on the Loose*

While preparing for the 1990 U.S. census, the Bureau of the Census and the U.S. Geological Survey cooperated in developing a major new database, **TIGER.** This Topological Integrated Geographic Encoding and Referencing (TIGER) system developed the first comprehensive digital street map of the United States, which covers 100 percent of the U.S. land area. Every street and road in the nation is listed. Address numbers are grouped along sections of every street in the 345 largest urban areas. All railroads and hydrographic features, Indian reservations, military bases, and other census-important special data are included. Besides the 50 states, the District of Columbia, Virgin Islands, Guam, Puerto Rico, American Samoa, and Northern Marina Islands are covered.

TIGER files will also be used by the U.S. government to increase efficiency of moving things and people, an activity that involves a half-trillion dollars a year. Efficiency typically improves from about 5 percent to 15 percent. Five percent of a half-trillion dollars is still about 25 billion dollars, a good return for the $182 million investment by the government.

*John C. Antenucci, Kay Brown, Peter L. Croswell, Michael J. Kevany, and Hugh Archer, *Geographic Information Systems,* Van Nostrand Reinhold, New York, 1991; and "GDT News," 1988, Summer.

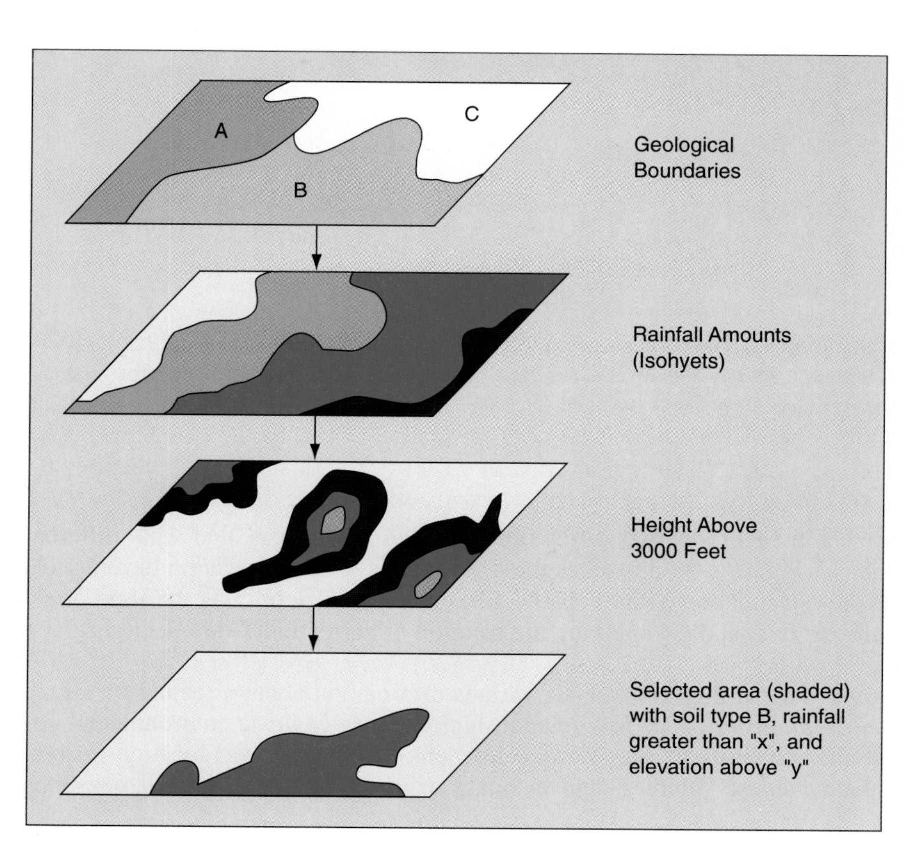

FIGURE 2-25 Conceptual view of a geographical information system (GIS). Each of the three maps would be an attribute (or property) of the area and would be a part of the database or separate databases. Each map would overlay each other digitally as one might overlay transparencies of physical maps. (Modified from Paul M. Mather, *Computer Applications in Geography,* John Wiley & Sons, New York, 1991, 207.)

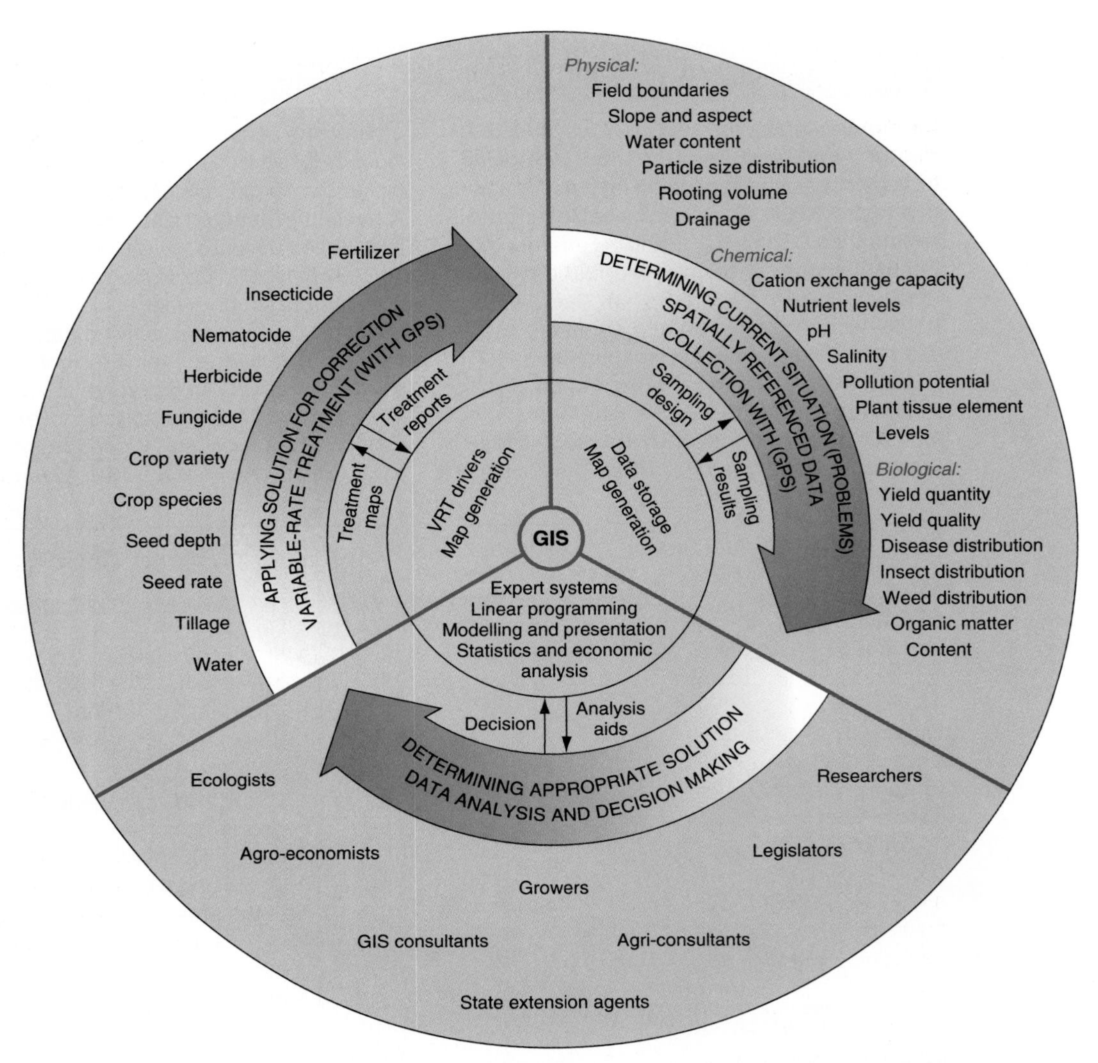

FIGURE 2-26 The components of precision farming showing the GIS as a central hub. (Courtesy of E. Lynn Usery, Stuart Pocknee, and Broughton Boydell, "Precision Farming Data Management Using Geographic Information Systems," *Photogrammetric Engineering & Remote Sensing* **61** (1995), No. 11, 1383–1391, University of Georgia.)

distributed, how one area's properties differ from those of other areas, and what different treatment is planned. Figure 2-26 illustrates the three phases of such precision farming: (1) determining the present situation with its spatial differences, (2) determining the appropriate solution for each soil area, and (3) applying the solution or correction. More detail is given in Chapter 13.

A GIS provides the ability to associate various information about a feature on a map and to create *new relationships.* The new relationships can help evaluate environmental impacts, determine the suitability of sites for development, identify the best location for certain uses, calculate harvest volumes, and evaluate many other spatial-related questions (Detail 2-5).

Detail 2-5 More Examples of GIS Uses*

Example 1. The United Nations Environment Programme (UNEP) was established in 1983 to coordinate global environmental management. One of the first hazards studied was **desertification.** Some of the data needed for the database were soil status, vulnerability of land to desertification, and animal and human population pressures.

The UNEP also began the Global Resource Information Database (GRID). The pilot project selected in 1987 was to develop a national environmental database for Uganda. The database helped model crop suitability and erosion potential.

Example 2. The South East Regional Research Laboratory (SERRL) at the University of London is primarily charged to the building and maintenance of an integrated settlement and infrastructure database for South East England. The database has twelve layers of cartographic data. The main layers are (1) settlement, (2) transport networks, (3) administrative boundaries, and (4) planning areas (green belts, areas of outstanding beauty, sites of special scientific interest, etc.).

Some uses already made of the database are as follows: (1) to evaluate accessibility of small- and medium-sized towns to the motorway and trunk road network; (2) updating urban/rural boundaries with remotely sensed data.

Example 3—A Planning Approach. Suppose that an East-coast state in the United States with several large population centers needs an assessment of potential sites for coal-fired electric generating stations. Help locate sites. These are some of the 15 to 25 criteria that might govern a suitable location:

- Cooling water will cause minimal disturbance to aquatic system nearby
- Sulfur-bearing smoke will not pass over urban areas nearby
- Scenic and recreation areas will not be degraded
- Station must be close enough to population centers to ensure labor supply
- Not built on high-grade (prime) agricultural land
- Transportation costs for fuel will be minimized
- Site should be flat and involve minimum earth moving
- Foundation material must be strong enough to support structures
- Access to site must not involve special difficulties (hills, rivers)
- Site must be accessible during all months of the year

A database with soils characteristics, rivers, land relief, distances, population distribution, weather (wind) directions, scenic areas, water flow volumes, aquaculture information, and much more is needed. Although costly to collect into a database, that database is then constantly available for future planning needs and for updating.

*Paul M. Mather, *Computer Applications in Geography,* John Wiley & Sons, New York, 1991.

Four Basic Elements of a GIS

The four basic elements of a GIS are (1) the computer *hardware* (the computer, other reading equipment), (2) the computer *software* (CD ROM disk or other programmed systems), (3) the *data* (databases from federal, state, or other sources, probably including some new data of your own), and (4) some *liveware* (the people who will be using the program and making decisions and observations about the new information). All of these are important. For example, *good data* must exist before it can be examined accurately and correctly analyzed in terms of its patterns and implications. A dictum in computer circles is, "Garbage in, garbage out." All parts of the GISs are important—an adequate computer, software to perform the desired functions, good and sufficient data, and competent people to use the available materials and equipment.

2:13 Global Positioning Systems (GPSs)

It is sometimes desirable to know exactly where one is in relation to the area around the site. Once the area has been separated into small units that differ from each other, where are they physically when you are in the field (Figure 2-27)? One way of locating where you are is by using widely dispersed survey markers tediously established from selected longitudinal and latitudinal established reference points, each reference point started without much reference to others on the Earth. As widely separated areas developed and surveys extended toward each other, the surveyors—approaching from separate, distant reference points—often did not merge with each others' lines. Now a new system, more universal and continuous worldwide, is available to make site location and repeat location precise and well referenced with other sites.

The Global Positioning System (GPS)

The **Global Positioning System** (**GPS**) is a $12 billion U.S. Defense Department constellation of 24 satellites positioned in very high orbits (11,000 mi) and circling the Earth twice daily. At any given time any place on Earth can have a line of sight to at least four satellites. Using trilateration, a minimum of three signals will each produce a surface, whose overlap on each other will locate the site in question, giving both longitude and latitude. Using a fourth satellite signal allows calculation of the altitude of the site.

How the GPS Works

Receivers get the signals from three or four satellites and measure the distance to the satellite by the *nanoseconds* of time it takes the returning radio signal to reach the receiver (186,000 mi/sec). With accurate atomic clocks in the satellites (accurate to 1 second per 30 years), the receiver then knows the distance of each satellite from the receiver. Using trilateration from the group of satellites, the computer calculates the receiver's position. Positions are updated as often as every second so that the receiver can compute navigational information such as bearing and speed. Data is always available from the satellites, even in stormy weather and for any location on Earth (Figure 2-28). However, the signals can be blocked by tall buildings, partially by dense forests, and by hills or other physical obstructions to the radio beams near the horizon.

FIGURE 2-27 Could you go out each year on these rolling Palouse loess hills (volcanic ash and wind-blown silt from southwestern Columbia basin) of eastern Washington and find the same spots each year? On hills such as these the soil will be different on the ridges, on the slopes, and in the bottoms; you would need to treat each differently. The global positioning system (GPS) would always identify your location within a few meters. If you set up well-marked reference points ahead of time (at least four), the accuracy can be known within a few centimeters. (Courtesy of USDA—Office of Governmental and Public Affairs; photo by Doug Wilson.)

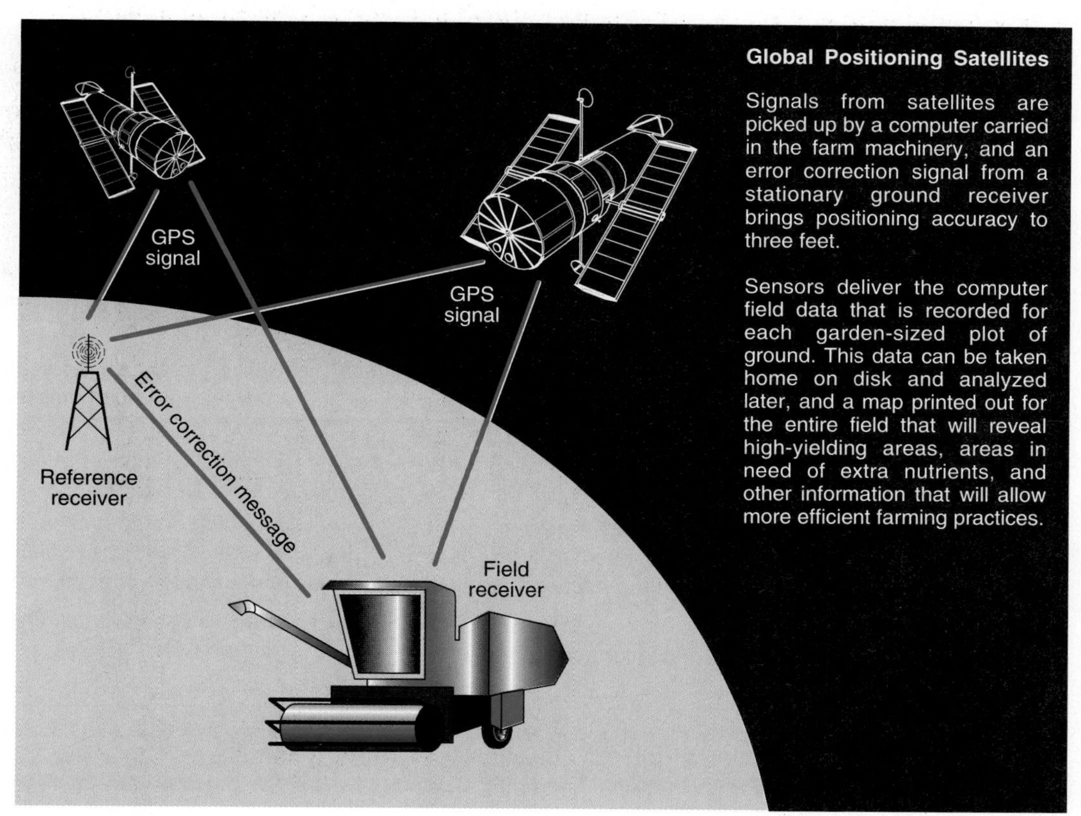

GPS signal
GPS signal
Error correction message
Reference receiver
Field receiver
Global Positioning Satellites
Signals from satellites are picked up by a computer carried in the farm machinery, and an error correction signal from a stationary ground receiver brings positioning accuracy to three feet.
Sensors deliver the computer field data that is recorded for each garden-sized plot of ground. This data can be taken home on disk and analyzed later, and a map printed out for the entire field that will reveal high-yielding areas, areas in need of extra nutrients, and other information that will allow more efficient farming practices.

FIGURE 2-28 An illustration of how the global positioning system works. A reference and stationary receiver provides stability and accuracy to improve the precision of a moving field receiver, such as a grain combine. (Courtesy of Jerry Leslie, "SDSU Research Aims at Making Precision Farming Pay Off," *South Dakota Farm and Home Research* **47** (1996), No. 3, 18–19.)

Accuracy of the GPS

Originally the system was intended for military use, and the Defense Department expected to be able to locate a site within 6 m to 8 m of its true location. To reduce the accuracy for enemy countries using the system so accurately, the signals are deliberately distorted, allowing accuracy to the public of about 25 m to 80 m of the true location. Users, however, have found that by using two receivers in conjunction, accuracies to within a few centimeters are possible with **differential GPS (DGPS)**.[16] In a study of precision and accuracy, others have observed that the actual distance from *true* may average 32 m to 39 m for 95 percent of measurements.[17] With 300 sequential fixes averaged (requiring just a few minutes), the mean distance was 25 m to 27 m. If *differential correction* was used, both precision (reproducibility) and accuracy (true value) were improved to 5 m to 7 m from the *true* value (95 percent were within 10 m to 15 m). With 300 replications, mean distances were less than 3 m from the true location. Manufacturers often advertise 1 m to 5 m accuracies. Location accuracies

[16]Anonymous, "Global Positioning System," *Compressed Air Magazine* **99** (No. 3), 1994, 38–41.
[17]Peter August, Joanne Michaud, Charles Logash, and Christopher Smith, "GPS for Environmental Applications: Accuracy and Precision of Locational Data," *Photogrammetric Engineering & Remote Sensing* **60** (No. 1), 1994, 41–45.

Detail 2-6 GPS—From Space Wars to Precision Geography*

Some of the uses made of Global Positioning Systems (GPS) range from critical uses to conveniences. Critical uses include the following:

- The improvement of airline guidance systems (will replace some radar at great cost reduction) to provide more accurate movement by providing exact latitude, longitude, altitude, and time. GPS will be especially useful in transatlantic flights when few land-based navigational aides are available. It is expected that it will allow planes to take off in zero visibility.
- Emergency medical teams can have their destinations pinpointed and their best routes outlined on dashboard maps.
- Stranded boaters on open seas can pinpoint their locations precisely.

Some less critical but just as commercially important to the public are the following:

- Surveyors can locate benchmarks for reference points better than previously.
- Detection of movements of only a few millimeters on an earthquake fault is believed possible.
- Delivery companies and rental car owners can develop automobile navigation systems that have dashboard maps indicating location and suggested or mandated routes.
- Mapping boundaries precisely of widely scattered sections of National Forest has already been done in Michigan.
- GPS is used to track animal herds in behavioral studies.

*Anonymous, "Global Positioning System," *Compressed Air Magazine* **99** (1994), No. 3, 38–41; and George V. Kinal, "Where on Earth Am I?", *BYTE,* February 1996, 175–178.

better than 1 m, even to 25 cm to 10 cm are claimed.[18,19] Individual GPS units weighing 1 kg to 2 kg advertise accuracies of 1 m to 5 m down to ±15 cm, with options of ±1.2 cm and ±0.4 cm.

A study to determine accuracy in North American forested areas listed accuracies of 6.6 m to 4.4 m under conifers and 3.9 m to 2.2 m under open skies.[20] To get better than about 15 m to 20 m accuracy, the ionospheric distortions must be eliminated. This is done primarily by the use of a differential GPS (DGPS) to get within a meter or less. This error is mostly from clock errors.[21]

The errors are relatively small and DGPS receivers can locate you within a few feet of the true location while you are moving. The errors are virtually beyond the control of GPS users. Errors come from (1) inaccurate orbital information, (2) errors in the satellite clocks—slight instability, (3) errors in the receiver clocks—large differences possible in various receivers, (4) ionospheric refraction that causes the GPS signal to bend and slow down *unpredictably,* and (5) obstructions (buildings, forests, mountains, any physical structures) that partially block the signal from the satellites.[22]

[18] Michael Dyment, "The Renaissance of the Surveying and Mapping Profession," *Professional Surveyor* **13** (1993), No. 5, Jul–Aug, 30–33.
[19] E. Lynn Usery, Stuart Pocknee, and Broughton Boydell, "Precision Farming Data Management Using Geographic Information Systems," *Photogrammetric Engineering & Remote Sensing* **61** (1995), No. 11, 1383–1391.
[20] Christopher Deckert and Paul V. Bolstad, "Forest Canopy, Terrain, and Distance Effects on Global Positioning System Point Accuracy," *Photogrammetric Engineering & Remote Sensing* **62** (1996), No. 3, 317–321.
[21] Reg Souleyrette, "Issues in Implementing GIS and GPS," *Professional Surveyor* **15** (1995), No. 7, 28–30.
[22] James Ferguson, "What Is It We Obtain from GPS?" *Professional Surveyor* **13** (1993), No. 4, 30–33.

How Valuable Is the GPS?

One commercial navigation company termed the GPS the *next utility,* becoming as basic to people as the telephone. It will affect most people in *some* way. It was used in the Gulf War so that troops and vehicles knew at any given moment where they were and where they needed to move to. In ocean areas, where few landmarks are available, the GPS enables ships and submarines to make accurate position fixes. Some of the other present and potential uses are given in Detail 2-6.

A unit installed on a car can make fairly accurate surveys of roads or railway alignments, even when traveling at 100 km/hr (62 mph). Costs for this information could be much lower and the convenience much greater than from using conventional means, at least so long as the Defense Department operates the satellite system without charge. GPS use will escalate in such areas as (1) tracking vehicles and shipments, (2) locating sites of equipment breakdowns or routes for emergency response, (3) resource management, (4) the military, (5) hunters, (6) hikers, and (7) well drillers. Fishermen in the Gulf of Mexico use the GPS regularly. Already, with high-resolution (high-priced) units, accuracy is comparable to the best conventional mapping. Low-resolution units look like hand-held calculators and cost less than $300.

> *There is nothing more difficult to take in hand, more perilous to conduct, or more uncertain in its success, than to take the lead in the introduction of a new order of things.*
>
> **—Nicolo Machiavelli,** *The Prince*

> *Rocks weathered to soil . . . are the ores which the living world mines for the elements that go into the make-up of palm and pine, of rice and redwood, of mice and men.*
>
> **—E. Epstein**

Questions

1. (a) What is a soil profile? (b) What is changed in a soil profile as the soil develops?
2. What is (are) the likely soil texture(s) of soils moderately developed from (a) ground moraine? (b) loess? (c) flat lowland river terraces? (d) glacial moraine outwash? (e) soil creep materials? and (f) dune landforms?
3. (a) Define a *landform.* (b) Define a *soil horizon.*
4. Describe in simple terms (a) an **A** horizon, (b) an **R** horizon, (c) a **B** horizon, and (d) an **E** horizon.
5. Describe some of the characteristics of a petrocalcic horizon.
6. Much amorphous silica in soils is quite soluble in alkaline solution. How can this help explain the presence of duripans below about a meter (about 39 in) in many arid-region soils?
7. What materials cement mineral particles together in various horizons?
8. Briefly define the following horizons: (a) mollic, (b) albic, (c) argillic, (d) calcic, (e) ochric, (f) oxic, and (g) duripan.
9. What letter horizon would likely be used for each of the following diagnostic horizons: (a) mollic, (b) argillic, (c) calcic, (d) umbric, (e) albic, (f) duripan, (g) ochric,

(h) natric, and (i) oxic? Keep in mind, some diagnostic layers may be in more than one letter horizon.

10. Describe, generally, the following landforms: (a) plateau, (b) alluvial fan, (c) delta, (d) river terrace, (e) floodplain, (f) eolian deposits, (g) loessial deposits, (h) ground moraine, and (i) colluvium.

11. Write horizon symbols for (a) a **B** horizon with large sesquioxide accumulation, (b) a **B** horizon with high accumulation of smectite clays, (c) a gleyed **C** horizon, (d) a **B** horizon with high carbonate accumulations, and (e) a cemented **C** horizon.

12. What is meant by two capital letters together, such as **EB** or **BA,** when referring to horizons? Be general.

13. (a) What is meant by a *soil individual*? (b) How is a *soil individual* defined? (c) Are all soil individuals the same size? Explain.

14. Describe briefly the kinds of uses of the personal computer in (a) soil mapping and (b) the modeling of what happens to certain attributes (pesticide flow, predicted erosion, water runoff) over an area.

15. Define (a) a GIS and (b) a database.

16. In what ways will the use of GIS help users of soil survey maps and other field data?

17. (a) Explain how soils from the same kind of landform (say a river terrace, but along different rivers) could be quite different from each other. (b) Consider the same question but considering soils from ground moraines in different areas or from loess in different areas.

18. (a) Why would one soil form an *ochric epipedon* but another soil would form a *mollic*? (b) Answer the same question comparing formation of *spodic* versus a *natric* horizon.

19. (a) Describe the GPS system and describe some of its uses. (b) What is a DGPS?

20. (a) List the five soil-forming factors. (b) Is erosion one of these factors? (c) How does each factor affect the final soil that is produced?

21. Explain how and why each of the following influences soil development: (a) aridity; (b) sand with a high percentage of quartz; (c) a high clay content; (d) long, cold periods each year; (e) yearly thin depositions of sediment; (f) yearly erosion losses; and (g) profile mixing.

22. (a) How is soil *quality* evaluated? (b) What is a good quality soil?

23. (a) How can soils be degraded? (b) How can their *health* be improved? (c) What causes degradation of soil?

Soil Physical Properties

The earth, like the body of an animal, is wasted at the same time that it is replaced. It has a state of growth and augmentation; it has another state which is that of diminution and decay.

—James Hutton

Every person has a right to his opinion. But no one has a right to be wrong with the facts.

—Bernard Baruch

3:1 Preview and Important Facts

PREVIEW

The **physical properties** of soils—texture, structure, density, porosity, water content, strength (consistency), temperature, and color—are dominant factors affecting the use of a soil. These properties determine the availability of oxygen in soils, the mobility of water into or through soils, and the ease of root penetration. (Soil water, a vital physical property, is described in detail in Chapter 4.)

The thousands of different soils exhibit multitudinous differences, such as varying amounts of stones and pebbles, coarse and fine sands, clays, clumps (aggregates) cemented by clays and organic matter, living and dead plant materials, the dark remnants of partly decomposed organic substances (humus), and animal life (ants, earthworms).

The soil, as seen by minute organisms such as bacteria and nematodes, is a rough terrain. Sands, silts, and clays are clumped together, and there are tortuous and irregular open pores of all sizes throughout the soil mass. In undisturbed soil (uncultivated or below tillage depth), pores may be lined with precipitated calcium carbonates, deposited layers of clay, iron oxides, or other minerals.[1] **Humus,** the residual, greatly altered organic matter in soils, coats

[1] Soluble substances in soil water can form insoluble materials as water is evaporated and used by plants. Calcium carbonate ($CaCO_3$), iron oxides (Fe_2O_3), silica (SiO_2), and minute particles of organic matter (humus) are some of the common insoluble materials formed in soils or moved into them.

many mineral particles, often cementing them together into relatively stable clumps called **aggregates.** Clay particles, which are smaller than sands and silts, are too tiny to be seen using the visible light microscope. Magnifications of about 25,000 times, using an electron microscope, are required to see individual clay particles, some of which are small enough to be attached to the surfaces of microscopic-sized organisms such as bacteria. This microscopic soil world is a fascinating one but still largely unexplored.

Texture is the physical property of particular importance to those using soils. *Texture* is the term used to indicate the proportions of sands, silts, and clays in each soil. The soil texture controls water contents, water intake rates, aeration, root penetration, and some chemical properties. **Soil structure** (aggregation) and **soil temperature** are other physical properties. **Bulk density,** the weight of a volume of soil, is related to **pore space** in the soil. Other physical properties include **soil color, soil air,** and **soil strength.**

Economically we can rarely afford to make great changes in soils' physical properties, but understanding the properties can improve our ability to manage soils better. We learn what is possible and what, if we try to do it, will only cause new problems.

IMPORTANT FACTS TO KNOW

1. The sizes of sands, silts, and clays—both in actual dimensions and in relation to some common objects (pinhead, thickness of a sheet of paper, etc.)
2. The meaning of textural classes, such as *loam, clay loam,* and *loamy sand,* in general terms (not exact percentages)
3. The similarities and differences between textural classes, such as *loam, clay loam,* and *loamy sand*
4. The manner of indicating in the textural class name the amounts of large fragments mixed in with sands, silts, and clays
5. The terms used for rock fragments, as size of the fragments increases
6. The meaning of the term *soil structure* and how formation of different soil structures is a natural process of soil development
7. The typical bulk density values of *average cultivated soils* and the limits of using bulk density as diagnostic values in studying soils
8. The importance of (a) pore space and (b) the pore sizes in soils
9. The method of measuring soil colors and the purpose for knowing a soil's color
10. The general basis used to classify soils for use in engineering and construction activities
11. How soil temperature can be altered by various factors, such as wetting or drying, plastic or vegetative mulches on the soil, graphite on snow, and the slope or aspect of the site

3:2 Soil Texture

Natural soils are composed of soil particles of varying sizes. The soil particle-size groups, called **soil separates,** are *sands* (the largest), *silts,* and *clays* (the smallest). The relative proportions of soil separates in a particular soil determine its **soil texture.**

Texture is an important soil characteristic because it greatly affects water intake rates (infiltration), water storage in the soil, the ease of tilling the soil, the amount of aeration (vital to root growth), and soil fertility. For instance, a coarse, sandy soil is easy to till, has plenty of aeration for good root growth, and is easily wetted—but it also dries rapidly and easily loses plant nutrients, which are drained away in percolating water. High-clay soils (more than 30 percent clay) have very small particles that fit closely together. Clays have few large pores, which means there are only tiny openings for water to flow into the soil. This makes high-clay soils difficult to wet, difficult to drain, and difficult to till.

Sizes of Soil Separates

The U.S. Department of Agriculture (USDA) has established limits of variation for the soil separates and has assigned a name to each size class (Table 3-1). This system has been approved by the Soil Science Society of America and is the one used in this book. Other particle-size classification systems are used in the United States and throughout the world.

Soil Textural Classes

Textural names are given to soils based on the relative proportions of each of the three soil separates: sand, silt, and clay. Soils that are preponderantly clay are called **clay** (textural class); those with high silt content are called **silt** (textural class); those with a high sand percentage are called **sand** (textural class). A soil that does not exhibit the dominant physical properties of any of these three groups (such as a soil with 40 percent sand, 40 percent silt, and 20 percent clay) is called **loam.** Note that loam does not contain equal percentages of sand, silt, and clay. It does, however, exhibit approximately equal *properties* of sand, silt, and clay.

The **textural triangle** (Figure 3-1) is used to determine the soil textural name after the percentages of sand, silt, and clay are determined from a laboratory analysis.

Because the soil's textural classification includes *only mineral particles* and those of less than 2 mm diameter, the sum of percentages of sand, silt, and clay equals 100 percent. Note that organic matter is not included in these percentages. Knowing the amount of any two percentages automatically fixes the percentage of the third. In reading the textural triangle, any two particle fractions will locate the textural class at the point where those two, and the third fraction, intersect.

Particle-Size (Mechanical) Analysis

The procedure used to separate a soil into various size groups from the coarsest sand, through silt, to the finest clay, is **particle-size analysis,** also called **mechanical analysis.** For mechanical analysis, the mineral matter less than 2 mm (0.08 in) in diameter is considered separately from the larger particles. All rocks, pebbles, roots, and other rubble are removed (and measured) by screening the finer soil parts through a 2 mm sieve before analysis. Humus is removed from the soil sample by destroying it with an oxidizing chemical (such as hydrogen peroxide) before particle-size separation is done. *Routine* (approximate) textural values are usually obtained without removal of organic matter. An example of determining soil textural class is given in Calculation 3-1.

Table 3-1 Soil Separates and Their Diameter Ranges

Soil Separate Name	*Diameter Range (mm)*	*Visual Size Comparison of Maximum Size*
Very coarse sand	2.0–1.0	House key thickness
Coarse sand	1.0–0.5	Small pinhead
Medium sand	0.5–0.25	Sugar or salt crystals
Fine sand	0.25–0.10	Thickness of a book page
Very fine sand	0.10–0.05	Invisible to the eye
Silt	0.05–0.002	Visible under a microscope
Clay	<0.002	Most are not visible even with a microscope

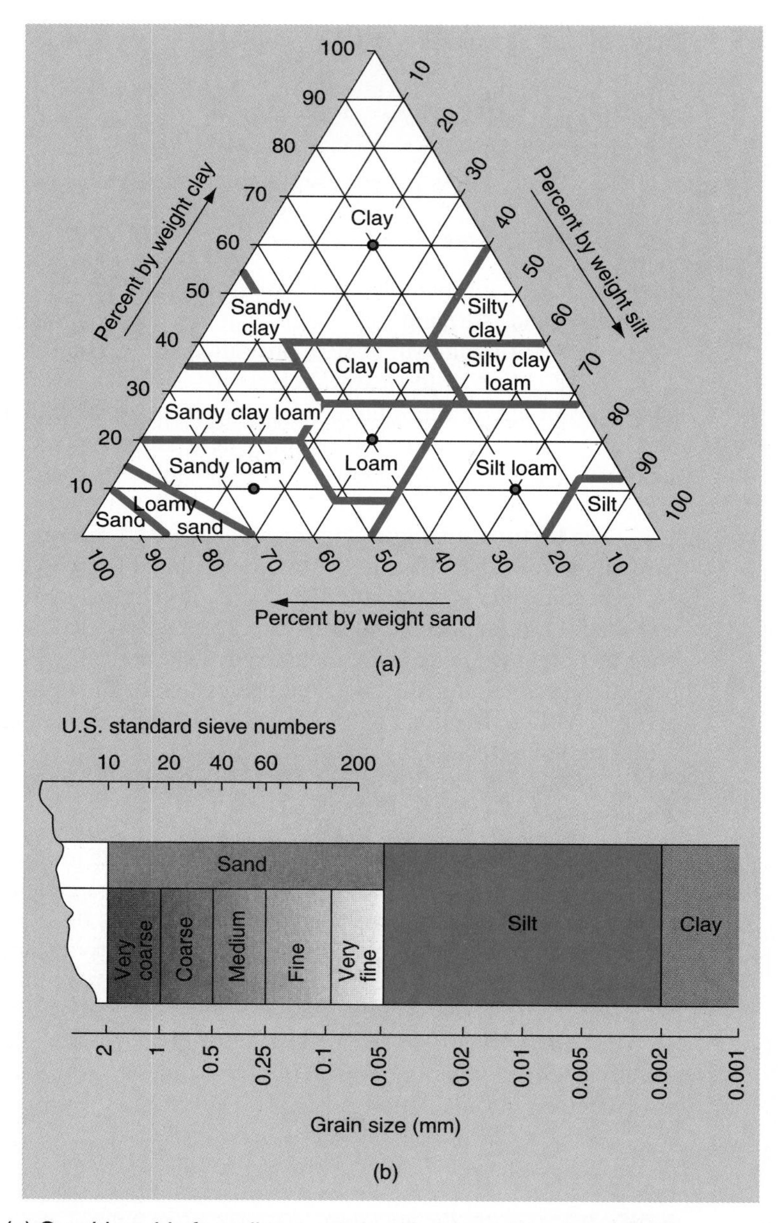

FIGURE 3-1 (a) Graphic guide for soil textural classification of the less-than-2-mm portion. The dots shown in several classes have the following percentages of size fractions:

Textural Class	*% Sand*	*% Silt*	*% Clay*
Clay (see center)	20	20	60
Silt loam	20	70	10
Sandy loam	65	25	10
Loam	40	40	20

(b) Separation of sand into various sand sizes and the U.S. standard sieve numbers used for these separations. A 10 mesh sieve has 10 wire divisions per inch in each of the two directions making up the screen.

Calculation 3-1 Determining Textural Class Names

Problem A sample of soil was screened and had the size separates in material smaller than 2 mm determined by particle-size (mechanical) analysis, with the following results:

$$\begin{aligned} \text{Sand content (2–0.05 mm diameter)} &= 140 \text{ g} \\ \text{Silt content (0.05–0.002 mm diameter)} &= 38 \text{ g} \\ \text{Clay content (<0.002 mm diameter)} &= \underline{22 \text{ g}} \\ \text{Total dry soil weight} &= 200 \text{ g} \end{aligned}$$

Determine the textural class name.

Solution Textural names consider only less-than-2-mm portion

$$\left(\frac{140 \text{ g}}{200 \text{ g}}\right)\left(\frac{100}{}\right) = 70\% \text{ sand}$$

$$\left(\frac{38 \text{ g}}{200 \text{ g}}\right)\left(\frac{100}{}\right) = 19\% \text{ silt}$$

$$\left(\frac{22 \text{ g}}{200 \text{ g}}\right)\left(\frac{100}{}\right) = 11\% \text{ clay}$$

1. Using the textural triangle (Figure 3-1), place the triangle with 100 percent clay at the top (apex) and read across, parallel with the base along the 11 percent line. Keeping this line in mind, turn the triangle so 100 percent silt is now at the top and read across, parallel to the new base of the triangle along the 19 percent line. The 11 percent clay and 19 percent silt lines intersect in *sandy loam.* The percentage sand value could have been used as easily as either clay or silt values, because the lines for all three fractions intersect at the same point. The content of organic matter is ignored. If the soil contains more than 15 percent (by volume) of particles larger than sand, a "coarse fragment" adjective is added to the textural name (e.g., *gravelly* sandy loam).
2. The correct complete name above is *sandy loam.*

The basis of particle-size separations is **Stokes law** of settling velocities: *The settling rate of a particle is the net difference between its downward force* (gravity) *against the buoyancy* (resistance to fall) *by surface friction and movement of the water.* It is assumed, because of the complex mathematical restrictions otherwise, that the particles are *smooth spheres.* This assumption is obviously incorrect and is a major weakness in the analysis. Typical fall rates will be 8711 D^2 cm^{-1} sec^{-1}, where D is the diameter of the particle. To calculate fall rates, enter the particle size in centimeters for D (and square it) and obtain the fall rates in cm/sec. Some typical values follow:

$$\begin{aligned} \text{medium sand (0.05 cm)} &= 22 \text{ cm/sec} \\ \text{fine sand (0.02 cm)} &= 3.5 \text{ cm/sec} \\ \text{medium silt (0.001 cm)} &= 0.087 \text{ cm/sec} \\ &= 0.52 \text{ cm/min} \end{aligned}$$

$$\begin{aligned} \text{coarse clay (0.0002 cm)} &= 0.00035 \text{ cm/sec} \\ &= 0.021 \text{ cm/min} \\ &= 1.26 \text{ cm/hr} \\ \text{fine clay (0.00002 cm)} &= 0.0000035 \text{ cm/sec} \\ &= 0.30 \text{ cm/day} \end{aligned}$$

3:3 Rock Fragments

Mineral particles larger than very coarse sand (>2.0 mm diameter) in soils are called **rock fragments.** They are classified by shape and size into *rounded* or *flat-shaped* particles. *Rounded* rock fragments are **gravel** (three sizes), **cobble, stone,** and **boulder;** *flat* rock fragments are **channer** (smallest), **flagstone, stone,** and **boulder** (Table 3-2).

The adjective describing rock fragments in soils is used as the first part of the textural class name under the following conditions:

- **<15% by volume:** No mention of rock fragments is used.
- **15–35% by volume:** The name of the dominant kind of rock fragment is used (e.g., *stony* loam).
- **35–60% by volume:** The word *very* precedes the name of the dominant kind of rock fragments (e.g., *very cobbly* sandy loam).
- **>60% by volume:** Add the word *extremely* in front of the coarse fragment name (e.g., *extremely gravelly* loam).

Of the more than 15,000 soil series described in the United States, about 17 percent are in soil groupings (*families*) containing 35 percent or more rock fragments by volume. Although rocks are a hindrance to cultivated crops, rocky soils (>90 percent rock fragments) are often productive for growing trees and other wildland plants.

Table 3-2 Terms for Describing Rock Fragments in Soils

Shape and Size	*Noun*	*Adjective*
Rounded, subrounded, angular, or irregular		
2–75 mm diameter	Pebbles	Gravelly
2–5 mm diameter	Fine pebbles	Fine gravelly
5–20 mm diameter	Medium pebbles	Medium gravelly
20–75 mm diameter	Coarse pebbles	Coarse gravelly
75–250 mm diameter	Cobbles	Cobbly
250–600 mm diameter	Stones	Stony
≥600 mm diameter	Boulders	Bouldery
Flat		
2–150 mm long	Channers	Channery
150–380 mm long	Flagstones	Flaggy
380–600 mm long	Stones	Stony
≥600 mm long	Boulders	Bouldery

Source: Soil Survey Manual, USDA—Soil Survey Division Staff, Oct. 1993, 142–144.

3:4 Soil Structure

Structure is the arrangement of sands, silts, and clays into stable (cemented) aggregates. **Aggregates** are secondary units or granules composed of many soil particles held together by organic substances, iron oxides, carbonates, clays, and/or silica. Natural aggregates are called **peds** and vary in their water stability; the word **clod** is used for a coherent mass of soil broken into any shape by artificial means, such as by tillage.

Two terms are often confused with *ped.* One is **fragment,** which consists of a piece of a broken ped, and the other is **concretion (nodule),** which is a coherent mass formed within the soil by the precipitation of certain chemicals dissolved in percolating (seepage) waters. Concretions are often small, like shotgun pellets, and they are sometimes referred to as **shot.**

Soil Structural Classes

Soil structural units (peds) are described by three characteristics: **type** (shape), **class** (size), and **grade** (strength of cohesion; see Table 3-3). Types of structure describe the ped shape with the terms **angular blocky, subangular blocky, columnar, granular, platy,** and **prismatic** (Figures 3-2 and 3-3).

Structure *classes* are the ped sizes, such as **very fine, fine, medium, coarse** (or **thick**), and **very coarse** (or **very thick**). Structure *grades* are evaluated by the distinctness, stability, and strength of the peds.

1. **Structureless:** Soils have no noticeable peds. It might be an unconsolidated mass such as noncoherent sand, called **single grain,** or it might be a cohesive mass, such as could occur in some loams or clayey soils, called **massive.**
2. **Structured:** The following structural grades are used:
   - *Weak:* Peds are barely distinguishable in part of the *moist* soil; only a few distinct peds can be separated from the soil mass.
   - *Moderate:* Peds are visible in place; many can be handled without breaking.
   - *Strong:* Most of the soil mass is visible as peds, most of which can be handled with ease without breaking.

Soil structure may exist as a compound structure in which large peds such as prisms or blocks may further fall apart into smaller blocks or smaller peds.

Soil structure influences many important properties of the soil, such as the rate of infiltration of water and air. Both granular (spheroidal) and single-grain (structureless but sandy) soils have rapid infiltration rates; blocky (blocklike) and prismatic soils have moderate rates; and platy and massive soil conditions have slow infiltration rates.

Genesis of Soil Structure

Structural peds form because of combinations of swelling–shrinking and adhesive substances. As a mass of soil swells (wets or freezes) and then shrinks (dries or thaws), lines of weakness (cracks) are formed. The soil mass between cracks—cemented by organic substances, iron oxides, clays, carbonates, and even silica—remains cohesive. Polysaccharides (chains of sugar molecules) are thought to be very common cements.

The amount of cracking is usually the minimum necessary to relieve the shrinking stress. The pattern of cracking forms mostly five- and six-sided shapes. Because the swelling–shrinkage vertically does not require cracks to form (the soil surface just sinks

Table 3-3 Types and Classes of Soil Structure

	Type (Shape and Arrangement of Peds)					
				Blocklike, Polyhedronlike, or Spheroidal		
	Platelike with the Vertical Dimension Greatly Less than the Other Two Dimensions; Faces Mostly Horizontal	**Prismlike with Two Dimensions Limited and Considerably Less than the Vertical; Vertical Faces Well Defined; Vertices Angular**		**Blocklike: Blocks or Polyhedrons Formed by the Faces of the Surrounding Peds**		**Spheroids of Polyhedrons Having Surfaces that Have Slight or No Accommodation to Faces of Surrounding Peds**
		Without Rounded Caps	**With Rounded Caps**	**Faces Flattened; Most Vertices Sharply Angular**	**Rounded and Flattened Faces and Vertices**	**Porous and Nonporous Peds**
Class	*Platy*	*Prismatic*	*Columnar*	*(Angular) Blocky*	*(Subangular) Blocky*	*Granular*
Very fine or very thin	Very thin platy; <1 mm	Very fine prismatic; <10 mm	Very fine columnar; <10 mm	Very fine angular blocky; <5 mm	Very fine subangular blocky; <5 mm	Very fine granular; <1 mm
Fine or thin	Thin platy; 1–2 mm	Fine prismatic; 10–20 mm	Fine columnar; 10–20 mm	Fine angular blocky; 5–10 mm	Fine subangular blocky; 5–10 mm	Fine granular; 1–2 mm
Medium	Medium platy; 2–5 mm	Medium prismatic; 20–50 mm	Medium columnar; 20–50 mm	Medium angular blocky; 10–20 mm	Medium subangular blocky; 20–50 mm	Medium granular; 2–5 mm
Coarse or thick	Thick platy; 5–10 mm	Coarse prismatic; 50–100 mm	Coarse columnar; 50–100 mm	Coarse angular blocky; 20–50 mm	Coarse subangular blocky; 20–50 mm	Coarse granular; 2–5 mm
Very coarse or very thick	Very thick platy; >10 mm	Very coarse prismatic; >100 mm	Very coarse columnar; >100 mm	Very coarse angular blocky; >50 mm	Very coarse subangular blocky; >50 mm	Very coarse granular; >10 mm

Source: Modified from *Soil Survey Manual,* USDA—Soil Survey Division Staff, October 1993.

FIGURE 3-2 Examples of structure types and crusting. (a) is massive and results from Pleistocene lake layering of different textures; although it appears to be a pedogenic development of platy structure, it is a deposition. (b) and (c) are platy and result from natural soil development: (b) is 1.3 cm (0.5 in) thick plates of a plowpan in sandy loam; (c) is at 122 cm (4 ft), caused by a fluctuating water table. (d) and (e) are subangular blocks: (e) has compound structure with the blocks within weak prisms (prism between arrows); (f) is a 10 cm × 23 cm (4 in × 9 in) prism from a clay loam. (g) shows two prisms 12.5 cm (5 in) long. (h) shows a dry farm silt loam soil in November, with winter wheat emerging; the soil has lost surface structure and formed a crust with a thin surface layer of massive (structureless) soil (notice upside-down piece of top soil—see arrow). The soil is fairly well structured below the surface crust. (Courtesy of Raymond W. Miller, Utah State University.)

down in shrinkage), the *prismatic* structures develop in early stages. In later development, especially in clayey soils, the cracking horizontally will develop and will form *blocky* peds. Prisms commonly break down further into smaller *blocky* peds (compound structure). These are mostly subsurface peds.

Platy structure requires force to separate horizontal soil layers. Frost heaving, fluctuating water tables, compaction (equipment or animals), and thin layering of different-textured alluvium or lacustrine material can aid the formation of plates.

Granular peds are mineral aggregates glued together mostly by organic substances, but so mixed by rodents, earthworms, frost action, and cultivation that all edges are rounded and the peds are small in size. Granular structure is limited to surface horizons unless buried by sediment.

Deterioration of Aggregates

Increasing exchangeable sodium (Na^+) most often speeds deterioration of structure. The sodium does not effectively neutralize the surface negative charge on soil particles. The result

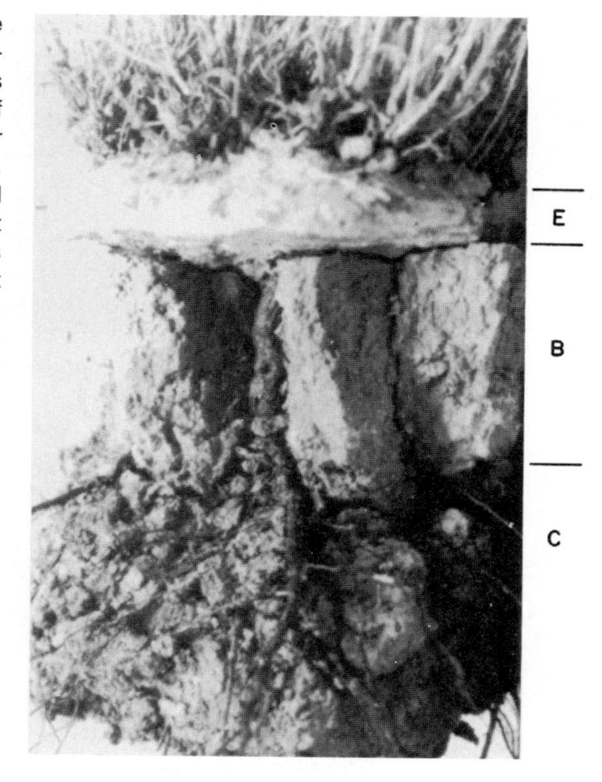

FIGURE 3-3 Soil that developed where considerable sodium occurred in the profile. Typically, the top 2.5–5 cm (1–2 in) is a silty white **E** horizon (massive leached of colloids) with a **B** horizon of prismatic or columnar structure at a shallow depth. Often the columns or prisms are coated with dark-colored humus. Massive parent material begins at a shallow depth. In this photograph parent material begins at 20 cm (8 in) deep. The example occurs in southeast central Utah with precipitation about 30 cm (12 in) annually. (*Source:* USDA—Soil Conservation Service; photo by John J. Bradshaw.)

is a repulsion of adjacent soil particles because of similar charges and the disintegration of structural peds (dispersion). As water flows into the dispersed soil, the dispersed clays and small organic colloids move with the water, lodge in the narrow necks of soil pores, and seal the soil. Soil with too much sodium can become almost impermeable to water, and it dries to hard crusts, inhibiting seedling emergence.

3:5 Particle Density and Bulk Density

Density is the mass of an object per unit volume. Approximate densities of some commonly known materials are as follows:

Material	*kg/m³*	*Mg/m³*	*g/cm³*	*lb/ft³*
Water	1000	1	1	62.4
Pine wood	700	0.7	0.7	44
Loose sand	1600	1.6	1.6	100
Quartz mineral	2600	2.6	2.6	162
Steel	7700	7.7	7.7	480
Lead	11,300	11.3	11.3	706

Water is the reference used in soil density measurements. Water weighs *one megagram per cubic meter (1 Mg/m³)* or *one gram per cubic centimeter,* which are convenient reference numbers to remember. Mineral soil densities, except for some volcanic materials, are *greater than* the density of water; organic soil densities are *less than* that of water.

Two density measurements—particle density and bulk density—are common for soils. **Particle density** is the density of the solid soil particles only; the measurement does not include water weight or pore (air) space. The dominant soil minerals—quartz, feldspars, micas, and clay minerals—average approximately 2650 kg/m^3 (2.65 Mg/m^3), the standard value used in calculations if particle density is not measured. Individual minerals have densities from 2000 kg/m^3 for bauxite (aluminum ore) to 5300 kg/m^3 for hematite (iron ore) or 7600 for galena (lead ore).

Bulk density, the density of a volume of soil as it exists naturally, includes any air space and organic materials in the soil volume. Because bulk density is calculated for the dried soil, water is not included in the sample weight. The bulk soil volume is assumed not to have changed by drying; only the water has been removed, leaving empty pores. One should not assume that there is no volume change during drying of soils with considerable amounts of swelling clays.

A loosened, fluffed soil with an increased total pore space will have a smaller weight per unit volume than the same soil after it is compacted.

Thus, bulk densities can be used to estimate differences in compaction of a given soil, such as that which might result after tillage with heavy equipment on soils. In conversation, clay soils are often called *heavy* soils and sandy soils are called *light* soils, terms that refer to the energy needed to till the soils. Actually, sands normally have greater bulk densities (are heavier) than clays. However, the observation that clay *seems* heavier (in weight) is not entirely wrong. The clay normally contains much more water, which is relatively heavy. However, by definition, this water weight is not involved in calculating the bulk density of the soil.

Unfortunately, because of the wide variety of soil textures and humus contents, bulk density alone is not an indication of soil suitability for plant growth. Soils of different bulk densities, because of different textures, may be equally good for plant growth. Table 3-4 illustrates some of these variations in bulk density that result from compaction, vegetation, and soil use. Figure 3-4 shows the changes in loblolly pine seedlings grown in three different soils with varied bulk densities due to different amounts of compaction. The average soil bulk density of cultivated loam is approximately 1100 kg/m^3 to 1400 kg/m^3. For good plant growth bulk densities should be below about 1400 kg/m^3 for clays and 1600 kg/m^3 for sands. In greenhouse potting mixtures the high amount of peat moss, vermiculite, or perlite used produces low bulk densities, such as 100 kg/m^3 to 400 kg/m^3 (See Chapter 21). Recall that all of these density values refer to the *dry* soil.

Bulk density values have various uses. For example, density values are needed to calculate total water storage capacity per soil volume. Secondly, soil layers can be evaluated to determine if they are too compacted to allow root penetration or to provide adequate aeration.

Measurement of bulk density commonly involves taking an undisturbed block of soil (clod or soil core), determining its volume, drying it, and weighing it. *Clods* can be coated with paraffin or liquid plastic and dipped into water to measure water displacement and, hence, to calculate volume. When soil cores are taken by a metal cylinder, the exact volume is determined by measuring the cylinder volume (Calculation 3-2). The formula for bulk density is as follows:

$$\text{bulk density} = (\rho_B) = \frac{\text{oven-dry soil mass}}{\text{soil volume}}$$

Some examples of the importance of bulk density to plant growth are given in Detail 3-1. The average weight of soil for a hectare (or acre) area per unit depth is calculated by multiplying the soil volume by its bulk density. A hectare-15 cm or acre-furrow-slice weight is

Table 3-4 Example of Bulk Densities of Soils as Affected by Texture, Compaction, Coarse Particles, and Other Features

Soil Treatment and Identification	Bulk Density ($Mg\ m^{-3}$)	Bulk Density ($kg\ m^{-3}$)	Pore Space (%)*
Rocky silt loam soil under aspen forest	1.62	1620	40
Loamy sand surface soil	1.5	1500	43
Decomposed peat (low particle density)	0.66	660	65(?)
Cotton field soil			
Tilled surface soil of a cotton field	1.3	1300	51
Trafficked interrows where wheels passed surface	1.67	1670	37
Traffic pan at about 25 cm (10 in) deep	1.7	1700	36
Undisturbed subsoil below traffic pan, clay loam	1.5	1500	43
Boulder clay in Suffolk, England			
3 years under grass pasture	1.13	1130	57
3 years in barley, no traffic	1.3	1300	51
3 years in barley, normal tractor traffic	1.63	1630	37
Long-term use, Nigeria, sandy loam			
Under bush 15–20 years	1.15	1150	57
3 years corn, last 2 years no-till	1.42	1420	46
5 years corn, conventional tillage	1.51	1510	43
Fragipan soil in Ohio, from glacial till			
Cultivated surface soil, silt loam	1.47	1470	45
Soil layer 30 cm (12 in) deep, loam	1.65	1650	38
Fragipan, 55 cm (21.5 in) deep, loam	1.76	1760	34
Beneath fragipan, 125 cm (49 in) deep, loam	1.85	1850	30
Paddy rice soil, Philippines, clay			
0–15 cm (0–6 in), average of 9 cores	0.66	660	75
15–30 cm (6–12 in), average of 9 cores	0.91	910	66
Grainfield fine sandy loam, Oklahoma, **Ap**	1.72	1720	35
Lower subsoil	1.80	1800	32
Miami silt loam, Wisconsin, **Ap**	1.28	1280	52
Lower subsoil	1.43	1430	46
Houston clay, Texas, **Ap**	1.24	1240	53
Lower subsoil	1.51	1510	43
Oxisol clay, Brazil, **Ap**	0.95	950	64
Lower subsoil	1.00	1000	62

*The percentage pore space was calculated using 2.7 for particle density except for the peat soil, which is estimated.

estimated for bulk densities of about 1300 kg/m^3. A hectare-15 cm volume of soil of bulk density 1300 kg/m^3 would weigh approximately 2,000,000 kg when oven dry. It can be calculated as follows:

$$1 \text{ hectare} = 10{,}000 \text{ m}^2$$

$$1 \text{ hectare-15 cm deep} = (10{,}000 \text{ m}^2)(0.15 \text{ m}) = 1500 \text{ m}^3$$

$$1 \text{ hectare-15 cm deep of bulk density } 1300 \text{ kg/m}^3 \text{ weighs } 1{,}950{,}000 \text{ kg}$$
$$\text{(often rounded off to 2,000,000 kg)}$$

An acre-7-in depth is estimated to weigh about 2,000,000 lb by a similar calculation. An acre-7-in depth is slightly deeper than 15 cm. The weight of one hectare-30 cm is estimated to be 4×10^6 kg (4 million kg).

FIGURE 3-4 Bulk density in relation to the growth of loblolly pine in Louisiana. Sand, loam, and clay soils were subjected to the following kinds of loosening or pressure, and the resulting soil material was used to grow loblolly pine seedlings: (1) soil loosened, (2) soil undisturbed, (3) soil subjected to static pressure of 3.5 kg/cm, (4) soil subjected to static pressure of 7.0 kg/cm, (5) soil subjected to static pressure of 10.5 kg/cm, (6) soil puddled (dispersed) plus same pressure as in treatment (5). (Courtesy of Louisiana State University.)

3:6 Soil Porosity and Permeability

Pore spaces (also called **voids**) in a soil consist of that portion of the soil volume not occupied by solids, either mineral or organic. Pores in soil that are the result of irregular shapes of primary particles and their packing are called **matrix pores.** Aggregation and the pushing forces of penetrating roots, worms, and insects, and of expanding gases entrapped by water are called **nonmatrix pores;** these actions change the sizes of pores, usually making them larger. Under field conditions all pore spaces are occupied by air, water, and roots.

Tortuous pathways is the term that best describes soil pores. The soil particles have irregular shapes and thus leave the spaces between them very irregular in size, shape, and direction. Sands have large and continuous pores. In contrast, clays—although containing more total pore space because of the minute size of each clay particle—have very small pores, which transmit water slowly.

Calculation 3-2 Calculating Soil Bulk Density

Problem A metal cylinder pushed into a loam soil is removed from the field and the soil it contains is dried in an oven. The measured data are as follows:

$$\text{Cylinder height} = 5.0 \text{ cm}$$

$$\text{Cylinder inside diameter} = 4.4 \text{ cm}$$

$$\text{Oven-dried soil weight} = 87.6 \text{ g}$$

Calculate the soil's bulk density.

Solution

1. The volume of the soil sample equals the volume of the cylinder. A cylinder's volume equals pi ($\pi = 3.14$) times the radius squared times the cylinder height ($V = \pi r^2 h$):

$$\text{volume} = \pi\left(\frac{\text{diameter}}{2}\right)^2 h$$

$$= (3.14)\left(\frac{4.4 \text{ cm}}{2}\right)^2 (5 \text{ cm}) = 76.0 \text{ cm}^3$$

2. The soil bulk density is soil dry weight divided by the volume of the soil. If measurements were in pounds and cubic feet, the calculations and answer would be in those units. In this example, the measurements are in grams and cubic centimeters. Thus,

$$\text{bulk density} = \frac{\text{soil mass}}{\text{soil volume}} = \left(\frac{87.6 \text{ g}}{76 \text{ cm}^3}\right)\left(\frac{1 \text{ kg}}{1000 \text{ g}}\right)\left(\frac{1{,}000{,}000 \text{ cm}^3}{1 \text{ m}^3}\right)$$

$$= 1150 \text{ kg/m}^3 \; (= 1.15 \text{ g/cm}^3)$$

Small bottlenecks in soil pores can fill with water and block air movement through them. Air exchange may be inadequate for plant root growth in some wet, clayey soils. The most rapid water and air movement is in sands and strongly aggregated soils in which aggregates act like sand grains and pack to form many large pores.

Nonmatrix pores are officially described according to their average diameter in millimeters as follows:

Very fine: <0.5 mm
Fine: 0.5–2 mm
Medium: 2–5 mm
Coarse: 5–10 mm
Very coarse: ≥ 10 mm

Water drains by gravitational force from pores larger than about 0.03 mm to 0.06 mm. In comparison, root hairs—the smallest plant roots—are between 0.0008 mm and 0.012 mm in diameter. In soils where most pores are smaller than 0.030 mm, attraction forces in the soil retain water within the fine pores, which results in a waterlogged soil and poor aeration. Thus,

Detail 3-1 Examples of Effects of Bulk Density on Growth

- Plant roots of field crops are hindered by soils high in bulk density, but in varying degrees. Tolerance to soil compaction is in this order:*

$$\text{alfalfa} > \text{corn} > \text{soybeans} > \text{sugar beets} > \text{dry edible beans}$$

- On two silt loam soils in Arkansas, subsoiling decreased soil bulk density by 16% and increased cotton lint yields by 13% in a year with normal rainfall and by 59% during a dry year.†
- Soil bulk densities greater than 1200 kg/m^3 (1.2 g/cm^3) on medium textured soils in the state of Washington were positively correlated with diseases of beans and peas.‡
- Subsoiling increased the yield of tobacco in North Carolina on soils with a bulk density of greater than 1630 kg/m^3 (1.63 g/cm^3) and/or a sand content greater than 73%. Responses were greater in dry weather.§

---

* A. J. M. Smucker, "Plant Root System Response to Compacted Soils," *Agronomy Abstracts* (1985).
†J. S. McConnell and M. H. Wilkerson, "Soil Compaction Effects on Cotton," *Arkansas Farm Research,* Mar.–Apr. 1987.
‡J. M. Kraft, "Compaction/Disease Interaction," *Agronomy Abstracts* (1985).
§M. J. Vepraskas, G. S. Miner, and G. F. Peedin, "Relationships of Soil Properties and Rainfall to Effects of Subsoiling on Tobacco Yield," *Agronomy Journal* **79** (1987), 141–146.

to the growing plant, pore sizes are of more importance than total pore space. The best balance of water retention (smaller pores) plus adequate air and water movement (larger pores) is in medium-textured soils, such as loams. Aggregation, or the lack of it, can modify this balance of large and small pores, which results from the soil texture. Soluble bicarbonates, silicates, and iron during wet periods will move in the soil and precipitate when drying occurs. Over many decades or centuries, pores become filled with the precipitates and often are cemented into hard layers (Figure 3-5).

The relative amounts of air and water in the pore spaces fluctuate continuously. During a rain, water pushes air from the pores, but as soon as soil water disappears by deep percolation (downward movement), evaporation, and transpiration (evaporation from plant leaf openings), air gradually replaces the water as it is lost from the pore space.

The percentage of a given volume of soil occupied by pore space may be calculated from the following formula:

$$\begin{aligned} E_p &= \%\ \text{pore space} \\ &= 100\% - \%\ \text{solid space} \\ &= 100\% - \left(\frac{\text{Bulk density}}{\text{Particle density}}\right)(100) \end{aligned}$$

or

$$E_p = 100\% - \left(\frac{\rho_B}{\rho_P}\right)(100)$$

A sample calculation is given in Calculation 3-3.

FIGURE 3-5 A cemented layer can be like semisoft rock. This Upton gravelly loam soil in New Mexico has a very shallow 30 cm (12 in) surface soil (**A** and **B,** the darker material top foot of depth) over a *hard indurated (lime-and-silica cemented) hardpan called* petrocalcic *or* duripan *(caliche),* the white material below the left 1 ft marker in the photo. This hardpan cannot be easily dug, even with a pick, and is often nearly impermeable to water movement through it. (*Source:* USDA—Soil Conservation Service; photo by Max V. Hodson.)

3:7 Soil Air

To survive, all living organisms require gaseous exchange, usually free oxygen (O_2). In soil, plant roots require O_2 for respiration and microorganisms need it for organic-matter decomposition. The most desired condition for growth of most plants is *well-aerated soil,* a condition in which oxygen exchange between soil air and atmospheric air is rapid. The factors influencing the rate of gaseous exchange include soil pore sizes and continuity, temperature, depth in the soil, wetting and drying of soil, and coverings (mulches) on the soil surface.

Composition of Soil Air

Atmospheric air has approximately the following composition of the gases that are also important in soils:

Nitrogen (N_2) = 79%
Oxygen (O_2) = 20.9%
Carbon dioxide (CO_2) = 0.035%
Water vapor (expressed as relative humidity) = 20–90%

Because O_2 is required in respiration and for organic-material decomposition, soil air is different from atmospheric air. Some O_2 is used and considerable CO_2 is produced. The differences are as follows:

Calculation 3-3 Calculating Pore Space Percentage

Problem A soil core was taken for determination of bulk density. The measurements were as follows:

Cylinder volume $= 73.6\ cm^3$ (see Calculation 3-2)

Dry soil weight $= 87.8$ g

Standard particle density $= 2650\ kg/m^3$ ($2.65\ g/cm^3$)

Calculate the percentage pore space.

Solution

1. $$\text{Bulk density} = \left(\frac{\text{soil weight}}{\text{soil volume}}\right) = \left(\frac{87.8\ g}{73.6\ cm^3}\right)\left(\frac{1\ kg}{1000\ kg}\right)\left(\frac{1{,}000{,}000\ cm^3}{1\ m^3}\right)$$
$$= 1190\ kg/m^3\ [= 1.19\ g/cm^3 = 1.19\ Mg\ m^{-3}]$$

2. $$\%\ \text{pore space} = 100\% - \left(\frac{\text{bulk density}}{\text{particle density}}\right)(100)$$
$$= 100\% - \left(\frac{1.19}{2.65}\right)(100)$$
$$= 100\% - 44.9\%$$
$$= 55.1\%, \text{ the soil's pore space percentage}$$

This soil would be a clay loam or other clayey soil. A sandy soil would have a bulk density closer to 1400 or 1500 kg/m^3 and a pore space percentage closer to 45%–50%.

Examples Several soil textures, bulk densities, and their pore space percentages are given as follows for cropped soils:

Soil Texture	*Bulk Density (kg/m^3)*	*Pore Space (%)*
Gravelly sand	1870	29.4
Coarse loamy sand	1680	36.6
Sandy loam	1510	43.0
Loam	1340	49.4
Clay loam	1260	52.5
Clay	1180	55.5

Soil Air	*Surface Soil (in %)*	*Subsoil (in %)*
Higher in carbon dioxide	0.5–6	3–10
Lower in oxygen	20.6–14	18–7
Higher in relative humidity	95–99	98–99.5

Rates of Oxygen Exchange

The **oxygen diffusion rate (ODR)** is the rate at which oxygen in the soil exchanges with oxygen in the atmosphere. Many large soil pores allow rapid air exchange (diffusion); small pores, or pores with bottleneck portions filled with water, decrease exchange rates. Diffusion of CO_2 gas through water is about 10,000 times slower than it is through air-filled pores. At a depth of 1 m (39 in) in the soil, the exchange rate may be only one-half to one-fourth as fast as in the top few centimeters. Exchange rates of air above 40×10^{-4} g/m^2 per minute seem to be fast enough for most plant roots and microbes. Root growth of some plants ceases when the rate reaches half of this value.

From these data it is easy to visualize reduced root growth occurring in deep clayey subsoils when they are wet because the water-filled portions of pores would block oxygen diffusion further. The small pores of clays usually have a slow ODR. Because of this problem some clayey soils are, in effect, *shallow soils,* as far as plant root growth is concerned.

Oxidation-Reduction Potential (Eh or Redox Potential)

All plants need O_2 for respiration. Respiration is necessary for life and growth. Most plants need the O_2 to be in the soil pores where roots are growing. However, a few plants, such as rice, can move O_2 *internally* from their tops, which are in the atmosphere, to their roots. Airflow is through internal large-diameter pores. By these means they are able to supply themselves O_2, even when growing in stagnant waters.

Most plants grow best in *oxidized* (aerated) soil because they need O_2. Free oxygen is the *primary acceptor of electrons* released during the oxidation of the carbon in organic materials. If the oxygen concentration is too low, certain elements or ions become the electron acceptors and are, in turn, reduced. The major elements and ions involved in this process, and comments about them, are given in Table 3-5.

Mostly undesirable things happen when the ODR is too low and the *redox value* drops: Toxic organic acids form, nitrate is denitrified, nitrogen is lost to the atmosphere, available sulfate is reduced to sulfide, and most plant roots cannot respire. Thus, plant root growth slows or stops. Aeration affects much more than O_2 concentrations.

Aeration and Energy for Plant Growth

Plants obtain energy from the sun, store it in chemical bonds (photosynthesis), and release and use the energy as they break these bonds (respiration). Splitting a glucose sugar (6-carbon sugar) into 2-carbon or 3-carbon fragments (fermentation) is common in **anaerobic glycolysis.** In the presence of O_2 **aerobic glycolysis plus respiration** (conversion of glucose carbon to carbon dioxide) releases and makes available much more energy, about 19 times more than

Table 3-5 Oxygen and the Ions in Soil Commonly Acting as Electron Acceptors and Their Ion Forms That Are Produced (When Oxygen Concentrations Are Low)

Form in Oxidized Soil	*Form in Reduced Soil*	*Approximate Eh Where Change Occurs (volts)**	*Comments about the Solubility of the Reduced Form Compared to Oxidized Form*
O_2	H_2O	0.38 to 0.32	Much more soluble
NO_3^-	N_2, N_2O	0.28 to 0.22	Lost to atmosphere
MnO_2	Mn^{2+}	0.28 to 0.22	Much more soluble
Fe_2O_3	Fe^{2+}	0.18 to 0.15	Much more soluble
SO_4^{2-}	S^{2-}	−0.12 to −0.18	Less soluble
CO_2	CH_4	−0.2 to −0.28	Lost to atmosphere

*Positive voltage values indicate oxidized soils. A typical soil is borderline and may be poorly aerated if the Eh is smaller than about 0.25 wet or 0.30 when at good moisture for plant growth. The values change with soil pH and location in the profile or even within a soil ped.

Source: Selected and modified data of W. H. Patrick, Jr., and C. N. Reddy, "Chemical Changes in Rice Soils," in *Soils and Rice,* International Rice Research Institute, Los Baños, Philippines, 1978, 361–379.

anaerobic breakdown alone (308 kcal vs. 16.2 kcal stored into adenosine triphosphate (ATP) bonds). The energy in the ATP energy bonds (the main mechanism for metabolic energy transfer) and released by the anaerobic glycolysis is 47.3 kcal, and the aerobic glycolysis plus respiration releases 686 kcal. Thus, much less energy flow exists in anaerobic conditions; anaerobic decomposition of organic matter is much slower than is aerobic decomposition.

Reduced aeration with deficient O_2 concentrations can be induced by various factors. Some of these are (1) waterlogging (ponding), (2) soil compaction of loams and clays, (3) very high contents of clay that swell when wet, thereby closing large pores, and (4) organic-matter decomposition by soil microorganisms that uses oxygen in soils that have a low ODR.

Some practices—deep ripping, drainage of excess water, and incorporation of bulky organic residues—are partly designed to aid soil aeration. Inadequate aeration should be expected in the following conditions:

- Poorly drained soils
- Soils of high clay content shortly after rainfall or irrigation
- Deep subsoils in clayey soils, especially if wet
- Highly compacted soils of fine texture
- Deeper portions of clayey soils having no structure (massive)

3:8 Rhizotrons—Seeing beneath the Soil Surface

Observing root growth and the activity of roots has long been a goal of many researchers. Root systems have been excavated in numerous ways to see what is actually happening down below. The National Soil Tilth Laboratory (NSTL) in Ames, Iowa has facilities that allow observation and the collection of solutions and soil air. There are separate temperature controls for above the soil surface and the root zone. These controlled chambers allow the repetition of experiments under identical temperature conditions and can vary those conditions to climates of many areas. One worker said they can simulate just about any soil and environment in the United States. Soil cores of intact soil from 40 cm to 50 cm diameter to 1 $m^2 \times 1.5$ m deep are gathered by pushing steel boxes into the soil, excavating them, lifting them by crane onto a truck, and driving them to the laboratory. Some of the largest columns weigh 3182 kg (7000 lb). Chemical movement, drainage, water balance, and soil gases can be monitored. A fiber-optic scope can measure and monitor root growth at certain depths. Microprocessors control temperature, moisture, and relative humidity for providing seasonal environmental changes.

3:9 Soil Consistence (Strength)

Soil consistence refers to attributes of soil material as expressed in degree of *cohesion* and *adhesion* or in *resistance to deformation on rupture.* Here, ". . . consistence includes: (1) resistance of soil material to rupture; (2) resistance to penetration; (3) plasticity, toughness, and stickiness of puddled soil material; and (4) the manner in which the soil material behaves when subject to compression."[2] A noncemented soil mass is evaluated under two moisture conditions: air-dry condition and field capacity water content. When a soil mass is cemented, soil strength is evaluated when air dry and again when wet after it has soaked in water for an hour. The descriptive terms for several of the consistence measurements are presented in

[2]Soil Survey Division Staff, *Soil Survey Manual,* USDA Handbook No. 18, October 1993, 172.

Tables 3-6 to 3-9. Not listed are *manner of failure classes* (brittleness to fluid) and *smeariness* (nonsmeary to strongly smeary).

3:10 Soil Color

For centuries people have recognized that in hot climates wearing white and pale-colored clothes is cooler than wearing dark-colored clothes, because more heat is absorbed by dark colors. Similarly, dark soils absorb more heat than do light-colored ones. Some black coal mining wastes and dark-colored oil–shale residues reach temperatures of 65.6°C to 70.0°C (150°F to 158°F), which are lethal to many plants that could otherwise grow in those soils. Although black soils having high humus content absorb more heat than light-colored soils, they also frequently hold more water. The water requires a relatively larger amount of heat than the soil minerals to raise its temperature. Evaporating water also requires considerable heat. The net result is that many dark soils are *not* warmer than adjacent lighter-colored soils because of the temperature-modifying effects of the soil moisture; in fact, they may be cooler except for a centimeter or so at the *dry* surface.

Table 3-6 Rupture Resistance Classes for Blocklike Specimens[a]

Test Description		Classes		
Operation	*Stress Applied*	*Moderately Dry and Very Dry*	*Slightly Dry to Wetter*	*Air Dried, then Submerged 1 Hr*
Specimen not obtainable	—	Loose	Loose	Not applicable
Fails under very slight force applied slowly between thumb and forefinger	<8N[b]	Soft	Very friable	Noncemented
Fails under slight force applied slowly between thumb and forefinger	8–20N	Slightly hard	Friable	Extremely weakly cemented
Fails under moderate force applied slowly between thumb and forefinger	20–40N	Moderately hard	Firm	Very weakly cemented
Fails under strong force applied slowly between thumb and forefinger (80N is about maximum force that can be applied by a person)	40–80N	Hard	Very firm	Weakly cemented
Cannot be failed between thumb and forefinger but can be between both hands or by placing on a nonresilient surface and applying gentle force underfoot	80–160N	Very hard	Extremely firm	Moderately cemented
Cannot be failed in hands but can be underfoot by full body weight (about 800N) applied slowly	160–800N	Extremely hard	Slightly rigid	Strongly cemented
Cannot be failed underfoot by full body weight but can be by <3J blow	800–3J	Rigid	Rigid	Very strongly cemented
Cannot be failed by blow of <3J	≥3J	Very rigid	Very rigid	Indurated

[a]Natural blocks of soil, about 25–30 mm dimension. If only smaller blocks are available, proportion the force needed (i.e., if only 15 mm block, half the force should be required to rupture it).
[b]Both force in newtons (N) and energy in joules (J) are used. The number of newtons is 10 times the kilograms of force. One joule is the energy delivered by dropping a 1 kg weight 10 cm.
Source: Slightly modified from Soil Survey Division Staff, *Soil Survey Manual,* USDA Handbook No. 18, U.S. Government Printing Office, Washington, DC 20402, revised October 1993, 174–175.

Table 3-7 Plasticity* Classes

Plasticity Classes	*Test Description*
Nonplastic	A roll 4 cm (1.6 in) long and 6 mm (1/4 in) thick that supports its own weight held on end cannot be formed.
Slightly plastic	A roll 4 cm long and 6 mm thick can be formed and, if held on end, will support its own weight. A roll 4 mm thick will not support its weight.
Moderately plastic	A roll 4 cm long and 4 mm thick can be formed and will support its own weight, but a roll 2 mm thick will not support its own weight.
Very plastic	A roll 4 cm long and 2 mm thick can be formed and will support its weight.

*Plasticity is the degree to which puddled <2 mm soil material is permanently deformed without rupturing by force applied continuously in any direction.
Source: Soil Survey Division Staff, *Soil Survey Manual,* USDA Handbook No. 18, October 1993, 178.

Table 3-8 Stickiness* Classes

Stickiness Classes	*Test Description*
Nonsticky	After release of pressure between thumb and forefinger, no soil material adheres to thumb or forefinger
Slightly sticky	After release of pressure, soil material adheres perceptibly to both digits. As digits are separated, material comes off one or the other rather cleanly.
Moderately sticky	After release of pressure, soil material adheres to both digits and tends to stretch slightly rather than pull completely free from either digit.
Very sticky	After release of pressure, soil material adheres strongly to both digits so it stretches when the digits are separated; material remains on both digits.

*Stickiness is the capacity of puddled <2 mm soil to adhere to other objects.
Source: Soil Survey Division Staff, *Soil Survey Manual,* USDA Handbook No. 18, October 1993, 179.

Table 3-9 Penetration Resistance Classes and Soil Toughness

Penetration Resistance Classes		***Toughness Classes***	
Classes	*Resistance* MPa*	*Classes*	*Criteria*
Small	<0.1	**Low**	Can reduce the specimen diameter at or near the plastic limit to 3 mm by <8N force (0.8 kg).
Extremely low	<0.01		
Very low	0.01–0.1		
Intermediate	0.1–2	**Medium**	Requires 8–20 N to reduce the specimen diameter at or near the plastic limit to 3 mm.
Low	0.1–1		
Moderate	1–2		
Large	>2	**High**	Requires >20 N (2 kg) to reduce the specimen diameter at or near the plastic limit to 3 mm.
High	2–4		
Very high	4–8		
Extremely high	≥8		

*Above 2 MPa, root penetration is greatly restricted; below about 1 Mpa, root restriction is small.
Source: Soil Survey Division Staff, *Soil Survey Manual,* USDA Handbook No. 18, October 1993, 179, 183.

Soil Color vs. Soil Properties

Soil color indicates many soil features. A change in soil color, compared to the adjacent soils, indicates a difference in the soil's mineral origin (parent material) or in soil development. White colors are common when salts or carbonate (lime) deposits exist in the soil. Spots of different color (**mottles**), usually rust colored, indicate a soil that has periods of inadequate aeration each year. Bluish, grayish, and greenish subsoils (**gleying**), with or without mottles, indicate longer periods each year of waterlogged conditions and inadequate aeration. Usually a soil horizon with a chroma ≤2 indicates a low level of aeration.

Within geographic regions darker colors usually indicate higher organic-matter contents. However, between contrasting climatic conditions, color is not a good indicator of organic matter content (for example, soils with high concentrations of dark-colored minerals may be dark colored, and some partially decomposed humus is darker in some environments than in others).

Munsell Soil Color Charts

Soil color determination is standardized and determined by the comparison of the soil color to **Munsell color charts.** These charts are similar to books of color chips found in paint stores in which many gradations of different color groups are shown (Figures 3-6 and 3-7). There are color charts for plant foliage and for other uses.

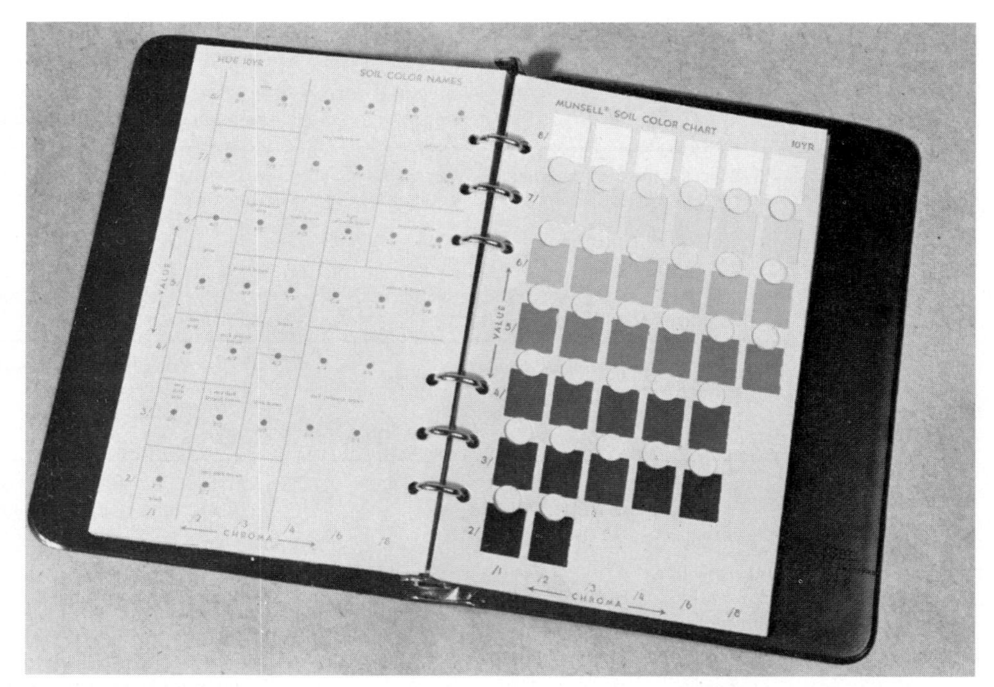

FIGURE 3-6 Sample page from the Munsell soil color charts, showing the hue 10YR. Color values appear vertically on the chart, and chroma (brilliance) is horizontal. Official color names at left are assigned to the specific color chips shown on the right page. Example: 10YR 4/3 is a dark brown. (Courtesy Munsell Color, 2441 N. Calvert St., Baltimore, MD 21218.)

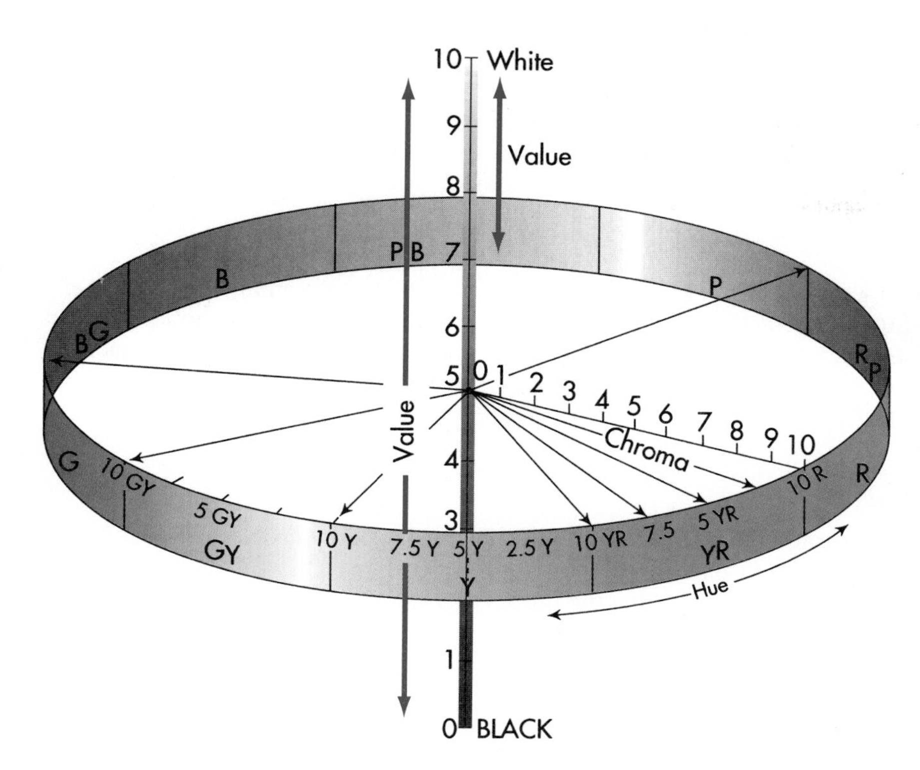

FIGURE 3-7 Illustration of the Munsell color scheme showing the hue terms and their subdivisions into 2.5, 5, 7.5, and 10 units. Chroma subdivisions range from 0 (neutral) to 10 (a very select portion of wavelengths, called *brilliance*). The value term ranges from 0 (black, no reflected light) to 10 (white, maximum light reflected). (Courtesy of Raymond W. Miller, Utah State University.)

Soil color notation is divided into the following three parts:

Hue: *The dominant spectral or rainbow color (red, yellow, blue, and green).*
Value: *The relative blackness or whiteness; the amount of reflected light.*
Chroma: *Purity of the color (chroma number increases and the color is more brilliant or bright as grayness decreases).*

To measure color and record it, match the soil color to the appropriate color chip and record the hue, value, and chroma in that order. Example recordings:

dark red	10R 3/6
reddish brown	5YR 6/3
black	7.5YR 2/0
light gray	10YR 7/1
yellow	5Y 7/8

Think of 10R = red, YR = orange, Y = yellow. Then, low value is blacker and high value whiter. The last number (chroma) has high numbers for more brilliance—less *muddy* color.

3:11 Soil Temperature

Soil temperatures vary from perpetual frost (**permafrost**) at a shallow depth in many soils in frigid Alaska and Siberia to warm, tropical areas, where daytime temperatures of the bare soil surface seldom fall below 40°C (104°F) on sunny days.

Relation of Soil and Air Temperature

Heat is a form of energy. Heat energy is lost from the Earth to space and is replaced by energy from the Sun. The temperature of the Earth and its atmosphere is a result of the *net* energy (R_N) of *incoming shortwave solar radiation* (R_S) minus the losses into space of (1) *reflected shortwave radiation* (albedo, R_R) and (2) *longwave infrared radiation* (emissivity, R_L).

$$R_N = R_S - R_R - R_L$$

Figure 3-8 illustrates the more complex energy flow, including *heat absorbance by the ground* (G), *heat absorbed by the air* (H), the *latent heat* (LE) used mostly to evaporate water, and *longwave radiation emitted from the Earth* (RL), which is lost as degraded heat energy. The more complete relationship follows:

$$R_N = G + H + LE = RL$$

The net heat absorbed by the Earth equals the heat lost in the form of far-infrared radiation. If this were not so, the Earth would be cooling or heating up gradually year after year.

Water has great influence on temperatures of soils and of the atmosphere (see Figure 3-8). Heating, cooling, and evaporation of water exert immense effects on the heat energy of an area. The evaporation process absorbs relatively large amounts of heat; condensation, likewise, releases large quantities of heat back to the system.

Detail 3-2 presents an example of relative numerical values for the energy balance in the atmosphere and in the Earth's surface. *Positive values are absorption of heat; negative values are losses of heat from that system.* Temperatures of an area of the Earth will be greatly modified by the water contents in that area.

The standard soil temperature is measured at a depth of 50 cm (19.5 in) or at the rock or hardpan contact if the soil is not 50 cm deep. The **average annual soil temperature (AST)** can be measured at a depth of 6 m (20 ft), although the value at about 3 m (10 ft) deep is usually quite close. The AST can be approximated by adding about 1°C (about 2°F) to the mean annual *air* temperature. This rule of thumb is not as accurate in arid, sunny regions as it is in other areas.

Soil temperatures resulting from the Sun's radiation change with depth and with the time of day. For example, the maximum daily temperature at deep soil depths is delayed, even by several hours, after the time when the air temperature reaches its maximum. Conversely on cool nights the deep soil layers do not cool as fast as surface layers because of the insulating effect of the overlying soil. Heat flow is slower in soil than in the atmosphere. The deeper the soil layer, the longer it takes a temperature change to reach it, and thus the less will be the actual temperature fluctuation from day to day or week to week (Figure 3-9). Daily temperature fluctuations seldom affect the soil deeper than about 30 cm to 40 cm (12 in to 16 in). Below about 1 m (3.3 ft) the soil changes slowly from season to season. The mean summer and mean winter temperatures at 1 m deep seldom differ by more than 5°C (9°F) in the subtropics; differences increase and decrease several degrees in temperate regions.

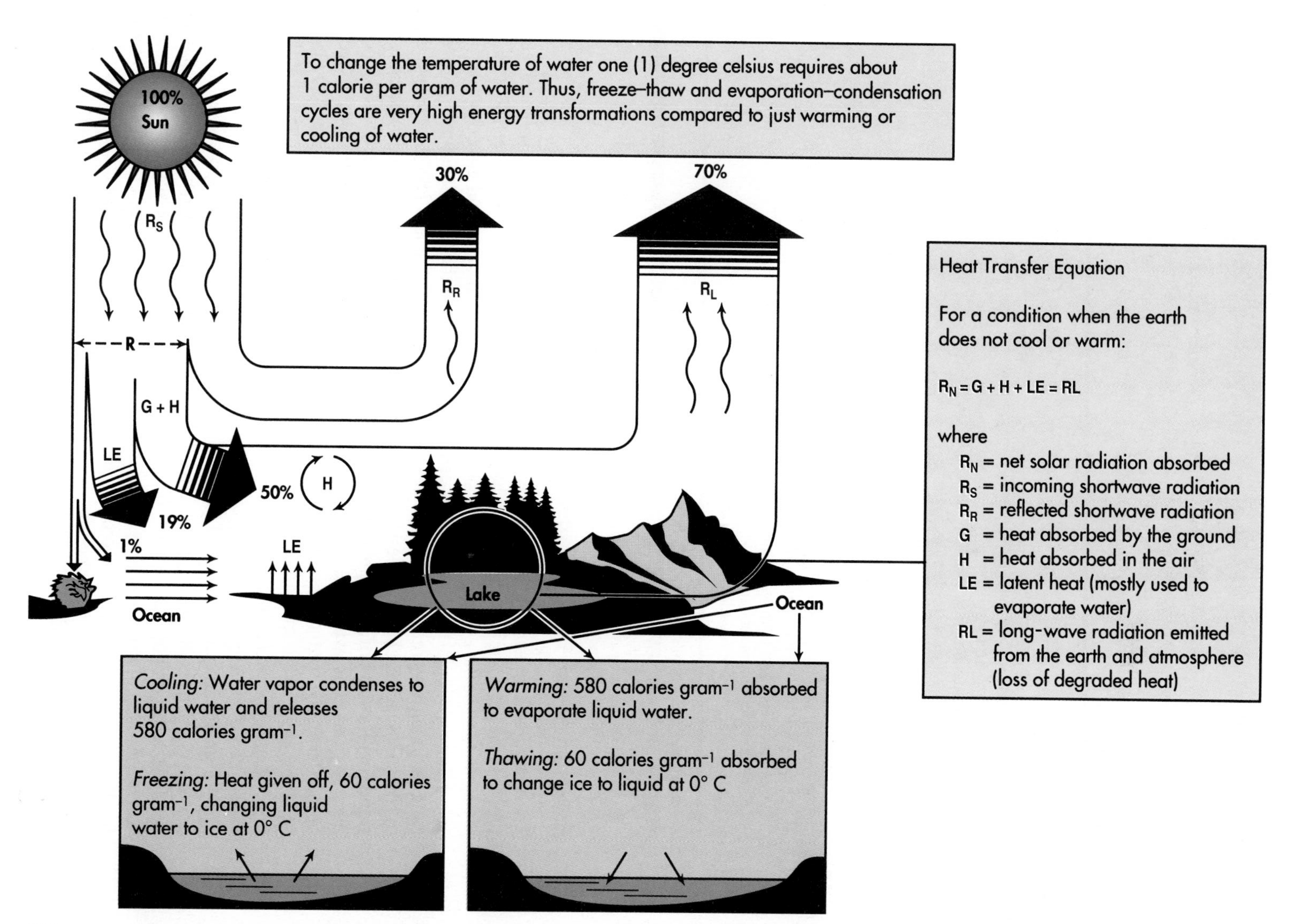

FIGURE 3-8 An illustration of short-wave incoming energy (Sun) and energy transformations in and from the Earth and its atmosphere, emphasizing the energy losses and gains modified by changes in phases of water. (Courtesy of Raymond W. Miller, Utah State University.)

Detail 3-2 Energy (Heat) Budgets

The Earth receives energy (short-wave) from the Sun and loses energy (long-wave and heat) to space. The overall Earth's balance is zero, but at any one location on the Earth that location may have a net energy gain (spring warming) or net energy loss (fall cooling). A typical yearly average for the Earth might be as follows:

*Gain and Loss**	*Sun's Short-Wave*	*Degraded Long-Wave*	*Net*
Space energy budget (= no change in Earth's temperature)			0
Atmospheric energy budget			0
Emission by water vapor to space	—	−38	
Emission by liquid water (clouds) to space	—	−26	
Absorbed sun energy by water, dust, ozone	+16	—	
Absorbed sun energy in clouds	+3	—	
Net heat from Earth (see last 4 items below)	—	+45	
Surface energy budget			0
Incoming solar radiation to ground	+51	—	
Emission of energy by Earth to space	—	−6	
Emission of energy by Earth to atmosphere	—	−118	
Absorption by Earth from water and CO_2	—	+103	
Sensible heat to atmosphere	—	−7	
Latent heat transfer (evaporation) to atmosphere	—	−23	

*Average long-term (year-round) average of 100% of incoming Sun's radiation.

Factors Influencing Soil Temperature

The temperature a soil attains depends on (1) how much heat reaches the soil surface (heat supply) and (2) what happens to that heat within the soil (dissipation of heat). The heat supply to the soil surface from external sources is reduced by organic soil coverings, which act as insulators. All plastic mulches make the soil hotter when the Sun is shining, but transparent plastic allows greater warming than opaque plastic. A Sun angle less than perpendicular to the soil surface also supplies less heat per unit of soil area (Figure 3-10). North-facing slopes in the northern hemisphere will receive less heat than will other directional slopes during winter months, when the Sun is low in the southern sky. (The reverse slopes have these properties in the southern hemisphere.) Rock gardens, or nonlevel lands, sloped to be closer to a right angle to the springtime Sun's rays will receive more heat per soil area and will warm faster than will flat (oblique-to-the-sun) surfaces. Light-colored soils will reflect more heat than will darker soils.

Some heat that is absorbed by soil is lost in evaporating soil water. If a soil's water content is high, much more heat is needed for temperature changes because the heat capacity (heat reservoir) of the water is three to five times more than for soil minerals. In a dark soil where the color is caused by large amounts of humus, the larger amounts of water held by the humus may offset the increased heat absorption due to the dark color. Although compacted soils have better heat conductivity than does loosened soil, they also have more material to heat per volume of soil. Thus a rock garden sloped south or west with sand soil (which holds less water) will warm fast in the spring after warming begins. It has less heat capacity to satisfy.

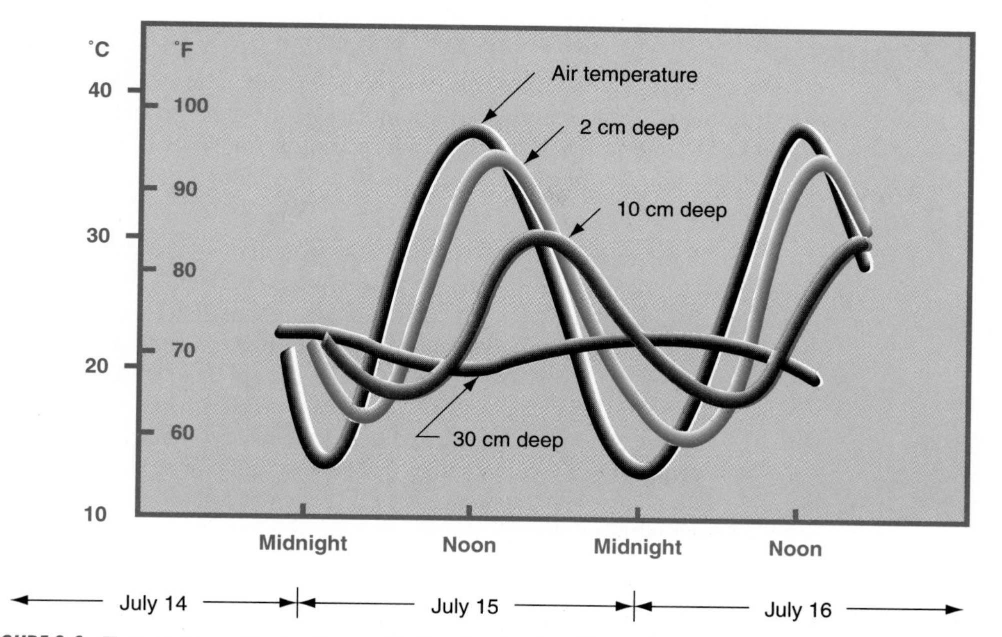

FIGURE 3-9 Temperature fluctuations with the depth of soil in northern Utah illustrating the reduced amplitude of change with deeper soil depth and the delay in time when maximums and minimums are reached. (Courtesy of Raymond W. Miller, Utah State University.)

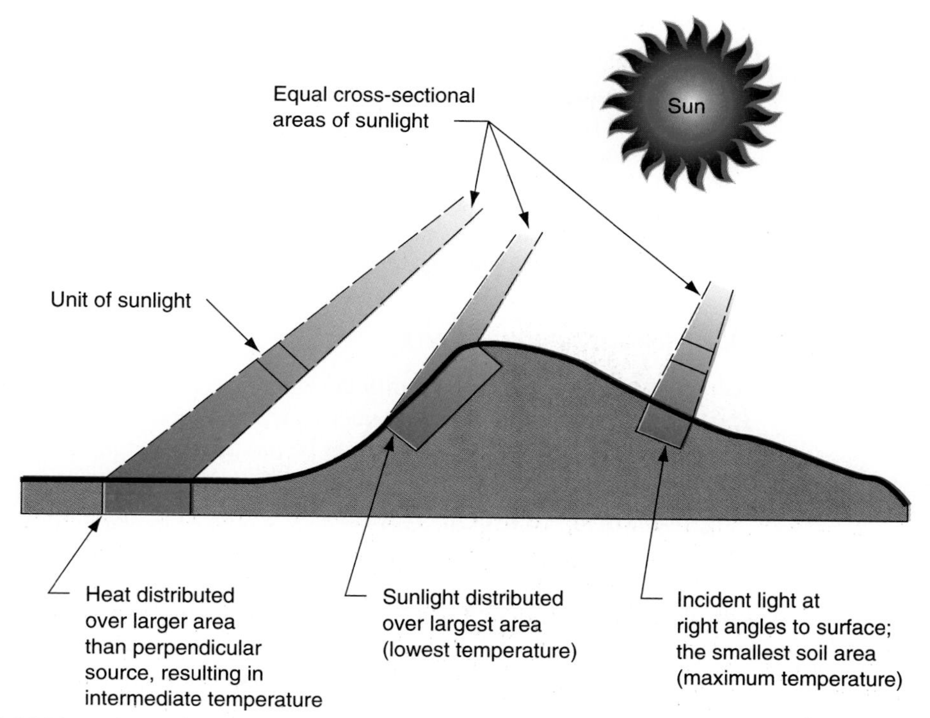

FIGURE 3-10 Effect of soil slope, aspect (direction of slope), and the Sun's position on the heat supplied to soil. (Courtesy of Raymond W. Miller, Utah State University.)

Living with Existing Temperatures

Generally, existing temperatures have just been endured and agricultural activities modified to fit them. Temperature restrictions can be mitigated usually only by expensive measures. Acknowledging the limitations and planning for them is wise soil and crop management. Some examples follow:

- To get maximum germination and growth of seeds, soil temperatures must be correct; 4°C–10°C (40°F–50°F) for wheat and peas; 10°C–29°C (50°F–85°F) for corn; 16°C–21°C (60°F–70°F) for potatoes; about 27°C (80°F) or above for sorghums and melons. Optimum emergence temperatures for other plants include the following: 8°C–11°C (46°F–52°F) for cabbage and spinach; 11°C–18°C (52°F–64°F) for beets and cauliflower; and 18°C–25°C (64°F–72°F) for asparagus, carrots, celery, endive, lettuce, onions, radishes, and tomatoes. Direct planting of onion seeds into cool spring soil (a recent change from transplants; done because of labor costs) produces late plants. When pregerminated seeds were planted, they emerged in 1.7°C–7.3°C soil within 7 days; regular seeds required 30 days. Maturity of onions from pregerminated seeds was 10–12 days earlier. Geranium seeds have most rapid germination at about 27°C (80°F). Sweet corn covered with clear plastic in Pennsylvania or other Midwest states has germinated 5 or 6 days earlier than uncovered seed rows.[3]
- When applying anhydrous ammonia in the fall, it is best to wait until the soil at a depth of 10 cm (4 in) is 10°C (50°F) or less. Below this temperature, conversion of ammonium to nitrate is slow and, therefore, leaching losses of nitrate will be minimal.
- Freezing and thawing of bare, saturated, fine-textured soils in cold areas, such as the intermountain West and the northern United States, may cause heaving and then death of shallow-rooted crops. The frost heaving of rock fragments requires many farmers to remove a new crop of rocks from plowed fields each year.
- Alternate freezing and thawing under conditions of moderate soil moisture improves the structure of cloddy soils, but this process with excess moisture destroys structure. Freezing and thawing of wet loessial and other silty slopes causes erosion as mudflows.

Modifying Temperature Effects

As technology enables us to modify the environment, we become less satisfied with tolerating existing undesirable climates. Conquering temperature, except in greenhouses, is not currently possible, but the effects in the field can be modified in limited ways, such as the following:

- The use of clear-plastic surface covers (flat or in canopy rows) in cold areas or cold seasons increases soil temperatures and permits successful plant growth during the cold period (Figure 3-11). The use of clear plastic mulches in Iowa and Alaska will mature sweet corn 4 to 8 days earlier than without it. In Orange County in southern California, plastic mulch speeds the growth (and protects berries from mold) of an early strawberry crop worth $15 million, produced on only 891 hectares (2200 acres).

---

[3]Rick Melnick, "Clear (Plastic) Options for Sweet Corn Production," *American Vegetable Grower* **44** (1996), No. 2, 17–18.

FIGURE 3-11 These clear-plastic strips are trapping the rays of the sun and warming the root environment, thus making the growth of this corn possible in Alaska. (*Source:* Lee Allen, D. H. Dinkel, and Arthur L. Brundage, University of Alaska Institute of Agricultural Sciences, *Agroborealis,* **1,** No. 2, Sept. 1969.)

- A black polyethylene mulch on the soil in a pineapple plantation in Hawaii was responsible for a 50% increase in the growth of pineapple. This growth increase was attributed to an increase in soil temperature of 1.5°C (2.7°F) during winter and not to an increase in soil moisture. Black mulches retain moisture and control weed growth (no light) but transmit less heat rays through the soil than do clear plastic mulches.
- In February early snow removal, using fly ash or graphite, has reduced winter wheat losses caused by snow mold in Idaho and northern Utah.[4,5] Snow-mold damage is increased with the length of time snow covers the land in spring. Sometimes snow mold kills 50%–70% of the grain; yields are reduced or replanting of spring grains is required. *Graphite* was applied in 1989 at rates of 13–20 kg/ha (at $0.45/kg) by airplane or snow-cat. Commercial applicators charged $15/ha. Graphite can be applied in liquid nitrogen and/or herbicide solutions. Treated snow usually melts 2–3 weeks earlier than does snow without graphite. In Utah small grain breeding nursery stock has been secured by yearly use to minimize snow-mold losses. It is being used on mountain pastures in April to speed plant growth on grazing lands (Figure 3-12).
- Mulches left on the soil through the winter act as insulation and retard warming in the spring. Incorporating straw stubble into the soil in the fall rather than leaving it on the surface as a mulch resulted in about 40% more growth of early corn in Minnesota.
- Early fruit bloom following a prolonged warm period is often followed by a freeze that can eliminate a year's production of apples or other fruits. Cooling by sprinkling

[4]T. A. Tindall, "Darkening Agents in a UAN Solution: Influence on Snow Mold and Yield of Winter Wheat," *Journal of Fertilizer Issues* **3** (1986), No. 4, 129–132.
[5]T. A. Tindall and S. A. Dewey, "Graphite-Nitrogen Suspensions with Selected Herbicides Applied to Snow Cover in Management of Winter Wheat," *Soil Science* **144** (1987), 218–223.

(a)

(b)

FIGURE 3-12 Distributing graphite on snow in Utah in early spring to remove snow cover. A fungus, called *snow mold,* becomes very active under the snow as the soil thaws and warms. Early snow removal can save a wheat crop. (a): Spreading. (b): Cleared areas (black strips) show where snow melted on research fields because of ash cover. (Courtesy of Dr. Terry Tindall, Utah State University.)

has maintained fruit buds in a dormant condition until flowering is safe. Spring sprinkling of lettuce to prevent frost damage is also a common practice.

- Specially contoured and planted rows can modify low spring temperature effects by protecting the young plants with soil and positioning them to receive more direct sunlight (greater warming), such as in many rock gardens.

3:12 Other Soil Physical Properties

Additional soil physical properties include plasticity stickiness, smeariness, and fluidity.

Plasticity: *The degree to which soil is permanently deformed, without rupture, by a force* applied continuously in any direction. Plasticity is determined in the field by rolling wet soil between the hands until a 3-mm (1/8-in) cylinder is formed. With continuous rolling while it dries, this 3-mm "wire" will break when the soil has dried to approximately the *plastic limit* (see following sections).

Stickiness: *The property of wet soil to adhere to another object.*

Smeariness: *A field term for clays that are thixotropic.* **Thixotropy** is the property exhibited by certain fine-textured clay gels of being "solid" upon standing but turning liquid when shaken or mixed. Many swelling clays used in well-drilling muds to lubricate the drill and lift cuttings to the surface are thixotropic.

Fluidity: *The property of nonthixotropic clay soils that flow under hand-pressure.*

3:13 Soil Physical Properties in Engineering Use

At least three systems of physical soil properties are used in engineering evaluation of soils for various construction purposes. Two of the most used are (1) the American Association of State Highway and Transportation Officials **(AASHTO)** and (2) the **Unified Classification System,** growing out of the Airfield Classification by the Corps of Engineers and the Bureau of Reclamation. The more used of these systems is the AASHTO, but there is increasing use of the Unified system. Although these two systems are in a sense *classification systems,* they are discussed in Chapter 3 because they are based mostly upon particle-size measurements and cohesion of the soil when it is moist or wet. Both systems involve particle-size groupings a bit different from the size fractions used in the USDA system.

In attempts to quantify and standardize procedures, those developing the AASHTO and Unified systems measured most particle-size groups by the amount of material that *passes through* or *is retained on* sieves of various sizes (Table 3-10). Material passing through a U.S. Standard No. 200 mesh sieve (200 wire-divisions per inch) is smaller than 0.075 mm (silts and clays, very fine sands). Material passing through a No. 40 mesh sieve is smaller than 0.425 mm (fine sands). Material passing through a No.10 sieve (10 wires per inch) is 2.0 mm or smaller (sand, silt, and clay), and a No. 4 sieve (4 wires per inch) retains material larger than 4.75 mm, which is coarse material (gravel and larger). Little separation of silts and clays is done because, to the engineers working with soils in construction, the rocks, gravels, and sands are of primary interest, although they must also have some

Table 3-10 Engineering Classification of Grain Sizes

Soil	*Diameter (mm)*	*Sieve Size*
Boulders	More than 300	12 in (about 31 cm)
Cobbles (rounded)	300–75	12 in–3 in
Gravel	75–4.76	3 in–Sieve No. 4 retains
Coarse	75–19	3 in–3/4 in
Fine	19–4.76	3/4 in–Passes Sieve No. 4
Sand	4.76–0.074	Sieve No. 4–sieve No. 200
Coarse	4.76–2.0	Sieve No. 4–sieve No. 10
Medium	2.0–0.42	Sieve No. 10–sieve No. 40
Fine	0.42–0.074	Sieve No. 40–sieve No. 200
Fines	Less than 0.074	Passes Sieve No. 200
(Silt size)	0.074–0.002	
(Clay size)	Less than 0.002	

Courtesy of Raymond W. Miller, Utah State University.

concern with the properties of silts and clays. Using these four sieves produces the following results:

Soil material *not passing* the No. 4 is gravel or coarser,
Soil material *retained* on the No. 10 is finer gravel (all passing through is <2 mm),
Soil material *retained* on the No. 40 holds the coarser sands, and
Soil material *retained* on the No. 200 is the finer sands.
Soil material *passing* through the No. 200 mesh sieve is silts and clays: it can only be separated by settling the soil suspended from water (mechanical analysis).

AASHTO System

The **AASHTO** classification system is based mostly upon three physical properties of a soil: (1) the mechanical analysis or particle-size fractions, (2) the liquid limit, and (3) the plasticity index. The particle-size analysis involves a particle-size grouping different from that of the USDA. Table 3-11 shows the AASHTO and three other engineering particle-size classifications. Notice that coarse sand in the *Unified system* has an upper size of 4.75 mm; its lower size of 2 mm is where USDA coarse sand begins. All fines in the engineering system are grouped together and include USDA clays, silts, and some very fine sands. It is obvious that the coarse fragments and sands are given considerable attention by engineers because of the uses they make of soils (subgrades, embankments, roadbeds, building support, dikes, etc.); see Figure 3-13.

Table 3-12 is used to simplify the determination of the group or subgroup in the **AASHTO** classification to which a soil belongs (see Calculation 3-4). Starting from the left column in Table 3-12, the first column into which the soil fits is its group and subgroup. For soils in groups **A-4, A-5, A-6,** and **A-7,** the ranges of the liquid limit and plasticity index are given in Figure 3-14.

Table 3-11 The Grain-Size Grouping for the USDA System and Three Widely Used Soils Mechanics Engineering Classification Systems

Classification System	Grain size, mm			
	100 — 10 — 1 — 0.1 — 0.01 — 0.001 — 0.0001			
USDA (1975)	Gravel	Sand	Silt	Clay
	75 — 2	0.05	0.002	
MIT (1931)	Gravel	Sand	Silt	Clay
	2	0.06	0.002	
AASHTO (1970)	Gravel	Sand	Silt	Clay Colloids
	75 — 2	0.05	0.002	
Unified (1953)	Gravel	Sand	Fines (silt + clay)	
	75 — 4.75	0.075		

Courtesy of Raymond W. Miller, Utah State University.

FIGURE 3-13 Engineers are concerned about the ability to compact soils, to have stable slopes and footings, and to avoid costly mistakes about unstable soil. This house in West Virginia was built on a leveled but not stabilized soil base. The Soil Conservation Service listed the soil and slope as having *severe* hazards. The evidence is clear and expensive. The house was torn down for salvage. (Courtesy of USDA—Soil Conservation Service.)

Liquid limit (LL): *The water content in the soil at which the soil will flow under a standardized agitation procedure.* A small amount of soil passing the No. 40 sieve (0.425 mm, smaller than medium sands; see Table 3-12) is mixed with water to a paste consistency, placed in a round-bottomed brass cup, and excess soil struck off to leave a maximum thickness of 10 mm (see Figure 3-15). The soil pat is divided into two segments with the grooving tool, and the crank on the tool is turned. The *water content at which 25 blows (cranks) cause the groove to close* is the **liquid limit.**

Plastic limit (PL): *The minimum water content at which the mixture acts as a plastic solid.* It is measured by rolling the wetted soil with the palm of the hand onto a frosted glass or other mildly absorbent surface into a thread or worm of soil 3 mm (1/8 in) diameter. This is repeated (soil gradually dries while being reworked several times) until the thread breaks up into short pieces as the rolling soil thread approaches the 3 mm diameter. This *water content when the thread breaks* is the **plastic limit.**

Plasticity index (PI): *The numerical difference between the liquid limit and the plastic limit.* It is the range of water content within which the soil exhibits the properties of a plastic solid; it is a measure of the cohesive properties of the soil. The term **silty** is applied to fine material having a **PI** of 10 or less. The term **clayey** is for fine material with a **PI** of 11 or greater, after rounding to the closest whole number. **Peats** and **mucks** were in the original classification as Group A-8 and usually fit Group A-7 but are left as Group A-8. They have *low density, high compressibility, high water content,* and *high organic-matter content.*

Table 3-12 Table to Determine AASHTO Classification of Soils

General Classification →	*Coarse: Granular Materials (35% or less of total sample passing No. 200)*							*Fine: Silt and Clay Materials (More than 35% of total sample passing No. 200)*			
Group Classification →	**A–1**		**A–3**	**A–2**				**A–4**	**A–5**	**A–6**	**A–7**
	A–1a	**A–1b**		**A–2–4**	**A–2–5**	**A–2–6**	**A–2–7**				
Percent passing sieve:											
No. 10 →	50 max.										
No. 40 →	30 max.	50 max.	51 max.								
No. 200 →	15 max.	25 max.	10 max.	35 max.	35 max.	35 max.	35 max.	36 min.	36 min.	36 min.	36 min.
Fraction passing sieve No. 40											
Liquid limit →				40 max.	41 max.	40 max.	41 max.	40 max.	41 min.	40 max.	41 min.
Plasticity index →	6 max.		NP	10 max.	10 max.	11 min.	11 min.	10 max.	10 max.	11 min.	11 min.
Usual constituent materials →	Stone fragments, gravel and sand		Fine Sand	Silty or clayey gravel and sand				Silty soils		Clayey soils	
Subgrade rating →	Excellent to good							Fair to poor			

Begin in the left column and proceed to the next column on the right until the soil characteristics fit the criteria in that column. Notice that group **A–3** is purposely out of letter sequence.
Modified after Duane L. Winegardner, *An Introduction to Soils for Environmental Professionals,* Lewis Publishers, New York, 1996.

Calculation 3-4 Determining AASHTO Group and Subgroup

Problem If a soil contains 65% of material passing a No. 200 sieve, has a liquid limit of 48, and has a plasticity index of 17, to what group does it belong?

Solution

1. Since more than 35% of the soil material passes the No. 200 sieve, it is a silt–clay material, so it must fit one of groups **A-4** to **A-7.** Examine the specified limits for group **A-4** first (see Table 3-12). The maximum value of the liquid limit in this column is 40, and the liquid limit of the soil being classified is 48. Therefore, it cannot be an **A-4** soil.
2. By proceeding to the right and comparing the actual soil properties with the values given in each successive column, it is found that the soil fits the specifications in the last column. The soil belongs to group **A-7.**
3. If it is desired to classify this soil into a subgroup, subtract 30 from the liquid limit of the soil (see Figure 3-14). The difference in this example is 18. Since the plasticity index is less than this difference, the soil will be above the diagonal line and is in the **A-7-5** subgroup. If the plasticity index had been greater than the liquid limit minus 30, the material would have been an **A-7-6** soil.

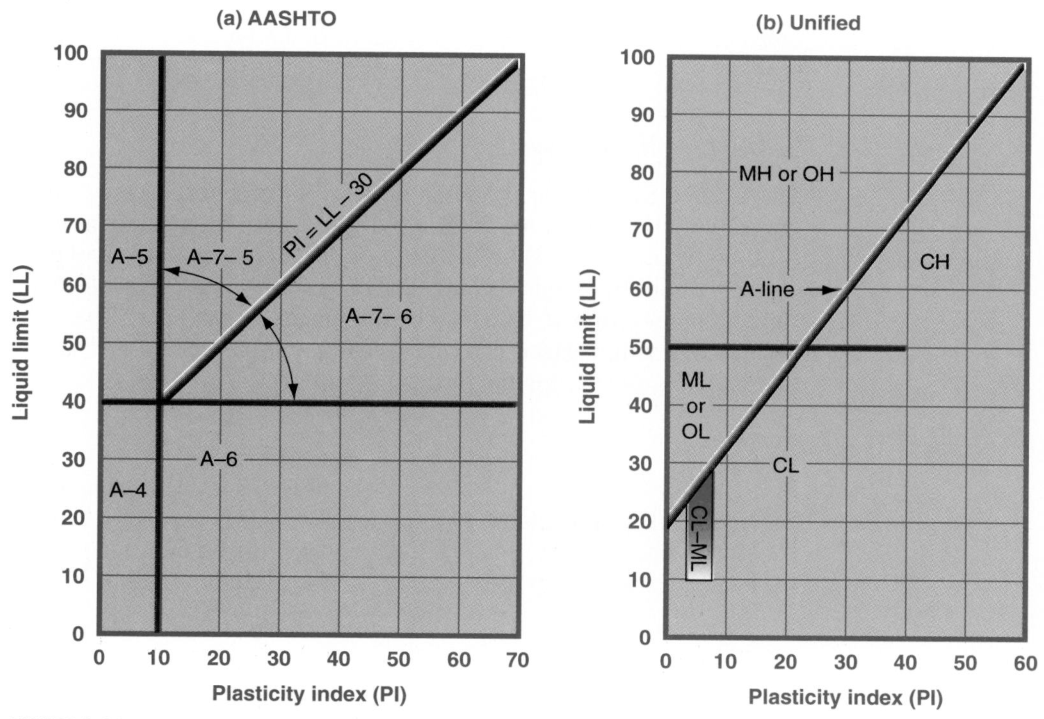

FIGURE 3-14 Visual soil identification using the liquid limit and plasticity index for the (a) AASHTO classification system and (b) Unified Classification system. (Courtesy of Raymond W. Miller, Utah State University.)

Check the fall distance of the cup in the liquid limit device and adjust, if necessary, so that the height of fall is exactly 1 cm. It is important that this measurement be made between the base and the point on the cup which comes in contact with the base. The grooving tool handle is a 1 cm rule.

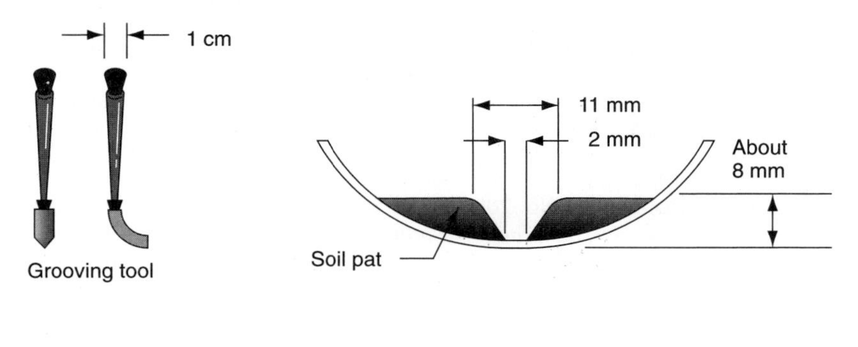

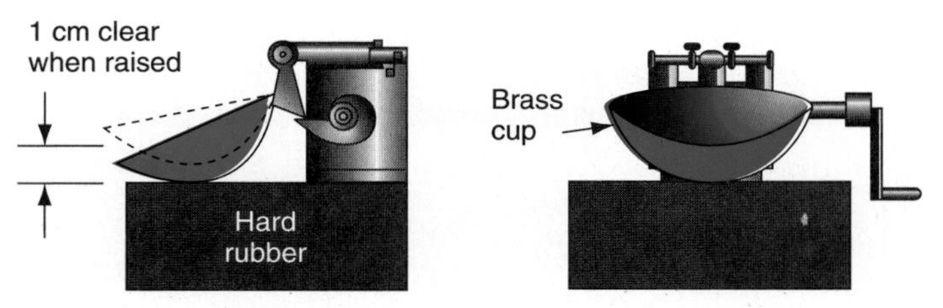

FIGURE 3-15 Equipment used to determine the liquid limit. The number of *cranks* is the number of *drops* of the bowl—one drop per revolution of the handle. (Courtesy of Raymond W. Miller, Utah State University.)

Unified Classification System

The Unified System was proposed because of the unsatisfactory nature of systems based on grain (particle) size alone. Soils are divided into three major groups: (1) **coarse-grained,** (2) **fine-grained,** and (3) **highly organic (peaty).** If more than 50 percent of the grain particles in the soil (by weight) is judged to be visually distinguishable by the naked eye, it is considered *coarse-grained.* By actual measurement if more than 50 percent is larger than No. 200 mesh (0.075 mm, coarser than fine sands), the soil is *coarse-grained.* All soils will be classified into 15 groups using the following symbols:

G: Gravel
S: Sand
M: Nonplastic or low plasticity fines
C: Plastic fines (clayey)
Pt: Peat, humus, swamp soils

O: Organic
W: Well graded (many sizes mixed)
P: Poorly graded (mostly a few fines)
L: Low liquid limit
H: High liquid limit

The **P** and **W** soils have very few fine materials. The **M** and **C** soils are *dirty* (have more than 12 percent material finer than 0.075 mm) but differ in *plastic* or *nonplastic* nature of the fines.

The soils are given double-letter symbols, such as **CH, GW,** and **SP,** to make up the 15 soil groups. The groups and their characteristics are shown in Figure 3-16. Notice that the soil classification is just that, a classification. It does not describe which materials are best for the purposes for which you wish to use the soil, although some tables are available to help suggest qualities. As *subgrade* material, usually the coarser fractions **(GW, GP, GM, GC, SW, WP, SM,** and **SC)** are better than those with more fines. The **P** (poorly graded) materials may

Major Divisions			Group Symbols	Description
Coarse Grained Soils Over 50% retained on 0.075 mm (#200) sieve	**Gravels** 50% or more retained on 4.75 mm (#4) sieve **Sands** <50% passes a #4 sieve	**Clean Gravels**	**GW**	Well-graded gravels and gravel-sand mixtures, all sizes, little or no fines
			GP	Poorly graded gravels and gravel-sand mixtures, few sizes, little or no fines
		Gravels with Fines	**GM**	Silty gravels, gravel-sand-silt mixtures
			GC	Clayey gravels, gravel-sand-clay mixtures
	Sands >50% passes a #4 sieve	**Clean Sands**	**SW**	Well-graded sands and gravelly sands, all sizes, little or no fines
			SP	Poorly graded sands and gravelly sands, few sizes, little or no fines
		Sands with Fines	**SM**	Silty sands, sand-silt mixtures
			SC	Clayey sands, sand-clay mixtures
Fine-Grained Soils 50% or more passes 0.075 mm (#200) sieve	**Silts and Clays** Liquid limit 50% or less		**ML**	Inorganic silts, very fine sands, rock flour, silty or clayey fine sands
			CL	Inorganic clays of low to medium plasticity, gravelly clays, sandy clays, silty-clays and lean clays
			OL	Organic silts and organic silty clays of low plasticity
	Silts and Clays Liquid limit greater than 50%		**MH**	Inorganic silts, micaceous or diatomaceous fine sand or silts, elastic silts
			CH	Inorganic clays of high plasticity, fat clays
			OH	Organic clays of medium to high plasticity
Highly Organic Soils			**PT**	Peat, muck and other highly organic soils

FIGURE 3-16 The Unified engineering soil classification system. (Modified after Irving S. Dunn, Loren R. Anderson, and Fred W. Kiefer, *Fundamentals of Geotechnical Analysis,* John Wiley & Sons, New York, 1980; and Duane L. Winegardner, *An Introduction to Soils for Environmental Professionals,* Lewis Publishers, New York, 1996.

lack enough fines of different sized materials for good binding of compacted soils. The **W** (well-graded) soils have many sizes of particles, which makes them compact into a tighter, hard mass. Some **M** (low plastic) materials may lack enough cohesion to bind the compacted soil to make a good subgrade.

Those soils with more than half of the material smaller than a No. 200 sieve (<0.074 mm) contain a lot of silt and clay and require the information about their **LL**s and **PI**s to classify them (see Figure 3-16). Generally in decreasing order from *good* to *very poor,* these finer-textured soils can be listed as subgrade soil material as follows:

ML > CL > OL > MH > CH > OH > Pt

The Unified Soil Classification System involves relatively few and inexpensive laboratory tests and provides a practicable basis for visual or field classification. However, the information given in this text can serve only as a starting point for describing engineering properties of soil masses. Shear strength, swelling, void ratios, and other measurements may be needed. You need, additionally, to take into account the characteristics of the soils *as they are found in nature.* One text suggests that classification systems must be used with caution[6]: "Blindly determining physical properties, such as compressibility, from empirical relationships and then using them in detailed calculations can lead to disastrous results."

Other Measurements

Damage to structures caused by soils that expand when wet and shrink when dry is one of the most costly of natural soil hazards in the United States. These soils are referred to as *expansive* (some say *expensive*) soils. The tendency for a soil to shrink and swell is of utmost importance for engineering purposes. To have adequate information for many decisions, the engineer will also need to measure additional soil factors: *Void ratio* (volume of voids—pore space—divided by volume of solids), *porosity* (volume of voids divided by total soil volume), *soil density, toughness index* (coherence strength of the kind of clay), *shrinkage limits* (large, even more than 10 percent for some swelling clays), *activity indices* (larger for swelling and sticky clays), and *resistance to shear* (a crucial but more difficult measurement). The number and quality of these properties will depend upon the size and kind of construction being done. Details of the measurement and interpretation of these properties require too much space to justify in this text. For further reading see Terzaghi, Peck, and Mesri[7] and Spangler and Handy.[8]

> *The light in the world comes principally from two sources—the sun and the student's lamp.*
>
> **—Bovee**

> *Despite all his accomplishments, man owes his existence to a thin layer of topsoil and the belief that it will rain.*
>
> **—Larry Neppl**

---

[6] Irving S. Dunn, Loren R. Anderson, and Fred W. Kiefer, *Fundamentals of Geotechnical Analysis,* John Wiley & Sons, New York, 1980, 32.

[7] Karl Terzaghi, Ralph B. Peck, and Gholamreza Mesri, "Soil Mechanics in Engineering Practice," John Wiley & Sons, New York, 3rd ed., 1996.

[8] Merlin G. Spangler and Richard L. Handy, *Soil Engineering,* Harper & Row, New York, 4th ed., 1982.

Questions

1. Using the textural triangle, determine the texture for the following three soils: (a) 10% sand, 15% clay, and 75% silt; (b) 41% clay, 40% silt, and 19% sand; and (c) 16% clay, 28% silt, and 56% sand.
2. State how the following are altered, in general, by increased clay content: (a) water infiltration, (b) pore size, (c) total pore space, and (d) bulk density.
3. In considerable detail (giving particle sizes, for example), state what each of the following names tells about that soil: (a) a gravelly loam, (b) a very cobbly loamy sand, and (c) an extremely stony sandy loam.
4. (a) How does soil air differ from atmospheric air? (b) What are the critical gases in the soil air for root growth?
5. The revised *Soil Survey Manual* uses the term *soil strength.* What is measured to indicate soil strength?
6. Of the colors 5YR 2/2, 5YR 6/2, 5YR 3/6, and 5Y 6/2, state which is (a) the blackest, (b) the least reddish, and (c) the most brilliant colored.
7. What are some of the practical uses of soil color information?
8. State how each of the following affects the warming or the cooling of soils: (a) transparent-plastic mulches, (b) opaque-plastic mulches, (c) organic mulches, (d) water content in the soil, and (e) soil color.
9. Explain how and why large bodies of water modify nearby land area temperatures.
10. How are freezing–thawing cycles major controls on heat flow in the immediate vicinity (within a few centimeters)?
11. By what mechanisms do soils lose heat to the surrounding atmosphere?
12. Define *liquid limit, plastic limit,* and *plasticity index.*
13. If the liquid limit is 37% and the plastic limit is 113%, what is the plasticity index?
14. Define the following terms in engineering classification systems usage: *well graded, poorly graded, nonplastic, fines,* and *subgroups* such as the **A-2-6** subgroup.

4 Soil Water Properties

If there is magic on this planet, it is in water.

—Loren Eisley

You never know the worth of water 'til the well runs dry.

—Benjamin Franklin

4:1 Preview and Important Facts

PREVIEW

Soils are water reservoirs with limited capacities and fascinating properties. Worldwide, insufficient soil water is the factor most limiting to plant growth. Historically we have compensated for lack of rainfall with irrigation, which allowed us to stabilize plant production through dry periods and to expand farm acreage into arid regions. With abundant irrigation water, often heavily subsidized by governments, knowledge of many aspects of soil–plant–water relations was thought to be unimportant. Now, with fewer irrigated acres in the United States than in 1978,[1] a thorough understanding of soil water and new ideas for soil water management are sorely needed.

Water is essential for all life. Water dissolves nutrients, making them available to microbes and higher plants. Water is the transpiration stream that carries these nutrients from the roots to the leaves of plants before evaporating from the stomata. Water is the fluid in which the contents of living cells are dissolved or suspended. Water is the source of hydrogen in the vital process of photosynthesis.

A typical wetted and drained loam soil might contain water in about 25 percent of its volume, and only about half of that water would be available to plants. The amount of soil water available to plants relates closely to soil texture. Silt loams often hold the most available water. Water entry into soils (infiltration) is rapid in sands but slow in clays, which cre-

[1]U.S. Department of Commerce, 1992 Census of Agriculture, Volume 3 Part 1: Farm and Ranch Irrigation Survey (1994), U.S. Government Printing Office, Washington, DC, 1996.

ates challenges to management for optimum crop production. Sandy soils usually hold small amounts of plant-available water and, therefore, sands require frequent rains or irrigations to supply adequate water. Clayey soils have slow infiltration and drainage of excess water. The retained water fills bottleneck pores and blocks adequate air exchange, which causes poor aeration and restricted root growth.

When fully wetted and drained, a soil is said to be at **field capacity.** When a soil is so dry that it can no longer absorb water fast enough to satisfy water losses, plants will wilt and soon die; this condition is called **permanent wilting point.** The water loss *out of* leaves into the air is called **transpiration.** The movement of water *into* the soil, **infiltration,** is regulated by soil texture and structure. Free water that moves into the soil by gravity flow is called **saturated flow.** After all free water has moved as far as it can, the films of water on the soil particles can still move slowly from thick films, where water is held less strongly, to areas with thin films, where water is more strongly adsorbed; this is called **unsaturated flow.**

It is important to understand the forces holding water in soils and to understand the mechanisms and limitations of water movement into and through soils. The close relationship of soil texture, soil structure, and water flow and retention needs to be understood.

IMPORTANT FACTS TO KNOW

1. The properties of soil and water that result in retention of water in soil
2. Terminology used for water: *water potential, matric potential, plant-available water, field capacity, permanent wilting percentage, mass water ratio,* and *volumetric water ratio* and *percentage*
3. The approximate amounts of water held in soils of different textures at field capacity and at the permanent wilting point
4. Why soil water is—or is not—available to plants
5. How to measure soil water contents (ratios and percentages) on a mass and volume basis
6. How to estimate soil water contents using modern instruments
7. How environmental factors affect evapotranspiration
8. Typical rates of water flow under saturated and unsaturated soil conditions
9. Some methods to increase the amount of water retained in soils

4:2 Water and Its Relation to Soil

Water, the most common substance on the surface of our planet, is peculiar stuff. Water has more anomalous properties than any other common substance, and it seems to defy the rules of chemistry. A molecule as small as water should be a gas but, instead, water is a liquid at room temperature. Water has an extremely high specific heat and its heat of vaporization is the highest of any known substance. Solid phase water (ice) is less dense than the liquid phase, causing ice to float. High surface tension gives water its tendency to bead up or form drops rather than spread over a greater area. The physical and biological processes on earth are inseparably linked to the peculiar properties of water.

Soil is a leaky water reservoir. When too much water is added, the excess runs off over the surface or into deeper layers. Why does the soil hold some of the water yet allow part of it to drain deeper? To understand the answer to this question, one must understand the force causing the attraction between soil and water.

Water is held in soils because of **hydrogen bonding** (Figure 4-1), which accounts for most of the anomalous properties of water. Oxygen and a few adjacent elements on the chemical periodic table (Appendix C, p. 694) exhibit strong electronegativity. This means that they retain more than an equal allotment of the shared electrons in a covalent bond and, therefore,

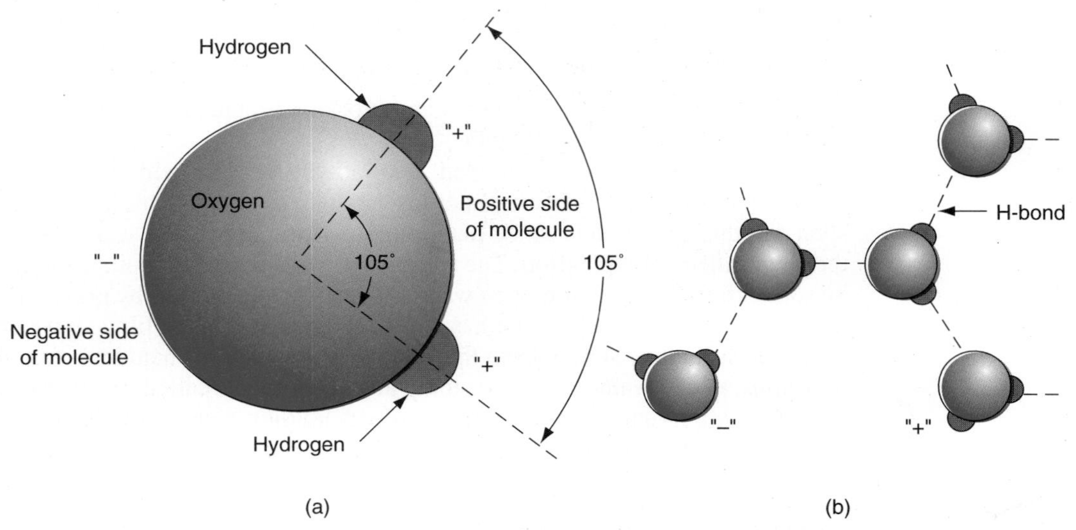

FIGURE 4-1 (a) Polar water molecule, H_2O. Because of the nonlinear positions of the hydrogens, water is polar. Water has one side that is more negative than positive and an opposite side (with two hydrogens) that is more positive than negative. *Polar* means there is no center of zero charge. In (b) the bonding of water to itself through H bonding is shown. The H of water in soils may bond to oxygen ions of soil mineral surfaces, thereby holding the water tightly to soil. The bonds in (b) become more rigid (less flexible) in colder temperature (ice or snow). A distance between molecules is shown for illustration purposes. In actual shape, molecules are adjacent to each other in close-fit orientation. (Courtesy of Raymond W. Miller, Utah State University.)

retain a partial negative charge. In soils, hydrogen bonding explains the attraction of positively charged hydrogens (H^+) of water to electronegative oxygens of other water molecules and to the oxygens on the surfaces of soil minerals and organic matter. This bonding of water to the surfaces of solid particles is termed **adsorption** or **adhesive bonding** (attraction of unlike molecules). The bonding of a water molecule to another water molecule is termed **cohesive bonding.** When fatty or oily substances that are low in oxygen coat soil particles, hydrogens of the water are not attracted to the coated surface (mostly carbons and hydrogens exposed). Such soils are called **water-repellent soils.** They are formed in nature under many plant covers and after forest fires, which tend to drive vaporized oils and resins into the soil where they coat the soil particles and cause them to resist wetting because of the C-H surface they expose to water.

Strong combined adhesion and cohesion forces cause water films of considerable thickness to be held on the surface of soil particles. Because the forces holding water in soil are surface-attractive forces, the more surface (the more clay and organic matter) a soil has, the greater is the amount of adsorbed water.

4:3 Terminology and Classifications for Soil Water

Adsorbed water in soils is less free to move than is water in a pool of water. Thus soil water has less **free energy** (has less ability to do work) than does water in a pool. Many concepts have been used to indicate the tendency for water to react, to move up or down, to change phases, or to enter a plant. Currently **soil water potential** is used to describe these tendencies.

Soil Water Potential

Soil water potential is defined as *the work the water can do as it moves from its present state to the* ***reference state,*** *which is the energy state of a pool of pure water at an elevation defined to be zero.* In most instances soil water has a potential less than zero, and that condition is indicated by giving the potential value a *negative sign.* ***Negative potential*** means that *work must be done on the water to remove it from the soil to a pool of water at the zero state.* The more tightly water is held by soil, the greater in absolute value is the negative number describing its potential.

The tendency of water to do work can be expressed in the following ways:

- **Suction or tension:** *Terms equivalent to negative pressure, used to avoid negative numbers when describing the status of water.* These terms were commonly used two decades ago, usually with units of *atmospheres* or *bars.* Although the terms *suction* and *tension* are conceptually acceptable, they are rarely used today.
- **Hydraulic head:** *Used extensively in engineering applications; usually negative numbers in the case of soil water.* *Hydraulic head* is the numerical equivalent of the height of a water column exerting a force equal to the attractive force between soil and water. Units are *length units* such as feet, meters, or centimeters.
- **pF scale:** *Logarithmic scale (similar to pH scale) of centimeters of hydraulic head.* This notation is used occasionally in engineering.
- **Water potential:** *The present, preferred system; values are usually negative, indicating that work must be done to the water to bring it to the zero (reference) state.* Water potential is usually expressed in units of energy per *mass* of water (*joules per kilogram*) or energy per *volume* of water. Because energy per volume equals pressure, pressure units such as *pascals* (and kilopascals, megapascals, etc.), *atmospheres* (now archaic), or *bars* are used.

Because all of these systems are used to some extent, and are found in some textbooks, one should be familiar with all of them. For scientific applications the *water potential* system using the SI units *megapascals* (*MPa*) and *kilopascals* (*kPa*) is now most widely used. Table 4-1 tabulates a brief summary of units and their equivalents used to report numerical values for water potential.

The **total water potential** is a summation of the following four components:

- **Matric potential:** *The effects of surface adsorption on the ability of water to do work.* Water adsorbed to surfaces of soil particles, or held in capillary pores by hydrogen bonding, is less free to move or react than is water in a pool. Work must be done on it to make it free. Where matric potential is a factor, values are always negative. In a saturated soil where water is free to flow, matric potential is not a factor and its value is, therefore, *zero.*
- **Solute (or osmotic) potential:** *The effects of dissolved substances on the ability of water to do work.* Water containing solutes (i.e., salts) is less able to do work than is pure water. For instance, it cannot pass as readily through a membrane, nor can it boil at the standard boiling point. Work (i.e., energy) is required to remove salts from water. Solute potential values are always *negative.*
- **Pressure potential:** *The effects of pressure from gases or overhead water on the ability of soil water to do work.* Pressurized water can flow further and faster and is,

Table 4-1 The Relation of the Pascal and Kilopascal to Other Units of Measure

The *pascal* is the S.I. unit for measurement of pressure

Pressure = force per area = energy per volume

1 pascal = 1 newton of force per square meter of area, or

= 1 newton-meter of energy per cubic meter of volume, or

= 1 joule of energy per cubic meter of volume (because 1 newton-meter = 1 joule)

Because the pascal is such a small unit for measuring the status of soil water, kilopascals are used.

1 kilopascal (kPa) Equals:	*Usage*
1000 pascals (Pa)	
0.001 megapascal (MPa)	
1 joule/kilogram (J/kg)	Used to express potential as energy per unit mass rather than unit volume. If 1 kg of water occupies 0.001 cubic meters, then 1 J/0.001 m^3 = 1000 J/m^3 = 1000 pascals = 1 kPa.
0.01 bar	A commonly used pressure unit. 1 bar ≈ 1 atmosphere pressure
0.00987 atmosphere	1 atmosphere = atmospheric pressure
10,000 $dyne/cm^2$	An older metric term for pressure
0.145 $pound/in^2$ (psi)	An English unit for pressure
10.2 cm of water	A unit of hydraulic head; a column of water 10.2 cm high in the Earth's gravitational field exerts a pressure of 1 kPa.
0.75 cm of mercury	Same concept as above, but adjusted for the greater density of mercury

therefore, more capable of doing work than nonpressurized water. Water can be under pressure due to gas pressures in the soil or to overhead water (i.e., not adsorbed, or free water, setting on top of the water in question). Overhead water is encountered under three conditions: (1) The water in question is located at a position in the soil profile that is below the water table, (2) the soil surface has *ponded* water (i.e., covered with stationary water), or (3) the soil surface is *flooded* (i.e., covered with flowing water). Pressure values are either *zero* or *positive.*

- **Gravitational potential:** *The effects of vertical position on the ability of water to do work.* The more elevated water is relative to a reference plane, the greater is its ability to do work. Gravitational potential is not affected by soil properties per se and cannot be measured on a soil sample brought to the laboratory; it is strictly a field phenomenon. Water can move spontaneously downhill, but work is required to move water uphill. Gravitational potential can be *positive* or *negative,* depending on whether the water in question is above (+) or below (−) the reference plane.

Water potential, as illustrated in the equations that follow,[2] is the sum of matric, osmotic (solute), and pressure potential but does not include gravitational potential. **Total water potential** is the sum of all four potential components.

[2]R. J. Hanks, *Applied Soil Physics,* 2nd ed., Springer-Verlag, New York, 1992, 24.

ψ_w	=	ψ_m	+	ψ_s	+	ψ_p
water potential		matric potential		solute potential		pressure potential
ψ_t	=	ψ_w	+	ψ_g		
total water potential		water potential		gravitational potential		

Effects of gravitational potential are usually small compared to the effects of other components, and when water contains few salts, the osmotic potential component may be insignificant. In other cases, such as where osmosis and phase changes are not important, osmotic potential can be ignored completely. In most field environments, other than wetlands or fields ponded from recent rains, the pressure component is zero. For these reasons, *matric potential is approximately equal to the total water potential* for most practical applications. However, in some instances such as flooded soils or soils in which one wishes salty water to enter a plant root (by osmosis), other components can be extremely important. In cases of nonflooded, nonponded, nonsalty soils where a less precise estimate will suffice, the following can be considered to be approximately true:

$$\psi_t \simeq \psi_w \simeq \psi_m$$

Water Potential Gradient and Water Flow

Soil water moves in response to a *water potential gradient.* **Water potential gradient** is the difference in total water potential between two locations in the soil, such as two depths in a soil profile. The gradient usually occurs when one location in the soil is drier than another because of evaporation or transpiration losses or because of recent wetting at one of the soil locations. *Water flows from areas of higher water potential (usually wetter areas) to areas of lower water potential (usually drier areas).* This water movement in soils at less than saturation water content, and in response to water potential gradients, is called **unsaturated flow.**

Water from rainfall or irrigation moves into and through *saturated* soil by gravity flow, often very rapidly. In a saturated soil, gravitational force (and pressure, if standing water is present) is more effective in causing water to move downward than is the matric attraction in causing the water to adhere to soil. This water movement is called **saturated flow.** Slower water movement occurs when the soil is not saturated and the water flow is influenced primarily by matric forces. This tendency for water to move to an area of lower matric potential (*unsaturated flow*) is a nondirectional force and results in such phenomena as water moving upward from an irrigation furrow into an elevated bed of soil.

The potential of different portions of water in a soil varies. Water that is adjacent to a particle surface has a lower potential (e.g., −800,000 kPa) than water on the outside portion of a thick water film (−10 to −30 kPa, perhaps loose enough for gravitational flow). Note that *water is held more tightly as the negative value increases in magnitude.* Figure 4-2 represents this concept of water potential as a function of distance from the particle surface.

If different molecules of water in soil are held with different degrees of force, what is the practical meaning of a measurement showing the soil water to have a potential of −1000 kPa? Such a reading means that the *most weakly held water* in that sample is at −1000 kPa; other portions of water in the soil are held more tightly and are therefore at lower potentials such as −1500 kPa or −800,000 kPa (such a large negative number would more appropriately be expressed as −800 MPa).

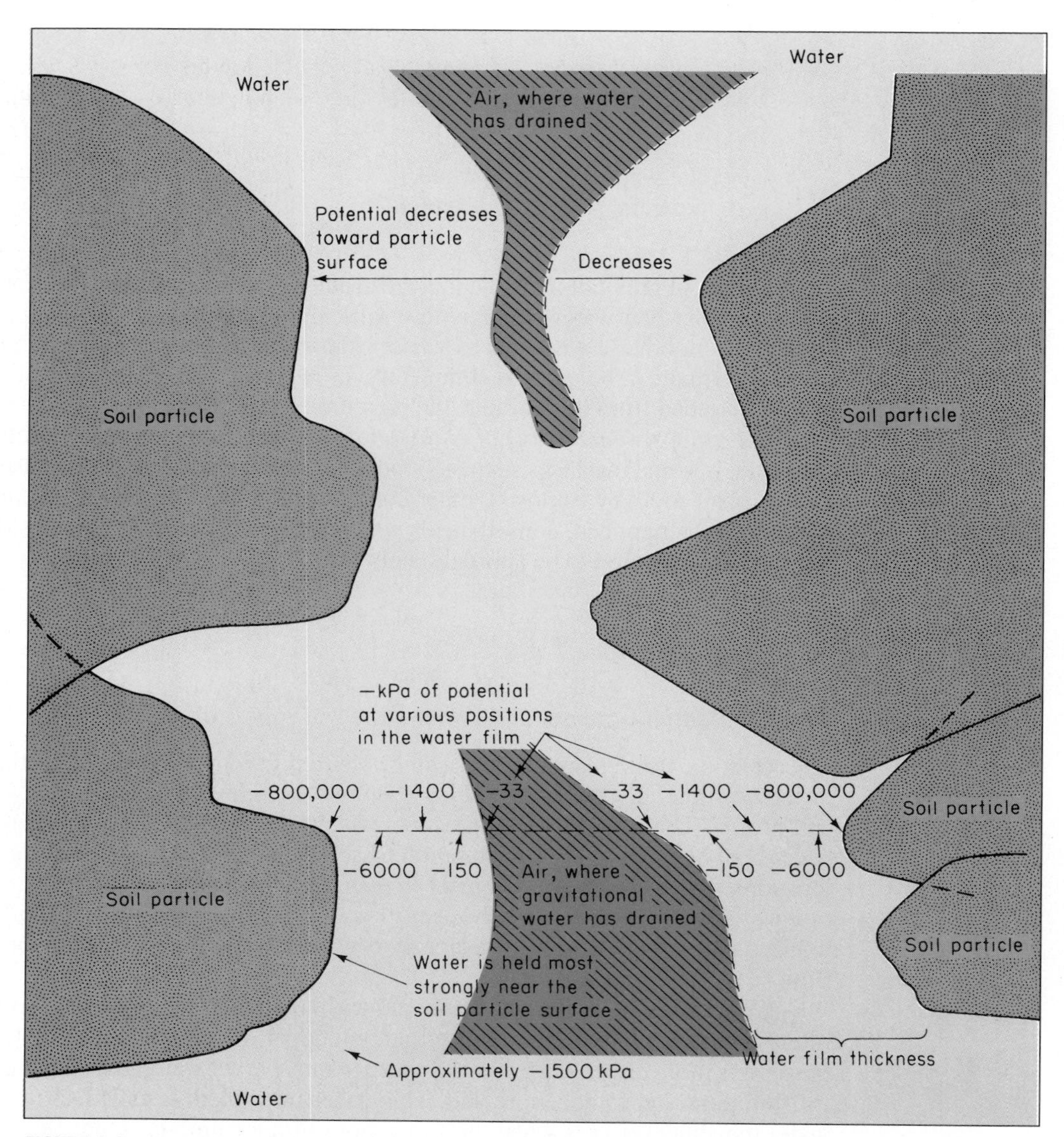

FIGURE 4-2 Cross-section of a soil pore and the solid soil particles that make up its walls, showing the increase in force with which water is held with changing distances from the soil particle surface. At some distance from the particle surfaces, water is held so weakly that the pull of gravity causes some of it to drain. The water layers depicted are arbitrary thicknesses that have been selected for illustration purposes only. (Courtesy of Raymond W. Miller, Utah State University.)

Soil Water Classification for Water Management

A useful classification of water potential is based on the availability of water to plants in the environment. Terms used include *gravitational* (*drainage*) *water, field capacity, permanent wilting percentage,* and *available water capacity.*

Gravitational water is water held more loosely than −33 kPa. This portion of water will drain freely from the soil by the force of gravity. In very sandy soils this reference value is considered to be −10 kPa rather than −33 kPa.

Field capacity is the amount of water in the soil when the water potential equals −33 kPa (or −10 kPa for sands). The field capacity is a measure of the greatest amount of water that a soil can store under conditions of complete wetting followed by free drainage. Field capacity values are used to determine the amount of irrigation water needed and the amount of stored soil water available to plants. Field capacity approximates the amount of water in a soil after it has been fully wetted and all gravitational water has drained away. For practical purposes the upper layer of a soil with normal drainage is considered to be at field capacity about one day after a heavy rain or irrigation ceases. Field soils reach this condition only momentarily because water continues to redistribute throughout the soil. Determining when the profile reaches an overall *average field capacity* is difficult to do precisely. The rate at which soils approach an average field capacity is slower with clayey soils than with sands. Figure 4-3 illustrates the rate at which two typical saturated soils drain to field capacity, assuming no evaporation or loss through plant uptake occurs during that time.

The **permanent wilting percentage** (also called **permanent wilting coefficient** and **permanent wilting point**) is the amount of water in the soil when the water potential is −1500 kPa. Water at the permanent wilting point (PWP) is held so strongly that plants are not able to absorb it fast enough for their needs. On a hot, dry day a plant such as corn may transpire excessively and temporarily wilt even when the water is only at −100 kPa or −200 kPa (the water is available but cannot be absorbed fast enough). In this instance, however, the plant recovers at night when transpiration slows. In contrast to such *temporary* wilting, the *PWP* indicates low water availability; in such conditions wilting plants do not recover except when additional water is added to the soil.

Plants, Wilting Point, and Available Water

Plants vary in their abilities to extract water from soil, but a large array of crop plants have root osmotic potentials between −1500 kPa and −2000 kPa and could be expected to wilt at soil water potentials near −1500 kPa. A recent study done to determine the lower limit of water available for cotton and sorghum confirmed that the use of −1500 kPa was appropriate

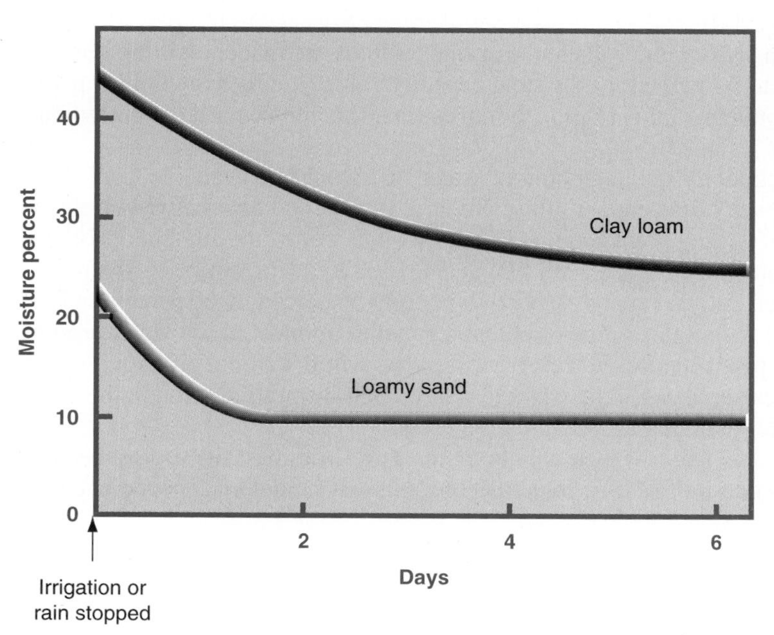

FIGURE 4-3 Rate at which two soils of different texture drain to field capacity. Field capacity is reached when the curve tends to level off to a constant moisture percentage. Drainage is faster in coarse-textured soils than in fine-textured ones. (Courtesy of Raymond W. Miller, Utah State University.)

FIGURE 4-4 Water is the medium of life. When irrigation water is absent or too costly, great effort is needed in dry areas to harvest crops. This Hopi Indian is dry-farming corn in a rainfall area of 200 mm to 300 mm (8 in–12 in). Ten to twenty kernels are planted in a hole 15 cm to 30 cm deep, covered, and packed. Outer leaves later are removed to allow inner plants to be productive. (Courtesy of USDA—NRCS, the Bureau of Indian Affairs and the Arizona Agricultural Experiment Station, *Soil Survey of Hopi Area, Arizona, Parts of Coconino and Navajo Counties.* Photo by Fred Kootswatewa, Hopi Tribe.)

and corresponded closely to the lower limit of adequate water availability to plants in the field.[3]

Some drought-tolerant and arid-zone plants such as cacti have special transpiration-resistant leaf coatings or other adaptations that permit them to survive long time periods when the soil is at or below −1500 kPa. All plants can absorb some water from soils drier than −1500 kPa, and some plants have the ability to extract some water even at potentials as low as −6000 kPa. Special adaptations are necessary for normal plants to survive in arid climates without irrigation (Figure 4-4).

Commonly, except for organic soils and extreme textures, permanent wilting percentage is about 40 percent to 50 percent of the field capacity value. If either the field capacity (FC) or PWP value is not known, this rule-of-thumb estimate should be used to approximate that water value.

Available water capacity is the amount of water that would be available *to plants* if the soil were at field capacity. It is equal to the difference in water content between field capacity and the permanent wilting point (Figure 4-5), or in other words, the amount of soil water held with a water potential between −33 kPa and −1500 kPa (for sands, between −10 kPa and −1500 kPa). Generally, available water will be from 50 percent to 60 percent of field capacity. The water held within this range of potentials makes up most of the water used by plants and is also called **plant-available water.** Most plants wilt if water at or below −1500 kPa is the only water present, because the water loss through transpiration through the stomata of the leaves is greater than that absorbed by the roots.

Although gravitational water is available to plants as it flows past the roots—because it is present only for short periods of time in permeable soils—it cannot be depended upon in

[3] M. J. Savage, J. T. Ritchie, W. L. Bland, and W. A. Dugas, "Lower Limit of Soil Water Availability," *Agronomy Journal* **88** (1996), 644–651.

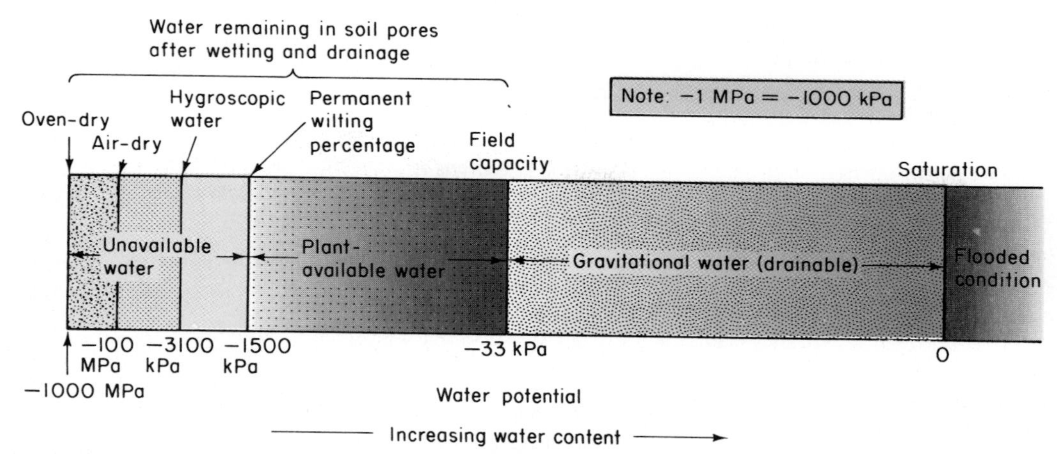

FIGURE 4-5 Soil water classifications and their water potentials. At water potential near −33 kPa, any additional water would be held loosely enough for gravity to overcome the adhesive forces and would drain away. Capillary water (that held by capillary action) remains; this is held at water potentials ranging from −33 kPa (field capacity) to −3100 kPa or more strongly (hygroscopic), depending on the pore sizes of the soil. Plants cannot survive with only capillary water held more strongly than about −1500 kPa (the point of permanent wilting). Soil water at the air-dry state is held by water potentials that vary from −100 MPa to −30 MPa, depending on humidity. (Courtesy of Raymond W. Miller, Utah State University.)

meeting plant water requirements. Some soils hold appreciable water at high water potentials (−5 kPa to −30 kPa) for a long enough time to consider it *plant-available water.* This occurs particularly in sandy soils that overlay either coarser materials (such as gravel) or slowly permeable subsoils (such as a clay or cemented layer).

Of course, not all available water will be used by plants. Some available water the plant could have used is lost by evaporation from the soil surface.

Capillary Water and Saturation Percentage

Two other water terms are often used—*capillary water* and *saturation percentage.* **Capillary water** is held tightly in small capillary pores by hydrogen bonding (i.e., matric forces). The height of rise (or strength of attraction) for water in a capillary tube is inversely proportional to the radius of the tube. In other words, the smaller the pore diameter, the greater the attraction. This capillary phenomenon explains the tendency for small pores to remain wet, whereas large pores drain easily. The **saturation percentage** is the water content of soil when all pores are filled with water. As a rule of thumb, saturation percentage is approximately *double the amount of water at field capacity.* The saturation percentage is approximated by the water in a *saturated paste* prepared for laboratory analysis of soluble salts.

4:4 Soils as Water Reservoirs

Which soils hold the most or the least water? Water in soil is held as films on particle surfaces and in small pores. Large pores (in sands and between large aggregates) allow water to drain by gravitational flow. Small pores (clays) retain water by the capillary phenomenon. Generally the more clay and humus in the soil, the larger the amount of water the soil can store (Table 4-2 and Figures 4-6 and 4-7). Not only is the quantity of water retained in clay soils large, but much of it is held very tightly by the clay surfaces. In fact, clay holds large amounts of water at both its field capacity and its permanent wilting point (Figure 4-6).

Table 4-2 Permanent Wilting Point, Field Capacity, and Plant-Available Water-Holding Capacity of Various Soil Textural Classes*

	Water per 30 cm of Soil Depth					
	Permanent Wilting Point		Field Capacity		Plant-Available Water Capacity	
Soil Texture	%	cm	%	cm	%	cm
Medium sand	1.7	0.7	6.8	3.0	5.1	2.3
Fine sand	2.3	1.0	8.5	3.7	6.2	2.7
Sandy loam	3.4	1.5	11.3	5.0	7.9	3.5
Fine sandy loam	4.5	2.0	14.7	6.5	10.2	4.5
Loam	6.8	3.0	18.1	8.0	11.3	5.0
Silt loam	7.9	3.5	19.8	8.7	11.9	5.2
Clay loam	10.2	4.5	21.5	9.5	11.3	5.0
Clay	14.7	6.5	22.6	10.0	7.9	3.5

Source: "Water," *Yearbook of Agriculture,* USDA, Washington, DC, 1955, 120.
*There is a variation in the amounts and kinds of sand, silt, and clay within any one textural group (such as within loam soils); therefore, there is also a variation in the water constants. For simplification an average value is given.

Smectites (swelling montmorillonitic clays; see Chapter 5) hold more water than a soil with similar amounts of nonswelling clays (kaolinite or sesquioxide clays).

Medium-textured soils have the unique combination of pores capable of holding large amounts of water, yet not so tightly that plants have difficulty extracting it. For this reason the largest *available water capacity* is found in silt loams and in other loamy soils (Figure 4-6). Soil humus, compaction (which alters pore size), and types of clay will affect available water capacity.

4:5 Soil Water Content

Soil water content can be measured or estimated in several ways. Methods include the classical gravimetric method to directly determine water content in the laboratory and the newer methods such as time domain reflectometry to quickly estimate water content in the field.

Measuring Water Content

The **gravimetric** method is a standard and reliable procedure used to directly measure mass water content. A soil is sampled, put into a container, weighed in the sampled (moist) condition, oven dried, and weighed again after drying. Drying is done at 105°C to 110°C (221°F to 230°F) until the sample weight is constant. This requires from several hours to several days, depending upon the size and wetness of the sample. A newer approach to gravimetric analysis allows drying the sample in a microwave oven.[4] Drying a 100 g sample to constant weight in this manner may require about 3 minutes.

The **mass water content** (θ_m) is the decimal fraction that equals the mass of water divided by the mass of oven-dried soil. The term **mass water percentage** (P_m) is frequently used to mean the mass water content × 100. Although in common usage, the term is not the

[4]ASTM, Standard test method for determination of water content of soil by microwave oven method D 4643-87, ASTM, Philadelphia, 1992.

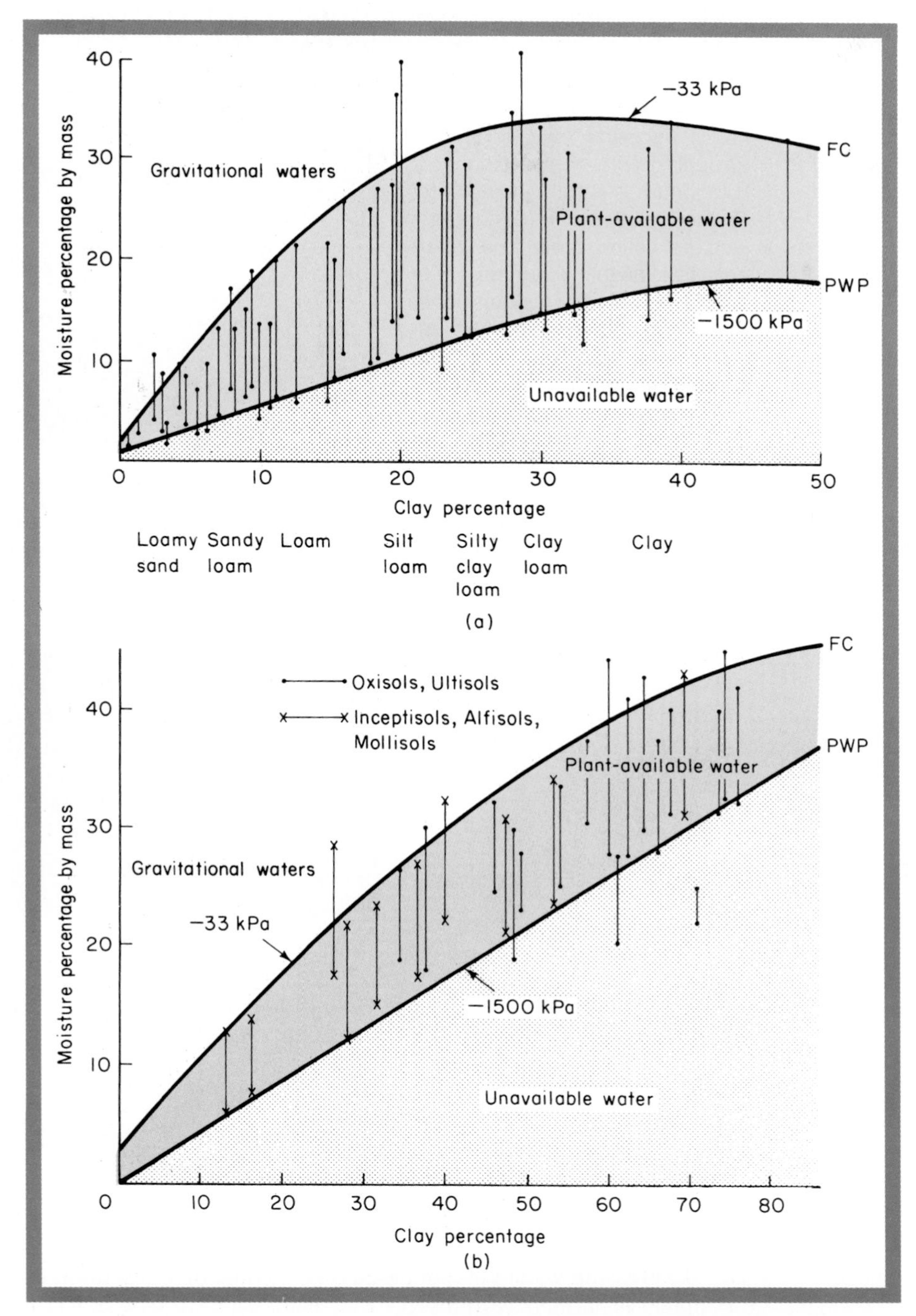

FIGURE 4-6 Actual soil moisture constants for selected soils. For each vertical line the upper end is the moisture content at −33 kPa (field capacity) and the lower end is the moisture content at −1500 kPa (permanent wilting percentage). The length of each line is the plant-available water percentage. The soils in (a) are from Utah, Nebraska, Iowa, California, and the northeastern United States (silicate clays). The soils in (b) are from Puerto Rico and the Virgin Islands and contain mostly sesquioxide and kaolinite clays. Notice the differences in clay percentages. Textural classes are approximate. (Courtesy of Raymond W. Miller, Utah State University.)

FIGURE 4-7 Under the same amount of annual rainfall (76 cm; 30 in) and approximately the same steepness of slope, the soil in (1) supports only xerophytic (drought-tolerant) plants such as Eriogonum, a shrub, and an herb, Phacelia. The soil in (2) supports western yellow pine. The difference is available water. In (1) the soil is a volcanic cinder cone, which has an available water capacity less than that of coarse sand, perhaps 1.25 cm/30 cm (0.5 in/ft) of soil depth. By contrast, the clay loam soil (2) has an available water capacity of perhaps 5 cm/30 cm (2 in/ft) of soil depth. (Courtesy of Michigan State University.)

same percentage as is usually calculated for grains and forage, because the denominator is not the mass of the whole (water + soil) but only the mass of *dry* soil:

$$\textbf{mass water content} = \theta_m = \frac{\text{mass of water}}{\text{mass of oven-dried soil}} \qquad \text{(Eq. 4-1)}$$

$$\textbf{mass water percentage} = P_m = \theta_m(100) \qquad \text{(Eq. 4-2)}$$

Expressing water content based on a volume of soil is often useful. The volumetric water content is defined as follows:

$$\textbf{volumetric water content} = \theta_v = \frac{\text{volume of water}}{\text{volume of soil}} \qquad \text{(Eq. 4-3)}$$

$$= \frac{\text{mass of water/density of water}}{\text{mass of oven-dried soil/soil bulk density}} \qquad \text{(Eq. 4-4)}$$

Using mass in g and density in g/mL, the density of water is 1, and Equation 4-4 simplifies to

$$\textbf{volumetric water content} = \theta_v = \text{(mass water content) (soil bulk density)} \qquad \text{(Eq. 4-5)}$$
$$= \theta_m\,(\rho_b)$$

One can determine, therefore, *volumetric water content* by measuring *mass water content* and multiplying that value by the soil *bulk density value.*

Unlike mass water content, volumetric water content can be converted to a true percentage. The formula is

$$\textbf{volume water percentage} = P_v = \theta_v\,(100) \qquad \text{(Eq. 4-6)}$$

Volumetric water content values can be used to estimate the amount of water in a field soil or the amount of water needed to wet the soil to a specified degree by rainfall or irrigation. See Calculation 4-1 for an example.

The term water *content* is a nonspecific term and *mass water ratio* and *volumetric water ratio* are sometimes used. The use of *ratio* does clarify what water *content* is meant. This book will use *content* because it is currently used in recent books and government publications.

Gains and Losses of Water

Water volume is used to determine changes in the amount of water in a soil. For example, how much irrigation water needs to be added, or how much water has been evaporated, or how deeply will a rainfall or irrigation wet a soil? The soil's water is usually given as a *depth of water* in a depth of soil, as though the water measured were pulled out of the soil, placed in a container that has the same cross-sectional area as the soil containing the water, and had its depth measured. Rainfall is measured similarly; for example, "rainfall was 1.3 inches" means there was rainfall equivalent to cover the area of the rain gauge to a depth of 1.3 inches.

The general equation for these kinds of calculations follows:

$$\frac{\text{Volume of water}}{\text{Volume of soil}} = \frac{\text{(area) (depth of water)}}{\text{(area) (depth of soil)}} = \frac{\text{depth of water}}{\text{depth of soil}} = \theta_v \qquad \text{(Eq. 4-7)}$$

Some simple algebraic rearrangement of Equation 4-7 reveals that

$$\text{depth of water} = \theta_v\,\text{(depth of soil)} \qquad \text{(Eq. 4-8a)}$$

Or, written in symbols, Equation 4-8a becomes

$$d_w = \theta_v(d_s) \qquad \text{(Eq. 4-8b)}$$

See Calculation 4-2 for sample calculations using this equation.

Calculation 4-1 Calculation of Soil Water Content Using Gravimetric Techniques

Problem A soil sample taken from a field was placed in a can, weighed, dried in an oven at 105°C (221°F), and reweighed. The measurements were as follows:

Moist soil plus can weight = 159 g

Oven-dried soil plus can weight = 134 g

Empty can weight = 41 g

Bulk density of the soil = 1400 kg/m^3

Calculate (a) the mass water content and percentage and (b) the volumetric water content and percentage of this soil when it was sampled.

Solution (a) The mass water content (θ_m) is the water mass divided by the dry soil mass. All calculations must involve only soil and water weights, so the weight of the can must first be subtracted.

$$\text{Moist soil only} = 159 \text{ g} - 41 \text{ g} = 118 \text{ g}$$

$$\text{Dried soil only} = 134 \text{ g} - 41 \text{ g} = 93 \text{ g}$$

Then,

$$\theta_m = \frac{\text{moist soil} - \text{oven-dried soil}}{\text{oven-dried soil}}$$

$$= \frac{118 \text{ g} - 93 \text{ g}}{93 \text{ g}} = 0.269$$

$$\text{Mass water percentage} = P_m = \theta_m (100) = (0.269)(100) = 26.9\%$$

(b) The volumetric water content (θ_v) is the fraction of the soil volume occupied by water. Using Equation 4-4 for θ_v yields the following:

$$\theta_v = \left(\frac{\text{weight of water}}{\text{weight of dry soil}}\right)\left(\frac{\text{bulk density of soil}}{\text{density of water}}\right)$$

$$= \left(\frac{25 \text{ g}}{93 \text{ g}}\right)\left(\frac{1400 \text{ kg/m}^3}{1000 \text{ kg/m}^3}\right) = 0.376 \text{ volumetric water content}$$

$$\text{volumetric water percentage} = P_v (100) = (0.376)(100) = 37.6\%$$

Compare the mass water percentages to the volumetric water percentage.

4:6 Instruments for Determining Water Content or Potential

Water content and matric potential are related uniquely for each soil. Water content can be measured by numerous ways: *pressure* (water release characteristic curve), *electrical resistance* (moisture blocks), *thermal dissipation blocks* (rate of heat dissipation), *slowed neutrons from a radioactive emitter* (neutron probe), *electromagnetic wave transmittance* (frequency domain reflectometry and time domain reflectometry), *vacuum–suction* (tensiometers), and *soil air relative humidity* (thermocouple psychrometers).[5]

---

[5] Thomas W. Ley, "An In-Depth Look at Soil Water Monitoring and Measurement Tools," *Irrigation Journal* **44** (1994), No. 3, 8, 11–14, 16–20.

Calculation 4-2 Water Volume Calculations

Problem Calculate these values: (a) the total water currently contained in the top 30 cm (12 in) of soil, (b) the depth to which 27.5 mm of rain would wet this uniform soil, and (c) the available water capacity of the top 30 cm of soil. The following soil information is given:

present mass water content (θ_m) $= 0.18$

θ_m at field capacity $= 0.23$

θ_m at permanent wilting point $= 0.09$

Bulk density (ρ_b) $= 1.3$ Mg/m^3 (or 1.3 g/cm^3)

Solution Equations 4-5 and 4-8b are involved in all three solutions. Equation 4-5 states that $\theta_v = \theta_m\ (\rho_b)$; Equation 4-8b states that $d_w = \theta_v\ (d_s)$.

(a) To calculate the total water depth in 30 cm of soil, first calculate θ_v according to the simplification described by Equation 4-5:

$$\theta_v = \theta_m\ (\rho_b) = 0.18\ (1.3) = 0.234$$

Then, using Equation 4-8b,

$$d_w = \theta_v\ (d_s) = 0.234\ (30\text{ cm}) = 7.02\text{ cm}$$

The top 30 cm of soil contains 7.02 cm of water.

(b) Rearrange Equation 4-8b: $d_s = d_w/\theta_v$. This equation implies that the *change* in d_s = the *change* in d_w/θ_v. The change in d_w = 27.5 mm of rain, expressed more conveniently as 2.75 cm.

Because the soil will wet in incremental layers to field capacity and will then drain deeper, the change in θ_v = (θ_v at field capacity − present θ_v) = (0.23) (1.3) − (0.18) (1.3) = 0.299 − 0.234 = 0.065. Now, the change in depth of wetted soil d_s = 2.75 cm/0.065 = 42.3 cm.

The soil is wetted 42.3 cm deep by 27.5 mm of rain.

(c) Available water capacity is the change in depth of water between field capacity and the permanent wilting point. Using Equation 4-8b and holding the depth of soil constant at 30 cm, we can write the following:

$$\begin{aligned}\text{available water capacity} &= (\theta_v \text{ at field capacity} - \theta_v \text{ at permanent wilting point})\ (30\text{ cm})\\ &= [(0.23)\ (1.3) - (0.09)\ (1.3)]\ (30\text{ cm})\\ &= (0.299 - 0.117)\ (30\text{ cm}) = 5.46\text{ cm}\end{aligned}$$

The available water capacity of the upper 30 cm of soil is 5.46 cm.

Water Release Characteristic Curve

The relation between water content and matric potential can be described by a **water release characteristic curve** (Figure 4-8). These characteristic curves can be constructed by placing a moist soil sample in a chamber set at a known pressure. The pressure forces water from the soil sample through a porous ceramic plate and out through a port in the chamber. When the soil water equilibrates with the pressure in the chamber, water no longer flows from the port. The soil matric potential is then assumed to be equal to the known pressure in the chamber. The soil water content is determined gravimetrically and the investigator then knows the moisture content at one particular matric potential value. Other samples from the same soil can be equilibrated to other pressures, resulting in a series of points representing different matric potential values and water contents. From this series of points one can construct the curve

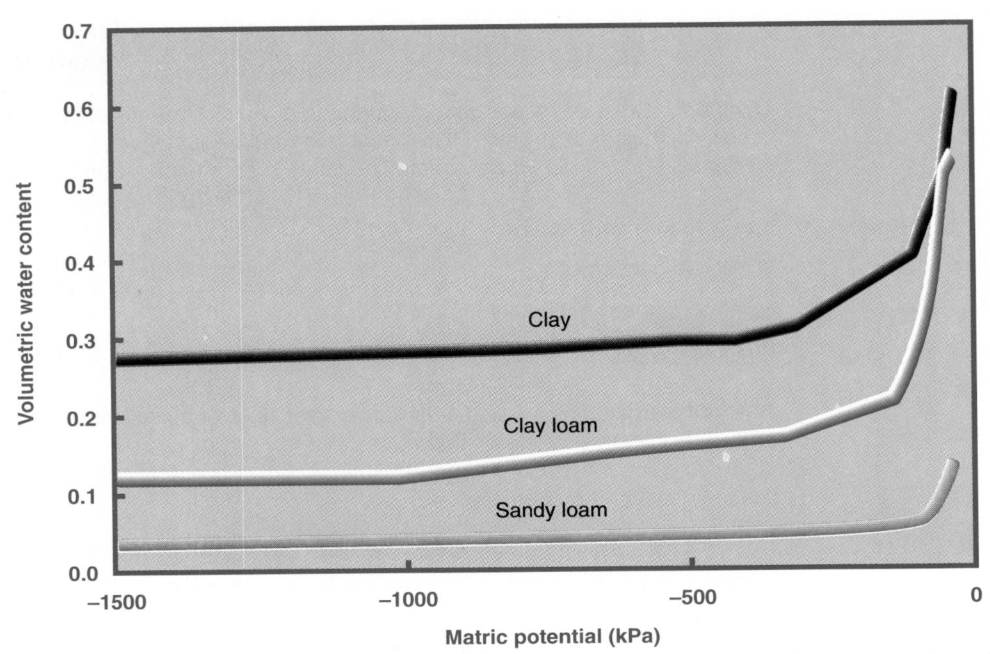

FIGURE 4-8 Generalized water release characteristic curves for three soils. As water content decreases, matric potential decreases (becomes more negative). Note that at any matric potential, the clay soil contains more water than the clay loam or sandy loam.

(as shown in Figure 4-8), which then allows one to estimate water potential from water content data, or water content from water potential data.

Porous Blocks

Porous blocks (of gypsum, ceramic, nylon, or fiberglass) are based upon the principle that electrical resistance between two electrodes in a porous block is proportional to water content and to concentration of the electrolyte. If the electrolyte is held approximately constant by the presence of a slowly soluble matrix such as gypsum or the soil's salt content, then resistance is approximately proportional to the water content of the blocks. In 1940 G. J. Bouyoucos introduced a gypsum block, inside of which were two electrodes a fixed distance apart. (For obvious reasons these blocks are also known as *Bouyoucos blocks, porous blocks, gypsum blocks,* and *resistance blocks.*) The blocks are buried in soil at various depths, and the resistance across the electrodes is measured as needed with a portable meter or continuously with a data logger (Figure 4-9). The blocks must be calibrated for each soil to relate electrical conductivity to either the *water content* or *water potential* for that soil. Improved moisture blocks are made of various materials intended to last longer than gypsum, which eventually dissolves. Gypsum blocks do not give accurate readings in wet soils but are useful where irrigation is needed and in drier settings, such as rangelands.

Thermal Dissipation Blocks

A ceramic block containing a small heater and temperature sensor allows measurement of the thermal dissipation of the block. Thermal conductance away from the block is proportional to the water content of the block and thus to soil water potential. Each block must be individually calibrated and they are somewhat expensive.

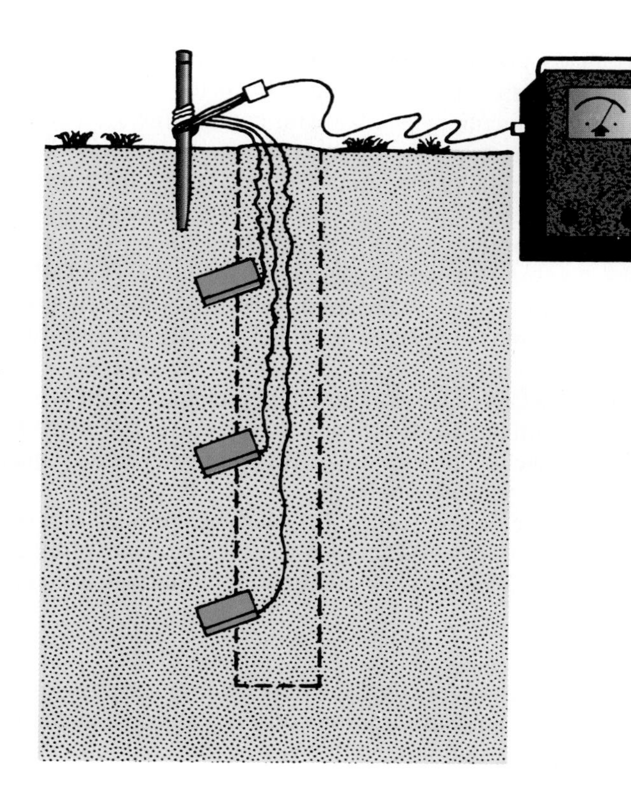

FIGURE 4-9 Electrical conductivity between two electrodes set a fixed distance apart inside a small block is an indirect measure of soil moisture. The blocks may be made of gypsum, nylon, fiberglass, or other porous material and can be buried in the soil at any desired depth; wires from each block extend to the surface of the soil for easy reading of conductivity. (*Source:* USDA, Agricultural Handbook 107, 1957.)

Neutron Scattering (Probe)

Neutron probes are used to estimate soil water contents. The probes look like flashlight cylinders attached by a long cord to a meter box. The probe contains radioactive material (radium–beryllium, americium–beryllium) that emits *fast-moving neutrons.* As the neutrons emitted from the probe collide with or deflect from equal-size hydrogen ions (of which water is a major source), they are slowed. A portion of the *slowed,* deflected neutrons are deflected back to the probe, where their return is detected. The rate of return of slowed neutrons can be calibrated quite precisely to the water content of a soil.

The neutron probe is used by lowering it into an access tube of plastic pipe or aluminum, which is about 5 cm (2 in) in diameter and as deep (long) as needed. The probe is suspended at each soil depth to be measured, and the neutron count is taken and converted by the meter box to volumetric moisture content. Although the neutron probe is easy to use and offers high precision, it is expensive to purchase and requires licensing and monitoring to handle radioactive materials. Mainly because of safety concerns and inconvenient recordkeeping required for the storage of radioactive material, use of the neutron probe is declining in favor of newer, safer technology.

Dielectric Constant Method

The dielectric constant (DC) of a nonconducting material indicates its ability to transmit high-frequency electromagnetic waves. This provides a way to measure water because dry soil has a DC of 2 to 5 but water's is about 78. Two approaches are used: *frequency domain reflectometry* and *time domain reflectometry.*

Frequency domain reflectometry (FDR) moisture probes use electromagnetic radiation reflected from the soil as a function of soil water content. The soil acts as a dielectric completing a capacitance circuit. High-frequency waves at about 150 MHz are pulsed through the capacitance circuit. Air space greatly affects the travel of the signal into the soil.

The equipment is commercially available for applications similar to the neutron probe but without the legal complications surrounding the neutron probe. Similar to the neutron probe, it is used by lowering the probe into access tubes, and the electronic data processor provides readings of volumetric water content in about 15 seconds per location (Figure 4-10).

Time domain reflectometry (TDR) is another technology used to provide safe, instantaneous readings of volumetric water content (Figure 4-11). The TDR propagates a high-frequency transverse electromagnetic wave along a cable connected to one or more probes inserted into the soil. The time between sending and receiving the return wave is measured. This travel time is a function of the dielectric constant of the material surrounding the probe. Because the dielectric constant of water is about 78 and that of soil is between 2 and 5, the presence of water greatly affects the travel time of the pulse. High water content slows the propagation velocity and yields a higher dielectric constant. The instrument converts travel time of the pulse to volumetric water content.

After a decade or more on the market, TDR has been refined and is enjoying a degree of acceptance. The greatest utility of TDR is for surface soil measurements into which the probe can be easily pushed. Accuracy is greater for sandy soils than in soils of heavier texture.

Tensiometers

Tensiometers are inexpensive, widely used instruments for measuring matric potential in the field (Figure 4-12). Tensiometers have a porous-clay cup attached to a tube filled with water. As the soil dries, water moves out through the porous cup, creating a vacuum (i.e., pressure

FIGURE 4-10 Moisture probe to determine soil water content by frequency domain reflectometry. Used for irrigation scheduling, this technology was intended primarily to replace neutron probes. As with the neutron probe, the moisture probe is lowered into access tubes from which field estimates of volumetric water content are obtained at various soil depths. (Courtesy of Troxler Electronic Laboratories, Inc.)

FIGURE 4-11 Soil moisture measurement system using time domain reflectometry (TDR). To obtain moisture readings, probes are pushed into the ground and can be left in place for data logging or remote sensing. (Courtesy of ESI Environmental Sensors, Inc., Division of Gabel Corporation.)

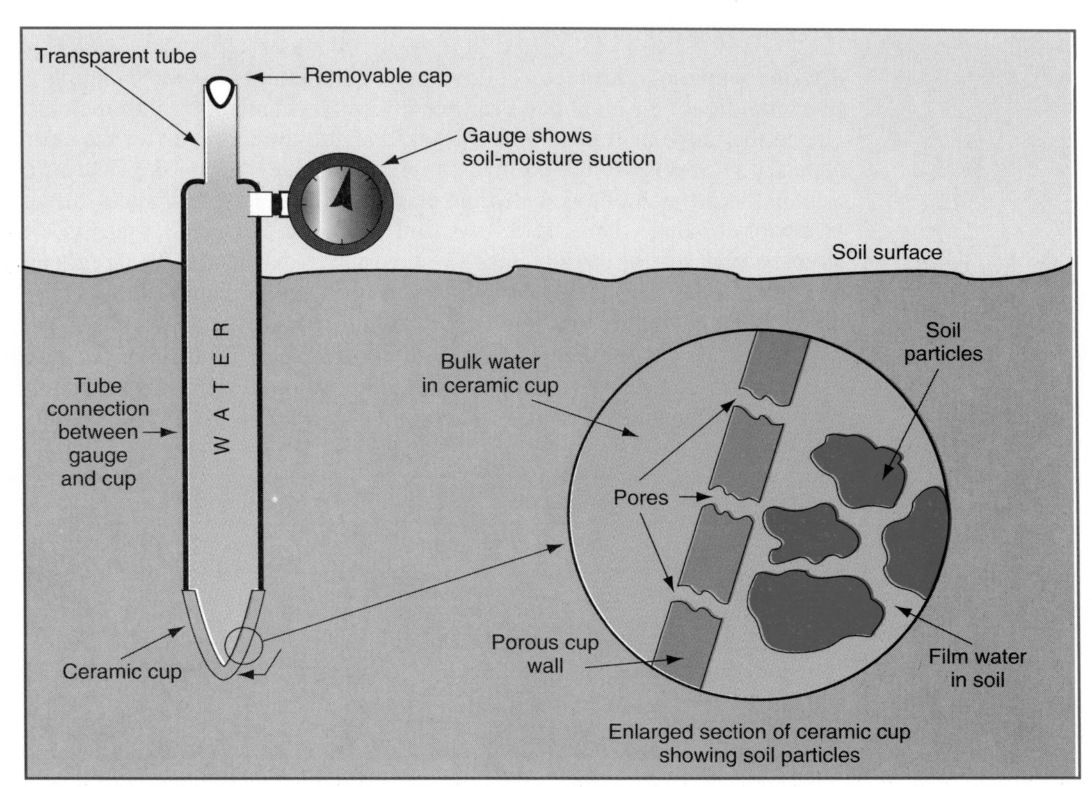

FIGURE 4-12 Sketch of a tensiometer inserted vertically into the surface of the soil and an enlarged schematic diagram of a section of the porous ceramic cup in contact with soil water. (*Source:* S. J. Richards and R. M. Hagan, "Soil Moisture Tensiometer," Extension Service Leaflet 100, California Agricultural Experiment Station, 1958.)

less than atmospheric pressure) on the water column. The tube is attached to a vacuum gauge, usually calibrated to read matric potential in centibars (1 centibar = 0.01 bar = 1 kPa).

Tensiometers can be used to schedule irrigation by placing an instrument at the depth of maximum root density. If desired, additional instruments could be placed at other depths. Using multiple instruments allows the user to determine whether water in the soil is flowing upwards or downwards. The user can add gravitational potential to the matric potential value for each depth, knowing that water will migrate toward the lower potential. As an example, a need for irrigation could be indicated by a reading of about −30 centibars (−30 kPa) in a sandy soil, or about −50 centibars in a loam and −70 centibars in a heavy clay. Automatic irrigation systems sometimes have a vacuum-sensing valve attached to permanently installed tensiometers. Sugarcane and potatoes are crops with which water control by tensiometer measurement has been successfully used. Orchards, ornamental nurseries, and turf farms use tensiometers extensively, particularly in California.

One can relate tensiometer gauge readings to soil water content by using a *water release characteristic curve* developed for each soil being measured. The principal limitation of tensiometers is that they do not measure matric potential values as low as the permanent wilting point. The actual range of effective measurement is only from 0 to −85 kPa. For irrigation scheduling this is not a severe drawback because irrigated crops are usually managed so that soil stays moist. Tensiometers are more useful for measuring moisture in sandy soils than in fine-textured soils because more of the plant-available water in a sandy soil falls within the readable range of tensiometers.

Thermocouple Psychrometers

Thermocouple psychrometers allow direct calculation of *water potential* by measuring relative humidity of the air above a sample in a small chamber. Psychrometers are reasonably precise in the range of −10 kPa to −7000 kPa and, therefore, cover the entire range from field capacity (−33 kPa) to the permanent wilting point (−1500 kPa). They are much more accurate on the dry end of the scale than on the wet end. Psychrometers are usually designed for laboratory rather than field use and are much more expensive than tensiometers. Psychrometers offer the advantage of measuring matric and solute potential simultaneously and can also be used to measure the water potential in plant tissue.

One study attempted to compare eight methods for their general suitability, which involved initial cost, accuracy, and six other factors (see Table 4-3). The *qualitative* evaluation lists neutron probe, TDR, and gravimetric sampling as the most accurate; the hand-push capacitance probe and tensiometer were fair to good. The TDR is the most expensive; the neu-

Table 4-3 A Qualitative Evaluation of Soil Water Monitoring Devices

Device	*NP*	*TDR*	*GS*	*ATCP*	*HPCP*	*TM*	*GB*	*SGB*
Initial cost	3	1	8	2	7	8	8	8
Field site setup needs	7	3	10	3	10	7	6	6
Getting a routine reading	8	8	1	8	4	10	8	8
Interpretation of reading	10	10	10	10	3	5	3	5
Accuracy	10	10	10	8	2	7	2	3
Composite rating*	49	49	52	48	42	47	41	47

*Two evaluations not included above were *maintenance* and *special considerations* (such as restrictions on use, ruggedness of equipment, speed, and so on). **Evaluation:** 1 is poorest, 10 is most favorable. **Methods:** NP = neutron probe, TDR = time domain reflectometry, GS = gravimetric sampling, ATCP = access-tube capacitance probe, HPCP = hand-push capacitance probe, TM = tensiometer, GB = gypsum block, and SGB = specialty gypsum blocks.
(*Source*: Modified from Thomas W. Ley, "An In-Depth Look at Soil Water Monitoring and Measurement Tools," *Irrigation Journal* **44** (1994), No. 3, 8, 11–14, 16–20.)

tron probe is rated poor on special considerations because of the regulatory obligations for radioactive material in its use.

4:7 Water Flow into and through Soils

The most rapid movement of water into and through soil is caused by the pull of gravity. *Saturated-flow* water moves because of water potentials larger (less negative) than −33 kPa. A much slower flow, *unsaturated flow,* occurs when the matric potential is low enough that the gravitational force is no longer strong enough to cause flow. Matric forces dominate at water potentials lower (more negative) than −33 kPa. Unsaturated flow can be in any direction. Soils with layers of different textures, such as alternating layers of sand and clays, may have water retained at potentials of −10 kPa to −30 kPa. In such cases, less water drains by gravity flow than in a uniformly textured soil.

Saturated Flow

Saturated flow is water flow caused by the pull of gravity. It begins with **water infiltration,** which is water movement into the upper layer of soil when rain or irrigation water is on the soil surface. When the soil profile is wetted, the movement of more water *through* the wetted soil is called **percolation.** Percolating water moving through the soil and substrata carries away dissolved nutrients and other salts. The process of removing soluble components in percolating water is called **leaching.**

Water infiltration is rapid into large, continuous pores in the soil as in coarse sands. It is reduced by anything that decreases either the size or amount of pore space or wetability, such as structure breakdown and pore clogging by lodged particles. The factors that control the rate of water movement into the soil include the following:

- **Percentage of sand, silt, and clay:** Sands permit rapid infiltration. Clays have slow infiltration that becomes even slower as the soil swells in response to the added water.
- **Soil structure:** Fine-textured soil with large water-stable aggregates (granular structure) has much greater infiltration rates than massive (structureless) soils. Blocky and prismatic structures are intermediate.
- **Organic matter:** The greater the amount of organic matter and the coarser it is, the more readily water enters the soil. Organic surface mulches are especially helpful for infiltration because they protect soil aggregates from breakdown by reducing the impact of raindrops and by continuing to supply the cementing agents, as the mulch decomposes, for aggregation.
- **Depth of the soil to hardpan, bedrock, or other impervious layers:** Shallow soils do not have the capacity and, therefore, do not permit as much total water to enter as do deep soils.
- **Amount of water in the soil:** Wet soils have slower infiltration rates than dry soils. This is because pores or cracks are fewer and smaller as clays become moist and swollen. Also, as dry, lower layers of a soil become wet, the matric potential gradient from top layers to lower layers disappears.
- **Soil temperature:** Warm soils take in water faster than cold soils. Also, frozen soils may or may not be capable of absorbing water, depending upon the porosity and water content when freezing took place.

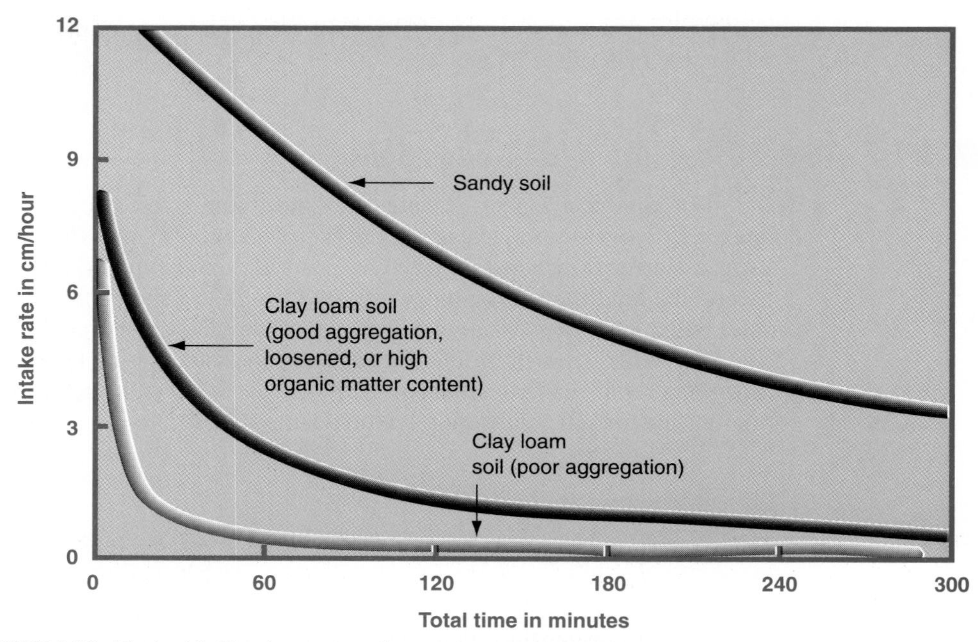

FIGURE 4-13 Typical infiltration rates of water into soils, illustrating the reduced intake rate with increased time of infiltration. The curves for clay loam soils may have various slopes (changes) in infiltration rates, as indicated by the white line, if particles begin to plug pores or soil swelling reduces pore sizes. When no more loose particles are moving and swelling has reached a maximum, porosity remains constant and the curves of intake rate level off. (Courtesy of Raymond W. Miller, Utah State University.)

- **Compaction:** Soil compaction, from causes such as vehicle traffic or heavy grazing, reduces pore space and slows infiltration.

A typical water infiltration (intake) curve is shown in Figure 4-13. Notice that with increased time and wetted depth, the infiltration rate decreases because (1) clays swell, which makes pores smaller and reduces water movement; (2) loose particles that flow with water become lodged in bottleneck pores and plug them; and (3) resistance to flow increases as water passes deeper through long pores. A dry, cracked, clayey soil may take up a large amount of water during the first hour but less in each succeeding hour, particularly if rainfall is dispersing the surface soil particles, causing them to move downward constricting pores.

The condition of a water-saturated soil that controls water flow through the entire root zone or through horizons of interest (usually the upper 152 cm [60 in]) is referred to as its **permeability.** Usually permeability is limited by the most compacted or clayey layer in the profile. Permeability is one of the most important factors determining the overall productivity of a soil or its suitability for development. Information on permeability is used in designing irrigation and drainage systems, terraces, and septic tank absorption fields. **Hydraulic conductivity,** as described in Detail 4-1, is the most commonly used indicator of a soil's permeability. For making and interpreting soil surveys, the USDA divides soils into the following eight *permeability classes*[6] based on permeability rates:

---

[6]Soil Survey Staff, Soil Conservation Service, "National Soil Survey Handbook," Section 618.35, U.S. Government Printing Office, Washington, DC, 1993.

Detail 4-1 Vertical Water Flow

The most common mathematical expression for the vertical water flow rate through soil is called **Darcy's Law.** Darcy stated that for any one soil, the quantity of water (*Q*) flowing through an area (*A*) over a given time (*t*) is directly proportional to the driving force. The driving force for water flow is the *water potential gradient,* i.e., the difference in water potential between two points, divided by the distance between the points ($\Delta\psi/L$). The proportionality constant relating the driving force to the flow rate is the *hydraulic conductivity* (*K*). Each soil has a unique number and size of pores, and therefore a unique *K* value. Darcy's Law is usually written as the equation shown on the left, but can be algebraically rearranged to solve for *K,* as shown in the equation on the right.

$$\frac{Q}{At} = \frac{K\Delta\psi}{L} \qquad K = \frac{QL}{At\Delta\psi}$$

where

- Q = water quantity; typical unit: cm^3
- A = area through which water passes; typical unit: cm^2
- t = time; typical unit: sec
- K = hydraulic conductivity; typical unit: cm/sec
- $\Delta\psi$ = difference in water potential between points X and Y; typical unit: cm
- L = distance between points X and Y; typical unit: cm

Some typical *K* values are given in the following examples. The 10 to the negative exponent means to move the decimal point that many places to the left. For example, 1.4×10^{-4} equals 0.00014. The *K* value indicates the soil's permeability to water flow. Small decimal values (e.g., 10^{-2}, 10^{-1}) indicate fast flow rates, while large negative exponent values (e.g., 10^{-9}) mean very slow rates.

Material	*K (cm/s)*
Gravel	1.5×10^{-1} to 2.0×10^{-2}
Sand	1.2×10^{-1} to 2.0×10^{-3}
Loam	1.7×10^{-4} to 1.7×10^{-7}
Clay	2.5×10^{-8} to 1.0×10^{-9}
Superstition sand (Arizona)	1.8×10^{-3}
Sarpy loam (Colorado)	1.4×10^{-3}
Millville silt loam (Utah)	4.7×10^{-4}

1. **Extremely slow:** <0.025 cm/hr (<0.01 in/hr).
2. **Very slow:** 0.025–0.15 cm/hr (0.01–0.06 in/hr).
3. **Slow:** 0.15–0.51 cm/hr (0.06–0.2 in/hr). Soils less permeable than *slow* are considered severely limited for camp sites, picnic areas, and playgrounds, and for tillage of agricultural fields.
4. **Moderately slow:** 0.51–1.5 cm/hr (0.2–0.6 in/hr). Soils less permeable than *moderately slow* are considered severely limited for septic tank fields and for irrigation.
5. **Moderate:** 1.5–5.1 cm/hr (0.6–2.0 in/hr).
6. **Moderately rapid:** 5.1–15 cm/hr (2.0–6.0 in/hr). Soils with permeability greater than *moderately rapid* are severely limited for septic tank fields or wastewater irrigation because they are poor filters.
7. **Rapid:** 15–51 cm/hr (6.0–20.0 in/hr).
8. **Very rapid:** >51 cm/hr (>20 in/hr).

In humid regions percolation below the upper soil horizons, or below the rooting depth of plants, is common (Figure 4-14). Excess water moves through the profile, dissolving soluble ions and carrying them into groundwater. The water also moves small soil particles (clays and organic colloids) downward until they lodge in pores or adsorb to ped surfaces. The leached soil layers lose most soluble salts and much of their adsorbed (exchangeable) calcium, potassium, magnesium, and sodium. Acidic hydrogen ions (primarily from carbonic acid formed from carbon dioxide dissolved in water) and soluble $Al(OH)_2^+$ ions replace these adsorbed basic cations. This percolation (leaching) causes humid-region soils to gradually become more acidic. Some general principles regarding the leaching of important soil ions follow:

- **Calcium (Ca):** the ion in largest amount in most leaching waters
- **Magnesium, Sulfur, Potassium (Mg, S, K):** next largest amounts, but amounts depend upon soil composition
- **Nitrogen (N):** leaches readily but concentrations are low in cropped soils unless nitrogen fertilizer was recently added
- **Phosphorus (P):** very little leached because soil phosphorus has low solubility

Water flow rates through tubes (uniform pores) increase with the square of the radius (r^2) of the tube; thus, doubling the radius of water-filled pores increases water flow rate about four times. However, because soil pores have irregular shapes and changing sizes, this simple flow rate comparison is inaccurate. Nevertheless, the general concept is still valid—larger pores have much greater flow rates than smaller pores. Unsaturated water movement is rapid through fine sand or well-aggregated loams (medium-sized pores) and slower through very fine and poorly aggregated clayey soils (very small pores).

Gravitational water partly drains from large water-filled pores, and the water is replaced by large air pockets that greatly reduce water flow rates. Water does not flow appreciably from small water-filled pores into large air-filled pores, just as a wet sponge suspended in the air does not drip into the air around it unless *gravitational* water exists at the edge. Water flow from a fine-textured layer (such as clay loam) into a coarse-textured one (sand) is slow, ex-

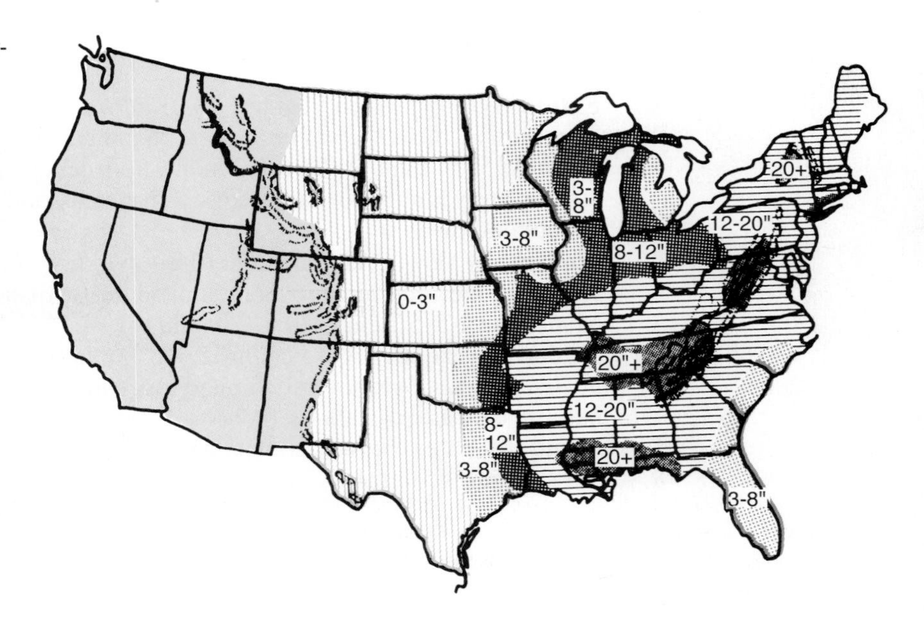

FIGURE 4-14 Average annual percolation. (*Source:* L. B. Nelson and R. E. Uhland, "Factors That Influence Loss of Fall Applied Fertilizers and Their Probable Importance in Different Sections of the United States," *Soil Science Society of America, Proceedings* **19** [1955], No. 4.)

cept at water potentials near 0 (free-flowing water). Even portions of gravitational water may flow very slowly from one layer to another in a stratified soil (soil with layers of contrasting textures). Because of this phenomenon nearly all *stratified* soils hold more water—as much as 50 percent to 60 percent more plant-available water—than if the profile were of a single uniform texture. Stratified field soils usually hold more water than laboratory measurements of −33 kPa values would suggest.

Unsaturated Flow

Unsaturated flow is the flow of water held with water potentials lower (more negative) than about −33 kPa. *Water will move toward the drier region (lower potential, or toward the greater pulling force).* In a soil with *uniform* texture and structure, water moves from wetter to drier areas. The water movement may be in any direction (Figure 4-15). The rate of flow increases as the water potential gradient (the difference in potential per unit of distance between the wet and dry zones) increases and also as the size of pores increases. Unsaturated flow may be as rapid as several centimeters per hour when the soil is wetted to near field capacity. When the soil is near the permanent wilting point, the flow is slower, such as a millimeter per hour or less.

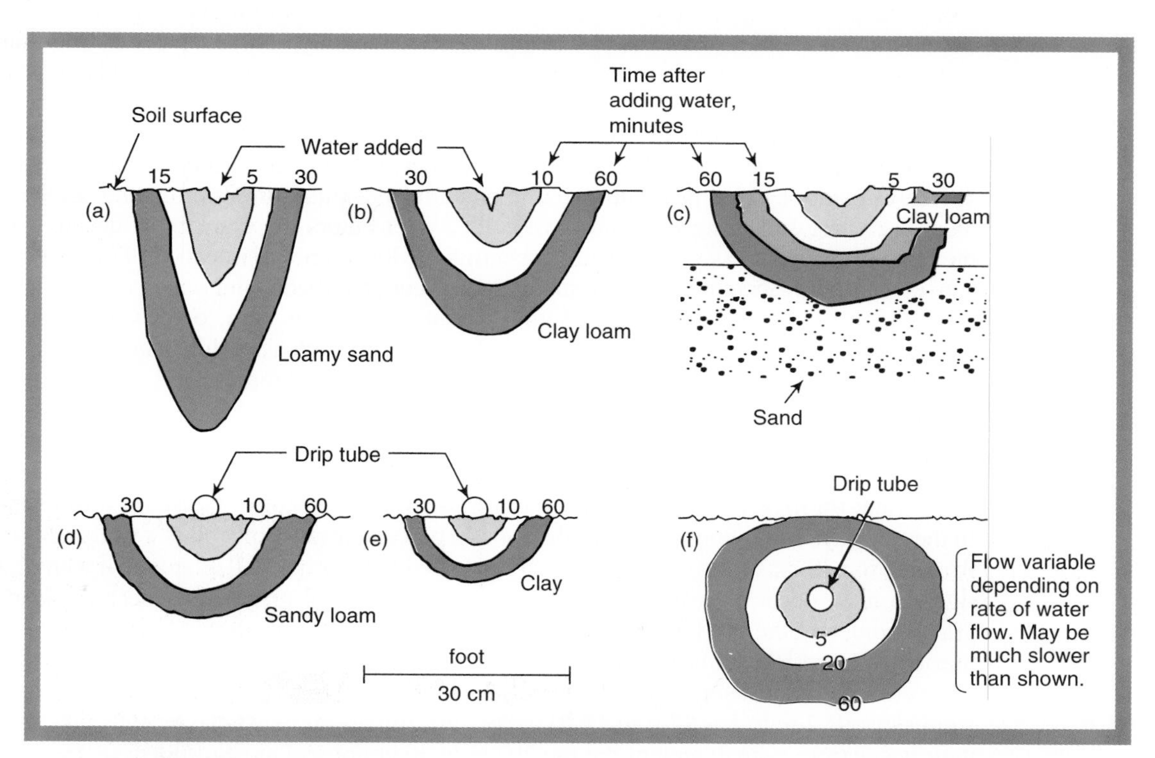

FIGURE 4-15 Typical wetting patterns expected for saturated and unsaturated flow as influenced by texture and location of the water source. *Saturated flow,* which always has unsaturated flow occurring at the same time, (a) is mostly vertical in coarse-textured soils, (b) has more lateral movement in fine-textured soils than occurred in (a), and (c) has a temporary inhibition to flow when sand is below a finer-textured soil. In (d) and (e) flow is still mostly saturated flow from drippers. In (f) water movement (unsaturated flow upward) is in all directions. Because of differing pore sizes, the speed of flow varies with texture and water content. (Courtesy of Raymond W. Miller, Utah State University. See also Walter H. Gardner, "How Water Moves in the Soil," *Crops and Soil* Magazine, Nov. 1979, 13–18.)

4:8 Water Uptake by Plants

The storage and release of water in the soil is only part of the soil water story. Other vital questions are these: (1) How do plants absorb water? (2) From what soil layers is water obtained? and (3) When is water most needed?

Water Absorption Mechanisms of Plants

Although a number of mechanisms are known to affect root absorption of water, more than 90 percent of the total water absorption is by passive absorption. **Passive absorption** is the pulling force on soil water by the continuous water column up through the plant cells as water is lost by transpiration. The plant can be thought of as a wick losing water at one end by transpiration while dipped in water at the other. Water encounters much resistance as it moves through the permeable membranes of the root, through porous cell walls, xylem tubes, and finally, through leaf cells. Plants grown hydroponically (in nutrient solution rather than soil) actually transpire more rapidly when their roots are cut off because water then need not pass membranes to passively enter the xylem.

Root extension is important to the absorption mechanism. Plant root systems are dynamic; some roots die as new ones grow and expand into new areas of the soil, encountering more soil water as roots extend. Because the water movement to roots by *unsaturated flow* occurs only over short distances of a few millimeters per day, root extension is important in aiding the plant's absorption of water, particularly when the soil root zone holds no gravitational water.

Active absorption requires that the plant expend energy to absorb water. The selective accumulation of soluble ions by the plant cells increases a plant's soluble salt content (i.e., decreases osmotic potential). This osmotic potential aids plants in water uptake but the plant must use energy to bring ions into root cells. Water enters the plant roots because of net movement toward areas of high salt concentration (low osmotic potential). For most plants active absorption accounts for very small amounts of absorbed water.

Many plants can also obtain some moisture from fog, rain, or dew by absorption through the leaf stomata. A few plants called **epiphytes,** such as orchids and Spanish moss, absorb water from the air. Most epiphytes grow in rain forest biomes.

Depths of Water Extraction

Most water for irrigated crops is extracted from shallow soil depths, where most roots reside. In the few days following wetting of the soil, 40 percent or more of a plant's water may come from the top 30 cm (12 in). Golf greens are irrigated with only small amounts of water but irrigation must occur almost daily because of shallow-rooted grasses and because sand and gravel comprise the base of the green. This coarse-textured base holds little water and few nutrients, but resists deformation by people walking on it. In contrast to the grasses, mesquite trees growing in arid soils can remove appreciable water from great distances laterally and from subsoils 2 m to 3 m (7 ft to 10 ft) deep. Turf grasses on moist clayey soils take most of their water from a small area of the top 20 cm to 30 cm of soil most of the time.

As the plants grow and the soil dries between wettings, an increasing portion of the water taken up by the plant comes from deeper soil layers until three-fourths of the water absorbed in a given day by deep-rooted plants may come from depths below 60 cm. Keeping the top 30 cm of soil moist is important to plant growth because that is the region where root densities and nutrient concentrations are greatest.

Soil wetness and adequate aeration are competing conditions in many soils. Usually clayey and compacted soils are likely to have poor aeration when wet. The larger and more continuous the pores, the more likely the soil will have good air exchange. Some examples are shown in Figure 4-16.

When Plants Need Water Most

Plants are obviously in need of water when they begin to wilt, but by the time wilting is visible, plant growth has already been reduced significantly. Wilting usually is due to insufficient soil moisture. However, plants wilt on well-moistened soils on warm, dry days because the water loss by evapotranspiration is greater than the absorption rate of water by plant roots. Although plants usually recover from such temporary wilting during the cooler evening or night periods, some growth loss has usually occurred. Many small-grain crops can exhibit temporary wilting during later growth without extensive loss in grain yield. However, drying to a condition of wilt will reduce *vegetative* growth of nearly all plants.

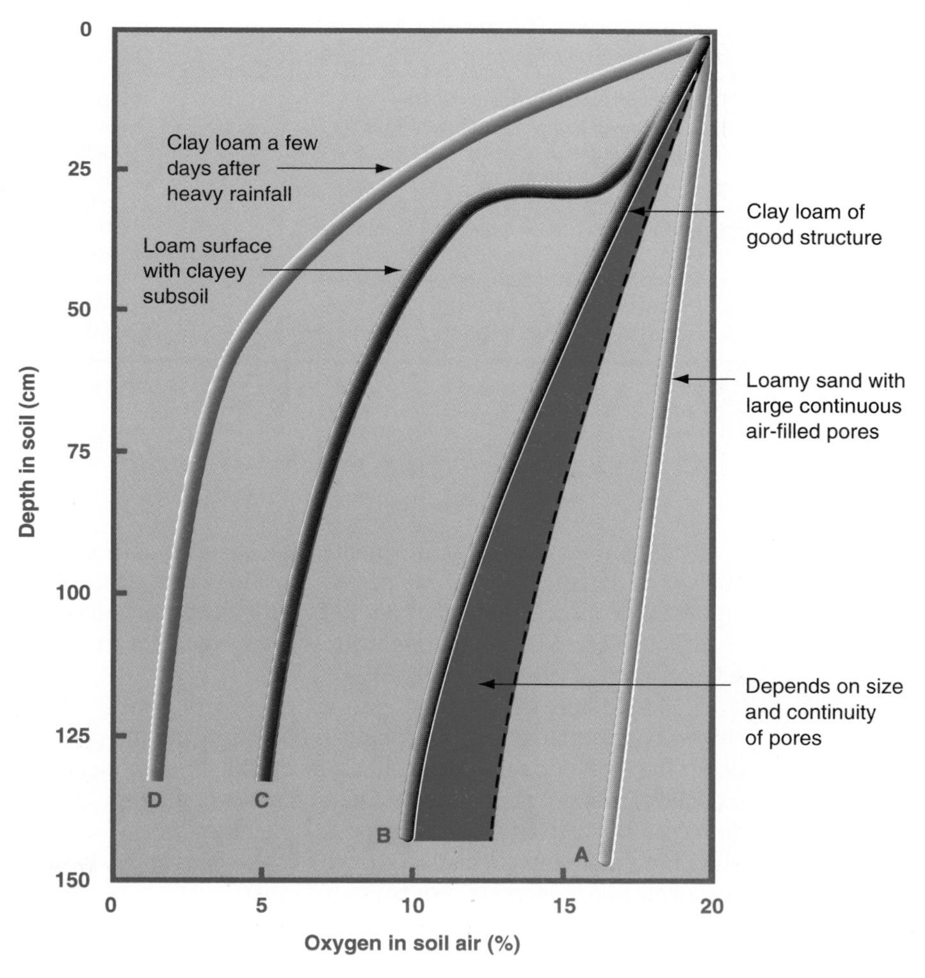

FIGURE 4-16 Typical curves for changes in oxygen content with changes in soil depth for four soil conditions: (A) a moist, loamy sand with large, continuous air-filled pores; (B) a clay loam with smaller and less continuous pores; (C) a loam surface **A** horizon with a clayey **Bt** horizon and clay loam subsoil layers; and (D) a clay loam soil in which water fills many pores after being wetted, thus blocking oxygen exchange to deeper soil depths. A clayey or wet soil becomes, in essence, a shallow soil because of poor aeration in its deeper layers. (Courtesy of Raymond W. Miller, Utah State University.)

Table 4-4 Effects of Water Stress (Dryness) on Crop Yields

Condition or Time of Water Deficiency (Stress)	*Yield Reduction (%)*
Five weeks prior to ear emergence	
Wheat grain	70
Total wheat plant tops	52
Stress 20 days during male meiosis (chromosome numbers halved)	
Total corn plant tops	29
Corn grain	0
Stress 20 days during grain filling	
Total corn plant tops	30
Corn grain	47
Sorghum grain, emerging grain head, reached −12.9 bars	17
Sorghum grain, milk through soft dough stage, reached −12.4 bars	10

*Growth Stage When Water Stress (Dryness) Reduces Yields the Most**	*Crops that Fit this Category*
1. Need high water levels constantly	Castor bean, cauliflower
2. From flower initiation to early stages of developing fruit or seed	Apricots, barley, beans, citrus, corn, cotton, oats, olives, peanuts, peas, small grains, soybeans, sunflower, tomatoes, wheat
3. During rapid fruit or tuber growth to harvest	Cherries, lettuce, peaches, potatoes, strawberries, turnips, watermelon
4. During head formation and enlargement	Broccoli, cabbage, lettuce
5. During tillering, prior to heading	Oats, sorghum
6. Period of maximum vegetative growth rate	Alfalfa, sugarcane, tobacco

*Food and Agricultural Organization of the United Nations, "Crop Water Requirements," Irrigation and Drainage Paper 24, FAO, Rome, 1975.

Sometimes damage to the plant can occur with only a few dry days. Drought stress for 10 days during the 30 to 40 day period after flowering reduces germination of the cottonseed produced to only 20 percent of normal.[7] Drought stress before or after this period has almost no effect on germination. Drought stress seems to affect germination most during the critical stage of protein filling in the seed.

Most plants have critical periods of growth during which insufficient water (**water stress**) is most damaging. For plants producing seed, the period from flowering to fertilization is the most critical. If drought occurs during this period, grain or seed yield will be greatly reduced. Table 4-4 shows yield reduction of several crops caused by drought. Similar reductions also apply to fruit and fiber-producing crops.

For maximum vegetative growth (alfalfa, sugarcane, pastures), water is most important during rapid size increase. For fruits and tubers, water is needed during fruit enlargement. For grain crops (oats, wheat), water is especially critical during tillering and the pollination and fertilization periods.

[7] E. L. Vigil, "Deadliest Days of Drought Stress," *Agricultural Research* **37** (1989), No. 7, 19.

4:9 Consumptive Use and Water Efficiency

The amount of water needed for plant growth includes the water lost in two ways: (1) the **transpiration** loss and (2) the **evaporation** loss from the soil. The two losses are often combined and called **evapotranspiration (ET). Consumptive use (CU)** is the quantity of water lost by evapotranspiration plus that contained in plant tissues. Consumptive use is practically the same as ET, because the water in plant tissues is only about 0.1 percent of the total consumptive use for slow-growing plants and up to about 1 percent for plants efficient in water use. Accurate measurements of water loss often require elaborate setups or expensive instrumentation (Figure 4-17).

Evapotranspiration

Evapotranspiration, the water lost by evaporation from soil and transpiration from plants, *increases* when the air is dry (low relative humidity), warm, moving (windy), and if the surface soil is near field capacity. Evaporation *decreases* when the opposite conditions occur: Air is motionless and has high relative humidity; temperatures are cool and the soil is dry (Figure 4-18). The maximum ET loss that would occur if the soil were kept near field capacity, is referred to as **potential ET (ET_p)**.

Evapotranspiration involves a large amount of water, but usually less than the amount of water that would evaporate from ponded water covering an equal area under similar atmospheric conditions. The amount of ET water loss can be estimated by using a weather station **Class A evaporation pan.** The water loss from mature plants is typically about 50 percent to 90 percent as much as the loss from a free water surface with the same area (Table 4-5). The amounts of these daily losses of water range from perhaps 0.1 cm per day during early growth in cool weather to more than 1.3 cm (0.5 in) per day during hot, dry weather (Table 4-6).

7

Water Use Efficiency

The amount of water required (transpiration + plant growth + evaporation from soil + drainage loss) to produce a unit of dry weight material (a kilogram of corn, for example) is a measure of **water use efficiency (WUE).**[8] How efficiently can water be used by plants? The **transpiration ratio (TR)** is the amount of water used per unit of dry matter produced. The transpiration ratio is the weight of water transpired divided by the weight of dry plant material produced (only above-ground plant portions are used—except for root crops, such as sugarbeets and potatoes). Transpiration ratios range from about 200:1 to 1000:1, with 300:1 to 700:1 being most common values. Hot, dry, irrigated regions typically have ratios of 600:1 to even 1000:1.

Transpiration ratios may be used to compute water needed for irrigation of particular crops. For example, if a field of alfalfa produces 10,000 kg per hectare and has a transpiration ratio of 500:1, the water used would be 500 × 10,000 kg = 5,000,000 kg of water needed to produce the crop, which equals 50 hectare-cm of water. (Using English units, this same field would produce 4.46 tons/acre and would require about 21 acre-inches, or 560,000 gallons/acre, of water.)

---

[8]Engineers may use a different definition: the net amount of water added to the root zone as a fraction of the amount of water taken from some source. Source: Daniel Hillel, ed., *Optimizing the Soil Physical Environment toward Greater Crop Yields,* Academic Press, New York, 1972, 90.

(a)

(b)

FIGURE 4-17 Consumptive use of water and leaching losses are determined with precision by research studies using lysimeters, which are sealed containers of various sizes with access or drainage tubes. (a) The 20,430-kg (45,040-lb) soil core in this lysimeter in Pawnee, Colorado, was lowered into a metal-lined hole onto large balance beams to weigh water losses and gains. (b) The lysimeter in place. Lines shown are power and other electrical lines to an adjacent trailer laboratory for making various recordings and measurements. (*Source:* International Biological Programs Grassland Biome Study, Colorado State University. Photos by Alan Brooks.)

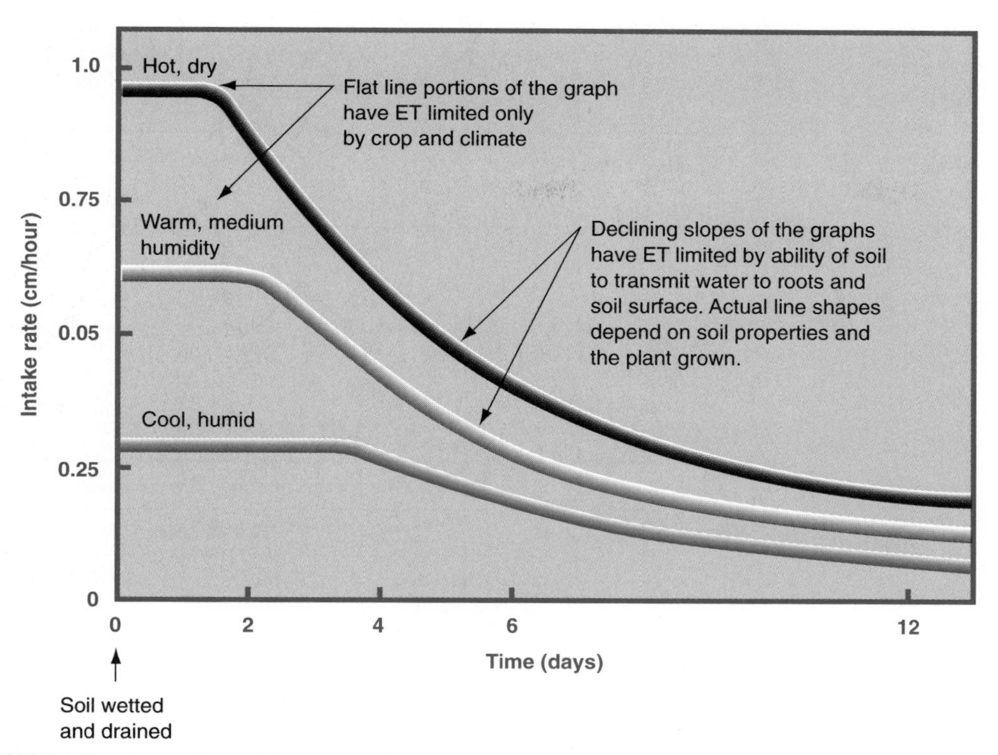
1.0
0.75
0.05
0.25
0
Intake rate (cm/hour)
Hot, dry
Flat line portions of the graph have ET limited only by crop and climate
Warm, medium humidity
Declining slopes of the graphs have ET limited by ability of soil to transmit water to roots and soil surface. Actual line shapes depend on soil properties and the plant grown.
Cool, humid
0
2
4
6
12
Time (days)
Soil wetted and drained

FIGURE 4-18 Illustration of the rates of evapotranspiration (ET) as affected by climatic conditions and by the ability of the soil to supply water to the plant. (Courtesy of Raymond W. Miller, Utah State University.)

4:10 Reducing Water Loss

Reducing water losses from a local area (farm, watershed) can only be accomplished by altering the conditions effecting the losses: evapotranspiration, runoff, and percolation through soil into groundwater. Movement of water by runoff and percolation is not always a loss, as the water may be useful elsewhere.

Reducing Evapotranspiration

Evapotranspiration losses can be reduced by selecting plants with greater water use efficiency, by reducing total growth, by shading or cooling the area, or by using a moisture barrier.

Mulches (straw, peat, gravel, formed aggregates, transparent and opaque plastic sheeting, asphalt) act as barriers to moisture movement out of the soil. Most mulches also keep the soil temperature cooler by shading the soil. Mulches lighter in color than the soil will also lower soil temperatures by reflecting more incoming solar radiation than the soil would reflect. Except for nurseries or landscaping purposes, mulches have not been considered economically practical for the sole purpose of evaporation control. Most often they are used for multiple purposes including weed control, soil erosion control, temperature control, evaporation control, and protection against rotting for fruit or berry crops that would otherwise touch the wet soil. Early-season irrigation in the San Joaquin Valley of California accentuates the area's drainage problem. Clear-plastic mulch placed over soils planted with cotton in

Table 4-5 Water Required for Evapotranspiration (ET) of Selected Crops Given as a Percentage of Maximum ET (the Amount of Water Evaporated from Class A Weather Station Pan)[a]

	Percentage of Free Water Evaporation in:			
Crop	*January*	*April*	*August*	*November*
Alfalfa	55	85	90	70
Citrus	50	55	55	50
Deciduous orchard	15	50	65	15
Grapes	15	40	60	25
Pasture grass	40	65	70	50
Pangola grass[b]	115	100	95	100
Plantain (cooking banana)[b]	80	85	75	110
Sugarcane[c]	75	50	75	90

ET as a Percentage of Maximum ET Values[d]

	For Western U.S. Areas[d]			
Crop	*Seedling Stage 20 Days after Planting*	*Has Developed 50% of Full Cover*	*When the Full Plant Cover Is Reached*	*50 Days after Reaching Full Cover (Maturing)*
Alfalfa	41	79	100	100
Beans	21	51	107	59
Corn	21	49	49	68
Pasture (full cover)	87	87	87	87
Peas	22	51	105	20
Potatoes	12	41	91	38
Small grains	17	51	104	19
Sugar beets	12	41	91	90

Sources: Selected and modified data from (1) J. E. Christiansen and G. H. Hargreaves, "Irrigation Requirements from Evaporation," *International Commission on Irrigation and Drainage, Seventh Congress,* 1968, R. 36, Question 23, p. 23.593. (2) M. E. Jensen, ed., *Consumptive Use of Water and Irrigation Water Requirements,* Report of ASCE Committee on Irrigation Water Requirements, American Society of Civil Engineers, New York, 1973.

[a]The depth of water used by the plant is estimated by multiplying the percentage shown, in decimal form, times the depth of water evaporated from a Class A weather station pan.

[b]South America (Colombia), tropical area with rainy weather May to August, very dry January to March.

[c]Planted in March, harvested a year later in April, with maximum growth October to February.

[d]These values assume that pan evaporation equals potential (maximum) evapotranspiration. This may not always be true; pan values may need adjustment first. Pan values vary such that potential evapotranspiration is 0.7 to 1.1 times the pan evaporation data.

March saved enough water for germination and avoided a preseason irrigation.[9] For growing Pima cotton this plastic mulch increased yields and profits. Petroleum mulches sprayed in thin strips above the seed row have also increased moisture retention and seed germination in California.

A common misconception in the use of surface mulches is that they always result in a final moisture saving. A mulch will reduce evaporation loss in the first days or weeks following wetting, but over a *long* drying period, the water loss from a mulch-covered soil approaches the same amount as that from uncovered soil. Barrier-type mulches (plastic sheet-

[9]E. Fereres and D. A. Goldhamer, "Plastic Mulch Increases Cotton Yield, Reduces Need for Preseason Irrigation," *California Agriculture* **45** (1991), No. 4, 25–28.

Table 4-6 Some Consumptive Use Rates of Various Plants in Various Locations and Situations

Plant	Use Condition and Location	Consumptive Use Rate	
		mm/Day	in/Day
Cotton	Peak use (blooming), July, Salt River Valley, Arizona	10.1	0.40
Oranges, navel	July, Orange County, California	11.7	0.46
Potatoes	Most active growth, warm, dry	6.1–8.1	0.24–0.32
Safflower	Peak growth, May, Mesa, Arizona	12.7	0.50
Soybeans	Peak use, Nebraska	11.7	0.46
Sugar beets	Peak use, Davis, California	7.6	0.30
Wheat, winter	Late fall, seedling stage	1.0–2.0	0.04–0.08
	Dormant winter period	0.2–1.0	0.008–0.04
	Peak growth, windy, arid	8.1–9.1	0.32–0.36

Source: J. S. Robins, J. T. Musick, D. C. Finfrock, and H. F. Rhoades, "Grain and Field Crops," Chapter 32; D. W. Henderson, "Sugar, Oil, and Fiber Crops, Part III—Oil Crops," Chapter 33; and J. R. Stockton, J. R. Carreker, and M. Hoover, "Sugar, Oil, and Fiber Crops, Part IV—Irrigation of Cotton and Other Fiber Crops," Chapter 33; all in *Irrigation of Agricultural Lands,* R. M. Hagan, H. R. Haise, and T. W. Edminster, eds., No. 11 in the Agronomy Series, American Society of Agronomy, Madison, WI, 1967.

ing, asphalt covers) are much more effective than porous ones and will greatly reduce evaporative water loss for a long time (Figure 4-19). Any shading cover (plant residues, seed hulls, gravel, rocks, wood chips) will reduce evaporative water losses for several days, and if those days are during the sprouting and new-growth periods, the moisture saving can be critical to the survival of the plants.

One of the distinctive practices of dryland grain production is fallow. **Fallow** is the practice of leaving land unplanted, commonly in alternate years, to accumulate a little extra water in the fallow year to be used in the next growing year. During fallow the fields are clean-cultivated to hinder weed development and so save water that would have been lost by transpiration.

The water saved by fallowing is small, but the conservation of even a small fraction of the annual rainfall can mean the difference between a profitable and unprofitable yield of grain. Approximately 10 cm (4 in) of *available water* are required to produce wheat plants from seed to maturity. Each additional 2.5 cm of water available in the soil produces an

FIGURE 4-19 Plastic mulches can conserve water and warm the soil earlier. These plastic covers are having holes punched for transplanting peppers near Irvine, California. Buried drip system lines help keep the water costs down. The plastic mulch allows early warming, reduces water evaporation, and prevents low-hanging fruit from touching a wet soil and rotting. The system is also used for strawberries and tomatoes. (*Source: Ag Consultant* Magazine, Aug. 1987, 4–5. Photo by Parry Klassen.)

additional 280 kg/ha to 470 kg/ha (4 bu/a to 7 bu/a) of wheat. Of course, yield increases also depend upon climate and soil characteristics such as fertility.

Plant selection can greatly influence ET. Sorghum uses less water than corn on the same site. Short-season crops such as lettuce and beans can be used to reduce water needs. Even watershed areas can have vegetation altered to save water. About 4 miilion hectares (10 million acres) of land in California are in foothill areas, receiving 38 cm to 51 cm (15 in to 20 in) of annual precipitation and supporting mainly low-value trees and shrubs (brush). Likewise throughout the Western states, shrubby species such as sagebrush occupy large land areas. These shrubs are usually deep rooted and use all the available moisture in the soil. Furthermore, their leaves and stems intercept a large amount of precipitation that is evaporated directly back into the atmosphere.

Research on the conversion of such brushland to grassland has demonstrated the following:

- Grasses root less deeply than trees and shrubs.
- Grasses become dormant earlier in the fall and thus leave more stored water in the soil for early spring growth.
- Grasses intercept less precipitation, thus permitting more water to enter the soil.
- Grasses protect the soil from erosion better than do shrubby species.
- Grasses permit more runoff water to be caught behind dams for use by municipalities and farms. On some watersheds replacing trees and shrubs with grasses has saved as much as 5 cm (2 in) more water in runoff per year.

Forest trees transpire large amounts of water into the atmosphere through their leaves or needles. Forests also intercept considerable quantities of rain or snow and permit it to be evaporated back into the atmosphere before the water ever reaches the soil. By removing part of the forest cover, more water reaches the soil and less is transpired. Clear cutting (total harvesting) of hardwoods in West Virginia resulted in only about one-third as much water loss by evapotranspiration as occurred in uncut forests. Although infiltration of water into the soil would likely increase as a result of a practice such as clear cutting, one must consider other consequences. The additional runoff from clear cutting may or may not be useful, and soil erosion is almost certain to increase.

Field-Scale Water Conservation

Irrigation techniques can be improved to conserve water. Runoff losses in furrow irrigation can be reduced by cutting down the size of the stream for the duration of soaking after the water has reached the furrow end. Surge flow reduces deep percolation. Land leveling increases surface irrigation water efficiency by allowing better water control. Irrigating more frequently, but only wetting alternate furrows each time, encourages maximum water use and minimum evaporation and drainage loss. The greatest water conservation, particularly on sandy soils, is with drip and sprinkler irrigation. Both drip and sprinkling irrigation can be efficiently adapted to hilly or irregular slope areas, shallow soils, sandy soils, and slowly permeable clay soils. Conversely, windy sites with very low relative humidity are not conducive to efficient use of conventional sprinklers because so much water evaporates before reaching the plants.

Reuse of wastewater is receiving considerable attention. Wastewaters include water from municipal water treatment plants, industries, and irrigation tailwaters. Wastewaters are typically of low quality because of their salt or sediment content, but they are inexpensive, reliable sources with availability largely unaffected by drought.

FIGURE 4-20 In Israel, terraces of both ancient and modern construction support vegetation on slopes that would otherwise be too dry and erodible. (Courtesy of Duane T. Gardiner, Texas A&M University-Kingsville.)

In dry areas **level terraces** with the lower end closed maximize water retention. In the 510-mm (20-in) rainfall belt of Texas, impounded water in closed-end terraces resulted in a 6-percent increase in yield of cotton compared with cotton grown on land not terraced, based on average yields over a 26-year period.

Conservation terraces are small, one-field watersheds that collect runoff on sloped areas for use on an adjacent level area. Terraces constructed in Israel, an arid nation, are still functional (Figure 4-20). Conservation terraces typically have about two to four times as much area in watershed as in the planted level terrace area. The crop on the flat water-receiving area should be of sufficiently high value to justify the cost if much soil needs to be moved in preparation of the area. A similar application of this principle is the use of catchment basins with collection, cultivated land ratios of 20:1 to 30:1, as used in the very dry areas of Israel, where with only 150 mm to 200 mm (6 in to 8 in) annual rainfall, water capturing supplies enough water to grow fruit trees, pine trees, and grain crops.

Water is more critical than energy. We have alternate sources of energy. But with water, there is no other choice.

—Eugene Odum

Questions

1. How does hydrogen bonding hold water to soil mineral surfaces?
2. Define (a) *water potential* and (b) *matric potential.*

3. Using kilopascals, what is the equivalent of a water potential value of −1 bar?
4. Which water in a film of water on a soil particle is held most strongly to the soil: the water near the particle or the water at the outer edge of the film? Explain.
5. What is the difference between *available water capacity* and *field capacity*?
6. Define (a) *saturation percentage* and (b) *permanent wilting percentage.*
7. Give an approximate volumetric water percentage for the wilting point of a typical (a) loamy sand, (b) loam, and (c) clay soil.
8. Calculate the mass water ratio of a soil if a sample of the soil at field capacity plus the can containing the sample weigh 248 g, the dry soil plus the can weigh 233 g, and the can weight is 141 g.
9. If a soil at field capacity has a mass water ratio of 0.24 and a bulk density of 1.3 g/cm^3, calculate its volumetric water percentage.
10. (a) How much water (in cm of water per 30 cm depth of soil) does the soil described in Question 9 contain at field capacity? (b) About how many centimeters of plant-available water will the soil hold in the top 90 cm? (Assume that the soil is uniformly wet to a depth of 90 cm.)
11. Describe how tensiometers are used to determine the soil water status for a crop.
12. In terms of forces and flow rates, describe the difference between *saturated flow* and *unsaturated flow*.
13. List factors that influence rates of water infiltration into soils.
14. Describe the phenomenon of *passive water uptake* by plants.
15. (a) Define *evapotranspiration.* (b) Define *consumptive use.* Estimate a daily consumptive use rate for soybeans in Nebraska.
16. (a) How do the losses by ET compare to water losses from a pan of water at the same location and time? (b) Why is knowledge of this relationship useful?
17. If an average transpiration ratio is about 450:1, why is this ratio for some crops higher and other crops lower than this?
18. How can ET water losses be reduced?

Soil Colloids and Chemical Properties

Clay minerals are alumino silicates that predominate in the clay fractions of soils in the intermediate to advanced stages of weathering.

—Garrison Sposito

The whole of nature is a trillion, trillion chemical machines . . .

—H. H. Seliger and W. D. McElroy

5:1 Preview and Important Facts

PREVIEW

The chemical properties of soils are determined by the colloids of soils. What is a colloid? What are the colloids in soils? A **colloid** is any solid substance whose particles are very small; thus, its surface properties are relatively more important than its mass (weight). Most colloids are smaller than a few *micrometers* (microns) in diameter (0.0001 cm). In Greek *colloid* means *glue.* The smaller particles, with their enormous surface-to-surface contact, would stick things together. Because of their large surface-to-mass ratio, soil colloids settle slowly from suspension.

The chemical properties of soils include mineral solubility, nutrient availability, soil reaction (pH), cation exchange, and buffering action. Chemical properties of soils are more influenced by the colloids (clays and humus) than by equal weights of the larger silt and sand particles. The predominant soil colloids in soils are *clays* and *humus.* **Humus** is a mixture of residues left after partial decay of organic substances in and on top of soils. Clays and humus are the sites of most chemical reactions in soils.

The various kinds of clays possess different properties. Clays have negatively charged sites in their lattices and attract and hold positively charged ions (cations) at those negatively-charged sites on the clay surface. Montmorillonite and vermiculite are swelling–shrinking clays that absorb large amounts of positively-charged ions (cations) and water. Kaolinite is the pottery-making clay; it has no swelling characteristics but also little cation adsorption capacity. Many other clays exist, including the very residual, nonswelling, slightly reactive sesquioxide clay.

The quantity of cations that can be held by a given amount of soil is the **cation exchange capacity** of that soil. The exchangeable cations are not easily removed by leaching action until they are replaced (exchanged) by other cations. The process of a solution cation replacing an adsorbed cation is called **cation exchange.**

Plants often excrete hydrogen ions (H^+) to the soil solution, which can replace nutrient cations from their exchange sites, thus allowing the nutrients to move to the plant root and be absorbed. If the absorbed ions on the colloids are mostly $Al(OH)_2^+$ and H^+, the soils are acidic. The pH is a measure of acidity or alkalinity: **soil pH** ranges from about pH 3.5 (extremely acidic) to pH 7 (neutral) to about pH 11 (extremely alkaline [or basic]). Most crop plants grow well between pH 5.5 and pH 8.5. In strongly acidic soils, soluble aluminum and manganese can reach toxic levels and microbial activity is greatly reduced.

IMPORTANT FACTS TO KNOW

1. The definition, identity, and properties of soil colloids
2. The various meanings of the word *clay*
3. The contrasting properties of montmorillonite, kaolinite, and metal oxides and the climates in which each becomes the dominant clay
4. The origins and general properties of clay minerals
5. The nature of the active groups in humus colloids and the variation in organic colloid composition
6. The causes and action of cation exchange and the soil ions mostly involved in cation exchange in soils
7. The meaning of *soil pH, soil acidity,* and the *ions causing acidity*
8. Some general amounts of calcium and potassium held on the cation exchange sites per hectare-30 cm deep
9. The effect of the soil cation exchange capacity (CEC) on the amount of lime needed and on the available potassium, magnesium, and calcium held in soils
10. The meaning of *basic cation saturation percentage* (BCSP)

5:2 Soil Clays

Clays, the active mineral portion of soils, are colloidal, and most clays are crystalline. The term *clay* has three meanings in soil usage: (1) It is a particle-size fraction composed of any mineral particles less than 2 microns[1] in effective diameter; (2) it is a name for a group of minerals of specific composition ; and (3) it is a soil textural class. Many materials of the clay-size fraction—such as gypsum, carbonates, or quartz—are small enough to be classified as clay on the basis of size but are not clay minerals. In contrast, some clay mineral particles may reach sizes of 4 or 5 μm—double the upper size limit of the clay-size fraction. *Clay,* as discussed in this chapter, includes only *clay minerals* of specific crystalline or amorphous composition and of any size up to several micrometers (Figure 5-1).

The Origin of Clays

Prior to the x-ray study of mineral compositions, clays were incorrectly thought to be just smaller particles of primary minerals, such as small particles of quartz, feldspars, micas, hornblende, or augite. Now clay minerals are known to have specific compositions that are not very similar to the primary minerals, except to micas. **Clay minerals** are mostly newly formed crystals, reformed mostly from the *partial dissolution of minerals,* leaving behind a

[1]One micrometer (μm), or micron, equals 0.001 mm; 1.0 mm equals 0.04 in. Thus, 25,400 μm equals 1.0 in.

FIGURE 5-1 Microscopic evidence that primary minerals and secondary (clay) minerals are crystalline. The regular shapes indicate definite crystal patterns. The large crystals are mica, a primary mineral (1), and the small crystals are kaolinite, a secondary clay mineral (arrow). Mexico. (Magnification 26,300 times.) (Courtesy of R. L. Sloane, University of Arizona.)

certain quantity of clay. There is also some precipitation from concentrated solutions. Clay types are determined by the minerals involved (amounts of silica, alumina, and cations produced in soil water) and the acidity of the leaching waters.[2] Clays are *secondary minerals.*

Laboratory syntheses of clays have proven that the kind of clay formed is determined by the proportions of the different ions, silica, and alumina in the solution during formation. Removal of some of these products by leaching will reduce clay formation rates and alter the kind of clay formed. Soils of regions that have hot, moist climates but are not leached excessively because of poor drainage have large amounts of the primary minerals dissolved, which then recrystallize to clays. These hot, humid, tropical soils tend to be high in clay content, even to depths of 5 m to 20 m (16.4 ft to 65.5 ft). Other nearby soils may have large portions of the weathered primary minerals flushed away. Some clays apparently form from slight alteration (selective solubility and rebuilding) of some primary minerals, particularly from the micas (biotite and muscovite), to form vermiculite and hydrous mica.

The composition of clays in a given soil can be a complex mixture of clays from various origins. Clays have been forming for thousands of centuries in some soils. Erosion moves some of those clays and redeposits them as alluvium in oceans and as sediments on land. Thus, a soil may have clays from several origins:

1. **Inherited clays:** Deposited as clays in sediments that are now forming a soil; perhaps the clays were formed in a different climate eons of time ago
2. **Modified clays:** Changed by further weathering (degradation) of moved and deposited clays

---

[2]Note that the terms **silica** and **alumina** are not the elements silicon and aluminum. Silica is the SiO_2 unit from mineral dissolution, and alumina is the Al_2O_3 unit actually built up structurally in multiples of the silica and alumina units, organized in small groups of the tetrahedral and octahedral sheet pattern of clays.

3. **Transformed clays** (aggradation): Accumulated from highly weathered clays deposited in alluvium that now provide soluble silica and other clay-forming materials
4. **Neoformed clays:** New clays formed entirely from crystallization of clays from soluble ions in solution and the dissolution of minerals in the developing soil

Nature of Clays

Most clays are crystalline; they are composed of definite, repeating arrangements of atoms. The majority of clays are made up of planes of oxygen atoms; silicon and aluminum atoms hold the oxygen planes together primarily by **ionic bonding,** which is the attraction of positively and negatively charged atoms (Figure 5-2).

Three or four planes of oxygen atoms with intervening silicon and aluminum ions (or other cations, depending on the clay) make up a **layer.** One clay particle is composed of many layers stacked like a deck of cards. A clay particle is called a *micelle.* A few clays have irregular structure; the oxygen and other atoms have regular orientation *only in small sections.* These *microcrystalline* clays are called *amorphous* materials. Structural details are given in Detail 5-1.

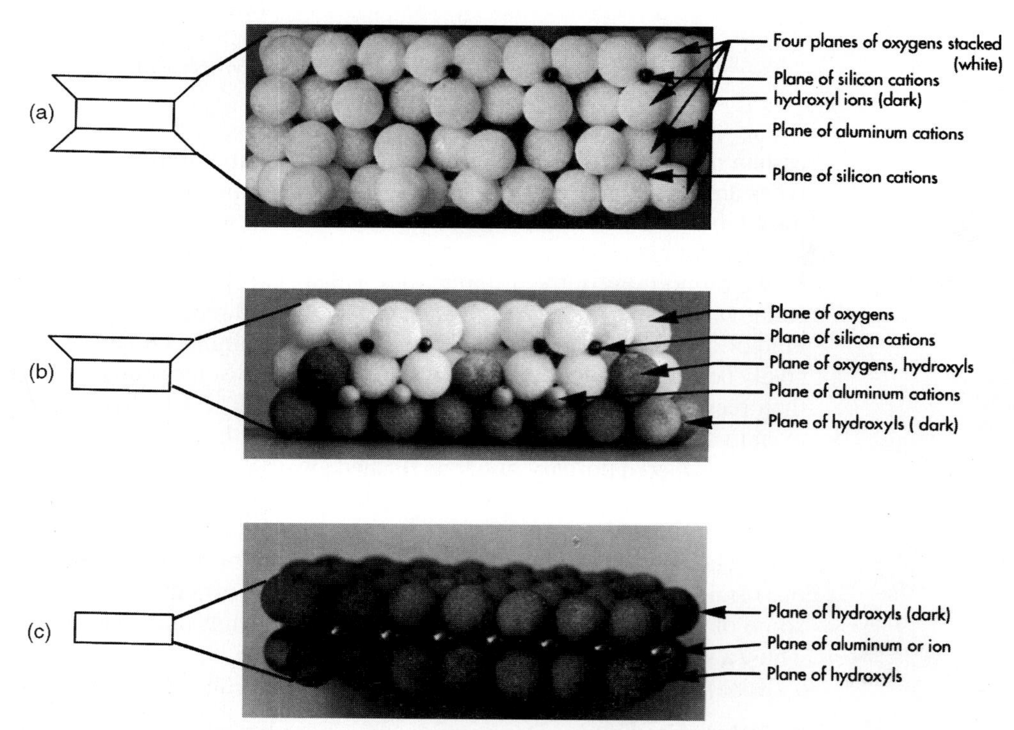

FIGURE 5-2 Representation of the basic orientation of atoms in clay minerals. The unit lattice of (a), typified by montmorillonite, has four planes of mostly oxygen ions with positively charged silicon (Si^{4+}) or of aluminum (Al^{3+}), mostly in the small spaces between oxygen ions, holding the oxygen ions together. Where only three oxygen ion planes occur (b), as in kaolinite, one plane is hydroxyls (OH^-, dark-colored gray). Many of the aluminum and iron clays (sesquioxides) have only planes of hydroxyls (c) or a mixture with oxygen ions and hydroxyl ions, which are about the same size. Clay minerals are mostly oxygen ions (70%–85% by volume). In addition to aluminum and silica, the ions of iron, zinc, magnesium, and potassium occur in the structural lattice of some clays (see Figure 5-5). (Courtesy of Raymond W. Miller, Utah State University.)

Detail 5-1 Silicate Clays: Structural Details

When a cation such as Si^{4+} or Al^{3+} is surrounded in a close fit by oxygens and has planes fitted on the exposed surfaces, the names of the geometric shapes are used to indicate the spacial relationships. Thus, the Si^{4+} surrounded by 4 oxygens forms a tetrahedron (4 sides when planes cover the surfaces) and the unit has **tetrahedral** coordination. The 6 oxygens needed to fit snugly around the larger Al^{3+} form an octahedron (8-sided figure) and the Al^{3+} is in **octahedral** coordination.

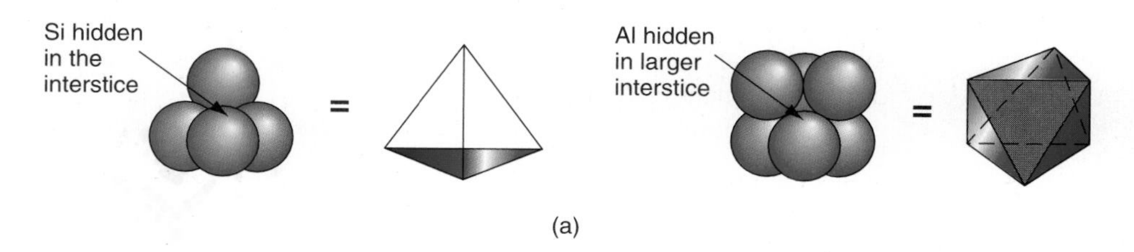

(a)

In Figure 5-2 the planes of oxygen held together by Si^{4+} are tetrahedrally oriented and are referred to as a **(silica) tetrahedral sheet.** One of these planes of oxygen plus a second plane of oxygens (or hydroxyls) are held together by Al^{3+}, are octahedrally oriented, and are referred to as the **(alumina) octahedral sheet.** One silica sheet per one alumina sheet is a 1:1 lattice. With 2 silica sheets per 1 alumina sheet, the lattice is a 2:1 type. Vermiculite and chlorite can be referred to as 2:1:1 or 2:2 lattice types.

For a different look at the structure, an expanded view of structure (A) of Figure 5-2 is shown below.

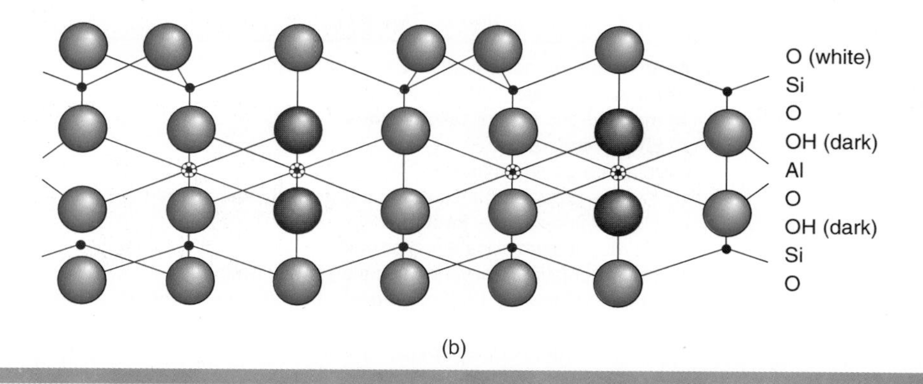

(b)

Clays have a net negative charge, which will attract and hold positively charged ions (**cations**).[3] Common soil cations include potassium (**K^+**), sodium (**Na^+**), ammonium (**NH_4^+**), calcium (**Ca^{2+}**), magnesium (**Mg^{2+}**), hydrogen (**H^+**), aluminum dihydroxide (**$Al[OH]_2^+$**), and aluminum (**Al^{3+}**). The amounts of these positive cations held by clays vary with the kind of clay. Plant roots can use some exchangeable cations as nutrients. Leaching

[3]Atoms react with other atoms to form compounds through interactions of their electrons. The tendency for an element to attract or to give up electrons is indicated by "+" or "−" symbols. If the element has lost or lacks some electrons, it will have a "+ charge" indicating the number of electrons in which it is deficient. The positively charged ions are attracted to negatively charged ions or sites. Any atom or group of atoms with "charge" is an *ion.* Positively charged ions are *cations* (pronounced cat′-eye-ons); negatively charged ions are *anions* (an′-eye-ons).

by rainfall over several years removes many of the common cations which are replaced by H^+ or $Al(OH)_2^+$ ions. Exchangeable cations must be replaced by other cations *as they are removed* (exchanged). More cations from weathering minerals or hydrogen from the dissolving of carbon dioxide in water are able to *exchange* for cations already on the exchange sites. (**Acidic cations** are mostly H^+, the hydroxy aluminum ions and aluminum Al^{3+} ions.)

The nomenclature of clays is complex, and no single, complete system is agreed upon by clay mineralogists. The outline in Table 5-1 illustrates the kinds of criteria used to group clays; clays are indicated for some of the categories. The complete system is much more detailed and extensive than Table 5-1 shows. The number of possible sites in the octahedral sheet with cations in them are referred to as *dioctahedral* (two cations, Al, for each three possible sites) and *trioctahedral* (three cations, Mg, for each three possible sites). See Detail 5-2 for chemical formulas.

The common clays and the different clay structures are shown diagrammatically in Figure 5-3.

Charge on Clays

Why do clays have a charge? A small amount of the charge comes from ionizable hydrogen ions, but most is from isomorphous substitution. **Ionizable hydrogen ions** are hydrogens from hydroxyl ions on clay surfaces. The —Al-OH or —Si-OH portion of the clay surface ionizes the H and leaves an unneutralized negative charge on the oxygen (—Al-O^- or —Si-O^-). The extent of ionized hydrogen depends on solution pH; more ionization occurs in more alkaline (basic) solutions.

Table 5-1 Brief Outline of Clay Minerals

1. **Amorphous** (nonoriented, small crystal units)
   a) Alumino-silicates: **allophane, imogolite**
   b) Iron hydrous oxides: **ferrihydrites**
   c) Silica: **opal, glass**
2. **Crystalline Layer Silicates**
   a) *Kandite group,* 1:1 lattice (one tetrahedral, one octahedral sheet per layer)
      (1) Alumino-silicate: **kaolinite**
      (2) Magnesium silicate: **serpentine**
      (3) Water layers between kaolinite sheets: **halloysite**
   b) *Smectite group,* 2:1 lattice (two tetrahedral, one octahedral sheet per layer)
      (1) No isomorphous substitution
         (a) Alumino-silicate: **pyrophyllite**
         (b) Magnesium-silicate: **talc**
      (2) Various isomorphous substitutions (swelling)
         (a) Alumino-silicate: **montmorillonite**
         (b) Magnesium-silicate: **saponite**
   c) *Hydrous mica group,* 2:1 lattice, K between layers
      (1) Alumino-silicate: **illite** or **hydrous mica**
      (2) Swelling mica-like clays: **vermiculite**
   d) *Chlorite group,* 2:2 or 2:1:1 lattice (two tetrahedral and two octahedral sheets of different composition)
      (1) Many **chlorites,** mostly nonswelling clays
3. **Crystalline Chain Silicates**
   a) Alumino-magnesium silicates, fibrous: **attapulgite**
4. **Crystalline Iron, Aluminum, and Titanium Oxides**
   a) *Metal oxides of iron:* ***goethite*** [$FeOOH$]
   b) *Metal oxides of aluminum:* ***gibbsite*** [$Al(OH)_3$]
      boehmite [$AlOOH$]
   c) *Metal oxide of titanium:* ***anatase*** [TiO_2]

Detail 5-2 Composition of Clays

Except for a few amorphous clays (*allophane, imogolite, ferrihydrite,* and others), soil clays are crystalline, with compositions that can be described by chemical formulas, as is shown in the following. In the formulas, the *M* shown in brackets at the end of some formulas represents the charges of *adsorbing cations* that are needed to neutralize charges caused by isomorphous substitution. The subscripts *x, y,* and *z* indicate variable amounts of substitution, and each letter indicates a particular numerical fraction less than 1.0.

Kaolinite	$Al_2(Si_4O_{10})(OH)_8$
Pyrophyllite	$Al_2(Si_4O_{10})(OH)_2$
Muscovite	$KAl_2(AlSi_3O_{10})(OH)_2$
Illite	$K(Al_{2-x-y}Fe_xMg_y)O_{10}(Si_{4-z}Al_z)(OH)_2 \cdot [x+y+z=M]$
Talc	$Mg_3\,(Si_4O_{10})\,(OH)_2$
Montmorillonite	$(Al_{2-x}Mg_x)(Si_{4-y}Al_y)O_{10}(OH)_2 \cdot [x+y=M]$
Saponite	$Mg_6(Si_{8-x}Al_x)O_{20}(OH)_4$
Chlorite	$Mg_3(Si_{4-x}Al_x)O_{10}(OH)_8$ $(Mg_{6-x-y}Al_xFe_y)$
Vermiculite	$(Mg_{3-x-y}Al_xFe_y)(Si_{4-z}Al_zO_{10})(OH)_2$ $(4.5\ HOH)$
Sesquioxides	$Fe_4O_{6-x}(OH)_x$ and $Al_4O_{6-x}(OH)_x$

These formulas illustrate the dominance of oxygen in minerals and its bonding to aluminum and silicon. Common substituting ions include Al^{3+} for Si^{4+}, and Fe^{3+}, Fe^{2+}, Zn^{2+}, and Mg^{2+} for Al^{3+}. Particular clays will vary from the example compositions shown above. Clays not listed include allophane, an amorphous aluminum silicate; imogolite, a microcrystalline allophane; and ferrihydrites, amorphous iron hydrous oxides.

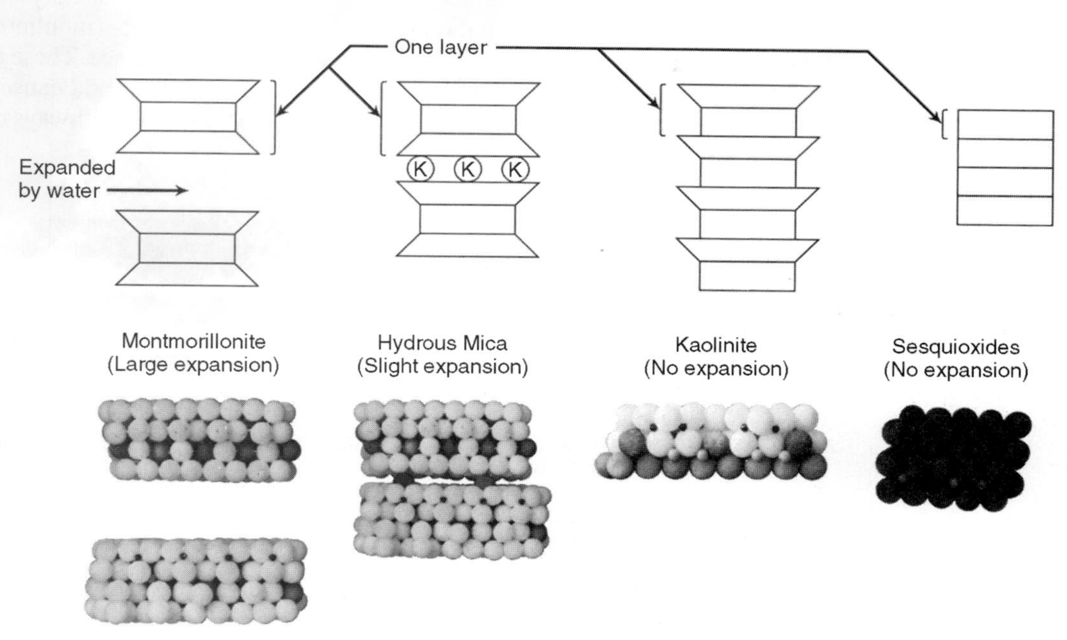

FIGURE 5-3 Schematic drawing of clay minerals. The shapes for sheets are identical to those given in Figure 5-2. The lattice layers of montmorillonite within a single clay particle can be greatly expanded by water. In contrast, hydrous mica with strong bonding by potassium ions has limited swelling in some layers, probably where potassium has been removed by leaching. Kaolinite, the potters' clay, has no swelling. Sesquioxides also have low or no swelling, and they usually have irregularly stacked sheets, making them x-ray amorphous. The coordination is octahedral, as suggested in the photo, but the crystallinity is quite poor. (Courtesy of Raymond W. Miller, Utah State University.)

Isomorphous substitution, the second source of charge on clay particles, is the substitution during clay genesis of one cation for another of similar size and usually with lower positive valence. In clay structures (having ionic bonds), certain cations fit into certain mineral lattice sites because of their convenient size and charge (see Table 5-2). Only during clay formation, as minerals undergo dissolution, can other ions of similar *size* and *charge,* if they are present, occupy some of those sites. The Si^{4+} dominates in tetrahedral sites, replaced sometimes by Al^{3+}. The Al^{3+} in octahedral sites is replaced by one or more of the following: Fe^{3+}, Fe^{2+}, Mg^{2+}, or Zn^{2+}. These ions, in high concentrations in the solution, *substitute* because of similar *size* and *positive* charge. Notice that most substitutions are by ions with lower charge (less positive) than the ones being replaced. Because the total negative charge from the structural oxygens remains unchanged, the lower positive charge of the total substituted cations results in an excess negative charge at that location in the structure.

The excess negative charge from isomorphous substitution and ionized hydrogens create *electron excess* sites that attract cations from the surrounding solution (see Figure 5-4), but these cations do *not* become a part of the clay structure; they are held somewhat loosely. The negatively charged locations are called **cation exchange sites.** Other cations in solution can compete with and often replace originally adsorbed cations. The total amount of the negative sites is referred to as the soil's **cation exchange capacity (CEC).** The cation exchange sites adsorb and hold mostly Ca^{2+}, Mg^{2+}, and K^{+} in soils of pH greater than about 6 or 7. In more acidic soils the sites are mostly occupied by $Al(OH)_2{}^{+}$, or Al^{3+} ions. Lesser amounts of other cations (Na^{+}, $NH_4{}^{+}$, Zn^{2+}, and other cations) occur on some of the exchange sites.

Besides the charge deficit that causes cation exchange, the *amount* of cation exchange also depends on the surface of the colloid exposed to the soil solution. Ionic attraction for cations is effective only at distances of a few oxygens' thicknesses. Clays whose layers are spread apart to allow the soil solution to pass between clay layers (montmorillonite, vermiculite) have accessible exchange sites also along this internal surface. These clays have weak bonding between layers so that water can spread the layers apart and cause swelling of the clay particle. Such swelling clays are poorly suited for supporting highways or buildings (see Vertisols, Chapter 7).

Table 5-2 Ionic Radii, Charge, and Coordination Numbers for Ions Common in the Crystal Lattice of Soil Clays*

Ionic Radii for Common Soil Ions (nm)

Nonhydrated				Hydrated	
Si^{4+}	0.042	Na^{+}	0.074	Na^{+}	0.790
Al^{3+}	0.051	Ca^{2+}	0.099	Ca^{2+}	0.960
Fe^{3+}	0.064	Mg^{2+}	0.066	Mg^{2+}	1.08
Fe^{2+}	0.074	K^{+}	0.133	K^{+}	0.530
O^{2-}	0.140				

Ionic Radii, Coordination Numbers, and Crystal Geometries

r_{Cation}/r_{Anion}	Coordination Number	Crystal Geometry
0.155–0.255	3	Trigonal planar
0.255–0.414	4	Tetrahedron
0.414–0.732	6	Octahedron
0.732–1.0	8	Close-fit lattice

*Based on nonhydrated sizes. Water molecules of hydration are stripped away when crystal lattices form. The free ions in water or soil solution will be hydrated and effectively be larger ions.

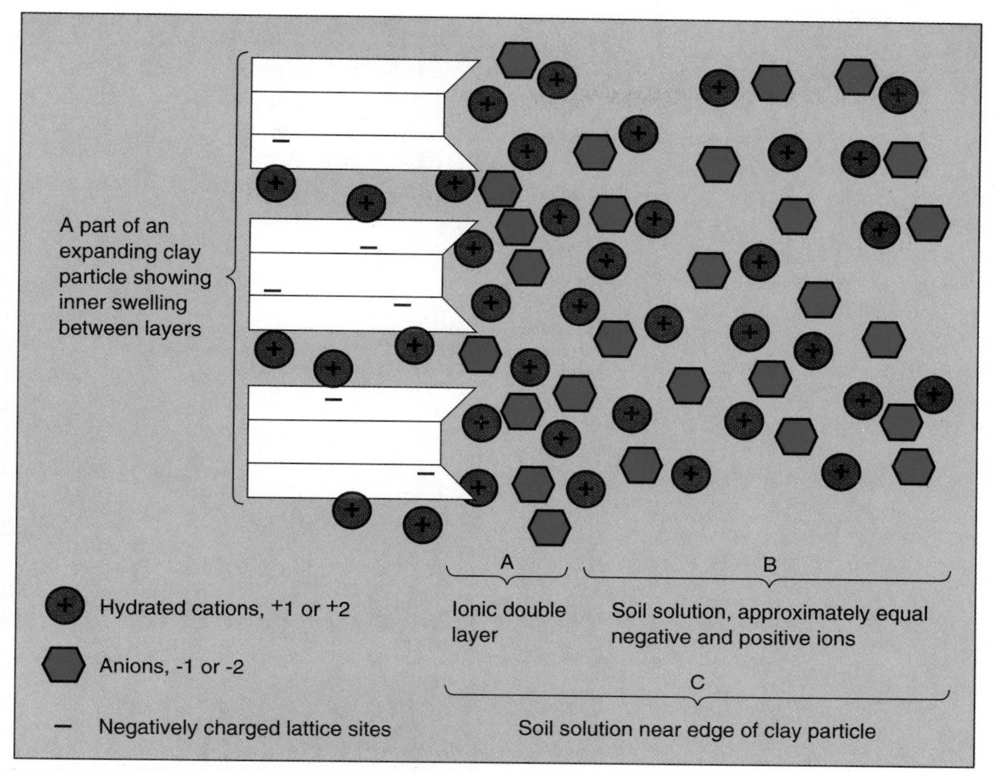

FIGURE 5-4 Illustration of the movement, mostly of cations, to the outer surfaces and interlayer areas as might occur in montmorillonite, vermiculite, and some expanded interlayers of illite. Adding a new cation, such as ammonium fertilizer, puts ammonium ions on many of the adsorption sites. (Courtesy of Raymond W. Miller, Utah State University.)

Other clays (kaolinite, chlorite) have tightly bonded layers and do not swell when wet. Kaolinite is used to make pottery, tile, and other fired-clay items because it does not shrink, crack, or deform when kiln-baked. Kaolinite has a low cation exchange capacity because of its relatively large size and lack of isomorphous substitution.

Silicate Clays

If clay minerals are examined under a high-powered microscope, evidence of their crystalline structure can be seen (Figure 5-5). Each crystalline clay is like a partial deck of magnetic cards—that is, the clay is thinner in one dimension than in the other two dimensions. Each card represents a *layer,* and the layer is nearly an exact replication of every other layer in that clay. The many cards adhere together to make up one clay particle called a **micelle.** Each layer is made up of two oxygen sheets in some clays, of three or four in others, and of six in still others (chlorites). A few hydroxyl ions (OH^-), which are almost the same size as oxygen ions, replace some of the oxygen ion sites. In some clays up to one-fourth of the silicon ion positions (tetrahedral sites) are substituted with aluminum. Similarly, other ions—such as magnesium, zinc, and iron—with atomic sizes similar to that of aluminum may fit into aluminum sites (octahedral sites) in place of aluminum.

Amorphous silicate clays (allophane) are mixtures of silica and alumina that have not formed well-oriented crystals; they lack crystallinity. Even other weathered oxides (iron oxide) may be a part of the mixture. Typically, these clays occur where large amounts of

(a)

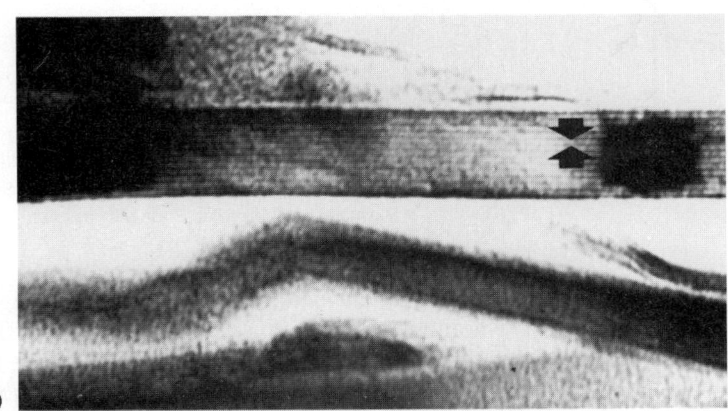

(b)

FIGURE 5-5 (a) This electron micrograph of Utah Dickite, a kaolinitelike clay mineral, is magnified 2720 times. Each thin portion comprises many clay particles stacked. A soil sample would most often have the individual particles less oriented, like decks of cards thrown into a pile along with many large irregularly shaped silt and sand particles of all sizes. In (b) a magnification of more than 700,000 times makes it possible to see layers of chlorite clay. The distance between arrows is about 1.4 nm (14 angstroms), the thickness of about six oxygen planes. Chlorite is a 2:2 or 2:1:1 sheet clay. (*Sources:* (a) B. F. Bohor and R. E. Hughes, "Scanning Electron Microscopy of Clays and Clay Minerals," *Clays and Clay Minerals* **19** (1971), 49–54; (b) J. L. Brown and M. L. Jackson, "Chlorite Examination by Ultramicrotomy and High Resolution Electron Microscopy," *Clays and Clay Minerals* **21** (1973), 1–7.)

weathered products existed but have not had the conditions or time for good crystal growth. Amorphous clays are common in soils forming from volcanic ash.

Amorphous clays are not well characterized but do exist in many soils in varying amounts. Their properties are often quite unusual, such as a great affinity for phosphorus. Almost all of their charge is from ionization of H^+ of accessible hydroxyl ions (OH^-); such sites can attract a positive ion or lose the H^+ into solution.

Kandites (kaolinite, nacrite, halloysite) are residues from extensive weathering in high-rainfall, acidic soils. Kaolinite, the most common clay of this group, has only one sheet of silica tetrahedra per sheet of alumina octahedra per layer. Thus, it is a 1:1 lattice clay. Almost no substitution of Al^{3+} for Si^{4+} or Mg^{2+} for Al^{3+} has occurred in kaolinite, so the net negative charge (cation exchange capacity) is low. However, each layer has one plane of oxygens (O^{2-}) replaced by hydroxyls (OH^-), which results in strong hydrogen bonds (—H—) to the oxygens in the adjacent layer. Kaolinites have such strong hydrogen bonding between layers that they do not allow water to penetrate between the layers. Kaolinite has almost no swelling.

Kaolinite is predominant in the clay fraction in acidic, humid, warm, well-drained soils, such as in the southeastern United States and in the humid soils in the tropics and subtropics, if they are well drained and have weathered for several hundred thousand years. As minerals weather, much of the silica and basic cations are dissolved and slowly leached from the profile. Less silica present in the remaining solution means that the proportion of alumina will be high enough to result in kaolinite formation rather than the higher silica-containing montmorillonite.

Smectites (montmorillonite, saponite) are the swelling, sticky clays.[4] They are often referred to as *2:1 lattice,* or *expanding-lattice,* clays. The *2:1* refers to the number of silica sheets (two) per alumina sheet (one) per clay layer. In montmorillonites, water easily penetrates between planes of adjacent oxygens, causing the individual layers of clay particles to separate and swell. If the solution has mostly sodium as the cations, the clay may swell three to ten times its dry volume and become like a gelatinous mixture. **Bentonite,** an impure deposit of montmorillonite or other swelling clay, is used to seal earthen ponds (see Figure 5-6), to spread water over fires when the gelatinous slurry is dropped by planes, to act as solution stiffeners or gels in well-drilling muds, and to act as thickeners in paints and lipstick. *Bentonite* can refer to any deposit of highly colloidal swelling clay.

Montmorillonite is most common in soils that have had little or no leaching. Leaching removes relatively more silica than it does alumina. Thus, leaching (water moving down through the soil) reduces the relative amount of silica available for montmorillonite clay formation. Soils of the arid regions, poorly drained soils, and soil developed from alkaline parent rocks such as limestone have mostly montmorillonite clays.

Hydrous mica and **illite** are obsolete terms, but we have no new names with which to replace them. The term *mica-like* is sometimes used. A third term—*fine-grained mica*—is also used for this poorly-defined group of minerals. In this text illite will be used. **Illite** has a gross structure similar to that of montmorillonite—a 2:1 lattice clay of silica-to-alumina sheets. However, these clays also have large potassium ions holding adjacent layers together so tightly that water cannot penetrate between layers. Thus, illite has slight to moderate swelling, depending upon how many of the planes of potassium ions have weathered out, allowing some clay layers to be separated and the clay to expand, somewhat like montmorillonite.

Because illite is quite similar to the structure of the primary micas, it is believed to form by limited alteration of primary micas. Indeed, some mineralogists do not consider illite a clay mineral because it is so similar to regular mica. Illite is found in soils still high in primary minerals (not extensively weathered). Both illite and montmorillonite may occur in similar slightly leached environments.

Vermiculite clay minerals are similar in structure to illite but without the interlayer potassium ions. (The Latin *vermiculari* means *to breed worms,* from the curled wormlike shapes formed upon heating and expanding vermiculite.) Vermiculite has the layers held weakly together by hydrated magnesium (six water molecules in octahedral coordination with Mg^{2+}). Thus, vermiculite has swelling but not as much as montmorillonite; it has a high cation exchange capacity.

---

[4]The clay called *montmorillonite* was first described near Montmorillon, a town in France. Its name came from the town. *Illite,* which in this text is referred to as *hydrous mica,* was named by geologists from the Illinois Geological Survey. The term *kaolinite* comes from the Chinese *kauling,* meaning *high ridge,* the name of the hill from which the first kaolinite was shipped to Europe for making fine china dishes.

(a)

(b)

FIGURE 5-6 (a) Pond near Winthrop, Washington, built on Owhi fine, sandy loam, a porous soil. The pond is sealed with a high-sodium montmorillonite clay (bentonite) that has swelled and now seals all pores, inhibiting much water loss. The pond was 15 years old at the time the European swans and peafowl were photographed. (b) Nonswelling clays can be used for bricks, tiles, and ceramics (china dinnerware and vases). Kaolinitic clays are found in tropical climates where weathering has been long and intensive, as in this site in Guatemala. The clay is molded, dried, and sometimes fired to harden the bricks or tile. (*Source:* (a) USDA—Soil Conservation Service, photo by William Brewster; (b) Raymond W. Miller, Utah State University.)

Chlorite clays are common in some soils. They are often called *2:2 lattice* (or *2:1:1 lattice*) clays because they are similar to the unit lattice of vermiculite, except the hydrated Mg in chlorite is a firmly bonded, complete, magnesium hydroxide octahedral sheet. Thus, a layer of chlorite has two silica tetrahedra, an alumina octahedra, and a magnesium octahedra sheet (2:2 or 2:1:1). Chlorites do not swell when wetted and have low cation exchange capacities.

Sesquioxide Clays (Metal Oxides and Hydrous Oxides)

Under conditions of extensive leaching by rainfall and long-time intensive weathering of minerals in humid, warm climates, most of the silica and much of the alumina in primary minerals are dissolved and slowly leached away. The remnant materials, which have lower solubilities, are sesquioxides. **Sesquioxides (metal oxides)** are mixtures of aluminum hydroxide, $Al(OH)_3$, iron oxide, Fe_2O_3, or iron hydroxide, $Fe(OH)_3$. (The Latin word *sesqui* means one and one-half times.) *Sesquioxide* refers to the clays of iron and aluminum because their formulas can be written $Al_2O_3 \cdot xH_2O$ and $Fe_2O_3 \cdot xH_2O$, one and one-half times more moles of oxygen than for Al or Fe. However, TiO_2 and MnO_2 are also often included in sesquioxides. Sesquioxide clays can range from amorphous to crystalline.

Small amounts of these clays exist in many soils, even in soils with a relatively low extent of weathering. However, they will be the predominant clay and occur in large percentages only in soils formed in the humid, hot, well-drained soils of tropical areas where intense weathering has occurred for perhaps hundreds of thousands of years. Iron oxide and iron hydrates commonly color the soils various shades of red to yellow, respectively.

Sesquioxide clays do not swell, are not sticky, and have many differences from the silicate clays. The sesquioxides form stable aggregates and may coat soil particles. Soils with 30 percent to 40 percent sesquioxide clays may absorb water almost as if they were fine sands. During World War II in some tropical islands of the Pacific, vehicles moved through sesquioxide muds nearly hub deep, something quite impossible in sticky, montmorillonitic clays. Mixtures of sesquioxides and kaolinite often have low stickiness.

The high percentage of iron and aluminum hydrous oxides furnish an enormous surface for adsorption of phosphorus. In addition to adsorbing to sesquioxides, phosphates can form insoluble phosphates with soluble iron and aluminum ions. Soils with large sesquioxide contents have a high phosphorus adsorption capacity. This often results in lower phosphorus efficiency of added fertilizers.

5:3 Organic Colloids (Humus)

Humus is a temporary intermediate product left after considerable decomposition of plant and animal remains—*temporary* because the organic substances remaining continue to decompose slowly. The humus is often referred to as an *organic colloid;* it consists of various chains and rings of linked carbon atoms. Like clay colloids, humus has a negative charge. Unlike most clays the charge on humus arises solely from ionization of hydrogens from R–OH groups. Acid and other functional groups on these large organic molecules release H^+ ions, leaving a negatively charged site where the H^+ had been.

Humus is amorphous, dark brown to black, nearly insoluble in water, but mostly soluble in dilute alkali (NaOH or KOH) solutions (Figure 5-7). It contains about 30 percent each of the nitrogen-rich proteins, the slow-to-decompose lignins, and complex sugars (polyuronides). The polyuronides comprise much of the varied organic substances that cement soil aggregates together. Humus has about 50 percent carbon, oxygen, 5 percent nitrogen, and lesser amounts of sulfur, phosphorus, and other elements. It also has a cation exchange capacity on a dry-weight basis many times greater than that of clay colloids. On a weight basis, a few percent humus exerts much greater influence than does several times more percent clay, especially for properties such as the cation exchange capacity.

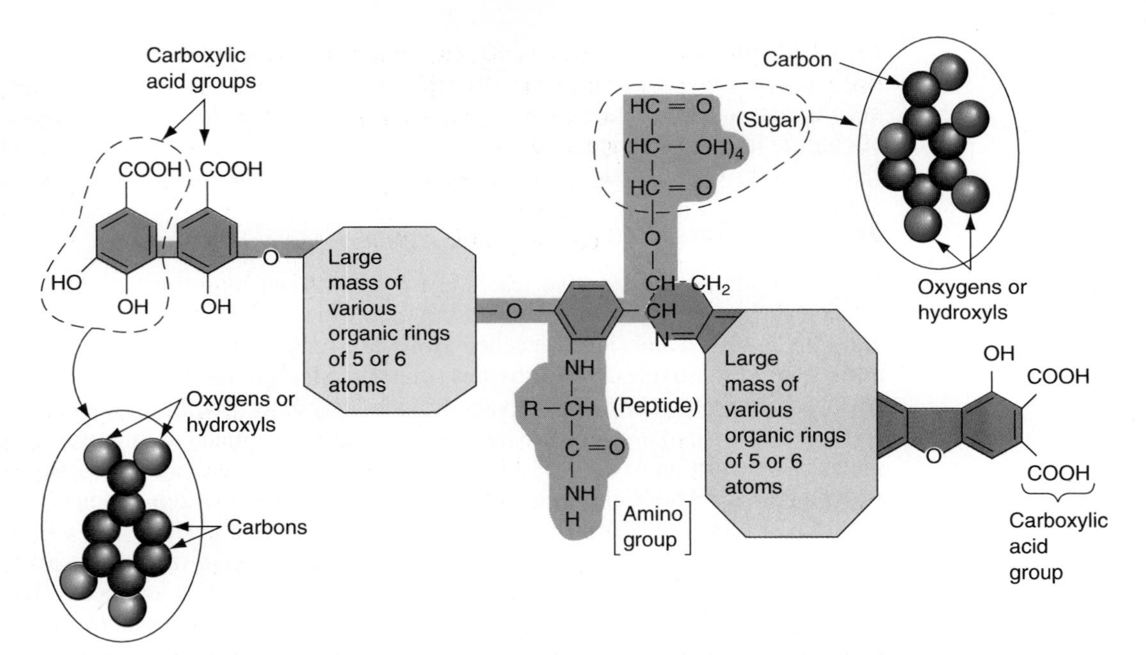

FIGURE 5-7 Schematic illustration of what a *portion* of a molecule of humus might look like. One molecule of humus would be from 10 to more than 100 times greater than the portion shown. Humus is more porous and stringy than are clay particles. The humus can attach to clays through bonding, thus helping to hold small particles into larger masses (aggregates). All humus molecules will be different from each other and are constantly changing as they are attacked by microbes and further decomposed. Distinctive features of all humus are (1) the many active groups, mostly **carboxylic acid groups** and **hydroxyl (—OH) groups;** (2) the large portion of humus that consists of rings of carbon atoms; and (3) the appearance of humus as a very large, knobby cord folded around and over itself somewhat. (Courtesy of Raymond W. Miller, Utah State University.)

5:4 Cation Exchange

Soil colloids generally have a net negative charge and will attract and hold positively charged ions (**cations**) to their surface near where the negative charge originates in the clay lattice. Other cations in the soil solution that approach the held cation might be able to replace it, thus, *exchanging* for it. *This replacement of one adsorbed cation by another cation free in solution is called* ***cation exchange.***

To illustrate cation exchange by an equation, represent the colloid (clay or humus) by any of several symbols, such as a rectangle, parallelogram, or capital *X.* When the equation is completed, the total charges on the left side of the equation equal the charges on the right side:

$$\underset{2\,K\quad 6\,[Al(OH)_2^{+}]}{\overset{12\,Ca\quad 4\,Mg}{2\,Na\bullet\boxed{-42\text{ charge}}}} + \underset{\text{(Fertilizer)}}{9\,K^{+} + 9\,Cl^{-}} \rightleftharpoons \underset{11\,K\quad 5\,[Al(OH)_2^{+}]}{\overset{10\,Ca\quad 3\,Mg}{\boxed{-42}}} \begin{array}{l} +2\,Ca^{2+} \\ +H^{+} + 2Na^{+} \\ +Al(OH)_3^{0} \\ +Mg^{2+} \\ +9\,Cl^{-}\end{array}$$

Adsorbed cations resist removal by leaching water but can be replaced (exchanged) by other cations in solution by **mass action** (competition for the negative site by the large *num-*

ber of ions present). Thus, when a potassium fertilizer with its cation, K^+, is added to the soil, many of the numerous potassium ions replace other cations already adsorbed to the exchange sites. Cation exchange takes place on the surfaces of clay and humus colloids as well as on the surfaces of plant root cell walls.

The cations most numerous on exchange sites in soils are calcium (Ca^{2+}), magnesium (Mg^{2+}), hydrogen (H^+), sodium (Na^+), potassium (K^+), $Al(OH)_2{}^+$ and aluminum (Al^{3+}).[5] The proportions of these cations on the colloid surfaces are constantly changing as ions are added from dissolving minerals or by additions of lime, gypsum, or fertilizers. Losses by plant absorption or by leaching also change cation proportions.

Cation Exchange

The cation exchange sites in soils secure cations but keep them exchangeable and thus somewhat available to plant roots. Any water moving through the soil will lose many of its soluble cations to the soil and pick up those cations that were replaced from the exchange sites by the cations being adsorbed. Cation nutrients—such as potassium (K^+), ammonium ($HN_4{}^+$), and calcium (Ca^{2+})—will not move far in the soil before large portions of them are readsorbed to the exchange sites and some of those adsorbed cations are replaced, sending the replaced ions back into the solution.

The mechanism of cation exchange is illustrated in Detail 5-3. Different ions in solution move at different speeds through a soil. However, for every cation that adsorbs, the cation originally on that site moves out into solution as it is replaced. The percolating water constantly carries some cations with it, as well as many anions. Certain cations (calcium) and anions (phosphates, for example) may form *insoluble salts* and precipitate out of solution. This precipitation may reduce the amount of ions moving with the percolation water. Table 5-3 lists the leaching losses of some common soil ions from various soils. Udolls, Udalfs, and Aquepts are soils of humid climates or wet conditions (see Chapter 7). Large losses of calcium occur because large amounts of calcium exist in most soils from the various weathering minerals. Plants absorb nitrate as it is made available, so plant-covered soil loses less nitrogen than similar barren soils. Less sulfur is needed by plants, so plant cover does not affect sulfate losses dramatically. Nonorganic soils have little loss of phosphates. Chlorides would easily leach with water.

The strength of adsorption increases as (1) the valence of the cation increases, (2) the cation's hydrated size is smaller, and (3) the strength of the site's negative charge increases. Thus, adsorption strength of cations increases in approximately the following order:

$$Na^+ < K^+ = NH_4{}^+ < Mg^{2+} = Ca^{2+} < Al(OH)_2{}^+ < H^+$$

[5]Aluminum forms octahedral coordination with six water molecules or hydroxyls, which have essentially the same size. As the soil water solution becomes less acidic (has more OH^- ions), one aluminum-held water molecule ionizes an H^+, which is less attracted to the oxygen of the water molecule held to the aluminum; the ionization leaves a hydroxyl ion attached to the aluminum, which neutralizes some of the Al^{3+} charge. The aluminum ion-hydroxyl unit becomes successively less positively charged by such ionization. At different pHs the following are the appropriate forms of aluminum in soil solution (with water molecules shown):

$Al(H_2O)_6{}^{3+}$	Predominant from below about pH 4.5
$Al(H_2O)_4(OH)_2{}^+$	Predominant between about pH 4.5 and 6.5
$Al(H_2O)_3(OH)_3{}^0$	Predominant between pH 6.5 and 8.0
$Al(H_2O)_2(OH)_4{}^-$	Predominant between pH 8.0 and 11

Notice the noncharged hydrated aluminum that exists at the near-neutral pH that will precipitate as the noncharged hydroxyl form; it is usually written as $Al(OH)_3$. The $Al(H_2O)_5(OH)^{2+}$ form does not exist in dominant amounts, so it has been omitted from this list. See Detail 8-1 for more information.

Detail 5-3 Chromatographic Ion Flow in Soils

In its broadest sense **chromatography** refers to processes that permit the separation of components in a mixture. This is possible in soils as a consequence of differences in rates at which the individual components of that mixture migrate through the stationary soil under the influence of mobile water. As soluble ions in water percolate through soil, some of these ions will adsorb to the cation exchange sites (and anions to anion exchange sites), replacing ions already adsorbed. $Al(OH)_2^+$ and H^+ ions would be adsorbed strongly, Ca^{2+} and Mg^{2+} less strongly, and K^+ and Na^+ least strongly. Some H_3BO_3 would adsorb to anion exchange sites (few sites), but NO_3^- and Cl^- ions would adsorb weakly and move nearly at the rate of water flow.

Chromatographic movement of ions is the flow of different ions through the soil at different speeds, partly because of adsorption differences. What happens? Suppose that a solution of water containing calcium, sodium, boric acid, and chloride is added onto the soil surface. Immediately, some of the calcium and sodium ions replace ions already adsorbed on the cation exchange sites. The replaced ions join the remaining ions still in solution and move deeper into the soil. Continuously, the ions in solution encounter new exchange sites and compete with each nearby adsorbed ion to occupy its site. The downward rate of movement of each ion will be different and will depend on (1) the rate of water flow, (2) the strength with which the ion is adsorbed to the site, and (3) the number and nature of competing ions in the solution (which compete with it for the site). The most strongly adsorbed ion will tend to move the most slowly down through the soil (it may appear not to move). Eventually some ions will drain out the lower part of the soil. The relative rates of movement should be, approximately (where H_3O^+ is hydrated H^+), as follows:

Cations: Slowest movement; $H_3O^+ = Al(OH)_2^+$

Faster movement; $Ca^{2+} = Mg^{2+} < K^+ < Na^+$

Anions: (Slow) $H_2PO_4^- < H_3BO_3 < SO_4^{2-} < NO_3^- = Cl^-$ (Fast)

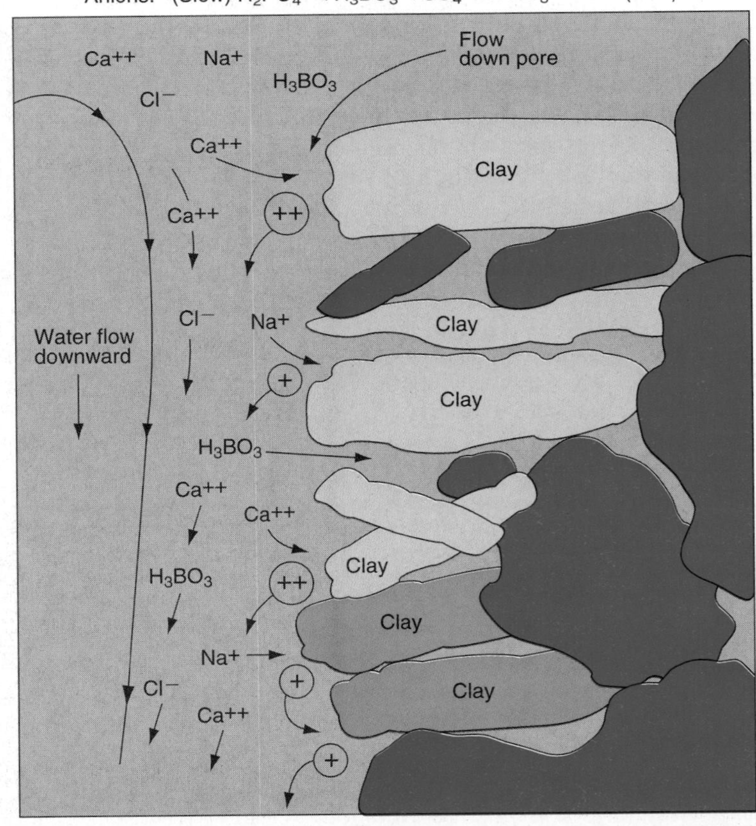

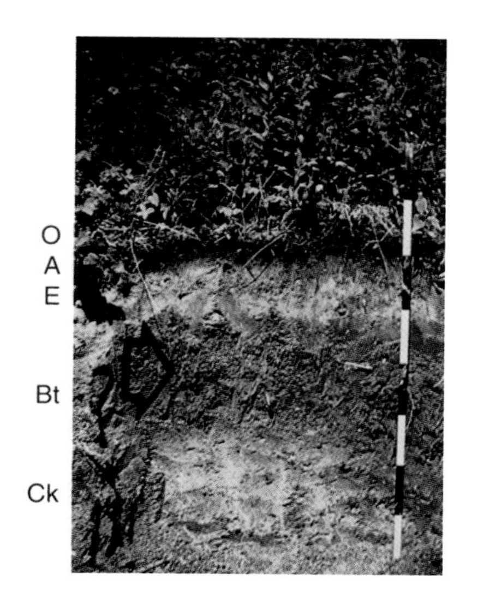

1 SOIL ORDER:
ALFISOL

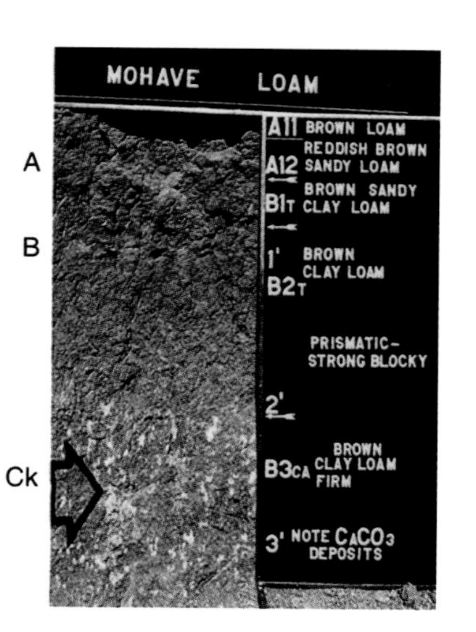

2 SOIL ORDER:
ARIDISOL

3 SOIL ORDER: ENTISOL

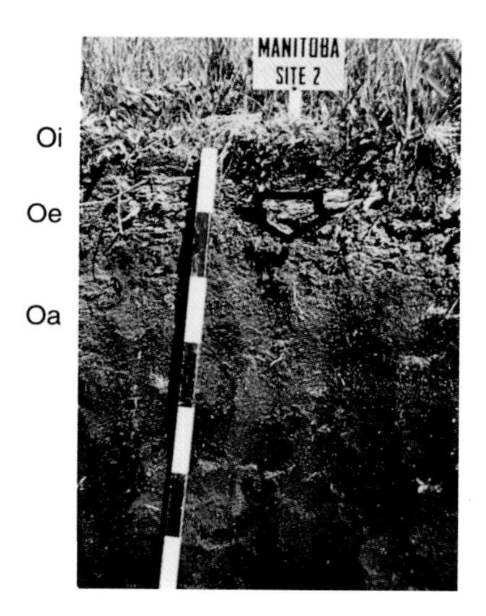

4 SOIL ORDER:
HISTOSOL

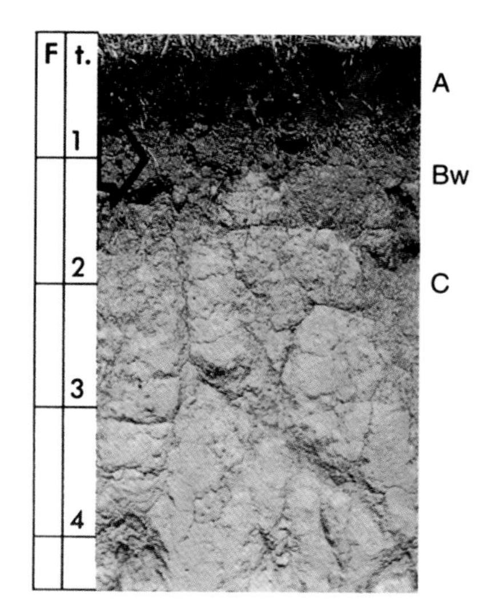

5 SOIL ORDER:
INCEPTISOL

Note: *Corresponding table appears on pages 230–231.*

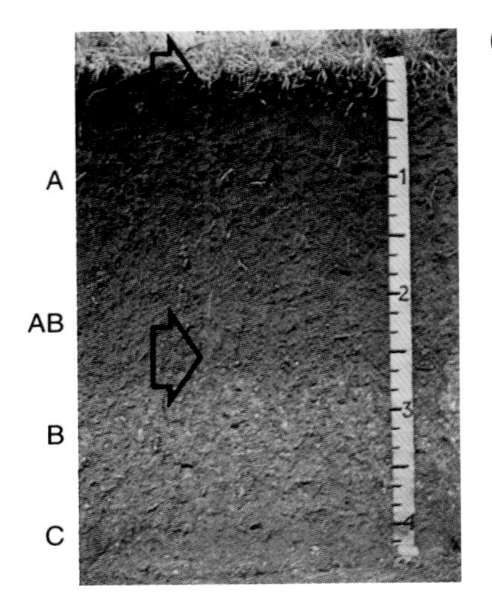

6 SOIL ORDER: MOLLISOL

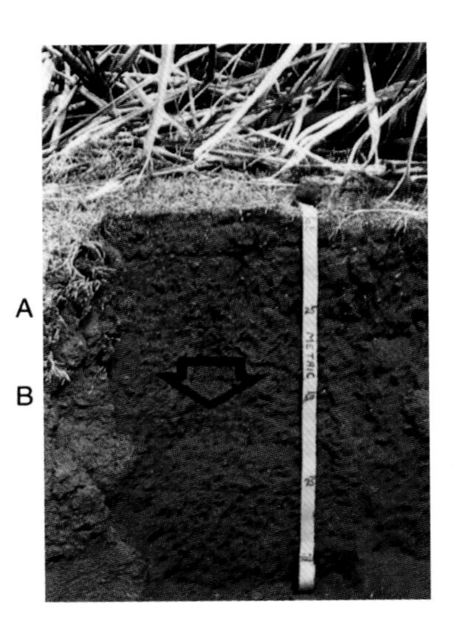

7 SOIL ORDER: OXISOL

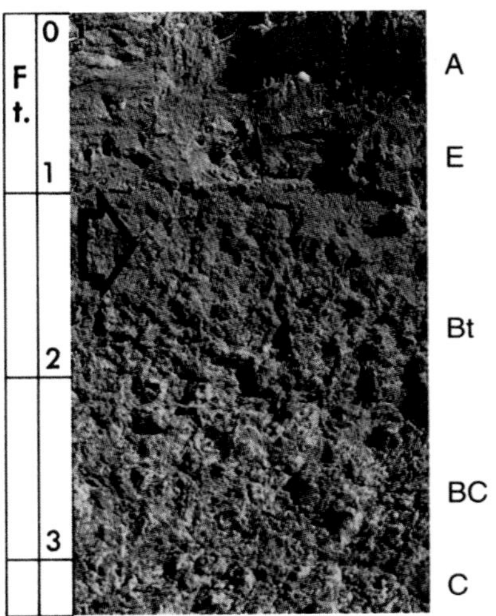

8 SOIL ORDER: SPODOSOL

9 SOIL ORDER: ULTISOL

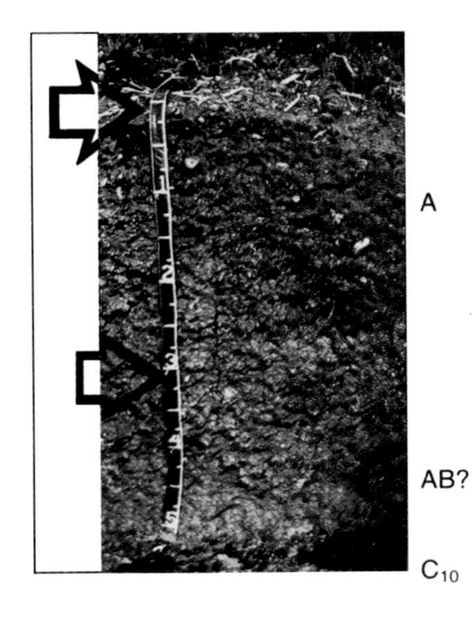

10 SOIL ORDER: VERTISOL

11 SOIL ORDER: ANDISOL

Table 5-3 Leaching Losses of Cations and Anions under Various Conditions

Condition	Loss per Year in kg ha^{-1}				
	Ca	*Mg*	*K*	*N*	*S*
Soil in Scotland (6-year average)	56	17	10	8	—
Uncropped well-drained (3.5-year average)					
Udoll in Illinois	101	52	1	86	12
Udalf in Illinois	42	21	3	—	3
New York Udalf, 10-year average					
Barren soil	446	71	81	77	59
Grass-covered soil	291	56	69	3	49
New York Aquept, 15-year average					
Barren soil	362	46	72	48	39
Rotation of crops	280	30	64	7	37
Canadian soil, 5-year average					
Grass under orchard	51	19	19	—	—
Clean-cultivated under orchard	374	104	45	—	—

When phosphorus was measured, only a trace was reported in leaching waters.

Source: Selected data from N. L. Brady, *The Nature and Properties of Soils,* 10th ed., Macmillan, New York, 1990, 420.

Water pollutants—such as heavy metals (lead [Pb^{2+}], cadmium [Cd^{2+}], mercury [Hg^{2+}]), and many others—can be disposed of in soil. Instead of leaching downward to pollute ground and surface waters, the heavy metals are almost all adsorbed to the cation exchange sites. They may finally form insoluble hydroxides, carbonates, or other minerals, but cation exchange helps initially to retain them and cleanses the percolating wastewater.

Liming a soil to correct the adverse acidity is a cation exchange reaction. When lime is added to neutralize acidic soils, most of the exchangeable H^+ and $Al(OH)_2{}^+$ are neutralized to alter the soil pH. The amount of lime required for neutralization is determined directly or indirectly by the amounts of exchangeable H^+ and $Al(OH)_2{}^+$ that must be replaced by calcium or other cations.

A sodium saturation percentage of more than 10 percent to 20 percent on exchange sites can cause poor soil structure and reduced aeration when the salinity is low. If, however, the salinity is moderate to high, the soil remains strongly flocculated. The effect of sodium to cause breakdown of aggregates (soil dispersion) is discussed in a later chapter.

Cation Exchange Capacity

Cation exchange capacity (CEC) is the quantity of exchangeable cation sites per unit weight of dry soil. It is measured in *centimoles$_c$ of cations per kilogram of dry soil* ($cmol_c/kg$) (see Detail 5-4). The term *centimoles$_c$* is used rather than weight units because the *number* of negative sites in a given soil sample does not change, but the *weights* of the cations that may be adsorbed to those sites at one time do change. One centimole$_c$ of cation X occupies the same number of cation exchange sites as one centimole$_c$ of cation Y (a different cation). (If chemical *weight* units were used, one gram of cation X would *not* occupy the same number of cation exchange sites as would one gram of cation Y.) Measuring the CEC in centimoles$_c$ of cations per kilogram of material keeps the CEC value the same regardless of what cations occupy the sites.

Representative cation exchange capacities of soils in the United States are shown in Tables 5-4 and 5-5. Note the very low CECs of sands and loams compared to those of clays. CEC values for the various colloids are given in Table 5-6.

Detail 5-4 Units for Cation Exchange Capacity (CEC)

As the world's nations try to agree on a uniform system of weights and measures, the United States is gradually converting to a metric system. The use of customary U.S. units (pounds, feet, gallons, etc.) is being eliminated, and metric units (kilograms, meters, liters, etc.) are replacing them [see the conversion unit tables in Appendix C].

Cation exchange capacity is reported in chemical terms. The unit used prior to 1982 was *milliequivalents of exchangeable cations per hundred grams of soil* (meq/100 g). The metric system no longer uses the term *equivalents;* now **moles** is the accepted chemical unit. All the calculations and concepts of *equivalents* are still mentally used by those who learned them, but the notation must be written differently. The old *equivalent* is represented by *moles* (+) or *mole*$_c$, which indicates a monovalent ion portion. For example, to write 6 meq/100 g in the newer metric system, do one of the following:

Old Nomenclature	***New Accepted Nomenclature***
6 meq/100 g	= 60 mmol$_c$ kg^{-1} (where mmol$_c$ = millimoles of charge)
	= 6 cmol$_c$ kg^{-1} (= centimoles)
	= 6 cmol (+) kg^{-1} of soil
	= 60 mmol (+) kg^{-1} of soil

Any of these five methods could be used to report the 6 meq/100 g. Also, the solidus (/) can be used in place of $^{-1}$—such as writing the second one above as 6 cmol$_c$/kg, or the first one as 60 mmol$_c$/kg.

The amounts of exchangeable cations in most soils are surprisingly large. In depths of normal rooting (60 cm to 90 cm), the amounts of exchangeable cations range from hundreds to thousands of kilograms per hectare (see Calculation 5-1 and Table 5-5). In Table 5-5 the high CEC of the soil called *Humod* is due to a high humus content; the high CEC of the Andept is due to its volcanic-ash origin (amorphous clays) plus its high humus content.

Exchangeable K^+ plus soluble K^+ are the major immediate sources of potassium to plants. A measurement of exchangeable K^+ allows an estimate of whether or not fertilizer potassium needs to be added. As an example, a measurement of less than 250 kg/ha of exchangeable K^+ to a depth of 30 cm (about 222 lb/a per 1-ft depth) indicates that a crop having a high potassium requirement would need added potassium fertilizers.

Is the cation exchange capacity (CEC) always the same in a given soil? The answer is yes—approximately—if the soil pH, humus, and clay contents remain the same. However, the soil's CEC does change as these properties change (see Detail 5-5).

Estimating CEC and Exchangeable Cations

The measurement of CEC is time-consuming and often not measured because of the high cost. One method used to estimate CEC is simple and straightforward (see Calculation 5-2). First, estimate or measure the clay and humus percentages. Second, assign an *average* CEC value to each 1 percent of clay or humus. To do this, divide the average CEC value for that material (= 100% material) by 100. This is the CEC for 1 percent of the material. Third, multiply the percent of the material by the CEC for 1 percent of it. Fourth, add up the CEC contributions for the clay and humus. In general, each percentage of humus contributes about 1.5 cmol$_c$/kg to 2.0 cmol$_c$/kg of CEC, and each percentage of clay contributes about these amounts in cmol$_c$/kg: Montmorillonite = 0.6–1.0, kaolinite = 0.02–0.08, and sesquioxides = 0.03. Because soil contents of illite, chlorite, or vermiculite are usually quite small, their contribution is small and difficult to estimate.

Table 5-4 Generalized Relationship between Soil Texture and Cation Exchange Capacity

Soil Texture	*Cation Exchange Capacity (centimoles$_c$ per kg of soil) (Normal Range)*
Sands	1–5
Fine sandy loams	5–10
Loams and silt loams	5–15
Clay loams	15–30
Clays	>30

Table 5-5 Typical Amounts of Exchangeable Cations of a Variety of Soils and the Amounts in Soil in kg/ha–30 cm or lb/a–ft Depths

	Typical Centimoles$_c$ of Cations per Kilogram of Soil					
Soil (Names of Suborders)	*CEC*	*Ca^{2+}*	*Mg^{2+}*	*K^+*	*Na^+*	*H^+, Al^{3+}*
Psamment (sandy soil, pH 6.4, Kansas)	5.2	1.9	1.2	0.3	tr	1.8
Argid (sandy loam, pH 6.3, California)	4.2[a]	2.2	1.1	0.8	0.2	1.8
Udoll (silt loam, pH 6.7, Illinois)	25.4	17.1	3.1	0.4	0.1	4.7
Ustert (clay loam, pH 6.4, Texas)	28.6[a]	23.0	4.3	0.8	0.3	4.5
Humod (**Oe** layer, pH 3.6, Alaska)	105.7	5.8	6.5	0.5	1.3	91.6
Aquult (sandy loam, pH 3.5, North Carolina)	24.0[a]	2.7	0.6	0.06	0.02	20.6
Orthox (clay, pH 4.9, Puerto Rico)	26.5	8.1	2.1	0.6	0.1	15.6
Aqualf (Si.cl.l.,[b] pH 5.4, Ohio)	37.8	8.6	4.3	0.6	0.0	4.3
Andept (Si.cl.l.,[b] pH 5.3, Oregon)	103.6	6.7	1.2	0.4	0.4	94.9

Amount of Exchangeable Cations for 1 cmol$_c$/kg of Soil CEC

Ion on the Exchange Complex	*Approximate kg/ha for 30-cm Depth*	*Approximate lb/a for 1-ft Depth*
Ca^{++}	800	800
Mg^{++}	480	480
K^+	1560	1560
Na^+	920	920
H^+	40	40

Source: Profile data from Soil Survey Staff, *Soil Taxonomy,* Agriculture Handbook 436, USDA—Soil Conservation Service, 1975, 754.

[a]Often the sum of cations does not equal the measured CEC because of pH of determination, solubility of certain minerals, and the adsorption strength of some of the cations involved.

[b]Silty clay loam.

Table 5-6 Representative Cation Exchange Capacities of the Common Soil Colloids Responsible for Most Soil Cation Exchange Capacity

Soil Colloid	*Cation Exchange Capacity (cmol$_c$/kg of Colloid)*
Humus	100–300
Vermiculite	80–150
Smectites (montmorillonite)	60–100
Illite	15–25
Kaolinite	2–8
Sesquioxides	0–3

Calculation 5-1 Calculating Exchangeable Cations in Soils

As an example of the large quantity of exchangeable ions in soils, the following problem demonstrates the method of calculating amounts of exchangeable cations (see Table 5-5).

Problem In laboratory analyses the following values were obtained for exchangeable potassium. Calculate whether potassium fertilizer will be needed for a corn crop in each soil.

Soil Description	*cmol of K/kg Soil*
Well-weathered Paleudult sandy loam, Johnston Co., North Carolina, pH 4.9	0.22
Highly weathered Orthox sandy clay loam, Belém, Brazil, pH 4.2	0.06
Arid Paleargid sandy loam, Cochise Co., Arizona, pH 6.6	0.78

Solution

1. Whether or not potassium fertilizer is needed depends on the crop grown, its expected yield, and the soil test correlation for the area. For this problem assume fertilizer is needed for the corn if the test for K^+ in the top 30 cm (1 ft) depth is less than 240 kg of exchangeable K^+/ha–30 cm.

2. **Soil weight.** Usually an average soil weight per hectare–30 cm is used for all soils in routine analysis. A hectare area 30 cm deep has a volume of about 100 m × 0.3 m;

$$100 \text{ m} \times 100 \text{ m} \times 0.3 \text{ m} = 300 \text{ m}^3.$$

If a bulk density of 1400 kg/m^3 is used as an average value, the weight of that soil would be

$$(3000 \text{ m}^3)\ (1400 \text{ kg/m}^3 \text{ bulk density}) = 4{,}200{,}000 \text{ kg of soil/ha–30 cm}$$

The approximate weight value of 4 million kg/ha–30 cm is often used to approximate average soil weights. *Remember it.* Use the weight of 4 million kg/ha–30 cm as the weight in this calculation.

3. **Weight of K^+.** To convert centimoles of K^+ to weight units,

$$1 \text{ cmol}_c \text{ of } K^+ = \left(\frac{1 \text{ cmol}_c \text{ K}}{\text{valence of K}}\right)\left(\frac{\text{atomic wt of K}}{1 \text{ mol K}}\right)\left(\frac{1 \text{ mol K}}{100 \text{ cmol K}}\right)$$

$$= \left(\frac{1}{1}\right)\left(\frac{39 \text{ g}}{1}\right)\left(\frac{1}{100}\right) = 0.39\text{g}$$

$$= 390 \text{ mg of } K^+ \text{ per cmol}_c \text{ of } K^+$$

4. **Kilograms of K^+ per hectare–30 cm** is calculated using the 4 million kg/ha–30 cm approximation of average soil weight.

$$1 \text{ cmol}_c \text{ } K^+/\text{kg}^{-1} \text{ soil} = 390 \text{ mg K per kg of soil (see step 3)}$$

$$\text{kg of K per ha–30 cm} = \left(\frac{390 \text{ mg K}}{1 \text{ kg soil}}\right)\left(\frac{1 \text{ kg}}{1{,}000{,}000 \text{ mg}}\right)\left(\frac{4{,}000{,}000 \text{ kg soil}}{1 \text{ ha–30 cm of soil}}\right)$$

$$= 1560 \text{ kg of } K^+ \text{ per hectare–30 cm depth}$$

Calculation 5-1, Cont'd Calculating Exchangeable Cations in Soils

Thus, 1 cmol of K^+/kg of soil is equivalent to 1560 kg of K^+ per hectare taken to a 30-cm depth. (This is the way the values in Table 5-5 were derived for the 30-cm depth of soil.) For the Paleudult soil from North Carolina with 0.22 $cmol_c$ K^+/kg,

$$\left(\frac{1560 \text{ kg K}}{\text{ha–30 cm}}\right)\left(\frac{0.22 \text{ cmol}_c}{1 \text{ cmol}_c}\right) = 343 \text{ kg K}^+ \text{ per ha–30 cm}$$

For the Orthox soil from Brazil with 0.06 $cmol_c$ K^+/kg,

$$\left(\frac{1560 \text{ kg K}}{\text{ha–30 cm}}\right)\left(\frac{0.06 \text{ cmol}_c}{1 \text{ cmol}_c}\right) = 94 \text{ kg K}^+ \text{ per ha–30 cm}$$

For the Paleargid from Arizona, with 0.78 $cmol_c$ K^+/kg,

$$\left(\frac{1560 \text{ kg K}}{\text{ha–30 cm}}\right)\left(\frac{0.78 \text{ cmol}_c}{1 \text{ cmol}_c}\right) = 1{,}217 \text{ kg K}^+ \text{ per ha–30 cm}$$

5. The soil from North Carolina is above the 240 kg K^+/ha–30 cm value previously set as the critical value. Additional K is not needed for this soil. Obviously the Orthox from Brazil is quite low in potassium; corn would respond to the addition of appreciable potassium fertilizer. The Arizona soil, typical of most arid-region soils, has adequate potassium available in it.

The kinds and proportions of the various cations on the CEC are discussed later. In general, the calcium ions dominate the CEC except in quite acidic soils. The $Al(OH)_2{}^+$ ion dominates in soils of pH about 4 to 6. Below pH 4 more Al^{3+} and H^+ occur on the exchange sites, but Al species and Ca^{2+} are the two dominant cations in most soils.

Importance of Cation Exchange

Cation exchange is an important phenomenon in soil fertility, in causing and correcting soil acidity and basicity, in changes altering soil physical properties, and as a mechanism in purifying or altering percolation waters. *The plant nutrients calcium, magnesium, and potassium are supplied to plants in large measure from exchangeable forms.* In fact, the usual test to predict a soil's ability to furnish potassium to the plant is a measure of its exchangeable potassium content. Cation exchange is very important in soils because of the following relationships:

- The exchangeable K is a major source of plant K.
- The exchangeable Mg and Ca are often major sources of plant Mg and Ca.
- The amount of lime required to raise the pH of an acidic soil is greater as the CEC is greater.
- Cation exchange sites hold Ca^{2+}, Mg^{2+}, K^+, Na^+, and $NH_4{}^+$ ions and slow their losses by leaching.
- Cation exchange sites hold fertilizer K^+ and $NH_4{}^+$ and greatly reduce their mobility in soils.
- Cation exchange sites adsorb many metals (Cd^{2+}, Zn^{2+}, Ni^{2+}, Pb^{2+}) that might be present in wastewaters. Adsorption removes most of them from the percolation water, thereby cleansing the water that drains into groundwaters or surface waters.

Detail 5-5 Variable Cation Exchange Capacities

The cation exchange capacity of a soil changes with a change in pH (acidity or basicity). The majority of the negatively charged exchange sites are from isomorphous substitutions and are a permanent charge. However, sesquioxides (metal oxides) and kaolinite have only a few lattice sites with isomorphous substitution. The various hydroxyls of clays, humus, and organic acids ionize H^+ into the soil water solution, thereby producing negatively charged cation exchange sites on these soil particles as the pH rises.

In acidic solutions (high H^+ concentration), fewer H^+ ions ionize off the R—OH. In basic solutions fewer H^+ are in solution and more R—OH sites are ionized to $R—O^-$. Nonionized sites do not react as exchangeable sites.

In more strongly acidic solutions, the H^+ hinders ionization of the R—OH groups. At lower pH, there will be *fewer* pH-dependent CEC sites ionized, which allows them to react as exchange sites.

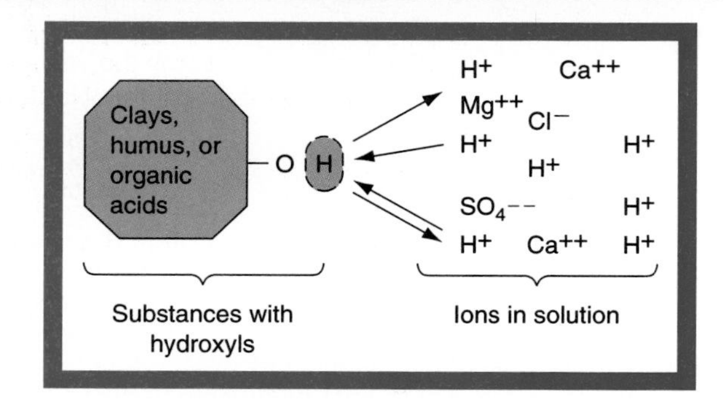

As much as 10%–40% of the soils' CEC may be from pH-dependent sites. Most CEC from humus is pH-dependent. As the pH rises, pH-dependent CEC also increases.

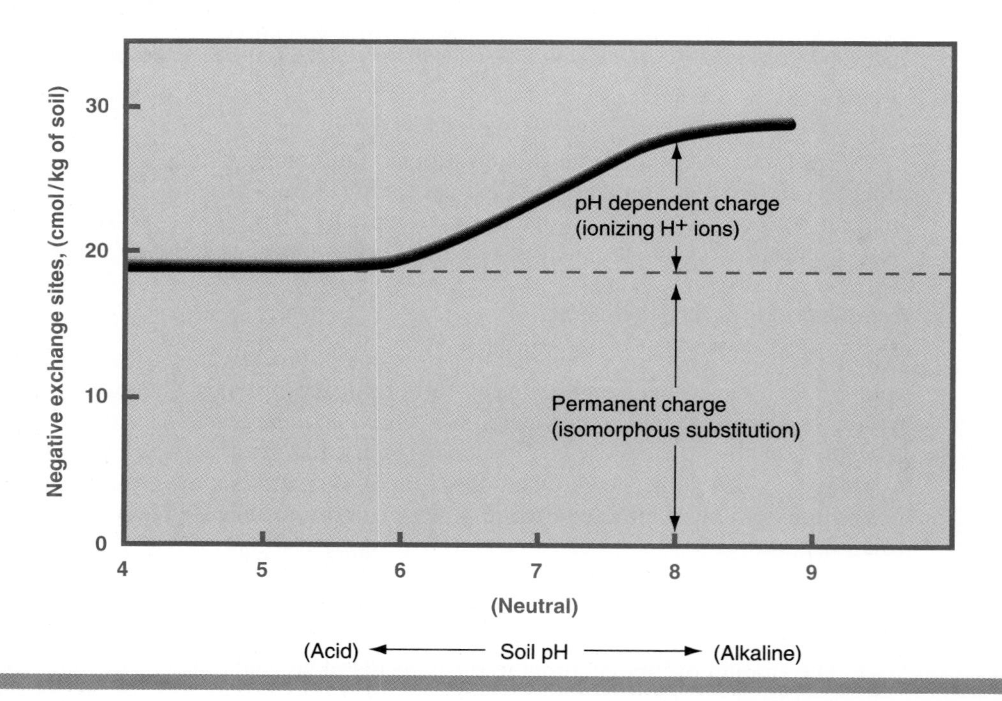

5:5 Anion Exchange and Adsorption

Note that the discussion of cation exchange capacity has considered primarily the plant nutrients calcium, magnesium, ammonium, and potassium. **Anions**—the negatively charged nutrients such as sulfate, nitrate, phosphate, molybdate, borate, and chloride—are *not held* on cation exchange sites. Also, few of the other micronutrients are held there; they form

Calculation 5-2 Estimating Cation Exchange Capacities

Estimating cation exchange capacity is done using the data of Table 5-6 and assuming that all cation exchange is on clay and soil humus. First, from the range of CEC values given, select an *average* CEC value for the humus and for each clay type. Second, divide the value from Table 5-4 by 100 to get the CEC value contributed by 1% of that material.

Problem From data of soils selected from Table 5-3, estimate the CEC for each horizon described below:

Soil	*Predominant Colloid*	*Clay (%)*	*Humus (%)*	*Measured CEC ($cmol_c$/kg)*
Sandy Entisol Ap	Montmorillonite	2.6	0.5	5.2
Subhumid-area Vertisol Ap	Montmorillonite	36.0	1.7	28.6
Spodosol **Oe** layer	Organic	24.0	85.0	105.7
Oxisol Ap	Kaolinite and metal oxides	73.1	4.7	26.5

Solution Using average values from Table 5-6, assume that each 1% of humus (10 g of humus/kg of soil) contributes 2 $cmol_c$ of CEC per kilogram of soil (divide the 200 $cmol_c$/kg for pure humus by 100%). In a similar manner, 1% of montmorillonite contributes about 0.8 $cmol_c$ (80 $cmol_c$/kg average for 100% clay). Similarly 1% of kaolinite contributes about 0.05 $cmol_c$ and 1% metal oxides contribute about 0.02 $cmol_c$ of CEC.

Sandy Entisol Ap: With 0.5% humus it should have 2 $cmol_c \times 0.5 = 1.0$ $cmol_c$/kg of CEC from humus. Assuming that the clay is montmorillonite, the 2.6% clay contributes 2.6×0.8 $cmol_c = 2.08$ $cmol_c$/kg. Adding the CEC from clay and from humus, this soil has an estimated CEC of $1.0 + 2.1 =$ **3.1 $cmol_c$ per kg of soil.**

Subhumid-area Vertisol Ap: The Ustert, a swelling soil, has montmorillonite clay, so its 36% clay contributes $36\% \times 0.8$ $cmol_c$ CEC/kg $= 28.8$ $cmol_c$/kg. Its 1.7% humus contributes $1.7\% \times 2.0$ $cmol_c$ of CEC/kg $=$ 3.4 $cmol_c$/kg, for a total of **32.2 $cmol_c$ per kg of soil.**

Spodosol Oe horizon: This organic layer with 85% humus has 2 $cmol_c$/kg $\times$ 85% humus $=$ 170 $cmol_c$ of CEC/kg contributed by humus. The 4.1% kaolinite clay provides 0.05 $cmol_c$/kg $\times$ 4.1% $=$ 0.21 $cmol_c$ of CEC per kg of soil.* The total is **170.2 $cmol_c$ per kg of soil** for its estimated CEC.

Oxisol: This soil of 73% clay probably has mixed kaolinite and sesquioxide clays (assume half of each). The soil would have (73%/2) $\times$ 0.05 $cmol_c$ CEC/kg for kaolinite and (73%/2) $\times$ 0.02 $cmol_c$ CEC/kg for metal oxides. This is a total of 2.5 $cmol_c$/kg of CEC from its clay. It also has 4.7% $\times$ 2.0 $cmol_c$/kg $=$ 9.4 $cmol_c$/kg from humus. The total estimated CEC is $2.5 + 9.4 =$ **11.9 $cmol_c$ per kg of soil.**

When comparing estimated values to measured values, there will be considerable error in some soils. This may be due to a poor choice of the average CEC values, the extent of humus decomposition, an incorrect estimate of the kind of clay, or the presence of an unusual colloid in that soil. It seems that extremes in percentage clay or humus produce less accurate estimates. However, using an estimation for more common soils that have 10%–25% clay and 1%–6% humus can provide a fair approximation of the soil's CEC. Consider that clays in humid climates are mostly kaolinite and hydrous mica, and soils in climates with less than 600 mm (24 in) annual precipitation are montmorillonite plus hydrous mica or vermiculite.

Source: Courtesy of Raymond W. Miller, Utah State University.

*The 24% kaolinite is 24% of the mineral portion, which is only 15% of the whole soil. So on a whole-soil basis, the soil has only (24%) $\times$ (15%) or 4.1% kaolinite.

Detail 5-6 The pH Defined

Soil pH is the *negative logarithm of the active hydrogen ion (H^+)* concentration in solution. When water (HOH) ionizes to H^+ and OH^- (a neutral solution), both H^+ and OH^- are in concentrations of 10^{-7} mole* per liter. The brackets [ ] mean concentration.

$$HOH \rightleftharpoons H^+ + OH^-$$

$$\frac{[H^+][OH^-]}{[HOH]} = 1 \times 10^{-14},$$

$$[H^+] = [OH^-] = 1 \times 10^{-7}$$

Thus, the negative logarithm of $[H^+]$ is 7, or pH 7. When the H^+ concentration is greater (more acidic), such as 10^{-4} mole per liter, the pH is lower (e.g., pH 4). In basic solution the OH^- concentration exceeds the H^+ concentration. The product of the H^+ and OH^- concentrations equals 10^{-14} mole per liter. When H^+ is 10^{-5}, OH^- is 10^{-9}, for example. Thus, a large concentration of $[H^+]$ of 10^{-2} is a pH of 2. When the soil is very alkaline, such as $[OH^-]^{-3}$, the $[H^+]$ is $10^{-14} - [OH^-]^{-3} = 10^{-11}$, and the pH is 11. High pH means low acidity. (The weight and volume units in chemistry are expressed in the centimeter-gram-second system, and conversion to U.S. units is not applicable here.)

*A mole is one molecular weight of ion or of the molecule.

hydroxides or carbonates of low solubility and are available to plants directly from the small amounts solubilized into the soil solution.

Anion exchange sites are either positively charged sites or ligand-exchange sites. Multicharged positive ions, such as iron and aluminum, have hydroxyls (OH^-) that can be exchanged with sulfate (SO_4^{2-}), phosphate ($H_2PO_4^-$ or HPO_4^{2-}), molybdate (MoO_4^{2-}), and some other anions. This is sometimes called *ligand exchange.* The highest **anion exchange capacities (AEC)** occur in amorphous silicate clays (volcanic-ash soils) and in iron and aluminum hydrous oxide clays; lesser exchange occurs in kaolinite. Phosphates, particularly, can be held firmly onto anion exchange sites.

Anion exchange capacities are generally low, usually only a *few tenths* of a $cmol_c$/kg of soil, in contrast to much larger cation exchange capacities. In soils with high percentages of iron oxides, values of a few $cmol_c$/kg have been measured. A few rare values as high as 40 $cmol_c$ of sulfate/kg of soil have been reported.[6]

5:6 Soil Reaction (pH)

The term *pH* is from the French *pouvoir hydrog'ne,* or *hydrogen power.* **Soil reaction (pH)** is an indication of the acidity or basicity of the soil and is reported in pH units (see Detail 5-6). At pH 7, hydrogen ion concentration (H^+) equals the hydroxyl ion concentration (OH^-). Thus, pH 7 is considered the neutral point. Below pH 7 the soil is increasingly acidic; above pH 7 the soil is increasingly alkaline (basic). Because the pH scale is logarithmic, the H^+ concentration, which is the substance measured when determining pH, has a tenfold change between each whole pH number. Thus, a soil of pH 5 has 100 times more H^+ in solution than is in a soil solution with a pH of 7. See Figure 5-8 for some pH relations among various materials and environments.

[6]H. Gebhardt and N. T. Coleman, "Anion Adsorption by Allophanic Tropical Soils: II. Sulfate Adsorption," *Soil Science Society of America, Proceedings* **38** (1974), 259–262.

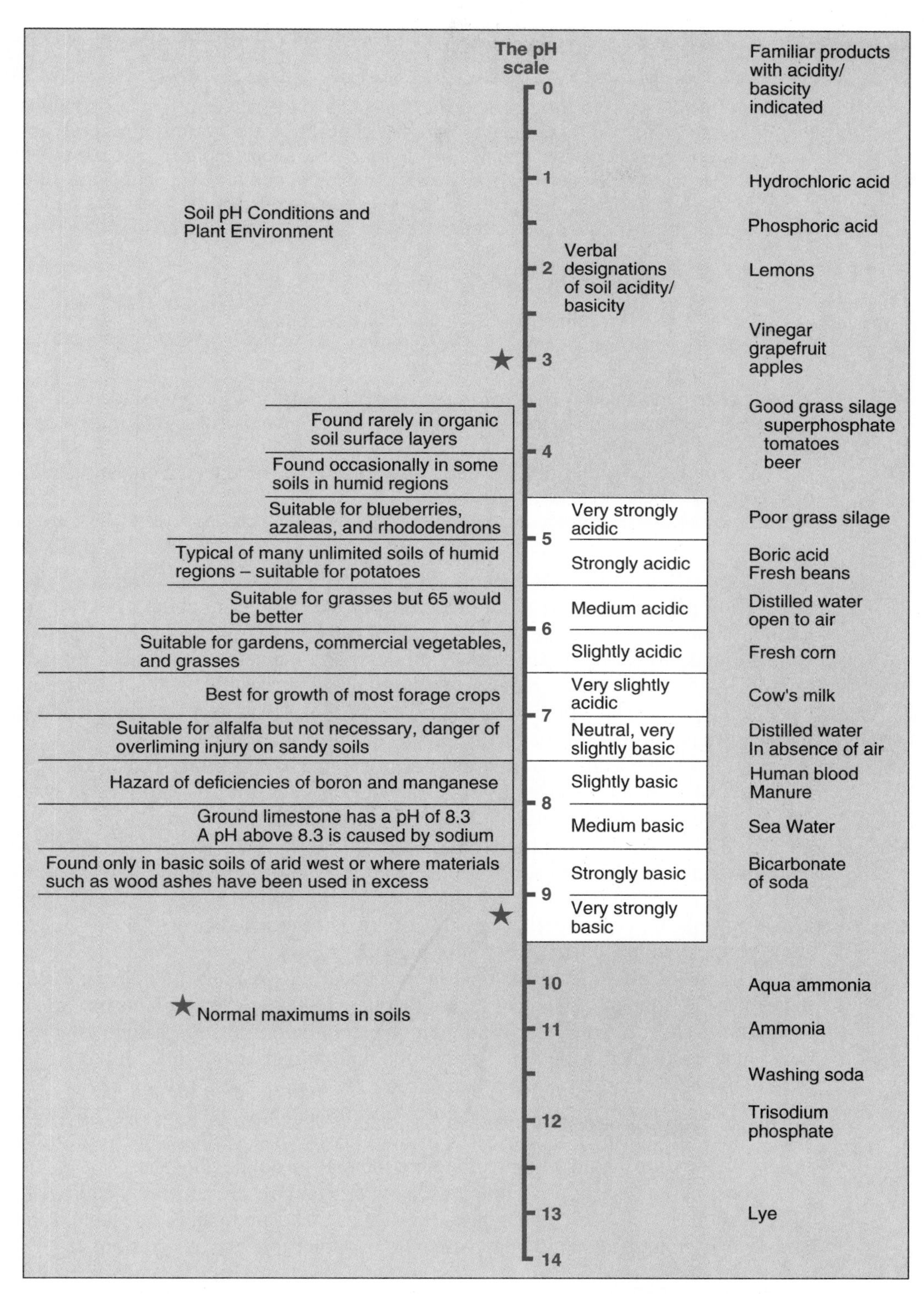

FIGURE 5-8 The entire pH scale ranges from 0 to 14, but soils under field conditions vary between pH 3.5 and 10.0. Few soils have pHs outside this range. In general most plants are best suited to a pH of 5.5 on organic soils and a pH of 6.5 on mineral soils. (*Source:* Adapted from Winston A. Way, "The Whys and Hows of Liming," University of Vermont Brieflet 997, 1968.)

Detail 5-7 Free vs. Bound Hydrogen Ions

The term **free H^+ in solution** refers to hydrogen ions in the solution. This is also the form of H^+ that is measured as pH. A much larger amount of acidity is the **bound** or **reserve acidity.** This refers to exchangeable cations that will, during reactions, supply more H^+ into the solution. These bound acidic cations include H^+ (in very acidic soils), and iron and aluminum ions (mostly as hydroxyl forms, e.g., $Al(OH)2^+$). As H^+ in solution is neutralized, the pH rises. This causes another of the waters attached to the aluminum to ionize a hydrogen and form an insoluble hydroxide.

$$Al(H_2O)_4(OH)_2{}^+ + OH^- = Al(H_2O)_3\,(OH)_3 \downarrow + HOH$$

(exchange aluminum) (basic solution) (precipitates) (water)

Neutralizing the solution H^+ would cause the pH to rise and some nonionized R—OH groups on humus and clays to ionize their H^+, leaving an R—O^- charged site. Thus, the cation exchange capacity also increases.

The amount of bound acidity is usually many times (10–1000 times) greater than the amount of free acidity in solution at any one time.

If the adsorbed ions on the colloids are mostly $Al(OH)_2{}^+$ and some H^+, the soils are acidic. **Soil pH** ranges from about pH 3.5 (extremely acidic) to about pH 11 (extremely alkaline [or basic]). Most crop plants grow well between pH 5.5 and pH 8.5. Strongly acidic soils are undesirable because soluble aluminum and manganese can reach toxic levels and microbial activity is greatly reduced. Strongly alkaline (basic) soils have lower micronutrient availability, except for boron, chloride, and molybdenum; iron, zinc, manganese, and macronutrient phosphorus may be deficient in soils of high pH.

Soils become acidic as rainfall leaches away the basic cations (Ca^{2+}, Mg^{2+}, K^+, Na^+) and many of them are replaced by H^+ from carbonic acid (H_2CO_3) formed in water by dissolved carbon dioxide. Eventually, $Al(OH)_2{}^+$ replaces H^+ (see Detail 5-7).

Soil basicity, although more difficult to alter than soil acidity, may be just as undesirable for plants. Nonleached soils or those high in calcium (low rainfall areas) may have pH values near 8.5. With increased exchangeable sodium, soils may reach values of over pH 10. Plants on soils of pH greater than about 9 usually have reduced growth or even die. However, some plants (halophytes) are tolerant of high pH and/or salt.

The major effect of a basic pH is to reduce the solubility of iron, zinc, copper, and manganese. Also, phosphate is often not readily available to some plants because of its precipitation in the soil solution by calcium or precipitation on *solid* calcium carbonate. Iron deficiency, associated with wet, clayey soils high in carbonates, has long been recognized (although not well understood); it is referred to as *lime-induced iron chlorosis.* Solutions of high pH have low solubilities of iron, zinc, manganese, and copper; the addition of phosphorus often further decreases the availability of those metals at the root surface or just inside the root, presumably by precipitating them as insoluble phosphates.

Most micronutrient problems caused by high soil pH are solved by adding special fertilizers, such as water-soluble **chelates** (key′-lates). Chelate ligands are stable, soluble complexers of the metal ions (although susceptible to microbial decomposition).

Importance of Soil pH

The soil pH is easily determined and provides various clues about other soil properties. The soil pH greatly affects the solubility of minerals. Strongly acidic soils (pH 4 to 5) usually dis-

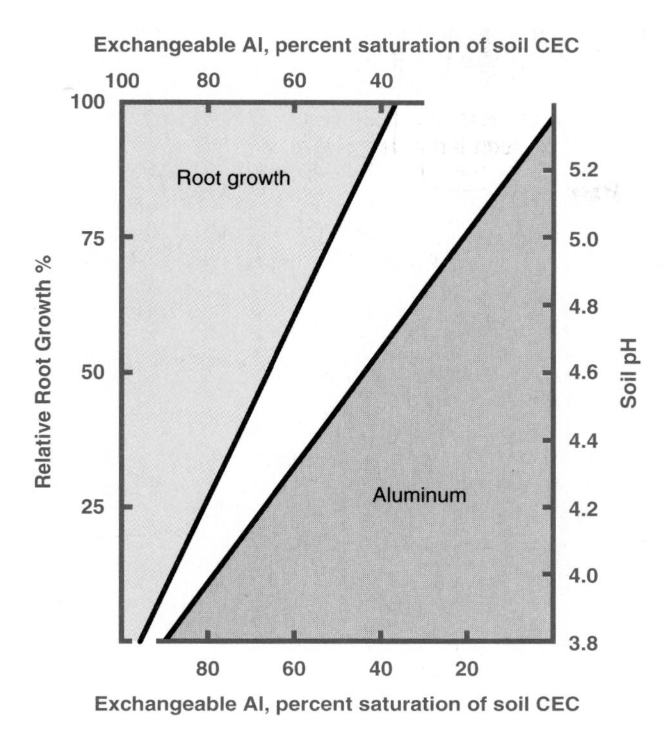

FIGURE 5-9 As a soil becomes more acidic (pH becomes lower), more soluble aluminum is available for absorption by plants. The more aluminum that is absorbed, the greater is its toxic effect and the greater will be the reduction in the root growth of corn sorghum. (*Source:* Redrawn from Eduardo Brenes and R. W. Pearson, "Root Responses of Three Gramineae Species of Soil Acidity in an Oxisol and an Ultisol," *Soil Science* **116** (1973), 295–302. © The Williams & Wilkins Co.)

solve high, even toxic, concentrations of soluble aluminum and manganese (Figure 5-9). Azaleas, tea, rhododendrons, cranberries, pineapple, blueberries, and several conifer timber species tolerate, or may even require, a strong acidity. In contrast, beans, barley, and sugarbeets do well only in slightly acidic to moderately basic soils because of a high calcium demand or inability to tolerate soluble aluminum. Alfalfa does best in a neutral or slightly basic pH because that environment is most hospitable to the beneficial bacteria that inhabit alfalfa roots. Most minerals are more soluble in acid soils than in neutral or slightly basic solutions. Soluble $Al(OH)_4^-$ may exist at high pH values. On mineral soils most agricultural crops do best in slightly acidic soils (pH 6.5); on organic soils, about pH 5.5.

The soil pH does influence plant growth by its effect on activity of beneficial microorganisms. Most nitrogen-fixing legume bacteria are not very active in strongly acidic soils. Bacteria that decompose soil organic matter, releasing nitrogen and other nutrients for plant use, are hindered by strong acidity. Fungi usually tolerate strong acidity better than do other microbes.

Basic Cation Saturation Percentage (BCSP)

The cations commonly adsorbed on exchange sites of soil colloids can be divided into acid-forming cations (aluminum and hydrogen) and base-forming cations (calcium, magnesium, potassium, sodium, and most others). The proportion of basic cations, in percentage, to the total cations on the cation exchange complex is the **basic cation saturation percentage (BCSP),** also referred to simply as *base saturation percentage.* For example, if a soil has a cation exchange capacity of 16 $cmol_c$/kg and 4.2 $cmol_c$ of those cations are aluminum plus hydrogen, the remaining cation exchange sites always contain the basic cations. To calculate

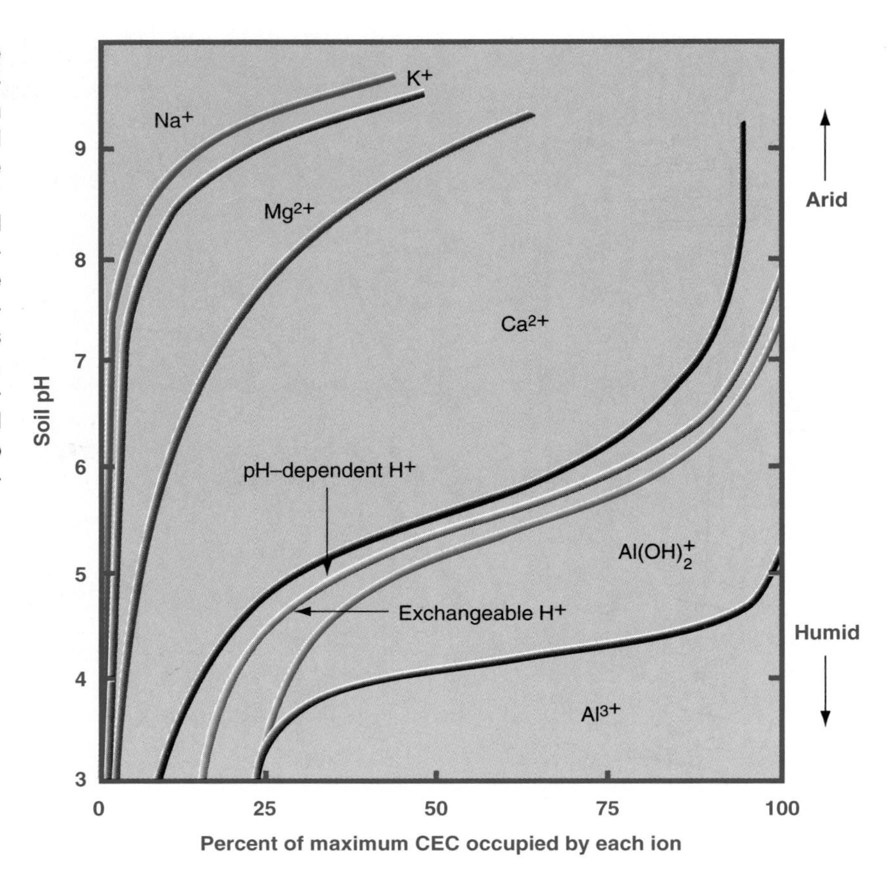

FIGURE 5-10 Generalization of the soil pH and the approximate proportions of various cations on the cation exchange sites for a clay loam soil and 3%–5% organic matter. The amount of H^+ on pH-dependent H^+ sites (ionizable organic matter and metal hydroxides) varies with the organic matter and metal hydroxide contents. A major portion of the exchangeable ions in acidic soils is some form of hydrated aluminum. Only a small portion of aluminum occurs as $Al(OH)^{2+}$, but $Al(OH)_2^+$ still predominates at those pH values, so $Al(OH)^{2+}$ is not shown in the diagram. (Courtesy of Raymond W. Miller, Utah State University.)

the basic cation percentage, the basic cations = 16.0 − 4.2 = 11.8 $cmol_c$/kg. The basic cation saturation percentage is (11.8/16) (100) = 73.7%. The more acidic a soil is, the lower is its percentage of basic cation saturation (Figure 5-10). At pH 7 or higher, most soils are essentially 100-percent basic cation saturated (very little exchangeable hydrogen and soluble aluminum). Compare proportions in Table 5-5.

Basic cation saturation does not provide much information that is not already available from pH values, but it does provide numerical values of the amount of exchangeable hydrogen and aluminum ion species, and this aids in predicting the amount of lime needed to neutralize the soil acidity.

5:7 Buffering in Soils

Most soils can resist appreciable pH changes when large amounts of a material—either strongly acidic or basic—are added, such as an acid-forming or base-forming fertilizer. This ability to resist a change in pH is the **buffering capacity** of the soil. The buffering capacity increases as the cation exchange capacity increases. To exhibit buffering, the soil must remove hydrogen ions (H^+) of added acids or neutralize the hydroxyls (OH^-) of added bases. This occurs by cation exchange and neutralization:

$$\text{colloid-H} + NH_4OH \text{ (aqueous ammonia)} \rightleftharpoons \text{colloid-}NH_4 + HOH \text{ (water)}$$
$$\text{colloid-2H} + CaCO_3 \text{ (lime)} \rightleftharpoons \text{colloid-Ca} + CO_2 \uparrow \text{(gas)} + HOH$$

In these two examples neither the aqueous ammonia fertilizer nor the lime greatly changes the soil pH because the *base* is *neutralized* to either exchangeable NH_4^+ or Ca^{2+} plus the neutral water. A *small* pH change does actually occur, and the pH will gradually change as more and more base is added.

A similar buffering reaction occurs with an added acid:

$$\text{colloid-Ca} + 2H_2CO_3 \text{ (carbonic acid)} \rightleftharpoons \text{colloid-2H} + Ca\,(HCO_3)_2$$

Carbon dioxide dissolves in water to form weak carbonic acid (H_2CO_3), which ionizes to produce free H^+ (see Detail 5-7). The free H^+ goes on the exchange sites and the calcium replaced forms calcium bicarbonate that is slightly basic. Eventually, the bicarbonate will precipitate as lime ($CaCO_3$) and leave neutral water again or will be leached from the soil.

The concept of buffering can be broadened to include a *resistance to change in the concentration of any ion in the solution* that is also adsorbed to the colloid. If calcium in solution is precipitated, additional exchangeable calcium will be exchanged off exchange sites into solution so that only a slight change in soluble calcium will occur. (The amount of exchangeable ions on the colloid are partly determined by concentrations of cations in solution.) Conversely, adding soluble calcium will force much of that calcium onto exchange sites, thereby replacing other exchangeable ions, again lowering the soluble calcium to a low concentration. Exchangeable ions are always in dynamic equilibrium with soluble ions in the soil solution.

Buffering action is effective in controlling soluble concentrations of H^+, Al^{3+}, Ca^{2+}, Mg^{2+}, K^+, and Na^+. The larger the CEC of the soil, the greater will be the buffering capacity. Soils high in humus and/or clay, particularly montmorillonite or vermiculite clays, will have high buffering. Organic soils and clays have higher CEC values and are more strongly buffered than are related sandy soils.

The role of metallic impurities is to increase the disorder of silica, . . . this is the genesis of clay minerals.

—G. Millot

Questions

1. (a) What are *colloids?* (b) How large are they? (c) What substances are the colloids in soils?
2. (a) What are *clay minerals?* (b) Are they inherited in some soils? (c) What are three different meanings of the word *clay?*
3. From what materials do clays form in (a) neoformation and (b) degradation?
4. Do clays contain basic cations (a) *in* their structures? (b) *adsorbed* to their structures?
5. (a) What charge [+ or −] do most clays carry? (b) What causes the charge?
6. How does (a) montmorillonite differ from kaolinite? (b) allophane differ from sesquioxides? (c) hydrous mica differ from vermiculite?
7. In what climatic conditions are each of the various clays expected to form? Explain for sesquioxides, kaolinite, and montmorillonite.
8. (a) What is the nature of *organic colloids?* (b) To what extent are all organic colloids similar?
9. (a) Define *cation exchange.* (b) Illustrate cation exchange by an equation.

10. (a) Which ions are involved in cation exchange? (b) Which ions are dominant in strongly acidic soils? (c) For which items or reactions is cation exchange important?
11. List some typical CEC ranges for (a) pure clays, (b) humus, and (c) soils.
12. How does the soil pH alter CEC?
13. (a) What is the basis used for estimating a soil's CEC? (b) Explain why estimates might be used. (c) How accurate are such estimates?
14. (a) What kinds of soils might have appreciable AEC? (b) Which clays have the largest AEC values?
15. (a) What is *soil pH?* (b) Give the extreme pH values found in most soils.
16. In what ways does soil pH adversely affect soil properties and plant growth?
17. (a) Define *BCSP.* (b) Is the BCSP high or low in arid-region soils?

Organisms and Their Residues

In nature nothing is wasted, for everything is part of a continuous cycle. Even the death of a creature provides nutrients that will eventually be reincorporated in the chain of life.

—Denis Hayes

6:1 Preview and Important Facts

PREVIEW

Soils appear to be inert masses of minerals but, in fact, they teem with organisms and plant roots. Soil organisms mix and aerate soil, fix atmospheric N_2, decompose dead organic substances, oxidize and reduce many elements, and recycle nutrients. Imagine a world with all the plant and animal bodies ever grown laying around on the surface! Truly, two of the most important processes on earth are **decomposition** (recycling elements) and **photosynthesis** (storing energy from the sun back into chemical bonds in plants).

Some organisms cause horrendous damage by reducing or destroying crop yields and by spreading animal disease. For example, in the United States parasitic nematodes (small roundworms) in the soil cause an estimated plant loss of about $2 billion annually; soil-borne diseases, such as *Phythium* root rot of cereals and *Fusarium* wilts of fruits and vegetables, cause an estimated $1 billion damage annually; and various large larvae (grubs, cutworms, wireworms, root maggots, and sod webworms) destroy several million dollars worth of crops annually. These losses are much greater in areas of the world where herbicides, fungicides, and insecticides are not commonly used.

Although the detrimental soil organisms are many, the beneficial organisms exceed them in effect. **Fungi** and **bacteria** are the most important organic matter decomposers. Decomposers produce **enzymes,** which reduce the **activation energy** necessary to break (or build) the bonds of the substance. Hundreds of different enzymes are needed; generally each enzyme breaks (or builds) only one kind of bond. Other organisms, **algae,** fix atmospheric nitrogen which then becomes part of the soil organic matter. **Actinomycetes** fix N_2 and help in the decomposition of organic materials. **Protozoa** are also decomposers and prey on bacteria.

The residue left after the initial stages of decomposition in soil is called **humus** or **soil organic matter.** Soils have varying amounts of humus, mostly in the top 15 cm to 40 cm of depth. The more arid the climate and the less the vegetation grown, the less humus there will be in the soil. Even a small amount (0.5 percent to 4 percent of humus) is important. Humus and actively-decomposing organic residues comprise the soil's **nitrogen reservoir;** they also furnish large portions of the phosphorus and sulfur that plants use and some of most nutrients; they protect soils against erosion forces; they supply the cementing substances to form desirable aggregates; and they loosen up the soil to provide better aeration and water movement.

Soil organic matter is constantly undergoing change and must be replenished continually to maintain soil productivity. Soil organic matter can be supplemented by the addition of organic amendments, such as animal manures, municipal sewage sludge and septage, logging and wood-manufacturing refuse, industrial organic residues, and food-processing residues. In many countries residues are used for purposes other than to improve the soil, such as using straw in mud for building structures and using manure for fuel (Figure 6-1).

IMPORTANT FACTS TO KNOW

1. The beneficial vs. nonbeneficial roles of animals, including earthworms, in soils
2. The differences between the *bulk soil* of an area and the *soil of the rhizosphere,* particularly in pH, microbial activity, and nutrient availability
3. The importance of fungi in soils and the many roles they play, particularly in decomposition, production of toxins, and associations with plant roots such as mycorrhizae
4. The importance of bacteria in soils and their role in decomposition and N_2 fixation, and their autotrophic and pathogenic roles. The sizes of bacteria in comparison to the sizes of soil pores and clay particles. The growth rates of bacteria.
5. The role of N_2 fixation and its importance in the ecology of various land areas, for agricultural crops, and for the world, in general
6. The process of N_2 fixation in nature
7. The general conditions favoring the growth of microbes and the methods and limitations involved in controlling unwanted microbes

FIGURE 6-1 Animal manures are excellent soil amendments, but very little is applied to cultivated fields in the tropics and subtropics. Only about 20% of the collected animal manures in India is applied to fields; the other 80% is made into manure cakes for fuel used to cook meals. (Courtesy of Roy L. Donahue.)

8. The unique nature of viruses and viroids, particularly whether or not they are living organisms, how they cause damage, and why they are so difficult to control
9. The nature of a biological reaction, its cause, benefits, and limitations
10. The benefits of soil organic matter, particularly (a) as a source of nitrogen, phosphorus, and sulfur; (b) in soil aggregation; and (c) as a contributor to soil cation exchange capacity
11. The values and properties of (a) animal manures, (b) crop residues, (c) sewage sludge, and (d) composts, particularly in a LISA program

6:2 General Classification of Soil Organisms

There is no single accepted classification for microorganisms. A comprehensive system that includes plants, animals, and microorganisms has five *kingdoms:* **Animalia, Plantae, Fungi, Protista,** and **Monera** (Detail 6-1). This list and the following discussion minimize discussion of the large animals and emphasize mostly the microorganisms in soils. Table 6-1 lists some details about the soil organisms in numbers and biomass.

6:3 Animalia: Rodents, Worms, and Insects

The large animal life—*Animalia* (formerly, *macrofaunae*)—that inhabit the soil range in size from large burrowing animals, such as badgers, down to mites (tiny arachnids barely visible to the eye).

Burrowing Animals

Large burrowing animals (e.g., moles, prairie dogs, gophers, mice, shrews, rabbits, badgers, woodchucks, armadillos, chipmunks) aerate the soil and alter its fertility and structure, but they also eat and destroy vegetation, which makes them more detrimental than beneficial from an agricultural point of view.

Earthworms

Earthworms are important soil organisms. Earthworms feed on animal and plant residues and on most deciduous leaves but not on waxy and resinous conifer needles. The ingested organic matter and fine-textured soil are excreted as small granular aggregates, which resist rupture by raindrop impact and provide abundant and readily available plant nutrients. However beneficial they might be, earthworms can spread plant diseases.

Earthworms aerate and stir the soil, which allows better water infiltration and easier root penetration. Some earthworm species burrow to 6 m deep (20 ft) in the soil, but most live in the common root zone, which averages 2 m (6.6 ft) in depth in the temperate regions. Earthworms prefer moist, well-aerated, warm (70°F, or 21°C) soils with soil pH between 5.0 and 8.4, plenty of palatable organic matter, low salt concentrations but high available calcium, fairly deep soil of medium or fine texture, and soil undisturbed by tillage.

Earthworms are hindered by heavy farm machinery and by sandy, salty, arid, and acid soils. Other unfavorable conditions include cold or hot and bare or barren soils as well as the existence of mice, mites, moles, millipedes, and strong insecticides.

Arthropods and Gastropods

Arthropods are joint-footed invertebrate organisms. Arthropods of importance in soil include mites, millipedes, centipedes, and insects such as spring-tails, proturans, diplurans, beetles, flies, ants, and termites. They feed mostly on decaying vegetation and help to aerate

Detail 6-1 Partial Outline of the Classification of Living Organisms with Emphasis on the Organisms Most Active in Soils*

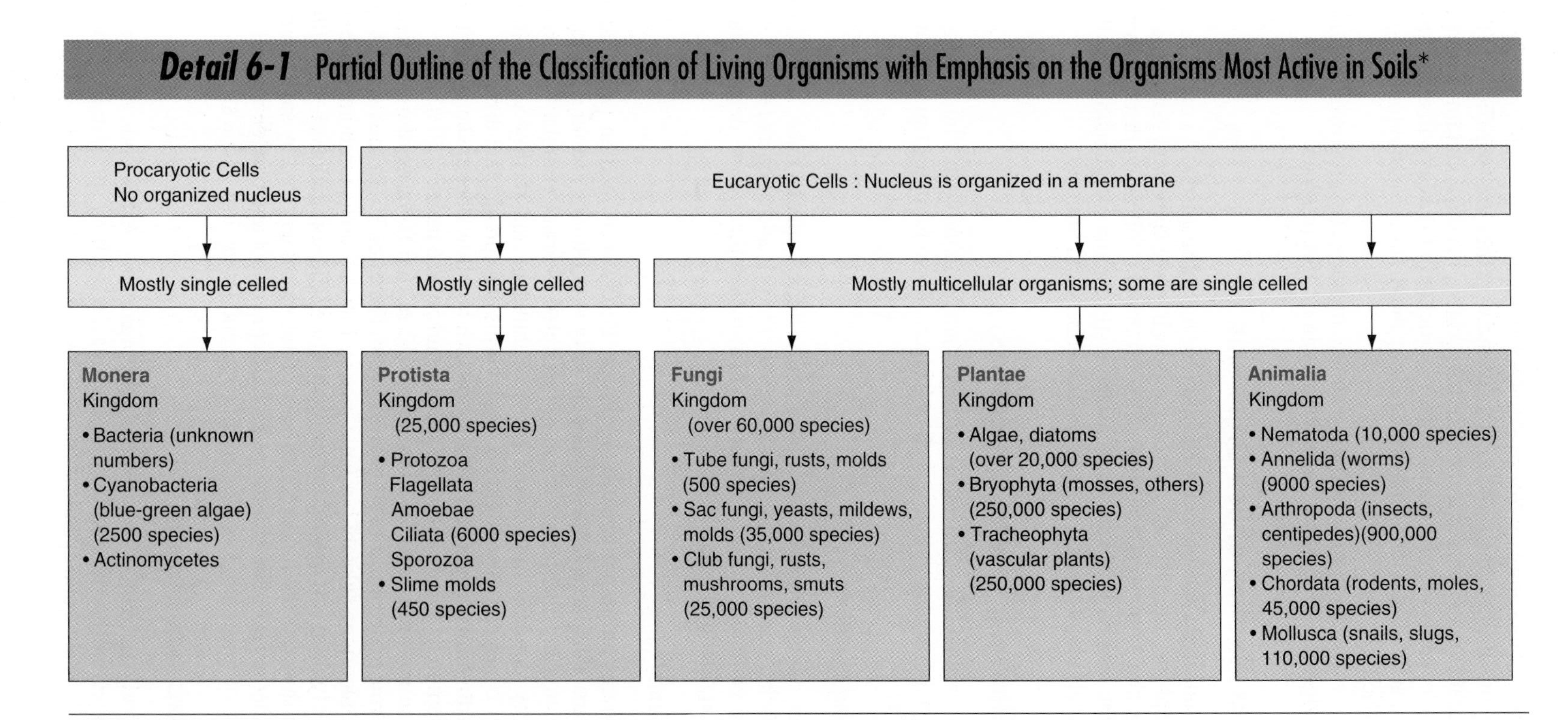

Source: Raymond W. Miller, Utah State University, from data by Edward W. Wilson, et al., *Life on Earth,* Sinauer Associates, Sunderland, MA, 1973.

*Most subkingdoms listed are common names within the five kingdoms but are not phyla names in most instances.

Table 6-1 Approximate Numbers, Biomass, and Carbon/Nitrogen Ratios of Microorganisms Common in Soils

Organism	*Number per Gram of Soil**	*Biomass per Hectare–15 cm (kg) Live Weight*	*Carbon/Nitrogen Ratio of the Organism*
Bacteria	10^7–10^9	400–5000	5/1 to 8/1
Actinomycetes	10^6–10^8	300–4000	5/1 to 8/1
Fungi	10^5–10^6	1000–10,000	8/1 to 20/1
Algae	10^4–10^5	50–600	5/1 to 9/1
Protozoa	10^4–10^5	15–200	5/1 to 10/1
Nematodes	10–50	10–100	5/1 to 10/1
Earthworms	—	100–1000	5/1 to 10/1
Other soil life	—	15–200	5/1 to 15/1

*One gram equals about 1/5 teaspoonful of soil.

the soil with their burrows; however, many species also can be pests because they are phytophagous (from the Greek *phyto,* plant, and *phagos,* to eat).

Ants and termites can radically change soil structure and till the soil; the net result can be beneficial or harmful, usually the latter (Figure 6-2).

Slugs and snails are important members of the **gastropods** (belly-footed organisms); they feed on decaying vegetation but will eat and damage living plants. In infested areas counts as high as 243,000 slugs per hectare with a live weight of 450 kg/ha (401 lb/a) have been reported.

Nematodes

Nematodes are microscopic, unsegmented, threadlike (which is the meaning of *nema*) worms; they are classified according to their different feeding habits. *Omnivorous* nematodes live mainly on decaying organic matter and are the most common of the soil nematodes. *Predaceous* nematodes prey on soil bacteria, fungi, algae, and even on other nematodes. *Parasitic* nematodes infest plant roots, causing the conspicuous knots that give visible proof of their presence. Parasitic nematodes are so prevalent that nearly all field and vegetable crops and trees are infected. The entry of nematodes into a plant allows easy entry for other pathogens (disease organisms), which may cause even more extensive damage than the nematodes themselves. Sugar beets and corn are particularly susceptible; an infested, untreated field can result in crop losses of more than 50 percent. Nematodes rank as the primary pest of soybean roots.

Nematodes are controlled by chemical fumigants (nematocides, which are expensive), hardwood bark, rotation with more-resistant plant species, or the use of resistant plant varieties (Figure 6-3). Control of nematodes by fumigation increased marketable yields of cantaloupes in California 50 percent to 100 percent.[1] Hardwood bark mixed with soil is a newer, lower-cost technique for the control of plant-parasitic nematodes. So far this control measure has been demonstrated with scientific accuracy on container-grown tomato and forsythia plants.

Nematodes can be beneficial, such as in control of white grub on golf courses, sod farms, and lawns (Detail 6-2). Nematodes help keep microorganisms under control, also.

[1]Erik Likums, "Cantaloupes vs. Nematodes," *Agricultural Research* **28** (1980), No. 9, 15.

(a)

(b)

FIGURE 6-2 Termite activities are a common disturbance on large areas in tropical climates. (a) Termite mounds in Guinea-Bissau on cultivated land and (b) a cross-section of a termite mound in northern India.
(Photos by permission of (a) Raymond W. Miller, Utah State University, and (b) Julian P. Donahue.)

FIGURE 6-3 Many plants are susceptible to injury by parasitic nematodes, and crop losses often exceed 50%. Control may be by specific chemicals, hardwood bark, rotation with resistant plant species, or selection of more-resistant varieties. (a) A field of sugar beets in Idaho infested with the nematode Heterodera schachtii and controlled, on the left, by an organophosphate nematocide. (b) Nematodes on blue spruce in a Wisconsin greenhouse were controlled, on the left, by steam sterilization. (c) Nevada: Ranger alfalfa variety is susceptible to nematodes, but Lahontan variety is resistant. (*Source:* USDA—Forage and Range Research Station, Logan, Utah. Photos by Gerald Griffin.)

6:4 Plantae: Plants and Algae

The **Plantae** organisms obtain energy from the sun and can exist as stationary life. Algae and diatoms are examples of *micro*plantae; the *macro*plantae range from the 250,000 species of Bryophytes (mosses) to the 250,000 species of Tracheophytes (vascular plants).

Plants

Most familiar plants have root systems, some of which are often 30 percent to 50 percent of the total plant mass—thousands of kilograms of biomass per hectare. Older roots develop thick, protective coverings of a mucilage. In contrast, **root hairs** are single cells of the root surface; they have thin walls that allow water and nutrient absorption.

Detail 6-2 Nematodes—The Bad and the Good[a,b]

Normally nematodes are considered pests that damage crops. The Golden nematode (*Globodera rostochiensis*), for example, has an enormous appetite for potato plant roots and can wipe out entire crops. The female can produce as many as 500 eggs that are able to persist up to 20 years or more in fields. Traces of chemicals used for nematode control were found in Long Island groundwater over a decade ago and stopped that pesticide use in many areas. Development of potato varieties resistant to the grub have slashed nematode populations by about 90 percent each year.

A good nematode (*Steinernema glaseri*) helps control the white grub, the larvae of the Japanese beetle that inflicts millions of dollars worth of damage annually on golf courses, cemeteries, sod farms, and lawns. The grubs feast on turf roots, killing the turf. The presence of the grubs attracts racoons, skunks, and birds; these animals dig up the turf in looking for grubs.

The nematodes follow the carbon dioxide trail released through the breathing vents of the grubs. The parasitic nematode then enters the larva through the breathing vents—or, in Japanese beetle larvae, who have protective plates over their vents, the nematode enters through the grub's mouthparts. Inside the grub the nematode releases a bacterium in the grub bloodstream that kills the grub within 24 hours. The nematodes killed 50 percent of the grubs in trial plots.

[a]S. M. Hays, "Golden Nematodes Are Anything But," *Agricultural Research* **44** (1996) No. 4, 16–17.
[b]D. Lyons-Johnson, "Nematode Takes on Japanese Beetle Grubs," *Agricultural Research* **44** (1996) No. 4, 12–13.

Roots exude, or secrete, many substances, including at least 18 amino acids, 10 sugars, 10 organic acids, various proteins, growth substances, growth inhibitors, microbe attractants, and repellents.[2] As early as 1832, Decandolle ascribed the problem of "soil sickness" to the toxic exudates produced by certain crop plants; yet it is only recently that extensive research has been able to identify many of these chemical excretions. The quantities of exuded material vary from about 2 percent of the "organic matter flowing into the root system" from plant tops to values of 7 percent exuded from plant roots in sand cultures. Amounts approaching 15 percent of the total radioactive carbon added were recovered in the rooting media when sorghum was given radioactive $^{14}CO_2$ for 48 hours.[3,4]

The products around a root in the soil constitute a gold mine of microbial substrates: carbohydrates and other chemicals mentioned earlier plus dead and sloughed root cells. As a result the environment surrounding the root (for a distance of 1 mm or so) will teem with microbes, each performing its function(s) of humus decomposition, N_2 fixation, or other activity. Some organisms will also be producing their own protective chemicals to help them survive. The area in the soil near the roots is called the **rhizosphere.** This portion of soil may be quite different in chemical properties from the bulk of the soil. The rhizosphere may also be one to two pH units more acidic than the neutral or alkaline soil as a whole.

[2]M. G. Hale, L. D. More, and G. J. Griffin, "Root Exudates and Exudation," in *Interactions between Non-pathogenic Soil Microorganisms and Plants,* Y. R. Dommergues and S. V. Krupa, eds., Elsevier Science Publishing, New York, 1978, 163–204.
[3]A. Haller and H. Stolp, "Quantitative Estimation of Root Exudation of Maize Plants," *Plant and Soils* **86** (1985), 207–216.
[4]Chung-Shih Tang, "Continuous Trapping Techniques for the Study of Allelochemicals from Higher Plants," Chapter 7 in *The Science of Allelopathy,* A. R. Putnam and C. S. Tang, eds., 1986, 114.

Algae

Soil **algae** are microscopic organisms that carry on photosynthesis. The main groups are green algae, yellow-green algae, and diatoms. What were once called *blue-green algae* have been reclassified into the Monera kingdom as *cyanobacteria* (also called *blue-green bacteria*). Algae are not important as decomposers of organic matter, but they are producers of fixed nitrogen and new photosynthetic growth. In soils kept moist and fertile, algal growth on the soil can produce considerable amounts of organic material—hundreds of kilograms per hectare annually.

6:5 Fungi: Molds, Mushrooms, Yeasts, and Rusts

Fungi are organisms without the ability to use the Sun for energy; they live on dead or living plant or animal tissue. Fungi are a curious assortment of one-celled organisms (*yeasts*), multicellular filamentous *molds, mildews, smuts,* and *rusts,* and the well-known *mushrooms,* to name a few. "As with other microorganisms, the classification of the fungi is difficult, sometimes ambiguous, vehemently debated, and subject to constant revision."[5]

Many microorganisms are beneficial (Detail 6-3), but not all.

Fungal Organic Matter Decomposers

One of the first visual evidences of decomposition of some materials is the appearance of fungal **mycelia,** a vegetative mass of threadlike branching filaments (**hyphae**). Molds on bread, on cheeses, on many rotting foods, and in forest litter exhibit mycelia. Fungi are vigorous decomposers of organic matter and readily attack cellulose (woody materials), lignins, gums, and other complex compounds, even in quite acidic conditions (unusual for most living organisms). Fungi also compete with economic plants for nutrients released from organic-residue decomposition, particularly nitrogen, phosphorus, and sulfur. Fungi also secrete substances that aid in the formation of water-stable soil aggregates.

Deleterious Fungi

Some fungi exist as predators on living cells. Hyphae of fungi can penetrate protozoa and, when they become immobile, slowly digest them. Even nematodes can be ensnared in mycelia and be devoured.

Although relatively few types of soil fungi are predatory, their importance is major. Various fungi cause *smuts* and *rusts* on grains and lawn grasses, many of the *wilts, powdery* and *downy mildews, leaf spots, cankers, scabs, clubfoot, blight, black wart, takeall, leaf curl,* and some *root rots.* The *amanita* mushrooms look like edible mushrooms but are very poisonous to people.

Cucumber fruit rot (belly rot) is a hazard in warm, humid climates, such as in the southern United States. All management control measures attempt to encourage drying (drainage, defoliants), incorporating surface plant debris (plowing) and allowing more open space (lower population density). The use of composted hardwood bark in place of sphagnum peat in pot culture has suppressed poinsettia crown and root rots. *Potato blight* in Ireland caused the potato famine of 1845 to 1849. The famine caused a mass migration to the United States in 1847 to 1854.

Even the fungal decomposers are not all innocuous. A recent flurry of interest has centered on **aflatoxins,** toxins produced by molds growing on grains (mostly *Aspergillus flavus*);

---

[5] R. M. Atlas, *Microbiology: Fundamentals and Applications,* 2nd ed., Macmillan, New York, 1988, 314.

Detail 6-3 Beneficial Organisms—New Finds

Scientific discoveries are often unusual and helpful. The ideas that follow are new discoveries about microorganisms and the chemicals they produce.

- The fungus *Trichoderma harzanium* produces an antimold substance with an aroma like that of a blend of coconut and celery (also found in natural peach essence). The extract kept nearly all of 6000 refrigerated New Zealand kiwi fruit safe from mold for a year with no flavor or quality loss. Half of the untreated fruit molded.[a]
- In a bad year aflatoxin mold can cost peanut growers millions of dollars. However, other *Aspergillus* fungi strains do not produce the toxin. Spreading these harmless strains in the soil crowds out the aflatoxin makers. Work is under way to produce a commercial product.[b]
- Called **Bt,** the bacterium *Bacillus thuringiensis* has been used for nearly 30 years. Its protoxins are harmless to people, livestock, and most beneficial insects (e.g., bees). In many caterpillars and other insects that chew on Bt-treated plants, Bt toxins rupture cell membranes in the midguts, causing starvation and dying. Until about 5 years ago the available strains were few; now more than 20 may be useful in commercial products. The Bt is applied as an insecticidal spray and sticks to plant leaves; it can last 1 to 4 days in soil. This is fortunate because the diamondback moth, which attacks cabbage and other vegetables, developed resistance to the early HD-1 strain.[c]
- Some biological control agents can be alternatives to pesticides for hard-to-control diseases, such as bacterial wilt of cucurbits. The *Erwinia tracheiphila* bacterium is insect-transmitted. Fungicides are not effective against bacterial diseases. Some *plant growth-promoting rhizobacteria* (PGPR) applied to cucumber as a seed treatment or root-drench application increases plant growth and seems to enhance the plant's own defense mechanisms against disease. It may be helpful to fight a wide spectrum of pathogens. The PGPR treatment may be thought of as a sort of vaccination against plant disease.[d]

[a]Hank Cutler, "Nature's Chemical Factory Yields a Kiwifruit Protector," *Agricultural Research* **44** (1996), No. 1, 23.
[b]Richard J. Cole, "Toxin to Get Competition From a Nontoxic Cousin," *Agricultural Research* **44** (1996), No. 2, 23.
[c]Phyllis A. Martin, "Banking on Bt," *Agricultural Research* **44** (1996), No. 3, 20–21.
[d]G. W. Zehnder, J. W. Kloepper, C. Yao, G. Wei, S. Tuzun, R. A. Shelby, O. L. Chambliss, and Jimmy Witt, "Root-Colonizing Bacteria Promote Cucumber Resistance to Disease and Insect Pests," *Highlights of Agricultural Research* (Alabama Agricultural Experiment Station) **42** (1995), No. 3, 20–21.

some of these are carcinogens (cancer-causing). One of them, Aflatoxin B1, is the most potent liver carcinogen known. In Hong Kong and Thailand a research team sampled grains and found that 80 percent of the peanut samples and 50 percent of the rice, corn, beans, and other cereals were infected with toxin-producing molds; one-third of those foods with mold were toxic to rats. In 1974, 106 people died and 297 became ill after eating contaminated corn; North Carolina alone estimated losses from contamination of corn to be nearly $32 million. Mycotoxins may produce birth defects, abortions, tumors, cancers, and other effects.

Mycorrhizae: The Root-Fungi Association

Mycorrhizae (my-koe-rye′-zee) means *fungus root.* Mycorrhizae are mutually beneficial (**symbiotic**) relationships between fungi and plant roots. Some fungi form a kind of sheath around the root, sometimes giving it a hairy, cottony appearance. The plant roots transmit substances (some supplied by exudation) to the fungi, and the fungi aid in transmitting nutrients and water to the plant roots. The fungal hyphae may extend the roots' lengths 100 fold. The hyphae reach into additional and wetter soil areas and help absorb many nutrients, particularly the less mobile nutrients such as phosphates, zinc, copper, and molybdenum, for transmission to the plant. Because they provide a protective cover, mycorrhizae increase the plant seedling's tolerance to drought, to high temperatures, to infection by disease fungi, and even to extreme soil acidity. The mycorrhizae do not fix N_2.

Two kinds of mycorrhizae are described by the fungal growth habits. **Ectomycorrhizal** fungi (*ecto,* outside) sheath the host root but penetrate only the outer cell layers of the root cells walls (Figure 6-4); the hyphae of **endomycorrhizal** fungi (*endo,* inside) actually penetrate into host cells. Some hyphae are from *vesicular-arbuscular mycorrhizae* (VAM). VAM are particularly helpful in phosphate absorption and they are the most common form found on plants, although they are the least visually obvious mycorrhizae.

When nursery soils are sterilized, or nonnative plants are grown in them, it is often beneficial to inoculate the soil with the appropriate fungus. For years pines transplanted from the

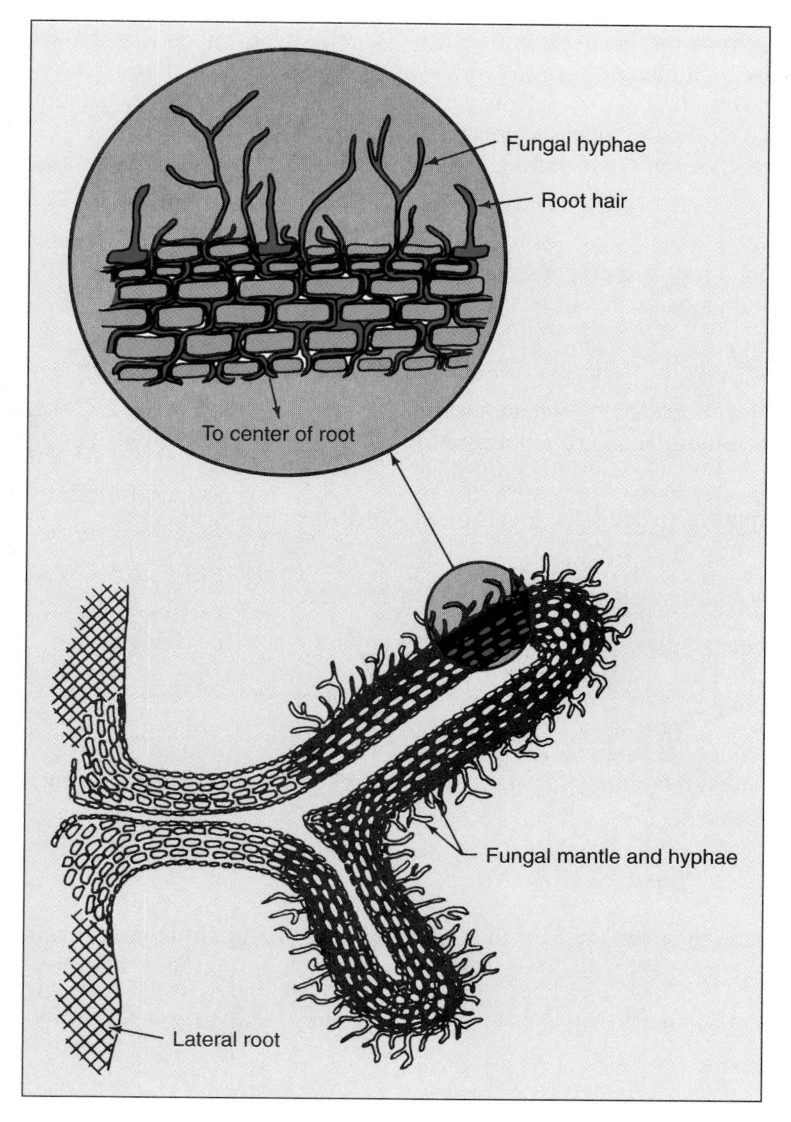

FIGURE 6-4 Ectomycorrhizae surrounding a root and penetrating around cells (replacing the middle lamella between cells). Root exudates stimulate the fungi. The fungus mat helps in nutrient absorption and water uptake; it also buffers the root against harsh soil environment. (Courtesy of Raymond W. Miller, Utah State University.)

United States to Puerto Rico would grow only a few inches, turn yellow, show severe phosphorus deficiency, then usually die. When pine-growing soil from the United States was used to inoculate the Puerto Rican soil, the treated pines grew to 2.4 m in 3 years, whereas the noninoculated pines were only 30 cm tall.

The practice of sterilizing greenhouse-nursery soil with methyl bromide gas led to plant zinc deficiencies, partially due to the unwitting destruction of zinc-absorbing mycorrhizae. Inoculated peach tree seedlings had higher levels of zinc than those that had only been fertilized with zinc chelate. Fumigated nursery soil produced severely stunted citrus trees due to low phosphorus absorption resulting from reduced mycorrhizae. The addition of several hundred kilograms of phosphorus per hectare corrected that problem.

Studies on *Glomus* (VAM) have added the fungi as an aid to increase phosphorus uptake from the soil.[6]

The greatest growth responses to mycorrhizal fungi are probably to plants in highly weathered tropical soils that are low in basic cations, are acidic, are low in phosphorus, and may have toxic levels of aluminum. Plants that have coarse or limited root systems should benefit the most. The mycorrhizal association improves on these soil conditions and helps protect such root systems from the hostile environment. Benefits from mycorrhizae have been shown for many crops, including pasture and forage legumes, corn, wheat, barley, tomatoes, many vegetables, herbaceous and tree fruits, onions, citrus, grapevines, sweetgum, pineapple, coffee, tea, cocoa, oil palm, papaya, cassava, and rubber trees.

The fungi–plant root symbiotic association, called *mycorrhizae,* is summarized in the following statements:

- Mycorrhizae, or fungus-root associations, are expected on most vascular plants and are probably present on most plants to some extent.
- The major aid from the fungi seems to be in increasing phosphorus uptake from low-phosphorus soils. This effect may be caused by several actions: (a) exploration of a larger volume of soil is increased by fungal hyphae, (b) organic acids exuded by fungi increase phosphorus solubility, and (c) altering of the concentrations of cations in solution by fungi, thereby increasing the phosphate solubility.
- An increase of phosphorus to adequate levels by mycorrhizae may be necessary before N_2-fixing bacteria in low-phosphorus soils can nodulate the roots.
- Mycorrhizal associations also aid zinc and copper absorption.
- Protection of roots against plant pathogens may be a common benefit. Colonization of roots by mycorrhizae has reduced root pathogen damage by *Fusarium* on tomato and the stunting of cotton.
- Mycorrhizal fungi reduce stress due to drought. They probably are protective to roots against many other stress conditions (drought, adverse pH, pathogens, low nutrition, and adverse temperatures).

6:6 Protista

Protozoa and slime molds are the main phyla of the **Protista** kingdom. A **protozoan** is a unicellular organism without a true cell wall. Protozoa ingest bacteria, fungi, other microbes,

[6] J. M. Barea, "Vesicular-Arbuscular Mycorrhizae as Modifiers of Soil Fertility," *Advances in Soil Science* **15** (1991), 1–40.

nematode larvae, and eggs, and even smaller protozoa.[7] The protozoan classes are named for their particular methods of movement in the soil water: *amoeboid* (moving by pseudopodia), *flagellar* (moving by whiplike units), and *ciliate* (moving by hairlike cilia that wave). The protozoans are numerous in soil; they help to control other microbes but also cause critical diseases (e.g., malaria). Protozoan digestion of bacteria and fungi influences microbial populations and hastens the recycling of plant nutrients.

Although protozoa cause few serious plant diseases, they are the cause of many animal problems. Some of these are *sleeping sickness, Clogas's disease, severe diarrhea, amoebic dysentery, Texas cattle fever,* and *malaria.*

6:7 Monera: Soil Bacteria and Actinomycetes

The soil microorganisms **bacteria** and **actinomycetes** belong to the **Monera** kingdom. Both decompose organic matter, although actinomycetes are not as effective as bacteria or fungi. Actinomycetes are the source of numerous beneficial antibiotics, but some produce musty tastes and odors in municipal water systems. Bacteria are arguably the most vital link in many of the Earth's ecological cycles.

Soil Bacteria

Bacteria are unicellular microorganisms; the name comes from the Greek word for *rod,* designating the usual bacterial shape. The number of bacteria in the soil usually exceeds that of all other microorganisms, although fungi may exceed bacteria in weight.

One gram of soil (about 0.2 teaspoon) can contain 100 million bacterial cells, 1 million actinomycetes, and 5 m (16 ft) of fungal mycelia. All of this is less than 0.05 percent of the dry-soil weight. Typical bacteria are a few micrometers long; one cm (0.4 in) of distance could have 10,000 to 100,000 bacteria end to end. As small as bacteria are, most clay particles are *much smaller.*

Some bacteria populations double in number in as little as 30 minutes, especially when the soil has an abundance of organic residues. However, most bacteria require several hours or days to double in the natural environment.

Bacteria are classified by nutritional patterns, oxygen needs, and symbiotic relationships (Table 6-2). **Autotrophic** (self-nutritive) bacteria manufacture their food by the synthesis of inorganic materials, such as plants do in photosynthesis. **Heterotrophic** (other, or different, nutrition) bacteria derive their food (carbon) and energy directly from organic substances. Fungi, protozoa, animals, and most bacteria are heterotrophs. Autotrophic bacteria are further defined as **photo-** or **chemo-** to designate their energy source.

Autotrophic Bacteria

Autotrophic bacteria obtain their nutritive carbon from carbon dioxide and, for energy, specific groups can oxidize ammonium, nitrates, sulfides, sulfur, ferrous iron, manganous ions, hydrogen gas, and carbon monoxide. The oxidation transforms nitrites, sulfides, and carbon monoxide into useful nitrates, sulfates, and carbon dioxide. Other oxidations eliminate toxic forms of carbon and manganese. These oxidation and reduction reactions can also be detrimental, causing rust, metal oxidation, and loss of nitrogen (denitrification).

Probably the most important groups of autotrophic soil bacteria are those that oxidize ammonium to *nitrites* (a toxic, transitory form of nitrogen) and then to *nitrates.* These *nitrifying* organisms achieve maximum growth under the following conditions:

[7] K. M. Old and J. F. Derbyshire, "Soil Fungi as Food for Giant Amoebae," *Soil Biology and Biochemistry* **10** (1978), 93–100.

Table 6-2 Subdivisions of Bacteria on the Basis of Their Reactions, Energy Requirements, and Other Properties

Divisions based on method of obtaining nutrition and energy:

Photoautotrophs Energy from sunlight; nutritive carbon from carbon dioxide
Photoheterotrophs Energy from sunlight; carbon from organic matter
Chemoautotrophs Energy from oxidation of inorganic substances such as nitrogen, iron, or sulfur; carbon from carbon dioxide
Chemoheterotrophs Energy and nutritive carbon from organic matter

Divisions based on oxygen requirement:

Aerobic Require free gaseous oxygen source; includes most active decomposers; most populous bacteria
Anaerobic Can use electron acceptors other than oxygen such as NO_3^- or SO_4^{2-}; do not require free oxygen
Facultative anaerobes Can be either aerobic or anaerobic in presence or absence of O_2

Dinitrogen fixers divided on presence or absence of symbiotic relationships:

Symbiotic N_2 fixers Associated with a host plant; both host and bacteria benefit; fixes N_2 from atmosphere
Nonsymbiotic N_2 fixers Exist as free bacteria without a host but fix N_2

- An abundance of proteins to release ammonium
- Adequate aeration
- A moist but not overly wet soil
- A large amount of calcium
- Optimum temperature between 20°C and 40°C (68°F to 104°F)

The bacterial nitrification process is shown diagrammatically as follows:

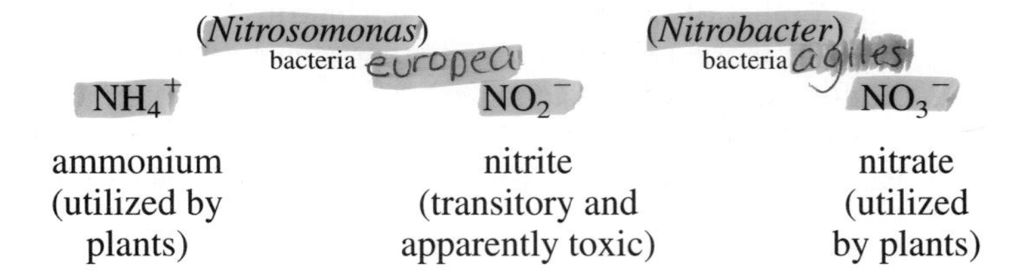

Nitrification is of great concern because of NO_3^- lost from soils by flowing with water into groundwaters. Although high levels of nitrate in waters may be beneficial to plants being irrigated, large concentrations of nitrate in drinking water or foods can cause severe health problems to babies of mammals. **Eutrophication,** an increase in nutrient concentrations in water, is evident by the increased algal growths; these masses of algae, when dead and decomposing, require oxygen that otherwise could be used by aquatic life.

Several groups of autotrophic bacteria can either oxidize or reduce carbon monoxide to carbon dioxide, or reduce it to methane, all of which are gases. These bacteria are literally life giving. The world's population adds 220 million tons (200 million Mg) of carbon monoxide to the atmosphere each year. With no conversion to carbon dioxide or methane, the atmosphere would become lethal to animal life within a few years. This bacterial conversion is anaerobic (without free oxygen).

Heterotrophic Bacteria

Heterotrophic bacteria are those that depend upon organic matter for their nutrition; most soil bacteria are in this group. Heterotrophic bacteria include both nitrogen-fixing and non-nitrogen-fixing groups. The nitrogen-fixing bacteria are further subdivided into *symbiotic* and *nonsymbiotic;* the symbiotic bacteria are commonly associated with plants in the legume family. Heterotrophic bacteria that do not fix nitrogen are the most prevalent soil bacteria and account for much of the decomposition of organic materials.

Symbiotic Bacteria[8,9]

The heterotrophic bacteria that fix atmospheric dinitrogen gas in plant root nodules (**symbiotic bacteria**) have a mutually helpful relationship with their host plants. The plant roots supply essential minerals and newly synthesized substances to the bacteria. Eventually the plants benefit from the atmospheric nitrogen that the bacteria fix as the bacteria either release some of it into the soil, where plant roots can use it, or use some of it to build protein for new bacteria.

The ancient Greeks knew that **legumes** (pod-bearing plants such as peas, beans, alfalfa, and clovers) have a beneficial effect upon both the companion crop and whatever crop was planted next in the same soil. It was not until 1838 that Boussingault, a French chemist, demonstrated that the beneficial effect was due to the fixation of atmospheric nitrogen in the legume **root nodules.** In 1879 Frank proved that artificial inoculation with specific bacteria resulted in legume root nodule formation and that the bacteria in these nodules fixed atmospheric nitrogen.

Symbiotic bacteria begin by infecting root hairs, causing an invagination (enclosing-like sheaths) inward through several cells. Surrounding plant cells proliferate rapidly, perhaps because of auxin hormone produced by the infecting bacteria. As the bacteria enter the nodule cells, they form enclosing membranes and produce methemoglobin and oxygen-carrying pigment (the nodule may be pink in cross-section). The hemoglobinlike material may be an oxygen sink or trap to keep the bacteria in an anaerobic environment, which is necessary for N_2 fixation. Other nonsymbiotic bacteria include both aerobic and anaerobic organisms that are capable of fixing N_2.

The fixation of the inert dinitrogen is accomplished by the enzyme **nitrogenase.** This enzyme lowers the **activation energy** (energy needed to cause the reaction to occur). The fixation progresses in reduction stages from dinitrogen (**$N{\equiv}N$**) through uncertain intermediates **$HN{=}NH$** and **$H_2N{-}NH_2$** to produce **$2NH_3$**. Finally, the amide group (**$-NH_2$**) is transformed into organic compounds such as amino acids. All of this takes place while the nitrogen is bonded to the enzyme(s).

The lifespan of a single bacterium may be only a few hours, and the bodies of a portion of the bacterial population are continuously dying, being decomposed, and releasing ammonium and nitrate ions. This nitrogen can be used by the host plant. In legumes a few entire nodules are sloughed during the season. Most of the nitrogen fixed is excreted by the bacteria; this nitrogen can be made available to the host plant and to other plants growing nearby. The amount of available nitrogen that leguminous crops add to the soil is phenomenal (see Table 6-3). Figure 6-5 shows nodulation on field peas; nodulation is more complete when all plant nutrients except nitrogen are available in sufficient quantities to the leguminous host

---

[8] David R. Benson, Daniel J. Arp, and R. H. Burns, "Cell-Free Nitrogenase and Hydrogenase from Actinorhizal Root Nodules," *Science* **205** (Aug. 1979), 688–689.

[9] John D. Tjepkema and Lawrence J. Winship, "Energy Requirement for Nitrogen Fixation in Actinorhizal and Legume Root Nodules," *Science* **209** (July 11, 1980), 279–280.

Table 6-3 Approximate Quantities of Nitrogen (N_2) Fixed by Selected Legumes, Nonlegumes, and Soil Systems under Ideal Environmental Conditions

Plant/Condition	*Amount of N_2 Fixed Each Growing Season*		
	kg/ha	*Range (kg/ha)*	*lb/a*
Legumes (symbiotic)			
Sesbania*	540	505–581	
Popinac*	450	110–548	
Alfalfa	224	128–600	200
Red clover	129	117–154	115
Kudzu	123		110
Soybean	112	157–200	100
Cowpea	101		90
Peanut	45		40
Bean	45		40
Lupines	—	150–169	—
Nonlegumes (symbiotic)			
Alder, red	168	40–300	150
Buckthorn	67		60
Casuarina	62		55
Sweet gale	9		8
Lichens	—	39–84	—

*Sesbania (*sesbania rostrata*); Popinac (*Leucaena leucocephala*).

FIGURE 6-5 Nodules on root of a field pea plant grown at the Matanuska Research Farm, Alaska. The pea seed was inoculated with a commercial Rhizobium bacterial culture when planted in early June. Photo taken September 13, near end of the growing season. (*Source:* L. J. Klebesadel, "Biological Nitrogen Fixation in Natural and Agricultural Situations in Alaska," *Agroborealis* [Jan. 1978], 11. Photo courtesy of researchers at the University of Alaska Agricultural Experiment Station.)

plants. The same bacterial species will not inoculate all legumes (a species that inoculates lupines will not inoculate lespedezas or trefoil); nevertheless, a few bacterial species can react symbiotically with several similar legumes. The best known symbiotic bacteria belong to the genus **Rhizobium.** Symbiotic heterotrophic bacteria specific to the crop to be grown are frequently added, or inoculated, in a dried, powdered form to the crop seed to ensure that nitrogen-fixing organisms are present (Figure 6-6). Applied rates are often about 5 g of inoculum per kg of seed (1 part inoculum to 200 parts of seed). Heavier applications of inocu-

FIGURE 6-6 All legume seed should be inoculated before planting to ensure sufficient bacteria for maximum nodule formation. Left: soybeans that were not inoculated. Right: soybeans that were properly inoculated. (Courtesy of Nitragin Co.)

lum mixed into peat granules trickled into soil as the seed is planted is an alternative technique to encourage nodulation.

Some plants that have been found to have symbiotic relationships with various N_2-fixing bacteria, including blue-green bacteria (**cyanobacteria**), are *Digitaria* (grass species), corn, sorghum, pearl millet, water fern (with blue-green bacteria), the tropical herb *Gunnera macrophylla* (with blue-green bacteria), Douglas fir (the bacteria are found on the needles), and white fir. In flooded Philippine rice fields, *Anabaena spiroides* and *Aulosira fertilissima* help to maintain the nitrogen level of the soil by utilizing atmospheric nitrogen.

In the early 1970s two Brazilian scientists reported a symbiotic bacterium associated in N_2 fixation with several tropical grasses and corn. Since then bacteria of the genus *Klebsiella* have been shown to be associated in N_2 fixation with numerous grasses, winter wheat, and Kentucky bluegrass.[10] However, the hope that common crops might be enabled to fix N_2 or to use less nitrogen—and thus reduce the need for extensive nitrogen fertilizers—seems still to be many decades away.[11]

Free-Living N_2-Fixing Heterotrophic Bacteria

Unlike symbiotic bacteria, the **free-living** nitrogen-fixing organisms do not need a specific host plant.

The anaerobic *Clostridium* are usually more abundant in soils than the aerobic *Azotobacter.* In aerobic conditions in tropical soils, the acid-tolerant N_2-fixer *A. beijerinckia* is most abundant. *Clostridium* develop best in poorly drained, acid soils; *Azotobacter* are more abundant in well-drained, neutral soils. The amounts of atmospheric nitrogen fixed are

[10] L. V. Wood, R. V. Klucas, and R. C. Shearman, "Nitrogen Fixation (Acetylene Reduction) by *Klebsiella pneumoniae* in Association with Park Kentucky Bluegrass (*Poa pratensis* L.)," *Canadian Journal of Microbiology* **27** (1981), 52–56.

[11] A. Quispel, "A Critical Evaluation of the Prospectus for Nitrogen Fixation with Non-legumes," *Plant Soil* **137** (1991), 1–11.

variable, but under ideal conditions the total nitrogen fixed varies from 1 kg/ha to 15 kg/ha (about 1 lb/a to 14 lb/a) per year.

Bacterial Diseases

Bacteria cause numerous plant diseases, although not as many as do fungi. Some of these are *wildfire* of tobacco; *blight* of soybeans, rice, and peas; *moko* of bananas; *wilt* of carnations, corn, and tomatoes; *slippery skin* of onions; *gumming* of sugar cane; *soft rot* of fruit; *leafy gall* of ornamentals; *pox* of sweet potatoes; *cane gal* of raspberries; aster *yellows;* and corn *stunt.* Human diseases are also of serious concern and include bacterial food poisoning (such as *salmonella*), *typhoid fever, Rocky Mountain spotted fever* from ticks, *parrot fever, rheumatic fever, cholera,* and *dental caries.*

6:8 Actinomycetes

Actinomycetes are taxonomically and morphologically related to both fungi and bacteria but are usually classified with bacteria. They are characterized by branched mycelia, similar to fungi, and resemble bacteria when the mycelia break into short fragments. Actinomycetes aid in decomposition of organic matter, especially cellulose and other resistant organic molecules. Like fungi, actinomycetes aid in the development of water-stable soil structure by secreting nonwater-soluble gummy substances. In the 1950s and 1960s actinomycetes attracted worldwide attention after the discovery that they produce useful antibiotics. Nearly 500 antibiotics have been isolated from actinomycetes, the most common of which are Streptomycin, Aureomycin, Terramycin, and Neomycin.

In the late 1970s actinomycetes were found to form symbiotic N_2-fixing relationships with a diverse selection of plants. Definite actinomycete-plant symbiosis has been established for many plants,[12] including boysenberry, soapberry, alder, bayberry, sweet fern, sweet gale, New Jersey tea, coffeeberry, buffaloberry, bitterbrush, mountain mahoganies, Australian pine, European autumn, Russian olive, and flooded rice. The great interest in symbiotic actinomycetes is partly due to the fact that actinomycetes infect at least seven different botanical families, whereas *Rhizobia* bacteria infect only the family *Leguminosae* (with a few exceptions).[13] Fixation in red alder (a tree) as much as 168 kg of nitrogen per hectare (150 lb/a) per year has been measured.

6:9 Soil Viruses and Viroids

Viruses are unique substances, many of which cause serious plant and animal diseases (*potato leaf roll, tobacco mosaic, raspberry ringspot, cacao mosaic, foot-and-mouth disease,* and *bovine leukemia*). The word *virus* was used by ancient Romans to mean *poison venom and/or secretion.* In early medical use it meant *any microscopic etiologic agent of disease.* **Viruses** are nonliving nucleic acids often surrounded by a protein coat. Most viruses are ribonucleic acid (RNA) material—genetically reproducible, nucleoprotein substances—and have a protective coat of a different protein that must be shed before the virus can be replicated. Currently some scientists list viruses as *acellular microorganisms,* meaning they are *without cells and cannot carry out physiological functions on their own;* yet they store genetic information that, inside a living organism cell, can be replicated. Viruses are capable of only a few life functions.

---

[12] D. L. Hensley and P. L. Carpenter, "The Effects of Temperature on N_2 Fixation (C_2H_2 Reduction) by Nodules of Legume and Actinomycete-Nodulated Woody Species," *Botanical Gazette* **140** (supplement) (1979), S58–S64.

[13] Peter Del Tredici, "Legumes Aren't the Only Nitrogen-Fixers," *Horticulture* (Mar. 1980), 30–33.

Viral-like substances have been classified into the following three kinds of materials that carry genetic material that can be replicated:[14]

- **Prion:** Various nucleic acids without a protective coat
- **Viroid:** No protective coat around RNA (naked RNA)
- **Virus:** Has a protective protein coat around its RNA or DNA; sometimes even a lipid-containing envelope surrounds the nucleic acid

Viruses and viroids cause many plant diseases in addition to those mentioned earlier. Some of these are *carnation latent virus, beet yellows virus, citrus tristerza virus, mosaics of cucumber, turnip yellow, tomato bushy stunt, rice dwarf virus, tomato spotted wilt, lettuce necrotic yellows,* and *wound tumors.* The spread of viruses can be controlled to some extent by removing the host carriers, some of which are nematodes, fungi, and roots of plants.

Enzymes in the soil act as catalysts and decompose viruses. Few viruses overwinter in soil but are commonly hosted and disseminated by nematodes or fungi; otherwise, common plant viruses in soils may survive only about 1 to 4 weeks. There is almost no chemical control in the field for virus infections. Denaturation (decomposition, high salt, high temperature, etc.) inactivates the virus but kills the infected organism as well.

6:10 Optimum Conditions for Microbial Activity

Microbes are in constant competition for organically combined carbon and other nutrients. Their ability to procure these growth materials depends on temperature, moisture, soil acidity, soil nutrient levels, suitable energy source, and the competition by other microbes, among other factors. Competition is intense. Generally, microbes are most numerous in soils with moisture near field capacity, with near neutral pH, with a high nutrient content, and at temperatures near 30°C (86°F).

Optimum Soil Water and pH for Soil Microorganisms

The water content near or just greater than the field capacity (wet, but adequately aerated) is near optimum for most microorganisms. Dryness kills many microbes, and many others tolerate dryness by developing resistant strains or entering into a dormant stage. A few anaerobes, because they are hindered by free-oxygen gas, grow best at saturation conditions.

The optimum soil pH is near 7. This is the pH of microbial cytoplasm (the cell material). Bacteria and actinomycetes are usually less tolerant of acidic soil conditions than are fungi, and few grow well at soil pH below 5. One exception is the sulfur-oxidizing genus **Thiobacillus,** which produces sulfuric acid; it tolerates soil pH even down to 0.6. Many fungi survive in forested and organic soils with pH as low as 3.0. Localized microenvironments near roots or decomposing residues can produce locales of lower pH than that of the soil as a whole, differences of as much as one or two pH units.

Optimum Temperatures and Other Conditions for Microorganisms

Microbial activity accelerates rapidly as temperature rises. Biological (enzymatic) reaction rates nearly double as temperature increases from 10°C to 20°C (50°F to 68°F). Plants and most microbes are nearly dormant at freezing (Figure 6-7). There are exceptions: Some microbes are

[14] R. M. Atlas, *Microbiology: Fundamentals and Applications,* 2nd ed., Macmillan, New York, 1988, 807.

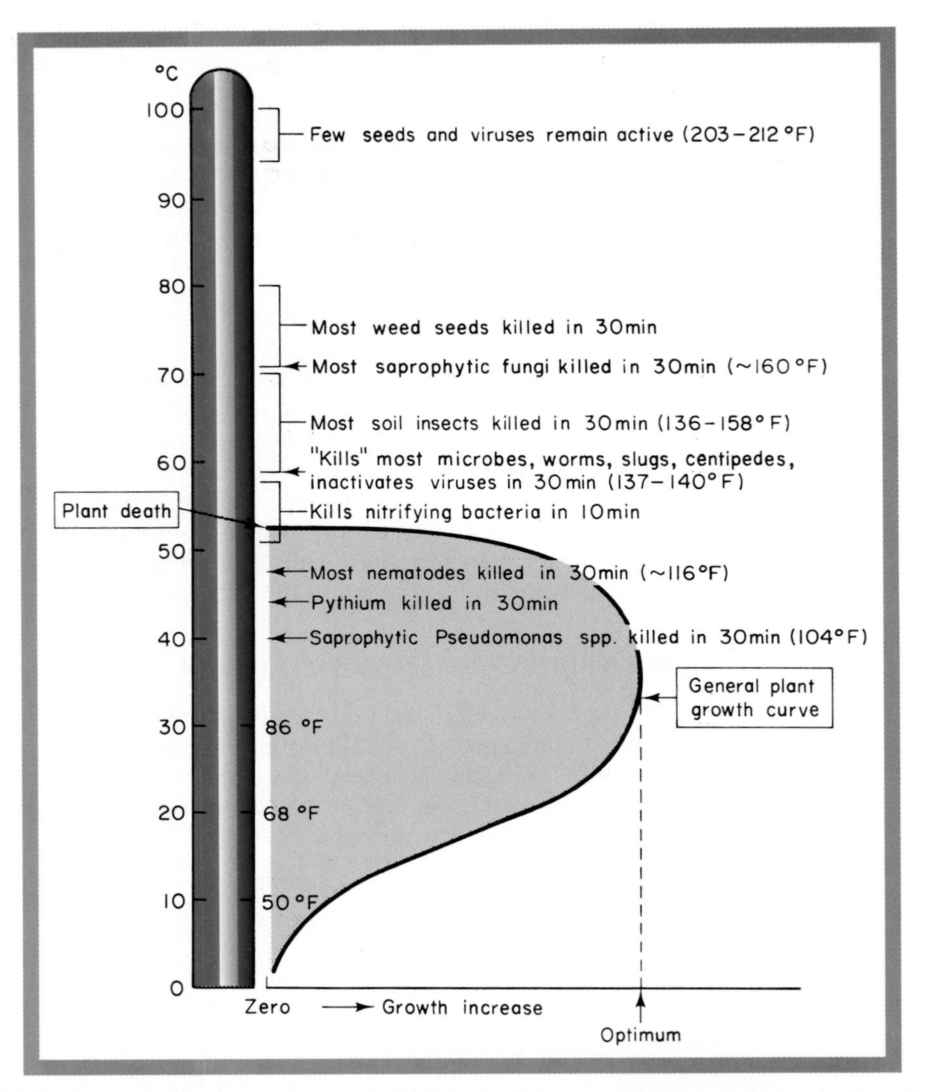

FIGURE 6-7 Approximate temperatures at which many living organisms and viruses in the soil are killed or inactivated. The plant growth curve is only approximate for many common crops, typically corn. (*Source:* Drawn by Raymond W. Miller from data of K. F. Baker and R. J. Cook, *Biological Control of Plant Pathogens,* W. H. Freeman, San Francisco, 1974.)

are able to tolerate very cold temperatures (psychrophiles, or cold lovers), and others tolerate relatively high temperatures (thermophiles, or heat lovers). The majority of soil bacteria and actinomycetes, however, have optimum activity temperatures similar to those of the mesophiles (middle group). The temperature tolerances of these three general groups are as follows:

- **Psychrophiles:** Can grow at temperatures below 5°C (41°F), but have optimum temperatures near 15°C to 20°C (59°F to 68°F)
- **Mesophiles:** Grow slightly near 0°C (32°F) and show little if any growth above 40°C (104°F); many die at this higher temperature; optimum temperature usually between 25°C and 37°C (77°F to 99°F)

- **Thermophiles:** Can tolerate 45°C to 75°C (113°F to 167°F), with optima between 55°C and 65°C (131°F to 149°F), such as some composting microbes

Microorganisms have high nutritive needs, especially for nitrogen, phosphorus, sulfur, and calcium. The carbon source (from organic substances) is most easily attacked when it is plentiful and from succulent young plants.

Organisms compete for nutrients with other species and even attack each other. Protozoa consume other microbes, particularly bacteria; one amoebic cell may devour several thousand bacteria per cell division (once or twice a day). Some microbes produce **antibodies,** substances that retard or kill other nearby organisms and acquire more nutrients by eliminating the competition. Even viruses can be detrimental to microbes. **Bacteriophages** are viruses that parasitize bacteria and cause their death and partial breakdown. Optimum growth conditions for microbes imply not only adequate moisture, temperature, pH, and nutrients, but also an absence or low level of microbial and viral enemies.

6:11 Influencing Microorganisms

People have not been very satisfactory in attempts to regulate the kinds and amounts of microorganisms in soils. The rate of change can be very rapid (hours) for many organisms, both in numbers and in interaction effects on other microorganisms. Some influence is possible, but don't expect to find easy microbial controls.

Encouraging Beneficial Organisms

The best approach to encouraging beneficial organisms is to maintain optimum conditions for them. Fortunately, most conditions good for plant growth also favor growth of beneficial microorganisms. Unfortunately, it does *not* follow that good conditions for plants and beneficial microorganisms are poor conditions for harmful ones. It can be as important to make conditions unfavorable for harmful microorganisms as it is to promote favorable conditions for beneficial ones; the beneficial ones may become expendable in the more desperate need to eradicate the harmful. Ways to encourage an active, and usually good, microbe population include the following actions:

1. Inoculate the soil with the desired symbiotic organism (e.g., *Rhizobia* and fungi for mycorrhizae) in a soil where the host plant has not grown for many years or has never grown. Some *Rhizobia* can survive 10 years in soil without a host plant, but inoculation is so simple and inexpensive that taking a chance on the microbe's presence is not worth the risk.
2. Lime the soil, for most crops, to values above pH 6. Do not apply too much lime.
3. Minimize soil fumigation or sterilization, which kills both harmful and beneficial biota. Many greenhouse mixtures must be sterilized because the harmful microbes are too hazardous in the optimum conditions in greenhouses.
4. Maintain as large a soil organic-matter level as is practical.
5. Try to avoid all contamination. Do not carry contaminated soil from infected fields to clean ones on uncleaned equipment. Burn or remove infested plants; do not discard by burial in the field or on compost piles.
6. Avoid causing stress conditions, such as drought, salt accumulation, waterlogging, or excess fertilizer additions.

Controlling Harmful Microorganisms

Controlling harmful soil microorganisms is more difficult than controlling above-ground pests because of the problem of distributing biocides (killing chemicals) within soils. Gases heavier than air (e.g., methyl bromide) can be injected into preplant soil with some success, but such chemicals are not generally usable on turf grasses because of phytotoxicity to the grass. Soluble solid biocide materials can be applied either to the soil and moved by irrigation or applied in solution to most soil areas to depths of 15 cm (6 in) or more.

Unfortunately, few chemical treatments are specific. Fungicides kill most fungi and nematocides kill most nematodes, the good along with the harmful. The circumstance should be desperate that calls for their heavy use. Chemical pesticides are about the only extensive control available, but caution and restraint in their use are essential.

Other methods of control are based on introducing natural enemies (viruses, insects, microbes) or in changing the soil conditions that favor the harmful growth. Keeping the pH below about 5.2 controls potato scab actinomycetes. Rots (fungi) in sweet potatoes can be ameliorated by careful maintenance of a moderately to strongly acid soil. Many wilts and rots are favored by hot, wet, limed soils. Reducing surface litter (mulch) or allowing adequate interirrigation drying may be enough to reduce the damage to acceptable levels. For effective control, knowledge of the conditions favoring each problem is needed.

Because different predicaments are caused by different organisms, it is difficult to generalize. Some sample recommendations for general management are itemized in the following list:

1. Always start with clean, disease-free plants and use varieties resistant to diseases known to be a problem in your area.
2. Maintain careful sanitation practices. Diseases from fungi, bacteria, and viruses are easily spread on digging or pruning tools, clothing that contacts plants, mud on shoes, chewing and sucking insects, and wind. Soil sterilization between pot or greenhouse bench crops may be essential. Control of insect vectors (carriers) is important.
3. Minimize mechanical damage to plants on stems or leaf tissue. These damaged areas become entry points for the omnipresent disease spores and reproductive bodies of disease organisms.
4. Carefully control water. Most fungi and bacteria require high moisture conditions. Excessively frequent sprinkling that leaves the foliage wet for long time periods favors disease. Often, allowing surface drying between irrigations and exposure to bright sunlight where plants tolerate it will decrease some bacterial and fungal diseases. Good ventilation (for drying) may also help.
5. Control soil acidity. Some wilts are particularly prevalent if soil is basic or has too much lime (calcium carbonate) in it. Controlling pH values near pH 6 are preferred.
6. Control infestations immediately.

6:12 Composition of Organic Matter

The number of organic substances is immense, and they are as variable in composition as they are numerous; yet all organic matter is composed of about 45 percent to 50 percent carbon with lesser amounts of oxygen and hydrogen plus small quantities of nitrogen, phosphorus, sulfur, and many other elements. Carbon atoms joined together into carbon chains of many lengths and linkages with **H**s attached are the basic skeleton of organic compounds. The relatively few remaining elements fill out the skeletons to make different groups of organic-matter substances called *proteins, lignins, carbohydrates, oils, fats, waxes,* and many other materials. **Humic substances** are the colloidal, amorphous, polymeric, dark-brown materials

Detail 6-4 Soil Humus

The residue left after extensive decomposition of organic materials in soils is called **humus.** It is extremely variable in composition and is quite resistant to further microbial decomposition. In most formulas written to try to depict humus, the major ingredients are (1) many active chemical functional groups exposed to the surrounding solution for reaction with other substances in the solution and (2) a very large cross-linked and folded molecule with molecular weights in the hundreds of thousands of grams per gram molecular weight. An example of one structure is shown in Figure 5-7.

As examples of the resistance of humus to decay, some of the carbon atoms stay for hundreds of years in the soil. Radioactive carbon dating values are the average age of all carbon atoms in the material (fresh material = zero age). Because some of the soil carbon is plant residues recently deposited, some of the humus carbon must be much older than the average age. Some reported values[a] are as follows:

Material	*Average Age of Carbon in the Organic Matter*
Frozen peat, Barrow, Alaska, 2.1 m (6.9 ft) deep	25,300 yr ± 9%
Arctic Brown soil humus	3000 yr ± 4.3%
Surface soils, Iowa and North Dakota	210 yr–440 yr ± 120 yr
Cheyenne grassland soil, 0 cm–15 cm (0 in–6 in) deep	1175 yr ± 100 yr
Manured plum orchard, Cheyenne soil	880 yr ± 75 yr
Bridgeport soil, Wyoming cultivated land	3280 yr
Carbon of substances adsorbed to clay, virgin soil	6690 yr
Humus in continuous wheat plots	1895 yr
Farmed slope (25%–40%), 31 cm–36 cm deep, Peru[b]	610 yr ± 60 yr
Farmed terrace, 112 cm–123 cm deep, Peru[b]	1610 yr ± 70 yr
Abandoned terrace, 110 cm–150 cm deep, Peru[b]	3520 yr ± 80 yr

The mechanisms hindering decomposition of soil carbon to CO_2 are many. Some of them are (1) adsorption of humus to clays, which hinders access of enzyme sites to the molecule's bonds, (2) lignin content, which reacts with proteins forming complex, hard-to-attack substances, (3) tannin–protein complexes, (4) insolubility of phenolic–amino acid substances, and (5) carbohydrate–amino acid derivatives of low accessibility to microorganisms and enzymes.

Source: Raymond W. Miller, Utah State University.

[a]F. E. Allison, *Soil Organic Matter and Its Role in Crop Production,* Elsevier Science Publishing, New York, 1973, 157–158.

[b]J. A. Sandor and N. S. Eash, "Ancient Agricultural Soils in the Andes of Southern Peru," *Soil Science Society of America Journal* **59** (1995), 170–179.

called humus (Detail 6-4). Soil **humus,** the complex array of substances left after extensive chemical and biological breakdown of fresh plant and animal residues, makes up 60 percent to 70 percent of the total organic carbon in soil. Because of its complexity humus is often divided by solubility separations into *fulvic acid, humic acid,* and *humin.* Both **fulvic acid** and **humic acid** are soluble in dilute sodium hydroxide solutions, but humic acid is larger and will precipitate out (be insoluble) when the solution is made acidic. **Humin** is the portion of humus that is insoluble in dilute sodium hydroxide.

The nature of soil humus is extremely complex. In addition to humin, humic acid, and fulvic acid, some of the other specific substances comprising soil humus are sugar amines, nucleic acids, phospholipids, vitamins, sulfolipids, and **polysaccharides**—the chains of sugar molecules that help to cement soil aggregates together. All of these substances are large complex molecules or portions of intricate mixtures. They are residual materials from plant tissues and substances synthesized or degraded as microbes decompose organic materials.

6:13 Decomposition of Organic Matter

When organic substances are manufactured by plants, the process of photosynthesis stores energy from the Sun in the plant's chemical bonds. When the substances are decomposed, the stored energy is again released as the bonds are broken. However, the decomposition process has an energy barrier, called the **activation energy,** which must be overcome. When wood is burned the activation energy is the elevated heating provided by the starting flame. In nature only a few reactions, such as lightning, are available to provide this heat energy. Most biological processes require a means to reduce this activation energy so that the process can occur in natural conditions; enzymes, which lower the activation energy between chemical bonds, provide this means.

Enzymes and the Biological Reaction

An **enzyme** is a substance that is able to lower the activation energy enough to allow the breaking or formation of a particular bond in a given natural environment (Detail 6-5). Such enzyme-influenced reactions are called **biological reactions.** The enzyme makes splitting the bond easier, but the process does not consume or destroy the enzyme. When one reaction is completed, the portions of the severed molecule diffuse into the solution and the enzyme is the same as it began and can be involved to split another similar bond. An activator, or any substance that is not consumed or changed in a process, is termed a **catalyst.** Enzymes are catalysts that aid decomposition (or building) of organic materials.

There is a different enzyme for breaking each kind of bond. Each enzyme is given a name descriptive of the particular reaction it does, plus the ending **-ase.** Table 6-4 lists a number of common enzymes and their particular reactions.

Enzymes are produced by plants, animals, and microorganisms, and some are functional even when outside the living cell. Many different enzymes are produced by a single organism, and many organisms produce the same enzymes.

Free enzymes in the soil have several fates. First, they may function for a while before they themselves are decomposed by other enzymes and undergo further chemical breakdown. Second, they may become **denatured** (inactivated permanently). Denaturation can occur because of substances or reactions that change the protein shape; this may result from a high concentration of soluble salts or high temperature in the environment. Third, enzymes may become inactive because some reaction blocks access of the active parts of the enzyme to the substrate—such as the adsorption of enzymes onto humus colloids or clay particles, by the bonding of enzymes to each other, or by bonding to other chemicals that do not separate off as does the usual compound.

When enzymes have been experimentally added to soils, they have existed free and active only for a short time (hours or days); yet periods of temporary inactivity (e.g., adsorption to humus colloids) can allow enzymes to remain active for years. Measured activity of a selected enzyme (*phosphatase*) in peats that had existed in frozen conditions for at least 9000 years was found to be greater than such activity measured in many fresh soils.[15] Active phos-

[15] T. W. Speir and D. J. Ross, "Soil Phosphatase and Sulphatase," Chapter 6 in *Soil Enzymes,* R. G. Burns, ed., Academic Press, New York, 1978, 209–210.

Detail 6-5 Example of Enzyme Action

The two substrates (materials involved in the process) are (1) a chain of amino acids (polypeptides) to be broken into shorter units and (2) water molecules. The *carboxypeptidase* enzyme in I has a configuration that allows electron bonding (weakens nearby bonds) at sites **A, B, C,** and **D.** In II, as the polypeptide chain attaches to the enzyme, the H from site **B** goes to the N in the polypeptide chain, as shown by the arrow. Other bonds are altered (weakened) as shown by dotted lines. Water supplies the H to regenerate the enzyme site at **B** and supplies OH to form the carboxyl in the polypeptide residue near site **A.** The two portions of the polypeptide chain are now split apart and leave the enzyme; the enzyme is rejuvenated and ready to split another similar bond.

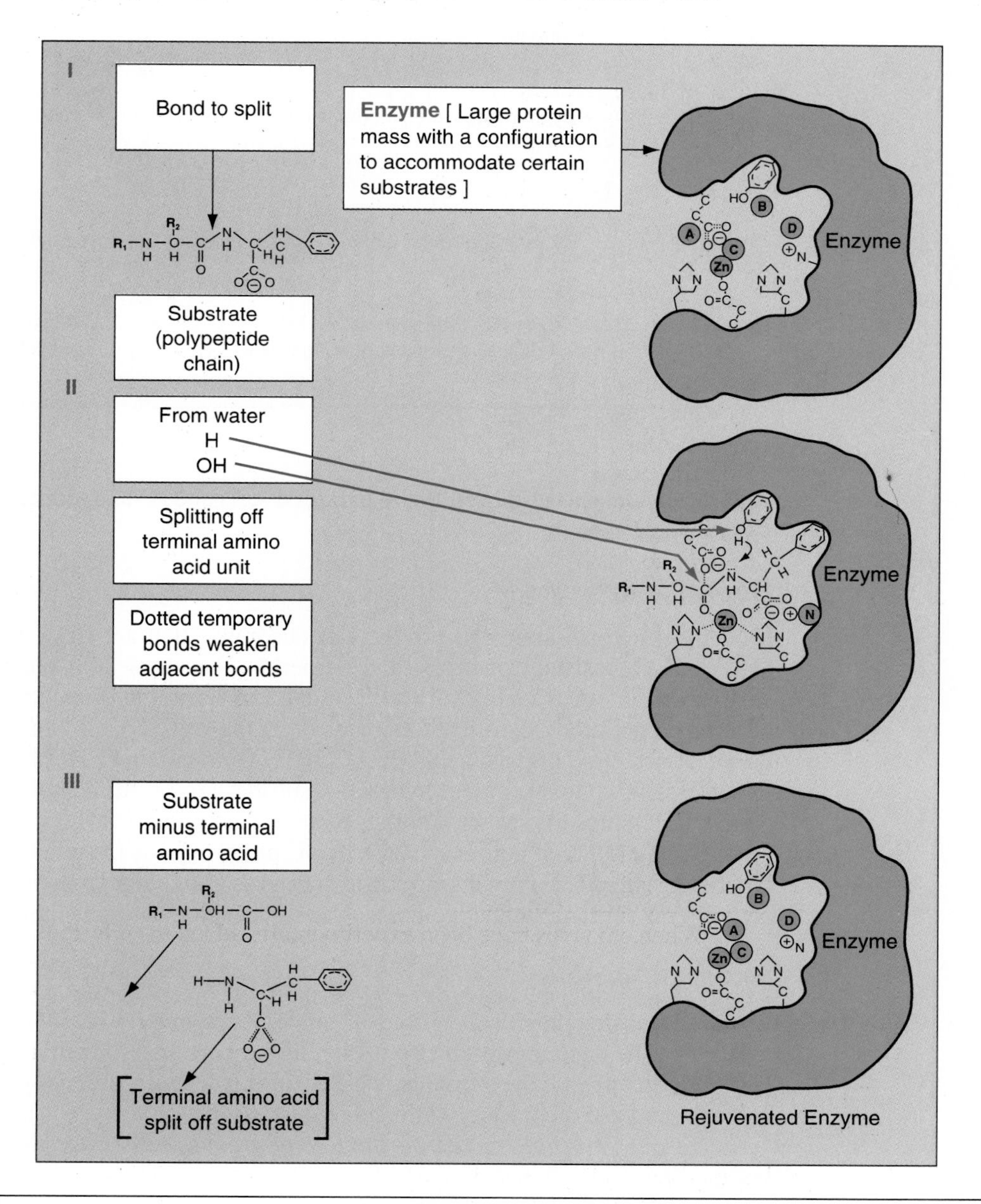

Source: Raymond W. Miller, Utah State University; drawn from details in E. O. Wilson, et al., *Life on Earth,* Sinauer Associates, Stamford, CN, 1973, 122–130.

Table 6-4 Some Common Enzymes and the Reactions They Influence

Enzyme	*What It Does*
Cellulase	Breaks celluloses (cell-wall fibers, wood), which are chains hundreds of sugar units long, into those component sugars. Important in organic-matter decay.
Urease	Breaks down urea (H_4N_2CO) to water, carbon dioxide, and ammonia. Makes animal urine and urea fertilizer nitrogen more available to plants. It requires the presence of Ni^{2+} ion.
Phosphatase	By involving water it breaks the humus—O—P(=O)—$(OH)_2$ bond to produce humus—OH and H_3PO_4, which helps to decompose humus, making phosphorus available to plants.
Sulfatase	By involving water, it breaks the humus—O—S(=O)(=O)—OH bond to produce humus—OH and H_2SO_4.
Protease	By involving water, it breaks the bond linking two amino acids [R_1—(NH)—C(=O)—R_2] to form separate amino acids (R_1—NH_2 and HO—C(=O)—R_2) or parts of proteins. This is a digestive process of living tissues. R_1 and R_2 are distinct portions of organic material.

phatase enzyme has also been found in buried soils and in lake sediments covered over about 13,000 years ago.

Products of Decomposition

In well-aerated soils the end products of decomposition are CO_2, NH_4^+, NO_3^-, $H_2PO_4^-$, SO_4^{2-}, H_2O, resistant residues, and numerous other essential plant nutrient elements in smaller quantities. If soil is not well aerated, less desirable products result. For example, in anaerobic conditions significant amounts of methane (CH_4), also called *swamp gas,* are produced. Also, some organic acids (R—COOH), ammonium (NH_4^+), various amine residues (R—NH_2), the toxic gases hydrogen sulfide (H_2S), dimethyl sulfide, and ethylene ($H_2C{=}CH_2$), plus the resistant humus residues, are produced.

One example of nutrient release by decomposition is the release of nitrogen. Table 6-5 illustrates estimated amounts of nitrogen released annually from soils of different textures and different organic-matter contents. Usually decomposition is more rapid in sandy soils.

Decomposition Action

The decomposing organisms excrete a variety of enzymes to begin breakdown of the organic materials. The decomposition rate is directly proportional to the numbers of microbes present. Microbes absorb the nutrients released during decomposition—particularly nitrogen and carbon—and use them for growth and reproduction.

Nitrogen most often controls the rate of organic matter decomposition; it is needed to build proteins in new bacterial and fungal populations. The nitrogen content in the microorganisms and in organic material is given as the **carbon:nitrogen ratio (C:N ratio).** A wide

Table 6-5 Nitrogen Released from Soil Organic Matter from Three Soil Textural Classes during the Growing Season

Soil Organic Matter (% of total soil weight)	Nitrogen Released (kg/ha)*		
	Sandy Loam	*Silt Loam*	*Clay Loam*
1	50	20	15
2	100	45	40
3	—	68	45
4	—	90	75
5	—	110	90

*Soils in Southern regions release more nitrogen and soils in Northern regions release less nitrogen than shown. When a good legume stand has been turned under, the crop that follows may have an additional 45 kg–56 kg (88 lb–123 lb) of available nitrogen per hectare.

Table 6-6 Approximate Percentages of Organic Carbon and Total Nitrogen and the C:N Ratio of Common Organic Materials Applied to or Growing on Arable Soils

Organic Material	*Organic Carbon (C) (%)*	*Total Nitrogen (N) (%)*	*C:N Ratio*
Crop residues			
Alfalfa (very young)	40	3	13:1
Clovers (mature)	40	2	20:1
Bluegrass	40	1.3	30:1
Cornstalks	40	1	40:1
Straw, small grain	40	0.5	80:1
Sawdust	50	0.1	500:1*
Soil microbes			
Bacteria	50	10	6:1
Actinomycetes	50	8.5	6:1
Fungi	50	5	12:1
Soil humus	50	4.5	12:1

*Some sawdust may reach C:N ratios of nearly 800:1.

organic-carbon:total-nitrogen ratio indicates a material relatively low in nitrogen content. Table 6-6 lists some common organic materials and their carbon and nitrogen contents by weight.

Bacteria, requiring 1 g of nitrogen for each 5 g to 6 g of carbon (C:N ratio of 5:1 or 6:1), are heavy users of nitrogen. If straw with its low nitrogen content (C:N ratio of 80:1) is incorporated into a soil low in nitrogen, bacteria will multiply slowly because the straw is a low nutrient food for the decomposing microorganisms. The process of decay can be hastened by adding more nitrogen (usually from fertilizers) to supply microbial needs. Bacteria (or fungi) will use any available nitrogen in the soil. Plants growing in a nitrogen-deficient soil are deficient in nitrogen because the soil microorganisms, which are more abundant and in more intimate contact with the nitrogen, are able to use most available nitrogen before it can become accessible to plant root surfaces. The same is true for phosphorus and other nutrients.

As decay of organic materials progresses, much of the carbon released escapes into the atmosphere as carbon dioxide (CO_2). This narrows the C:N ratio in the organic matter (which

Detail 6-6 Decomposition Rates of Organic Materials

All plant residues contain materials easy to decompose and difficult to decompose. A mixture of residues from Austrian and Scots pines and English oak (*Pinus nigra, p. sylvestris,* and *Quercus robur*) had the following composition and rates of decomposition.

Original Litter	*Portion of Whole (%)*	***Percentage Lost by Decomposition by:*** *1st Year*	*2nd Year*	*5th Year*	*10th Year*
Sugars	15	99	100	—	—
Cellulose	20	90	100	—	—
Hemicellulose	15	75	92	100	—
Lignins	40	50	74	97	100
Waxes	5	25	43	77	95
Phenols	5	10	20	43	70
Whole litter materials		55.1	79.6	87.1	98.2

Within the first year just over half of the material is decomposed. By the end of the second year 80% is gone, and at the end of 5 years 13% of the original matter is still not lost. When organic residues are mixed in soils, some adsorption to clays slows decomposition rates of some materials (often proteins) even more than is shown above.

includes microbe bodies), because only a little of the nitrogen is lost while large quantities of carbon are expelled into the atmosphere.

Eventually the easily decomposable residues are gone and only materials that are decomposed more slowly remain (Detail 6-6). The food and energy source is now in short supply and more of the bacteria and fungi die. Their bodies, having a high nitrogen content, are decomposed by other living microorganisms, releasing carbon dioxide and liberating some nitrogen to the soil solution. Most of this released nitrogen is available to growing plants.

Dense populations of microorganisms inhabit the upper soil surface and have handy access to the soil nitrogen sources. Plant residues with C:N ratios of 20:1 or narrower have sufficient nitrogen to supply the decomposing microorganisms and also to release nitrogen for plant use. Residues with C:N ratios 20:1 to 30:1 supply sufficient nitrogen for decomposition but not enough to result in much release of nitrogen for plant use the first few weeks after incorporation. Residues with C:N ratios wider than 30:1 decompose slowly because they lack sufficient nitrogen for the microorganisms to use for increasing their numbers, which causes microbes to use nitrogen already available in the soil.

If environmental conditions are favorable, the rate of decomposition of plant residues is most rapid during the first two weeks after incorporation into the soil. This pattern is given diagrammatically in Figure 6-8 and shows the radical differences in the amounts of nitrogen released during decomposition of a narrow C:N ratio material (alfalfa) and a wide ratio material (oat straw).

Whichever conditions favor microorganism growth will favor fast decomposition rates:

- **Temperature:** Cold periods retard plant growth and organic matter decomposition. Continuous *cold* temperatures lower soil humus because little plant material is grown. Continuous *warm* temperatures aid high plant production but also promote faster decomposition.

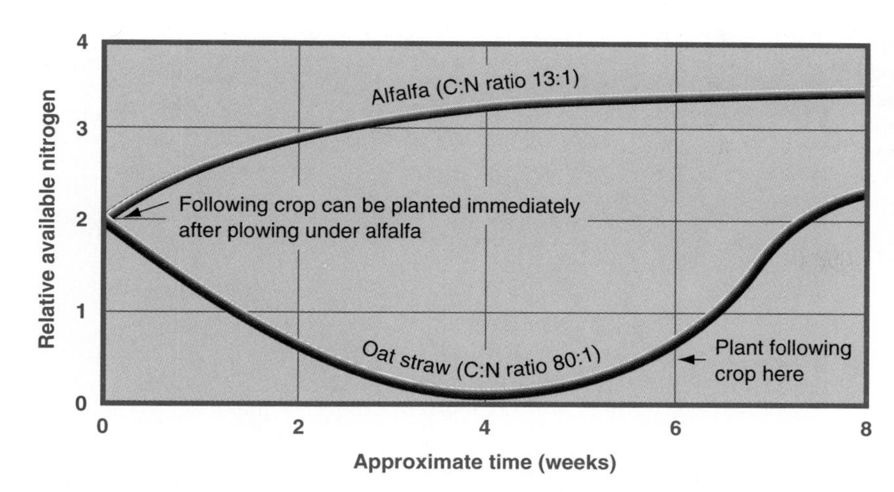

FIGURE 6-8 Relationship between the carbon-to-nitrogen (C:N) ratio of young alfalfa and mature oat straw and the time of planting a following crop after their incorporation in the soil. Notice that if the available soil nitrogen increases, there will not be as much nitrogen deficiency as with the oat straw.

- **Water:** Extremes of both arid (dry) and anaerobic (waterlogged) conditions reduce growth of most plants and microbial decomposition. In general a cool, wet climate results in the greatest amount of soil organic matter. The wetness encourages plant growth while the cool temperature slows decomposition.
- **Soil texture:** Soils higher in clays tend to retain larger amounts of humus, other conditions being equal. Most organic substances adsorb to clay surfaces by many kinds of bonds.

Plants also need **suitable soil pH** and **adequate nutrients.** Other decomposition inhibitors include toxic levels of elements (aluminum, manganese, boron, selenium, chloride), soluble salts, shade, and organic phytotoxins.

6:14 Effects of Soil Organic Matter

The list of positive effects of soil organic matter is so varied and extensive that it makes one think of extravagant claims printed on old-time patent medicine labels. However, soil organic matter can be harmful, too. Is it good to have a lot in the soil? How much is best?

Benefits of Soil Organic Matter

Organic matter provides many benefits. When organic gardeners praise organic matter and condemn mineral fertilizers, one reason is because there are many things provided by organic matter that are not provided by mineral fertilizers. Most of the claims of benefits from organic matter are true (Figure 6-9).

- Organic matter is the source of 90%–95% of the nitrogen in unfertilized soils.
- Organic matter can be the major source of both available phosphorus and available sulfur when soil humus is present in appreciable amounts (about 2% or more).
- Organic matter supplies directly, or indirectly through microbial action, the major soil aggregate-forming cements, particularly the long sugar chains called **polysaccharides.**
- Organic matter contributes to the cation exchange capacity, often furnishing 30%–70% of the total amount. The large available surfaces of humus have many cation exchange sites that adsorb nutrients for eventual plant use and temporarily

FIGURE 6-9 Composted organic wastes are often the only nutrients added to cropped fields. This compost is added onto a commune field near Peking, and it is unloaded in piles, later to be spread uniformly over the fields. (*Source:* International Rice Research Institute, Manila, Philippines. Photo by Randolph Barker.)

adsorb heavy-metal pollutants (lead, cadmium, etc.), which are usually derived from applied wastewaters. Adsorption of pollutants helps clean contaminated water.

- Organic matter commonly increases water content at field capacity, increases available-water content in sandy soils, and increases both air and water flow rates through fine-textured soil. The latter effect is probably due mainly to soil aggregation, which produces larger soil pores.
- Organic matter acts as a chelate. A ligand is any organic compound that can bond to a metal (usually iron, zinc, copper, or manganese) by more than one bond and form a ring or cyclic structure by that bonding. The ligand–metal is called a **chelate** (key′-late). The *soluble* chelates help mobilize micronutrient metal ions, increasing their availability to plants and mobility in soils.
- Organic matter is a carbon supply for many microbes that perform other beneficial functions in soil (e.g., free dinitrogen fixers, denitrifiers).
- When left on top of soil as a mulch, organic matter reduces erosion, shades the soil (which prevents rapid moisture loss), and keeps the soil cooler in very hot weather and warmer in cold weather.
- Most soils above the Arctic Circle and below the Antarctic Circle depend on a thick layer of organic matter to stabilize them. When this layer is destroyed by fire or construction activities, soils may become warmer by 9°C (20°F). In summer this causes melting of permafrost and results in very severe surface and pothole (vertical) soil erosion.
- Humus buffers the soil against a rapid change in acidity, alkalinity, salinity, and degradation by pesticides and toxic heavy metals.
- Humus reduces the crystallization and hardening of plinthite (laterite) in soils in the humid tropics, which are rich in soluble iron and aluminum. Humate complexes with iron and aluminum reduce crystallization. Organic matter also reduces hardening by maintaining more uniform soil temperature and soil moisture.

Mulch tillage leaves organic matter as a mulch. Mulch tillage is used on grain lands where wind or water is likely to cause extensive erosion of exposed soil. A coarse surface mulch increases the percentage of water that seeps into the soil. On one bare Texas soil, a 7 cm rain wetted the soil to a depth of 39 cm; the addition of a 36 Mg/ha (16 t/a) straw mulch increased the water penetration to 75 cm (29 in), even though the straw itself absorbed some water.

Surface organic matter can insulate soil, retarding heat flow between the atmosphere and the soil. In hot summers this benefits some plant roots, but in cool areas it slows soil warming in the spring. In cold areas one should leave small amounts of plant residues on the soil surface to allow the maximum rate of warming in the spring, leaving only enough residue to control soil erosion.

Some rather curious effects may emanate from soil humus. The release of various vitamins, hormones, amino acids, and other substances exhibits unexpected effects on plants. An extract from canola plant pollen (called *brassinolide*) in concentrations of one-billionth of a gram causes cells to elongate and, when applied on the eyes of seed potatoes, has increased potato yields 24 percent.[16]

Triacontanol, an alcohol occurring in alfalfa, is a plant growth regulator. As small an amount as a single milligram (portion of one drop) per hectare stimulates the growth of corn, tomatoes, lettuce, and rice.[17]

The list of growth substances produced by microorganisms in soils is very long; it includes indole acetic acid (IAA), vitamins, gibberellic acid, and numerous unidentified auxins (growth promoters). In large amounts some of these substances can produce abnormal growth in plants. In one study several of these growth substances were effective in helping roots penetrate a compacted subsoil nearly as well as where the subsoil was not compacted.

Organic Matter as a Source of Nutrients

Are soil organic materials good sources of nutrients (fertilizers)? Whether or not organic materials are good as sources of plant nutrients depends on their furnishing (1) the needed nutrients (2) in adequate amounts (3) at the right time (4) in an efficient (low-loss) manner. The major lack from organic matter is the inadequate amount of nutrients supplied during stages of peak plant demand. Decomposition rates increase as the weather warms, which provides maximum plant growth conditions. To some extent nutrients are released in larger amounts as crop needs increase. However, this is generally correct only in temperate areas where crops are planted in the cool spring and reach high nutrient demands in the early warm periods of summer. In the tropics or subtropics, where less distinct cold–warm cycles regulate planting and plant-growth dates, nutrient release rates do not correspond as well with plant nutrient needs. Late-planted crops also lack a good need–supply relationship.

Whether or not adequate nutrient amounts are released depends on the amount and kind of organic matter, soil texture, and crop needs. Because these factors vary, no one soil condition can supply optimum nutrients for the many different crop situations.

Perhaps the major advantages of soil organic matter as a nutrient source are its conservation of nutrients against leaching losses and its continual release of nutrients. Because only small portions of organic matter decompose in a few days, heavy rainfall or irrigation is able to leach only small amounts of solubilized nutrients. Immediately after a leaching action, further organic matter decomposition continues and supplies new quantities of the many nutrients in the humus.

[16]Judy McBride, "Pushing Plants to Full Potential," *Agricultural Research* **28** (1979), No. 2, 14–15.

[17]Stanley K. Ries, "Regulation of Plant Growth with Tricontanol," *Critical Reviews in Plant Science* **2** (1985), No. 3, 239–285.

Organic materials should not be added indiscriminately to soil in large amounts in an effort to supply adequate nitrogen and/or phosphorus. There may be attendant problems that are worse than the nutrient deficiencies. The production of toxins (allelopathic substances, soluble salts from manures, and heavy metals—e.g., cadmium, mercury, zinc, and lead in sewage sludge) are all accumulative in the soil.

Detrimental Effects of Soil Organic Matter

Numerous plants contain or produce phytotoxins (plant toxins). The phytotoxins make such residues undesirable organic matter, at least temporarily. Unfortunately, the problem of phytotoxin production cannot always be avoided; the decomposition of many plant residues produces such toxins. This production of toxins is a form of allelopathy.

The term *allelopathy* was introduced in 1937 by Hans Molisch (1856–1937) in Germany. **Allelopathy** is any *beneficial or harmful* effect of the chemicals produced by one plant on another plant. Most literature until recently reported only harmful effects. Either toxic or beneficial chemicals, called **allelochemicals,** may be in the plant and exuded by roots or leached from the foliage or roots. In other instances the chemicals may be released after death of the plant or produced by hydrolysis or other enzymatic breakdown processes of the plant materials. For example, acetic acid is produced in the cool, wet, anaerobic decomposition of wheat straw and is toxic to barley seedlings planted into the wheat stubble.[18,19] Table 6-7 lists just a few of the observed allelopathic effects of various plants on themselves or other plants.

The allelochemicals now identified include many organic substances broadly categorized as *alkaloids, benzoxazinones, cinnamic acid derivatives, coumarins, cyanogenic compounds, flavonoids, polyacetylenes, quinones, and terpenes.* Root leachates and soil extracts, which contained *homovanillic acid,* suppressed the growth of young citrus trees in old citrus orchards. Leachate from ryegrass (*Lolium perenne*) suppressed growth of many woody ornamental plants.

In 1928 *juglone* (5-hydroxy napthoquinone) was identified as a potent allelochemical from black walnut trees. More recently many quinones, formed by oxidation of various phenolics, are believed to provide some resistance to plant pathogens. Marigold releases four thiophenes and two benzofurans into the rhizosphere to produce the allelochemical activities of this plant.

As an example of the extent of the study of allelochemicals, Putnam states, "Numerous families and species of plants (perhaps as many as 2000) have been found to be cyanogenic."[20] These include only the cyanide-containing substances that liberate HCN (cyanide gas) upon hydrolysis. Some of the plants having these substances are sorghum, johnsongrass, and peach.

A number of weed species appear to impose detrimental influences on other plants. Older farmers have often been convinced that weeds have an unfair advantage; perhaps this is true. Many of these chemical-bearing weeds are some of the aggressive perennials we constantly struggle to control, such as nutsedges, quackgrass, and johnsongrass. About ninety weed species have alleged allelopathic attributes, but evidence for some of them is a bit

---

[18] Joan M. Wallace and L. F. Elliott, "Phytotoxins from Anaerobically Decomposing Wheat Straw," *Soil Biology and Biochemistry* **11** (1979), 325–330.

[19] M. J. Krogmeier and J. M. Bremner, "Effects of Water-Soluble Constituents of Plant Residues on Water Uptake by Seeds," *Agronomy Abstracts* (1986), 181.

[20] Alan R. Putnam, "Phytotoxicity of Plant Residues," Chapter 14, 285–314, in Paul W. Unger (ed.), *Managing Agricultural Residues* (1994), Lewis Publishers, Ann Arbor, MI.

Table 6-7 Examples of Allelopathy*

Plant Producing the Allelochemical and the Damage Done	*Allelochemical*
Datura stremonium (Solonaceae or jimsonweed) produced alkaloids scopolamine and hyoscyamine that are extremely toxic to sunflowers and many cereals. Cyanide can be released during decomposition also.	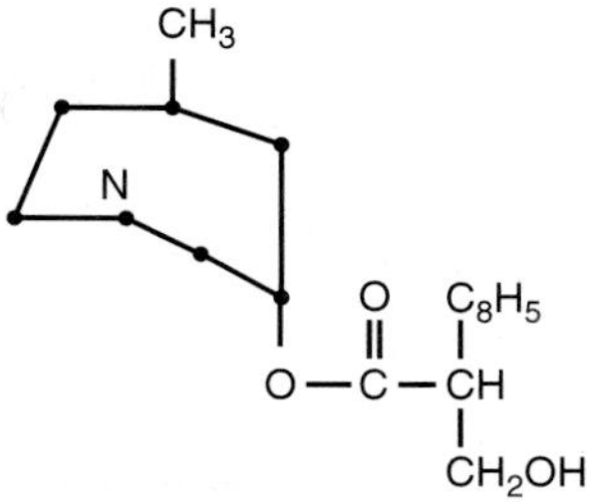 Hyoscyamine
Coffea arabica (coffee) and *Camellia senensis* (tea) produce caffeine and other alkaloids (methylated xanthine) that decompose slowly, allowing a buildup in soils that causes degeneration of old plantations of that same plant.	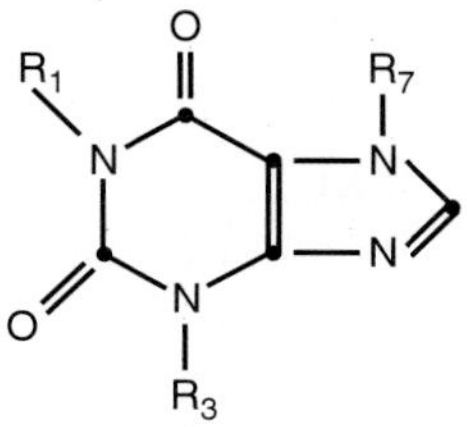 Methylated Xanthine
Many cereals, including rye, corn, and wheat, produce the glycosides that break down into benzoxazinones and then into DIBOA, which has antifungal properties and is toxic to the weed velvetleaf.	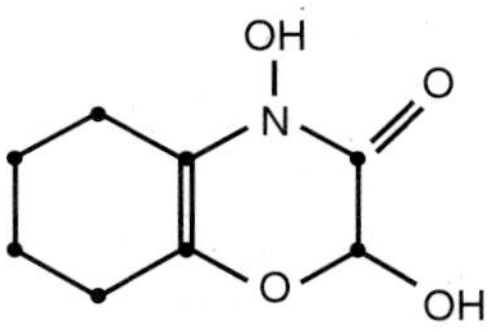 DIBOA
Peach (*Prunus persica*) roots (and unripe fruits and seeds) produce large quantities of amygdalin, which is hydrolyzed to HCN gas and benzaldehyde, both extremely inhibitory to young peach roots.	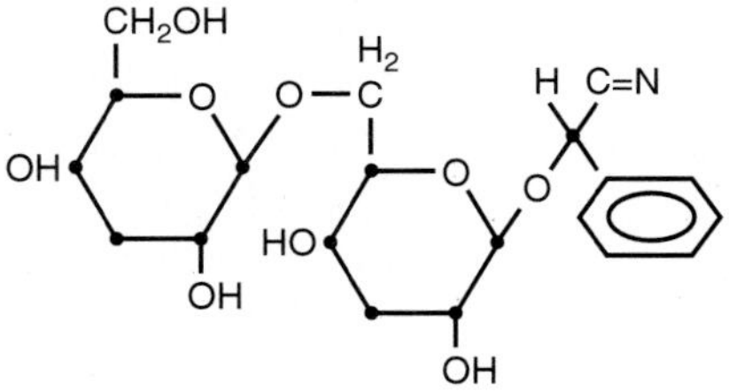 Amygdalin
Agropyron repens (quackgrass) produces flavones (tricin) that inhibit seedling root growth of many plants and inhibit nodulation on several legumes.	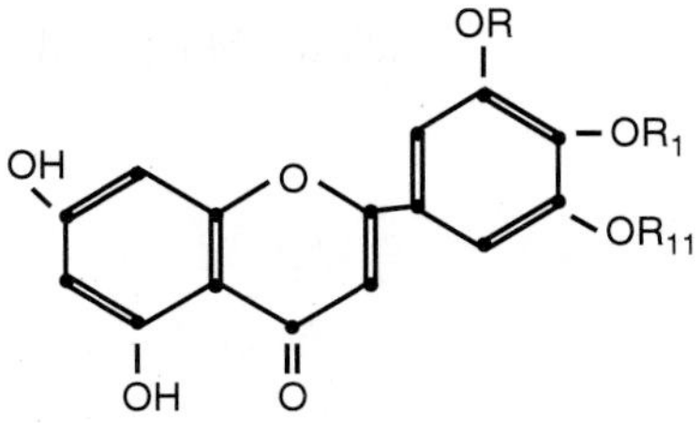 Tricin

Continued.

Table 6-7 Examples of Allelopathy* (*Cont'd*)

Plant Producing the Allelochemical and the Damage Done	*Allelochemical*
Ambrosia artemiscifolia (ragweed) produced many sesquiterpenes (β-bisabolene) that are strongly inhibitory to crop seeds during their germination.	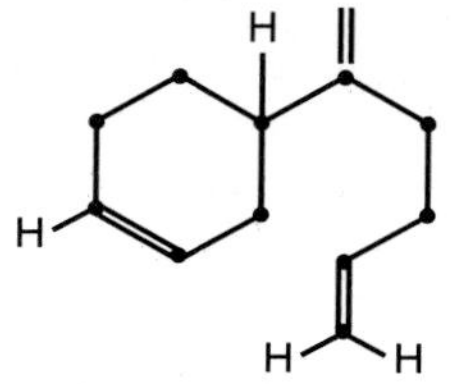 β-bisabolene

*Selected examples from Alan R. Putnam, "Phytotoxicity of Plant Residues," Chapter 14, pp. 285–314, in Paul W. Unger (ed.), "Managing Agricultural Residues," Lewis Publishers, Ann Arbor, MI, 1994.

meager (Detail 6-7). More recently interest and research has begun to study *positive* impacts for employing some of these plant residues effects. Some researchers believe there is considerable promise in using some of these residues for *selective pest control.*

Economical Levels of Soil Organic Matter

How much soil humus should be maintained in soils? Wide ranges in organic-matter contents in soils (Table 6-8) ensue from the effects of climate, plant cover, and management variations that occur in different soils because of management, location, texture, vegetative cover, and soil origin. Uncultivated soils are higher in total soil organic matter (both *on* and *in* soil) than those soils are after years of cultivation. The obvious exceptions include arid soils that supported almost no plants before cultivation but due to irrigation have an increase in humus content, resulting from moisture and plant growth rather than from tillage.

Plants can grow well in mineral soils without any organic matter at all and also in organic soils, if fertilizers are added. Commercial greenhouses and nurseries use mixed potting media having as much as 30 percent to 100 percent peat moss or other organic materials. The optimum level of organic matter that should be maintained in soil has never been determined, nor is it any single value for all soils.

Generally, organic matter in the soil improves plant growth. To a farmer, however, the bottom line for deciding a good level of organic matter to maintain depends on *cost* and *convenience.* It is usually too costly and inconvenient to purposely build up organic matter in soils by means other than incorporating plant residues and disposing of available manures, plant residues, or sewage sludge onto nearby soil. A practical rule of thumb is to *use all available crop residues* by incorporating them into the soil (rather than burning them), apply the minimum fertilizer economically feasible to produce maximum plant size for larger harvests, and add other suitable residues that might be available or need dumping.

6:15 Organic Waste Materials

Many organic materials are available to be added to soil. These are usually beneficial to the soil, which provides a means for useful disposal of wastes. The materials include animal manures, sewage sludges, septage (septic tank wastes), wastewaters, composts, and food-processing wastes.

Detail 6-7 Allelopathy in Action

A weed called *dyer's woad (Isatis tinctoria)* is a spreading nuisance and serious weed problem in several Western states.* Used in Europe for thousands of years to make blue dye, yellow-flowered dyer's woad uses allelopathy to aid its spread. When seed pods fall and rot, they exude a toxin into the soil that kills the roots of nearby grass plants. The only known resolution is for the toxin to be washed away by rainfall. Optimistically, the toxin, when identified, could be synthesized and used as an environmentally safe herbicide. In the meantime, the meter-tall dyer's woad spreads to new areas, competitive over weeds and crops alike because of its allelopathic toxin and its big, leafy rosettes that shade nearby plants.

*J. A. Young, "Dyer's Woad Wages Chemical War in the West," *Agricultural Research* **36** (1988), No. 7, 4.

Table 6-8 Organic-Matter Contents of Surface Soils from a Variety of Locations, under Various Plant Cover, and in Various Climates

Soil and Site Description	*Organic-Matter Content (%)*
Arizona–Nevada soils, 280 mm (11 in) rainfall yearly	
Aridisol (loamy coarse sand)	0.60
Aridisol (fine sandy loam)	1.64
Aridisol (silty clay), pH 8.4	1.19
India, 1000 mm–1500 mm (39 in–58.5 in) rainfall, cultivated	
Acid sulfate paddy soil, pH 6.2	0.75
Alluvial paddy soil, pH 6.3	2.88
Oxisol (clay), pH 5.2	5.52
Hawaii, cultivated soil	
Oxisol (clay) from basalt, pH 5.9, 1100 mm (43 in) rainfall	0.7
Oxisol (clay) from basalt, pH 5.6, 2400 mm (94 in) rainfall	6.0
Inceptisol (loam) from volcanic ash, pH 5.1, 2500 mm (98 in) rainfall	12.0
Iowa, U.S., 800 mm (31 in) precipitation, cultivated	
Entisol (silt loam)	2.59
Mollisol (loam)	2.83
Mollisol (silty clay)	4.9
Nebraska, U.S., 450 mm–800 mm (18 in–31 in) precipitation, cultivated	
Mollisol (silty clay loam) (33% clay), pH 5.2	3.8
Entisol (loamy fine sand) (6% clay), pH 5.9	1.4
Liberia, Africa, 2000+ mm (78+ in) rainfall, virgin soils	
Ultisol (sandy clay loam), pH 4.9	3.45
Ultisol (sandy loam), pH 4.3	2.28
Turkey, 900 mm (35 in) rainfall, cultivated	
Inceptisol (clay) (61% clay), calcareous	1.1
Vertisol (clay) (54% clay), calcareous	0.9
Mollisol (loam) (20% clay), 65% lime	4.3
Entisol (sandy loam) (6% clay), from volcanic ash	0.6

Continued.

Table 6-8 Organic-Matter Contents of Surface Soils from a Variety of Locations, under Various Plant Cover, and in Various Climates (*Cont'd*)

Soil and Site Description	*Organic-Matter Content (%)*
Costa Rica, low elevation, 1500 mm–2000 mm (58.5 in–78 in) rainfall	
Ultisol (clay) forested, pH 5.5, low elevation	3.74
Inceptisol (clay loam) forested, pH 5.2, from volcanic ash	3.97
Inceptisol (silty clay) forested, pH 4.1, from volcanic ash	18.94
Inceptisol (silty clay) forested, pH 6.2, from volcanic ash	10.25
Ultisol (clay) forested, pH 5.1, low elevation	4.34
Santa Barbara, California, 480 mm (19 in) rainfall	
Mollisol (loam), pH 7.3	7.85
Mollisol (very fine sandy loam), pH 7.8	11.3
Michigan–Indiana soils, cultivated, 750 mm–1000 mm (29 in–39 in) rainfall	
Mollisol (sand) 2.9% clay, pH 6.6, imperfectly drained	6.15
Alfisol (sand) 3.4% clay, pH 7.5, well drained	1.81
Mollisol (loam) 19.8% clay, pH 6.3, imperfectly drained	8.24

Animal Manures

How important are animal manures as additives to land areas? Thirty-nine percent of all livestock in the United States are fed in confinement and safe, environmental disposal of their manure is an important public concern. The most pragmatic means of disposing of manure is by adding it to soil, where its nutritive value can be recycled by growing plants. Manures are widely used for several reasons: First, the manure is produced by animals and must be disposed of somewhere. Environmental concerns require better storage and more timely usage of the material now than in the past (Figure 6-10). Second, it is a good fertilizer and soil conditioner, even with some possible problems discussed in Chapter 18.

Manure Composition and Use

Cattle manures (dry basis) average 3 percent N, 0.8 percent P, 2 percent K, 25 percent organic carbon, plus varying amounts of other elements essential for plant growth. Barnyard manure, because of water content, has a low fertilizer grade (Table 6-9). Manures differ in composition, partly as a result of the differences among the kinds of feeds that animals consume (see

FIGURE 6-10 Using manure for nutrient and soil improvement. The manure is held in a 3200-gal storage tank (stores manure from 125 cows for 30 days). When it is environmentally suitable to apply it (when it can be incorporated into the soil), the manure is spread on the fields. (*Source:* USDA—Soil Conservation Service.)

Table 6-9 Typical Composition of Selected Animal Manures (Dry-Weight Basis)*

Constituent	Beef/Dairy (%)	Poultry (%)	Swine (%)	Sheep (%)
Nitrogen (N)	2–8	5–8	3–5	3–5
Phosphorus (P)	0.2–1.0	1–2	0.5–1.0	0.4–0.8
Potassium (K)	1–3	1–2	1–2	2–3
Magnesium (Mg)	1.0–1.5	2–3	0.08	0.2
Sodium (Na)	1–3	1–2	0.05	0.05
Total soluble salts	6–15	2–5	1–2	1–2

*Data were obtained from many sources.

Table 6-9). In comparison with chemical fertilizers, all manures supply relatively small quantities of plant nutrients per unit of dry weight. However, the *micronutrient* content in manures is usually higher than in chemical fertilizers to which micronutrient fertilizers have not been intentionally added. Manures have high soluble salt contents (6 percent to 15 percent of the total dry weight in beef or dairy cattle manure). This is partly because salt often is fed to livestock to increase appetite and to reduce kidney stones; much of the salt is voided in the manure. If wisely used, manures are good fertilizers, although they are low in phosphorus.

Sewage Sludges

Sewage sludges are the solids settled out in sewage treatment plants and are much like livestock manures (see Table 6-10). They are especially well adapted for application to turf grasses on golf courses, in cemeteries, and around public buildings. Commercial nurseries are also potential users of sludge. Larger quantities of sludge are expected to be used for the revegetation of soils drastically disturbed by surface mining and construction. Forests, pastures, and rangelands are also suitable sites for applying large amounts of sewage sludges and their wastewaters.[21]

Although variable in chemical composition, sludges contain about 4.0 percent nitrogen (N), 2.0 percent phosphorus (P), and 0.4 percent K on a dry-weight basis. They also contain toxic elements and heavy metals, such as boron, cadmium, copper, mercury, nickel, lead, selenium, and zinc. There is real concern that toxic quantities of the metals will be detrimental to plants, animals, and people. Cadmium is of special concern because it is relatively soluble, mobile in plants, and highly toxic.

Pathogenic (disease-producing) organisms present in some sewage wastewaters and sludges may cause cholera, diarrhea (amoebic and bacterial), hepatitis, pinworms, poliomyelitis, and tapeworms. This is an example where natural, organic systems of agriculture may not be advantageous. Composting can eliminate many of these pathogens.

Composting[22,23]

Composting is the microbial decomposition of piled organic materials into partially decomposed residues, which are called *compost* or *humus.* Composting is not a common process in nature because there are usually no piles of organic materials to decompose. Where thick

[21]Alex Hershaft and J. Bruce Truett, *Long-Term Effects of Slow-Rate Land Application of Municipal Wastewaters,* EPA-600/57-81-152, Environmental Protection Agency, Washington, DC, 1981.

[22]Jerry Maniac, Marjorie Hunt, and editors of *Organic Gardening Magazine, The Radial Guide to Composting,* Rodale Press, Emmaus, PA, 1979.

[23]Staff of *Organic Gardening Magazine, The Encyclopedia of Organic Gardening,* Rodale Press, Emmaus, PA, 1978, 235–248.

Table 6-10 Human Effluents in the United States: Total Produced and Their Major Plant Nutrient Content (in Thousands of Tons, Dry-Weight Basis) (Estimated for 1975 and Predicted for 1990)[a]

	Total Produced		Major Plant Nutrients[b] Nitrogen (N)		Phosphorus (P)		Potassium (K)	
Human Effluent	*1975*	*1990*	*1975*	*1990*	*1975*	*1990*	*1975*	*1990*
Sewage sludge	4300	5418	172	217	86	108	17	36
Septage	694	972	18	25	11	15	3	4

Source: Adapted from U.S. Environmental Protection Agency, *Composting of Municipal Solid Wastes in the United States,* EPA, Washington, DC, 1971.

[a]Tons × 0.9072 = metric tons.

[b]Based on typical sewage sludge composition of 4.0% N, 2.0% P (4.6% P_2O_5), and 0.4% K (0.5% K_2O) and on septage composition of 4.4% N, 1.6% P (3.7% P_2O_5), and 0.4% K (0.5% K_2O) (dry-weight basis). Organic-carbon percentage of both sludge and septage is estimated at 25% (dry-weight basis) because of the amounts of nonorganic materials (soil, plastics, precipitated salts) in sewage.

accumulations of organic matter do occur naturally (e.g., in stagnant water or under cold, acid forests), conditions are far from optimum for the decomposition of the residues, unlike those of well-maintained compost piles. The process of enzymatic decomposition of organic matter by microbes in composting is similar to that which occurs as plant remains in and on the soil decompose. Various nutrients (nitrogen, phosphorus, sulfur, calcium, and others) are released as the less complex compounds decompose; the more complex residues continue to decompose at slower rates for months and years.

Composting requires conditions that are favorable for microbial growth. The process can be either anaerobic or aerobic, but it is much faster and less odoriferous if done aerobically. Anaerobic composting may also produce plant-toxic organic acids, ethylene, and others. For optimum decomposition rates, the materials must be kept moist and warm, and sufficient nutrients must be available. To destroy pathogens effectively, exact temperatures for determined lengths of time must be reached inside the composting mixture, such as about 71°C (160°F) for an hour or more. The complete destruction of plant pathogens by composting is nearly impossible because the outside of the compost heap is cooler than the required killing temperature. The composting produces heat and compost piles have actually ignited themselves from the heat produced.

Practically any organic substance can be composted, but some are less suitable than others. Large pieces of shells, wood, cornstalks, greases, oils, pesticides, large masses of wet materials, weeds with seeds, diseased plants, newspaper, and pine and other conifer needles are not generally good for composting. Some of these materials can be used, though, if they are first shredded and well mixed with other substances.

Large amounts of low-cellulose-content materials, such as food wastes and animal tankage, are difficult to compost because they form large masses of gelatinous, anaerobic material in piles. The addition of fibrous materials (sawdust, straw, shredded paper, leaves, wood chips, mature plant materials) to gelatinous or dense masses improves composting action by increasing aeration.

Good composting usually requires several weeks of aerobic decomposition at elevated temperatures, high moisture, and high nutrient levels. Commercial operations can speed the natural process by continually mixing the compost and adding heat from an external source when necessary. Some facilities composting municipal garbage turn out a good compost in a month. When mixing is infrequent and little is done to assist decomposition, the compost may require several months to reach finished conditions. Composting is such an inexact process that there are no certain criteria for determining when the compost materials are ready for use. Although the principle remains the same, methods of composting vary widely from the small, backyard, enclosed compost bin to those of large-scale operations (Figure 6-11).

Adding Nutrients and Water to Composts

If composted materials are known to be deficient in particular plant nutrients, the missing ingredients can be added to the compost. Nitrogen, phosphates, and lime are the most frequently added. Legumes, seed meals, manures, and animal slaughter wastes (dried blood, fish scraps, tankage) are commonly recommended organic sources.

Water is essential for all enzymatic processes, so composts must be kept moist to ensure rapid decomposition. For small compost piles, plastic or other moisture barriers can be used to cover composts during dry, hot periods.

Composted materials often are not sources of large amounts of available nutrients. Fresh organic residues that are high in nutrients (young legume plants, most manures, vegetable wastes) can be spread directly on the soil (and incorporated) without composting because they rapidly decompose and release nutrients that plants can absorb. Low-nutrient materials (sawdust, mature weedy plants, straw) should be composted first, when it is practical to do so. Composts should be added to soils in manners similar to adding humus or animal manures. They are generally low-analysis materials, commonly with less than 2 percent of nitrogen, phosphorus, or potassium.

Composts and Public Health

Public health problems encountered in compost materials are viruses, pathogens, and parasite eggs in the wastes. Composting would seem to favor the growth of these harmful organisms, but during decomposition the compost can heat up enough that only thermophilic (heat-tolerant) microbes are active in the interior portions. At temperatures near 65°C to 75°C (149°F to 167°F), most hazardous pathogens are destroyed in a few days but may not be killed at the edges of the compost. If temperatures are as low as 55°C (131°F), the time necessary to destroy pathogens lengthens to weeks or even months. Pathogens are likely to be in sewage sludges and some municipal garbage.

Composting does not remove nonorganic materials such as lead, mercury, cadmium, arsenic, cyanide, strong acids, and other inorganics; nor does it prevent the formation of

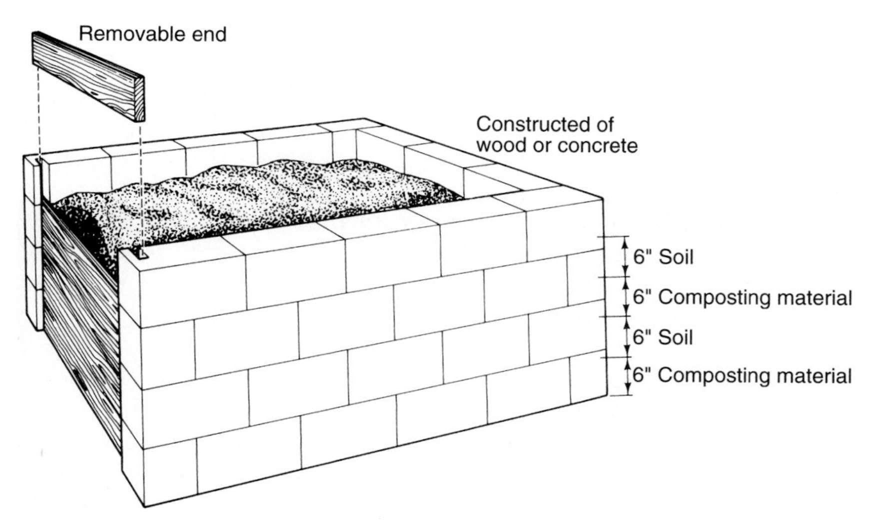

FIGURE 6-11 Well-constructed backyard type of compost bin made from concrete blocks. (Courtesy of Vocational Instructional Services, Texas A&M University.)

mycotoxins—substances that are toxic to plants or animals and formed by fungal action during decomposition of organic materials. Also, *teratogenic* (deformity-causing), *carcinogenic* (cancer-causing), or other harmful organic substances may be produced naturally by the composting process.

Phytotoxins (substances toxic to plants) and bad odors are produced during early stages of decomposition of some organic materials, particularly in anaerobic conditions. These can often be dissipated by evaporation of the substances at the elevated temperatures while in the composting bed.

6:16 Low-Input Sustainable Agriculture

The high cost and anticipated limited future supplies of petroleum and other agricultural inputs in the world have caused new interest in an agricultural program with reduced inputs. Fixation of N_2, crop rotations, and fewer passes with equipment over the land will reduce energy inputs.

What Is Low-Input Sustainable Agriculture?

Low-input sustainable agriculture (LISA) is farming with a minimal use of chemicals and energy. LISA is like organic farming in many respects—but with important differences. LISA will reduce—but not totally eliminate—chemical and fertilizer use. Organic farmers also restrain from the use of most chemicals and fertilizers, but they do it because the materials are not *natural* or because they believe the materials are health hazards or harmful to the soil quality. Some ancient terraced land is low-input sustainable agriculture (Figure 6-12). Many terraces in the Andes of Peru have been farmed for more than 1500 years. Alfalfa protects the soil (in rotations) and provides animal fodder; the animals produce manure to supplement the ashes added as fertilizers. The steep terrain made terracing essential for control of soil erosion.

Agriculture in the 1960s and 1970s was expensive in its energy use. Fields were traversed many times each season to plow, disc, harrow, smooth, furrow, plant, apply fertilizers and pesticides, and harvest. Energy was cheap and yields were high. Suddenly, energy costs increased. Some overworked, barren soils continue to erode extensively. Soil erosion is still severe in some areas, and it must be controlled. We should take a new look at the desirable levels of tillage and the economics of water, fertilizer, pesticides, and tillage inputs that we use to produce crops. LISA is sustainable agriculture—permanent, *forever* use of the land for production.

FIGURE 6-12 Bench terraces in the Andes Mountains of Peru exist at an elevation of 9000 to 12,000 ft. These in Colca Valley are part of about 1 million ha of terraces in the Andes that produce potatoes, corn, quinoa, several grains, and alfalfa. Builders, some 1500 years ago, constructed hundreds of miles of 2.5-m-tall rock-and-soil terrace walls and hundreds of miles of irrigation canals. Only 58 percent of the terraced area is now farmed by 3000 people. It is believed that 60,000 once farmed the area. (Courtesy of Charles G. Mahaffey, Augustana College, Rock Island, IL.)

Table 6-11 Comparison of Yields between Conventional Farming and Organic Farming Methods

Comparison	Yield Organic/Conventional*
Aggregate of crops and livestock production	
Illinois Amish vs. conventional	0.62
Pennsylvania Amish vs. conventional	
Old-order Amish	1.03
Nebraska sect	0.56
Wisconsin Amish vs. conventional	0.70
Wheat yields, 20 New York and Pennsylvania farms	0.78
Midwest, 14 organic vs. 14 conventional farms	0.87
Corn yields, 5 organic vs. 5 conventional farms	0.85
Soybean yields, 3 organic vs. 3 conventional farms	1.07
Wheat yields, 4 organic vs. 4 conventional farms	0.91
Miscellaneous crop yields, 18 farms	0.97
Virginia, four gardens of 7 vegetables	0.20
Estimated national average if exports were such as to require use of most cropland	
Wheat	0.47
Corn	0.50
Other feed grains	0.30
Soybeans	0.49
Cotton	0.44

Sources: (1) Council for Agricultural Science and Technology, "Organic and Conventional Farming Compared," CAST Report 84, CAST, Ames, Iowa, 1980, 32. (2) R. C. Lambe and J. G. Petty, " 'Chemical' Garden Out-Yields 'Organic' Garden," *Agri-News Newspaper* **4,** No. 2, Feb. 1973, 1, 3.

*Values of 1.0 mean yields of both methods are equal; less than 1 means organic farms produce less than conventional. Most yields were farmers' estimates, which can be good but may be inaccurate as well.

Low-Input Sustainable Agriculture vs. Conventional Farming

Which compromises allow us to evolve the best balance of production costs with the need to minimize our use of limited natural resources, particularly energy reserves? Any low-input sustainable agriculture system usually involves (1) better management of organic materials as nutrient reservoirs and (2) less chemical pesticide and fertilizer use. It is essential to schedule crop rotations involving N_2 fixers to build nitrogen soil supplies. Mixtures of crops or selected rotations will help reduce weeds and other pests with minimal pesticide use. Crop varieties resistant to certain diseases will minimize pesticide needs. Many of these principles have been studied for decades by organic farmers and numerous Amish peoples in the midwestern United States. Agriculture, before the machinery age, was low input (there were few fertilizers and pesticides, and equipment was small and pulled by horses).

Agriculturalists in the new age of LISA will try to marry the best practices of the organic farmer with the inexpensive but helpful inputs of better crop varieties, smaller quantities of more effective pesticides and fertilizers, reduced tillage, effective crop rotations, and integrated pest management. These changes will probably result in lower crop yields. But if input costs are lower, profits may still be higher. Many areas where reduced tillage is used have maintained yields near those with conventional management. It will require greater attention to many details of production to keep yields high with the techniques of LISA.

Some comparisons of crops grown under organic and conventional farming methods are shown in Table 6-11. Amish farms are considered here because they are similar in operation to organic farms (few chemicals are used) and have been established for a long time. It is expected that well-managed organic farms could produce as effectively as many conventional farms, but this is not often observed. Organic farms more often yield only 60 percent to 90 percent as much as conventional practices. Modeled projections, as given in the last five items of Table 6-11, suggest that conversion to organic methods on a wide scale is likely to produce less than half as much of several important world crops.

Organic farms usually produce almost as much or more energy than is used in their crop production. In contrast, conventional farms overall tend to produce only 30 percent to 60 percent as much energy as is used to produce the crops. Conventional farming currently requires a net energy input; in its present form it is not self-sustaining.

6:17 Decision Case Studies

What is a *decision case?* **Decision cases** have been called a *documentation of reality.* They describe *actual,* not simulated or contrived, situations that required a decision.[24] A limited number of cases are in the literature, but more will be published and classes may make their own concerning some local, resolved issue. The College of Agriculture at the University of Minnesota has attempted to encourage and lead in the use of case situations in their classes.[25] Students, in decision cases, are given specific facts from which decisions have to be reached and which have been considered by working people in real situations. This develops an atmosphere of curiosity, interest, and debate which can serve as a catalyst for student learning. Sometimes there is no right answer, but that can intensify debate and learning.

For those interested, a few example decision case studies will be referenced in the chapters pertinent to their subject. In this chapter there are two examples for consideration:

1. S. R. Simmons, D. Putnam, and D. Otterby, "Mueller Farm: Lupin as an Alternative Crop for On-Farm Protein Production, *Journal of Natural Resources and Life Sciences Education* **21** (1992), No. 1, 9–14.
2. R. Kent Crookston and Melvin J. Stanford, "Dick and Sharon Thompson's 'Problem Child': A Decision Case in Sustainable Agriculture," *Journal of Natural Resources and Life Sciences Education* **21** (1992), No. 1, 15–19.

In nature there are neither rewards nor punishments—there are consequences.

—Robert G. Ingersoll

Questions

1. (a) In which ways can earthworms be beneficial to the growth of plants and yet be nonessential? (b) What are the practical solutions to a nematode infestation?
2. (a) Describe the nature of the *rhizosphere.* (b) Does the rhizosphere have a different pH and/or microbial activity than is in the bulk soil? Explain.
3. How important are fungi as (a) organic matter decomposers? (b) pathogens? (c) toxins to plants? (d) toxins to animals and people?

---

[24] Steve R. Simmons, R. Kent Crookston, and Melvin J. Stanford, "A Case for Case Study," *Journal of Natural Resources and Life Sciences Education* **21** (1992), No. 1, 2–3.

[25] Department of Agronomy and Plant Genetics, 411 Borlaug Hall, University of Minnesota, St. Paul, MN 55108.

4. (a) Define *mycorrhizae.* (b) How extensive are mycorrhizae in nature?
5. Of what value are autotrophic bacteria?
6. (a) How are *nitrogenase, Rhizobium,* and *activation energy* interrelated? (b) What are *legume nodules?* (c) How much nitrogen is fixed by alfalfa, soybeans, and peanuts?
7. (a) What are *actinomycetes?* (b) What beneficial functions do they perform?
8. (a) Are *viruses* living substances? (b) How do viruses cause damage?
9. For the majority of soil microorganisms, what are the optimum conditions of water, temperature, pH, and nutrients?
10. What management practices (a) encourage growth of beneficial microbes? (b) help control harmful organisms?
11. What are the origins and the general composition of soil humus?
12. (a) What is an *enzyme?* (b) Enzymes lower activation energy. What does this mean?
13. (a) Define a *biological reaction.* (b) How are biological reactions related to organic-matter decomposition?
14. Answer the following questions about decomposition of soil organic matter. (a) What initiates the breakdown? (b) Which products result? (c) What is the term for the solid residue that is left?
15. Although soil organic matter will supply some of all plant nutrients, what is the nutrient for which it is most important? Explain why this is so important.
16. What is the relationship between *soil aggregation, organic-matter decomposition,* and *polysaccharides?*
17. Briefly discuss some beneficial and detrimental effects of soil organic materials.
18. Evaluate animal manures as a soil amendment (a) to supply nutrients and (b) that may contain materials detrimental to crops or soil.
19. (a) List the conditions and materials for making good compost. (b) Does old compost release more nutrients than plowed-under green-manure crops?
20. (a) Define *LISA.* (b) Why is the concept of LISA important?

Soil Taxonomy

The ends of scientific classifications are best answered when the objects are formed into groups respecting which a greater number of general propositions can be made, and those propositions more important than could be made respecting any other groups into which the same things could be distributed.

—John Stuart Mill

7:1 Preview and Important Facts

PREVIEW

Soil taxonomy[1] is a classification of soils. It is a grouping of soils into categories based on each soil's morphology (appearance and form).

The first complete U.S. taxonomic classification was published in 1938[2] and modified in 1949.[3] In the 1950s and early 1960s a new system was developed through a series of enlarged publications called *approximations.* In 1961 a comprehensive *Seventh Approximation* was printed and distributed worldwide to soil taxonomists for suggestions. In 1965 the U.S. Soil Conservation Service (currently the Natural Resources Conservation Service—NRCS) officially began the use of this system and published it in December 1975.[4] Although this U.S. system has applications worldwide, many countries—among them France, Canada, China, Brazil, and Russia—use their own systems.

In 1994 the U.S. soil classification system (**Soil Taxonomy**) organized all soils into 11 **orders,** 60 **suborders,** 294 **great groups,** 2366 **subgroups,** and many more **families** and **series.**[5] Each series is subdivided into mapping units, which are called **phases of series;**

---

[1] Taxonomy is derived from the Greek word *taxis,* meaning *arrangement.*

[2] M. Baldwin, Charles E. Kellogg, and J. Thorp, "Soil Classification," in *Soils and Men,* USDA Yearbook of Agriculture, U.S. Government Printing Office, Washington, DC, 1938, 979–1001.

[3] J. Thorp and Guy D. Smith, "Higher Categories of Soil Classification," *Soil Science* **67** (1949), 117–126.

[4] Soil Survey Staff, *Soil Taxonomy: A Basic System of Soil Classification for Making and Interpreting Soil Surveys,* Agriculture Handbook 436, Soil Conservation Service, USDA, Washington, DC, 1975.

[5] Soil Survey Staff, *Keys to Soil Taxonomy,* 6th Edition, USDA Soil Conservation Service, 1994.

these mapping units are not subdivisions in the classification system. There had been ten soil orders originally, but because of the large number of volcanic soils and their uniqueness, scientists separated out an eleventh order, which was called Andisols (Detail 7-1 and Figure 7-1).

Presently, there are eleven orders but a twelfth order—*Gelisols*—is proposed (see Detail 7-2). Gelisols have already been used in the FAO World map (inside the front cover) and cover much of the very cold area soils. In general **Gelisols** are soils that have *permafrost* within the 100 cm to 200 cm depth (the choice of 100 cm or 200 cm is yet to be determined). When greater detail is known and more soils are studied, new orders (as well as new suborders, new great groups, etc.) may be defined and described.

IMPORTANT FACTS TO KNOW

1. The construction of a soil great group name
2. A one-sentence definition each of *soil taxonomy, pedon, soil series, soil order, diagnostic horizon,* and *formative element*
3. In general terms, the description of these soil moisture regimes: aquic, udic, ustic, xeric, and aridic
4. The reason why Entisols and Inceptisols may include very productive soils as well as very nonproductive soils
5. The somewhat unique profile features of many Aridisols and the likely productivity of Aridisols
6. The distribution of Mollisols worldwide and their general fertility
7. The major problems in cropping Histosols, Vertisols, and Spodosols
8. The importance of Alfisols to world agricultural production
9. The unusual features of Andisols
10. The fertility problems of Ultisols and Oxisols and the reason why many of these soils are productive
11. The composition of plinthite and why it causes problems in cultivation of some Oxisols

7:2 U.S. System of Soil Taxonomy

To select and classify *individual soils,* a minimum soil volume of about 1 m^2 to 10 m^2 and as deep as roots grow, called a **pedon** (ped′-don), is chosen to represent a soil individual. Most *soil individuals* are present in larger than these minimum volumes and those multipedon areas are called **polypedons** (many pedons).

The six categories in the U.S. system of soil taxonomy are, in decreasing rank, *order, suborder, great group, subgroup, family,* and *series. Soil phase* is a utilitarian convenience term that is outside of the system of soil taxonomy.[6]

The six categories are briefly explained and defined below (names are considered as proper nouns and the top four categories should always be capitalized).

Order is the most general category in the system. *All soils belong to one of the eleven soil orders* (twelve if the tentative Gelisols is included). Five of the soil orders exist in a wide variety of climates: **Histosols** (organic soils), the undeveloped

[6]The six categories in the U.S. system of soil taxonomy were inspired by the six categories in the worldwide system of plant and animal taxonomy: *phylum, class, order, family, genus,* and *species.*

Detail 7-1 Brief Characterizations of the Soil Orders of the United States

Soil Order[a]	General Features
Gelisols	Gelisols have permafrost within 100 cm (depth of 100 cm or 200 cm still debated).
Entisols	Entisols have no profile development except perhaps a shallow marginal **A.** Many recent river floodplains, volcanic ash deposits, unconsolidated deposits with horizons eroded away, and sands are Entisols.
Inceptisols	Inceptisols, especially in humid regions, have weak to moderate horizon development. Horizon development has been retarded because of cold climates, waterlogged soils, or lack of time for stronger development.
Andisols	Andisols are soils with more than 60% volcanic ejecta (ash, cinders, pumice, basalt) with bulk densities below 900 kg/m^3. They have enough weathering to produce dark **A** horizons and early-stage secondary minerals (allophane, imogolite, ferrihydrite clays).[b] Andisols have high adsorption and immobilization of phosphorus and very high cation exchange capacities.
Histosols	Histosols are organic soils (peats and mucks) consisting of variable depths of accumulated plant remains in bogs, marshes, and swamps.
Aridisols	Aridisols exist in dry climates and some have developed horizons of lime or gypsum accumulations, salty layers, and/or **A** and **Bt** horizons.
Mollisols	Mollisols are mostly grassland, but with some broadleaf forest-covered soils with relatively deep, dark **A** horizons; they may have **B** horizons and lime accumulation.
Vertisols	Vertisols have a high content of clays that swell when wetted. Vertisols require distinct wet and dry seasons to develop because deep wide cracks when the soil is dry are a necessary feature. Usually, Vertisols have only deep self-mixed **A** horizons (top soil falls into cracks seasonally, gradually mixing the soil to the depth of the cracking). These soils exist most in temperate to tropical climates with distinct wet and dry seasons.
Alfisols	Alfisols develop in humid and subhumid climates, have precipitation of 500 mm–1300 mm (about 20 in–50 in), and are frequently under forest vegetation. Clay accumulation in a **Bt** horizon and available water much of the growing season are characteristic features. A thick **E** horizon is also common. They are slightly to moderately acid.
Spodosols	Spodosols are typically the sandy, leached soils of cool, coniferous forests. Usually **O** horizons, strongly acidic profiles, and well-leached **E**s are expected. The most characteristic feature is a **Bh** or **Bs** of accumulated organic material plus iron and aluminum oxides.
Ultisols	Ultisols are strongly acid, extensively weathered soils of tropical and subtropical climates. A thick **E** and clay accumulation in a **Bt** are the most characteristic features.
Oxisols	Oxisols are excessively weathered; few original minerals are left unweathered. Often Oxisols are more than 3 m (10 ft) deep, have low fertility, have dominantly iron and aluminum oxide clays, and are acid. Oxisols develop only in tropical and subtropical climates.

[a]Orders are arranged in approximate sequence from undeveloped soil to increased extent of profile development or increased extent of mineral weathering. The 1981 letter horizons are used.

[b]Allophane is a rapidly formed amorphous aluminosilicate clay; imogolite is slightly more crystallized. Ferrihydrite is rapidly formed amorphous iron hydrous oxide.

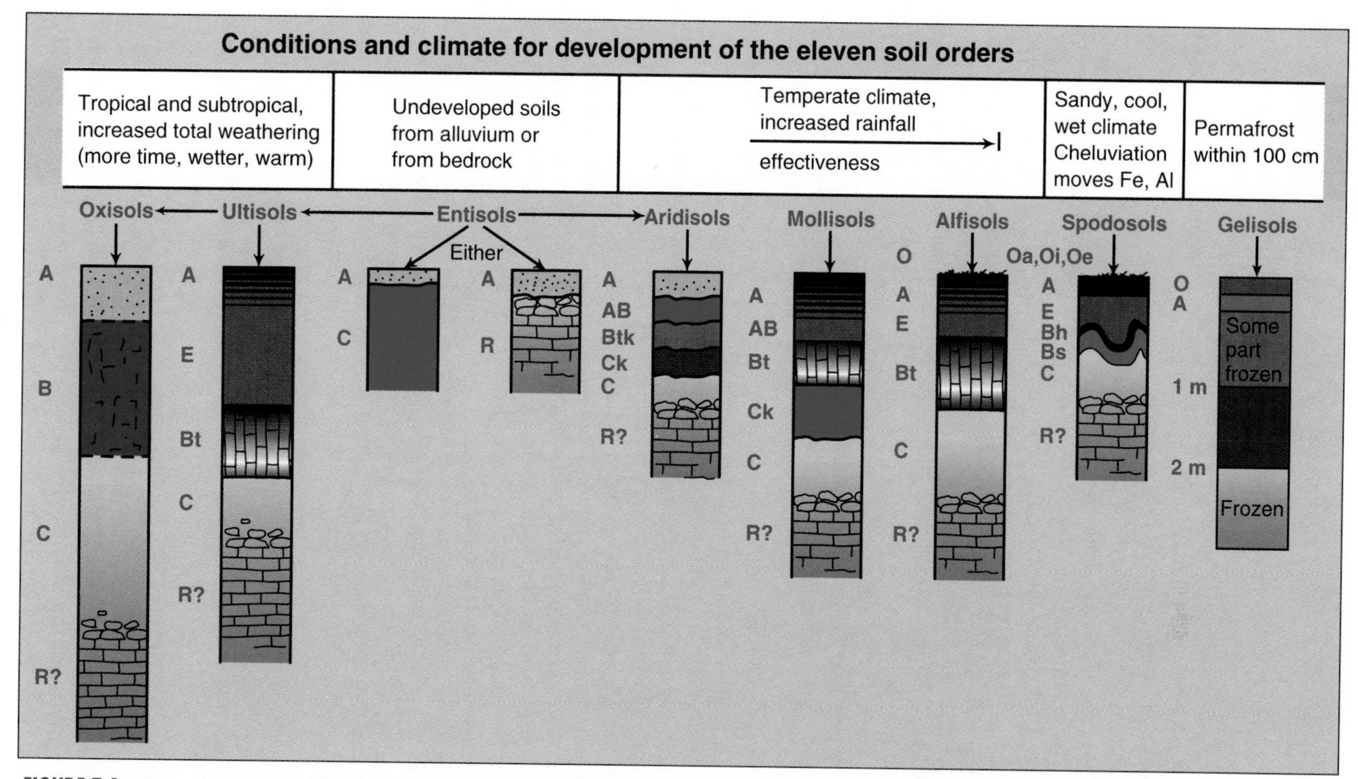

FIGURE 7-1 An outline of 8 of the 11 soil orders (and proposed 12th order) suggesting typical climatic environments and horizon sequences. Question marks indicate horizons that are found in some of these soils but are not found in all soils of that order. (Courtesy of Raymond W. Miller, Utah State University.)

Detail 7-2 Gelisols—A Twelfth Order?

The Soil Survey Division of the Natural Resources Conservation Service of the USDA is currently contemplating a new soil order for the soil classification system. **Gelisols**—soils of very cold areas with permafrost somewhere within 100 cm (or possibly 200 cm)—is being considered. Details are still being formulated. The term *Gelisols* is derived from the Greek *gelid,* meaning very cold. The taxonomy system already has a *pergelic* soil temperature regime, *Gelicryands,* and *Gelic* sub-units in the FAO/UNESCO map ledger (see World map inside front cover).

The *root* of the order is derived as with other orders and is **el,** from G**el**isol. In the Canadian soil classification, permafrost-affected soils are termed *Cryosols,* but the *root* of Cryosols (cry) would be awkward to use for suborder and lower levels because it does not start with a vowel. A characteristic of such cold soils is cryoturbation—oddly disturbed and mixed soils. Macroscopic features include irregular or broken horizons, deep incorporation of organic matter, oriented stones, and granular, platy, and vesicular structures in surface horizons. Massive soil (lack of structure) is common from freeze-thaw pressures and desiccation.

The three proposed suborders are as follows:

- **Histels:** Organic soils with permafrost within the designated depth
- **Turbels:** Mineral soils affected by cryoturbation (mixed, disturbed)
- **Haplels:** Mineral soils with minimal horizons and little or no cryoturbation

Some Gelisols, in the upper portion above permafrost, may have intergrades to mollic, umbric, aquic and other related soils having diagnostic horizons. The developers of the new order are proposing a revision in the soil temperature subclasses to allow classification of Gelisols. Three temperature classes (mean annual soil temperature, MAST) in Gelisols are proposed:

- **Hypergelic soils:** MAST at 50 cm deep colder than −9°C
- **Pergelic soils:** MAST between −4°C and −9°C
- **Subgelic soils:** MAST between +1°C and −3.9°C

The lower-case letter to indicate a *cryoturbated horizon*—chosen mostly because most single letters are already in use—is **jj.** Thus, a cryoturbated A1 horizon would be **A1jj.** Dry permafrost may be different from ice-cemented permafrost. Since **"f"** is already used for *frozen,* **fm** is proposed for *ice-cemented permafrost* and **f** alone for horizons with *dry* permafrost. If the horizon has more than 75 percent water, designate the horizon with **Wfm.**

Source: Written from review materials dated June 20, 1996, supplied by Paul Reich of the Natural Resources Conservation Service, USDA, Washington, DC.

Entisols, the slightly developed **Inceptisols** and volcanic material **Andisols,** and the swelling-clay **Vertisols.** The pending order, Gelisols, requires permafrost somewhere within the top 1 m (2 m?). The other six orders are mostly products of time and of the soil-forming factors, particularly the microclimate in which each developed (Figure 7-2).

Mollisols are usually naturally fertile soils, slightly leached, that occur in semiarid to subhumid climates, originally under grasses or broadleaf forests. **Alfisols** are fertile soils in good moisture regimes; they are usually productive, nonirrigated lands. **Ultisols** are leached, acidic, and of low-to-moderate fertility. They are

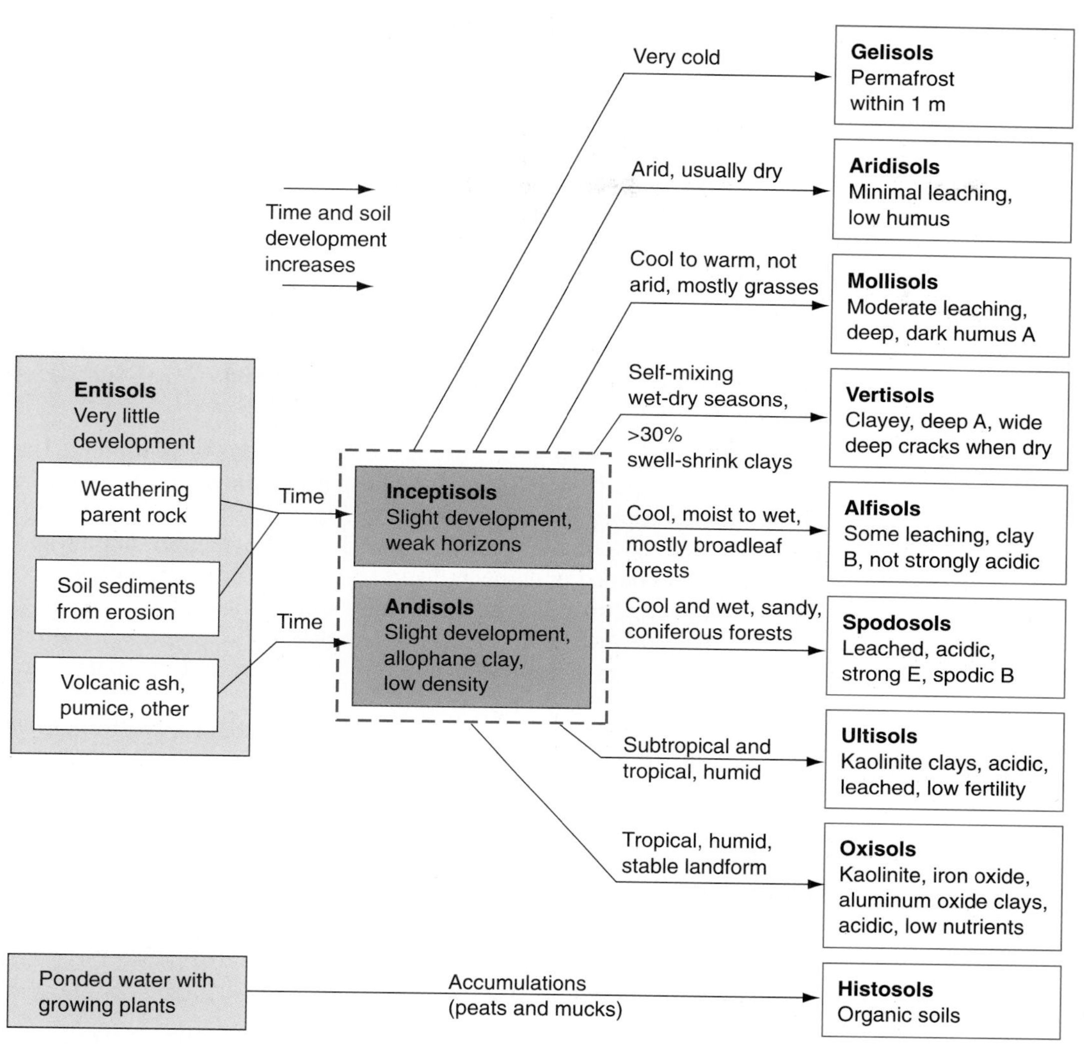

FIGURE 7-2 Schematic outline of the eleven soil orders in the U.S. Soil Classification System, illustrating the general conditions under which each order exists and listing some of the properties of each order. The order, Andisols, fits as an early-stage intermediate soil that can develop into other soils. A twelfth order, Gelisols, is proposed. (Courtesy of Raymond W. Miller, Utah State University.)

probably the most productive soils because they exist in areas with long frost-free periods and warm climates. When irrigated, **Aridisols,** the arid-region soils, are often very productive. In contrast, the infertile **Oxisols,** formed in hot, wet tropics, often exist in climates excellent year-round for plant growth. Unfortunately, Oxisols must be left in evergreen forests or carefully fertilized and limed for good yields of cultivated crops. The acidic, sandy **Spodosols,** found in cool climates, are some of the poorest soils for cultivation; they are used for cool-season crops, such as some grains, potatoes, tree nurseries, and pastures, but they need lime and fertilizer. (See Detail 7-1 and Figure 7-1 for additional facts about the soil orders.)

The most extensive soils in the United States are Mollisols (nearly 25 percent). The worldwide distribution of soil orders is Aridisols (more than 23 percent), Alfisols (more than 13 percent), Inceptisols (16 percent), Ultisols (8 percent), and Oxisols (nearly 9 percent). The other six orders are present in lesser amounts.

Nearly one-fifth of the world's land area is mountainous and has various amounts of the soil orders so intermixed as to be, on large-scale maps of the United States, uncategorized except as "soils of mountains," "miscellaneous," "salt flats," "rocklands," and "icefields."

Suborder Suborders within a soil order are differentiated largely on the basis of additional soil properties and horizons resulting from differences in *soil moisture, soil temperature,* and *dominating effects of chemical or textural features.* New suborders are periodically being considered. Suborders are distinctive to each order and are not interchangeable to other orders. Many suborders are given in the last sections of this chapter.

Great Group Soil great groups are subdivisions of suborders. The great groups found in the United States have been established largely on the basis of *differentiating soil horizons* and *soil features.* The differentiating soil horizons include those that have accumulated clay, iron, and/or humus and those that have pans (hardened or cemented soil layers) that interfere with water movement or root penetration. The differentiating soil features include the self-mixing (expansion and contraction) properties of clays, soil temperature, and major differences in content of calcium, magnesium, sodium, potassium, gypsum, and other salts.

(It is important to note that the 1938 U.S. classification system included a category called *great soil groups,* with names such as *chernozem, podzol, solometz,* and *laterite.* These are not the same as **great groups** in the current system.)

Subgroup Each soil great group is divided into three kinds of subgroups: one representing the central (**typic**) concept of the soil group; a second, which has properties that intergrade toward other orders, suborders, or great groups; and a third (extragrade), which has properties that prevent its classification as *typic* or as an intergrade to another soil category.

Family Soil families are separated within a subgroup primarily on the basis of *soil properties important to the growth of plants or behavior of soils when used for engineering purposes.* The soil properties used include texture, mineralogy, reaction (pH), average soil temperature, the area's moisture regime, permeability, thickness of horizons, structure, and consistency. More than 5000 families have been identified in the United States.

Series Each family contains several similar soil series. Each of the more than 15,000 soil series in the United States has narrower ranges of characteristics than the soil family. The name of the soil series has no pedogenic (soil formation) significance but represents a prominent geographic feature such as the name of a river, town, landmark, mountain peak, or area near where the series was first recognized. Soil series are differentiated on the basis of *observable* and *mappable soil characteristics.* Soils in the same series have about the same color, texture, structure, consistency, thickness, and reaction (pH) and have the *same number and arrangements of horizons and similar chemical and mineralogical properties.* Soils of a given series have about the *same kinds, thickness, appearance, and properties of horizons.*

Examples of series names of soils follow:

Name of Series	*Origin of Name*
Jay	City of Jay, Oklahoma
Houston Black	Houston County, Texas
Merrimac	Merrimac River in New Hampshire
Houghton (peat)	Houghton Lake, Michigan

Soil phase is a term used in association with soil series but not considered a classification category of the system.[7] **Phases of soil series** are mapping units. Their major use is to delineate soil areas for practical uses, such as farming and municipal or county zoning. Phases have differences in *surface soil texture,* the *solum thickness,* the *percentage slope,* the *stoniness,* the *saltiness,* the *extent of erosion damage,* and other conditions. Mapping units are **polypedons** that may have *small* portions of other polypedons included in them. Examples of phases are Jay silty clay loam, 0 to 3 percent slope; Houston Black clay, 2 percent to 5 percent slope, eroded; Houghton muck; and Dixie loam, 0 to 1 percent slope, shallow, rocky.

In addition to the U.S. system of soil taxonomy, other systems include those of France, Canada, Brazil, and the former USSR. Utilitarian soil maps of the world with a common legend in English, French, Spanish, and Russian have been printed on a scale of 1:5,000,000. They were prepared and published by continent by the Food and Agriculture Organization of the United Nations (FAO) in cooperation with the United Nations Educational, Scientific, and Cultural Organization (UNESCO). As might be expected, taxonomy *keys*—printed and as computer software—are being developed.[8] So far only a limited number of keys have been made available, perhaps partly because the system has been changed since its publication in 1975. Detail 7-3 is an abbreviated key to soil orders that illustrates the technique of separating out the soil orders. A booklet "Keys to Soil Taxonomy" is published by the Soil Survey Staff.[9]

7:3 Constructing Taxonomic Names

To construct suborder and great group names, words were coined from many roots in Latin and Greek, and a few others from English, German, French, and Japanese. All names lower than the order (but not the *series* or *family* name) contain a **root** (portion of the order name) as part of the new name; so the soil order to which the soil belongs is always known. The **root** is the syllable that precedes the vowel connecting the ending *-sol.* For example, the root of Aridisol is *id;* that of Spodosol is *od.* **Formative elements** are used to suggest properties of the soil (see Table 7-1). An example of naming is given using the fourth most extensive U.S. soil order, Ultisols, which make up 12.8 percent of the land surface of the United States. The example illustrates the use of formative elements and their origin and meanings in naming **taxa** (categories). This Ultisol is shown in color photo 9 of Table 7-2. Table 7-2 lists eleven of the soil orders in color with descriptions also given.

---

[7]The word *phase* may also be used in soil taxonomy as a subdivision of a soil order, suborder, great group, subgroup, or family. Its most common use, however, is as a subdivision of a series in delineating a soil mapping unit. *Phase of soil series* has replaced *soil type* as a mapping unit.

[8]Computer software *Keys to Soil Taxonomy* and a users' manual (1987) are available from the Department of Soil Science, North Carolina State University, P.O. Box 7619, Raleigh, NC 27695-7619. This listing does not suggest an endorsement of the software by the authors or publisher but is intended as information to interested persons [*Source: Agrotechnology Transfer* **2** (1986), 16].

[9]Soil Survey Staff, *Keys to Soil Taxonomy,* USDA—Natural Resources Conservation Service, seventh edition, U.S. Government Printing Office, Washington, DC, 1996.

Detail 7-3 Key to Soil Orders

This key is an abbreviated guide to distinguish between soil orders. Match the properties of the soil to the listed key *in sequence.* The first listing that it fits is the soil order to which the soil belongs.

1. Soils that have permafrost within the top 1 m—**Gelisols**
2. Soils that are organic in (a) more than half the thickness of the upper 81 cm or (b) at least in 2/3 of the soil depth to lithic or paralithic contact—**Histosols**
3. Mineral soils without a plaggen, argillic, or kandic horizon but with an illuvial spodic horizon within 2 m of the surface—**Spodosols**
4. Other soils that have andic soil properties in 60% or more of the soil from the surface to 60 cm or lithic or paralithic contact, formed from mostly volcanic ejecta with bulk densities below 0.95 Mg/m^3—**Andisols**
5. Other soils with (a) an oxic horizon top within 150 cm *or* (b) 40% or more clay within 18 cm and a kandic horizon with weatherable mineral properties of an oxic horizon, with upper boundary within 100 cm of soil surface—**Oxisols**
6. Other soils with more than 30% clay to 50 cm *and* cracks that open and close periodically (usually smectite-type clays)—**Vertisols**
7. Other soils that have (a) an aridic soil moisture regime, (b) an ochric or anthropic horizon, *and* (c) one or more of argillic, calcic, cambic, gypsic, natric, petrocalcic, petrogypsic, *or* a salic or duripan—**Aridisols**
8. Other soils with an argillic (or kandic) horizon, base saturation $<$35%, and enough moisture for crops in most years—**Ultisols**
9. Other soils that have (a) a mollic horizon and (b) a high base saturation $>$50% throughout the soil to at least 180 cm or to a lithic or paralithic contact, whichever comes first—**Mollisols**
10. Other soils with (a) an argillic, a kandic, or a natric horizon *or* (b) a fragipan with clay films $>$1 mm thick in some part. Usually enough moisture for crops in most years—**Alfisols**
11. Other soils that have (a) no illuvial clay horizon but do have (i) a cambic horizon or (ii) aquic conditions within 50 cm or (iii) within 150 cm the start of a calcic, petrocalcic, gypsic, petrogypsic, placic, duripan, fragipan, or oxic starting within 200 cm, or sulfuric horizon within 150 cm; *or* (b) a histic, a mollic, plaggen, or umbric horizon or ESP $>$15% decreasing below 50 cm—**Inceptisols**
12. Other soils with weak or no diagnostic horizons—**Entisols**

Order **Ult***isols* (Latin, *ultimus,* last = well weathered). The root for *Ultisols* is the syllable *ult,* followed by the vowel which connects *I* to the ending *sols.*

Suborder **Ud***ults* (Latin, *udus,* humid, = adequate rainfall) consists of the *root* preceded by a formative element.

Great group **Pale***udults* (Greek, *paleos,* old = excessive development) is the suborder preceded by a formative element.

Subgroup **Aquic** *Paleudults* (Latin, *aqua,* water = overly wet).

Family **Fine-silty over clayey, siliceous, thermic** *Aquic Paleudults* **family.** *Fine-silty* means "a high content of silt and clay"; *over clayey* means "the subsoil is over

35 percent clay"; *siliceous* means ">90 percent silica materials"; *thermic* means "an annual soil temperature between 59° F and 72°F (15° C and 22°C)."

Series **Sawyer,** named after the city of Sawyer, Oklahoma. The soil occurs in several southern states.

Throughout the remainder of this chapter, there are many examples of complete soil names.

Although it seems possible to have hundreds of combinations of the formative elements combined with soil order roots, conditions in nature exclude many of them. For example, it is unlikely to find lime in a soil in very wet (acidic soil) regions, so a *Calciudult* is unlikely to occur. Many of the formative elements referring to accumulation of soluble salts, gypsum, or lime are unlikely to occur in Ultisols, Oxisols, and Spodosols. Try to think of some other unlikely combinations, as an exercise in looking over the formative elements and learning their meanings. In various tables and figures in the rest of this chapter, many names are given as examples, particularly the eleven soils described in Table 7-2.

7:4 Soil Moisture Regimes

The **moisture regimes** represent an attempt to indicate the extent of naturally available water in the soil depth of maximum root proliferation (soil control section). The **soil control section** for moisture regimes is the depth between where 2.5 cm (1 in) of water would wet that dry soil and where 7.5 cm (2.4 in) of water would wet it. In clayey soils, this soil control layer is from about 8 to 25 cm (3.1 to 9.8 in) deep; for sands it is from about 20 to 60 cm (7.8 to 23.4 in) deep. For greater detail, refer to the Glossary under *Soil moisture regimes.* Many soil suborders have formative elements indicating the moisture regime in which they occur (Detail 7-4). The moisture regimes used are the following:

Aquic conditions Usually wet with anaerobic saturation for a period long enough to produce visual evidence of poor aeration (mottling and gleying [grey coloring])

Udic Usually has adequate water throughout the year

Perudic Extremely wet, percolation in all months not frozen

Ustic Deficient in water, rains in the cropping season; most of the water available comes during the summer season

Xeric Deficient in water and with a dry cropping season; most of the precipitation comes in the winter time of year

Aridic Very water deficient; long dry periods; short wet periods

Torric Same criteria as aridic, but used at specified locations in the classification system

Aquic conditions has replaced what was called *aquic moisture regime* because *aquic conditions* are adaptable to more situations. The periods of saturation may be only a few days, but *aquic conditions must involve* **reducing conditions.** Aquic conditions include any of the following: (1) *redoximorphic features* (wetness-caused mottles), (2) *redox concentrations of Fe and Mn,* (3) *redox depletions of Fe and Mn* leaving low-chroma wetness mottles, and (4) *reduced matrix* that changes color when exposed to air (oxygen). Extended periods of

Table 7-1 Formative Elements Used in Names of Great Groups (and Some Suborders) and the Adjectives Used as Extragrades for All Levels, Particularly Subgroups

Formative Element	*Derivation*	*Mnemonicon*	*Connotation*
	Formative Elements in Names of Great Groups		
Acr	Modified from Gr. *akros,* at the end	Acrolith	Extreme weathering
Agr	L. *ager,* field	Agriculture	An agric horizon
Alb	L. *albus,* white	Albino	An albic horizon
And	Modified from *ando* (black soil, Japanese)	Ando	Ando-like, volcanic ash
Aqu	L. *aqua,* water	Aquarium	Aquic conditions
Arg	Modified from argillic horizon; L. *argilla,* white clay	Argillite	An argillic horizon
Bor	Gr. *boreas,* northern	Boreal	Cool
Calc	L. *calcis,* lime	Calcium	A calcic horizon
Camb	L. *cambriare,* to exchange	Change	A cambic horizon
Chrom	Gr. *chroma,* color	Chroma	High chroma, bright color
Cry	Gr. *kryos,* icy cold	Crystal	Cold
Dur	L. *durus,* hard	Durable	A duripan
Dystr, dys	Modified from Gr. *dys,* ill; dystrophic, infertile	Dystrophic	Low basic cation saturation
Eutr, eu	Modified from Gr. *eu,* good; eutrophic, fertile	Eutrophic	High basic cation saturation
Ferr	L. *ferrum,* iron	Ferric	Presence of iron oxide accumulation
Fibr	L. *fibra,* fiber	Fibrous	Least decomposed stage
Fluv	L. *fluvus,* river	Fluvial	Flood plain, alluvium
Frag	Modified from L. *fragilis,* brittle	Fragile	Presence of fragipan
Fragloss	Compound of *fra(g)* and *gloss*		Tongued fragipan
Gibbs	Modified from *gibbsite*	Gibbsite	Presence of gibbsite in sheets or nodules
Gloss	Gr. *glossa,* tongue	Glossary	Tongued, portions of horizon penetrate below it
Gyps	L. *gypsum,* gypsum	Gypsum	Presence of a gypsic horizon
Hal	Gr. *hals,* salt	Halophyte	Salty
Hapl	Gr. *haplous,* simple	Haploid	Minimum horizon development
Hum	L. *humus,* earth	Humus	Presence of humus
Hydr	Gr. *hydor,* water	Hydrophobia	Presence of water
Luv	Gr. *louo,* to wash	Ablution	Illuvial
Med	L. *media,* middle	Medium	Of temperate climate
Nadur	Compound of *na(tr)* and *dur*		Sodic duripan
Natr	Modified from *natrium,* sodium		Presence of natric horizon
Ochr	Gr. base of *ochros,* pale	Ocher	Presence of ochric epipedon
Pale	Gr. *paleos,* old	Paleosol	Excessive development, usually very old
Pell	Gr. *pellos,* dusky		Low chroma, dull color
Plac	Gr. base of *plax,* flat stone		Presence of a thin pan
Plagg	Modified from Ger. *plaggen,* sod		Presence of plaggen epipedon
Plinth	Gr. *plinthos,* brick		Presence of plinthite

Psamm	Gr. *psammos,* sand	Psammite	Sand texture
Quartz	Ger. *quarz,* quartz	Quartz	High quartz content
Rhod	Gr. base of rhodon, rose	Rhododendron	Dark red color
Sal	L. base of *sal,* salt	Saline	Presence of salic horizon
Sapr	Gr. *sapros,* rotten	Saprophyte	Most decomposed stage
Sider	Gr. *sideros,* iron	Siderite	Presence of free-iron oxides
Sombr	F. *sombre,* dark	Somber	A dark-colored horizon
Sphagn	Gr. *sphagnos,* bog	Sphagnum	Presence of sphagnum, peat moss
Sulf	L. *sulfur,* sulfur	Sulfur	Presence of sulfides or their oxidation products
Torr	L. *torridus,* hot and dry	Torrid	Torric moisture regime
Trop	Modified from G. *tropikos,* of the solstice	Tropical	Humid and continually warm
Ud	L. *udus,* humid	Udometer	Udic moisture regime
Umbr	L. base of *umbra,* shade	Umbrella	Presence of umbric epipedon
Ust	L. base of *ustus,* burnt	Combustion	Ustic moisture regime
Verm	L. base of *vermes,* worm	Vermiform	Worm-worked or mixed by animals
Vitr	L. *vitrum,* glass	Vitreous	Presence of volcanic glass
Xer	Gr. *xeros,* dry	Xerophyte	A xeric moisture regime

Adjectives in Names of Extragrades and Their Meaning

Abruptic	L. *abruptum,* torn off	Abrupt	Abrupt textural change between horizons
Aeric	Gr. *aerios,* air	Aerial	Aeration
Anthropic	Modified from Gr. *anthropos,* man	Anthropology	An Anthropic epipedon
Arenic	L. *arena,* sand	Arenose	Sandy epipedon between 50 cm and 1 m (19.5 in and 39.4 in) thick
Cumulic	L. *cumulus,* heap	Accumulation	Thicker-than-normal epipedon
Epiaquic	Gr. *epi,* over, above, and *aquic*		Surface wetness, standing water
Glossic	Gr. *glossa,* tongue	Glossary	Tongued horizon boundaries
Grossarenic	L. *grossus,* thick, and L. *arena,* sand		Thick sandy epipedon >1 m (39.4 in) thick
Hydro	Gr. *hydor,* water	Hydroponics	Presence of water, waterlogged
Leptic	Gr. *leptos,* thin		A thin soil
Limnic	Modified from Gr. *limn,* lake	Limnology	Presence of limnic layer, lake bottom origin
Lithic	Gr. *lithos,* stone	Lithosphere	Presence of a shallow lithic (rock) contact
Pachic	Gr. *pachys,* thick	Pachyderm	Thicker-than-normal epipedon
Paralithic	Gr. *para,* beside, and *lithic*		Presence of a shallow paralithic (hardpan) contact
Pergelic	L. *per,* throughout in time and space, and L. *gelare,* to freeze		Permanently frozen or having permafrost
Petrocalcic	Gr. *petra,* rock, and *calcic* from calcium		Presence of a petrocalcic horizon
Petroferric	Gr. *petra,* rock, and L. *ferrum,* iron		Presence of a petroferric contact (ironstone)
Plinthic	Modified from Gr. *plinthos,* brick	Plinthite	Presence of plinthite
Ruptic	L. *ruptum,* broken	Rupture	Intermittent or broken horizons, discontinuous
Superic	L. *superare,* to overtop	Superimpose	Presence of plinthite at the surface
Terric	L. *terra,* earth		A mineral substratum, used with Histosols
Thapto	Gr. *thapto,* buried		A buried soil

Source: Soil Survey Staff, *Soil Taxonomy: A Basic System of Soil Classification for Making and Interpreting Soil Surveys,* Agriculture Handbook 436, USDA, Washington, DC, 1975, 89–90.

***Note:** Figure numbers refer to photographs appearing in the color insert between pages 224 and 225.*

Table 7-2 Soil Profiles in Color Representing the Eleven Soil Orders: Their Taxonomy and Diagnostic Horizons (Listed Alphabetically)

Figure No.	Order	Suborder	Great Group	Subgroup	Family*	Series
1	Alfisols	Boralfs	Cryoboralfs	Typic Cryoboralfs	Loamy, mixed, frigid	Waitville
2	Aridisols	Argids	Haplargids	Typic Haplargids	Fine-loamy, mixed, thermic	Mohave
3	Entisols	Psamments	Ustipsamments	Typic Ustipsamments	Mixed, mesic	Valentine
4	Histosols	Hemists	Cryohemists	Terric Cryohemists	Sphagnic, borustic, euic	Grindstone
5	Inceptisols	Ochrepts	Eurochrepts	Typic Eurochrepts	Coarse-silty, mixed, thermic	Natchez
6	Mollisols	Udolls	Hapludolls	Typic Hapludolls	Fine-loamy, mixed, mesic	Clarion
7	Oxisols	Torroxs	Torroxs	Typic Torroxs	Clayey, kaolinitic, isohyperthermic	Molokai
8	Spodosols	Orthods	Haplorthods	Typic Haplorthods	Sandy, mixed, frigid	Kalkaska
9	Ultisols	Udults	Paleudults	Aquic Paleudults	Fine-silty over clayey, siliceous, thermic	Sawyer
10	Vertisols	Usterts	Pellusterts	Udic Pellusterts	Fine, montmorillonitic, thermic	Houston Black
11	Andisols	Udands	Hydrudands	Hydrous Isohyperthermic Acrudoxic Hydrudands	Loamy, mixed isohyperthermic	Hilo

*The terms given in the table refer to only the family prefix because the family name also includes the subgroup names: loamy, mixed, frigid Typic Cryoboralf.

Figure No.	Phases of Series	General Diagnostic Horizons or Differentiating Features		Specific Diagnostic Horizon Indicated by Arrows on Photos	Location of Soil Profile
		Epipedon	Endopedon		
1	Waitville sandy loam	High base saturation	Argillic, ochric, and calcic horizons	Argillic endopedon	Manitoba, Canada
2	Mohave loam, 1%–5% slopes	Ochric horizon (light colored)	Calcic, natric, or gypsic horizon	Calcic endopedon	Arizona
3	Valentine fine sand	—	Sandy to a depth of 1 m	Sandy to a depth of 1 m	Nebraska
4	Grindstone peat, deep phase*	Histic horizon	—	Histic epipedon	Manitoba, Canada
5	Natchez silt loam 7%–40% slopes	Ochric horizon (light colored)	Cambic horizon	Cambic endopedon	Mississippi
6	Clarion loam 0%–3% slopes	Mollic horizon (dark colored)	Cambic horizon	Mollic epipedon	Polk County, Iowa
7	Molokai silty clay loam 2%–12% slopes	—	Oxic horizon	Oxic endopedon	Molokai, Hawaii
8	Kalkaska sand, 0%–40% slopes	—	Spodic horizon	Spodic endopedon	Osceola County, Michigan
9	Sawyer loamy sand, 2%–8% slopes	Low base saturation	Argillic (clay) horizon	Argillic endopedon	Dooly County, Georgia
10	Houston Black clay, 1%–3% slopes	>30% clay in all horizons, wide and deep cracks, moist chromas of less than 1.5 in upper foot		50%–60% clay, moist chroma of 1 in upper 38 in (1 mi)	Travis County, Texas
11	Hilo silty clay loam, 0%–10% slopes	Weak melanic	Cambic	Cambic endopedon	Mauna Kea, Hawaii

Courtesy credits: Figures 1, 4, Canada Department of Agriculture; Figures 2, 3, 5, 7, 10, USDA—Soil Conservation Service; Figure 6, Roy W. Simonson; Figure 8, Eugene P. Whiteside; and Figure 9, David A. Lietzke, both at Michigan State University.

*When no slope is indicated, it is 0%–1%.

Detail 7-4 Using Soil Moisture Regimes

Photo 10 in Table 7-2, which shows color photos of soils, is a Vertisol. At the suborder level it is dry enough in winters to be called a **Ustert** (*ust* means water deficient, mostly in the nongrowing [winter] period). At the subgroup level it is called a **Udic Pellustert** because, although it is water deficient in winter, it is not very dry and is almost wet enough to be in the **udic** (wet year-round) moisture regime.

seepage, the soil aspect (the compass direction), and infiltration all influence the soil's water content.

Because moisture regimes indicate when and how long the soil's *major root zone* is wetted, the precipitation patterns are only part of the cause for the moisture regime. For example, in Western U.S. mountains, a sandy ridge could have a dry summer (*xeric* regime), while a poorly drained valley bottom clay could have an *aquic* regime, kept waterlogged by seepage. A high-elevation mountain slope facing north may have enough coolness, periodic seepage, snow cover, and light summer rains to have an *udic* regime.

7:5 Terminology for Family Groupings

Family groupings are based on one or more of the following categories: (1) mineral particle classes, (2) the soil's mineralogy, (3) soil temperature classes, (4) sometimes the available rooting depth, and (5) less frequently the pH, lime, sand particle coatings, and/or permanent cracks. Only the first three properties (which are the most commonly involved) are discussed below.

Particle-Size Classes

The intent of using particle-size classes is to separate soils of quite different textures in the control section into separate families. The **control section** is usually from the top of the **Bt** to a depth of 50 cm (about 20 in); if no **Bt** exists, the control section is often between 25 cm and 100 cm deep (about 10 in to 39 in) or to rock if less than 100 cm. The seven groups follow:

Fragmental Mostly made up of stones, cobbles, gravel, and very coarse sands without enough fine particles to fill voids larger than 1 mm

Sandy-skeletal More than 35 percent of material is coarser than 2 mm diameter with enough sand (see *Sandy,* below) to fill voids larger than 1 mm

Loamy-skeletal Same as *Sandy-skeletal,* except loam fills the voids (see *Loamy* below)

Clayey-skeletal Same as *Sandy-skeletal,* except clay fills the voids (see *Clayey* below)

Sandy All material is sand or loamy sand, except the very fine sand size fraction.

Loamy All material is between *sandy* and *clayey.*

Clayey Material is more than 35 percent clay. (*Fine clayey* is 35 percent to 60 percent clay; *very fine clayey* is more than 60 percent clay.)

Soil Mineralogy Classes

For most of the root zone soil, terms are used to indicate dominant minerals, such as **carbonatic** (carbonates), **serpentinic** (serpentine minerals), and **siliceous** (more than 90 percent silica). The most common term is **mixed,** meaning that no single kind of mineral makes up more than 40 percent of the soil. *Clayey* soils with more than 50 percent of the clay of one type may be called **montmorillonitic, kaolinitic, illitic, oxic,** and so on, to indicate the dominant clay. *Mixed* is used with clayey soils, also. Other terms may be used for special materials (e.g., **cindery** for more than 60 percent of volcanic ash or cinders). The same *control section* is used as for particle-size classes.

Soil Temperature Classes for Family Groupings

Temperature classes are based on *mean and annual soil temperature,* the *average summer temperature,* and the *difference between mean summer and mean winter temperatures.* Mean annual soil temperature (MAST) is determined by measurement at 50 cm deep to represent the major root zone (5 cm to 100 cm). MAST can be estimated in most of the United States by adding 1°C to the mean annual *air* temperature. A single reading at 10 m (32.8 ft) deep is within 1°C of the annual average value. Also, if the average summer temperature of the top 100 cm is used, the average temperature at different depths is calculated by adding 0.6°C for each 10 cm (4 in) above 50 cm (20 in) deep and subtracting the same amount for each 10 cm deeper than 50 cm. The soil temperature classes for temperate region soils are defined in terms of mean annual soil temperature as follows:

- **Pergelic:** 0°C (32°F); permafrost present (unless dry)
- **Cryic:** 0°C–8°C (32°F–47°F); summer soil temperature *below* about 15°C
- **Frigid:** 0°C–8°C (32°F–47°F); summer soil temperature *above* about 15°C
- **Mesic:** 8°C–15°C (47°F–59°F)
- **Thermic:** 15°C–22°C (59°F–72°F)
- **Hyperthermic:** >22°C (72°F)

Tropical region soils, characterized by having a difference in temperature between mean summer and mean winter of less than 5°C (9°F), are grouped into the following mean annual soil temperature classes by adding **iso-** in front of the correct temperature name: **isofrigid, isomesic, isothermic,** and **isohyperthermic.** These four *iso-* temperatures are common in many tropical climates at different elevations.

Examples of Family Names

The family name contains several words, usually describing texture, mineralogy, and soil temperature. A correct sequence for writing family and subgroup names is not established. Because the U.S. Taxonomy system develops from the most specific to the most general, some people prefer to write the names with the sequence from left to right: *series, family,* and then *subgroup.* The reverse order can also be used. Six examples of names follow:

Colby series (Colorado) The fine-silty, mixed, calcareous, mesic family of Ustic Torriorthents

Minoa series (New York) The coarse-loamy, mixed, mesic family of Aquic Dystric Eutrochepts

Sharpsburg series (Nebraska) The fine, montmorillonitic, mesic family of Typic Argiudolls
Houston Black series (Texas) The fine, montmorillonitic, thermic family of Udic Pellusterts
Nebish Series (Minnesota) The fine, loamy, mixed, frigid family of Typic Eutroboralfs
Makaweli series (Hawaii) The fine, kaolinitic, isohyperthermic family of Oxic Haplustox.

7:6 U.S. Soil Classification: A Brief Look at Orders

The eleven (or twelve) soil orders, which include all the world's soils, are a diverse group. A very generalized outline of these orders in a soil development scheme was shown previously in Figure 7-2.

The orders are discussed in the following sections in a development sequence beginning with organic parent materials (*Histosols*). These are followed by *Entisols,* the least developed mineral soils, and then the *Inceptisols* and *Andisols* of intermediate development. The remaining orders are discussed roughly in order of extent of profile leaching and weathering, beginning with the dry area *Aridisols.* Next follow the soils with increasingly more effective water: *Mollisols* and *Alfisols. Vertisols* tend to occur in many areas where wetting and drying cycles exist, so they are listed after Alfisols. *Spodosols* are inserted next at this point, although they are unique in requiring sandy profiles and strong leaching. *Ultisols* and *Oxisols* are discussed last because they are very extensively weathered and low in primary materials. These weathering conditions are usually found only in tropical and subtropical climates having wet cycles and year-round mineral weathering. Weathering may be 300 percent to 800 percent greater in these conditions than in temperate regions. *Gelisols,* a proposed order, is not discussed.

Other land areas in the United States are designated as "miscellaneous land types" and total 4.5 percent of all land areas in the fifty states.

7:7 Soil Order: Histosol (Greek, *histos,* tissue); organic soils

The suborders are as follows:

Fibrists (Latin *fibra;* fiber): largely undecomposed plant fibers
Folists (Latin, *folia;* leaf): mostly leaf mat accumulations
Hemists[10] (Greek, *hemi;* half): half are recognizable plant fibers
Saprists (Greek, *sapros;* rotten): unrecognizable fibers due to prolonged decomposition

Properties and Classification of Histosols

Histosols are organic soils. For soils formed in or under water, the required distinguishing feature for a Histosol is a minimum *organic carbon* content of 12 percent when the mineral portion has no clay, increasing proportionately to 18 percent as the mineral portion increases

[10]Suborder illustrated by Figure 7-3.

to 60 percent or more clay. For soils formed on a lithic (shallow to rock) contact, organic carbon content must be 20 percent or more. Organic soils have higher water retention, high cation-exchange capacities, the usual nutrient deficiencies (particularly nitrogen, potassium, and copper), and low bulk densities (as low as 0.1 Mg/m^3, or 6.3 lb/ft^3).

Histosols are common in wet, cold areas, such as Alaska, Finland, and Canada and in wet areas having stagnant marshes and swamps, such as Ireland.

The classification of Histosols is based upon both their physical properties (kind of organic material, stage of decomposition, thickness, bulk density, water-holding capacity, temperature, permeability) and their chemical properties (presence of carbonates, bog iron, sulfates and sulfides, the pH, basic cation saturation, and carbon:nitrogen ratio). The extent of decomposition is estimated by the portion of the material recognizable as to source (leaves, plant stem, kind of plant) (Figure 7-3). In the 1938 U.S. classification system, Histosols were known as **Bog** soils.

Management of Histosols

Some management problems of Histosols are unique. When drainage is developed, the soil organic matter decomposes rapidly and drastically reduces the soil volume. Land surfaces have dropped in elevation (**subsided**) as much as 3.7 m (12 ft) in a 40-year period after drainage of the organic soil (Figure 7-4). The organic matter will burn; setting brush fires, draining fuel from motorized equipment, and smoking pose real hazards. For seed bed preparation the soil may need to be rolled by 25,000-pound packers to firm it rather than plowed to loosen it. It is difficult when planting to get good seed–water contact.

Histosols are used for many valuable vegetable crops. Costs for drainage and the large amounts of lime usually required, if the soils are acidic, make it economically necessary to grow high-value crops.

Distribution of Histosols

Histosols occur in large bodies in Florida and Georgia and in numerous areas in the Northeast and Great Lakes states, totaling 0.5 percent of all soils in the United States. Outside the

FIGURE 7-3 This *Grindstone Series* is mapped as Grindstone peat (*order:* Histosols; *suborder:* Hemists), deep phase. It has a 30-cm organic-material horizon (organic fiber) made of fibrous, spongy, sphagnum peat moss; is medium acid; and has a 76% identifiable fiber content. With further depth, fibers of mosses and herbaceous mixed residues decrease gradually from 54% at the 30-cm depth to 45% at 121 cm. Clay layers exist below 121 cm. Infiltration rate is rapid when saturated. Manitoba, Canada. Scale is 15-cm intervals. Terminology is that of Canada and is not identical to the new U.S. system. (*Source:* Canada Department of Agriculture.)

FIGURE 7-4 Subsidence, the sinking of the land, occurs on organic soils when they are drained and the organic matter rapidly decomposes, drastically reducing the soil volume. The 2.7-m deep concrete post in the photo was placed in this Terra Ceia (Saprist) soil down to limestone bedrock in 1924, with its top flush with the soil surface. The photo, taken in 1972, shows an average subsidence of about 2.5 cm per year. Florida. (Courtesy of Victor W. Carlisle, University of Florida.)

United States the largest continuous body of Histosols is in Canada, south of Hudson Bay; another large area is located in northwestern Canada. Histosols occur on all continents, but in areas too small to be shown on the soil map of the world. The worldwide extent of Histosols is only 1.2 percent of all soils, and they rank last in total area of all soil orders.

7:8 Soil Order: Entisols (recent soils), without pedogenic horizons

The suborders are as follows:

Aquents (Latin, *aqua;* water): overly wet part of the year
Arents (Latin, *arare;* to plow): horizon from plowing
Fluvents (Latin, *fluvius;* river): alluvial deposits
Orthents[11] (Greek, *orthos;* true): loamy or clayey textures
Psamments (Greek, *psammos;* sand): sandy profiles

Entisols are soils of slight soil development. No developed (pedogenic) horizons exist except possibly an **Ap** (plowed) or weak **A** from slight organic-matter accumulation. Entisols are identified by the absence of distinct pedogenic (naturally developed) horizons. An **ochric epipedon** may exist.

The lack of soil development does not imply that Entisols are simple, identical soils. They range from deep sand to stratified river-deposited clays and from recent volcanic-ash deposits (or erosion-exposed surfaces) to dry, arid lake beds.

---

[11]Suborder illustrated by Figure 7-5.

FIGURE 7-5 This *Colby Series* is a silt loam that grades from a 1.0-cm grayish-brown **A** into a 10-cm-thick **AC** horizon and finally into a parent material with lime accumulation (**Ck**) at 20 cm deep. The fairly dry climate where this soil is found does not have enough precipitation to leach lime out of the profile, and the soil pH is moderately alkaline. Infiltration rate is moderate when saturated. The complete taxonomy: *Order*—Entisols; *Suborder*—Orthents; *Great Group*—Torriorthents; *Subgroup*—Ustic Torriorthents; *Family*—Ustic Torriorthents, fine-silty, mixed, calcareous, mesic; *Series*—Colby, Kansas. *Torr* = very arid climate, *orth* = typical, *ent* = Entisols order. Scale is in feet (feet × 30.5 = cm). (Courtesy of Arvad Cline, USDA—Soil Conservation Service.)

Properties and Classification of Entisols

The various conditions creating Entisols indicate some of their profile characteristics and their classification. The absence of distinct pedogenic horizons in Entisols may be due, for example, to the following:

- Presence of a parent material too inert to develop soil horizons, such as quartz sands. These sand deposits could have existed for centuries with little profile development.
- Formation of the soil from a parent material that dissolves almost completely with very little residue, such as some rare limestones composed mostly of carbonates.
- Insufficient time to develop horizons, as in recently deposited volcanic ash, river terrace alluvium, or other recent alluvial deposits.
- Ecological conditions not conducive to horizon formation, as is true of soil on the moon, in very dry deserts (Figure 7-5), and in permafrost areas.
- Occurrence on steep slopes where the rate of surface erosion equals or exceeds the rate of soil profile formation or where the deposition covers soil too frequently for development.

Entisols are a soil order that includes the former **Azonal** soils and a few **Low Humic Gley** of the 1938 soil classification system.

Management of Entisols

Many recent alluvial deposits may be excellent and productive soils. Much of deposited alluvium is topsoil eroded and deposited over many years, even centuries. The soil texture can vary from gravels through sands to silts and clays. Some Entisols may exist on river terraces above flooding levels, while other areas may be subjected to periodic flooding. Well-sorted beach sand deposits, deposited in geologic ages past, may be inert and infertile. Thus, some Entisols may be ancient river terraces of excellent texture, high fertility, deep soils, and highly productive. Poorly drained and flooded low delta areas may be wet, poorly aerated clays (such as the Mississippi River delta) and be difficult to cultivate. Recent volcanic-ash deposits (many areas, including the St. Helens blast in 1980) may be erosive because of the high silt and fine-sand contents.

Distribution of Entisols

Entisols are widely distributed in the United States and include river floodplains, rocky soils of mountainous areas, barren islands of the East and Gulf coasts, and beach sands. They occupy about 8 percent of all U.S. soils. Entisols are on all continents, occupying about 11 percent of the world land surface; they rank sixth in area among the soil orders. Rainfall may be high or low, and vegetation may be forest, savanna (grass, shrubs, occasional trees), or grass. Precipitation and vegetation are, therefore, not diagnostic of the soil order. Some Entisols are excellent agricultural soils (floodplains, volcanic ash), but others are quite poor soils.

7:9 Soil Order: Inceptisols (Latin, *inceptum;* beginning, inception), pedogenic horizons

The suborders are as follows:

Aquepts (Latin, *aqua;* water): overly wet a part of the year
Ochrepts[12] (Greek base of *ochros;* pale): light-colored surface
Plaggepts (modified from German, *plaggen;* sod): human-made organic surface
Tropepts (modified from Greek, *tropikos;* of the solstice, tropical): uniform year-round temperature
Umbrepts (Latin, *umbra;* shade): acidic, dark-colored surface

Soils of the **Inceptisols** order are weakly developed. They lack sufficient development to fit into other orders but have more development then Entisols.

Properties and Classification of Inceptisols

Inceptisols are more weathered and developed soils than Entisols. As the suborder Ochrepts name indicates, some Inceptisols have thin or light-colored **A** horizons. Other Inceptisols have dark surface horizons, either naturally developed (Umbrepts) or formed by humus additions by people (Plaggepts). The lack of profile and horizon development may be due to several factors, including the following:

- The deposit may be recent (few decades). The weathering processes (clay movement, lime movement, weathering) have been active too short a time to develop strong horizons.
- The parent material may be resistant to weathering, thereby slowing development. Some quartz sands may be Inceptisols.
- Erosion (or deliberate horizon mixing as in rice-paddy puddling) may be just fast enough to remove the developing soil before strong horizons can form. Slow, periodic

[12]Suborder illustrated by Figure 7-6.

depositions (as in floodplains or by wind erosion) would also be effective in hindering development of a pedogenic horizon.

- Wetness, cold, and other conditions slowing translocation and weathering in the soil allow Inceptisols to exist longer than in better aerated or warmer soils (Figure 7-6).

Many rice paddy soils and poorly drained soils are Inceptisols. Inceptisols were officially known in the United States before 1965 as **Brown Forest, Low Humic Gley,** and **Sol Brun Acide.**

Management of Inceptisols

Inceptisols are as variable as the Entisols. Many are very fertile; others are productive because of excellent climates or textures plus good management. Some Inceptisols may be overly wet or exist in the cold regions of the world (Alaska, Siberia). Many of the soils used for rice production in Asia and the Pacific Islands are large areas of Inceptisols. Waterlogging and soil mixing and puddling of these soils for centuries have hindered profile development. People of many of these areas have tilled the soil, added large amounts of organic wastes and manures as fertilizers, and in other ways produced crops, but they have hindered profile development.

Distribution of Inceptisols

Inceptisols in the United States occur mostly in the Middle Atlantic and Pacific states, accounting for about 16 percent of all soils in the 50 states. They develop in many climates. On a global basis, Inceptisols occupy about 16 percent of the land surface; they rank third in area of the soil orders. The largest area of Inceptisols is in mainland China; extensive areas also are on islands of the East Indies, the Northern cold regions, and along the Amazon River basin.

FIGURE 7-6 This *Minoa Series* is a very fine, sandy loam with a 25-cm-deep plow layer (**Ap**), dark colored, and medium acid. Yellowish and brown mottles occur at 25 cm–56 cm in the granular structure, which indicate both some soil development (**B**) and periodic water tables. Mottles become more distinct with depth (waterlogging is more severe) through the **BC** to parent material (**C**) at 81 cm deep. Infiltration rate is slow when saturated. The complete taxonomy: *Order*—Inceptisols; *Suborder*—Ochrepts; *Great Group*—Eutrochrepts; *Subgroup*—Aquic Dystric Eutrochrepts; *Family*—Aquic Dystric Eutrochrepts, coarse-loamy, mixed, mesic; *Series*—Minoa. New York State. *Aqu* = excess water much of the time, *dystr* = low basic cation saturation, *eutr* = high basic cation saturation, *ochr* = presence of ochric epipedon, *ept* = Inceptisols order. Scale is in feet (feet × 30.5 = cm). (*Source:* USDA—Soil Conservation Service.)

7:10 Soil Order: Andisols (Japanese, *Ando,* black soil; high volcanic ash content)

Suborders: (Udands eliminated; no longer a suborder)

Aquands (Latin, *aqua;* water): overly wet a period of the year
Cryands (Greek, *kryos;* icy cold): very cold, cryic or pergelic temperatures
Torrands (Latin, *torridus;* hot and dry): torric (dry) moisture regime
Ustands (Latin, *ustus;* burnt): ustic (dry winter) moisture regime
Vitrands (Latin, *vitrum;* glass): presence of volcanic glass
Xerands (Greek, *xeros;* dry): xeric (dry summer) moisture regime

Andisols are mostly soils that are weakly to moderately developed. The majority of Andisols form from volcanic ejecta (ash, cinders, pumice, or some basalts). Andisols must have enough soil development to distinguish them from Entisols but not so much as to mask the influence of the unique parent material. After extensive weathering, Andisols may become soils of other orders. Most Andisols have been developing for less than 5000–10,000 years.

Andisols must have *andic soil properties* in a cumulative thickness of 35 cm (13.7 in) or more within the top 60 cm (23.4 in) of the soil.[13] The **andic properties** include (1) low bulk density, (2) potential for wind erosion, (3) amorphous clays, (4) high macroporosity with rapid drainage, and (5) low soil strength when mechanically disturbed. The **melanic** epipedon (at least 6 percent organic carbon through a 30-cm depth) was defined for these soils.

Properties and Classification of Andisols

Mineral transformations in Andisols are in early stages forming largely amorphous clays (allophane, imogolite, ferrihydrite). **Imogolite** is a slightly crystalline allophane, and **ferrihydrite** is an amorphous iron hydrous oxide. Andisols exist in the total range of climates from cold to hot and from wet to dry. **Andic** properties must exist in at least 35 cm of the top 60 cm of soil. This allows for some minimal mixing with other soil materials.

The criteria for defining an *Andisol* are complex but can be approximated by the following items: (1) more than 60 percent by volume of volcaniclastic materials coarser than 2 mm, *or* (2) a bulk density of <900 kg/m^3 (lighter than water), *or* (3) considerable volcanic-glass content. In addition Andisols should have (4) a very high phosphate retention (high oxalate-extractable Al and Fe). Finally, (5) the soils will contain mostly early-stage minerals, particularly amorphous clays. There will be few crystalline clays (smectites, kaolinite, vermiculite) and minimal translocation of colloids.

Management of Andisols

Many Andisols are among the most productive soils of the world when managed well. The highest measured rice yield (about 10 Mg/ha) was produced on volcanic-ash soil in the Philippines. Many volcanic soils of Hawaii are highly productive.

[13]R. L. Parfitt and B. Claydon, "Andisols: The Development of a New Order in Soil Taxonomy," *Geoderma* **49** (1991), 181–198.

Andisols tend to have large amounts of humus (7%–12% organic carbon contents in many soils). The amorphous allophane clays have very high cation exchange capacities (often 150 $cmol_c$/kg, which is higher than for montmorillonite). Unfortunately these soils also rapidly adsorb or precipitate phosphorus. The efficiency of added fertilizer phosphorus is often less than 10 percent, compared to 10 percent to 30 percent in most soils. This phosphorus problem is caused by the high contents of amorphous Al and Fe clays.

Andisols hold large amounts of water, but when they dry these soils may be difficult to rewet, and the dry soils are often loose and dusty. Erosion is a concern even though the aggregates are often quite resistant to disintegration by rainfall.

Distribution of Andisols

Andisols are extensive in many islands of the Pacific Ocean (Hawaii, Indonesia, The Philippines, Japan, the Aleutians). Chile has large areas of Andisols, as does New Zealand, Ecuador, east central Africa, and Central America. Additional areas are found in Spain, France, Italy, the United States, and many other countries. Almost 2 percent of the United States and of the World soils are Andisols. Prior to 1965 Andisols were known as **Ando-** and some of several other suborders and great groups.

7:11 Soil Order: Aridisols (Latin, *aridus;* dry, dry more than 6 months a year)

The suborders (Orthids eliminated; six suborders recently added) are as follows:

Argids[14] (Latin, *argilla;* white clay): clay accumulation as argillic or natric horizons
Calcids (Latin *calcis;* lime): has a calcic horizon
Cambids (Latin *cambiere;* to exchange): has a cambic horizon
Cryids (Greek, *kryos;* icy cold): very cold, cryic or pergelic temperatures
Durids (Latin *durus;* hard): has a duripan
Gypsids (Latin *gypsum;* gypsum): has a gypsic horizon
Salids (Latin base *sal;* salty): salty soils

A long dry period and only short periods of wetness in the upper soil dominate the soil-forming processes of **Aridisols.** A lack of water reduces leaching of basic cations, retards mineral weathering, and may allow accumulation of soluble salts.

Properties and Classification of Aridisols

The increase of suborders from two to seven for Aridisols indicates the great variety that can exist in arid-region soils. In temperate areas annual rainfall for Aridisols may be as high as 300 mm to 350 mm (about 12 in–14 in) per year. Some of the soils will have carbonates and/or soluble salts throughout the profile. Other soils may have some shallow leaching to about 15 cm to 40 cm deep and no salt accumulation. In some soils layers of carbonate accumulation exist. In some of these soils, deep, thick carbonate and silica-cemented layers occur. These are believed by some to have developed in ancient times during a period with a

[14]Suborder illustrated by Figure 7-7.

wetter climate. Some of these argillic horizons may also seem to be too thick or strongly developed for the present dry climates.

The aridity restricts plant growth, resulting in low soil organic-matter contents in the soils (0.5%–1.5%). The soils may be of almost any texture. Leaching is slight, basic cation saturation is about 100 percent, and primarily minerals make up most of the soils forming from parent rock.

Some of the profile features occurring frequently in Aridisols are (1) lime layers, (2) salt or gypsum accumulations, (3) low organic-matter accumulation, and (4) a calcareous profile. Some soils have lime-cemented hardpans (duripans or caliche); others have clay accumulation **Bt** (argillic) horizons (Figure 7-7). The **Bt** horizons may be very old (other climates) or weathered in place in some soils. Natural vegetative cover commonly is various scattered desert shrubbery, such as sagebrush, rabbitbrush, mesquite, shadscale, creosote bush, and shortgrasses in hot deserts. Desert regions also occur in *cold*-climate areas; the soils are commonly Aridisols and dry Entisols, and the vegetative cover differs from those of *hot*-climate areas.

Aridisols include the former great soil groups **Desert, Reddish Desert, Sierozem, Solonchak,** a few **Brown** and **Reddish Brown,** and associated **Solonetz** of the 1938 soil classification system.

Management of Aridisols

Many Aridisols are among the most productive soils when they are irrigated and fertilized. Many of the productive valleys of California, Arizona, Utah, New Mexico, and western Texas are mostly Aridisols. The soils' low humus contents make addition of nitrogen essential. In contrast, a lack of leaching allows potassium accumulation. Potassium deficiency is rare, but it is found in sandy soils and a few shallow soils or in soils developed from low-potassium parent material. Soil pH values are about 7 to 8.5. Deficiencies of zinc and iron—and, to a lesser extent, manganese and copper—are common.

Irrigation is essential. If the soils are not already salty, irrigation waters may add enough salt to develop salty soils. Sunny climates make production good if fertilizer needs are met.

FIGURE 7-7 This *Dixie Series* is gravelly, loam rangeland soil with a 15-cm **A** of dark brown granular loam. Its pH is 7.2; it has about 20% gravel. A clay accumulation (**Bt**) occurs at 23 cm–38 cm. Between 38 and 91 cm, a weakly to strongly lime-cemented hardpan (*calcic* endopedon, *duripan,* or caliche) and labeled as a **Ck** is too hard for roots to penetrate unless it is fractured (see arrow). Below 91 cm is soft, massive, calcareous, very gravelly loam of pH 7.9. Infiltration rate is slow when saturated. Scale is in feet (feet × 30.5 = cm). The complete taxonomy: *Order*—Aridisols; *Suborder*—Argids; *Great Group*—Haplargids; *Subgroup*—Xerollic Haplargids; *Family*—Xerollic Haplargids, fine-loamy, mixed, mesic; *Series*—Dixie. Utah. *Xer* = dry summers, *oll* = Mollisol-like, *hapl* = minimum horizon development, *arg* = argillic horizon, *id* = Aridisol order. (*Source:* USDA—Soil Conservation Service.)

Distribution of Aridisols

Aridisols in the United States are located primarily in the Western Mountain and Pacific states in areas of low rainfall where scattered grasses and desert shrubs dominate the vegetation. They make up more than 11 percent of all soil orders in the United States. Worldwide, Aridisols rank *first* in area of all soil orders at more than 23 percent (Figure 7-8). The Sahara Desert is shown on the map of Figure 7-8 as Aridisols, as are large areas of central and southern Asia, Australia, southern Africa, and southern South America.

7:12 Soil Order: Mollisols (Latin, *mollis;* soft), organic-rich surface horizons

The suborders are as follows:

Albolls (Latin, *albus;* white): have a leached **E** (albic) horizon
Aquolls (Latin, *aqua;* water): overly wet a portion of the year
Borolls (Greek, *boreas;* northern): cold areas
Rendolls (from *Rendzina;* Polish word for *noise* in plowing dark, limey, clay soils): high lime percentage in parent material
Udolls[15] (Latin, *udus;* humid): adequate water most of the year
Ustolls (Latin, *ustus;* burnt): dry many months, but water in summer
Xerolls (Greek, *xeros;* dry): dry summer, some leaching in winter

Mollisols are dark-colored soils of grasslands and some hardwood forests. Their distinguishing feature is a deep, dark-colored, surface **A** horizon (**mollic epipedon**) that is thick, dark in color, strong in structure (due to the presence of organic matter), and more than 50 percent saturated with basic cations (mostly calcium). In the profile picture the mollic epipedon extends from the surface to a depth of 55 cm (21.5 in), as shown by arrows on the scale (Figure 7-9).

Properties and Classification of Mollisols

Mollisols have a large number of suborders. Suborder names indicate the wide distribution of Mollisols in cold, temperate, and humid-to-semiarid climates. Some form on parent material high in calcium carbonate (40 percent or more lime). Others are well leached (Albolls) or exist in wet conditions (Aquolls).

Mollisols were known formerly as **Rendzina, Prairie, Chernozem, Chestnut, Brunizem,** and associated **Solonetz** and **Humic Gley** in the 1938 soil classification system.

Management of Mollisols

Mollisols, formed under grasses or some broadleaf forests, tend to be some of the most fertile soils. They have higher humus and, thus, higher nitrogen (and some other nutrients) than the Aridisols have. Mollisols in the wetter climates (Udolls) do not need irrigation and have been highly productive soils for centuries, even though they may be acidic and need lime additions.

---

[15]Suborder illustrated by Figure 7-9.

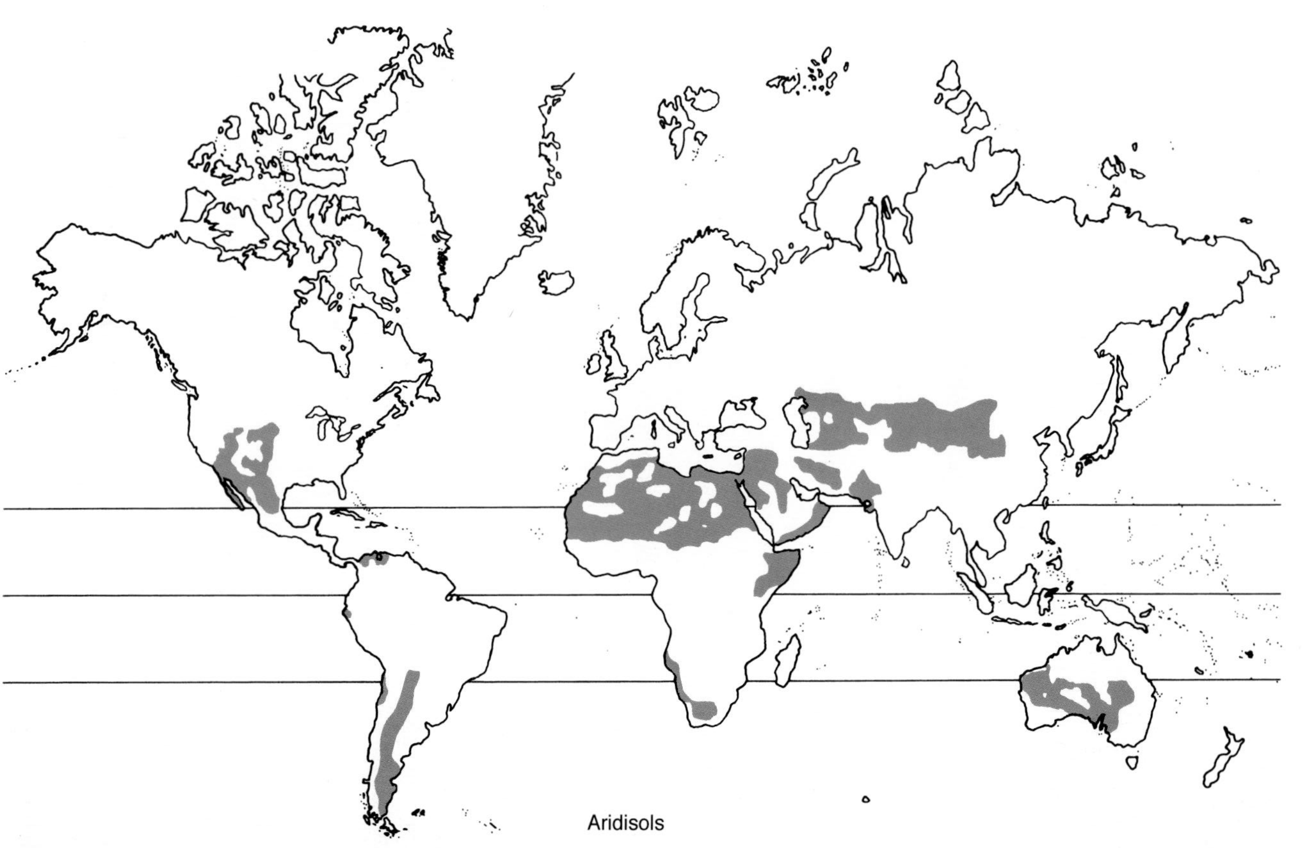

FIGURE 7-8 World distribution of Aridisols. Occupying 23% of the world land area, Aridisols are the most extensive soil order; they occur in areas of low rainfall. (Courtesy of Raymond W. Miller, Utah State University.)

FIGURE 7-9 This *Sharpsburg Series* is a silty clay loam with a dark surface soil to 28 cm (**Alp. Al2**), granular and subangular blocky structure, friable, and slightly acid. With further depth the profile grades through a transition zone (**AB**) to a thick clay accumulation layer (**Bt** argillic) from 43 cm–112 cm. This clayey layer is strongly acid, moderately dark, with prismatic structure breaking to blocky structure. Some deep wetness is indicated by a few mottles. The mollic epipedon is about 55 cm deep. The parent material begins at 124 cm deep. Infiltration rate is moderate when saturated. The complete taxonomy: *Order*—Mollisols; *Suborder*—Udolls; *Great Group*—Argiudolls; *Subgroup*—Typic Argiudolls; *Family*—Typic Argiudolls, fine, montmorillonitic, mesic; *Series*—Sharpsburg, Iowa, *Ud* = seldom has drought, moisture well distributed; *montmorillonitic* = has more than 50% of clay as montmorillonitic clay. Scale is in feet (feet × 30.5 = cm). (*Source:* USDA—Soil Conservation Service.)

The subhumid areas, such as the Great Plains of the United States, have the well-known *Chernozem* soils (of the old classification) with their black surface soils, often 60 cm to 80 cm (about 23 in–31 in) deep. Dryland wheat and sorghum are grown on many of these soils and on even drier Mollisols.

Because about one-fourth of U.S. soils are Mollisols, it is obvious that there are enormous variations in climates and crops grown on these soils. The drier Mollisols grow dryland grains but are also irrigated for many kinds of crops.

The higher humus content in Mollisols than in Aridisols and the limited leaching, except in the wetter suborders, make most Mollisols quite fertile soils without fertilization. Very little—if any—lime is needed, except in wetter climates. Only Alfisols, perhaps, have a higher *natural* productivity. The soils' textures, depths, climates, and lack of inhibiting conditions are valuable.

Distribution of Mollisols

The largest contiguous body of Mollisols in North America is in the Great Plains, extending northward into Canada and southward almost to the Gulf of Mexico. Other areas occur in the intermountain region in the West and Northwest. Mollisols are the most extensive of the U.S. soil orders, totaling nearly 25 percent of U.S. soils. The largest body of Mollisols in the world is in central Europe, central Asia, and northern China (Figure 7-10). A second large area is in Argentina and adjoining countries in South America. Worldwide, Mollisols rank seventh in area and total about 4 percent of all soils of the world. Precipitation is subhumid to semiarid in regions varying from north temperate to alpine (mountainous) to tropical.

7:13 Soil Order: Vertisols (Latin, *verto;* turn), self-swallowing clays

The suborders are as follows:

Aquerts (Latin, *aqua;* water): reducing, waterlogged conditions for some time period
Cryerts (Greek, *kryos;* icy cold): cold-area soils
Torrerts (Latin, *torridus;* hot and dry): in tropical areas, quite dry season

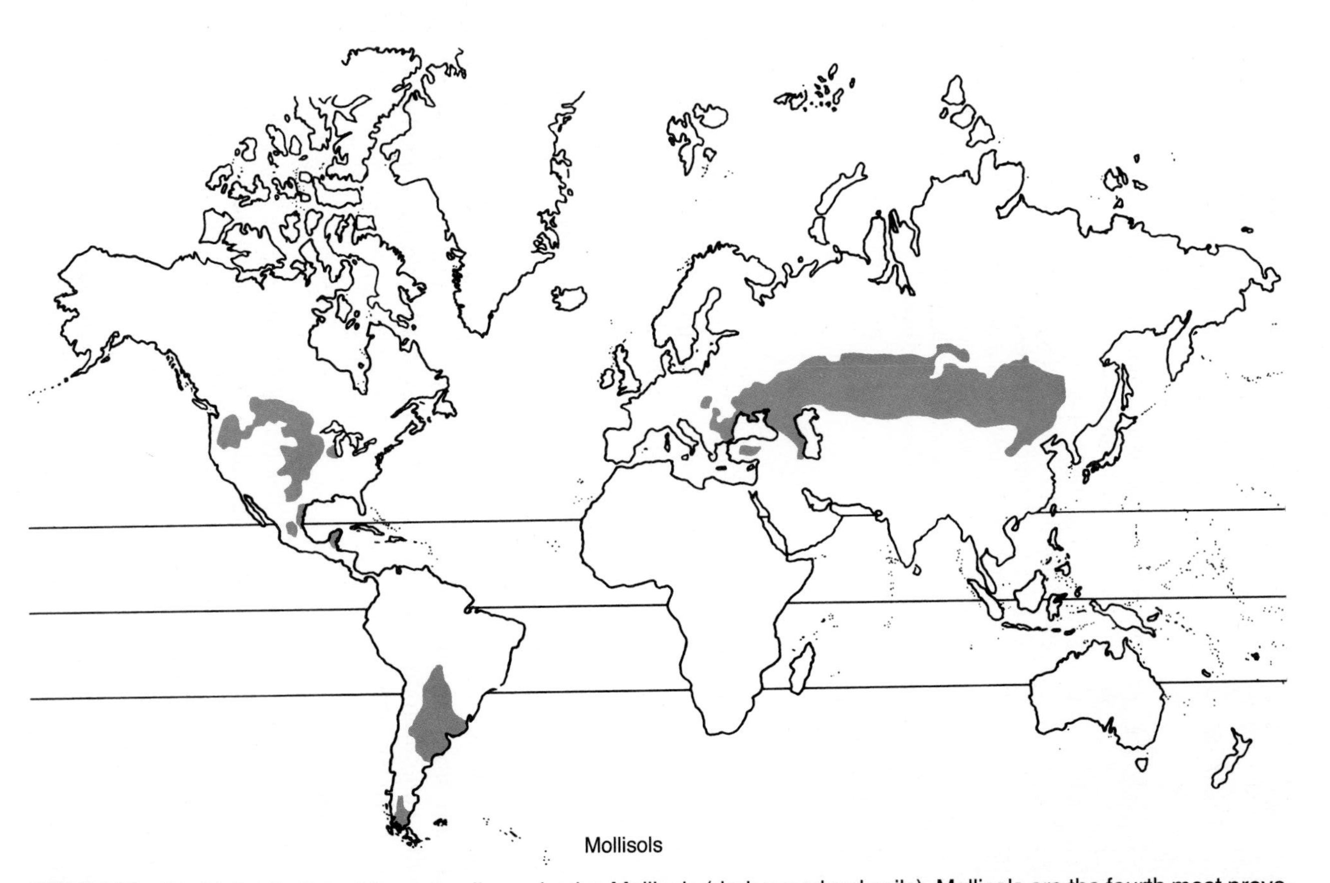

FIGURE 7-10 World distribution of the naturally productive Mollisols (dark grassland soils). Mollisols are the fourth most prevalent soil order, covering 4.1 percent of the world land surface, and the majority are in cultivation. Some of the Mollisols in drier climates can be profitably irrigated. (Courtesy of Raymond W. Miller, Utah State University.)

Uderts (Latin, *udus;* humid): adequate water most of the year
Usterts[16] (Latin, *ustus;* burnt): dry many months, adequate water most summers
Xererts (Greek, *xeros;* dry): dry summer, some leaching in winter

Vertisols are *self-mixing* soils, with more than 30% swelling clays. Vertisols develop from parent materials high in limestones and marl or from basic rocks, such as basalt. Vertisols expand and contract more than the soils of any other order because of their high swelling-clay contents. These *expansive* clays (high-swelling and sticky) are difficult to cultivate and are poor support for roadbeds and buildings. The swelling and shrinking churns and mixes the upper horizons, disrupting horizon development in most of the **A** and **B** horizons. They are resilient soils resisting degradation, but they may contribute to pollution through water flow through their cracks or in selective adsorption of certain organic solvents.[17] **Gilgai** (wavy soil surface) often develops on Vertisols.

Properties and Classification of Vertisols

By definition, Vertisols have more than 30 percent clay (mostly montmorillonitic) and 30 $cmol_c$/kg cation exchange capacity, and when dry they have wide, deep cracks more than 1 cm (0.4 in) wide *at a depth of 51 cm (20 in).* Vertisols occur in humid, semiarid climates with noticeable wet and dry cycles. The vegetation is dominantly tall grasses and scattered trees and shrubs.

Vertisols have profiles of deep **A** horizons, often more than 1 m (39 in) deep and usually no **B** horizon. Vertisols also show evidence of vertical and angled mass movement during swelling by the presence of wavy soil surfaces (**gilgai** micro relief), **slickensides** (smoothed pressure surfaces on peds), and wedge-shaped (parallelepiped) compound subsoil aggregates tilted at an angle from the horizontal. *Vertical movement* of 3.7 cm (1.4 in) at a depth of 91 cm (3 ft) due to dislocation during wetting–drying cycles was measured for a Vertisol similar to that in Figure 7-11. The light-colored horizon below 91 cm (3 ft) in the Vertisol in Figure 7-11 contains more than 50 percent calcium carbonate.

Vertisols were known as **Rendzinas** in the 1938 soil classification system and later as **Grumusols.**

Management of Vertisols

Cultivation of Vertisols is difficult. When wet the soils are very sticky; when dry they are very hard. Thick, hard soil crusts can inhibit seedling emergence. Plowing when too dry produces large clods and requires enormous power. When wet and cracks swell shut, the soils have low permeability to water. Often, about the only water the soil absorbs is what immediately fills the dry cracks. Thus, wetting may be shallow, only 40 cm to 60 cm (16 in–23 in), even with prolonged soil-water contact time. Traffic on the soil when it is wet is almost impossible and will compact the soil and leave deep tracks.

Vertisols generally are quite fertile, with high cation-exchange capacities and relatively high humus contents to great depths (>1 m); however, aeration of the wetted soil is quite poor. This restricts root growth, and penetration and rooting is often *shallow.* During drying cycles cracking soil breaks many roots, retarding plant growth. Infiltration rates can begin at

---

[16]Suborder illustrated by Figure 7-11.
[17]C. E. Coulombe, L. P. Wilding, and J. B. Dixon, "Overview of Vertisols: Characteristics and Impacts on Society," *Advances in Agronomy* **57** (1996), 290–377.

FIGURE 7-11 This *Houston Black Series* is a clay with the top 97 cm divided into three **A** horizons (a mollic epipedon), which differ from each other in structure mostly; the deeper two horizons differ in color also. Wormcasts, lime and lime concretions, and iron–-manganese concretions are found in the profile. The layers at a depth of 97 cm–264 cm are called **AC** horizons. They have many features of the **A** horizons but are much lighter in color, lower in organic matter, and mottled. Infiltration rate is very slow when saturated. The complete taxonomy: *Order*—Vertisols; *Suborder*—Usterts; *Great Group*—Pellusterts; *Subgroup*—Udic Pellusterts; *Family*—Udic Pellusterts, fine, montmorillonitic, thermic; *Series*—Houston Black. San Marcos, Hays County, Texas. *Pell* = low chroma, *ert* = Vertisols order, *thermic* = annual average temperature between 15°C and 22°C (59°F–72°F). Scale is in feet (feet × 30.5 = cm). (Courtesy of Alan Anderson, USDA—Soil Conservation Service.)

10 cm/hr while cracks are being filled, then slow to less than 0.2 cm/hr after the top soil layer wets and swells.

Wetted Vertisols have low support strength. Roads and railroads on Vertisols often become deformed or misaligned by the weight and vibration of traffic during wet soil conditions. Some soil mechanical engineers refer to the *expansive* Vertisol as *expensive* soils because they require extra engineering and building costs. Telephone poles and other vertical poles lean after years of wetting–drying cycles and windstorms during wet periods (Figure 7-12).

Distribution of Vertisols

Vertisols occur primarily in central and southeastern Texas and, to a lesser extent, in Alabama, Mississippi, and California, totaling 1 percent of all soil orders in the United States. Outside the United States, large areas of Vertisols occur in India, Australia, and eastern Africa. Ranking ninth in area worldwide, Vertisols make up more than 2 percent of the total land surface of the world.

7:14 Soil Order: Alfisols (*al*—aluminum, *fer*—ferric iron), movement of Al, Fe, and clay into the B horizon

The suborders are as follows:

Aqualfs (Latin, *aqua;* water): seasonally saturated with water
Boralfs[18] (Greek, *boreas;* northern): cold area
Udalfs (Latin, *udus;* humid): adequate water most of year
Ustalfs (Latin, *ustus;* burnt): dry in winter, most of its water in summer
Xeralfs (Greek, *xeros;* dry): dry in summer, some leaching in winter

[18]Suborder illustrated by Figure 7-13.

FIGURE 7-12 Cracks in this Vertisol in southern India range from 2 cm (0.8 in) to more than 5 cm (2 in) wide. Cracking tears plant roots and speeds drying of the soil. (Courtesy of Raymond W. Miller, Utah State University.)

Alfisols are soils with high enough effective precipitation to move clays downward and form an **argillic** (clay accumulation) horizon. The basic cation saturation percentage is relatively high (>50 percent), and they are usually fertile. These soils usually form under various forests and brush cover.

Properties and Classification of Alfisols

Alfisols occur in many forests receiving marginal amounts of precipitation or in wetter areas where high-lime parent materials retard development of strong acidity. Some of the most notable features of Alfisols are as follows:

- Translocated clay in a **Bt** (**t** is for the German *ton;* clay) clay accumulation horizon (argillic), shown in Figure 7-13, occurring 23 cm to 74 cm (9 in–29 in) in depth (between arrows).
- A medium-to-high supply of basic cations, such as calcium and magnesium, which is evidence of only *mild* leaching. This is in contrast to Ultisols, which have had *severe* leaching.
- Water is available for good plant growth for three or more warm-season months.

Alfisols were formerly classified as **Gray-Brown Podzolic, Noncalcic Brown, Gray Wooded, Degraded Chernozem,** and associated **Half-Bog** and **Planosol** soils in the 1938 classification system.

FIGURE 7-13 This *Nebish Series* is a loam and has a thin 7.6 cm surface layer (**A**) with a 15 cm leached, light-colored layer (**E**) beneath it. Structure is weak; pH is neutral. The clay accumulation layer (**Bt** argillic) is clay loam texture, occurs (between arrows) from the 23-cm to 74-cm depth, has blocky structure, and is slightly acid. A transition layer (**BCt**) grades into parent material (**C**) of loam texture at 84 cm. The parent material is not strongly leached because it is still moderately basic in pH. Infiltration rate is moderate when saturated. The complete taxonomy: *Order*—Alfisols; *Suborder*—Boralfs; *Great Group*—Eutroboralfs; *Subgroup*—Typic Eutroboralfs; *Family*—Typic Eutroboralfs, fine-loamy, mixed, frigid; *Series*—Nebish. Minnesota. *Bor* = cold area, *frigid* = less than 8°C (47°F) average annual soil temperature. (Courtesy of William M. Johnson, USDA—Soil Conservation Service.)

Management of Alfisols

If relief and climate are favorable, many Alfisols produce well when converted to cropland. Most are leached of lime to at least half a meter or deeper and are slightly to moderately acid in the surface horizon. Leaching can be severe enough to form a leached horizon (**E**). Often the clayey accumulation is not favorable to plant growth, particularly if the surface is eroded, exposing the clay as a surface layer.

Without the assistance of irrigation or fertilization, Alfisols are probably the most naturally productive soils used for crops. They include many of the fertile soils of the corn belt in Indiana, Ohio, Michigan, and Wisconsin, as well as woodland-covered soils in Texas and Colorado. Alfisols occupy extensive portions of England, Europe, north central Russia, Ghana, and many other regions that were prominent agricultural areas in early colonization and developmental periods.

The moderate acidity in the upper soil profile may require addition of lime for the best growth of many crops. Although some of these soils exist in areas with adequate precipitation for good crop production without irrigation, long, dry periods can occur and supplemental irrigation systems are sometimes used. Addition of fertilizers (nitrogen, phosphorus, potassium) usually increases yields.

Distribution of Alfisols

As shown on the Soil Map of the United States (inside back cover), Alfisols occur in large bodies in the North Central states and in the Mountain states and make up about 13 percent of the total land area of the 50 states. Alfisols occupy more than 13 percent of the land surface of the world (ranking second in area) and occur on all continents (Figure 7-14). Humid and subhumid climates and tall grasses, savannas, and oak–hickory forests characterize the climate and native vegetation where Alfisols occur.

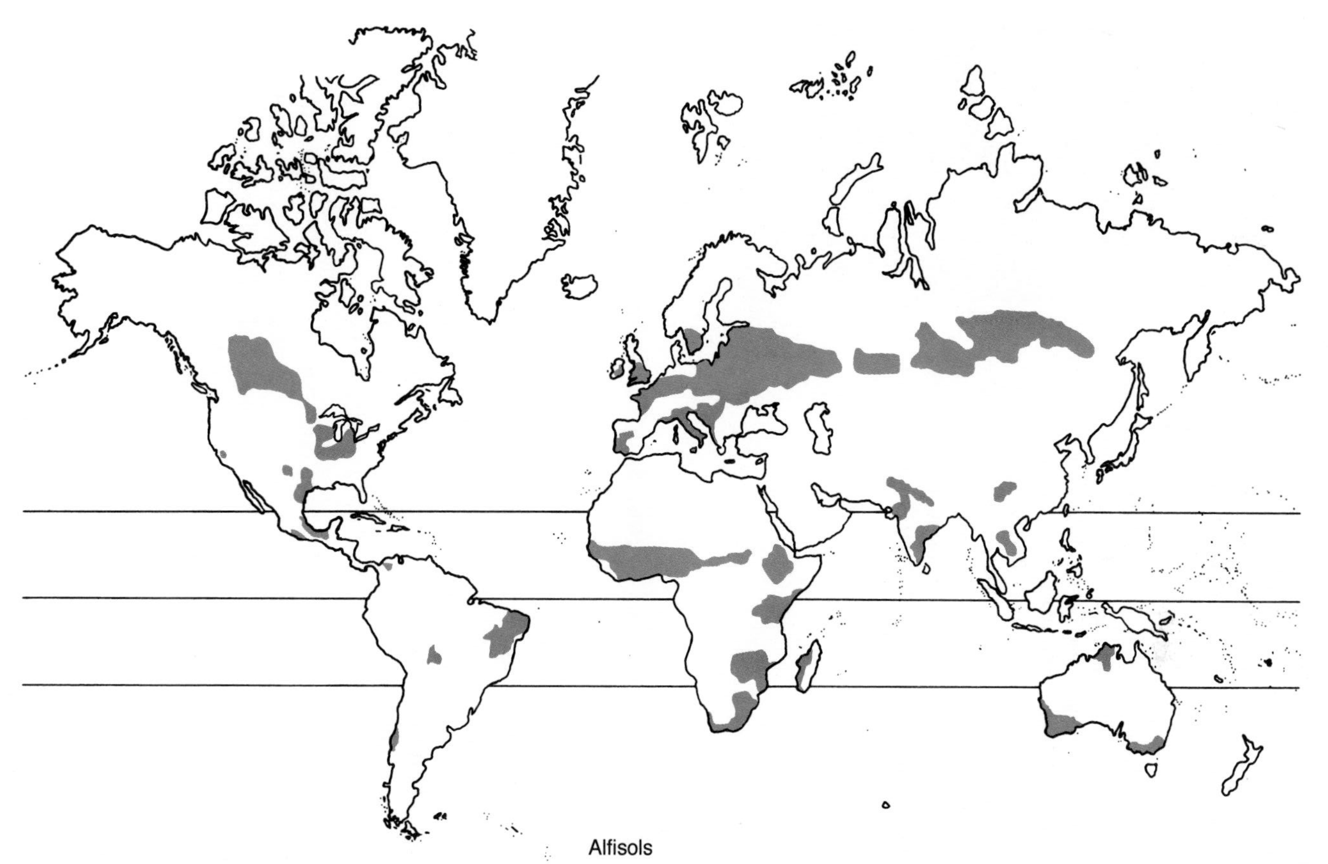

FIGURE 7-14 World distribution of Alfisols. Occupying 13.2% of the world land area (second largest soil order), Alfisols occur mostly under forests, especially of the broadleaf type. (Courtesy of Raymond W. Miller, Utah State University.)

FIGURE 7-15 This *Adams Series* is a sandy loam found in northern temperate and subarctic latitudes that has a strongly acid humus-and-roots layer (**Oe** or **Oa**) over a 10-cm-thick, gray, leached layer (**E,** albic). Below the **E** is a very strongly acid, dark-brown humus layer (**Bh**) 5 cm thick. The next 51 cm is brown sand with amorphous iron and aluminum oxide deposits (**Bs** or spodic). Parent material, a strongly acidic light-yellow sand, occurs at 66 cm deep. Infiltration rate is rapid when saturated. The complete taxonomy: *Order*—Spodosols; *Suborder*—Orthods; *Great Group*—Haplorthods; *Subgroup*—Typic Haplorthods; *Family*—Typic Haplorthods, sandy, mixed, frigid; *Series*—Adams. East Middlebury, Vermont. *Od* = Spodosol order. (Courtesy of A. R. Midgley, Vermont Agricultural Experiment Station.)

7:15 Soil Order: Spodosols (Greek, *spodos;* wood ash), gray color of E horizon

The suborders are as follows:

Aquods (Latin, *aqua;* water): seasonally saturated with water
Cryods (Greek, *kryos;* icy cold): very cold soils
Humods (Latin, *humus;* earth): humus accumulation is extensive
Orthods[19] (Greek, *orthos;* true): central concept, with humus and iron oxide accumulation

Spodosols, by definition, have high sand content. The high rainfall and easy leaching produces translocation of humus and/or sesquioxide colloids into a **spodic B** horizon. These soils usually occur in cold, wet climates, but are also found in Florida. Spodosols are usually found under acidic conifer forests or other vegetation that develops acidic soils and produces organic chemicals that mobilize iron, aluminum, and humus colloids.

Properties and Classification of Spodosols

Spodosols have moderately to strongly acidic sandy profiles with an ashy white upper horizon over a dark-brown **B** horizon and yellowish subsoils: The typical profile is (1) black organic litter and humified layer, (2) the white leached **E** layer called **albic** horizon, and (3) a thin deposition of **Bs** or **Bh** layer of iron oxides and/or humus colloids (Figure 7-15). These soils are typical of sandy soils under cool, wet, conifer, and deciduous forests. They are usually well leached; basic cation saturation percentage is very low. The unique features of

---

[19]Suborder illustrated by Figure 7-15.

Spodosols are the strongly acidic, leached, white **albic** horizon (**E**) and the sandy **B** horizons, often cemented with humus and noncrystalline aluminum and/or iron, which is a black or brown color (**spodic** horizon). The albic horizon may be missing (mixed with **Oa** and **B**) in cultivated soils. Clayey soils never develop into Spodosols because of the slow infiltration of clays.

There are many variations of Spodosols. A deep, 1.8-m (6-ft), iron-pan Spodosol occurs under a pigmy forest only 1.4 m to 3 m (4.6 ft–9.8 ft) tall in coastal California. In contrast, very shallow Spodosols, only 46 cm (18 in) deep, are found in Alaska. Spodosols were known in the 1938 soil classification system as **Podzol, Groundwater Podzol,** and **Brown Podzolic.**

Management of Spodosols

Cultivation of Spodosols requires considerable fertilization and lime additions to make them less acidic and more productive. Some acidic Spodosols are used without lime additions for blueberry cultivation in coastal North Carolina, but most cultivated Spodosols are used for crops adapted to the cooler North Central U.S. climate, such as grains, potatoes, strawberries, raspberries, pastures, and silage corn.

The typically cold climates of Spodosols restrict their uses to a relatively few crops. Spodosols also typically occur in areas anciently glaciated during the Pleistocene time and often have nonlevel relief and sometimes contain gravel and stones.

Distribution of Spodosols

Spodosols exist in large tracts in New England, at high elevations in the Middle Atlantic states, and in the northern part of the Great Lakes states, constituting 5 percent of all soil orders in the United States. Between latitudes 42° N and 60° N, Spodosols are common in Canada, northern Europe, and northern Asia. They rank eighth worldwide and occupy less than 4 percent of the world's land surface.

7:16 Soil Order: Ultisols (Latin, *ultimus;* last, highly leached, clay accumulation in B horizon)

The suborders are as follows:

Aquults (Latin, *aqua;* water): seasonally saturated with water
Humults (Latin, *humus;* earth): high-humus surface accumulation
Udults[20] (Latin, *udus;* humid): adequate water most of year
Ustults (Latin, *ustus;* burnt): dry many months, some water in summer
Xerults (Greek, *xeros;* dry): dry during summer, some leaching in winter

Ultisols are the warm, humid-area soils too low in basic cation saturation (acidic) to be Alfisols or Mollisols but not weathered enough to be Oxisols. Usually Ultisols have developed in humid climates, tropical-to-subtropical temperatures [average annual temperature more than 8°C (47°F)], and a forest or forest-plus-grass (savanna) vegetation. Many of the red-clay hills of the southeastern United States are the exposed **B** horizons of eroded Ultisols.

---

[20]Suborder illustrated by Figure 7-16.

Properties and Classification of Ultisols

Ultisols often have an **umbric** epipedon (dark-colored, strongly acidic **A**). The common characteristic profile features are a thick clay accumulation layer (**Bt**), which is moderate to strongly acidic, often a surface horizon dark with humus (**A**), and typically a leached layer (**E**). Intensive weathering forms the clayey **B,** often a reddish soil and having a basic cation saturation <35 percent. Appreciable quantities of primary minerals (feldspars, micas) are still present and possibly some illite or montmorillonite clays, although kaolinite dominates the clay fraction of most Ultisols. Base saturation in Ultisols decreases with depth because easily leached bases are gone, and the present sources of bases are fertilizer and organic residues on the soil surface.

Ultisols were formerly classified as **Red-yellow Podzolic, Reddish Brown Lateritic,** and associated **Planosol** and **Half-Bog** in the 1938 soil classification system.

Management of Ultisols

With a high level of management, Ultisols can be some of the world's most *productive* soils. They exist in areas that are frost-free for long periods and also in humid areas with enough rainfall for crops or with adequate water reserves for irrigation (Figure 7-16); however, their *nutrient reserve,* although better than that of Oxisols, is relatively low to moderate. Both fertilization and liming are necessary in continuous cultivation to produce moderate-to-high yields. Optimum yields on these areas require good management of fertilizer and liming alternatives and crop selection; insects are abundant and fungus diseases are prevalent in the warm, humid climate. Southern pines mixed with hardwoods are common as natural cover on these soils; timber production is profitable on many Ultisols.

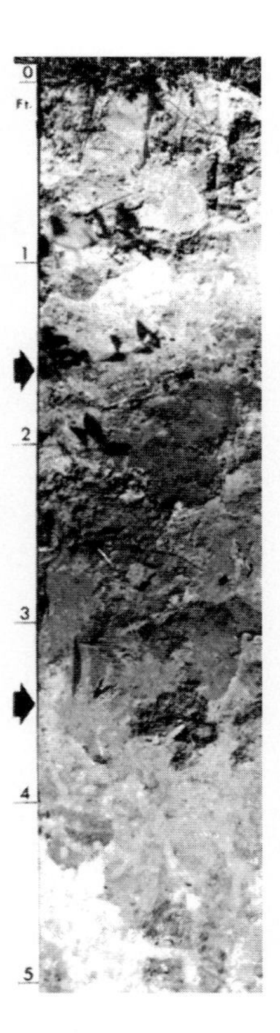

FIGURE 7-16 This *Ruston Series* has a moderately acidic, fine sandy loam plow layer 10 cm thick (an **Ap**), a leached (**E**) layer 30 cm thick, and a strongly acidic, 28-cm-thick, sandy clay loam (**Bt** or argillic) below the upper arrow. This younger soil above 109 cm is developing on an ancient, strongly acidic, deep **E′B′** profile (starting at the lower arrow). The ancient **Bt** lower boundary is at 234 cm deep (below bottom of the picture). The infiltration rate is moderate when saturated. The profile sequence from the soil surface to the depth of the photo is **Ap, E, B1t, B2t, 2E′, 2Bt′.** The complete taxonomy: *Order*—Ultisols; *Suborder*—Udults; *Great Group*—Paleudults; *Subgroup*—Typic Paleudults; *Family*—Typic Paleudults, fine-loamy, siliceous, thermic; *Series*—Ruston. Louisiana. *Pale* = old, excessive development, *ud* = seldom has drought, *ult* = Ultisols order, *thermic* = annual soil temperature is 15°C–22°C (59°F–72°F), *siliceous* = more than 90% of sand and coarse silt are silica minerals such as quartz and opal. (*Source:* USDA—Soil Conservation Service.)

Ultisols, without added fertilizers, often become worn-out soils, which is exactly what happened in the cotton fields in the southeastern United States in the mid 1800s. Humus is quickly decomposed in the year-round warm climate, and soon the supply of nitrogen from humus has declined. The high free-iron and aluminum hydrous oxides often keep soluble phosphorus at low values. The extensive leaching in the humid climate reduces plant-available soil potassium, making it commonly deficient in these soils. Ultisols are used for intensive crop production of a wide variety of crops.

Distribution of Ultisols

Ultisols are located mostly in the southern Atlantic states, the eastern South Central states, and in the Pacific states, mostly in subtropical climates. They account for almost 13 percent of the land area of the United States. Central America, South America, western Africa, southeastern Asia, and Australia all have large areas of Ultisols. Ranking seventh in area worldwide among the eleven soil orders, Ultisols occupy a little more than 8 percent of the world's land area.

7:17 Soil Order: Oxisols (French, *oxide;* oxides), very highly oxidized throughout profile

The suborders (Humox and Orthox are eliminated; Perox and Udox are added) are as follows:

Aquox (Latin, *aqua;* water): seasonally saturated with water
Perox (Latin, *per,* throughout; *udus,* humid): perudic moisture regime
Torrox (Latin, *torridus;* hot and dry): dry for half the year, arid seasons
Udox (Latin, *udus;* humid): Udic (wet) moisture regime
Ustox (Latin, *ustus;* burnt): dry winter, some water in summer

Oxisols are the most extensively weathered of all the soils. They are typically found on old landforms in humid, tropical, or subtropical climates. They are usually yellowish to bright red in color and are weathered to several meters in depth. Many changes have been made in the classification system for separations and criteria for Oxisols.

Properties and Classification of Oxisols

Oxisols have lost much of their silica (because primary minerals have weathered), and they are rich in the more residual iron and aluminum hydrous oxide residues, which have very low solubility. Sesquioxide and kaolinite clays dominate.

The unique feature of an Oxisol is the presence of the **oxic endopedon** and often that of **plinthite** (Detail 7-5 and Figure 7-17). Weathering is accelerated by high year-round temperatures and moisture, so the highly weathered Oxisols typically occur in continuously hot and humid, tropical and subtropical areas, usually under hardwood forests.

Oxisols have developed on old upland, medium-to-fine textured parent materials that have weathered into crystalline, kaolinitic-type clays with a net *negative* charge and amorphous iron and aluminum oxides with a net *positive* charge. Clay in the *oxic horizon often has a net positive charge* that is mostly a pH-dependent charge.

Weatherable minerals in Oxisols are either absent or present only in trace amounts. There are, therefore, small reserves of basic cations in these soils, mostly those on the

Detail 7-5 Oxic Horizons, Plinthite, Laterite

Oxic horizons, which are diagnostic for Oxisols, are usually unexciting in appearance but may be yellow to bright brownish-red in color. They are so excessively weathered that only a small percentage of primary minerals (feldspars, micas, ferromagnesium minerals) are left. These horizons consist dominantly of mixtures of materials resistant to weathering and solubility, such as hydrated oxides of iron and aluminum, titanium oxides, a small amount of quartz sands, and kaolinite.

Plinthite (from the Greek, *plinthos;* brick) is material high in iron and aluminum oxides found in many Oxisols. Wet anaerobic conditions solubilize reduced iron; drying allows it to solidify as oxidized and low-solubility iron oxides. After many decades plinthite may harden irreversibly to **ironstones,** which are cemented masses of material originally termed **laterite** (from the Latin, *later;* brick), which has been discarded by the U.S. soil taxonomy system because of its imprecise meaning.

When the softer plinthite is cut into building blocks (bricks), the wetting–drying cycles over several seasons will harden the bricks until they can be used for building (Figure 7-17). Clearing forests exposes the soil to erosion and more frequent wetting–drying cycles of the exposed plinthite; this speeds its conversion to ironstone. Optimum conditions necessary for *maximum* formation and hardening of plinthite into ironstone are as follows:

1. A fairly level land surface (for reduced erosion) situated at the foot of a seepage slope to accumulate iron in seepage water
2. An adequate supply of soluble iron from incoming seepage or as a residue from weathering
3. Alternating wet and dry seasons of approximately equal duration and sufficient rain continuously during the rainy period to saturate the zone of iron segregation to form temporarily anaerobic conditions in which iron is readily solubilized

Soils with ironstone are usually highly leached and strongly acidic and have low fertility. Nitrogen is the first nutrient to be limiting. Phosphorus is strongly adsorbed to iron oxides and will be a problem to keep available. Zinc, molybdenum, and copper are also usually low. The hardened ironstone layer restricts root elongation, decreases available water during dry periods, and causes poor drainage and excess water during rainy periods.

Ironstone exists as large rock layers or as smaller nodules (gravel) in soil. Some of this gravel is dug out from its pits and used as an all-weather surface material for gravel roads. Recovery of the ironstone (softening it) requires many decades of vegetative cover and rainfall. Complete recovery is unlikely to occur even after centuries of excellent conditions.

limited cation exchange complex, in plant tissue, and in very deep, less-weathered soil layers. This fact explains one of the principal causes of the usual rapid decline in crop yields when Oxisols are cleared and cultivated.

There is little *translocation* of clay in Oxisols; **Bt** horizons are, therefore, usually faint, diffuse, or absent. Clays (sesquioxides and kaolinite) in Oxisols are usually not dispersible by shaking in distilled water in contrast to clays in other soil orders. Clays often will settle in water like fine sand. High permeability and low erodibility are further characteristics of Oxisols. Poor management has resulted in large erosion losses. Oxisols were known as **Laterites** in the 1938 soil taxonomy system and as **Latosols** from 1950 to 1965.

Management of Oxisols

Oxisols have unique management requirements. Except for the nutrients cycled in organic matter, the soils are very low in nutrients. When the covering native vegetation is burned in

FIGURE 7-17 Hardened plinthite (ironstone, laterite) makes a good building stone, which gets harder with cycles of wetting and drying. Southern India. (Scale in upper photo is feet.) (Photos by Roy L. Donahue.)

shifting cultivation, nutrients from the ash are temporarily added to the soil, but they are used in a year or two by the cultivated crop or lost due to leaching and declining additions of organic matter. Some exposed and eroded soils (but not all Oxisols) may gradually harden and become impossible to cultivate due to formation of ironstone from plinthite. Cultivated crops require careful fertilization for optimum yields. Added phosphorus has low efficiency because it readily forms insoluble iron and aluminum phosphates, which are not very available to plants. With adequate nitrogen, phosphorus, and potassium, the common crops of the area—bananas, sugarcane, coffee, rice, and pineapples—are productive. Soils containing significant amounts of plinthite must be kept covered by vegetation (in timber, coffee, or other tree crops, and/or organic mulches) to hinder drying and irreversible damage.

Oxisols may be highly productive for carbohydrate and oil crops (both products are mostly carbon, hydrogen, and oxygen; all are derived from air and water rather than from soil minerals) but are less productive of protein foods (which require large amounts of nitrogen and sulfur). Because most Oxisols occur near the equator, where daylight seldom exceeds 12.5 hours daily throughout the year, corn production is less than in temperate regions, such as the U.S. Midwest, where there are 14 to 15 hours of daylight during much of the growing season.

Distribution of Oxisols

The map of the United States inside the back cover does not include Oxisols in the legend because their occurrences, which are only in Hawaii, Puerto Rico, and the Virgin Islands, are too small in area to be shown on this map scale. Oxisols occupy 0.01 percent of the soils of the United

States but are very extensive in tropical South America and Africa (Figure 7-18; also see Figure 7-19). On a world basis they rank fifth and total nearly 9 percent in area of all soil orders.

7:18 Other Soil Classification Systems[21]

There are many soil classification systems—French, South African, Australian, Canadian, Russian, and still others. Some of these are limited mostly to soils of that country and do not attempt a comprehensive coverage of world soils. None are equated simply to terms in any other classification. The Food and Agricultural Organization (FAO) of the United Nations has prepared a world map with described classification units. The FAO world soils are given in Table 7-3 with approximate comparisons to the Canadian system, the 1975 U.S. system, and the 1938 U.S. system. This comparison provides an acquaintance with taxonomic names and approximate relationships of the systems.

FAO System

The FAO soil classification system is worldwide, but it is *not* a system of units grouped into higher categories. The units are designed as the legend of a soil map of the world. The soil map has about 5000 **units.** These units relate most closely to *great groups* in the U.S. system. The FAO system uses the U.S. system of *diagnostic horizons,* although they are sometimes more simplified in definition. Only a selection of the total FAO mapping units are given.

Canadian System

The Canadian system has six categories: *order, great group, subgroup, family, series,* and *type.* The three highest categories are pedogenic and based on the soil's morphology. The great groups are based on diagnostic horizons present. The subgroups have an *orthic* (typic) subgroup and *intergrades,* just as in the U.S. *subgroups.* The system is very limited in extent, mostly to Canadian soil areas.

Comparisons of the United States, FAO, and Canadian Systems

A tabulation of the FAO system is given as the basis for comparing the four systems: FAO, Canadian, the 1975 U.S. system, and the 1938 U.S. system. These comparisons are only approximate because the systems are very different. The *great group* of the U.S. 1975 system is most accurately related to the first *subunit* level of the FAO system. The meanings of most of the FAO subunit names and adjectives are identifiable from the *formative elements* given in Table 7-1. A few terms not given in Table 7-1 follow.

Orthic: central concept of that soil
Calcaric: shallow to lime (2 cm–25 cm)
Gelic: permafrost within 200 cm
Gleyic: hydromorphic (anaerobic)
Luvic: leached, clay moved downward
Solodic: <6% Na in the CEC
Takyric: clay texture, massive crust, dry
Thionic: sulfuric material or horizon
Vertic: Vertisol-like properties

The FAO, United States, and Canadian systems are compared in Table 7-3.

---

[21] E. A. FitzPatrick, *Soils: Their Formation, Classification and Distribution,* Longman Publishers, 1980.

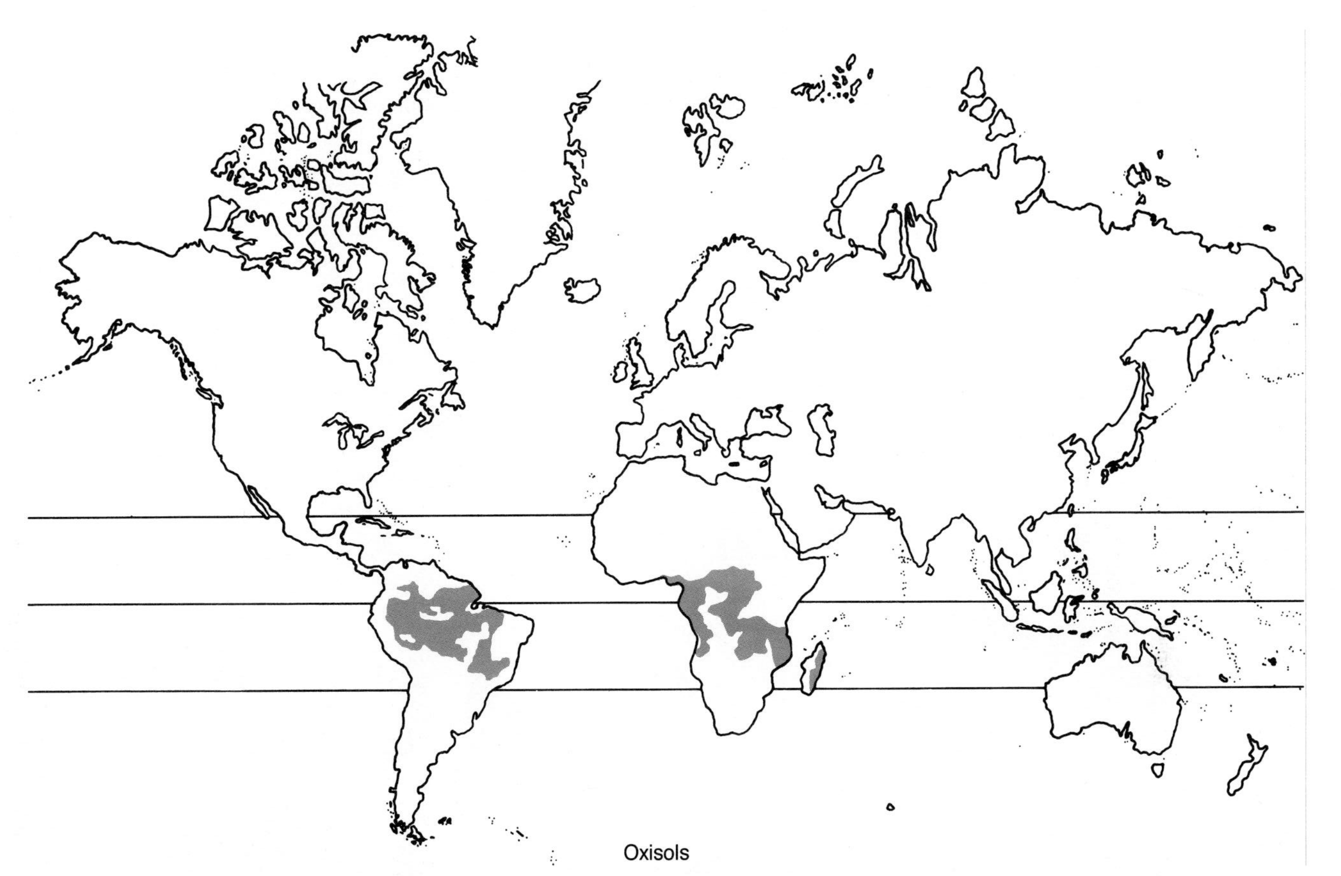

FIGURE 7-18 World distribution of Oxisols (excessively weathered soils). Although only 8.5% of the world land area, they are the fifth most extensive soil order. Oxisols are usually infertile, and some may harden irreversibly because of drying cycles. (Courtesy of Raymond W. Miller, Utah State University.)

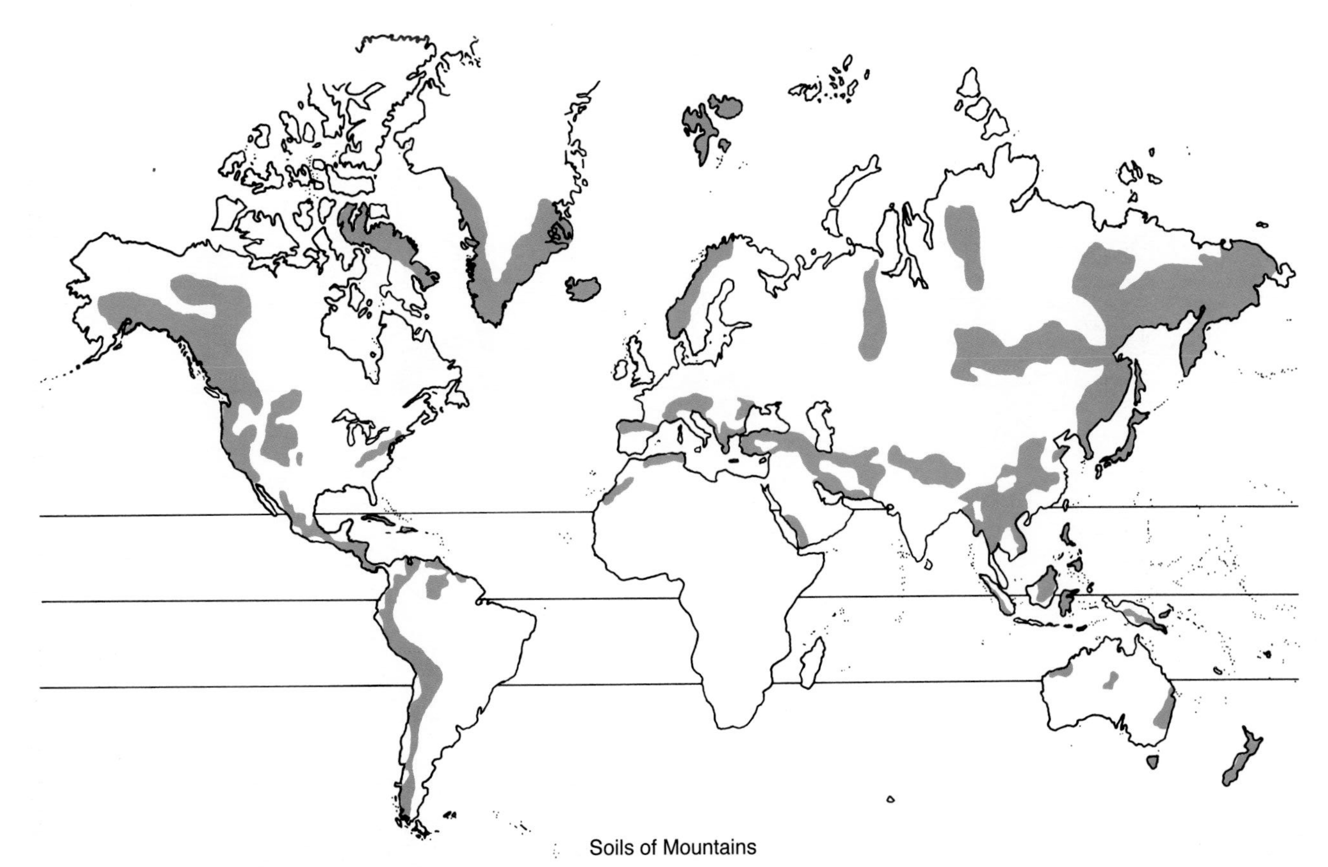

FIGURE 7-19 Soils of Mountains are intermixed regions of mountain slopes and valleys and include many kinds of soils. Nearly 20% of the world land area is in such mountainous regions. (Courtesy of Raymond W. Miller, Utah State University.)

Table 7-3 A Comparison of the FAO, Canadian, and Two U.S. Systems of Soil Classification Names and Taxonomy Terms

		U.S. Systems	
FAO System and Name Meanings	*Canadian*	*1975*	*1938*
ACRISOLS Latin *acris* = very acidic, low base status. *Subunits:*[a] Orthic, Ferric, Humic, Plinthic	—	**ULTISOLS** Hapl—ults[b] Pale—ults Hum—ults Plinth—ults	**Red-Yellow Podzolic** soils
ANDOSOLS Japanese *an* = black, *do* = soil. *Subunits:* Ochric, Mollic, Humic, Vitric	**Brown Wooded, Acid Brown Forest** soils	**ANDISOLS** Several suborders and great groups	**Andosols**
ARENOSOLS Latin *arena* = sand. *Subunits:* Cambic, Luvic, Ferralic, Albic	—	**Psamments** Several subgroups	**Sands**
CAMBISOLS Latin *cambiare* = change. *Subunits:* Eutric, Dystric, Humic, Gleyic, Golic, Calcic, Chromic, Vertic, Ferralic	**Brown Forest, Brown Wooded** soils	**INCEPTISOLS** Many **Ochrepts**	**Brown Earths**
CHERNOZEMS Russian *chern* = black, *zemlja* = earth. *Subunits:* Haplic, Calcic, Luvic, Glossic	**Rego Blacks,** Many **Black** soils	**MOLLISOLS** Several **Borolls**	**Chernozems**
FERRALSOLS Latin *ferrum* = iron and aluminum. *Subunits:* Orthic, Xanthic, Rhodic, Hemic, Acric, Plinthic	—	**OXISOLS** Most suborders	**Latosols**
FLUVISOLS Latin *fluvius* = river (alluvial deposits). *Subunits:* Eutric, Calcaric, Dystric, Thionic	**Regosols**	**Fluvents**	**Alluvial** soils
GELOSOLS Greek *gelid* = very cold, permafrost in part	**Cryosols?**	**Gelisols**	No equivalent
GLEYSOLS Russian *gley* = mucky soil mass. *Subunits:* Eutric, Calcaric, Dystric, Mollic, Humic, Plinthic, Gelic	**Gleysols**	**Aquents, Aquepts, Aquolls**	**Gley** soils
GREYZEMS English *grey* and Russian *zemlja* = earth. *Subunits:* Orthic, Gleyic	**Dark Grey**	**MOLLISOLS Borolls, Aquolls**	**Chernozems**
HISTOSOLS Greek *histos* = tissue. *Subunits:* Eutric, Dystric, Gelic	**Organic soils**	**HISTOSOLS**	**Bog** soils
KASTANOZEMS Latin *castaneo* = Chestnut, Russian *zemlja* = earth. *Subunits:* Haplic, Calcic, Luvic	**Brown** and **Dark Brown** soils	**MOLLISOLS Ustolls, Borolls**	**Chestnut** soils
LITHOSOLS Greek *lithos* = stone shallow to rock. *Subunits:* none	—	**Lithic** subgroups	**Lithosols**
LUVISOLS Latin *luo* = to wash, Illuvial clay layer. *Subunits:* Orthic, Chromic, Calcic, Vertic, Ferric, Albic, Plinthic, Gleyic	**Luvisolic Grey-Brown Podzolic Grey Luvisols**	**ALFISOLS** Many suborders	**Gray Brown Podzolic** soils

Continued.

Table 7-3 A Comparison of the FAO, Canadian, and Two U.S. Systems of Soil Classification Names and Taxonomy Terms (*Cont'd*)

FAO System and Name Meanings	*Canadian*	*U.S. Systems 1975*	*U.S. Systems 1938*
NITOSOLS Latin *nitidus* = shiny, shiny ped surfaces. *Subunits:* Eutric, Dystric, Humic	—	**Paleudalfs,** many **Udults, Tropohumults**	**Latosols**
PHAEOZEMS Greek *phaios* = dusky, Russian *zemlja* = earth. *Subunits:* Haplic, Calcaric, Luvic, Gleyic	**Rego Dark, Grey** soils	**Udolls** and **Aquolls**	**Brunizems**
PLANOSOLS Latin *planus* = flat, level, poorly drained. *Subunits:* Eutric, Dystric, Mollic, Humic, Solodic, Gelic	**Gleysolic, Gleysol, Solods**	**Pale—alfs, Albaquults, Aqualfs, Albolls**	**Planosols**
PODZOLS Russian *pod* = under, *zola* = ash, white layer. *Subunits:* Orthic, Leptic, Ferric, Humic, Placic, Gleyic	**Podzolic, Podzols**	**SPODOSOLS Orthods, Humods, Aquods**	**Podzols**
PODZOLUVISOLS From Podzol and Luvisol. *Subunits:* Eutric, Dystric, Gleyic	—	**MOLLISOLS Udalfs, Boralfs, Aqualfs**	**Gray-Brown Podzolic** soils
RANKERS Austrian *rank* = steep slope, shallow soils. No *Subunits*	—	**Lithic Haplumbrepts**	**Regosols**
REGOSOLS Greek *rhegos* = blanket, thin soil. *Subunits:* Eutric, Calcaric, Dystric, Gelic	**Orthic Regosols, Cryic Regosols**	**Orthents, Psamments**	**Regosols**
RENDZINAS Polish *rzedzic* = noise, stoney soil. No *Subunits*	**Calcareous Rego black** soils	**Rendolls**	**Rendzinas**
SOLONCHAKS Russian *sol* = salt, affected by salt. *Subunits:* Orthic, Mollic, Takyric, Gleyic	**Saline** subgroups	**Salids**	**Solonchaks**
SOLONETZ Russian *sol* = salt, affected by salt. *Subunits:* Orthic, Mollic, Gleyic	**Solonetzic**	**Natr—alfs Nadurargids**	**Solonetz**
VERTISOLS Latin *verto* = turn, self-mixing. *Subunits:* Pellic, Chromic	—	**VERTISOLS Pell—erts Chrom—erts**	**Grumosols**
XEROSOLS Greek *xeros* = dry areas. *Subunits:* Haplic, Calcic, Gypsic, Luvic	—	**ARIDISOLS Calcids Gypsids —argids**	**Desert** soils
YERMOSOLS Spanish *yermo* = desert areas. *Subunits:* Haplic, Calcic	—	**ARIDISOLS Cambids Argids**	**Desert, Red Desert, Gray Desert** soils

[a]Subunits are the adjective plus the name, as Orthic Acrisol and Ferric Acrisol; only adjective is given.
[b]Hapl—ults indicates that many of the —ults suborders (with Hapl for great group) are involved.
Source: Modified and selected from E. A. FitzPatrick, *Soils: Their Formation, Classification, and Distribution,* Longman Publishers, 1980.

7:19 Soil Maps: The United States and The World

A generalized soil map of the United States is given on the inside back cover of this book. The map was prepared mostly to show distributions of soil orders.

The soil map of the World is shown on the inside front cover of this book. When all the percentages given for the 12 soil orders are totaled for the world, they add up to about 96 percent. The other 5 percent to 6 percent is barren land. Some soil areas are so intermixed that they are referred to on the scale of the World map only as **Soil in Areas with Mountains** (about 19 percent of total) and **Icefields** plus **Rugged Mountains** (1.6 percent). These areas plus the soil orders total 101 percent because of rounding. Most of the *Soils in Areas with Mountains* occur in western Canada, in intermountain United States, a strip from northern Mexico to Panama, the Andes of western South America, Norway, northeastern Russia, Japan, central Asia, and most of interior China and southeast Asia (Figure 7-19). These mountain areas contain small valleys of various soil orders typical of the climate and other soil-forming factors in which they occur and with many soil orders on the mountain slopes—all areas too small to be shown separately on a map of this scale.

When we try to pick out anything by itself, we find it hitched to everything else in the universe.

—John Muir

Questions

1. Define (a) *soil taxonomy,* (b) *soil classification,* (c) a *pedon,* and (d) a *polypedon.*
2. Define (a) a *root* of an order, and (b) a *formative element.*
3. Soils of which order are the most extensively weathered?
4. Which soil order has the *least developed* soils?
5. State which soil order probably fits soils on these landforms: (a) recent deep ash deposits, such as that from Mt. St. Helens' eruption in 1980, (b) newly exposed ground moraine, (c) deep alluvium in arid Nevada, (d) old landforms of tall grass prairies of the Great Plains, and (e) the well-weathered, old, red soils of Georgia (not Oxisols).
6. Why might many Alaskan soils and wet, tropical-island, rice-paddy soils be listed as *Inceptisols?*
7. (a) How are *xeric* and *ustic* moisture regimes different? (b) How do *udic* and *aquic* differ?
8. (a) How is temperature categorized? (b) What does *iso* mean in these names?
9. (a) Can Histosols have nitrogen deficiency? (b) What are some other problems in using Histosols?
10. (a) Are Entisols good soils, poor soils, or some of both? (b) Can they contain weathered products? Explain. (c) Are Entisols unweathered materials? (d) Describe a nonproductive Entisol.
11. (a) What is unique about Vertisols? (b) What special management problems exist with Vertisols in whichever climate they are found?
12. (a) Describe the origin and characteristics of plinthite. (b) How does plinthite become ironstone? (c) How may plinthite complicate cultivation in those Oxisols containing it?
13. To what extent are potassium and lime needed in Oxisols and Ultisols?
14. How can some Inceptisols be (a) very *young* soils? (b) very *old* soils? (c) Where might some quite productive Inceptisols be found? (d) How do Inceptisols and Andisols differ? Explain.

15. (a) What climatic areas have extensive areas of Aridisols? (b) What are some expected properties in various Aridisols? (c) What do all Aridisols need for high production?
16. (a) How good are Mollisols for crops? (b) How should Mollisols be managed? (c) Are Mollisols important soils for the United States? Explain.
17. (a) Explain the reasons Vertisols *invert* themselves. (b) Are drying cycles needed for this inversion? (c) Why don't Vertisols have argillic horizons, generally? (d) What is the general definition of *Vertisols?*
18. (a) Why are many Alfisols the most naturally productive soils? (b) Why might they be more productive than most Mollisols? (c) To what extent does leaching occur in most Alfisols?
19. Do Alfisols need (a) lime? (b) fertilizers? (c) Where are most Alfisols?
20. (a) Why are Spodosols quite poor soil for growing a variety of crops? (b) Where do most Spodosols occur?
21. (a) How important are (1) irrigation, (2) fertilization, and (3) liming in Ultisols? (b) In which climatic areas are most Ultisols? (c) What clays do they have?
22. Locate on the U.S. map the dominant soil orders, area-wise, of (a) the Midwest, (b) the Southeast, (c) the Great Plains, (d) the arid Southwest, and (e) the wet, cold Northeast.
23. Locate on the world map the large areas of (a) Mollisols, (b) Aridisols, (c) Alfisols, and (d) Ultisols.
24. (a) What causes *subsidence* in Histosols? (b) How does subsidence alter management? (c) Why do Histosols usually need drainage?

Acidic Soils and Their Management

I cannot say whether things will get better if we change; what I can say is they must change if they are to get better.

—G. C. Lichtenberg

The best never let a little rain stand in their way.

—Gene Kelly

8:1 Preview and Important Facts

PREVIEW

The two ions H^+ and Al^{3+} are considered to be the most important causes of toxicity to freshwater biota and to organisms in soil solutions. Soils in humid regions become acidic as acidic rain water flows through them, replacing basic cations (Ca^{2+}, Mg^{2+}, Na^+, K^+) from cation exchange sites with H^+, Al^{3+}, and $Al(OH)_2{}^+$ ions. Soils high in humus and clay have the potential of developing highly buffered acidic soils, because these colloids have larger amounts of cation exchange sites than do silts and sands. The acidity of soils controls the plant types that grow well on those soils. Some plants tolerate the acidity and survive or compete better than other plants.

In the production of crops, **lime** is added to acidic soils to raise their pH (lessen their acidity). Columella, a Roman philosopher, recorded the use of lime in A.D. 45 to enhance plant growth; however, Edmund Ruffin, a Virginia farmer–scientist from 1825 to 1845, may have been the first person to apply lime on the soil specifically to correct a condition that he called *soil acidity.* Today lime is one of the most common agricultural soil amendments in humid regions.

Crop response to added lime does not occur immediately, perhaps not for several weeks or months, and large amounts are usually needed to be effective. Crops respond to liming because it improves microbial activity; it can correct calcium, magnesium, and molybdenum deficiencies; it lessens aluminum, manganese, and iron toxicities (they precipitate); and it makes phosphorus more available and potassium more efficient in plant nutrition. Ammonium fertilizers and urea make soil more acidic; the continued use of these fertilizers on acidic soils without also adding lime may decrease soil productivity.

IMPORTANT FACTS TO KNOW

1. The reason some soils become acidic but others remain alkaline
2. The reasons most plants grow poorly in strongly acidic soils
3. The substances that can be used as limes and the composition of the most commonly used agricultural lime
4. The neutralizing index of lime materials
5. The changes caused by lime addition to acidic soils
6. The pH value desired from a liming program
7. The yield increases expected from crops growing in limed soils
8. The typical amounts of lime needed to correct soil acidity
9. The effects of overliming and why those effects occur
10. The preferred ways to add lime to soils
11. The methods to acidify soils
12. The soils on which acidification might be practical

8:2 Why Some Soils Are Acidic

Most soils become acidic because of leaching. Carbon dioxide (CO_2) dissolved in water, excreted H^+ (H_3O^+) from roots, and organic acids produced during humus decomposition furnish H^+ in percolating water. As soil solution pH becomes acidic, aluminum hydroxides in-

(a)

(b)

FIGURE 8-1 (a) Soils are acid partly because ammonium nitrogen fertilizers are used on them. In (a) the center plot received 200 lb/a (224 kg/ha) of ammonium sulfate per year for 21 years. In (b) the center plot received for 21 years—the same as in (a)—but in addition it received 230 lb of limestone per acre (258 kg/ha) per year for the same period. (*Source:* Alabama Agricultural Experiment Station.)

teract and some soluble hydrated $Al(OH)_2^+$ ions form (Detail 8-1). As percolating water moves these $Al(OH)_2^+$ and H^+ ions through the soil, many absorbed basic cations (Ca^{2+}, Mg^{2+}, Na^+, K^+) are replaced by these acidic cations. Such a leached soil becomes more acidic after decades or centuries of leaching. At pH values below about 4.7, appreciable amounts of Al^{3+} ions will exist in solution and on cation exchange sites.

Most soils in areas receiving 500 mm (20 in) or more rainfall yearly will develop acidity. In areas with higher rainfall, strong acidity is expected with pH values of 4.5 to 5.5 common. Extreme acidity values are pH 3.0 to 3.5. The sources of H^+ ions are as follows:

- Carbonic acid from carbon dioxide dissolved in rainfall
- Carbon dioxide from humus decomposition and root respiration dissolved in water
- Oxidation of NH_4^+ from fertilizers (see Figure 8-1)
- Excreted H^+ ions by plant roots
- Acid rain (sulfur and nitrogen oxide pollutants)
- Crop removal of the basic cations (Ca, Mg, K, Na) and excretion of H^+ by roots

Details of the mechanisms of soil acidification by natural processes are given in Detail 8-2. Equations representing soil acidification follow. The *(−65)* is an arbitrary charge used to indicate a total negative charge for that colloid. The dots at each cation represent the valences for that cation attracted to the negative sites of the colloid.

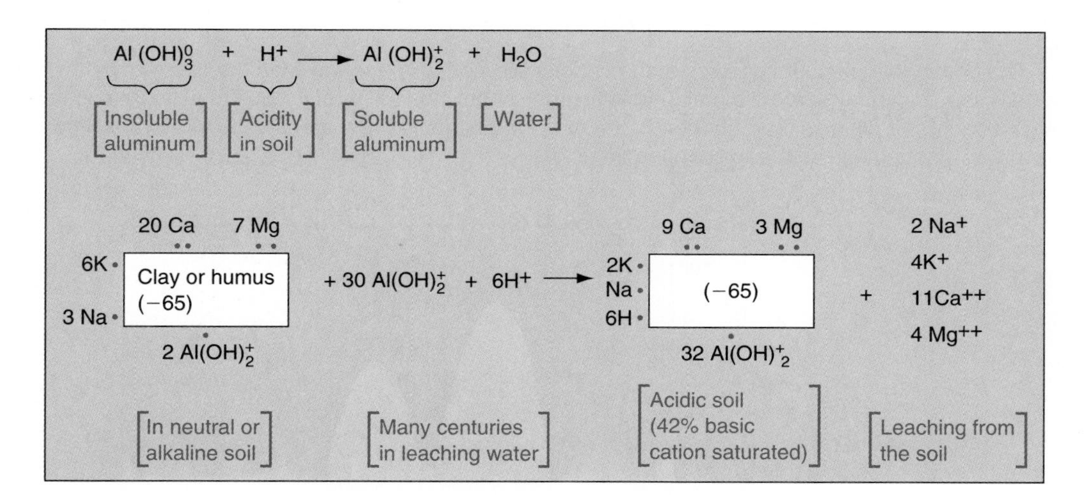

Acidification in nature requires removal of some cations yearly for many centuries of time and high rainfall for leaching water to wash out basic cations. Acidic soils in the United States are found primarily in the high-rainfall areas of the states east of the Mississippi River, the Pacific coastal soils of the Northwest, and some of the mountain areas. For example, Logan in northern Utah averages about 400 mm (15.6 in) of rainfall annually and seldom are soils there more acidic than pH 6.5. Mountains within 50 km (31 mi) will often have 750 mm to 900 mm (29 in–35 in) of precipitation, and soils are acidic (pH 5 to 6).

8:3 Ecological Relation of Soil Acidity

In nature acidic soils remain acidic; nature does not add lime to adjust the pH. Plants that live on acidic soils must tolerate the acidity that is there. The conditions in the acidic soils can be summarized as follows:

Detail 8-1 Aluminum Ions in Soil Solution

In water solutions, aluminum always forms octahedral coordination with some combination of water molecules and hydroxyl ions, much as it does with six oxygens or hydroxyls in clay minerals. Aluminum's natural coordination is with six of these molecules (octahedral coordination); the size of hydroxyls, oxygen, and water are similar. If the soil is not too strongly acidic, one or more of the water molecules ionize, releasing hydrogen, H^+, to the solution, increasing the solution acidity. The possible chemical forms simply indicated are as follows:

$$[Al(H_2O)_6]^{3+} \longrightarrow [Al(H_2O)_5(OH)]^{2+} + H^+ \quad \text{(pH about 4.5–5) (limited)}$$

$$Al(H_2O)_5(OH)^{2+} \longrightarrow [Al(H_2O)_4(OH)_2]^{+} + H^+ \quad \text{(pH about 5–6.5)}$$

$$Al(H_2O)_4(OH)_2{}^{+} \longrightarrow [Al(H_2O)_3(OH)_3]^{0} + H^+ \quad \text{(pH about 6.5–8.5)}$$

Most often the hydration is not indicated, and the ions are simply written $\mathbf{Al^{3+}}$, $\mathbf{Al(OH)^{2+}}$, $\mathbf{Al(OH)_2{}^{+}}$, or $\mathbf{Al(OH)_3}$. The water molecules are ignored.

Iron has similar reactions and products in soil solutions, but pH values differ.

Aluminum and iron tend to form larger units than described above, mostly of hydroxides.[a] These may be rings or clumps of Al^{3+} and OH^-, such as

$$[Al_6(OH)_{12}]^{6+} \quad \text{and} \quad [Al_{10}(OH)_{22}]^{8+}$$

At pH near 8 and reaching a maximum concentration near pH 10, two new soluble aluminum species, described simply as $[Al(H_2O)_2(OH)_4]^-$ and $[Al(H_2O)(OH)_5]^{2-}$, occur.[b] These may cause toxicity in excessively basic soils. As soils are leached, some basic cations are exchanged off the CEC and the exchange sites are increasingly occupied by aluminum species.

The dominant aluminum species at the various solution pHs are shown in the diagram below.[c] Notice that $Al(OH)_2{}^+$ will be the dominant *soluble* aluminum form from about pH 4.7 to pH 7.5. Thus, most exchangeable acidity is the $Al(OH)_2{}^+$ ion. At pH less than about 4.5, the Al^{3+} ion and increasing amounts of H^+ ions will be in solution and on the cation exchange sites as the exchangeable acidity.

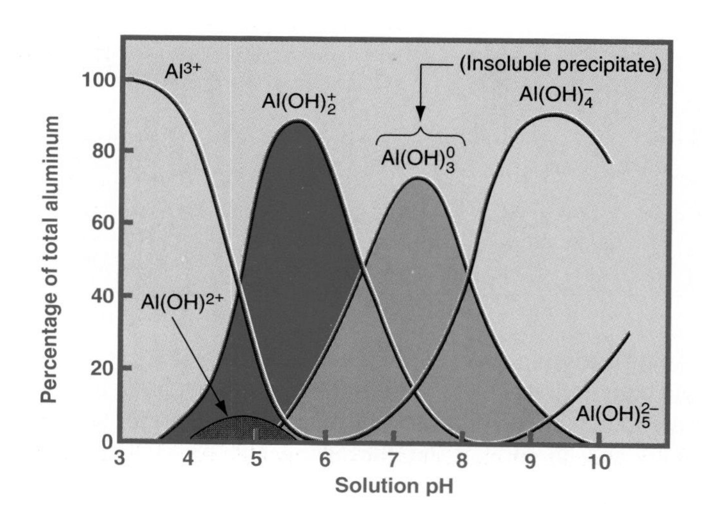

---

[a]P. H. Hsu and T. F. Bates, "Fixation of Hydroxyl-Aluminum Polymers by Vermiculite," *Soil Science Society of America, Proceedings* **28** (1964), 763–769; and H. L. Bohn, B. L. McNeal, and G. A. O'Conner, *Soil Chemistry,* John Wiley, New York, 1979, 200.

[b]P. R. Hesse, *A Textbook of Soil Chemical Analysis,* Chemical Publishing Co., New York, 1971, 341.

[c]Modified by Raymond W. Miller after G. M. Marion, D. M. Hendricks, G. R. Dutt, and W. H. Fuller, "Aluminum and Silica Solubility in Soils," *Soil Science* **121** (1976), 76.

Detail 8-2 Sources of Solution Acidity in Nature

The hydrogen ions in soil water, which over centuries work to produce acidic soils, are from these six sources:

1. Carbon dioxide of the atmosphere dissolves in raindrops.

$$CO_2 + HOH \longrightarrow H_2CO_3 \longrightarrow HCO_3^- + H^+$$

2. *Carbon dioxide* from decomposing organic matter and root respiration dissolves in water to form weak carbonic acid.

$$CO_2 + HOH \longrightarrow H_2CO_3 \longrightarrow HCO_3^- + H^+$$

These acidified waters percolate through the soil to gradually cause soil acidity. Percolating waters continuously move small amounts of H^+, which replace solubilized basic cations of calcium, magnesium, potassium, sodium, and other elements. The replaced basic elements are leached from the root zone.

3. *Ammonium-containing fertilizers* are oxidized by bacteria to form nitrate and hydrogen ions (Figure 8-1). For each NH_4^+ cation oxidized, two H^+ result.

$$NH_4^+ + 2O_2 \xrightarrow{\text{(nitrifying bacteria)}} NO_3^- + H_2O + 2H^+$$

This reaction applies to any source of NH_4^+, including urea after hydrolysis and the mineralization of NH_4^+ from organic materials.

4. Some hydrogen ions are *excreted by plant roots,* which are exchanged for other nutritive cations. The amounts of hydrogen ion released lower the pH as much as 1.2 pH units in the soil near roots.

5. *Acidic rain* falls when airborne sulfur oxides (mostly sulfur dioxide, SO_2) and nitrogen oxides (mostly nitric oxide, NO) are converted to sulfuric acid (H_2SO_4) and nitric acid (HNO_3) through oxidation and are then dissolved into raindrops. Rainwater made acidic by these strong acids may have a pH as low as 2. The original sources of these polluting oxides are the burning of fossil fuels—such as wood, coal, and petroleum products—and from forest and range fires. Acidity in rain has been enough to eradicate fish from many lakes. Soils are less affected than surface waters because most soils are highly buffered. Although forest fires or SO_2 emissions from geologic phenomena are natural contributors to acid rain, for the most part acidic rain is not considered a natural phenomenon.

6. *Crop removal* helps make soils more acidic by depleting the reserves of calcium, magnesium, and potassium. For example, a yield of 13 metric tons per hectare (6 t/a) of alfalfa removes 45 kg (100 lb) of calcium (Ca) and 9 kg (20 lb) of magnesium (Mg).

- Acidic soils are leached soils; the more leached they are, usually the more strongly acidic they are.
- The most strongly acidic soils have (1) few basic cations (Ca, K, Mg, Na) available for root absorption; (2) high exchangeable Al, H, and Mn; and (3) low contents of the more easily leached nutrients (S, B, Zn, Mo, Cl).
- Levels of toxic Al and Mn ions increase as pH decreases.
- Many microbial processes, including N_2-fixation, are slowed down by strong acidity.

Plants that grow well in acidic soils often have a low requirement for basic cations, particularly of Ca and K; are able to tolerate low nutrient availability in general; and tolerate the Al and Mn levels that occur in acidic soils. Some plants more tolerant of strong acidity include tea, pineapples, blueberries, coffee, and azaleas. Some of the proposed mechanisms to tolerate potent concentrations of aluminum are exclusion mechanisms such as immobilization of aluminum in the cell wall, selective permeability to aluminum at the plasmalemma (membrane), a plant-induced pH barrier in the rhizosphere, and exudation of aluminum as chelates. Additionally, *internal* tolerance mechanisms might include compartmentation of aluminum in the vacuole, chelation in the cytosol, aluminum binding proteins, and evolution of Al-tolerant enzymes.

Where temperatures are warm year around and rainfall is high (tropics, subtropics), many soils are extensively weathered (Ultisols, Oxisols) and are both acidic and very low in basic cations. Their primary mineral storehouse is mostly gone. Without many nutrients, plants in such environments grow slowly and contain low essential nutrient levels in their tissues.

8:4 Composition of Lime

Liming materials are usually the carbonates, oxides, hydroxides, and silicates of calcium and magnesium. More than 90 percent of the agricultural lime used is impure, crushed calcium carbonate (powdered limestone); the next most used are carbonates of calcium plus magnesium (dolomitic lime); a much smaller quantity is composed of calcium oxide or calcium hydroxide. Wood ashes, burned limestone (forming CaO), and marl (soft calcium carbonate with clays) are also used sometimes on acid soils to increase plant growth. In the building trades *lime* refers to calcium oxide, a caustic powder that is slaked in water for use in various brick-laying mortars; although this material is usable as agricultural lime, it is caustic and dusty, making handling difficult.

The most commonly used liming materials include the following:

- **Calcic limestone ($CaCO_3$):** Ground fine for use (also called **aglime**).
- **Dolomitic limestone [$Ca{\cdot}Mg(CO_3)_2$]:** A ground limestone high in magnesium (also called **aglime**). Although state laws vary, the average composition of dolomitic limestones sold in the United States is about 51 percent $CaCO_3$, 34 percent $MgCO_3$, and 15 percent soil and other impurities.
- **Quicklime (CaO):** Burned limestone.
- **Hydrated** (slaked) **lime [$Ca(OH)_2$]:** Lime from quicklime that has changed to the hydroxide form as a result of reactions with water.
- **Marl ($CaCO_3$):** Lime from the bottom of small freshwater ponds, which has accumulated by precipitation from drainage waters high in lime.
- **Chalk ($CaCO_3$):** Results from soft limestone deposited long ago in oceans.
- **Blast furnace slag ($CaSiO_3$ and $CaSiO_4$):** A by-product of the iron industry. Some slags contain phosphorus and a mixture of CaO and $Ca(OH)_2$. This product is called *basic slag* and is used primarily for its phosphorus content.
- **Ground oystershell, wood ashes,** and **by-product lime:** Miscellaneous sources from paper mills, sugar-beet plants, tanneries, water-softening plants, fly ash from coal-burning plants, and cement-plant flue dust.
- **Fluid lime:** The suspension in water of *any* suitable liming material that has a fineness of $<$60-mesh ($<$250 μm).

Gypsum ($CaSO_4 \cdot 2H_2O$) *is not a lime,* but it is sometimes added to soil to supply calcium, which might alleviate somewhat the toxicity from soluble aluminum, a primary action of lime.

Chemical Guarantees of Lime

There are several methods of expressing the relative chemical value of lime, the most common of which is the **Calcium Carbonate Equivalent,** sometimes known as the **total neutralizing power.** If a lime is chemically pure calcium carbonate (calcite), the calcium carbonate equivalent is 100. If all the lime is in the calcium carbonate form but it is only 85 percent pure, the

Calculation 8-1 Calculation of the Calcium Carbonate Equivalent

Problem What is the calcium carbonate equivalent of 100 kg of pure CaO?

Solution Use the Periodic Table in Appendix C to obtain atomic weights for the different materials. The calculation is as follows:

$$\% \text{ CaCO}_3 = \text{grams of pure CaCO}_3 \text{ that is equal to 100 g of the CaO used}$$

Add the gram atomic weights:

$$\text{CaO gram molecular weight} = 40 + 16 = 56 \text{ g}$$

$$\text{CaCO}_3 \text{ gram molecular weight} = 40 + 12 + 48 = 100 \text{ g}$$

Therefore, 56 g of CaO neutralizes the same amount of acid as does 100 g of $CaCO_3$. If we represent soil acidity by H^+, the following equations show the neutralization process:

$$\text{CaO} + \text{HOH} + 2\text{H}^+ \longrightarrow \text{Ca}^{2+} + 2\text{HOH}$$

$$\text{CaCO}_3 + \text{CO}_2 + \text{HOH} + 2\text{H}^+ \longrightarrow \text{Ca}^{2+} + \text{H}_2\text{CO}_3$$

The H_2CO_3 breaks down into CO_2 gas and water. We can now write the relationship between CaO and $CaCO_3$ from the molecular weights above as the following:

$$\frac{? \text{ g CaCO}_3}{100 \text{ g CaO}} = \frac{\text{Mol. wt. of CaCO}_3}{\text{Mol. wt. of CaO}}(100) = \frac{100 \text{ g}}{56 \text{ g}}(100)$$

$$? \text{ g of CaCO}_3 = (100 \text{ g CaCO}_3)\left(\frac{100 \text{ g}}{56 \text{ g}}\right) = 179 \text{ g}$$

The 179 g is the grams of $CaCO_3$ equivalent to 100 g of CaO = the calcium carbonate equivalent of CaO.

---

calcium carbonate equivalent is 85. Limestone is seldom pure; it formed in ocean bottoms with clay and silt sediments. Any lime can be calculated to its calcium carbonate equivalent by the use of atomic and molecular weights.[1] (See Calculation 8-1 and Table 8-1.)

Physical Guarantees of Lime

The neutralizing power of liming material is determined by the rate of solubility of the chemical compounds in the lime. Calcium oxide is more soluble than calcium carbonate, and calcic limestone is more soluble than dolomitic limestone; calcium silicate is the least soluble of all these liming materials. Fine lime particles react faster in the soil than do coarse particles (Figure 8-2).

No U.S. laws govern a commercial supplier's physical guarantee of lime; this regulation is currently being left to the states. Most states have lime laws, enforced by each state department of agriculture. In general, the physical guarantees of lime in the respective states may be averaged roughly in the following way: *Eighty-five percent of the lime particles must pass through a 16-mesh (1.18-mm) sieve and 30 percent must pass through a 100-mesh (150-μm) sieve.*[2] *Liquid lime* (suspension of about 50:50 lime to water) is often even finer.

---

[1]The "Model Agricultural Liming Material Bill" proposes the following minimum calcium carbonate equivalents: quicklime 140, hydrated lime 110, ground limestone 80, blast furnace slag 80, and ground oystershell 80. Reference: *Abstract of State Laws and ACP Specifications for Agricultural Liming Materials,* 3rd ed., National Limestone Institute, Fairfax, VA, 1977, 75.

[2]Sieves are designated by the number of openings per linear inch. An 8-mesh (2.36-mm) sieve has 8 openings per linear inch, and a 60-mesh (250-μm) sieve has 60 openings per linear inch.

Table 8-1 Lime Conversion Factors

To Convert from This Material* (Column A)	To Each of These: *Ca*	*CaO*	*$CaCO_3$*
	Multiply Column A by:		
Calcium (Ca)	1.00	1.40	2.50
Calcium oxide (CaO)	0.71	1.00	1.78
Calcium hydroxide [$Ca(OH)_2$]	0.54	0.78	1.35
Calcium carbonate ($CaCO_3$)	0.40	0.56	1.00
Magnesium (Mg)	1.65	2.31	4.12
Magnesium oxide (MgO)	0.99	1.39	2.48
Magnesium hydroxide [$Mg(OH)_2$]	0.69	1.00	1.72
Magnesium carbonate ($MgCO_3$)	0.48	0.67	1.19
Dolomite, pure ($CaCO_3 \cdot MgCO_3$)	0.43	0.63	1.09

*Calculated using the following atomic weights: calcium—40.08, oxygen—16.00, carbon—12.01, and hydrogen—1.00. For example, to convert from Ca to CaO equivalent, divide the molecular weight of CaO by the atomic weight of Ca = 56.08/40.08 = 1.40. Therefore, to convert Ca to CaO, multiply the kilograms (pounds) of Ca by 1.40 to obtain equivalent kilograms (pounds) of CaO.

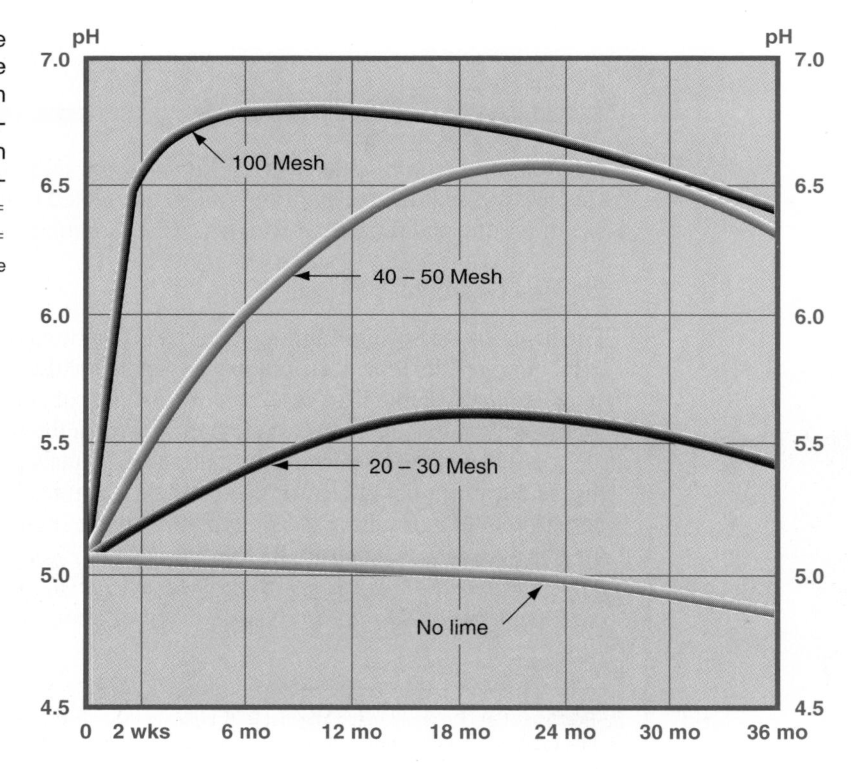

FIGURE 8-2 The finer the limestone, the more quickly it reacts with the soil to raise the pH and the more quickly some calcium becomes available to the plant. A satisfactory fineness is 85% through a 16-mesh (1.18-mm) sieve and 30% through a 100-mesh (150-μm) sieve. *Note:* 20 mesh = 850 μm, 30 mesh = 600 μm, 40 mesh = 425 μm, 50 mesh = 300 μm. (*Source:* The Fertilizer Institute and Purdue University.)

Calculation 8-2 Effective Calcium Carbonate (The Neutralizing Index)

An effective technique for assessing a liming material is to consider both chemical and physical factors together. First, determine the calcium carbonate equivalent as explained in Calculation 8-1, then use the sieve analysis (which expresses fineness). Together these calculate the **effective calcium carbonate,** also called **neutralizing index.**

Problem Calculate the neutralizing index of this lime:

Percent calcium carbonate equivalent: 90%

Sieve analysis: Retained on 8-mesh sieve	=	10%
Retained on 60-mesh sieve	=	20%
Passing 60-mesh sieve	=	70%
Total		100%

Solution Calculating the **fineness factor:**

1. Greater than 8-mesh sieve size lime is presumed too coarse to neutralize soil acidity within 3 years after application and so has zero effectiveness. Therefore, 10% of sample > 8 mesh × 0 effectiveness = 0.
2. Lime that passes between 8- and 60-mesh sieve is presumed 50% as effective as finer lime: 20% sample × 0.50 effectiveness = 10.
3. Presuming < 60-mesh lime is 100% effective: 70% × 1.00 effectiveness = 70.
4. Total of fineness factors = 80. Effective calcium carbonate (neutralizing index) = percent calcium carbonate equivalent × fineness factor = 0.90 × 80 = 72.

Sources: Adapted from L. S. Murphy and Hunter Follett, "Liming—Take Another Look at the Basics," *Solutions* (Jan.–Feb. 1978), 53, 54, 56, 58, 60, 62, 64–67; K. T. Winter, et al., "Liming Has No Miracles," *Solutions* (Mar.–Apr. 1980), 12, 14, 16, 24, 28, 30, 32, 34; and K. A. Kelling and E. E. Schulte, "Liming Materials—Which Will Work Better?" *Solutions* (Mar.–Apr. 1980), 52–54, 56, 60.

One unique technique for evaluating the *effectiveness* of liming materials is to calculate *a neutralizing index,* which is the calcium carbonate equivalent multiplied by an arbitrary fineness factor. Such a **neutralizing index** gives a better evaluation of the quality of the material than would either factor by itself (Calculation 8-2). Fine materials and a large calcium carbonate equivalent is more effective than a coarse, less pure lime. A good lime should have a neutralizing index of 70 or higher.

8:5 Reactions of Lime Added to Acidic Soils

Strongly acidic soils greatly restrict the growth of most plants. Notable exceptions are blueberries, cranberries, watermelons, white potatoes, tea, and pineapples; these crops do well on strongly acidic soils. By contrast, alfalfa and sweet clover yield their maximum harvest only when the soils are neutral to slightly basic.

On strongly acidic soils the majority of crop plants produce yields less than their potential when grown on soils less acidic for one or more of the following reasons:

- **Aluminum toxicity** is perhaps the most important cause of reduced plant growth (Figure 8-3; see also Detail 8-1).

(a)

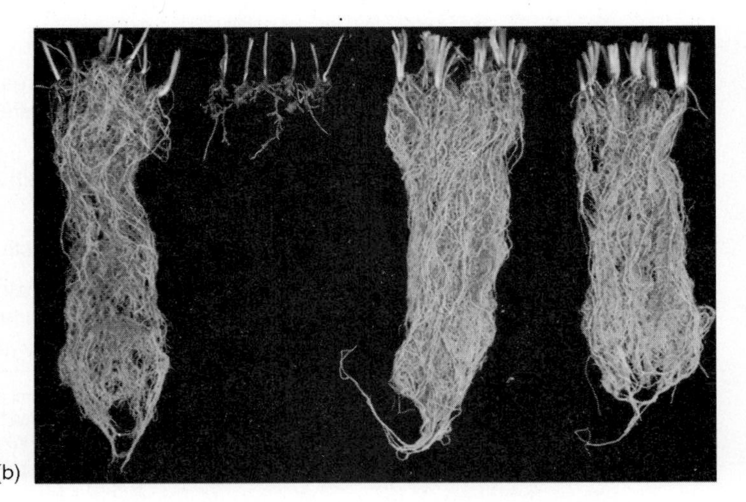

(b)

(c)

FIGURE 8-3 Excess soluble aluminum, which occurs in soils of pH about 5 or lower, is toxic to most plants. In (a) and (b) the two wheat cultivars on the left, Thorne and Redcoat, respectively, at pH 4.3 show different tolerances to acidity. The same cultivars on the right at pH 5.8 (after adding 3000 ppm calcium carbonate). (c) shows tall fescue grass grown in nutrient solutions containing, from left to right, 0, 2, and 4 ppm soluble aluminum. (*Source: Agronomy Journal,* vol. 66, by permission of the American Society of Agronomy; photos by A. L. Fleming, J. W. Schwartz, and C. D. Foy.)

- **Reduced microorganism activity**
- **Manganese toxicity**
- **Iron toxicity** in a few soils
- **Calcium deficiency**
- **Magnesium deficiency**
- **Molybdenum deficiency,** especially for legumes and the cabbage family
- **Nitrogen, phosphorus, and/or sulfur deficiency** because of very slow organic-matter decomposition

The addition of lime raises the soil pH and eliminates two major problems of acid soils: excess (toxic) soluble aluminum[3] and slow microbial activity. The process of changing pH by the addition of lime is illustrated by the equations in Figure 8-4 and is a reversal of acidification. The adsorbed acidic aluminum ions are replaced with calcium ions from the lime; the released H^+ is neutralized by the carbonates or hydroxides.

Liming also has the following benefits in addition to relieving toxic Al levels and speeding microbial processes:

- The raised pH reduces *excess* soluble manganese and iron (as possible toxins) by causing them to form insoluble hydroxides.
- Calcium and magnesium, deficient in many acidic soils, are added if the lime is *dolomitic* (contains both calcium and magnesium carbonates) rather than *calcic* lime (only calcium carbonate).
- Lime makes phosphorus in acidic soils more available. In strongly acidic soils iron and aluminum combine with the fertilizer phosphates to make insoluble compounds. *Liming* reduces the soluble iron and aluminum; therefore, less phosphorus will form insoluble iron and aluminum phosphates.

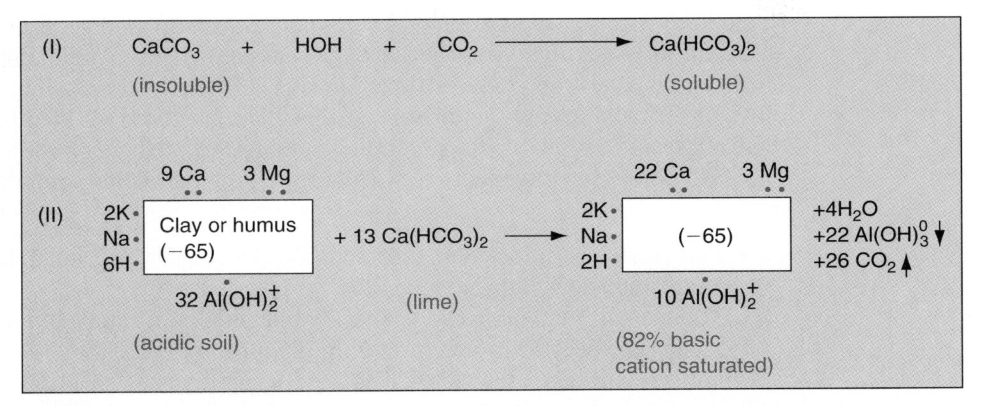

FIGURE 8-4 Lime (calcium carbonate) added to soil dissolves in the solution as calcium bicarbonate. The calcium in the lime exchanges for the exchangeable acidity [$Al(OH)_2^+$ and H^+ ions]. The $Al(OH)_3$ formed in the higher pH solution is insoluble and precipitates. Any H^+ exchanged is neutralized by forming dilute carbonic acid with bicarbonate, which is unstable, giving off gaseous carbon dioxide and water. (Courtesy of Raymond W. Miller, Utah State University.)

[3]F. Ahmad and K. H. Tan, "Effect of Lime and Organic Matter on Soybean Seedlings Grown in Aluminum-Toxic Soil," *Soil Science Society of America Journal* **50** (1986), 656–661.

- Lime makes potassium more efficient in plant nutrition. When potassium is plentiful, plants absorb more of it than they need. Lime reduces the excessive uptake of potassium. Economically, the practice of liming is desirable because the plant absorbs more of the cheaper calcium and less of the costlier potassium. Because calcium is often deficient in animal rations and potassium in excess, it is desirable to increase the percentage of calcium in the plant.
- Lime increases the availability of nitrogen by creating a more favorable environment for microbes, which hastens the decomposition of organic matter (the soil bacteria are more active at pH levels less strongly acidic).
- Lime on acidic soils increases plant-available molybdenum.
- Liming an acidic soil above pH 6.5 reduces the solubility and plant uptake of potentially toxic heavy metals—such as cadmium, copper, lead, nickel, and zinc—from added sources such as sewage sludge.

8:6 Crops, Lime, and Soil

On mineral soils below pH 5, most crops respond to the judicious use of lime. Adding lime changes the soil's pH, improves the microorganisms' activities, and increases the availability of nitrogen and phosphorus. Adding excess lime is not desirable and is additional cost.

Benefits of Correct Liming

On adequately fertilized soils in Maryland, the proper amount of lime applied on soil with a pH of 5.6 increased corn yields valued at $3.95 for each dollar spent on lime. In Puerto Rico sugarcane yielded 50 metric tons per hectare (22.3 tons per acre) when exchangeable aluminum provided more than 70 percent of the exchangeable ions; adding lime until the exchangeable aluminum was less than 30 percent increased the yield more than four times. Soybean yields in Brazil went from 1753 kg/ha (1565 lb/a) with no liming to 3960 kg/ha (3536 lb/a) with the application of 28 Mg/ha (12.5 ton/a) of lime. Raising the pH to 5.2 was sufficient for these high yields because the soils are Oxisols. Although price changes in lime, fertilizers, and the crop produced determine the net benefit derived, lime is a profitable soil additive on the most strongly acidic soils. Although liming to pH 6.5 is commonly recommended, there are specific soils, crops, or cropping systems that are better adjusted to lower pH values.

The following three facts about liming soil are particularly important:

- Phosphorus additions plus lime additions frequently give much larger increases in yields than does lime alone.
- Toxic levels of soluble and exchangeable aluminum can be almost eliminated by raising the pH to 5.2–5.5; further liming—from pH 6.0 to 6.5—usually, although not always, still increases yields. The beneficial effects of raising the pH from 5.3 to 6.5 may be due to an increase in biological activity, which increases the available nitrogen, molybdenum, and other nutrients.
- Adding excess lime (raising pH higher than 6.0 or 6.5) may require addition of those plant nutrients that become less available at higher pH. Some crops (small grains, sugar beets, alfalfa) are not injured by overliming, especially on fine-textured, humid-region soils.

To arrive at a satisfactory solution to the problem of how much lime to apply, the pH requirements of the proposed crop as well as the actual pH of untreated soil should be considered. The *soluble and exchangeable acidity,* particularly the exchangeable acidity, must be

Table 8-2 Relative Tolerance to Acidity of Selected Plants

Very Low Tolerance to Acidity	*Low Tolerance to Acidity*	*Moderate Tolerance to Acidity*	*High Tolerance to Acidity*
Least acid tolerant			
Alfalfa	Blackberry	Alsike clover	Azalea
Asparagus	Cabbage	Buckwheat	Blueberry
Barley	Cantaloupe	Oats	Coffee, arabica
Beans, field	Corn	Peanut	Cranberry
Cotton	Crown vetch	Potato	Kudzu
Kentucky bluegrass	Fescue grass	Raspberry	Lespedeza
Peas	Grain sorghum	Rice	common
Red clover	Some grasses	Rye	Korean
Soybean	Lettuce	Strawberry	Sericea
Spinach	Peanut	Vetch	Napier grass
Sugar beet	Sweet potato	Watermelon	Pineapples
Sunflower	Tobacco		Rhododendron
Sweet clover	Trefoil		Stylosanthes
	Wheat		(tropical legume)
	White clover		Tea
			Most acid tolerant

Sources: Modified from R. G. Hanson, "Corrective Liming of Missouri Soils," Science and Technology Guide 9102, *Agronomy* **11** (1977); and D. R. Christenson and E. C. Doll, "Lime for Michigan Soils," Michigan State University Extension Bulletin 471, 1979.

partially neutralized. Buffered solutions are used in soil tests to estimate the lime needed to neutralize portions of the exchangeable hydrogen ions.

Costs of lime vary with the lime quality and type plus transportation distances. In 1996 in central Arkansas *aglime* costs were $12 to $24 per ton spread, suspension lime was about $60 per dry ton spread, and pelletized lime was $90 to $100 per ton spread.[4]

The relative benefits of added lime to certain crops are listed in Table 8-2; alfalfa, barley, cotton, sugar beets, and sweet clover have the highest need for pH adjustment; corn, tobacco, and wheat have a moderate need; buckwheat, potatoes, rice, and rye do well in a low-pH soil. Blueberries, cranberries, and pineapples tolerate strongly acidic soils. If liming costs for a particular soil are high, a crop with a need for less lime sometimes can be grown as an economic alternative.

How Much Lime to Apply?

To evaluate the amount of lime to apply, both the pH requirement of the crop to be grown and the pH and buffer capacity (cation exchange capacity, CEC) of the cultivated soil should be determined. The lime need can be interpreted more accurately when the soil series is known, because the series defines soil texture, structure, mineralogy, and other root-zone characteristics such as humus content and permeability, which may affect the lime response.

A relationship of texture, cation exchange capacity, and buffer capacity (resistance to a change in ion concentration) is shown in Figure 8-5. The more clay and organic matter there is in a soil, the more lime is needed to change the pH, because the soil colloids contain large quantities of exchangeable aluminum and hydrogen ions due

[4]C. S. Snyder, J. H. Muir, and G. M. Lessman, "Spring-Applied Aglime Can Provide Immediate Soybean Response," *Better Crops* **80** (1996), No. 1, 3–4.

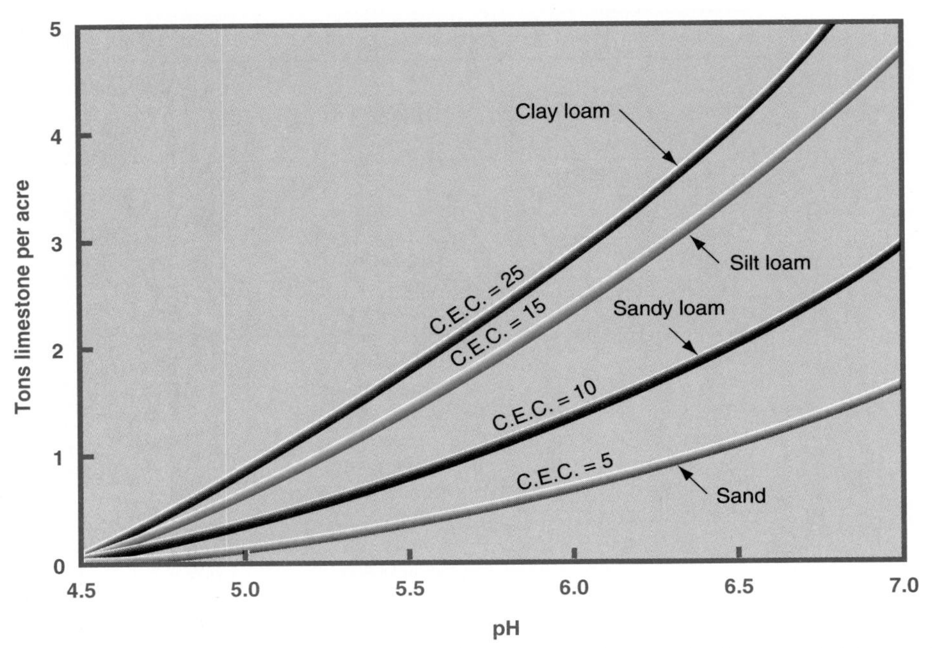

FIGURE 8-5 Approximate tons of limestone required to raise the pH of a 17-cm (7-in) layer of soil of four textural classes with typical cation exchange capacities (CEC) in milliequivalents per 100 g of soil. (*Sources:* Modified from D. R. Christenson and E. C. Doll, "Lime for Michigan Soils," Michigan State University Extension Bulletin 471, 1979; and R. G. Hanson, "Corrective Liming of Missouri Soils," Science and Technology Guide 9102, *Agronomy* **11** [1977].)

to their high cation exchange capacities. It is true that the greater the amount of organic matter in a soil, the lower the pH required for greatest plant nutrient availability (Figure 8-6). For example, one recommendation to encompass soils very high in organic matter is the following:

$<$10% organic matter, ideal pH to adjust to = 6.5

10% organic matter, ideal pH to adjust to = 6.0

20% organic matter, ideal pH to adjust to = 5.5

The amount of pH change desired and the type of clay present also cause variation in the amount of lime needed to change the pH. The relative amount of lime needed in soil of the same initial pH with the principal clay minerals is as follows:

vermiculite $>$ montmorillonite $>$ illite $>$ kaolinite $>$ sesquioxides (metal oxides)

Oxisols and Ultisols respond differently to pH change when they are limed because the predominant clay minerals are sesquioxides in Oxisols and kaolinite in Ultisols.

The general relationship between soil pH and plant nutrient availability is provided in Figure 8-6. The primary nutrients (nitrogen, phosphorus, potassium) and the secondary nutrients (sulfur, calcium, magnesium) are as available—or more available—at a pH of 5.5 and 6.5 for *organic* and *mineral* soils than at any other pH. Molybdenum, copper, and boron availabilities are also relatively high at pH 5.5 and 6.5, respectively. The micronutrients iron, manganese, and zinc are less available at a pH of 6.5 than at lower pH values.

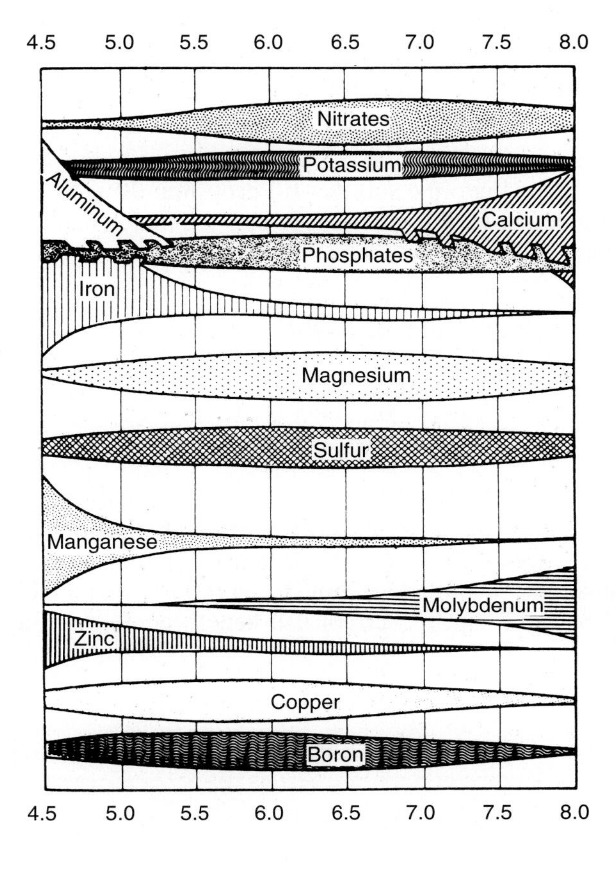

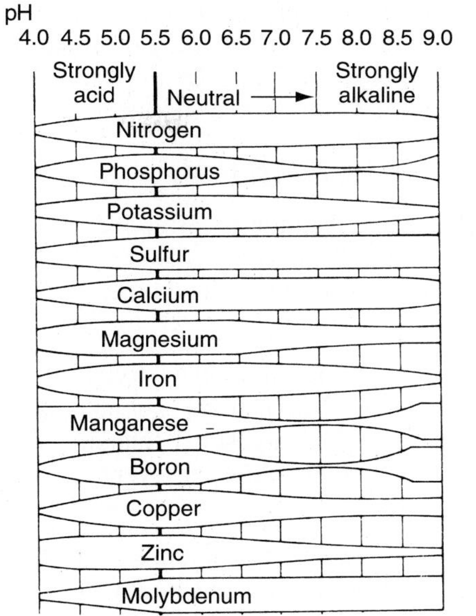

FIGURE 8-6 Theoretical relationship between soil pH and relative plant nutrient availability (the wider the bar, the greater the plant availability). (a) Organic soils. Note that the pH is about 5.5 for greatest plant availability of most nutrients in organic soils. (b) Mineral soils. Note that the pH is about 6.5 for greatest availability for the most nutrients. The lime requirement is one pH unit higher than for organic soils. Where elements are shown interlocking, the two elements at that pH combine to form insoluble compounds, which reduces phosphate solubility. (*Sources:* (a) Department of Crop and Soil Sciences, Michigan State University; (b) *Soils Handbook,* Kentucky Agricultural Experiment Station, Miscellaneous Publication 383, 1970, 28.)

Methods of Applying Lime

The most *efficient* way to use lime is to apply small amounts every year or two. However, frequent applications generally increase the cost. The *usual* liming practice consists of a compromise between that which is most effective and which is cheapest applied: *Add lime less often but in larger amounts.*

Lime can be applied to advantage at any stage in the cropping system, but it is usually best applied 4 to 12 months in advance of seeding a legume or a few months before planting a high-value crop that responds well to lime. Although it is usually recommended to add lime several months ahead of the crop that most needs lime added (to allow time for pH changes), good plant response in the southern United States has been observed when lime is applied in the spring, even at planting time.[5]

The usual method of applying lime is to spread it on the soil surface by a truck with a specially built, V-shaped bed and a spreading mechanism in the rear. Newly spread lime should be well mixed within the whole plow layer. On strongly acidic soils, where 6.7 Mg/ha

[5]C. S. Snyder, J. H. Muir, and G. M. Lessman, "Spring-Applied Aglime Can Provide Immediate Soybean Response," *Better Crops* **80** (1996), No. 1, 3–4.

to 13 Mg/ha or more of lime are required, half the amount may be applied before plowing and the other half applied and disked in after plowing. When not more than 4.5 Mg/ha are needed, the entire amount may be applied and disked in before seeding the legume or legume–grass mixture. When both surface soils and subsoils are strongly acidic, as with Ultisols, it sometimes pays to incorporate lime to a depth of about 30 cm (12 in).[6]

Liming No-Till Fields

Lime reaction is faster and greater when mixed into the soil. **No-till cropping** depends on rainfall for the downward migration of added lime. A lack of tillage plus the mulch of crop residues on no-till soil means the following:

- Soil microbial activity takes place at a much shallower depth.
- There is more root activity.
- The largest pH change and fertilizer buildup is in the surface (10 cm–20 cm) rather than in deeper layers.

Lime applications to no-till fields are not as effective as application to an equivalent cultivated soil, but liming no-till acidic soils is still worth the cost. Surface applications of lime in late winter for eight consecutive years in Virginia (in a medium-rainfall area) produced greater corn yields on a no-till area than on a tilled and limed area (Figure 8-7). Field research indicates that pH changes in the top 3 cm to 5 cm of soil benefit apple trees in Washington and wheat in Pennsylvania.[7,8]

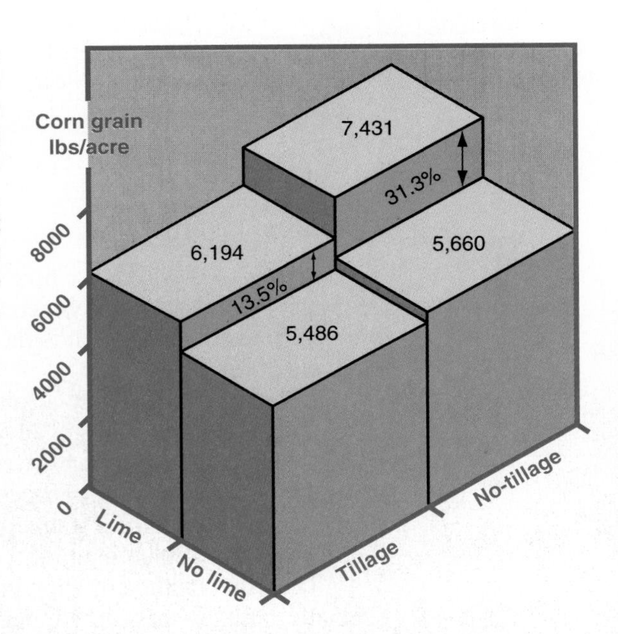

FIGURE 8-7 Surface-applied lime, which is a necessary method of addition in no-till operations, resulted in greater growth during 8 years on Frederick silt loam in Virginia than did lime incorporated by conventional operations. These results indicate that in at least some instances the surface application of lime will be a suitable method of application. (*Source:* Drawn from data by W. W. Moschler, D. C. Martens, C. I. Rich, and G. M. Shear, "Comparative Lime Effects on Continuous No-Tillage and Conventionally Tilled Corn," *Agronomy Journal* **65** [1973], 781–783.)

---

[6]B. D. Doss, W. T. Dumas, and Z. F. Lund, "Depth of Lime Incorporation for Correction of Subsoil Acidity," *Agronomy Journal* **71** (July–Aug. 1979), 541–544.
[7]Timothy J. Smith, "Time to Re-Apply Lime to Orchards in Washington?" *Better Crops* **80** (1996), No. 1, 8–9.
[8]Douglas B. Beegle, "Lime Needs Under No-Till Conditions," *Better Crops* **80** (1996), No. 1, 16–18.

Nitrogen fertilizers applied on the soil surface in no-till farming acidify the *surface* 2.5 cm (1 in) of soil. This reduces the effectiveness of certain herbicides, such as atrazine and simazine, and limits plant growth and yields.

Surface-applied lime induced alfalfa roots to grow more deeply, to 84 cm (33 in), even though the soil pH at this depth was 4.1 and exchangeable Al was 225 mg/kg.[9] This emphasizes the strong need for Ca^{2+} ions in order for alfalfa to develop properly and grow into soils with limited Ca^{2+}.

8:7 Lime Balance Sheet

When a soil has had its acidity lessened by lime, how often must lime be added and how much is needed to keep the soil pH suitable? The answers depend upon the rate of lime loss. Lime is neutralized or lost from the soil by activities such as the following:

- *Neutralization by acid-forming fertilizers (ammonium):* A rapid change
- *Neutralization by the acid formed by carbon dioxide dissolved in water (from air, respiration, and organic matter decomposition):* A slow continual process
- *Leaching:* A slow change
- *Removal in harvested or grazed crops:* A slow loss
- *Erosion:* As topsoil is eroded there is left more acidic subsoil needing lime

If compensating lime is not included, ammonium fertilizers may neutralize an average of 45.5 kg (100 lb) of field-applied lime per year. Each kilogram of nitrogen (N) from ammonium fertilizers generates soil acidity equal to 1.8 kg of pure calcium carbonate. Thus, to maintain soil pH, for each 100 kg of ammonium nitrogen (N) fertilizer applied, 180 kg of calcium carbonate equivalent must be applied. Each year in the United States more soil acidity is generated by nitrogen fertilizers than is neutralized by lime—by a ratio of 4:3.

The carbonic acid formed by carbon dioxide dissolved in water helps to solubilize, and thus leach, limestone:

$$\underset{\text{Carbonic acid}}{H_2CO_3} + \underset{\substack{\text{Calcium carbonate}\\\text{(limestone)}\\\text{(less soluble)}}}{CaCO_3} \longrightarrow \underset{\substack{\text{Calcium bicarbonate}\\\text{(more soluble and}\\\text{more leachable)}}}{Ca(HCO_3)_2}$$

In high-rainfall areas leaching losses may average 112 kg/ha (100 lb/a) per year. Harvested and grazed crops may remove the calcium equivalent of another 56 kg/ha to 224 kg/ha (50 lb/a to 200 lb/a) of lime. Erosion, if active, may remove another 45 kg/ha to 112 kg/ha (40 lb/a to 100 lb/a) of unreacted lime plus high-pH soil in the lost topsoil. These losses approximate 336 kg/ha to 560 kg/ha (330 lb/a to 500 lb/a) yearly. To maintain the desired pH in these conditions would require the application of an additional 2.2 Mg/ha (1 t/a) of lime every fifth year. This is in addition to the lime required to counteract the acidification due to N fertilizers.

8:8 Acidifying Soils

Some plants actually grow better—or have fewer problems—if the soil is moderately acidic rather than neutral or with a basic pH. It is possible to increase soil acidity, but, as with liming, the pH change may require months or longer. Elemental sulfur, iron, aluminum com-

[9] J. E. Rechcigl and R. D. Reneau, Jr., "Effects of Subsurface Acidity on Alfalfa in a Tatum Clay Loam," *Communications in Soil Science and Plant Analysis* **15** (1984), 811–818.

Table 8-3 Materials Added to Soils to Make Them More Acidic (Lower pH) and the Approximate Quantity Needed for Various Soils to Alter the pH

Amendment Added	*kg Equivalent to 1 kg of Sulfur*
Sulfur	1.0
Sulfuric acid (98%)	3.1
Iron Sulfate [$Fe_2(SO_4)_3 \cdot 7H_2O$]	8.7
Aluminum sulfate [$Al_2(SO_4)_3 \cdot 18H_2O$]	6.9

The pH Change Wanted	***kg of Sulfur Needed per Hectare****		
	Sand	*Loam*	*Clay*
8.5–6.5	2200	2800	3300
7.5–6.5	550	900	1100
7.0–6.5	100	170	350

Source: Selected and recalculated data from *Western Fertilizer Handbook,* Interstate Printers and Publishers, Danville, Ill., 1975, 231–232.

*Assumes noncalcareous soils and average organic matter for temperate-region soils. Soils high in organic matter and in swelling (montmorillonite clays) may have slightly higher requirements than shown.

pounds, and sulfuric acid are the most common materials used to acidify soils. Waste sulfuric acid, collected from scrubbing industrial smoke stacks (which prevent sulfur oxides from entering the atmosphere), is frequently utilized because it allows beneficial use of a waste by-product. The most effective materials are listed in Table 8-3. The reactions of soil acidification by additions of these materials follow:

$$2S + 3O_2 + 2H_2O \xrightarrow{} \text{Thiobacillus bacteria} \quad 4H^+ + 2SO_4^{2-}$$

$$Fe^{3+} + 3H_2O \longrightarrow \underset{\text{(insoluble)}}{Fe(OH)_3} + 3H^+$$

$$Al^{3+} + 3H_2O \longrightarrow \underset{\text{(insoluble)}}{Al(OH)_3} + 3H^+$$

The H^+ produced in each reaction can be consumed, as they are produced, by neutralization in the alkaline soils with bases such as $Ca(OH)_2$, $Mg(OH)_2$, NaOH, and the bicarbonates of these cations. Examples follow:

$$2H^+ + Ca(OH)_2 + SO_4^{2-} \longrightarrow H_2O + CaSO_4$$

$$Ca^{2+} + 2HCO_3- + SO_4^{2-} + 2H^+ \longrightarrow CaSO_4 + 2H_2CO_3$$

(H_2CO_3 is unstable and evolves CO_2 to leave H_2O)

Making soil more acidic (or less basic) may be desirable for lessening infection from pathogens (rots, scabs) or making micronutrient metals more soluble for better plant absorption. Sweet potatoes grown on Mississippi River terrace soils in Louisiana can be devastated by soil rot if the soil pH is higher than 6, but these particular soils have toxic amounts of manganese, which produce *crinkle-leaf* symptoms if the pH falls below 4.9.[10] The management of such soil is to lime carefully when needed to grow soybeans and various other crops but to keep the pH between pH 5 and 6 when used for sweet potatoes. Calcareous soils usually require too much amendment to neutralize the native lime to justify trying to acidify these soils, although growers on these soils sometimes deliberately select the most acid-forming of the N fertilizers.

> *Aglime is an important source of Ca for peanuts grown in the acidic soils of the Southeastern U.S. . . . due to [its] ability to supply Ca, whether it was applied for that purpose or to increase soil pH.*
>
> **—Gary J. Gascho (University of Georgia)**

Questions

1. Most soils, if all substances remained, would weather to produce neutral or alkaline soils. Explain, then, why so many soils are acidic and why some soils remain alkaline.
2. Show how (a) humus decomposition, (b) root respiration, and (c) ammonium nitrification produce acidity.
3. What are the two dominant ions on the cation exchange sites in soils of about pH 5.5?
4. Why are strongly acidic soils poor growing media for most plants?
5. (a) Define *lime.* (b) What actual material is most used as agricultural lime?
6. (a) Define the *neutralizing index.* (b) What are the important considerations in selecting a lime?
7. (a) In general terms, what does lime do when added to acidic soils? (b) How does liming improve the soil for growth of most plants?
8. Compare the importance of lime versus fertilizer.
9. (a) Which crops respond to lime? (b) Do all crops require liming of acidic soils? Discuss.
10. What is the preferred method of adding lime to soils? Why?
11. Considering the downward mobility of Ca^{2+}, is lime effective on no-till lands? Explain.
12. In relative terms, evaluate *clays, sands, low CEC,* and *high CEC* acidic soils for their need for lime.
13. (a) How can soils be deliberately acidified? (b) Which materials are used?

[10] L. G. Jones, R. L. Constantin, J. M. Cannon, W. J. Martin, and T. P. Hernandez, "Effects of Soil Amendment and Fertilizer Applications on Sweet Potato Growth, Production, and Quality," Louisiana Agricultural Experiment Station *Bulletin* **704,** 1977.

Salt-Affected Soils and Their Reclamation

Unless we change the direction we are headed, we might end up where we are going.

—Old Chinese proverb

9:1 Preview and Important Facts

PREVIEW

Salt is the savor of foods but the scourge of agriculture; in excess, salt kills growing plants. As early as 3500 B.C., the people of Mesopotamia farmed some of the richest land in the world—the Fertile Crescent of the Tigris and Euphrates Rivers in Turkey and Iraq. For nearly 5000 years they grew wheat and barley there, then only salt-tolerant barley; ultimately, the salt took over completely and nothing grew, and the land was abandoned. About 2100 years ago the Romans plowed the fields of conquered semiarid Carthage and applied salt to ensure that the Carthaginians could not reestablish their powerful metropolis; efforts to recolonize the area 24 years later failed because the salted fields were still unproductive.[1] In China now, more than 7 million hectares are classified as saline.

Soluble salts are those inorganic chemicals that are more soluble than gypsum ($CaSO_4 \cdot 2H_2O$), which has a solubility of 2.4 g per L (0.032 oz/gal) of water at 0°C. Common table salt (NaCl) has a solubility nearly 150 times greater than that of gypsum (357 g per L, or 47.7 oz/gal). Reclaiming salty soils requires leaching the soluble salts out of the soil. In theory removing salt is easy: The salts are dissolved in irrigation water and leached out of the soil profile. Reclamation of salt-affected soils requires (1) adequate internal drainage, (2) replacement of excess exchangeable sodium in sodic soils, and (3) leaching out of the soluble salts. Establishing internal soil drainage may be the most difficult requirement.

If the salt content is not too high, salt-affected soils can be used for crops with careful management. In managing salty soils, select salt-tolerant plants suitable for that soil, climate, and farm operation; if irrigating, irrigate frequently to keep salts diluted; irrigate in non-

[1]Moses Hadas, *Imperial Rome,* Time, New York, 1965, 38–39.

growing-season periods to leach salts partially downward; and plant seeds in the low-salt concentration areas of seedbeds. Nonirrigated soils have fewer options for management.

The area of salt-affected soils will increase as irrigation continues. Managing and reclaiming salt-affected soils is, indeed, an increasing worldwide concern.

IMPORTANT FACTS TO KNOW

1. The origins and composition of soluble salts
2. The kinds and properties of salt-affected soils
3. How soluble salt contents are measured and reported
4. The approximate salt contents that cause damage to plants
5. How damage to plants caused by soluble salts and excess exchangeable sodium is reduced
6. The principles used in the reclamation and management of salt-affected soils

9:2 Soluble Salts and Plant Growth

Soluble salts in water cannot be seen. Salts are solid when dry and sometimes can be seen on the surface of soils during drying conditions (Figures 9-1 and 9-2). Descriptive names have been given to various salty soils, such as *white alkali, black alkali, slick spots,* and *summer snow.* The white appearance of salts on the surface of some soils explains the names *white alkali* and *summer snow.* If soil has high exchangeable sodium, its pH will be about 9; soil humus colloids disperse, coloring puddles of surface water black, like puddles of oil. After drying, the soil has black crusts over its surface and ped faces (black alkali).

FIGURE 9-1 Toxic salt accumulation (white layer on ridge at arrow 1) has prevented the growth of all plants on this irrigated cotton field in southwestern Texas. The accumulation of excess salt could have been caused by a slight surface depression and/or a concentration of clay in the soil profile that reduced infiltration of irrigation water, as seen in the insert (arrow 2). (*Source:* Texas Agricultural Experiment Station, El Paso.)

FIGURE 9-2 Nonproductive area caused by salt accumulation (white areas) in northeastern Montana. The saline seep develops when water infiltrates into soil below root depth to an almost impermeable dense clay, which causes water to move laterally downslope. Eventually, at lower elevations on slopes, the water seeps laterally at a shallow depth below the soil surface and then to the surface by capillarity. As water evaporates on the surface, it leaves calcium, magnesium, and sodium sulfate salts. The use of deep-rooted crops in rotation and intensive cropping with minimum summer fallow so that soil water gets used up in areas above the low seep area have been recommended for reducing seepage movement. (*Source:* A. D. Halvorson and A. L. Black, "Saline-Seep Development in Dryland Soils of Northeastern Montana," *Journal of Soil and Water Conservation* **29** [1974], No. 2, 77–81; also *North Dakota Farm Research* **33** [1976], No. 4, 3–9.)

Most soluble salts in soils are composed of the cations sodium (Na^+), calcium (Ca^{2+}), and magnesium (Mg^{2+}) and the anions chloride (Cl^-), sulfate (SO_4^{2-}), and bicarbonate (HCO_3^-). Relatively smaller quantities of potassium (K^+), ammonium (NH_4^+), nitrate (NO_3^-), and carbonate (CO_3^{2-}) also occur, as do many other ions. In some soil solutions, soluble-salt concentrations are higher than in seawaters, which are 3 percent to 4 percent total salts.

The cations and anions that form soluble salts come from minerals as they weather. If precipitation is too low to provide leaching water, usually less than about 380 mm (15 in) annually, most or all of the soluble salts remain in the soil. As water evaporates from the soil surface, the soil salts move toward the surface but remain within or on the soil.

Large contents of soluble salts act osmotically to lower the water potential, making it harder for plants in salty (**saline**) soils to absorb water from the soil solution. A high exchangeable sodium percentage (more than 15% = **sodic soil**) causes the soil to disperse (aggregates break down), making the soil slowly permeable or impermeable to water. Irrigation can cause salt accumulation because all surface waters and groundwaters contain soluble salts.

Measuring Soluble Salts

Soluble salts are measured by electrical conductivity and the units of conductance (International System, or SI) are **decisiemens per meter (dS/m);** some scientists and recent literature still report in the older units of **mmhos per centimeter (mmhos/cm).** The relationships of the units of conductance are as follows:

Basic Unit	*Units Most Used in Soils*	*Old Units Still in Literature*
Siemens meter^{-1} **S m^{-1}**	Decisiemens meter^{-1} **dS m^{-1}**	Millimhos centimeter^{-1} **mmhos cm^{-1}**
	$1\ S\ m^{-1} = 10\ dS\ m^{-1} = 10\ mmhos\ cm^{-1}$	

The range of plant tolerance to salt is approximately as given below for the conductivity of the soil's **saturation paste extract:**

Extract Conductivity (dS m^{-1})	*Growth Reduction by Salt in Soil*
0–2	Few plants affected
2–4	Some sensitive plants affected (strawberries)
4–8	Many plants affected
8–16	Most crop plants affected
16+	Few plants can survive

Some approximate conversions from electrical conductivity to other relationships used to measure soluble salts are as follows:

$(dS\ m^{-1}) \times \mathbf{(640)} \cong$ Total dissolved solids (TDS) in mg L^{-1}
$(dS\ m^{-1}) \times \mathbf{(0.36)} \cong$ Osmotic potential of solution in bars
$(dS\ m^{-1}) \times \mathbf{(10)} \cong$ $mmol_c\ L^{-1}$ of total cations or of total anions

Effects of Salt Concentration

Although specific toxicities due to high concentrations of sodium, chloride, boron, or other ions can occur, salts usually reduce plant growth by the **osmotic effect.** High salt concentration increases the forces that hold water in the soil and require plant roots to expend more energy to extract the water. During a drying period salt in soil solutions may become concentrated enough to kill plants by pulling water from them (**exosmosis**). Salts move in water to the soil surface, where they accumulate, sometimes becoming visible as powdery, white salt crusts.

Salts are usually most damaging to *young* plants, but not necessarily at the time of germination, although high salt concentrations can slow or inhibit seed germination. Plant species have variable tolerances to the salt in soils, and the specific effects on different parts of a plant also vary (Figures 9-3 and 9-4). The minor differences in salt content at which any particular plant experiences growth reduction are often due to differences in the test conditions, plant age, or cultivar used; also, plant tolerances usually increase with maturity. Thus, it should be understood that the data in Figures 9-3 and 9-4 may vary.

The equation used to indicate yield loss assumes a straight-line decrease in yield with an increase in salt:

$$Yr = 100 - b(EC_e - a) \qquad \text{(Eq. 9-1)}$$

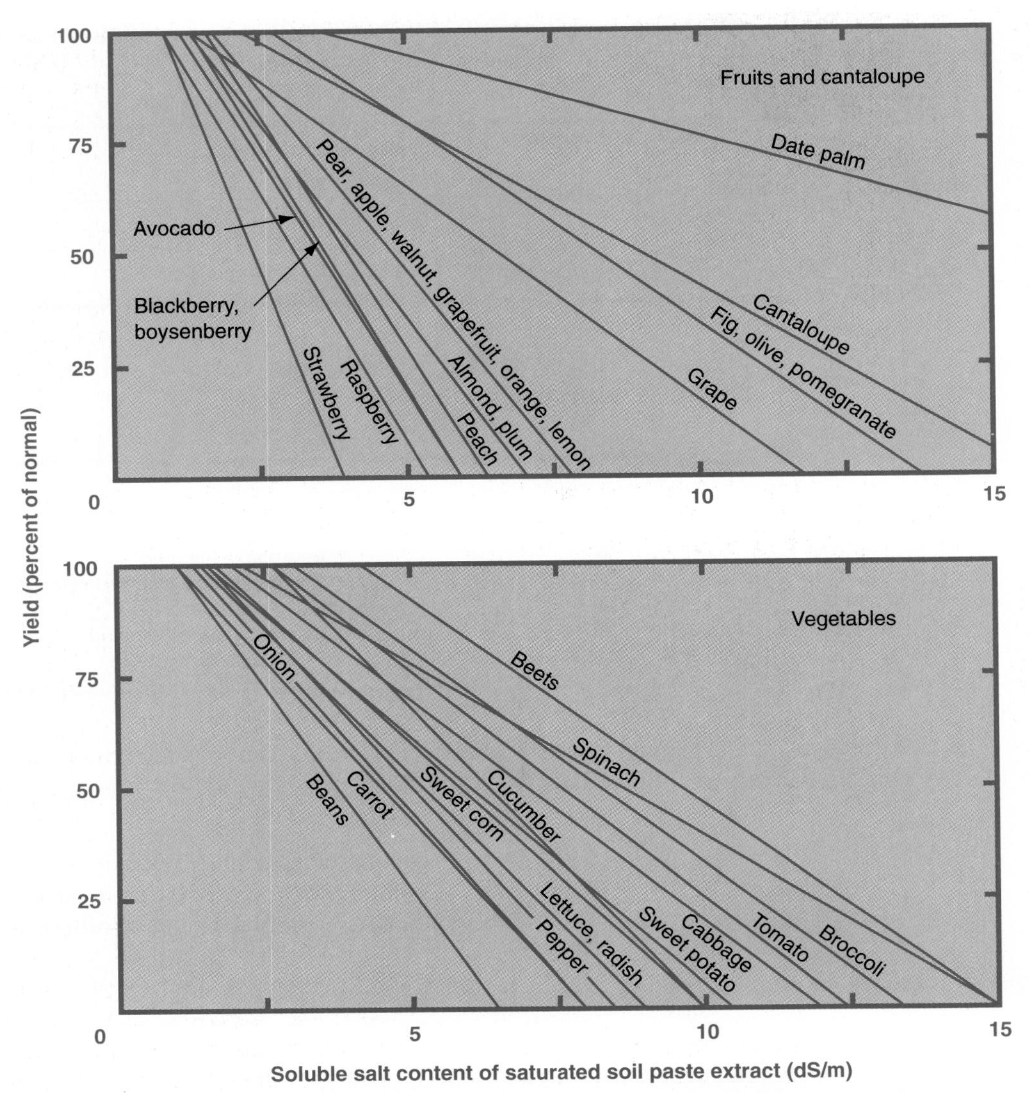

FIGURE 9-3 Yield reduction of fruits and vegetable crops by soluble salts in soils. Notice that this salt scale is expanded more than that of Figure 9-4. (*Source:* Drawn from data in *Soil Survey Investigations for Irrigation,* FAO Soils Bulletin 42, Food and Agricultural Organization of the United Nations, Rome, 1979, 72–74. Courtesy of Raymond W. Miller, Utah State University.)

where Yr is the relative yield compared to 100 percent, when the salt is no problem; b is the percentage yield loss per dS m^{-1} increase in salinity in excess of a, the threshold level of soil salinity at which yield decreases begin; and EC_e is the dS m^{-1} of the soil of interest (i.e., what is the yield loss in this soil at this EC_e?).[2]

Most plants are least affected by soil salts when in their *mature* stages. Plants in the germination and seedling stages may be quite sensitive to salt damage. The data for Figures 9-3

[2]J. D. Rhoades, "Overview: Diagnosis of Salinity Problems and Selection of Control Practices," Chapter 2, in Kenneth K. Tanji (ed.), *Agricultural Salinity Assessment and Management,* American Society of Civil Engineers, 345 East 47th Street, New York, NY, 10017, 1990, 20.

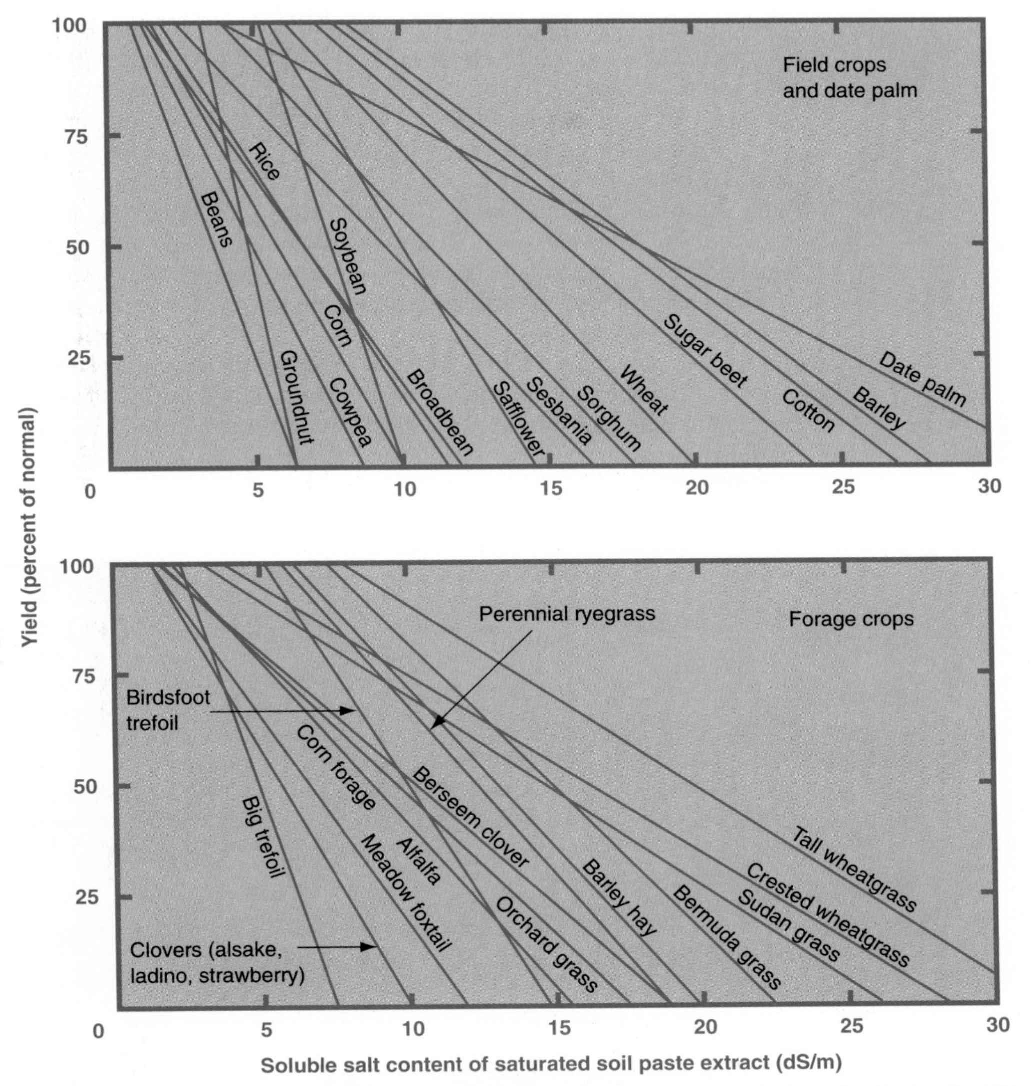

FIGURE 9-4 Yield reduction of field crops and forage crops by soluble salts in soils. Notice that this salt scale is double the salt range of Figure 9-3. (*Source:* Drawn from data in *Soil Survey Investigations for Irrigation,* FAO Soils Bulletin 42, Food and Agricultural Organization of the United Nations, Rome, 1979, 72–74. Courtesy of Raymond W. Miller, Utah State University.)

and 9-4 are for salt damage to plants *already germinated and in the later seedling stage;* crop yields may be greatly reduced for some crops at lower salt levels than shown if that salt level reduces the number of germinated seeds or number of seedlings that survive to older stages of growth.

The effect of salt on crop yield must also consider the productive part, which is important for yield. For example, salt levels for corn yields must be kept lower when producing corn for grain (Figure 9-3; a 50 percent yield loss at 6 dS m^{-1}) than when producing corn for forage (Figure 9-4; a 50 percent yield loss at 9 dS m^{-1}).

Both barley and cotton have considerable salt tolerance, but high concentrations of salt affect the *vegetative growth* (stems, leaves) more than the seed heads of barley or the bolls of cotton. Conversely rice grain yields are reduced before vegetative growth is affected.

However, because rice can be grown in *ponded water* (the most dilute condition possible for soil salts), the crop is often grown in high-salt soil in the early stages of reclamation (salt removal). Even if a soil has an electrical conductivity lower than 4 dS m^{-1} and is not termed *saline,* it may still have enough salt to lower the yields of salt-sensitive plants.

A common problem with salt-affected soil is a reduced permeability to water. The porosity of the soil becomes gradually altered; some soils can become completely impermeable. Soils become impermeable by two mechanisms: **swelling** of clays, which reduces pore sizes but is not caused by salt, and **dispersion** of the soil, so that aggregates break down, and smaller mineral and organic particles move with water and plug bottleneck pores, greatly reducing any flow through the soil. Dispersion is probably the most frequent cause of reduced infiltration or hydraulic conductivity.

Exchangeable sodium exerts its greatest effect on plant growth by favoring soil dispersion. As little as 10 percent exchangeable sodium in fine-textured (clayey) soils and 20 percent in sandier soils can cause dispersion damage. Colloid dispersion makes the soil less permeable, or even *im*permeable, and causes it to form hard surface crusts when dry (Figure 9-5).

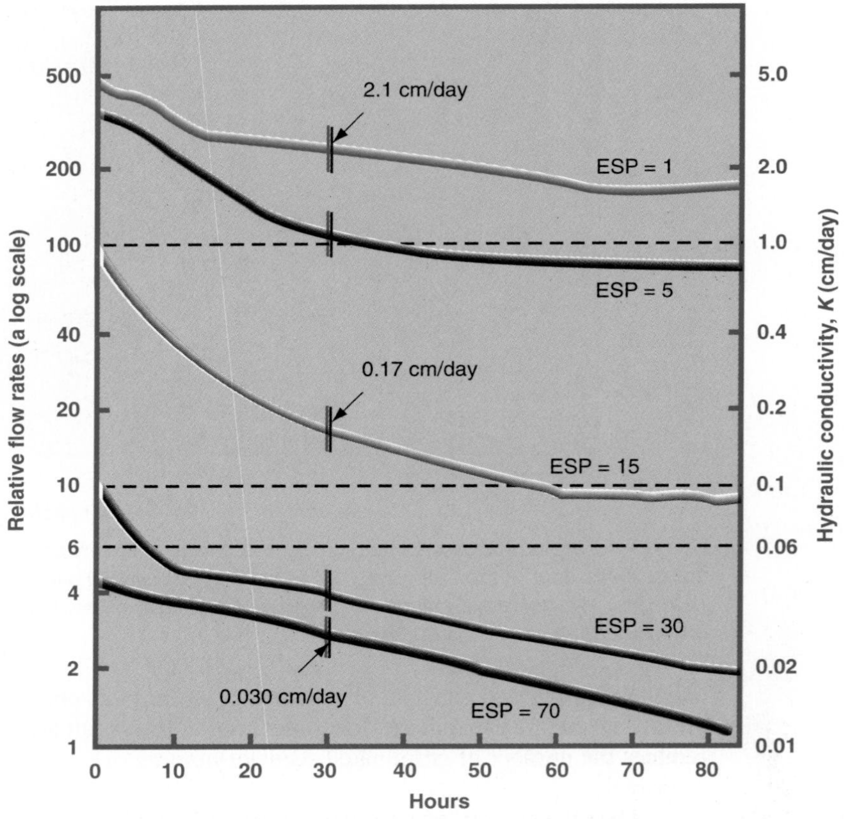

FIGURE 9-5 Example of the change in water flow (hydraulic conductivity) through a soil as influenced by the exchangeable sodium percentage. Each curve is for the same soil with a different exchangeable sodium percentage. Note that *water flow in all soils decreases with time as clays swell.* The swelling reduces pore sizes. Also, flowing water carries soil particles that lodge in pores, further reducing flow through the soil. Percolation of water through soil with an ESP of only 5 is 47 times faster at the end of 30 hours flow than flow in that soil with ESP of 70; it is more than 8 times faster than at an ESP of 15.

Exchangeable sodium percentages (ESP) as low as 2 percent to 5 percent have brought about reduction in the hydraulic conductivity of two Vertisols.[3] Not all soils have dispersion problems at the same exchangeable sodium percentage. Montmorillonite clays are the most easily dispersed. Some clayey soils disperse with only 9 percent to 10 percent exchangeable sodium. Kaolinitic soils, however, may be quite stable to even 25 percent to 35 percent exchangeable sodium. In many Oxisols and Ultisols, the kaolinite–metal oxide clay mixtures may be fairly stable even to values of nearly 45 percent to 50 percent exchangeable sodium. Soils low in clay content can tolerate higher exchangeable sodium percentages than can clayey soils because of their good permeability and large pores. Dispersion results in destruction of soil structure. The upper soil pores become filled with lodged dispersed particles, and both air and water exchange into and out of the soil are reduced. The hardened crusts can physically inhibit seedling emergence.

Effects of Specific Ions

Sodium and chloride are particularly toxic to woody ornamentals and fruit crops. Citrus, stone fruit, and blackberries are among the sensitive ones; grapes are quite sodium tolerant. Many deciduous fruits can be injured by as little as 5 percent exchangeable sodium. In woody ornamentals and fruit trees (on a dry-weight basis), as little as 0.5 percent to 1.0 percent chloride and 0.25 percent to 0.5 percent sodium can cause leaf-injury symptoms. Boron, at concentrations in plant tissue of 80 mg/kg to 300 mg/kg, is often toxic to plants.[4] Rapid absorption of sodium or chloride by the leaves of stone fruit and citrus trees makes some sources of water that might be satisfactorily added as surface irrigation *unsatisfactory* to use if sprinkler-applied. Strawberry leaves, which absorb sodium and chloride more slowly, and avocado leaves, which absorb little or none, can be sprinkle-irrigated safely with those same waters.

9:3 Saline and Sodic Soils

Salted soils are classified on the basis of two criteria: the *total soluble salt content* and the *exchangeable sodium percentage* (or, more recently, *sodium adsorption ratio*). Additional information is collected for some uses, such as *kinds of ions* and the *residual sodium carbonate content.*

Because ions in water conduct electrical current, electrical conductivity (EC) is a fast, simple method of estimating the amount of total soluble salts in a soil sample. To measure a soil's EC (electrical conductivity), a weighed soil sample is mixed with water to form a saturated paste or a more dilute mixture; the liquid is then removed by filtration or pressure or suction filtration; then the conductivity of the extract is measured.

Exchangeable Sodium Percentage (ESP)

The **exchangeable sodium percentage (ESP)** is the portion of total exchangeable cations that are sodium; so,

$$\frac{\text{Exchangeable Na}}{\text{Total exchangeable cations}}(100) = \text{ESP} \qquad \text{(Eq. 9-2)}$$

[3]Giuseppino Gescimanno, Massimo Iovino, and Giuseppe Provenzano, "Influence of Salinity and Sodicity on Soil Structural and Hydraulic Characteristics," *Soil Science Society of America Journal* **59** (1995), 1701–1708.

[4]J. D. Rhoades, "Overview: Diagnosis of Salinity Problems and Selection of Control Practices, Chapter 2 in Kenneth K. Tanji (ed.), *Agricultural Salinity Assessment and Management,* American Society of Civil Engineers, 345 East 47th Street, New York, NY 10017, 1990, 19–41.

The value of ESP is used extensively to indicate the hazard that the soil is or will disperse, thereby reducing its hydraulic conductivity (rate of water flow through it). An ESP value equal to or greater than 15 indicates a *sodic* soil. Soils with low hydraulic conductivity do not drain well; after a rain such soils often appear wet. These poorly drained soils often have considerable clay which is sticky because of the sodium. Such soils are called *slick spots*.

Sodium Adsorption Ratio

The **sodium adsorption ratio (SAR)** is used to estimate what the exchangeable sodium percentage of a soil is, or what it is likely to become, if the water of known SAR is used for years on that soil. The SAR has a good correlation to the *exchangeable sodium percentage* (ESP) and is much easier to calculate exactly or to estimate from a few simple analyses than is the ESP. The SAR defined, using ion concentrations of mmoles/L, is

$$\mathrm{SAR} = \frac{\mathrm{Na}^{+}}{\sqrt{\mathrm{Ca}^{2+} + \mathrm{Mg}^{2+}}} \qquad \text{(Eq. 9-3a)}$$

If values are in mmoles$_c$/L or milliequivalents/L, then

$$\mathrm{SAR} = \frac{\mathrm{Na}^{+}}{\sqrt{\dfrac{\mathrm{Ca}^{2+} + \mathrm{Mg}^{2+}}{2}}} \qquad \text{(Eq. 9-3b)}$$

The units of the actual SAR are $(\mathrm{mmoles_c}\ \mathrm{L}^{-1})^{\frac{1}{2}}$ and are ignored in typical usage (Calculation 9-1).

When the SAR is 13, the soil will probably lose permeability as salts are removed. An SAR of 13 replaces an ESP of 15 percent as the criteria for sodic soils in the United States. This relation, determined empirically, between the SAR and the ESP is approximately

$$\frac{\mathrm{ESP}}{100 - \mathrm{ESP}} = 0.015\ \mathrm{SAR} \qquad \text{(Eq. 9-4)}$$

For aggregated particles the minimum dispersion (maximum stability) is at the pH of lowest surface charge (point of zero charge). For layer silicates such as montmorillonite, the dispersion increases with increased pH. For iron oxide (and sesquioxides) the material can disperse at pH values below 6 or 7 and above 9. The greatest stability of sesquioxides is at a slightly alkaline pH (7 to 8).

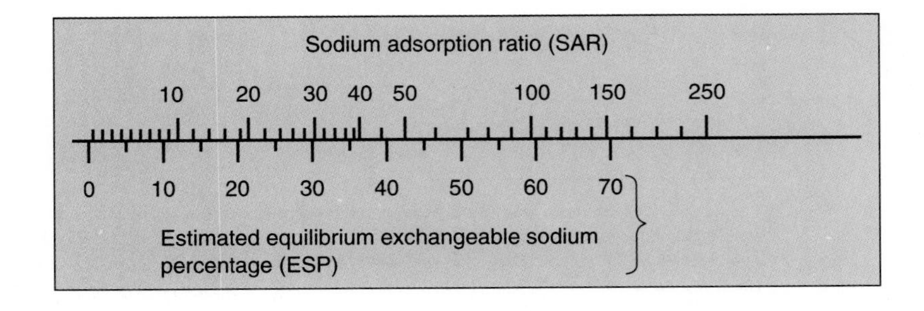

Calculation 9-1 Example Calculation of Sodium Adsorption Ratio (SAR)

To calculate the value of the SAR of a soil solution or water sample, simply insert the values in the following equation as $mmol_c/L$:

$$\text{SAR} = \frac{Na^+}{\sqrt{(Ca^{2+} + Mg^{2+})/2}}$$

Problem Calculate the SAR for soils 1, 2, and 4 from Table 9-2.

Solution The $mmol_c/L$ is calculated by dividing the mg/L of the cation by the ion's atomic weight and multiplying by the cation's valence (oxidation number). For example, calculate the $mmol_c/L$ for Ca^{2+} in soil 1 of Table 9-2. From the table, there are 1301 mg of Ca^{2+} and the gram atomic weight of Ca^{2+} (from chemical periodic table) is 40.08 g:

$$mmol_c \text{ of } Ca^{2+}\ L^{-1} = \left(\frac{1301 \text{ mg}}{L}\right)\left(\frac{1 g}{1000 \text{ mg}}\right)\left(\frac{1 \text{ mol } Ca^{2+}}{40.08 \text{ g}}\right)\left(\frac{2 \text{ mol}_c\ Ca^{2+}}{1 \text{ mol } Ca^{2+}}\right)\left(\frac{1000 \text{ mmol}_c}{1 \text{ mol}_c}\right)$$

$$= \left(\frac{2602 \text{ mmol}_c}{40.08 \text{ L}}\right) = 64.9 \text{ mmol}_c\ L^{-1} \text{ of } Ca^{2+}$$

for Soil 1:

$$\text{SAR} = \frac{15.26 \text{ mmol}_c/L}{\sqrt{\dfrac{65.0 + 34.2 \text{ mmol}_c/L}{2}}} = \frac{15.26}{\sqrt{49.6}} = \frac{15.26}{7.04} = 2.24$$

for Soil 2:

$$\text{SAR} = \frac{79.5 \text{ mmol}_c/L}{\sqrt{\dfrac{6.7 + 9.9 \text{ mmol}_c/L}{2}}} = \frac{79.5}{\sqrt{8.3}} = \frac{79.5}{2.88} = 27.6$$

for Soil 4:

$$\text{SAR} = \frac{29.2 \text{ mmol}_c/L}{\sqrt{\dfrac{1.1 + 0.33 \text{ mmol}_c/L}{2}}} = \frac{29.2}{\sqrt{0.71}} = \frac{29.2}{0.84} = 37.8$$

Source: Raymond W. Miller, Utah State University.

Residual Sodium Carbonate (RSC)

The **residual sodium carbonate (RSC)** value, used to indicate alkalinity of water, is defined by the equation

$$\text{RSC} = (HCO_3^- + CO_3^{2-}) - (Ca^{2+} + Mg^{2+}) \quad \text{(Eq. 9-5)}$$

where all units are *millimoles of charge per liter* ($mmol_c\ L^{-1}$). As soils dry, the calcium and magnesium ions precipitate as low-solubility carbonates. The carbonate–bicarbonate ions left

Table 9-1 Classification of Salt-Affected Soils[a]

Name for Soil	*Electrical Conductivity of Saturation Extract EC_e (decisiemens meter^{-1})*	*Sodium Adsorption Ratio (SAR)*
Normal soils	Less than 4	Less than 13
Saline soils[b]	More than 4	Less than 13
Sodic soils[c]	Less than 4	More than 13
Saline-sodic soils[d]	More than 4	More than 13

Source: Glossary of *Soil Science Terms,* Soil Science Society of America, Madison, WI, 1979, 14–15.
[a]Although the salt content division concentration is left at 4.0 decisiemens meter^{-1} (= 4dS m^{-1}), plants sensitive to salts may be affected by contents as low as 2.0 dS m^{-1}; salt-tolerant plants may not be affected below 8.0 dS m^{-1} salt content.
[b]Formerly called *white alkali* soils and *solonchak.*
[c]Formerly called *black alkali* because of dispersed black organic-matter coatings on peds and the soil surface.
[d]Formerly called *white alkali* or *black alkali,* depending on the visual appearance of the individual soil.

in the solution require considerable H^+ to neutralize them. The generally accepted scale of alkalinity hazard for water is as follows:

RSC $>$ 2.5	Hazardous
RSC = 1.25–2.50	Potentially hazardous
RSC $<$ 1.25	Generally safe

If a *hazardous* water is repeatedly added to soil, the soil will become alkaline and likely become a *sodic* or *sodic-saline* soil over time.

Salty Soil Classification

Salt-affected soils have one of three names: *saline, sodic,* or *saline-sodic* (Table 9-1).

- **Saline** soil has a saturation extract conductivity of 4.0 decisiemens per meter or greater and has an SAR less than 13.
- **Sodic** soil has an SAR of the saturation extract of 13 or more but has low salt content.
- **Saline-sodic** soil has *both* the salt concentration to qualify as *saline* and an SAR of 13 or more needed to qualify as *sodic.*

Table 9-2 lists some typical soluble-salt data for four salted soils: Soil 1 is saline, soils 2 and 3 are saline-sodic, and soil 4 is sodic. However, soil 1, the highest salt of the four soils listed, has only about 0.7 percent soluble salts; many soils are much saltier than this.

9:4 The Salt Problem and Salt Balance

Salt buildup is an existing or potential danger on *almost all* of the 17 million hectares (42 million acres) of irrigated land in the United States, and it is an increasing problem on nonirrigated semiarid and arid cropland and rangeland. Much of the world's uncropped land that might supply future food is in arid and semiarid regions where irrigation would be necessary.

Table 9-2 Soluble-Salt Constituents in the Saturation Extracts from Three Selected Salty Soils and One Sodic Soil, Showing the Relative Amounts of the Various Ions[a]

Soil 1 (pH 7.8; EC 7.6) Saline				*Soil 2 (pH 7.3; EC 9.2) Saline-Sodic*			
Cations		*Anions*		*Cations*		*Anions*	
Na^{+}	351	Cl^{-}	4329	Na^{+}	1828	Cl^{-}	2556
Ca^{2+}	1301	SO_4^{2-b}	452	Ca^{2+}	134	SO_4^{2-b}	965
Mg^{2+}	411	HCO_3^{-}	117	Mg^{2+}	119	HCO_3^{-}	146
K^{+}	41	CO_3^{2-}	0	K^{+}	20	CO_3^{2-}	0

Soil 3 (pH 8.1; EC 7.9) Saline-Sodic				*Soil 4 (pH 9.6; EC 3.2) Sodic*			
Cations		*Anions*		*Cations*		*Anions*	
Na^{+}	1661	Cl^{-}	75	Na^{+}	672	Cl^{-}	266
Ca^{2+}	278	SO_4^{2-}	4325	Ca^{2+}	22	SO_4^{2-}	221
Mg^{2+}	71	HCO_3^{2-}	183	Mg^{2+}	4	HCO_3^{-}	1141
K^{+}	23	CO_3^{2-}	12	K^{+}	160	CO_3^{2-}	0

Source: Soils 2, 3, and 4 selected and modified from J. D. Rhoades and Leon Bernstein, "Chemical, Physical, and Biological Characteristics of Irrigation and Soil Water," in *Water and Water Pollution Handbook,* vol. 1, L. L. Ciaccio, ed., Marcel Dekker, New York, 1971, 160.

[a]Values are milligrams of the ion per liter. To get millimoles$_c$ divide each value by the ion's atomic or polyatomic weight and multiply by the ion's valence. Soil 1 is saline, soils 2 and 3 are saline-sodic, and soil 4 is sodic. EC is electrical conductivity in dS m^{-1} (decisiemens per meter), a measure of salt content. Ion concentrations are given in milligrams per liter; multiply by 0.000134 for ounces per gallon. As an example: to get mmol$_c$ for sulfate in soil 1, divide 452 by weight of sulfate (= 96) and multiply it by valence of 2 = 9.4 mmol$_c$ of sulfate per liter.

[b]Actually, sulfate plus nitrate, calculated as only sulfate.

Continual application of water, all of which contains salts, will continually increase the soluble salts in soils unless the soils have periodic leaching. The need to move salt out of soils periodically will degrade water somewhere else where the salt goes.

Restrictions on Salty Water Disposal

The demand for water and water quality legislation, such as the Porter–Cologne Water Quality Act in California, promote the reuse of drainage water. Eventually, the water will concentrate enough salts so it is no longer suitable for agriculture. This salty wastewater must be recycled to groundwaters or streams. Eventually, this dumping of wastewaters will be permitted only after its salt content is lowered to an acceptable level—500 ppm in California.

Leaching Requirement

A **salt balance** exists when *outgoing* salt is equal to *incoming* salt. Successful salt management depends on adequate leaching. Leaching occurs whenever irrigation plus rainfall exceed the soil's water storage capacity, assuming that the soil has drainage. Leaching by drainage is essential to control salt content in soil.

The amount of extra water added to leach the soil is called the *leaching requirement.* The **leaching requirement (L_r)** is the *minimum leaching fraction that the crop can tolerate* *without yield reduction.* A salt balance is desired such that *salt added = salt removed.* This can be represented as follows:

$$\text{Salt in } \textbf{irrigation} + \text{salt in } \textbf{rainfall} - \text{salt } \textbf{removed in drainage} = 0 \qquad \text{(Eq. 9-6)}$$

$$D_iC_i + D_rC_r - D_dC_d = 0$$

where D = depth of water; C = salt content; and i, r, and d = irrigation, rain, and drainage, respectively. If irrigation plus rainfall = a = **added** water, and EC = electrical conductivity of the water, then

$$D_aC_a = D_dC_d$$

This can be rearranged and equated to the leaching fraction, L, as follows:

$$L = \frac{D_d}{D_a} = \frac{C_a}{C_d} = \frac{EC_a}{EC_d} \qquad L = L_r = \frac{EC_a}{EC_d} \qquad \text{(Eq. 9-7)}$$

Thus, the leaching requirement (L_r, to indicate *requirement* but is the same as L) can be defined as the *ratio of electrical conductivities of applied water to that of drainage water.* When the amount of water to wet the soil is calculated, the leaching fraction indicates the *additional* water that must be added for leaching. For example, if the L_r is 0.2 and an irrigation of 3.2 cm is planned, 0.2 × 3.2 cm = 0.64 cm additional water should be applied (Figure 9-6).

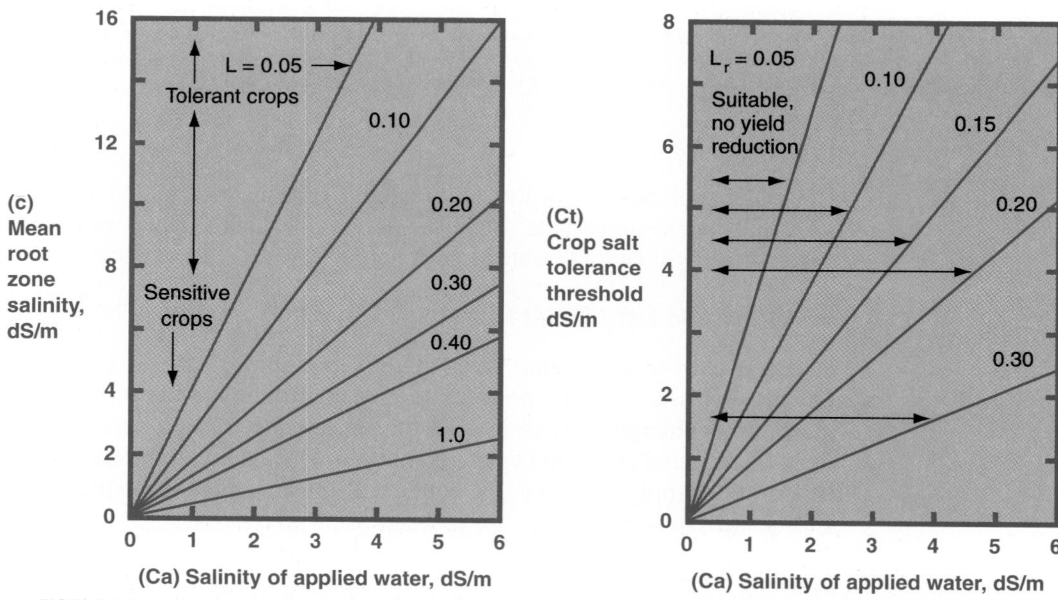

FIGURE 9-6 Mean root-zone salinity and crop-tolerance threshold values as a function of the leaching requirement and the salinity of applied water. Values above and to the left of each leaching fraction line indicate values of salinity in the root zone and values that meet the crop's salt tolerance without yield loss. For each water-salt value, extend a line up to the leaching fraction line and then direct it left, parallel to the base line, until it intersects the vertical axis. Read the value on the vertical axis. (Modified and redrawn from Glenn J. Hoffman, "Leaching Fraction and Root Zone Salinity Control," Chapter 12, in K. Tanji (ed.) *Agricultural Salinity Assessment and Management,* American Society of Civil Engineers, 345 East 47th Street, New York, NY 10017, 1990, 504–529.)

Calculation 9-2 Leaching Requirement Calculation

Problem Assume that an irrigation water has a conductivity of 108 mS m^{-1} (= 1.08 dS m^{-1} or 1.08 mmhos/cm). The field corn planted has a 50 percent yield reduction at a soil saturation extract conductivity of 6 dS m^{-1} (from Figure 9-4). Calculate the additional amount of water to add if the water needed to wet the profile is 6.35 cm (2.5 in).

Solution Substituting in the leaching requirement equation yields the following:

$$L_r = \frac{EC_a}{EC_d} = \frac{1.08 \text{ dS/m}}{6 \text{ dS/m}} = 0.18$$

This decimal (or fraction) is that fraction of the amount of water needed to wet the soil that must be added additionally. The total water needed is

$$6.35 \text{ cm} + (0.18)(6.35) = 7.49 \text{ cm } (2.95 \text{ in})$$

The leaching requirement in use is more complicated than just described. The salt content at different depths in the irrigated soil will vary in the amount of salt and the extent of leaching each irrigation. Additionally, the amount of concentration of salt between irrigations changes with the depth in soil as the soil water is used by plants. Even the frequency of irrigation (less drying between irrigations) can alter the salt content in root-zone water and thus modify the leaching requirement needed to keep the plant from yield reductions.[5]

Using surface-flow irrigation the effort to achieve adequate deep wetting at the lower end of the field means that enough excess water for leaching is already being added at the top. With increased use of saltier water and with less water available, more attention must be given to the leaching requirement than is given with common surface-flow irrigation which has overapplication. Notice that as irrigation water becomes saltier, the fraction L_r becomes larger, meaning that more water must be added for leaching to avoid salt buildup (see Calculation 9-2). The use of brackish water to grow salt-tolerant crops is proposed as one way to find more water in some water-scarce areas.[6]

Some scientists claim that the leaching requirement may be satisfactorily reduced to only 25 percent to 40 percent of the amount given in Equation 9-7.[7] However, the limit of the leaching requirement is not simple to determine. The salt sensitivity of the plant must be considered. For example, in California the more salt-tolerant wheat and sorghum can have a leaching requirement as low as 0.08 before growth reduction occurs with water containing 1350 ppm of salts. Lettuce, a more salt-sensitive plant, must have a leaching requirement of 0.20, two-and-a-half times more.[8]

In the Wellton–Mohawk project in southwestern Arizona, a leaching requirement (L_r) of 0.42 on soil using Colorado River water (150 metric tons of salt per 1000 cubic meters of water) caused salt removal in drainage water of 22.3 metric tons per hectare (9.95 t/a); when the L_r was reduced to 0.10 in an effort to comply with the international agreement to reduce the Colorado River water salt load, only 9.96 metric tons per hectare (4.44 t/a) of salt were removed.

---

[5]Parker F. Pratt and Donald L. Suarez, "Irrigation Water Quality Assessments," Chapter 11, in K. Tanji (ed.), *Agricultural Salinity Assessment and Management,* American Society of Civil Engineers, 345 East 47th Street, New York, NY 10017, 1990, 504–529.

[6]Don Gardner, "Irrigated Land May Include Widespread Use of Saline Waters," *Irrigation Age* **18** (1984), No. 5, 6–7.

[7]J. van Schilfgaarde, L. Bernstein, D. Rhoades, and S. L. Rawlins, "Irrigation Management for Salt Control," *Journal of Irrigation and Drainage Division* **100** (1974), No. IR3, 321.

[8]G. J. Hoffman, S. L. Rawlins, J. D. Oster, J. A. Jobes, and S. D. Merrill, "Leaching Requirement for Salinity Control: [Part I.] Wheat, Sorghum, and Lettuce," *Agricultural Water Management* **2** (1979), 177–192.

Obviously, the low leaching requirement results in more salt accumulation in time. The less even the water application is, the greater the L_r must be to keep all of the soil area low in salts.

9:5 Reclaiming Salty Soils

Three general rules to reclaim soils affected by salt are the following:

1. **Establish internal drainage.** For some soils drainage is already adequate. In other soils drainage might require the installation of drainage systems (tile lines, open ditches, etc.). Drainage may be impractical (too costly) or impossible (too flat, no near outlet, cost prohibitive, or illegal according to Environmental Protection Agency regulations). It must be possible to wet the root zone without excessive ponding or runoff; otherwise, water can never move *through* the profile and leach the soil.
2. **Replace excess exchangeable sodium,** if needed. This is necessary for some sodic and saline-sodic soils. The extent of this need varies with soil texture, kind of clay, quality of available water, extent of present damage, and so on. If soil is poorly permeable and the SAR is higher than 10, the soil probably has excessive sodium.
3. **Leach out most of the soluble salts.** Saline soils must have their salt content lowered, at least in part of the root zone. In sodic soils most of the *replaced sodium* must also be leached from the root zone. *Without leaching there is no lasting reclamation.* It is desirable to use good-quality irrigation water for leaching out salts, except when severe dispersion might require more salty water for leaching in early stages of reclamation.

Although these three general guides seem simple, often the physical problems and costs are great. Different situations require different approaches. When one considers economics, as land owners will certainly do, compromises and different techniques may be tried in each instance.

Reclaiming Saline Soils

Saline soils are relatively easy to reclaim for crop production if adequate amounts of low-salt irrigation water are available, internal and surface drainage are present, and salt disposal dump areas (sinks) are available. The main problem is to leach most of the salts downward and out of the root zone.

Frequently, saline soils have a high water table or a dense gypsum layer or are fine textured. These conditions reduce the movement of irrigation water downward and make it difficult to leach the salts to the desired depth below the plant root zone. In salty soils that have a high water table, artificial drainage is necessary before excess salts can be removed. Deep chiseling or deep plowing may be used on soils with impervious layers to open the soil for the desired downward movement of percolating salty water. This process is expensive and may need to be repeated several times.

Reclamation of saline soils, particularly when only rainfall or limited irrigation is used, can be hastened by the application of a surface organic mulch, as reported from the Rio Grande Valley of Texas. Because mulch slows surface evaporation, salt movement to the soil surface in evaporative water is decreased, and the *net downward movement* of salts is increased. Cotton gin trash and chopped woody plants are equally effective when applied at the rate of about 67 metric tons per hectare (30 t/a). With mulch the surface soil salt content decreases, whether the area receives only natural rainfall or supplemental sprinkler irrigation. Crop residues have also helped in North Dakota.

The quantity of water required to remove salts from the soil depends on many things: how deep the salts are to be leached, what percentage of the salts is to be removed, and how

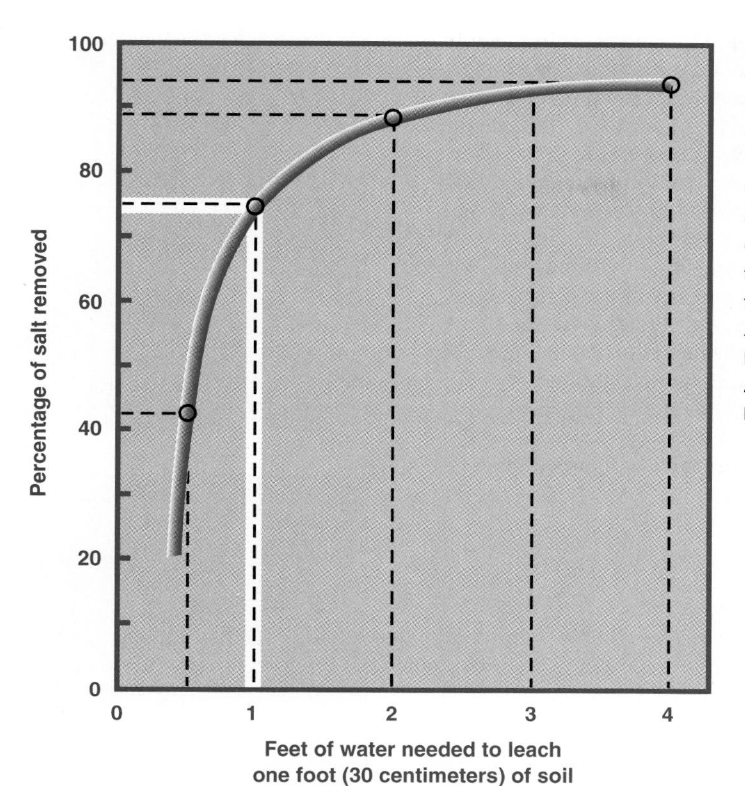

FIGURE 9-7 Estimated amount of normal (low-salt) water required to remove salts from saline soils. About 30 cm (1 ft) of water through 30 cm (1 ft) of soil removes nearly 80% of the salt. (*Source:* Data of R. C. Reeve; redrawn and modified from J. D. Rhoades, "Drainage for Salinity Control," in *Drainage for Agriculture,* J. V. Schilfgaarde (ed.), No. 17 in the Agronomy Series, American Society of Agronomy, Madison, WI, 1974, 433–461.)

the leaching is done (ponding constantly or intermittent sprinkling). A general guide is that with ponded water about 30 cm (12 in) of water are required to remove 70 percent to 80 percent of the salt for each 30 cm depth of soil to be leached (Figure 9-7). Intermittent water additions are more efficient and reduce the water applied to about 70 percent of that needed with ponding (continuous) leaching methods.

Soil boron may be sufficiently high to be a problem as a specific toxic element; its removal by leaching is slower than with soluble salts. Boron is weakly adsorbed by some soil constituents, and leaching may require three times more water than is needed to remove other soluble salts to the same extent.

Reclaiming Sodic and Saline-Sodic Soils

The reclamation of sodic soils may require a technique modified from that used for reclamation of saline soils. In sodic soils the exchangeable sodium is so great that the resulting dispersed soil is almost impervious to water (Figure 9-8). But even if water could move downward freely in sodic soils, water alone would not leach out the excess *exchangeable* sodium. The sodium must first be replaced by another cation and then leached downward and out of the root zone.

By cationic exchange reactions, calcium is used to replace sodium in sodic soils. Of all calcium compounds, calcium sulfate (gypsum, $CaSO_4 \cdot 2H_2O$) is considered the most convenient and cheapest for this purpose. Calcium solubilized from gypsum replaces sodium, leaving soluble sodium sulfate in the water, which is then leached out. The leaching waters themselves may rarely contain enough cations to replace the sodium. In a calcareous saline-sodic soil in India, leaching with a good low-salt irrigation water (SAR = 0.6, pH 7.5)

FIGURE 9-8 This sodic soil in North Dakota occurs in spots where high exchangeable sodium (16%–20% at a 25-cm [10-in] depth) on the clayey soil makes it impermeable to water. (a) These *gumbo* or *scab spots* hold ponded water for days. (b) A close-up shows the shallow surface soil (an **A**) and the rounded tops of *columns* making up the **Bt** clay horizon. The scale on the stake is 1 foot (30 cm) divided into inches. (*Source:* Fred M. Sandoval and G. A. Reichman, "Some Properties of Solonetzic (Sodic) Soils in Western North Dakota," *Canadian Journal of Soil Science* **51** [1971], 143–155. Photos by Fred Sandoval.)

(a)

(b)

was effective in removing exchangeable sodium without added gypsum. Without gypsum about 50 percent more water was required to remove the same amount of sodium as removed by the 35 cm (13.7 in) of water used with gypsum. The limiting factor in reclamation of sodic soils is usually permeability; *if the soil is permeable, gypsum addition is less critical to reclamation.*

If soil permeability when gypsum is added is still very low, $CaCl_2$ and H_2SO_4 have been used effectively (but they are costly). Sulfur, which oxidizes to sulfuric acid during several months, is used also in the reclamation of sodic soils and for lowering the pH of the soil. The relative required amounts of several amendments for reclaiming sodic soils are shown in Table 9-3.

When sulfur is added to the soil, *Thiobacillus* bacteria slowly oxidize the sulfur to sulfuric acid (H_2SO_4). The hydrogen ions of sulfuric acid can replace sodium ions on the soil cation exchange sites; or, if the soil contains lime ($CaCO_3$), the sulfuric acid may react to form gypsum, which then has the same effect as applied gypsum.

$$\underset{\text{(lime)}}{CaCO_3} + H_2SO_4 + H_2O \rightarrow \underset{\text{(gypsum)}}{CaSO_4 \cdot 2H_2O} + CO_2 \uparrow$$

Table 9-3 Estimated Efficiencies for Various Materials Used to Reclaim Sodic Soils Compared to Gypsum

Material	*Tons of Material Equivalent to 1 Ton of Gypsum*
Gypsum	1.00
Sulfuric acid	0.57
Sulfur	0.18
Lime-sulfur	0.75
Iron sulfate	1.62

Applications of about 40 Mg/ha (18 t/a) of gypsum in Nevada increased water infiltration and the depth of water penetration.[9] Three years after applying the gypsum, the water penetrated to a depth of 48 cm (19 in) in the soil receiving the gypsum and only 25 cm (10 in) into the soil that did not receive gypsum. This method reduced the exchangeable sodium percentage from 42 percent to 18 percent during a 3-year period on the gypsum-treated soil. At the same time the plot without gypsum increased in exchangeable sodium from 50 percent to 53 percent. Yields of hay increased by amounts of 0.1 Mg/ha to 2.3 Mg/ha (0.05 t/a to 1.02 t/a) per year as a result of the application of gypsum.

The **gypsum requirement (GR)** is the calculated amount of gypsum necessary to add to reclaim the soil (Calculation 9-3). Its numerical definition depends upon the weight used for the volume of soil reclaimed and the claimed gypsum efficiency. Using a soil bulk density of 1340 kg/m for an average soil, the gypsum requirement is described by the following two equations:

$$\text{GR} = \frac{\text{Mg of gypsum needed}}{\text{hectare of soil to some fixed depth}} \qquad \text{(Eq. 9-8a)}$$

$$\text{GR} = (\text{Na}_x)\ 4.50 \text{ Mg of gypsum per hectare-30 cm} \qquad \text{(Eq. 9-8b)}$$

where Na_x is the centimoles per kilogram of exchangeable sodium *to be replaced by calcium* from the added gypsum. When calculated in tons (2000 lb) per acre,

$$\text{GR} = (\text{Na}_x)\ 1.80 \text{ tons of gypsum per acre-foot} \qquad \text{(Eq. 9-8c)}$$

The conversion factors (4.50 or 1.80) change as soil bulk density, soil depth reclaimed, or gypsum efficiency change. The above values assume about 75 percent to 80 percent efficiency (about 25 percent more gypsum is added than is calculated by chemical formulas); (see Calculation 9-3).

A significant innovation in reclaiming sodic and saline-sodic soils is the initial use of *salty* water for leaching. Such a use appears paradoxical but is scientifically sound. A high salt content in water keeps sodic soils flocculated. The floccules have large pores between them and allow penetration of the leaching waters. Thus, the first water used for leaching may be moderately salty water. After most of the exchangeable sodium is removed by the calcium in the salts in the water or from gypsum additions, water of lower salt content can be used for final leaching.

[9] The confusion in using *tons* in the English, U.S., and metric systems has led to the currently used suggestion to indicate *metric ton* (= 1000 kg) by the symbol *Mg* (megagrams = 10^6 grams = 1000 kg).

Calculation 9-3 Gypsum Requirement Calculation

Problem A sodic soil has an average exchangeable sodium percentage (ESP) of 24 percent in the top 45 cm (17.5 in) and a cation exchange capacity (CEC) of 18 centimoles$_c$ per kg of soil. The average exchangeable sodium to be left in the top 30 cm (12 in) of soil is selected as 6 percent. Calculate the amount of gypsum and the amount of sulfur required to reclaim the soil in its top 30 cm.

Solution The gypsum requirement factor is calculated as follows, using a soil bulk density of 1340 kg/m^3:

$$GR = \left(\frac{1\text{ cmol}_c\text{ of Na to exchange}}{1\text{ kg of soil}}\right)\left(\frac{\text{mol}_c\text{ weight of gypsum}}{\text{mol}_c\text{ weight of sodium}}\right)\left(\frac{4 \times 10^6\text{ kg soil}}{1\text{ ha-30 cm}}\right)$$

$$\left(\frac{\text{kg of Na}}{1\text{cmol}_c\text{of Na}}\right)\left(\frac{1\text{ Mg}}{1000\text{ kg}}\right)$$

$$= \left(\frac{\text{Na}_x}{}\right)\left(\frac{85.5\text{ g gypsum}}{23\text{ g Na}}\right)\left(\frac{4 \times 10^6\text{ kg}}{1\text{ ha-30 cm}}\right)\left(\frac{0.00023\text{ kg Na}}{1\text{ cmol}_c\text{ of Na}}\right)\left(\frac{1\text{ Mg}}{1000\text{ kg}}\right)$$

$$= \text{Na}_x\ (3.42)\text{ Mg of gypsum per ha-30 cm depth}$$

where the (Na_x) is the cmol$_c$ of exchangeable sodium per kilogram of soil to be replaced by calcium from gypsum.* If the gypsum is not pure, or for other reasons is not 100 percent efficient, more than the amount calculated must be added. Experience suggests that the gypsum is only 75 percent to 80 percent efficient, so the amount added must be increased accordingly. Adding 30 percent more gives these approximate equations for the gypsum requirement (GR):

$$GR = (\text{Na}_x)\ 4.50\text{ Mg of gypsum per hectare-30 cm}$$

or

$$GR = (\text{Na}_x)\ 1.80\text{ tons of gypsum per acre-foot}$$

To solve the given problem, perform the following steps:

1. The cmoles$_c$ of Na needing replacement are calculated as the total exchangeable Na minus the exchangeable Na to be left.

$$(\text{CEC})\ (\text{ESP/kg}) = \text{cmol}_c\text{ of exchangeable Na per kilogram of soil}$$

$$= \left(\frac{18\text{ cmol}_c}{1\text{ kg}}\right)\left(\frac{24\%\text{ of sites have Na}}{100\%\text{ of exchange sites}}\right) = \left(\frac{432\text{ cmol}_c}{100\text{ kg}}\right)$$

$$= 4.32\text{ cmol}_c\text{ per kilogram of soil is exchangeable Na}$$

The cmol$_c$ of exchangeable Na to leave in soil is

$$= \left(\frac{18\text{ cmol}_c}{1\text{ kg}}\right)\left(\frac{6\%\text{ exchangeable Na to be left}}{100\%\text{ total exchange sites}}\right) = \left(\frac{108\text{ cmol}_c}{100\text{ kg}}\right)$$

$$= 1.08\text{ cmol per kg of soil to be left}$$

---

*In the SI metric system, millimoles (+) or mmol$_c$ [or centimoles (+) = cmol$_c$] is used in place of milliequivalents (meq). For sodium, 1 meq = 1 mmol because the valence of sodium is 1. One correct way to write the units is as follows, for millimoles per kilogram of soil:

$$\text{mmol}(+)\text{ kg}^{-1}\text{ or mmol}(+)\text{/kg or mmol}_c\text{/kg}$$

The "+" or subscript *c* is used to indicate an amount (an equivalent) that will *react with* 1 mole of a salt, base, or acid of a monovalent element.

Calculation 9-3, Cont'd Gypsum Requirement Calculation

The difference in total exchangeable Na and exchangeable Na to be left is the Na to be replaced = 4.32 − 1.08 = 3.24 cmol.

2. Putting the values from step 1 into the equation shown above for gypsum requirement,

$$GR = 4.50(Na_x) = 4.50\ (3.24\ cmol_c)$$
$$= 14.6\ \text{Mg per hectare of gypsum needed}$$

3. For the amount of sulfur needed, refer to Table 9-3. It shows that only 0.18 times as much sulfur by weight is needed compared to gypsum, so the needed sulfur is

$$14.58\ \text{Mg gypsum}\ (0.18) = 2.62\ \text{Mg of sulfur/hectare}$$
$$(= 1.05\ \text{tons of sulfur/acre})$$

Gypsum dissolves slowly, and use efficiencies, such as uneven spreading, are seldom 100 percent. The preceding equations assume that about 30 percent extra gypsum or sulfur is added. The equations finally used have already incorporated those corrections. Added elemental sulfur is even slower in reaction, often needing many months or longer to be oxidized in moist soil.

Source: Raymond W. Miller, Utah State University.

9:6 Salt Precipitation Theory

The elimination of salts and exchangeable sodium from soils by leaching is one satisfactory method of reclaiming salted soils, but the leached salts are washed into groundwater or streams, making those waters more salty, thus polluting them. More costly water, legal restrictions on disposal of wastewaters, legal restrictions on additions of salts to groundwaters, and excess salts in downstream river waters have all focused attention on a relatively new concept in managing salty soils: **precipitation of salts.** Briefly, this idea suggests that instead of leaching salts completely away into groundwaters, they can be leached to only 0.9 m to 1.8 m deep (3 ft to 6 ft) or a little deeper, where much of the salt would form slightly soluble gypsum ($CaSO_4{\cdot}2H_2O$) or carbonates ($CaCO_3$, $MgCO_3$) during dry cycles and not react any longer as soluble salts. Any water eventually reaching groundwater or surface water would be carrying less total salts than it contained initially.

A number of questions immediately arise about using the salt precipitation approach:

- How much salt will precipitate?
- How much will the remaining salt hinder plant growth?
- What are the techniques and hazards in using this method?
- What are the costs?

Amount of Salt that Precipitates

The amount of salt precipitating out will vary with the cation and anion composition of those salts. The ions precipitating will be mostly those of calcium, magnesium, carbonate, bicarbonate, and some sulfate. Estimates are that an average of about 30 percent of the total salts may eventually precipitate.

The Huntington Power Plant in Utah was faced in the 1970s with high costs to remove salt from waste, cooling waters. The plant managers considered adding the water to soil. Although soluble salt accumulation in the soil was expected, the accumulation was not as rapid as anticipated. One explanation is that a large portion of the added salt precipitates to relatively low solubility substances during drying cycles during the year. A desalting plant built to clean this wastewater, rather than dispose of it onto soil, was estimated to cost about $12 million. The project is still being monitored after 20 years of water application. Boron accumulation levels are now being monitored, because toxic levels may be accumulating.

Effects of the Remaining Soluble Salts

If only one-third of soil salt is actually removed from activity by precipitation, how will the remaining salt affect plant growth? The answer is, "Just as much as that amount of soluble salt ever did"; but the position of the salt in the root zone has been altered and the salt will be higher in sodium and chloride. The salt has been moved to the lower root zone, some of it even below the root zone. Most plant roots proliferate in the upper 30 cm to 60 cm from which they absorb water. Only for short times during hot summer days at maximum evapotranspiration losses will plants need much water from the deeper root zone; thus, if the salts affect only water uptake rather than exerting a toxicity, normal yields should be obtainable as long as the upper root zone is moist and sufficiently low in salts. Little effect on yield of corn and tomatoes occurred even when *two-thirds of the root zone* (lower part) was too salty. Alfalfa, normally tolerant of only 6 dS/m to 8 dS/m conductivity, yielded *normally* with its *lower* root system in solutions of 30 dS/m to 35 dS/m.

Hazards of the Technique and Its Costs

The third question concerning techniques and hazards appears to have obvious answers, but limited research has been done. The management technique is simply to *add less water, but to do it more carefully to ensure uniform depth of wetting.* The leaching requirement can be reduced by 60 percent to 75 percent of that calculated by the equation in Calculation 9-2 by more careful watering. Drip and sprinkler methods of applying water will work best for precise movement of salts to redetermined depths.

The hazards in this technique include careless water application, causing inadequate downward salt movement; and possible toxic effects on some plants of sodium, boron, and chloride. Soils with shallow water tables and seep areas are not suitable for this technique.

Salt controls will be expensive, and best management practices are still in developmental stages for many situations. Eventually, some salt must be leached beyond the root zones and will accumulate in groundwaters or surface waters. Until now the problem of removing excess salts from wastewaters has not been placed upon the individual farmer. Various states have regulations on releasing water having high salt contents back into water resources. Groundwaters or surface waters in California having more than 500 mg/L salts must have salts removed or be ponded on the farm. Salt removal may require desalinization, ponding to allow some salts to precipitate, or mixing with less salty water before disposal. Treatment of sewage wastewaters from large cities can cost millions of dollars annually for treatment.

9:7 Managing Salty Soils[10]

It is not always possible or practical to eliminate all salts from soil, but managing the soil *to minimize salt damage* is a necessary part of using salted soils. Although in sodic soils some of the exchangeable sodium must be removed, for slightly saline soils the control of water,

[10]FAO, "The Use of Saline Waters for Crop Production," *FAO Irrigation and Drainage Paper* **48,** Food and Agriculture Organization of the United Nations, Rome, Italy, 1992, 93–107.

the proper techniques of planting, and the choice of tolerant crops are essential for their successful use in crop production.

Water Control

Maintaining a high water content in the soil, near field capacity, dilutes salts and lessens their toxic and osmotic effects. If soil is irrigated lightly but frequently to keep it at a high moisture content during the salt-sensitive germination and seedling stage, plants are able to survive to the more-tolerant mature stage of growth.

Limited leaching done before planting (when water is usually more plentiful than later), or a light irrigation by sprinklers after planting, will move salts below the planting and early-rooting zone. Later, when the salt gradually moves upward with water, the plant will be more mature and more salt tolerant. Sprinkler-applied water after planting in the Imperial Valley of California increased lettuce germination 20 percent. Soils that develop hard crusts must not be irrigated after planting until the seedling emerges, or the crust must be sprinkled at the time of emergence to soften it so seedlings can break through (Figure 9-9).

The Imperial Valley of California has a severe salt problem. It has naturally salty areas, inadequate natural drainage, and irrigation being done with salty Colorado River water. Correct water management practices in combination with drainage for salt control are imperative. To keep the average salt content in the root zone as low as 8.0 dS m^{-1} (moderately high) requires 17 percent excess irrigation water to ensure leaching. One initial sprinkler irrigation on winter vegetable crops moves salt down below the seed germination zone; then conventional surface irrigation is used. Double rows and sloping beds are also used to reduce salt damage.

Planting Position

Salt moves with water, and some will accumulate in the surface soil or furrow ridge tops as water with its salt moves upward and evaporates. Figure 9-10 illustrates furrow cross-sections and planting positions used to avoid damage from salt accumulation zones. Notice that these methods avoid the centers of wide ridges and the tops of narrow ridges where salt will be most concentrated from furrow irrigation. Sprinkler methods of irrigation will eliminate some of these problems by washing salt deeper into the soil profile, even from ridge tops.

FIGURE 9-9 Where salt is a problem, lettuce may avoid some salt injury by being planted on the slope of the ridge, because salt concentrates on the ridge crest. The white areas are salt crusts. Notice the vacant spots in many of the rows in this Imperial Valley, California, field where plants have been killed by the salt. (Courtesy of Alvin R. Southard, Utah State University.)

FIGURE 9-10 Methods of preparing seedbeds and of planting to reduce the effects of soluble salts on plants. Shaded areas illustrate locations where salt would normally concentrate because of water flow and evaporation. Darkest shaded areas are highest in salt. Diagrams (c) and (d) are bed shapes of the same planting at planting time and after the plant is growing well, respectively. (Courtesy of Raymond W. Miller, Utah State University.)

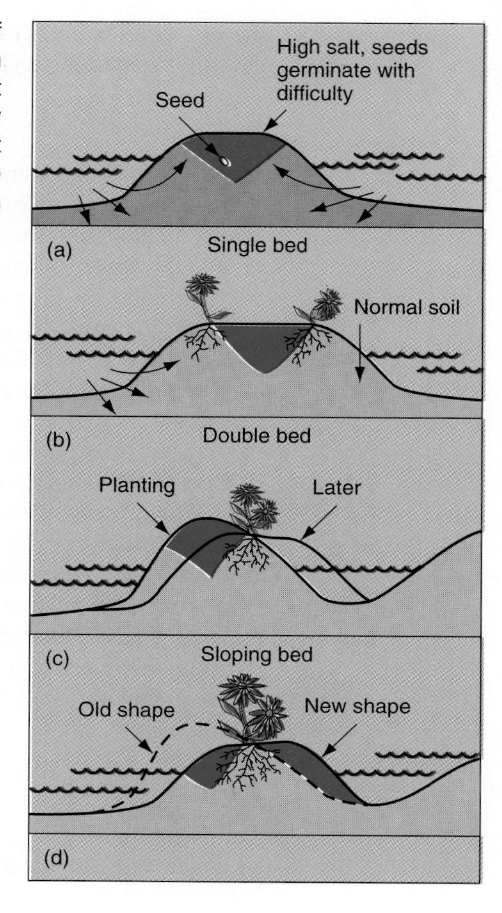

Choice of Crops

The choice of crops is based on tolerance to salt, adaptability to the climate or soil characteristics, and value of the crop in the individual crop activity. The chance of a crop failure or loss is less if an adequately salt-tolerant crop is selected. Obviously, the soil must have a suitable texture, adequate depth, and exist in a climate suitable to the crop. The purpose of farming is to make a profit, so a high-value crop is preferred. Yet, if a crop entirely different than is grown on the rest of the farm is selected, new seeders, harvesters, herbicides, and other equipment and farm programming may be needed for that crop.

Adding Soluble Silica—Effective or Not?

A recent attempt to reduce the amount of damage from soluble salts involves adding soluble silica.[11] Adding 1 mM of Si/L to a culture solution (hydroponics) having 120 mM of NaCl per liter yielded 18 percent more barley dry matter than if only the NaCl was in solution. The loss of ions from barley leaves was decreased when silica was present. Although the mechanism of silica on better (or less reduction in) yields is not understood, some researchers sug-

---

[11] Yongchao Liang, Qirong Shen, Zhenguo Shen, and Tongsheng Ma, "Effects of Silicon on Salinity Tolerance of Two Barley Cultivars," *Journal of Plant Nutrition* **19** (1996), 173–183.

gest that silica causes reduction in membrane permeability, reducing sodium uptake. Some studies have observed that added silica increases a plant's tolerance to salt, but they have not suggested a reason.

Using Brackish Water

Israel, a water-short nation, has pioneered the use of *brackish water*[12] (less salty than seawater). In 1975 brackish water made up 10 percent of Israel's fresh water use. Waters having an electrical conductivity of 3.7 dS m^{-1} and an SAR of 8.5 have been used on cotton and other salt-tolerant crops.[13] With brackish water, cotton produced the same plant size and more bolls than with lower-salt-content water. The soil salt buildup, which could result rapidly from using brackish water, can only be controlled by regular leaching. Water is applied at a rate that always exceeds evapotranspiration. Extensive drying of the soil between irrigations must be avoided (which would move salt upward to the surface). Mixing brackish and fresh water to get water of suitably low salt content to use for irrigation is being practiced extensively in Israel. Whether the practice is wise is still being debated.

Saline Seeps

A second unusual situation in the management of salt deposition on soil is the treatment of saline seeps in Montana and the Dakotas (see Figures 9-2 and 9-8). **Saline seeps** develop on several kinds of layered substrata, but all have water moving in a downslope direction within the soil (recharge area), usually above a low-permeability geologic layer, and probably dissolving more salts as it flows. The water seeps eventually to a soil surface at some lower elevation (seep area), where it is evaporated or transpired, leaving salt accumulations.

Preventing development of saline seeps requires preventing water seepage. This has been done by using the water on the recharge area, before it can seep away, usually by growing a plant cover to increase water use by transpiration.[14] Deep-rooted alfalfa on 80 percent of the recharge areas has hindered seep formation because there is less water to seep; wheat-fallow areas had inadequate retardation of water flow and hastened development of new seeps.

9:8 What Makes Plants Salt-Tolerant?

Scientists have tried for decades to understand why different plants tolerate different salt concentrations, their overriding hope being that a higher salt tolerance could some day be bred into plants. An Asian wild rice reportedly survives in soils with saturated soil paste extract values of 30 dS/m to 40 dS/m.[15] Mangrove forest trees tolerate high salt concentrations along many ocean coastal waters.

Limited information is available on the plant salt tolerance mechanisms. Some of the mechanisms so far identified would seem to be very plant specific and difficult to transfer

---

[12]Brackish water is water of salinity high enough to significantly restrict its direct use yet not prevent its use completely. In Israel so much citrus is grown that chloride content (toxic to citrus) is used to define brackish water. Brackish water has 400 mg/L–4000 mg/L chloride or about 1000 ppm–10,000 ppm salts.

[13]Marvin Twersky, Dov Pasternak, and Ilan Borovic, "Effects of Brackish Water Irrigation on Yield and Development of Cotton," in *International Symposium on Brackish Water as a Factor in Development,* Ben Gurion University of the Negev, and others, Beer-Sheva, Israel, 1975, 135–140.

[14]A. D. Halvorson and C. A. Reule, "Alfalfa for Hydrologic Control of Saline Seeps," *Soil Science Society of America Journal* **44** (1980), No. 2, 370–374.

[15]A. R. Bal and S. K. Dutt, "Mechanisms of Salt Tolerance in Wild Rice (*Oryza coarctata Roxb*)," *Plant and Soil* **92** (1986), 399–404.

genetically. Perhaps genetic-engineering techniques can be utilized. Some of the observed plant techniques to tolerate high-salt environments are as follows:[16]

- Accumulation of high levels of sodium and chloride in shoots is associated with concentrations (usually increases) of certain plant organic compounds in cells.
- Exclusion of salt ions from uptake by the root cells. Many important grains (wheat, barley, rye, triticale) seem to employ this ability.
- Excretion of absorbed salts from the plant by means of *salt glands* that either burst, dripping out the high-salt solution, or finally drop off the plant. Many **halophytes** (salt-loving plants) have this mechanism.

9:9 Monitoring Salts in the Field[17]

There is a need to monitor salt in the field. Although taking soil samples for laboratory analysis is relatively simple, there are limitations:

- Salt patterns in a field are not homogeneous and regular. Salt distribution and amount can change with each irrigation. Also, a large number of samples is needed, and the cost to prepare the saturated paste extracts is quite high.
- The laboratory values are not true field values. The laboratory sample has been dried and then rewetted with an amount of water double its field capacity and then extensively stirred.
- The process of using laboratory analyses is relatively slow. Lag time from sampling to the final data is usually several days, at best. A quick method is needed.

Five methods used for salt measurements in situ are the following:

- Vacuum extractors
- In-place sensors
- Bulk soil electrical conductivity
- The four-electrode salinity probe
- The newest method—electromagnetic induction—is fast and nondestructive.

Extractors remove soil water with mild suction, but only when the soil is at nearly field capacity. The salt content in a small volume of soil is measured since water from only a small volume of soil around each extractor is removed. The **in-place sensors** have the same limitation of measuring only a small localized portion of soil where the probe is located. Neither of these two methods is easily adapted to measuring large soil areas.

The bulk soil and the four-electrode probe measure the electrical conductivity and are used to measure larger soil volumes and obtain average salt values for each soil volume. The **bulk soil method** uses four metal stakes (called *probes,* but screw drivers will work) that are pushed into the soil in a line. As electrical current from a battery is passed through two of the probes, the resistance to current flow *through the soil* to the other two probes is measured. For a uniform soil the current penetrates a depth that is about one-third the distance between

---

[16]L. Gorham, R. G. Wyn Jones, and E. McDonnell, "Some Mechanisms of Salt Tolerance in Crop Plants," *Plant and Soil* **89** (1985), 15–40.

[17]J. D. Rhoades and D. L. Corwin, "Monitoring Soil Salinity," *Journal of Soil and Water Conservation* **39** (1984), 172–175.

the two outer probes. The measurement averages the soil salinity to this depth. Thus, the volume and depth of soil measured can be varied by the distance between probes; this distance can be selected by appropriate probe placement.

The four-probe method is fast and averages the salt content in a larger volume of soil than do the other methods; it also measures salinity in the soil in its normal physical state and water contents. However, *the readings are specific for each soil and require calibration for each soil* in order to interpret the readings obtained to some particular salt condition.

For more localized measurement of salinity in field conditions, the **four-electrode salinity probe** can be used. This single large probe has four annular metal rings molded into the end of a plastic probe. The rings are a fixed distance (a few centimeters) apart. The probe (about 15 cm long × 3 cm diameter, with a 1.5 m-long tube handle) is tapered slightly at the end. This allows the probe to be inserted to any depth into a hole of the probe's diameter (usually dug with a soil-coring tube). The electrical current is passed from a battery, and the measured salinity is an average inside a sphere of soil about 30 cm (1 ft) in diameter around the probe.

The major advantage of the four-electrode probe method is the ability, once calibration is done, to make many very rapid readings. Rapid monitoring of salt changes and variation over large areas could be quick and easy but each measurement is still from a very small volume of soil.

The **electromagnetic induction (EMI)** method is a high-tech tool that should be used with the GPS system (see Chapter 2).[18] The electromagnetic induction meter records conductivity readings which are proportional to the amount of salts in soil solution, all without contact with the soil. When EMI is combined with the GPS to locate and record position in the field, salt distributions can be quickly done. In southern Alberta, Canada, 6000 data points were collected on a 49-hectare field in 3 hours. Deliberate crossings of previous lines were done to check the readings. The 51 cross-over points verified that the readings were repeatable and accurate. The system can usually survey up to a hundred hectares per day.

Old alfalfa is more tolerant of salt-affected soil than young alfalfa, and . . . deep-rooted legumes show a greater resistance to such soils than the shallow-rooted ones.

—N. C. Brady

High salt concentrations in the rhizosphere generate stress through water deficits and ion toxicity. Exclusion of salt and osmotic adjustment both play major roles in tolerance of high salt environments.

—William G. Hopkins

Questions

1. (a) Can soluble salts be seen? Explain. (b) Why was the term *summer snow* probably used for some salty soils?
2. (a) Explain what each of these tells about salt in soil: *ESP, SAR,* and *EC* (electrical conductivity). (b) Why is *SAR* replacing *ESP*? (c) Give the units in which electrical conductivity is reported.

---

[18] Colin McKenzie, "Global Positioning Systems and Electromagnetic Induction—High Tech Tools for Salinity Mapping," *Better Crops* **80** (1996), No. 3, 34–36.

3. (a) What is the source of soluble salts? (b) What are the six most numerous ions of soluble salts?
4. (a) What is the effect of a high ESP? (b) How is ESP related to *slick spots*? (c) What is the relationship of SAR to ESP?
5. If calcium precipitates as insoluble carbonates, how will this change the soil SAR?
6. (a) Give definitions (the SAR, ESP, EC) for sodic, saline, and saline-sodic soils. (b) Could a nonsaline soil, by definition, have enough salt to damage some crops?
7. What is the major damage of high exchangeable sodium percentages on soil clays?
8. What does a high salt content affect that reduces plant growth?
9. Discuss a salt balance for soil.
10. (a) What are the three key requirements needed to accomplish reclamation of salt-affected soils? (b) How does reclamation of sodic soils differ from reclamation of saline soils?
11. (a) Why is gypsum sometimes used for reclamation? (b) What will happen if gypsum-treated soils are not adequately leached (the Na_2SO_4 stays in the soil)?
12. (a) Which soluble salt ions precipitate? (b) Which ions are left in solution? (c) Which factors limit the use of the salt precipitation management scheme?
13. (a) Why is waste-salt disposal (reclamation waters, irrigation runoff water, etc.) a serious problem? (b) Will the problem become more serious? Explain.
14. (a) In which ways can water control without profile leaching be used to reduce damage from soil salts? (b) Are any of these techniques practical?
15. What are some management techniques for reducing salt buildup in irrigated soils?
16. How is salt measured in situ (in field soil at normal water content)?

Nitrogen, Phosphorus, and Potassium

Nitrogen is the key to successful organic matter management and thereby to successful soil management.

—R. L. Cook and B. G. Ellis

In order for mineral nutrients to be taken up by the plant, they must, at some point, be taken across the plasma membranes to root cells. Nutrient uptake by roots is therefore fundamentally a cellular problem, governed by the rules of membrane transport.

—William G. Hopkins

10:1 Preview and Important Facts

PREVIEW

Plants are known to need at least 16 essential elements to grow, although more than 90 elements can be absorbed by plants. From the air and water, plants utilize hydrogen, oxygen, and carbon. The **macronutrients,** those absorbed in large amounts from soil and fertilizers, are nitrogen, phosphorus, and potassium (the three primary fertilizer nutrients). The secondary nutrients are calcium, magnesium, and sulfur. The **micronutrients,** those absorbed in lesser quantities (also called *trace elements*), are chlorine, copper, boron, iron, manganese, molybdenum, and zinc.

The principal soil storehouse for large amounts of the nutrient anions is soil organic matter. Decomposition of organic matter releases nutrients. Organic matter holds more than 95 percent of the soil nitrogen, often half or more of the total soil phosphorus, and as much as 80 percent of the soil sulfur. Boron and molybdenum reserves are stored in organic matter as well as adsorbed to iron and aluminum oxides and other solids through the solids' hydroxyl (OH) groups.

As nutrients are absorbed from the soil solution, they are replenished from several sources, such as exchangeable (adsorbed) ions on clay minerals and humus, the slow dissolution of soil minerals, and the decomposition of soil organic matter. Seldom is the rate of renewal for all essential elements from untreated soils fast enough to achieve maximum crop production. The supply storehouse for each of the macronutrients—nitrogen, phosphorus, and potassium—is very different. **Nitrogen** is released from decomposing organic materials, N_2-fixation by *Rhizobia* and a few other microorganisms, and from added fertilizers or organic wastes. **Phosphorus** comes from both decomposing organic matter and from limited slightly-soluble inorganic sources. Because of the low solubility of phosphorus minerals, keeping phosphorus available is difficult. **Potassium** is rapidly available only from soluble and exchangeable forms or from *fresh* plant materials. The only alternative, if *exchangeable* potassium is low, is to add fertilizer materials.

The differences that exist in the soil chemistry among the essential elements and the diverse roles of those elements in plants and animals are truly amazing. For example, nitrogen is supplied to plants from air-N_2 that is fixed and from decomposing organic matter, but there are *no mineral sources.* Phosphorus comes from both organic matter and mineral sources, but it has *no volatile form* at normal temperatures. Potassium, the third nutrient, is supplied *mainly from mineral sources;* it has *no volatile source,* and *organic sources* are of limited importance.

The role of nitrogen in plants is mostly structural (all amino acids and proteins contain nitrogen). Phosphorus occurs in membranes, as *esters,* and it is involved in a biological energy transfer system (adenosine triphosphate [**ATP**] and adenosine diphosphate [**ADP**]). Potassium, in contrast, does not occur in any known compound in the plant; it exists in the plant as a free ion, but it is active in many plant functions.

IMPORTANT FACTS TO KNOW

1. The general mechanism of nutrient movement to root surfaces and the process of nutrient absorption into root cells
2. The basic (simplified) cycles for nitrogen, phosphorus, and potassium
3. Soluble forms of nitrogen, phosphorus, and potassium
4. The general mechanism of N_2-fixation and its importance
5. The relative importance of soil organic matter as a nitrogen, phosphorus, and potassium reservoir
6. The process of nitrification and why it is acidifying
7. The mechanisms by which nitrogen is lost from soils (a) in water and (b) as gaseous forms
8. The conditions necessary to cause denitrification
9. The conditions necessary to initiate ammonia volatilization
10. Some important characteristics of these fertilizers: anhydrous ammonia, urea, ammonium nitrate, and ammonium phosphate
11. The nature of controlled-release nitrogen fertilizers, their value, and relative costs
12. The conditions that decrease phosphate absorption by roots
13. The properties of phosphorus and potassium that cause each to be so often deficient in soils
14. The amounts of nitrogen, phosphorus, and potassium (a) lost in leaching, (b) removed in crops, and (c) mineralized from soil humus

10:2 Elements Needed by Plants and Animals

The sixteen essential elements needed by plants are as follows:

- Carbon
- Oxygen
- Hydrogen
- Nitrogen
- Calcium
- Potassium
- Magnesium
- Phosphorus
- Sulfur
- Chlorine
- Iron
- Boron
- Manganese
- Zinc
- Copper
- Molybdenum

The elements cobalt, nickel, silicon, sodium, and vanadium are also needed by some plants. Elements not shown to be needed by all plants, but helpful in growth to some, are often called **beneficial elements.** (Some consider nickel—listed in this group rather than in the group of essential elements—to be an essential element.)

Humans and other animals require, in addition to the plant nutrients, the elements sodium, iodine, selenium, and cobalt. There is evidence that most mammals also need fluorine, chromium, nickel, vanadium, silicon, tin, arsenic, and cadmium. Table 10-1 lists these elements and their ionic forms in soil solution that are available for plant use.

Carbon, oxygen, and hydrogen are obtained from air and water. Only nitrogen, phosphorus, and potassium—the three most applied nutrients—are discussed in this chapter. The remaining ten nutrients are discussed in Chapter 11.

Nitrogen

Nitrogen ***is most often the limiting nutrient in plant growth;*** it is a constituent of chlorophyll, plant proteins, and nucleic acids. Nitrogen is utilized by plants as the ammonium cation or as the nitrate anion. Atmospheric dinitrogen (N_2) is made available by **nitrogen fixation,** which requires the action of specific microorganisms. Organic soil nitrogen is made available

Table 10-1 The 16 Plant Nutrients and 6 Additional Elements, Their Chemical Symbols, Content in Plants, and the Form(s) Common in Air, Water, and Soil and Available for Plant Uptake

Element and Symbol	*Portion of Plant* (%)*	*Ion or Molecule*
Carbon (C)	41.2	CO_2 (mostly through leaves)
Oxygen (O)	46.3	CO_2 (mostly through leaves), H_2O, O_2
Hydrogen (H)	5.4	HOH (hydrogen from water), H^+
Nitrogen (N)	3.3	NH_4^+ (ammonium), NO_3^- (nitrate)
Calcium (Ca)	2.1	Ca^{2+}
Potassium (K)	0.80	K^+
Magnesium (Mg)	0.42	Mg^{2+}
Phosphorus (P)	0.30	$H_2PO_4^-$, HPO_4^{2-} (phosphates)
Sulfur (S)	0.085	SO_4^{2-}
Chlorine (Cl)	0.011	Cl^- (chloride)
Iron (Fe)	0.0066	Fe^{2+}, Fe^{3+} (ferrous, ferric) = Fe(II), Fe(III)
Boron (B)	0.0045	H_3BO_3, (boric acid)
Manganese (Mn)	0.0036	Mn^{2+} = Mn(II)
Zinc (Zn)	0.0009	Zn^{2+} = Zn(II)
Copper (Cu)	0.0007	Cu^{2+} = Cu(II)
Molybdenum (Mo)	0.000005	MoO_4^{2-} (molybdate)
Cobalt (Co)	—	Co^{2+} = Co(II)
Nickel (Ni)	—	Ni^{2+} = Ni(II)
Selenium (Se)	—	SeO_4^{2-} (selenate)
Silicon (Si)	—	$Si(OH)_4$ (nonionized)
Sodium (Na)	—	Na^+
Vanadium (V)	—	VO_3^- (vanadate)

Source: Selected data and additions by Raymond W. Miller from B. G. Ellis and B. D. Knezek, "Adsorption Reaction of Micronutrients in Soils," 68; and L. O. Tiffin, "Translocation of Micronutrients in Plants," 204, both in *Micronutrients in Agriculture,* R. C. Dinauer, ed., Soil Science Society of America, Madison, WI, 1972.

*Estimates of element contents taken from oven-dried alfalfa. Various plants will have different values, particularly in content of potassium, nitrogen, calcium, phosphorus, and sulfur.

by **mineralization,** which is the microbial decomposition of organic matter that releases nitrogen first as amino acids (**aminization**) and finally as ammonium ions, **ammonification.** Some of the ammonium ions released are adsorbed on cation exchange sites.

Nitrification is the bacterial oxidation of ammonium cations to nitrate anions, which in turn are used by plants, lost by **leaching,** or **denitrified** by bacteria to volatile N_2 and N_2O gases. Ammonium also can be volatilized as ammonia gas. Significant amounts of soil ammonium and nitrate are used by microorganisms and become part of the soil's organic substances; this process is called **immobilization**. Nitrogen, immobilized in microbial bodies or other organic matter, is temporarily unavailable for plant use.

Phosphorus

Phosphorus ***is the second most often limiting nutrient.*** It is contained in plant-cell nuclei and is part of energy storage and transfer chemicals in the plant. Soils have low total and low plant-available phosphate supplies because mineral phosphate forms are not readily soluble. Phosphorus used by the plant is taken up as the HPO_4^{2-} and $H_2PO_4^-$ anions. Most soluble phosphates become fixed (precipitated or adsorbed to form insoluble compounds) before plants can absorb them from the solution. Organic phosphates are important—perhaps even major—phosphate sources in most soils.

Potassium

Potassium ***is the third most commonly added fertilizer nutrient.*** In soils potassium is released from weathering minerals and from cation exchange sites. Primary minerals containing potassium have very low solubility, so most potassium available to plants during a growing season is supplied from the soil's *soluble* and *exchangeable* potassium reservoir. Thus, most potassium deficiencies are in leached soils (humid climatic areas), where most soluble and much exchangeable potassium has been leached out. Remember that the cation, $\mathbf{K^+}$, is very soluble in water yet moves only slowly in soils because of adsorption to cation exchange sites.

Potassium is an enigma. First, it is one of the elements whose usual chemical compounds are highly soluble, yet its soil mineral forms—micas and orthoclase feldspar ($KAlSi_3O_8$)—are only very slowly soluble. Second, it is the most abundant metal cation (often up to 2 percent or 3 percent of dry weight) in plant cells, but soil humus furnishes very little potassium during decomposition. Third, decomposition of *fresh* plant residues supplies almost immediately whatever potassium the plant contained. Potassium occurs in plants only as a mobile soluble ion, K^+, and is not an integral part of any specific compound.

Potassium affects cell division, the formation of carbohydrates, translocation of sugars, various enzyme actions, the opening and closing of stomatal guard cells, the resistance of some plants to certain diseases, cell permeability, and several other functions. More than sixty enzymes are known to require potassium for activation. It is particularly important in plant control of water in the plant, probably in an osmotic role.

A deficiency of potassium causes white speckles in legume leaves and necrotic tissue to form on the edges of older leaves and bronzed, leathery leaves. Other symptoms may occur in the younger leaves of some plants.

10:3 Mechanisms of Nutrient Uptake

Prior to their absorption into root cells, nutrients reach the surface of roots by three mechanisms: mass flow, diffusion, and root interception. **Mass flow,** the most important of these mechanisms so far as quantity is concerned, is the movement of dissolved plant nutrients in flowing soil solution. **Diffusion** is movement by normal atomic dispersion from a higher

Detail 10-1 Uptake of Ions by Roots—Facts (and Fiction?)*

There is space within plants that is called *apparent free space* (AFS). It exists in cell walls between bundles of wall structural materials. This free space (estimated to be 10 percent to 25 percent of the plant volume) is within the plant but *outside* the living cell membranes. Ions in soil solution can enter rapidly by water flow and diffusion into this free space and even move to the plant transport systems to be transported through the plant xylem. Perhaps this is how the 90 or so elements (rather than just the *essential elements*) get into plants.

How ions move *through* cell membranes into cells is still only partially understood. Scientists believe that one way ions pass through membranes is through **ion channels,** which are made of proteins. Many or all of these *channels* can open or close their gates by shape change (see diagram).

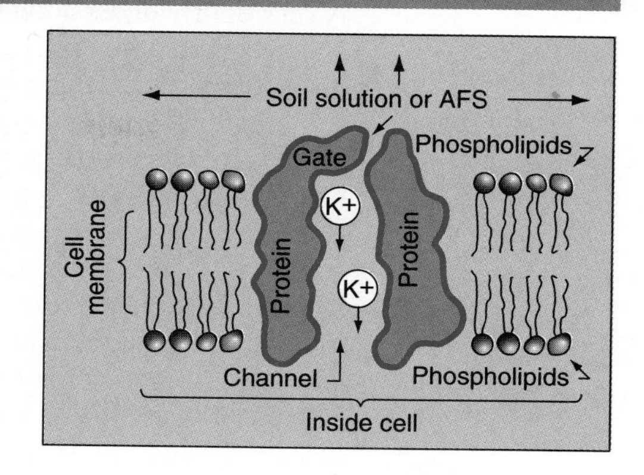

The ability of plants to concentrate soluble ions inside the plant, even when extracting them from dilute solutions outside the plant, has intrigued scientists for a long time. It is now known that cells develop an **electrochemical gradient** across the membrane from the negative charges of *carboxyl* groups on large immobile organic molecules in the cells. Even with a high interior K^+ concentration, if some of the positive ions start to migrate out of the cell, the developed electrical potential would soon attract them back into the cell.

A second passageway is by **carrier proteins,** which actually bond to ions and by shape change carry them across the membrane, like a back-and-forth change of shape as new ions are moved across the membrane. All of this sounds kind of like magic. The more we learn of life systems, the more spectacular it all seems.

*William G. Hopkins, *Introduction to Plant Physiology,* John Wiley & Sons, Inc., New York, 1995, 81–95.

concentration of a nutrient (such as near its dissolving mineral source) to areas of lower concentration. **Root interception** is the extension (growth) of plant roots into new soil areas where there are untapped supplies of nutrients in the soil solution.

All three supply processes are in constant operation during growth. The importance of each mechanism in supplying nutrients to the root surface varies with the chemical properties of each nutrient. Mass flow involves large amounts of water which flows to roots as the plant absorbs and transpires water. Mass flow is the dominant mechanism carrying nutrients to roots; it supplies about 80 percent of nitrogen, calcium, and sulfur to root surfaces. Diffusion is the dominant transport mechanism for phosphorus and potassium.

Absorption of Substances into Roots

The mechanisms of absorption into the root cells are not well understood. The cell walls are porous, and the soil solution can move through some or all of the cell walls, causing intimate contact of the soil solution with the outer membranes of many cells. For a nutrient to cross a cell membrane and pass into a cell, the molecule or ion must go through a passageway or bond with an ion-specific carrier. Perhaps both mechanisms are operating, one for particular ions, and another for other ions or molecules (water, sugars, etc.); see Detail 10-1.

The Role of Electrical Balance

A very poorly understood mechanism of electrical balance is also involved in ion absorption and accumulation. An electrical potential across the cell membranes has been measured by many scientists. As nutrient cations are absorbed, either some H^+ ions are excreted into the soil solution or more organic acid anions are produced inside the cell to provide negative charges to balance the absorbed positive charges (cations). Likewise, as some elements can be partially excluded from absorption, others can be preferentially absorbed, even against a *concentration gradient* (can be absorbed from a low-concentration soil solution and transferred to a higher concentration in the plant cell); see Figure 10-1. As nutrient anions are absorbed by the plant, more compensating cations are absorbed and/or HCO_3^- ions are excreted into the soil solution in order to maintain an electron balance in the cell. Perhaps the H^+ ions and HCO_3^- ions are excreted into the soil solution first to aid solubility of soil nutrients. There is still confusion about the mechanisms, but the processes involved are slowly being discovered and the mechanisms clarified.

Absorption through Stomata

Plants also absorb nutrients through small openings in leaves called **stomata.** Carbon enters almost entirely through the stomata as carbon dioxide, and the plant releases the oxygen (O_2) produced during photosynthesis out through the stomata. Hydrogen, as a part of water molecules, is absorbed through stomata, but this intake is usually small compared to the amount entering through the roots. Other nutrients are also absorbed through the stomata; soluble ions from fertilizer-enriched water from overhead sprinkler irrigation or other sprays are absorbed to some extent. Direct absorption through the leaves seldom exceeds a few kilograms per hectare per application of a nutrient solution. However, the small concentrations of micronutrients needed can be satisfactorily added by a nutrient foliar spray. Such sprays are widely used to supply fruits and berries with iron, zinc, manganese, molybdenum, boron, or copper.

10:4 Soil Nitrogen Gains and Transformations

Nitrogen is the key nutrient in plant growth. It is the most commonly deficient nutrient and is often the controlling factor in plant growth. Nitrogen is a constituent of plant proteins, chlorophyll (the green plant pigment important to photosynthesis), nucleic acids (DNA, RNA), and other plant substances. Adequate nitrogen often produces thinner cell walls, which results in more tender, succulent plants.

A deficiency of nitrogen causes poor plant yields. Typically not enough of the soil nitrogen is in a chemical form that can be utilized by plants. Most soil nitrogen is in organic matter. Large quantities of nitrogen, in the form of N_2 gas, reside in the atmosphere above the surface of the earth. This form of nitrogen, however, cannot be used by the majority of plants; the N_2 must first be changed by microorganisms into organic, then ionic, forms. The nitrogen cycle is shown in Figure 10-2. *Learn this cycle well.*

Nitrogen is a unique plant nutrient. Unlike the other essential elements, plants can absorb nitrogen in either the cationic form (ammonium ion, NH_4^+) or the anionic form (nitrate, NO_3^-), although only a small part of soil nitrogen occurs in these forms at any one time. Nitrate nitrogen is soluble and mobile in soils and is easily leached. Both nitrate and ammonium forms may be absorbed by microorganisms or converted to gaseous nitrogen forms (N_2 or NH_3, respectively) and lost to the atmosphere. A deficiency of nitrogen causes plants to grow poorly, spindly, and stunted.

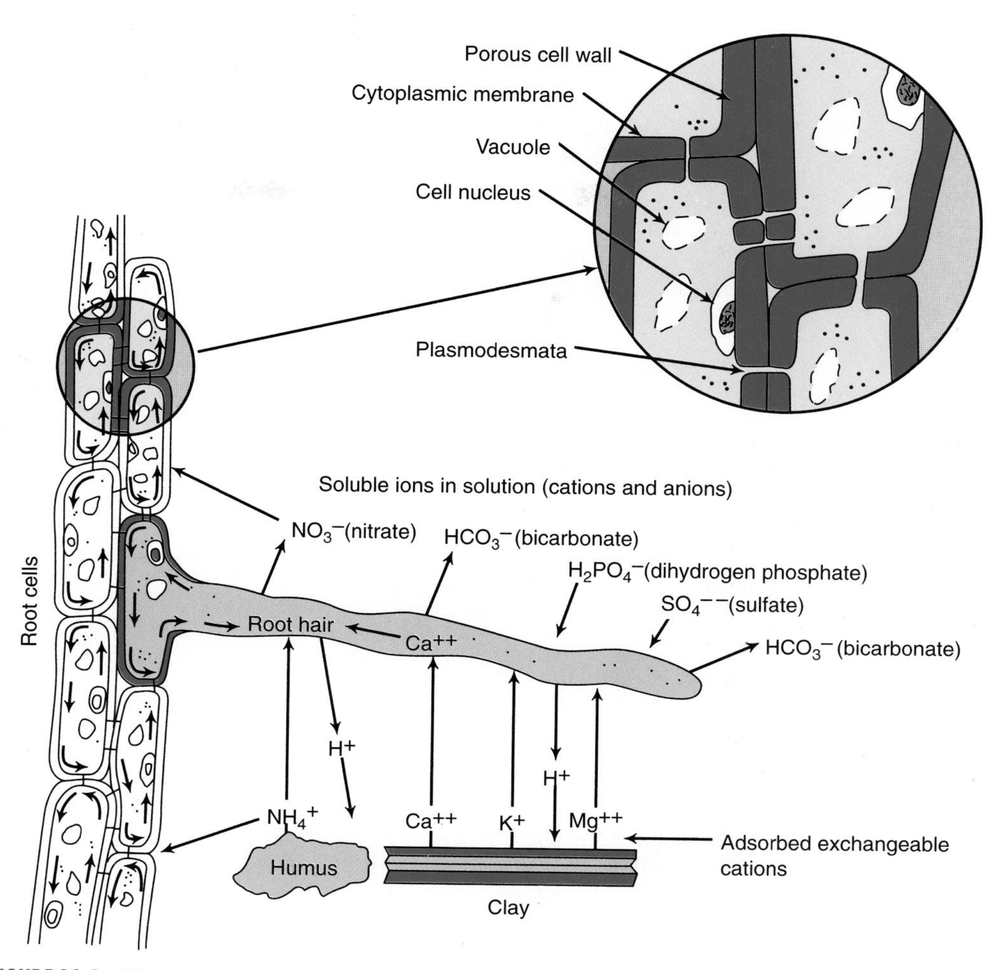

FIGURE 10-1 Diagrammatic scheme showing root structure and illustrating that a root hair absorbs nutrients from the soil solution and from adsorbed (exchangeable) ions on a clay crystal or humus colloid. A root hair is an extension of one of the epidermal (surface) cells of the plant root and is thought by some scientists to absorb nearly all the plant's water and nutrients. However, much evidence indicates that older and larger roots are also active in water absorption. Water can move within and through the cell walls and pore spaces between cells and thus furnish the cells with large amounts of contact between soil solution and the cell membranes enclosing the active cell protoplasm. Plasmodesmata are fine strand connections of cytoplasm between cells through which absorbed water and nutrients move. In the insert they are shown exaggerated in comparative size. Note that to maintain electrical balance within the plant, approximately equivalent amounts of HCO_3^- must be exuded to soil to balance the total *anion* uptake. Similarly, sufficient H^+ ions must be exuded to balance the total *cation* uptake.

Are both nitrate and ammonium ion forms equally good and equally available to the plant? Many studies have shown that selected plants seem to grow better with one form or the other. If the NH_4^+ form is absorbed by the plant, it does not have to be reduced again inside the plant to the amino form ($—NH_2$) as does NO_3^-, which saves the plant energy. Even the flavors or tastes of foods can be influenced by the form of the nitrogen plants use. For example, French beans fed ammonium ion were smaller, were three to ten times higher in amino

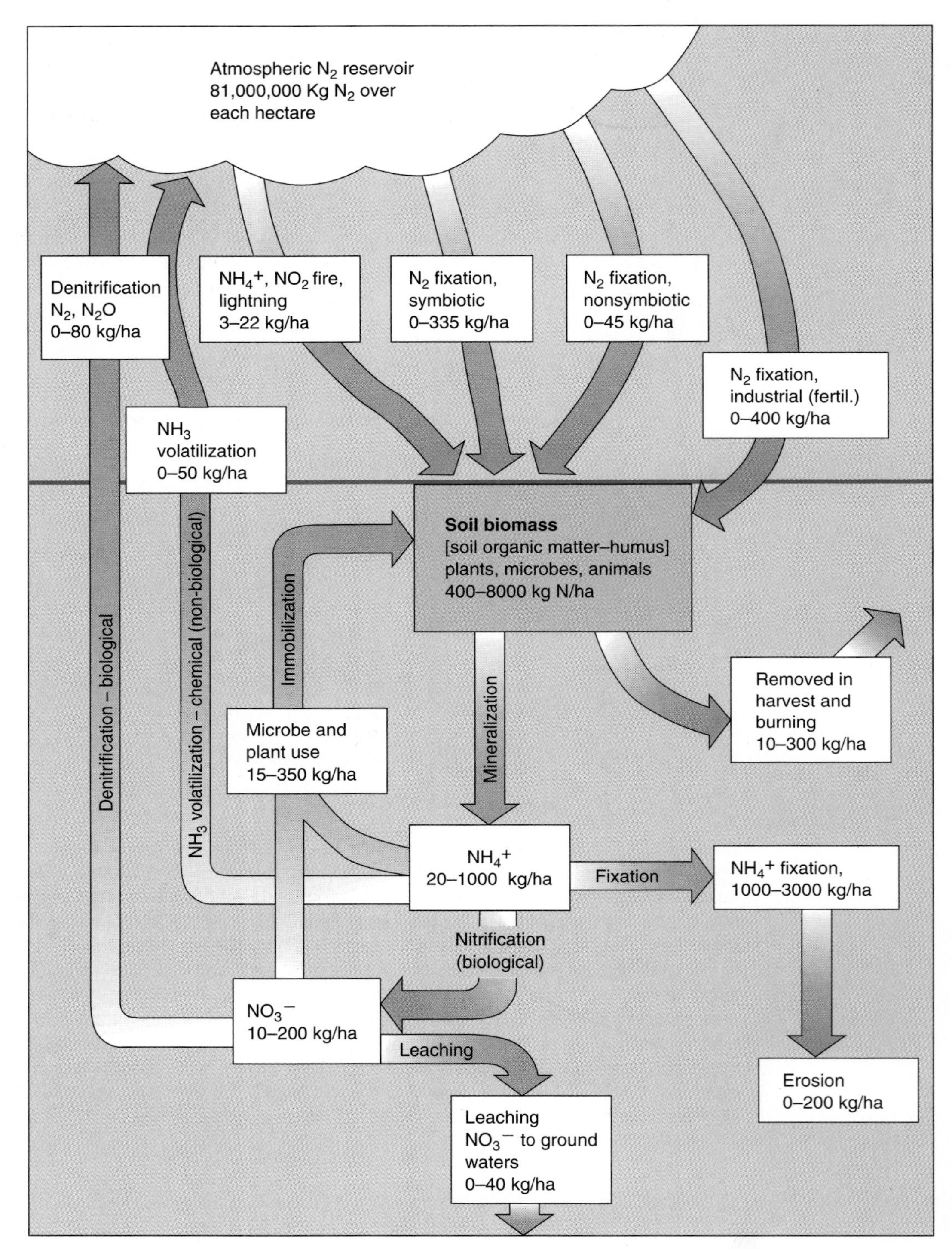

FIGURE 10-2 Nitrogen cycle, showing soil nitrogen changes, additions, and losses. (Courtesy of Raymond W. Miller, Utah State University.)

acid contents, and expended only half as much metabolic energy (ATP) per gram of material produced as did plants given nitrate ions. In contrast, beans supplied with nitrate ion, rather than ammonium ion, were 50 percent larger and had 10 to 30 times more organic acids.[1] Numerous studies, such as the two that follow, describe increased yields when both NH_4^+ and NO_3^- forms are available.

- NH_4^+ plus NO_3^- increased wheat yields 7%–47% in fourteen studies. Tillers also were increased by having both nitrogen forms, compared to having NO_3^- only.
- In hydroponic studies corn fertilized with equivalent amounts of NH_4^+ and NO_3^- averaged about 12.6 Mg/ha (200 bu/a). With only nitrate available, the corn averaged 11.3 Mg/ha (179 bu/a).

Fixation of Dinitrogen Gas (N_2)

A major source of soil nitrogen comes from **nitrogen fixation,** a microbial action in which the relatively *inert* dinitrogen (N_2) is taken from the soil air and changed into forms used by the plants. The nitrogen fixation by microorganisms is either *symbiotic* or *nonsymbiotic.* In **symbiotic fixation,** either bacteria or actinomycetes cause the formation of root nodules in certain host plants and then inhabit those nodules, where they fix nitrogen; see Detail 10-2.

The amounts of N_2 fixed by soil bacteria and actinomycetes vary enormously and are usually lower in soils that have high nitrogen levels or have had nitrogen fertilizers added. In Denmark three-year averages of two pea cultivars fixed 165 kg/ha and 136 kg/ha of N, and one field bean cultivar fixed 186 kg/ha, even though 50 kg/ha of N was added as fertilizer at planting.[2] Blue-green bacteria (originally called *blue-green algae*) fixed only 25 kg/ha. Alfalfas fixed from 128 kg/ha to 600 kg/ha, and nonlegume fixers fixed from about 40 kg/ha to 300 kg/ha. Only a few crops, such as alfalfa, can usually fix enough nitrogen to produce optimum growth without fertilizer N additions. In one study growing legumes with no added nitrogen fertilizer fixed 80 percent of their needs; at least 20 percent was supplied from decomposing soil humus. Some nonlegumes such as alder (*Alnus*) fix from 80 kg/ha to 170 kg/ha of N.

In free or **nonsymbiotic N_2** fixation, specific types of microorganisms exist independently in soil and in water, convert nitrogen (N_2) into body-tissue nitrogen forms, and then release it for plant use when they die and are decomposed. Nitrogen fixed yearly by nonsymbiosis varies from a few kg/ha to more than 45 kg/ha; 5 kg/ha to 8 kg/ha as an average.

Mineralization of Nitrogen

The major source of nitrogen in nonfertilized soils is the release of the amino groups ($—NH_2$) during decomposition of organic material. The overall conversion of organic nitrogen to the ammonium form is termed **mineralization.** Soil organic matter contains an average of about 5 percent nitrogen. Only about 1 percent to 3 percent of the total soil organic matter is decomposed yearly, but as much as 50 percent of the *fresh* residues may be decomposed. The decomposition rate is fast in warm, well-aerated, moist soils—such as most sands in wet summers—and it is slow in clays during cool seasons.

---

[1] S. Chaillou, J. F. Norot-Gaudry, C. Lesaint, L. Salsac, and E. Jolivet, "Nitrate or Ammonium Nutrition in French Bean," *Plant and Soil* **91** (1986), 363–365.
[2] E. S. Jensen, "Symbiotic N_2 Fixation in Pea and Field Bean Estimated by ^{15}N Fertilizer Dilution in Field Experiments with Barley as a Reference Crop," *Plant and Soil* **92** (1986), 3–13.

Detail 10-2 N_2-Fixation—Very Energy Consumptive*

Nitrogen usable by plants comes only from decaying organic substances, N_2-fixation, or added soluble nitrogen. N_2-fixation adds to a soil nitrogen supply that is usually inadequate for maximum plant growth. Unfortunately, this N_2-fixation is not *energy-cheap* and must be done in *anaerobic* (oxygen-free) environments.

Rhizobia bacteria stimulate a root hair to curl and enlarge. Within the curl the bacteria dissolve the cellulose cell walls and enter the root cells. The bacteria then form an *infection thread* through the root cortex cells into which the bacteria eventually are released. But, how does the bacteria keep an anaerobic environment? The infection thread actually *buds off,* keeping the bacteria in each *bud* contained in the small vesicles. The bacteria in each vesicle multiply until the larger **bacteroid,** surrounded by its oxygen-proof membrane, stops growing and starts fixing dinitrogen. This enlarged root area (*bulb on the root*), which can be many times the root diameter, is called a **nodule.**

The dinitrogen fixation is represented by the following equation, in which ATP and ADP are the high energy phosphate bonds before and after energy release, respectively:

$$8H^+ + 8e^- + N_2 + 16ATP \longleftarrow 2NH_3 + H_2 + 16\ ADP + 16P_i$$

The enzyme complex that accomplishes the conversion is **dinitrogenase.** This complex involves an Fe-protein and an MoFe-protein.

The energy needed is supplied by 16 moles of adenosine triphosphate (ATP), which releases large amounts of energy as one ester phosphate is split off leaving adenosine diphosphate (ADP). The total ATP needed is estimated to total 25 ATP for each molecule of N_2 fixed. Another estimate is that the fixed carbon (in consumed carbohydrate energy) requires about 12 grams of carbon to fix 1 gram of dinitrogen. Fixation of dinitrogen is energy expensive.

Nitrogen use from all sources is energy expensive. Absorption of NH_4^+ or NO_3^- from solution is energy expensive. Assimilating NH_4^+ consumes 2 percent to 5 percent of the plant's *total* energy production. For assimilation of nitrate, which must then be reduced to ammonium, the cost is nearly 15 percent of the *total* energy production. The ecology of a system is closely tied to the nitrogen in the system.

*William G. Hopkins, *Introduction to Plant Physiology,* John Wiley & Sons, Inc., New York, 1995, 101–122.

As an example of the amount of nitrogen released from soil organic matter, assume that 2 percent of the organic matter in a soil is mineralized. If the soil has an organic matter content of 4 percent, mineralization would produce about 80 kg of nitrogen as ammonium ion [4% humus in soil × 2 million kg of soil per hectare-15 cm × 5% N in the humus × 2% mineralized = 80 kg]. Some rates of mineralization are faster than the 2 percent per year described. Warm seasons, good moisture, or high organic-matter contents will increase this amount of released nitrogen. Adverse conditions and lower organic matter contents will decrease the nitrogen released. The 80 kg of nitrogen is about 25 percent to 35 percent of the nitrogen needed by a corn crop. In contrast, it is enough for many natural vegetation covers, such as forests and slower-growing plants. In natural systems the plant cover often grows as fast as nitrogen supplies permit. A larger soil humus content and a greater percentage decomposed (in warmer climates) would provide more released nitrogen.

Four examples,[3] all having high decomposition rates, are shown in Table 10-2. Rates of organic matter decomposition, amounts of organic matter in soils, and the nitrogen content in

[3]S. Ross, *Soil Processes: A Systematic Approach,* Routledge Publishers, New York, 1989, 59.

Table 10-2 Turnover Rates of Organic Carbon in Four Plant Cover Systems

Land Cover	*Soil Depth (cm)*	*Production Mg C/ha/yr*	*Total C in Soil Mg C/ha*	*Decomposed %/yr*
Continuous wheat	0–23	2.6	26	4
Continuous meadow	0–23	2.7–3.2	77	3
Tropical rain forest	0–30	9–10	44	11
Beech forest (cold)	0–30	7.1	72	3

Source: Modified from S. Ross, *Soil Processes: A Systematic Approach,* Routledge Publishers, New York, 1989, 59.

the organic matter all influence the amount of nitrogen released each year through decomposition. Notice that the warm, tropical forest has a high yearly decomposition rate of 11 percent of a high yearly biomass production. In ecosystems without outside sources of nitrogen, the growth rate and species present will often be related to the pool of nitrogen in the soil.

Nitrification of Ammonium

Nitrification is the oxidation of ammonium cations to nitrate anions by bacteria or other organisms. Nitrification is a rapid microbial transformation. Most small amounts of mineralized ammonium ions are nitrified within 1 or 2 days unless the soil is strongly acidic, cold, or waterlogged; these conditions slow nitrification markedly. Some ammonium ions are adsorbed temporarily to the negatively-charged cation exchange sites of clay or organic particles, some ammonium ions are fixed in clay lattices (**ammonium fixation**), and other ammonium ions are used directly by plants. Eventually, most of the ammonium ions in the soil are oxidized by *Nitrosomonas* bacteria to NO_2^- (nitrite) and then to NO_3^- by *Nitrobacter,* as follows:

$$2NH_4^+ + 3O_2 \xrightarrow[\text{bacteria}]{\textit{Nitrosomonas}} 2NO_2^- + 4H^+ + 2H_2O + \text{Energy}$$

$$2NO_2^- + O_2 \xrightarrow[\text{bacteria}]{\textit{Nitrobacter}} 2NO_3^- + \text{Energy}$$

Seldom do large amounts of nitrite (NO_2^-) accumulate, which is fortunate because it is toxic to living organisms, including plants. Note that the oxidation of each NH_4^+ to NO_3^- also releases $2H^+$, which is important in acidification of soils. About 1.8 kg of pure lime is required to neutralize the acidity produced by oxidation of 1 kg of urea-nitrogen or ammonium-nitrogen.

Nitrification is slowed by conditions unfavorable to the bacteria: dryness, cold, or toxic chemicals. Figure 10-3 illustrates two locations and their reported nitrification rates. Note that the desirable ammonium thiosulfate fertilizer, temporarily producing NH_3 plus SO_2, inhibits the second step involving *Nitrobacter* from producing an accumulation of toxic nitrate ion (NO_2^-) in the soil. In many of the soils nitrification was complete within 1 or 2 weeks after large amounts of ammonium had been applied.

Other Fixation Reactions Involving Soil Nitrogen

In addition to transformations previously mentioned, nitrogen can be **immobilized. Immobilization** is the incorporation of soluble nitrogen (mostly NH_4^+ and NO_3^-) into the

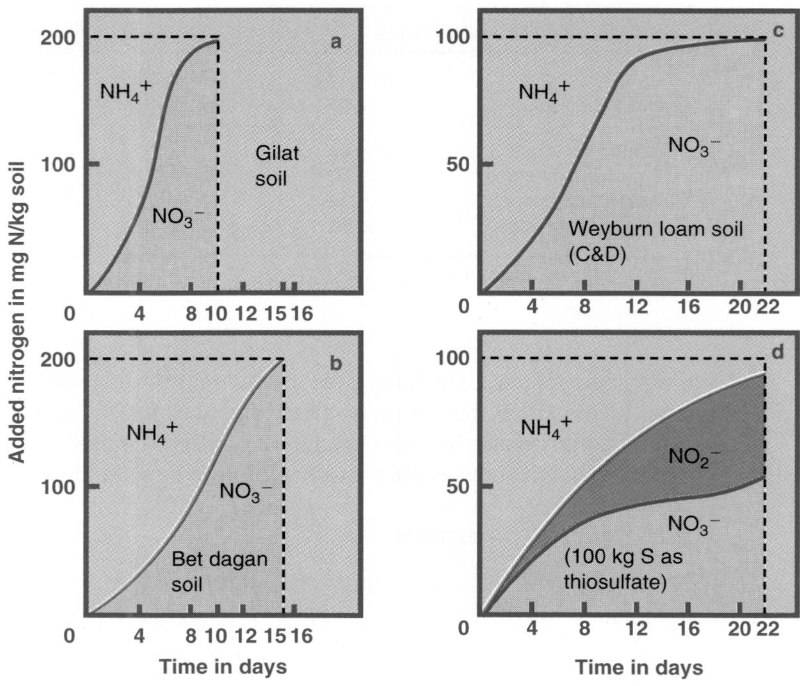

FIGURE 10-3 Rates of nitrification for three soils (a, b, c) and retardation of nitrification by ammonium thiosulfate (d). Soils (a) and (b) are in Israel; (c) and (d) (Weyburn loam) are in Alberta, Canada. (*Sources:* Aviva Hadas, Sala Feigenbaum, A. Feigin, and Rita Portnoy, "Nitrification Rates in Profiles of Differently Managed Soil Types," *Soil Science Society of America Journal* **50** [1986], 633–639; H. H. Janzen and J. R. Bettany, "Influence of Thiosulfate on Nitrification of Ammonium in Soil," *Soil Science Society of America Journal* **50** [1986], 803–806. Redrawn by Raymond W. Miller, Utah State University.)

bodies of plants or microbes. The nitrogen again becomes complex organic compounds in new organism bodies.

Ammonium ion can be fixed into soil clays, too. **Ammonium fixation** is the process by which certain clays bond NH_4^+ tightly between mineral lattices. Most of this fixed ammonium is neither exchangeable nor available to plants. Fixation is most common in vermiculite, illite, and montmorillonite clays, where the ammonium ions fit into locations between clay layers within the clay particle— similar to where potassium ions fit in illite. Ammonium fixation occurs in small amounts compared to the quantities adsorbed to cation exchange sites.

10:5 Nitrogen Losses from the System

Nitrogen losses occur as leaching losses in percolating water, removal in harvests or by animals, and as two gaseous losses. Although fixation and immobilization are losses from *available* forms, they still exist in the soil for possible conversion later to available forms again and are not considered as losses here. Harvest and animal losses are not discussed.

Leaching of Soil Nitrogen

Nitrate (NO_3^-) is the most readily leached form of nitrogen. Both ammonium (NH_4^+) and nitrate ions are very soluble in water, but the positively-charged ammonium ion is held to cation exchange sites and resists leaching. Leaching losses of nitrates are increased as the quantities of percolating water increase and when plant growth is not sufficient to absorb the nitrates as they are produced.

Losses of nitrogen from the soil covered with an actively growing crop are usually only a few kilograms per year unless large amounts of a soluble fertilizer were recently added. A corn-covered Udoll moist Mollisol in Missouri lost only 1.3 percent of added nitrogen in runoff but 30 percent of added nitrogen in groundwater.[4] On fields where fertilizer application is poorly timed for plant use or coincident with heavy rainfall and leaching, leaching losses can be 50 kg/ha to 80 kg/ha of nitrogen yearly. It is not unusual to find nitrogen losses exceeding 20 kg/ha on crops heavily fertilized.

Nitrification Inhibitors[5, 6, 7]

If oxidation of ammonium ions to the nitrate form is slowed or hindered, less nitrate is produced and less nitrogen is lost by leaching or denitrification. Several dozen chemicals have been tested for their ability to inhibit nitrification. The best known **nitrification inhibitor (NI)** is **Nitrapyrin** [2-chloro-6-(trichloromethyl)pyridine], also called **N-Serve®**. Dicyandiamide **(DCD)** and **ATC** (4-amino-1,2,4-triazole) are two other inhibitors. Potassium azide (KN_3) has been used extensively in Japan.

Nitrapyrin (like many inhibitors) inhibits the *Nitrosomonas* bacteria that convert ammonium to nitrite, the first step of nitrification. Nitrapyrin has the disadvantage of being a volatile liquid, and evaporates rapidly from sandy soils. DCD and ATC, in comparison, are easier to handle, are water-soluble, do not volatilize, and can be applied as coatings to granular fertilizers.

In a new fertilizer—ammonium thiosulfate—the thiosulfate acts as a nitrification inhibitor. The conversion of ammonium ion to the easily leached nitrate is slowed down (see Figure 10-3d). The retardation action occurs on the *Nitrobacter* organisms (the second step, nitrite to nitrate) rather than on the *Nitrosomonas*.

Nitrification inhibitors should be placed with the nitrogen to be most efficient. Rates of about 0.2 kg/ha to 2.0 kg/ha of active NI material or 0.5 percent to 1.0 percent of added fertilizer are satisfactory to retard nitrification, but results vary excessively with conditions of application and the soil. Some studies also show poor results with additions of Nitrapyrin of less than 2 kg/ha to 5 kg/ha (1.8 lb/a to 4.5 lb/a). Retention of added urea or ammonium nitrogen after a month may be 50 percent to 90 percent of the amounts added; after three months, retention as NH_4^+ may be only 20 percent to 30 percent.

Although NIs are used commercially, the results are not consistent, even in what seem to be duplicate circumstances. Even when oxidation of ammonium ions is inhibited, as shown by higher ammonium-to-nitrate ratios, there may not be consequent yield increases, indicating that other unidentified causes are more influential than just the conserved ammonium fertilizer. NIs have a gradually reduced effectiveness over time because of their volatilization, adsorption, leaching, or microbial breakdown. The effective life of NIs is also shortened by warmer temperatures and use on sandy soils, which increase volatilization and leaching losses above those occurring in finer-textured soils.

---

[4]D. W. Blevins, D. H. Wilkison, D. P. Kelley, and S. R. Silva, "Movement of Nitrate Fertilizer to Glacial Till and Runoff From a Claypan Soil," *Journal of Environmental Quality* **25** (1996), 589–593.

[5]T. F. Gutherie and A. A. Bomke, "Nitrification Inhibition by N-Serve and ATC in Soils of Varying Texture," *Soil Science Society of America Journal* **44** (1980), 314–320.

[6]George W. Bengtson, "Nutrient Losses from a Forest Soil as Affected by Nitrapyrin Applied with Granular Urea," *Soil Science Society of America Journal* **43** (1979), 1029–1033.

[7]R. J. Goos, B. E. Johnson, and W. H. Aherns, "New Uses for Ammonium Thiosulfate," *Solutions* **34** (1990), No. 3, 32–33.

Use of NIs can be detrimental to plant growth. Some of the observed effects are a toxic level of ammonium ion on plant roots (radishes) and a claimed reduction in cation uptake because NH_4^+, a cation, is absorbed in considerable amounts, but research has produced conflicting results. Nitrification inhibitors currently have some valid uses, such as retaining fall-applied ammonium fertilizers through the winter with minimum leaching losses, but few general-use recommendations have been validated by research.

There is increased interest in using NI materials to minimize the NO_3^- leaching into surface and groundwaters, thus avoiding pollution as well as avoiding nitrogen losses from the soil, but nitrates are a health problem where food supplies are adequate (see Chapter 18).

The search continues for new and better NIs. A recent one, **CMP [1-carbomoyl-3(5)methylpyrazol]** worked well in calcareous soil.[8] Without CMP, ammonium sulfate was nitrified in 3 weeks. With CMP, only 10 percent was nitrified after 3 weeks and 42 percent was nitrified after 8 weeks during warm growing conditions.

Gaseous Losses of Soil Nitrogen

In addition to nitrogen lost by leaching or made unavailable by immobilization and ammonium fixation, soil nitrogen can be lost through two mechanisms producing gaseous forms that escape into the atmosphere: denitrification and ammonia volatilization.

Denitrification is the change by bacteria of nitrate to a nitrogen gas (mostly N_2, some N_2O, and less of other oxides). Denitrification is a *biological process.* Usually denitrification is the *most extensive gaseous nitrogen loss.* When poor aeration limits the amount of free oxygen in the soil, a few specifically adapted bacteria are forced to use the nitrogen in NO_3^- as an electron acceptor (normally, O_2 is used). The end products of the process are dinitrogen gas (N_2) and/or nitrous oxide (N_2O); these gases volatilize from the soil into the atmosphere.

Denitrification is rapid. Even when conditions favorable for denitrification exist for only a day or less, appreciable losses of nitrogen as N_2 can occur. Estimates of total losses by denitrification on cropped lands average 10 percent to 20 percent of all nitrates formed from added fertilizers, and, in extreme conditions, can be as much as 40 percent to 60 percent of added nitrate–nitrogen. A loss of 10 percent to 15 percent is believed to be common.

Large denitrification losses result from (1) a lack of adequate free gaseous oxygen in the soil or solution; (2) an energy source of oxidizable organic matter (food) for the bacteria; and (3) warm, slightly acidic soils. Waterlogging for even a few hours in warm soil that contains decomposable humus may develop anaerobic conditions. In Ontario, Canada, killing grass cover with herbicide caused accumulation of nitrate, which caused a 20-fold to 30-fold increase in denitrification.[9] Even interiors of soil aggregates after wetting by rain or irrigation may develop anaerobic interiors. Spherical aggregates (granules) of a silty clay loam and of a silt loam from Iowa often had anaerobic interiors in aggregates larger than 20 mm (0.8 in) diameter. One granule of only 8 mm (0.31 in) diameter had an anaerobic zone (Figure 10-4).[10] All aggregates involved in denitrification also had a measured zone that was anaerobic. Poorly structured clayey soils, especially subsoils, may have large peds (50 mm to 100 mm, or 2 in to 4 in, diameter) that have poor porosity and are temporarily anaerobic after they are wetted.

Ammonia volatilization losses occur when ammonium is in a *basic* (high pH) solution. It is a *chemical, not biological,* process. The greatest losses occur from surface applica-

---

[8] P. I. Orphenos, "Inhibition of Ammonium Sulphate Nitrification by Methylpyrazol in a Highly Calcareous Soil," *Plant and Soils* **144** (1992), 145–147.

[9] Mario Tenuta and Eric G. Beauchamp, "Denitrification Following Herbicide Application to a Grass Sward," *Canadian Journal of Soil Science* **76** (1995), No. 1, 15–22.

[10] Alan J. Sexstone, Niels Peter Revsbech, Timothy B. Parkin, and James M. Tiedje, "Direct Measurement of Oxygen Profiles and Denitrification Rates in Soil Aggregates," *Soil Science Society of America Journal* **49** (1985), 645–651.

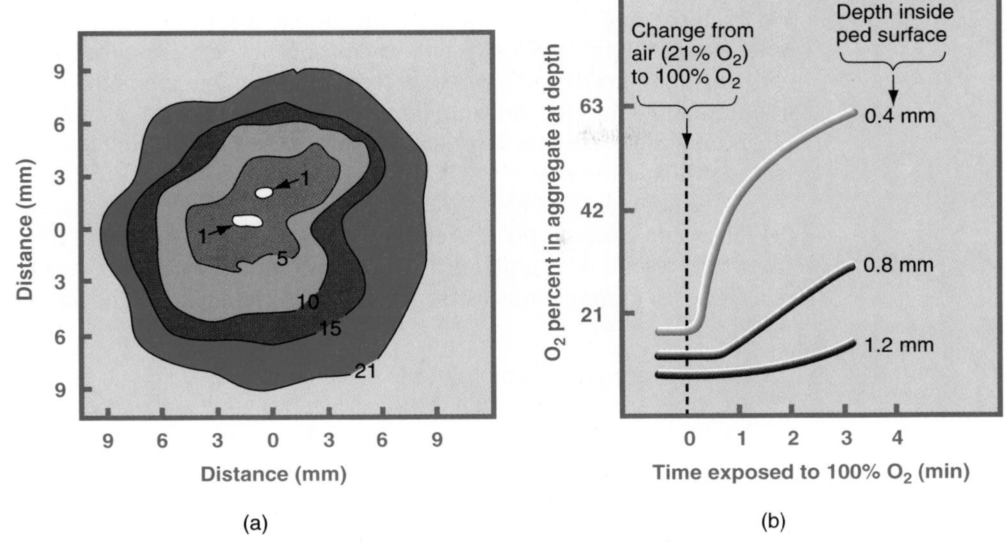

FIGURE 10-4 Typical O_2 characteristics in wetted spherical soil peds: (a) cross-section of a ped showing O_2 levels from the outside (21%) inward to centers having only 1%; (b) rate of O_2 concentration changes at three distances from the ped surface as air O_2 is increased. Pores filled with water would transmit almost no O_2 that was not already dissolved in the water. (*Source:* A. J. Sexstone, N. P. Revsbech, T. B. Parkin, and J. M. Tiedje, *Soil Science Society of America Journal* **49** [1985], 645–651. Redrawn by Raymond W. Miller, Utah State University.)

tions of any ammonium or urea fertilizer on calcareous (high carbonate content) soils. Small losses of ammonia also occur from nonfertilized soils. Ammonia volatilization losses from applied ammonium or urea fertilizers can be as much as 30 percent but usually are less than 10 percent. Studies have shown that large amounts of nitrogen (15%–30%) can be lost as ammonia even from corn leaves.[11]

To minimize volatilization, cover the fertilizer with soil or leach it in with irrigation or rainfall. Broadcasting on the surface may cause large losses, especially on a turf. Urea, plus the enzyme *urease,* in soil forms ammonium carbonate. In basic soils calcium hydroxide reacts on the ammonium carbonate to form ammonium hydroxide:

$$Ca(OH)_2 + (NH_4)_2CO_3 \rightarrow 2NH_4OH + CaCO_3 \downarrow \text{(precipitated)}$$

$$NH_4OH \text{ easily decomposes to } NH_3 \uparrow \text{(gas)} + H_2O$$

The loss of NH_3 gas from NH_4^+-forming fertilizers depends on the final pH of the NH_4^+ solution at the soil surface. Up to about pH 9, the ratio of NH_3 partial pressure in air versus that in water increases about ten-fold for each pH unit increase. Thus, the NH_3 lost from a solution of pH 8.2 will be much less than if the pH were 9 or higher.

As an NH_4^+ ion is added to calcareous soil, two reactions occur:

$$(NH_4)_2Y + CaCO_3 = (NH_4)_2CO_3 + CaY \qquad (Y = \text{various anions})$$

$$(NH_4)_2CO_3 + HOH = 2NH_3 \uparrow \text{(gas)} + 2HOH + CO_2 \uparrow \text{(gas)}$$

[11] D. D. Francis, J. S. Schepers, and M. F. Vigil, "Post-Anthesis Nitrogen Loss from Corn," *Agronomy Journal* **85** (1993), 654–663.

It has been determined that if the CaY salt is very *insoluble,* the NH_3 loss is greatest. For example, adding ammonium fluoride to the soil forms calcium fluoride (CaF_2), which is quite low in solubility. This system can reach a pH of 9 (from the large amount of ammonium carbonate produced) and have 30 percent to 40 percent ammonia loss in 1 hour.[12] Salts such as ammonium chloride and ammonium nitrate do not form insoluble calcium salts; thus, little ammonium carbonate is produced to raise the pH. The amount of ammonia lost depends on the soil solution pH.

Ammonia is unstable in water and is evolved in increasing amounts as the soil solution pH nears and exceeds pH 9. Actually, however, small ammonia losses have been reported from acidic soils with a pH of 4. If ammonium carbonate can form, the chance for loss is great. Losses of ammonia gas from soils can be summarized as follows:

- Losses are greatest on high-pH calcareous soils
- Losses are greatest when fertilizer is left on the soil *surface*
- Losses increase with higher temperatures, especially as surface soil dries out after being wetted (drying concentrates ammonia)
- Losses are greatest in soils of low cation exchange capacities
- Appreciable losses of urea applied on grass or pastures are possible because fertilizer on pastures is applied on the surface

Urease inhibitors slow conversion of urea to ammonium. Urea is hydrolyzed by the enzyme *urease* to ammonium carbonate. If urea is spread on a calcareous soil surface or on another alkaline soil and left several days to dissolve in the morning dew or from soil water, great losses of ammonia are likely. To slow this enzymatic action until the soluble urea can be leached or tilled into the soil where the NH_4^+ can be adsorbed onto soil particles, inhibitors of urease action are needed. The inhibitors are usually substances that react with the urease, thereby blocking the enzyme from reacting with urea. Typical inhibitors of urease include phenols, quinones, benzoquinones, various insecticides, and substituted ureas. Applied inhibitors are adsorbed to soil materials gradually, decomposed by microbial activity, or react with metals such as iron.

10:6 Nitrogen Balance

More than any other major plant nutrient, nitrogen in the soil is subject to a complex system of gains, losses, and interrelated reactions. Intelligent management demands a working knowledge of these relationships and their comparative magnitudes.

Losses and Gains of Nitrogen

Figure 10-5 depicts some values for nitrogen changes of a grazed clover/grass pasture. Tables 10-3 and 10-4 list amounts of nitrogen (and the relative percentage of change) in various soil-crop systems. For example, if solid nitrogen fertilizer materials are added before planting, enough fertilizer must be added to supply plant needs plus some for the expected losses by leaching, immobilization, denitrification, and volatilization. Knowing the expected losses helps to resolve the question of whether or not the increased fertilizer cost outweighs the convenience of an early application. Even more critical now is the requirement *not to exceed the fertilizer needed* so that contamination by leaching and runoff will occur.

---

[12] L. B. Fenn and L. R. Hossner, "Ammonia Volatilization from Ammonium or Ammonium-Forming Nitrogen Fertilizers," *Advances in Soil Science* **1** (1985), 123–169.

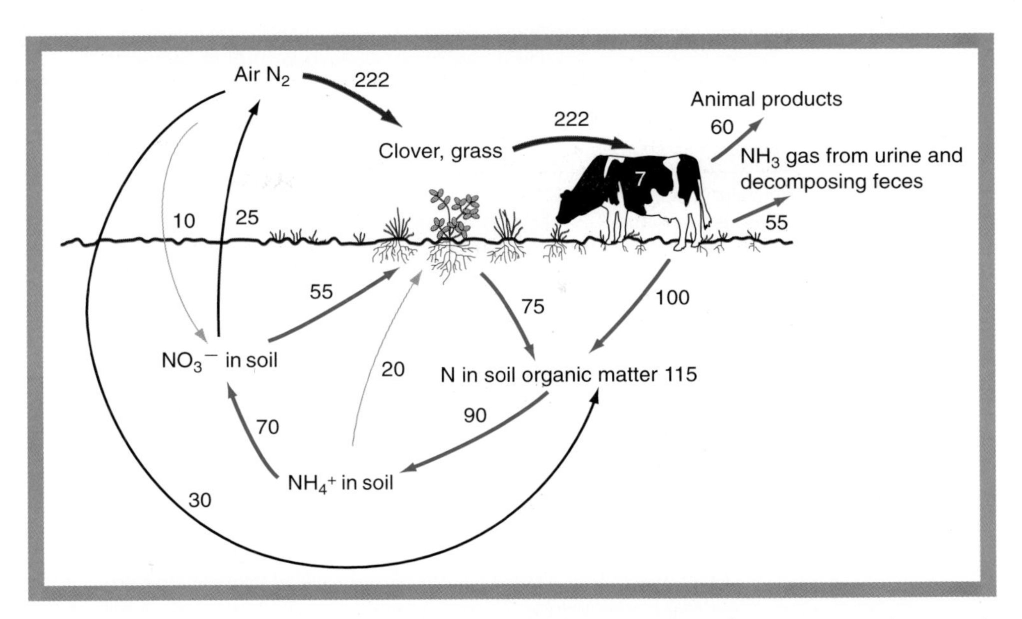

FIGURE 10-5 Example of nitrogen changes, in pounds per acre, in a clover-grass system used as pasture (grazed). Pounds per acre times 1.12 equals kg/ha. (*Source:* Redrawn from data by D. C. Whitehead, 1970, quoted by R. A. Date, "Nitrogen, a Major Limitation in the Productivity of Natural Communities, Crops, and Pastures in the Pacific Area," *Soil Biology and Biochemistry* **5** [1973], 10. Courtesy of Raymond W. Miller, Utah State University.)

Table 10-3 Examples of Nitrogen Gains, Losses, and Transformations (in kg/ha/yr) for Eight Different Cropping Systems[a]

Description of Change	*Grazed Bluegrass (N.C.)*	*Corn Grain (Ind.)*	*Soybean Seeds (Ark.)*	*Wheat (Kansas)*	*Irish Potatoes (Maine)*	*Cotton (Calif.)*	*Loblolly Pine (Miss.)*	*Douglas Fir (Wash.)*
Additions								
Added fertilizer	168	112	0	34	168	179	—	—
Irrigation, floodwater	—	10	—	—	—	50	—	—
Sediments added	10	10	10	6	6	3	11	10
N_2-fixation	—	—	123	—	—	—	8	—
Removals								
Harvested product	38	85	90	36	80	79	12	10
Denitrification	5	15	15	5	15	20	1	1
Volatilization of ammonia	98[b]	—	—	—	—	—	—	—
Leaching loss	—	15	10	4	64	83	1	1
Erosion and runoff	14	16	16	5	15	50	3	2
Recycling process								
Uptake from soil	151	126	120	56	145	127	20	35
Manure from grazing	60	—	—	—	—	—	—	—
Plant residues left	113	41	30	20	65	48	9	25
Mineralization from humus	48	50	15	28	65	48	6	—

Source: Data from M. J. Frissel, ed., *Cycling of Mineral Nutrients in Agricultural Ecosystems,* Elsevier Science Publishing, New York, 1978, 202–243.

[a]A dash means that no measurement was made or the item does not apply to the system.

[b]Losses from voided animal urine and feces as ammonia gas.

Table 10-4 Nitrogen Balance for Productive Cropping Systems*

Nitrogen Resources	Percentage of Original Source Changed	Amount of Nitrogen	
		lb/a	kg/ha
Initially in soil		3564	3992
Nitrogen available			
Mineralized N (from decomposition of organic matter to yield nitrates)	2% of initial N	71	80
Additional sources			
Fixation by bacteria and/or algae (free + symbiotic)		18	19
Addition from fertilizers		89	100
Precipitation (rain, dew)		7	8
Total available		185	207
Nitrogen losses			
Denitrification (bacterial change of nitrates to volatile N_2) and ammonia volatilization			
From fertilizer	15% of original fert. N	13	15
From mineralized nitrates	5% of total mineralized	4	4
Removed in crop plants			
Absorbed from fertilizer	55%	49	55
Absorbed from mineralized N	45%	32	36
Absorbed from fixation	50%	9	10
Absorbed from precipitation	100%	7	8
Leaching and runoff		4	4
Total losses		118	132
Nitrogen immobilized by microbial activity			
From fertilizer N not denitrified	45%	27	30
From mineralized N not denitrified	55%	36	41
From fixation of N	50%	9	10
Total immobilized N		72	81

Source: Modified and calculated from data by R. D. Hauck, "Quantitative Estimates of Nitrogen-Cycle Processes—Concepts and Review," in *Nitrogen-15 in Soil-Plant Studies,* Panel Proceedings Series of International Atomic Energy Agency, Vienna, 1971, IAEA-PL-341/6; 77.

*A hypothetical system, averaged from numerous studies using tracer ^{15}N; these values are an approximate guide for average soils.

In Table 10-3 eight crop systems are shown. Notice the appreciable nitrogen in the irrigation water regularly added to cotton, and N_2-fixation by soybeans, which eliminates the need for nitrogen fertilizer. Also, notice the variable loss of nitrogen by leaching, which is increased by high fertilizer additions and extensive leaching (when it rains frequently). Mineralization values are variable, and nitrogen turnover time in the two forested areas is much longer; in cultivated crops nitrogen is released faster. Most of the nitrogen in dense forests is recycled. Ammonia volatilization is not even shown in most of the soils in Table 10-3 because the soils are acidic and losses were expected to be low.

Models of Nitrate Movement

Numerous models are involved in trying to predict nitrate losses in percolating waters from rainfall and irrigation; see **Appendix A.**

10:7 Materials Supplying Nitrogen

The production of nitrogenous fertilizers has increased faster than that of any other type of chemical fertilizer. The principal nitrogenous fertilizer materials and their percentages of nitrogen are given in Table 10-5; note that the list does not include the many specialty and experimental materials.

Ammonia and Aqueous Nitrogen

Anhydrous ammonia (NH_3) is the principal nitrogenous fertilizer, with more than 90 percent of all nitrogenous fertilizers consisting of ammonia or compounds made from ammonia. **Ammonia** is a colorless gas containing one atom of nitrogen (atomic weight 14) and three atoms of hydrogen (atomic weight 1), so the percentage of N in pure NH_3 is as follows:

$$\%\text{ N} = \left(\frac{14}{14 + 3}\right)(100) = 82.35$$

Commercial grade ammonia is 99.5 percent pure and, therefore, contains 82 percent nitrogen.

Anhydrous ammonia is manufactured from atmospheric nitrogen and natural gas to supply hydrogen. Energy costs have increased the proportion of ammonia used compared to other nitrogenous forms because it is the *first industrial product* of the conversion of atmospheric nitrogen to usable form. Other nitrogen fertilizers require further processing and more energy. Urea, for similar reasons, has become the most popular *solid* nitrogenous material on the market.

Anhydrous ammonia is applied by injecting it into the soil behind chisels set about 12 cm (5 in) deep and using tractor- or truck-mounted pressure tanks. Anhydrous ammonia has physical properties somewhat like those of butane or propane gas; it is liquid when under tank pressure but gaseous at atmospheric pressure. Anhydrous ammonia is the *least expensive* nitrogen fertilizer.

Table 10-5 Principal Nitrogenous Materials

Material	*Nitrogen Content (N) (%)*
Anhydrous ammonia	82
Urea	45–46
Ammonium nitrate	33.5
Aqua ammonia	20–24
Ammonium sulfate	20–21
Diammonium phosphate (plus 46%–53% available P_2O_5)	18–21
Ammonium phosphate sulfate (plus 20% available P_2O_5)	16
Ammonium polyphosphate (having 34%–37% P_2O_5)	10–11
Sodium nitrate	16
Potassium nitrate (also containing 44% K_2O)	13
Urea-formaldehyde (slow-release materials)	30–40
Sulfur-coated urea	15–45
Organic products (animal manures, sewage sludge, meat meal, cottonseed meal, fish meal)	1–12

Many safety precautions must be observed when handling anhydrous ammonia. The most serious threat from anhydrous ammonia is *blindness.* Most accidents occur when the anhydrous ammonia is being transferred from one tank to another. Some precautions are as follows:

- Use and wear proper equipment. Wear goggles. Anhydrous ammonia will corrode copper, brass, and galvanized parts.[13]
- Keep flame away from mixtures of 16%–25% ammonia because they will burn.
- Keep away from ammonia if it escapes into the atmosphere. Ammonia is very soluble in water; living tissues, especially the eyes, are high in water content. Ammonia causes severe irritation of the eyes, nose, throat, and lungs. The skin can be easily burned; rubber goggles and gloves give some protection. Ammonia that contacts plants kills them because of the high concentration of NH_4OH produced in plant tissues.
- Have water available, even in a squeeze bottle; it is the best first-aid material.
- Store only in pressure tanks that are designed to withstand pressures of at least 17.5 kg/cm^2 (250 lb/in^2). Do not overfill tanks; leave room for vapor.
- Paint all ammonia tanks white to help reflect heat. Store tanks in a cool, shady place.
- Arrange for an inspection of all tanks at least once a year and check equipment for worn parts; see that fittings are tight.

Because anhydrous ammonia is difficult to handle, water solutions of ammonia, urea, ammonium phosphate, or nitrogen in other soluble mixtures are being more widely used each year. **Urea ammonium nitrate (UAN)** solutions are widely used. These solutions can be spread on the soil surface and cultivated in, or injected directly into, the soil.

Urea, Ammonium Nitrate, and Ammonium Sulfate

Urea [$CO(NH_2)_2$] is a synthetic organic fertilizer, now cheaper per pound of N than any other solid nitrogenous material; it contains about 45 percent N. Unlike anhydrous ammonia or ammonium salts, urea cannot be absorbed by plants until the nitrogen it contains is converted by the enzyme urease to ammonium.

Urea is readily soluble and leachable when it is first applied to the soil, but when it is changed to ammonium by the action of urease, it is held as exchangeable ammonium cations by clay and humus in a form readily available to plants. Under favorable warm temperatures and moist soil conditions, urea can be hydrolyzed to ammonium carbonate and then by bacterial action to nitrate within less than a week.

Biuret, a manufacturing contaminant of urea, is one possible hazard because it is toxic to sensitive plants, such as tobacco, in concentrations of more than 1 percent.

Urea is a popular nitrogen fertilizer because it is usually the cheapest solid nitrogenous fertilizer and is readily soluble in water, making it a convenient material for application in sprinkler water, as sprays, or as aqueous nitrogen solutions.

Ammonium nitrate (NH_4NO_3) is a good, relatively cheap source of solid nitrogen fertilizer, analyzing 33.5 percent N. Half the nitrogen content is in the ammonium form, and the other half is in the nitrate form. When added to a cool soil, the nitrate ions are immediately available for plant use and are mobile in the soil. The ammonium cations are adsorbed on the exchange sites in the soil and are available to plant roots, although they are not mobile. The ammonium ions nitrify rapidly to nitrate. Because NH_4NO_3 is explosive (if mixed with diesel fuel) and more expensive than urea, many dealers have discontinued it in favor of urea.

---

[13]W. Mueller, "The Nasty Fertilizer," *Agrichemical Age* **33** (1989), No. 8, 8, 9, 12.

Ammonium sulfate [$(NH_4)_2SO_4$] is manufactured mostly from recovered coke-oven gases; it contains 21 percent N. Because of its relatively high cost, it is less popular than ammonium nitrate and urea. For use on rice, however, ammonium sulfate is one of the best forms of nitrogenous fertilizers because the nitrogen, as the ammonium ion, is all potentially available to the plant. If the nitrate instead of the sulfate form were used, under the anaerobic conditions of a flooded rice field, the nitrate would be denitrified. This means only about half the N (the ammonium ions only) of ammonium nitrate can be utilized, compared to all of the nitrogen in the ammonium sulfate form. Ammonium sulfate is the most strongly acid-forming of all common fertilizers.

Organic wastes (plant residues, animal manures, sewage sludge, composts, food processing plant wastes) are all useful and can be considered as controlled-release fertilizers. Nutrient contents are low (1%–8% N, 0.2%–1% P, 0.5%–3% K). One Mg (megagram = metric ton) of dry material will release about 5 kg to 15 kg of nitrogen per hectare per season. The amount of nitrogen released does greatly vary with nitrogen content and rates of decomposition. Unfortunately, these materials may also carry weed seeds, diseases, soluble salts, heavy metals, and other problems. To reduce pollution and burial (landfill) problems, using them on soils has promise.

Controlled-Release Nitrogen Fertilizers

Crop recoveries of the usual fertilizer nitrogen are commonly 40 percent to 70 percent of that added to soil. In porous (sandy) soils and high rainfall areas, recovery can be even less because of more leaching. A less soluble, slow-release fertilizer reduces some of the nitrogen losses responsible for low nitrogen recoveries. Unfortunately, slow-release materials, which do reduce leaching losses, also dissolve too slowly at the time of high nitrogen need (in rapid plant growth periods, such as that just before the formation of grains for cereals). Slow-release fertilizers are expensive, perhaps two or three times the cost of regular nitrogen fertilizers. Some slow-release nitrogen sources and mixed fertilizers are described in Table 10-6—materials that have proven useful on slower-growing grasses, permanent pastures, and greenhouse plants.

Sulfur-coated urea is one of the popular controlled-release nitrogen fertilizers. Granular urea is spray-coated with molten sulfur; then a wax coating and, later, a clay film coating are applied to improve handling characteristics. Water solubility of *untreated urea* is 100 percent within a few minutes; in contrast, from one formula of *sulfur-coated urea,* only 1 percent of the coated urea dissolved about every 5 days.

Sulfur-coated urea pellets are *not* effective on flooded rice fields because an insoluble coating of iron sulfide (FeS_2) forms around each pellet, making them so slowly soluble that they are almost valueless.

Various nitrogen materials are made into polymers using urea and formaldehyde (or isobutylaldehyde) or are coated by polymer material. These produce materials of high percentage nitrogen and various rates of release. Materials such as Nutricote®, described in Table 10-6, can have almost any nutrient composition and yet the controlled release rates can range from relatively short to long time periods. Such materials are well adapted to nursery and greenhouse plants as well as to turf and pastures. Release rates of several slow-release nitrogen fertilizer materials are shown in Figure 10-6. In one study Nutricote® and Osmocote® materials produced better chrysanthemums than the treatments fertilized with liquid fertilizer.[14]

[14]James W. Boodley, "New Japanese Fertilizer Shows Dramatic Results," *Florists' Review* **168** (1981), No. 4354, May 14, 10–11.

Table 10-6 Some Slow-Release Nitrogen Sources and Mixed Fertilizers and Their Properties

Material	*Origin*	*Properties*	*Comments*
1. Natural organic sources	Manures, crop residues, sewage, sludge, composts	Low N (1%–3%); released by microbial decomposition; released more rapidly in warm, moist, fertile soil	Expensive to store and add; may have toxic salts, weed seeds, and heavy metals; relative cost is 5–6 × *
2. Polymers			
a. Isobutylidene diurea (IBDU)	Polymerization of urea with isobutylaldehyde	Hydrolyzes in water to urea; smaller particles hydrolyze fastest; slow in dry soil	Most is made available within 80–100 days; 8–20 mesh sizes suggested for turf; RC = 3–4 × *
b. Polyform UF	Polymerization of urea with formaldehyde	Two-thirds water soluble, 1/3 water insoluble; requires microbes for dissolution	A fast-release nitroform material; fastest in warm, moist conditions; RC = 3–4 × *
c. Ureaform® and Uramite®	Polymerization of urea with formaldehyde	One-third water soluble, 2/3 water insoluble; requires microbes for dissolution	Slower release nitroform materials; some residues even after one year sometimes; RC = 3–4 × *
d. Nutricote® type 100, 140, 180, 270, 360	High-quality mixed fertilizer coated with various polyolefin resins and release-controlling additive	Release rate is increased by higher temperature; type 100 and other numbers indicate the days during which controlled nutrient release occurs	Is used for supplying all three nutrients: N, P, K; not affected by pH, water content, or microbes
e. Osmocote®	Contains mixed fertilizer in an organic resin coating	Nutrients released as water vapor enters coating; nutrients diffuse out or burst coating	Coming in varying mixtures with N only or with various proportions of N, P, and K
3. Sulfur-coated urea (SCU)	Urea granules coated by molten sulfur, then wax, then clay conditioner	Releases urea as microbes decompose sulfur coating or water penetrates cracks in coating; about 20%–30% is released in water at 100°F during 7 days; acidifies soil	Least uniform release rate of slow-release materials; depends on sulfur thickness; can be crushed by equipment

Sources: Richard L. Duble, "Nitrogen, Fertilizers for Turfgrass," Parts One and Two, *Turf-Grass Times* **14** (1978), No. 1; 14–23, and **14** (1978), No. 2; 18–19; and James W. Boodley, "New Japanese Fertilizer Shows Dramatic Results," *Florists' Review* **168,** No. 4354, May 14, 1981, 10–11.
*Relative cost (RC) compared to cost for urea. (*4* × means 4 times more expensive than urea alone.)

10:8 Soil Phosphorus

Phosphorus is the second key plant nutrient; it is the second-most-often deficient nutrient. Phosphorus is an essential part of nucleoproteins in the cell nuclei; these molecules carry the inheritance characteristics of living organisms. Phosphate-to-phosphate *ester* bonds are plants' major energy storage and energy transfer bonds [**ATP** (adenosine triphosphate) and **ADP** (adenosine diphosphate)]. Phosphorus has roles in cell division, in stimulation of early root growth, in hastening plant maturity, in energy transformations within the cells, and in fruiting and seed production. For humans and other animals eating the plants, phosphorus is critical for growth of bones and teeth, which are mostly calcium phosphates (Figure 10-7).

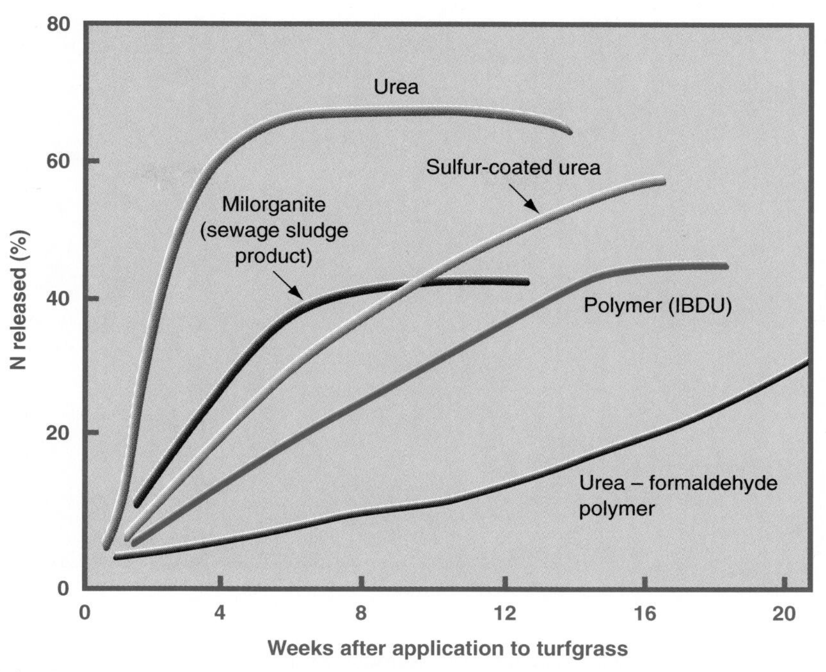

FIGURE 10-6 Examples of rates of nitrogen release from various slow-release nitrogen fertilizers added to turfgrass and compared to urea. (*Source:* Redrawn from Richard L. Duble (ed.), "Nitrogen Fertilizers for Turfgrass," *Turf-Grass Times* **14** [1978], No. 1; 14, 15, 21, 23.)

FIGURE 10-7 The forage available to these cows in Minnesota does not contain sufficient phosphorus because the soil is deficient in this mineral. The result is weakened animals that chew on bones or on the bark of trees. (*Source:* Extension Service, University of Minnesota.)

Plants differ in their ability to compete for soil phosphorus. Young plants absorb phosphorus rapidly, if it is available. Winter wheat absorbs about 70 percent of its phosphorus between tillering and flowering. A cold, wet spring usually results in a retardation of plant growth, often because of inadequate phosphorus absorption. The peak phosphorus demand for corn begins just 3 weeks into the growing season, and the root system is still relatively small then (Figure 10-8).

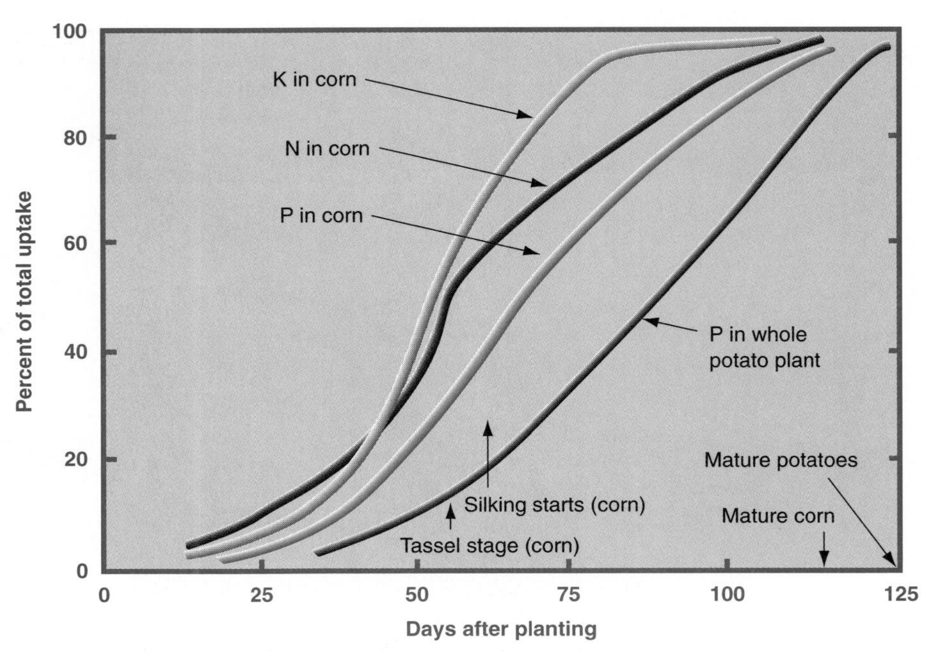

FIGURE 10-8 Uptake rates of phosphorus by corn and potatoes; uptake of nitrogen and potassium by corn shown for comparison. A straight line diagonally from 0 to 100% would indicate a uniform rate of uptake with time. The S-shaped curve is typical of plant growth: slow at both the beginning and end and rapid in midgrowth. Variable rates for different species of plants and different environments will alter the curve shapes.

10:9 The Phosphorus Problem

Getting adequate phosphorus into plants is a widespread problem. Why? The short answer is this: *Soil phosphates are insoluble.* But in reality nothing is *absolutely insoluble,* so this answer really means that the soil forms are of *very low solubility.* Added soluble phosphates will readily combine with cations in soil solution to form low-solubility substances. Some examples of soil solution concentrations of various nutrients are given in Table 10-7. Although plants need about one-fifth to one-tenth as much phosphorus as they do nitrogen and potassium, the concentration of phosphates in solution is only about one-twentieth—or less—as high as the concentrations of nitrogen and potassium.

Data of this type caused scientists to wonder if all or most of the nutrients used by the root could flow to the root in the water absorbed by the roots. This carrying of nutrients to the root in the absorbed water—**mass flow**—was discussed earlier. Scientists soon calculated that the amount of phosphorus dissolved in the soil solution and carried to the root in the amount of water absorbed by the plant was not enough phosphorus to supply even low needs by plants. It is now believed that most phosphorus and potassium are supplied to root surfaces by **diffusion** rather than by mass flow (Table 10-8). However, movement by diffusion is very slow, approximately 0.02 mm to 0.1 mm per hour for HPO_4^{2-}. Perhaps soil water turbulence by temperature gradients and flowing water that isn't adsorbed by all the roots that it passes may help explain the mechanism. Whatever the reason, *low solubilities of soil phosphates are the major problem in getting and keeping soil phosphates available to plants.* A phosphorus cycle is shown in Figure 10-9.

Table 10-7 Examples of the Soil Solution Concentrations of Macronutrients*

Nutrient	Fertile Indiana Alfisol: Total Available (kg/ha-20 cm)	Fertile Indiana Alfisol: Amount in Solution (mg/L)	Range in Average Soils (mg/L)	Needed for 9500 kg of Corn/ha (kg/ha)
NO_3^-	200	60	6–1240	190
NH_4^+	—	—	1.8–36	—
$H_2PO_4^- + HPO_4^{2-}$	100	0.8	0.1–1.9	40
K^+	400	14	3.9–39	195
Ca^{2+}	6000	60	2.0–100	40
Mg^{2+}	1500	40	1.2–60	45
SO_4^{2-}	100	26	9.6–960	22

Source: Combined and calculated data from Stanley A. Barber, *Soil Nutrient Bioavailability,* Wiley-Interscience, New York, 1984, 94–95.
*Values in the table are amounts of the element, not of the ion.

Table 10-8 Percentage of Each Macronutrient Supplied to Corn from a Fertile Alfisol Soil by the Three Supply Mechanisms

Nutrient	Approximate Percentage of Nutrient Supplied: Root Interception	Mass Flow	Diffusion
Nitrogen	1	79	20
Phosphorus	3	5	92
Potassium	2	18	80
Calcium	150	375	0
Magnesium	33	222	0
Sulfur	5	295	0

Source: Calculated percentages from data by Stanley A. Barber, *Soil Nutrient Bioavailability,* Wiley-Interscience, New York, 1984, 96.

10:10 Mineral Phosphorus

Making phosphorus available to plants is critical because (1) the supply of phosphorus in most soils is low, and (2) the phosphates in soils are not readily available for plant use. The total phosphorus in an average soil is approximately 0.05 percent by weight (400 kg/ha to 2000 kg/ha), of which only an infinitesimal part is available to the plant at any one time.

The original natural source of phosphorus is the mineral *apatite,* a calcium phosphate of low solubility and typical formula $Ca_5(PO_4)_3F$. **Apatite (rock phosphate)** and small amounts of other phosphates of iron, aluminum, manganese, and zinc make up the inorganic phosphate minerals. These rock phosphates, when powdered, can be used directly as low-quality fertilizers.

The soluble ion $H_2PO_4^-$ rapidly reacts in soil to form insoluble phosphates, a process loosely termed **phosphate fixation** (precipitation and adsorption). In acid soils the phosphate ions react with soluble iron and aluminum ions to form insoluble phosphates. Also phosphates adsorb to surfaces of insoluble iron, aluminum, and manganese hydrous oxides.

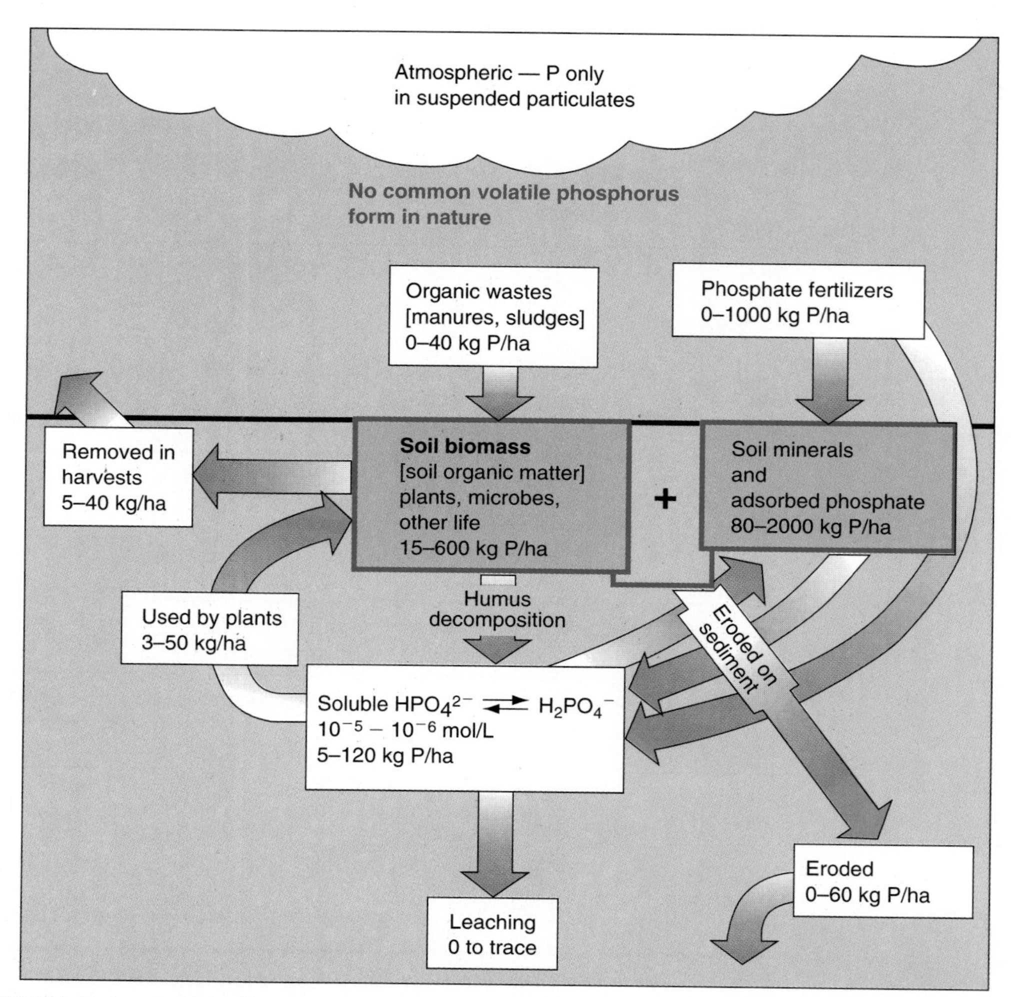

FIGURE 10-9 The phosphorus cycle showing gains, losses, and transformations in the soil. Depositions to form phosphate deposits in ocean bottoms not shown. (Courtesy of Raymond W. Miller, Utah State University.)

In alkaline soils low-solubility calcium triphosphate is formed. Soluble phosphate ions also adsorb on solid calcium carbonate surfaces. Phosphorus is most available at about pH 6.5 for mineral soils and pH 5.5 for organic ones (Figure 10-10). There is no efficient mechanism in the soil to retain $H_2PO_4^-$ or HPO_4^{2-} ions in large quantities as exchangeable anions. Thus, much of the phosphorus used by plants, other than that from applied phosphate fertilizers, is believed to come from organic phosphates released by decomposition of organic matter.

Reactions of Dissolved Phosphates

The description of phosphate reactions from an added superphosphate fertilizer pellet is informative. The monocalcium phosphate (MCP) in the pellet is water-soluble and dissolves readily in soil water. The soluble $H_2PO_4^-$ ion moves by diffusion outward from the concentration area of the dissolving pellet. The solution of phosphate may have a pH of about 0.6 to 1.4 and high concentrations of both P (3.4 molar–5 molar) and Ca (1 molar).

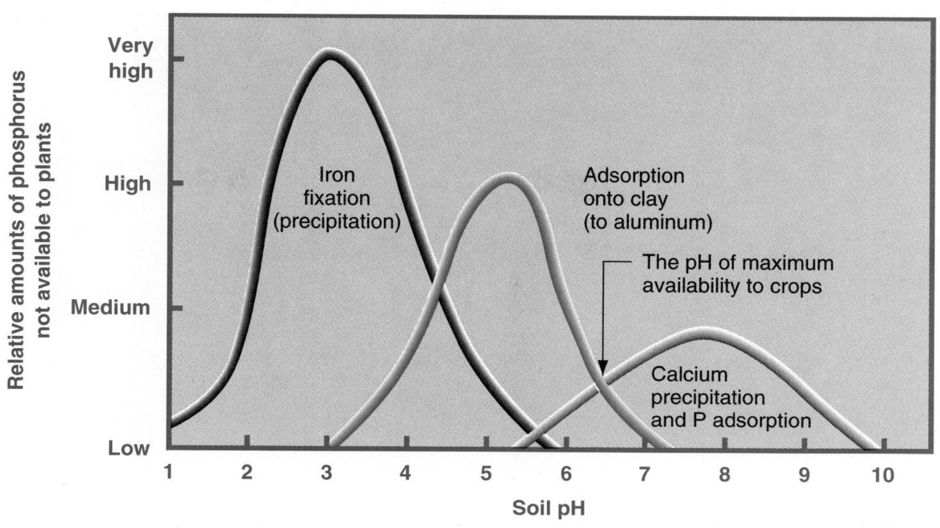

FIGURE 10-10 Soil phosphorus is precipitated and adsorbed to soil minerals, or made less available, by the formation of less-soluble phosphates of iron, aluminum (from clays), and calcium. At a soil pH below 5.5 (acidic), both iron and aluminum precipitate phosphorus. At a pH above 7.0 (basic), calcium precipitates phosphorus. Maximum phosphorus availability is at a pH of 6.5 for mineral soils and about 5.5 for organic soils and Oxisols. Addition of lime (a basic substance) to soils of pH 5.5 or less improves phosphorus availability to crops. (*Source:* George D. Scarseth, *Man and His Earth,* Iowa State University Press, Ames, 1962, 143.)

If the soil is *calcareous* with a high pH, various low-solubility calcium phosphates form as follows:

Phosphate Name	*Formula*	*Solubility in Water*
Monocalcium phosphate (MCP)	$Ca(H_2PO_4)_2 \cdot H_2O$	18 g/L
Dicalcium phosphate dihydrate (DCPD)	$CaHPO_4 \cdot 2H_2O$	0.316 g/L
Tricalcium phosphate (TCP)	$Ca_3(PO_4)_2$	0.02 g/L
(There are many additional phosphates of calcium)		

As the phosphate diffuses outward from the pellet, some of it precipitates, mostly as DCPD. Some of the $H_2PO_4^-$ ions diffuse farther and are adsorbed to solid lime and to iron or to aluminum hydrous oxides.

If the soil is *noncalcareous,* the soluble phosphate forms various insoluble aluminum phosphates, such as taranakites $[H_6K_3Al_5(PO_4)_8{\cdot}18H_2O$ and $H_6(NH_4)_3Al_5(PO_4)_8{\cdot}18H_2O]$ and Al–Fe phosphates $[H_8(Al,Fe)_3(PO_4)_6{\cdot}6H_2O]$, as well as some calcium phosphates.

The area of the fertilizer pellet has three zones: (1) the dissolving pellet; (2) the second zone of concentrated phosphate-causing precipitation; and (3) the outer zone, where phosphate adsorption is dominant.

Phosphorus Buildup in Soils

A Gilat soil in Israel had applications of organic materials and/or mineral fertilizers for 24 years (1961–1984). The additions caused large differences in phosphate reserves (Figure 10-11). The *control* (soil with no phosphorus additions) had enough phosphorus

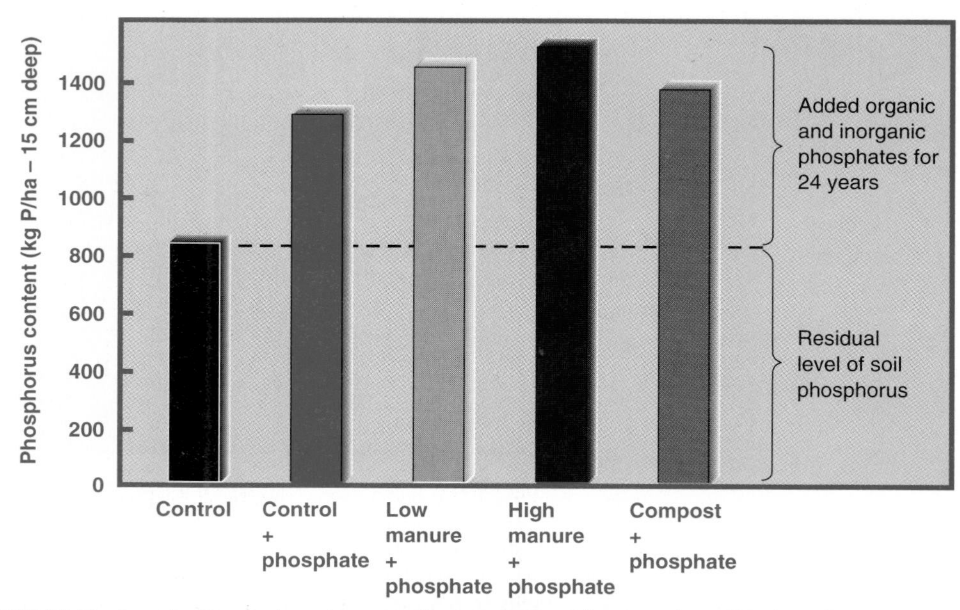

FIGURE 10-11 Levels of soil phosphorus before and after 24 years of application of manures, compost, and inorganic phosphate fertilizers. (*Source:* Unpublished data by Raymond W. Miller and Amos Feigin, 1985.)

for about 320 cropping years if each crop needed 40 kg of phosphorus and half that was left in the roots in a 120-cm (47-in) soil depth. This assumes that all depths to 120 cm could supply the same phosphorus content as in the surface soil. If most of the phosphorus was supplied by the top 30 cm of soil, phosphorus in the nonfertilized soil would be completely exhausted in 80 years. Supplies of phosphorus in soils are not usually enormous, and the low solubilities of phosphate compounds aggravate the problem. Phosphate mined for fertilizers is a resource that has finite limits. Some economists and scientists predict that world phosphorus reserves may be unable to supply affordable amounts even for the next century. The good news is that phosphorus levels in U.S. farmland are higher than they were a generation ago.

Phosphates in Anaerobic Soils

In anaerobic conditions (e.g., paddy rice), phosphates are more soluble than in aerated soils. When paddy rice is flooded, much of the iron becomes soluble (as ferrous iron) in anaerobic conditions and some iron phosphates may be solubilized.[15] Interestingly, the solubilized iron reprecipitates when the soil is drained (and reaerated). This fresh iron precipitate is somewhat disorganized (amorphous) and in one study did not adsorb added phosphate as strongly or as much as did the iron compounds before they solubilized in the anaerobic soil. A drained soil may, within 2 or 3 years, change back to the unwanted high phosphorus adsorption that it had before it was initially flooded.[16]

---

[15] R. A. Khalid, W. H. Patrick, Jr., and F. J. Peterson, "Relationship between Rice Yield and Soil Phosphorus Evaluated under Aerobic and Anaerobic Conditions," *Soil Science and Plant Nutrition* **25** (1979), No. 2, 155–164.

[16] R. N. Sah and D. S. Mikkelsen, "Sorption and Bioavailability of Phosphorus during the Drainage Period of Flooded-Drained Soils," *Plant and Soil* **92** (1986), 265–278.

10:11 Organic Soil Phosphorus

Phosphorus is a part of DNA and RNA and has a unique role in the adenosine triphosphate (ATP) energy storage bond. *Phosphatase* (an enzyme splitting off phosphates from complex organic molecules) in the plant can split off $H_2PO_4^-$ from organic phosphates. Some of the $H_2PO_4^-$ is exuded by the plant roots (along with some phosphatases). The phosphatases (also produced by bacteria and other organisms) continue to hydrolyze organic phosphorus from dead organic residues, making it available for absorption into living plants and microorganisms. Other enzymes active in the decomposition of organic matter release various organic molecules containing phosphates; these phosphate compounds are more soluble than either the complex organic phosphates or most phosphate minerals. Phosphates that move from soils, but which are not attached to erosion sediment, are believed to be mostly these soluble organic phosphates, which may comprise as much as half or more of the soluble soil phosphorus. However, large losses from dissolution of superphosphate have been recorded in Eastern Australia.[17] In South Dakota movement down to 30 cm deep from a heavy surface application of phosphate in the field was explained as movement through large macropores with the gravity drainage during the spring.[18]

It is not known conclusively if plants absorb the organic phosphates directly or whether the phosphatases found in normal soils split off the phosphates before the plants absorb them. Soils high in organic matter (4%–5%) and with conditions favorable to microorganisms (which speed decomposition) have better supplies of organic phosphates for plant uptake than do soils low in organic matter.

The rate of turnover of organic phosphorus is rapid in conditions favorable for microorganisms. Although critical concentrations are not well established, it is believed that when the carbon:organic-P ratio is about 200:1 or narrower, phosphorus is readily released (mineralized) into the soil solution. If ratios are 300:1 or wider (about 0.2% P in organic matter), the microorganisms use most of the phosphorus, immobilizing it into their cells instead of releasing it for plant use.

10:12 Managing Soil Phosphorus

Soil phosphorus availability is low in many sandy low-humus soils. Oxisols and other soils high in sesquioxide clays, iron oxides, and aluminum oxides often have high phosphorus adsorption capacities, rapidly adsorbing added soluble phosphates.

Soil pH, because of its influence on the presence and solubility of calcium, iron, and aluminum, and because of its effect on bacterial growth, influences available phosphorus. Optimum phosphorus availability in mineral soils is believed to be near pH 6.5, although some scientists question this conclusion. For example, on cotton at soil pH between 6.0 and 7.0, 11.2 kg/ha (10 lb/a) of fertilizer phosphorus was as efficient as double that amount at a soil pH of 5.0.

Phosphate deficiencies are usually remedied by application of phosphate fertilizers. Because phosphorus precipitates and adsorbs, phosphates are often placed in a band about 5 cm (2 in) to one or both sides of the seed and 5 cm below it to minimize contact with the soil but to be close to the first young roots. Even with these precautions, only about 20 percent of the phosphorus added is used by a crop during the season the phosphorus was applied.

---

[17]Nick R. Austin, J. Bernard Prendergast, and Matthew D. Collins, "Phosphorus Losses in Irrigation Runoff from Fertilized Pasture," *Journal of Environmental Quality* **25** (1996), 63–68.

[18]E. M. White, "Surface Banding Phosphorus to Increase Movement Into Soils," *Communications in Soil Science and Plant Analysis* **27** (1996), No. 9, 10; 2005–2016.

Table 10-9 Examples of Phosphorus Gains, Losses, and Transformations (kg/ha/yr) for Eight Different Cropping Systems*

Description of Change	Grazed Bluegrass (N.C.)	Corn Grain (Ind.)	Soybean Seeds (Ark.)	Wheat (Kansas)	Irish Potatoes (Maine)	Cotton (Calif.)	Loblolly Pine (Miss.)	Douglas Fir (Wash.)
Additions								
Added fertilizer	24	30	19	13	101	14	—	—
Irrigation, floodwater	—	tr	—	—	—	tr	—	—
Sediments added	tr	—	tr	tr	tr	tr	tr	tr
Removals								
Harvested product	—	15	10	7	10	13	1	2
Leaching loss	—	tr	tr	tr	5	tr	tr	—
Erosion, runoff	tr	3	3	3	tr	1	1	—
Recycling processes								
Uptake from soil	20	22	13	10	16	19	2	6
Manure from grazing	15	—	—	—	—	—	—	—
Plant residues left	—	7	3	3	6	6	1	5
Mineralization from humus	5	7	—	3	—	6	—	—

*A dash means that no measurement was made or does not apply to the system; *tr* means that a trace amount was measured.

Excess phosphorus is mostly retained in soil. These excess additions can cause problems, such as zinc deficiency. In susceptible plants, such as corn, beans, and flax, too much soluble phosphate causes zinc deficiency. Because zinc phosphate should still be soluble enough to supply adequate zinc, the problem may be *inside* the plant or from interactions not clearly understood.

For maximum phosphorus efficiency, the following procedures are recommended:

- Maintain the soil between pH 6.0 and 7.0, if practical.
- Promote as much relatively fresh organic matter in the soil as is economically practical, to release phosphorus as it decomposes.
- Band phosphorus fertilizers for row crops. Broadcast some or all phosphate fertilizer and incorporate it if the crop is not planted in rows (e.g., pastures) or if the soil phosphorus level is very low (roots need some phosphorus throughout the soil for good growth).
- Expect to need more phosphorus on aerobic soils than on anaerobic ones.

Some examples of phosphorus gains, losses, and transformations are shown in Table 10-9. Notice that much less phosphorus is needed than was needed of nitrogen, that almost no phosphorus is lost by leaching, and that relatively little data are given on the mineralization of phosphate (see Figure 10-9, the P cycle). The mineralization data shown suggest that the phosphorus available from mineralization is usually inadequate to meet crop needs.

10:13 Materials Supplying Phosphorus

The phosphorus ores used for fertilizer manufacture in the United States come mostly from Florida, where ancient oceans have left millions of years' accumulation of marine shell organisms. Extensive deposits of apatite, a phosphorus-bearing mineral, exist in the western United States and are used to some extent as a fertilizer.

Table 10-10 Phosphate Fertilizers in Common Use in the United States

Material	*Total P** *(%)*	*Available P** *(%)*
Rock phosphate (mined phosphate minerals)	13–17	0.8–2.2
Superphosphates		
Ordinary (sulfuric-acid treated) (85% of P is water-soluble)	9	9
Triple (phosphoric-acid treated) (most of P is water-soluble)	20–22	20–22
Concentrated (about same as triple)	24	24
Superphosphoric acid (polyphosphoric acid)	30–33	30–33
Wet-process phosphoric acid, crude (used in many solid mixed fertilizers)	13	13
Basic slag (also acts as a lime)	3.5–5.2	(?) 2–3
Bone phosphate (steamed bonemeal)	10–13	(?) 2–3

*To obtain P_2O_5 equivalent, multiply P content by 2.29. *Available phosphorus* is that which is soluble in a dilute ammonium citrate solution.

Phosphate Ores and Deposits

Rock phosphate ore is mined and ground to make the commercial **rock phosphate;** when mixed with sulfuric acid, it is made into **superphosphate,** 8 percent to 9 percent P and 48 percent gypsum (Table 10-10). If rock phosphate is mixed with phosphoric acid, it makes **triple superphosphate,** 20 percent to 22 percent P (40%–45% P_2O_5). (*Note:* Most phosphorus fertilizers are reported as calculated P_2O_5, even though none of it exists as this oxide. This is one of the customs in labeling fertilizers that some people have been trying to change.)

Mixed Nitrogen–Phosphorus Fertilizers

Other phosphorus fertilizer materials are **diammonium phosphate (DAP,** 46%–53% P_2O_5 and 18%–21% N), **monoammonium phosphate (MAP,** 48% P_2O_5 and 11% N), **ammonium phosphate sulfate** (20% P_2O_5 and 16% N), and **basic slag** (approximately 10% citrate-soluble P_2O_5). A fertilizer that is gaining favor is ammonium polyphosphate, which in liquid forms analyzes 10-34-0 and 11-37-0 (Table 10-11). The 10 is 10% total N, 37 is 37% P_2O_5, and 0 is 0% potassium calculated as K_2O. All fertilizers have these three numbers in this order. (See Chapter 12.)

When the price of sulfur (and, therefore, of sulfuric acid) increases, there will be renewed interest in using nitric acid to make several kinds of **nitric phosphates,** the most common of which contains 20 percent P_2O_5 and 20 percent N.

Many of the solid fertilizers used are mixtures (formulations) containing two or three of the elements nitrogen, phosphorus, and potassium (see Table 10-11). Many also contain sulfur. These are usually made from high-analysis liquids such as superphosphoric acid and gaseous ammonia, resulting in fertilizers with high concentrations of several nutrients.

10:14 Soil Potassium

Potassium is an element whose chemical compounds are usually highly soluble, yet its mineral forms—micas and orthoclase feldspar ($KAlSi_3O_8$)—are only very slowly soluble. The *total* amount of potassium found in most soils is sufficient to last several decades, even centuries; yet the low solubility of soil micas and feldspars (the soil minerals that contain potassium) supply only very small amounts of potassium during a growing season. Second, soil humus also supplies very little potassium for plant use. Decomposition of *fresh* plant residues releases

Table 10-11 Multinutrient Fertilizers in Common Use in the United States

Material	*Comments*	*Grade*
Ammonium polyphosphate	Used in liquid fertilizers	11-37-0 and 10-34-0
Ammonium phosphates	Water-soluble, good for fast-growing crops	11-48-0 to 18-46-0 to 21-53-0
Ammonium phosphate nitrate	Ammonium nitrate, anhydrous ammonia, and phosphoric acid; water soluble	30-10-0 27-12-0 22-22-0
Urea-ammonium phosphate	In developmental stage, damaging to germinating seeds	25-35-0 to 34-17-0
Ammoniated superphosphate	Well-known material	4-16-0 to 8-32-0
Nitric phosphates	No established advantage over other materials	12-35-0 to 17-22-0
Ammonium thiosulfate	Used in liquid fertilizers	12-0-0-26S

Source: Tabulated from details in U.S. Jones, *Fertilizers and Soil Fertility,* Reston Publishing, Reston, VA, 1979, 151–166.

whatever potassium the plant absorbed for growth, because the potassium occurs in plants only as a mobile soluble ion, K^+, rather than as an integral part of any specific compound. Potassium is known to affect cell division, the formation of carbohydrates, translocation of sugars, various enzyme actions, the resistance of some plants to certain diseases, cell permeability, and several other functions. More than 60 enzymes are known to require potassium for activation. It is particularly important in plant control of water (regulation of osmosis in the plant). Third, potassium is the most abundant metal cation (often up to 2%–3% of dry weight) in plant cells, but soil humus furnishes very little potassium during decomposition.

Forms of Soil Potassium

Most potassium used by plants in a given season comes from exchangeable K^+ and soluble K^+. In neutral and basic soils, soluble K^+ alone may be adequate to supply modest plant needs. In most soils, particularly acid ones, *exchangeable* K^+ *is the major source of potassium to plants.* The exchangeable K^+ accumulates as the mica and feldspars weather and as potassium in plant residues is released into the soil solution. Although some soils may contain as much as 2 percent total potassium (90,000 kg/ha-30 cm, or about 81,000 lb/a-ft), high-yielding crop plants depend on the *exchangeable* supply, which often is a *small* reservoir of readily available potassium. Exchangeable plus soluble potassium in the top 15 cm may be less than 100 kg/ha to 200 kg/ha in many acidic soils, a level that is inadequate or marginal for plant growth. About 170 kg/ha to 200 kg/ha (50 lb/a to 180 lb/a) is about the minimum amount considered necessary for a good cultivated crop.

Potassium Losses and Gains

Soluble or exchangeable potassium may be taken up in excess amounts by plants; this is called **luxury consumption.** Excess uptake may reduce Mg absorption in the plant; if the excess uptake results from overapplication of fertilizers, it is an expensive waste. Other than being absorbed by plants, soluble K^+ can be (1) immobilized into microbe bodies, (2) lost in leaching waters, and (3) entrapped between layers of illite and similar clays during drying.

The latter process, called **potassium fixation,** is similar to ammonium entrapment between clay layers.

In Table 10-12 potassium losses, gains, and other changes are shown for several crop systems. Notice the relatively high erosion and leaching losses. Although no potassium release is shown for mineralization of humus, the addition of undecomposed organic materials to soil does replace large amounts of potassium (dry, fresh plant material contains 1%–2% K). This potassium in the plant material can be used almost as fast as water flows through the material and then to roots. If the potassium is not used by plants, some of it attaches to cation exchange sites in the humus or on clays.

The potassium cycle is shown as Figure 10-12. It is a simple cycle. Microorganisms are involved directly only in decomposition of fresh plant and animal residues. There is only one ion form of potassium, K^+, and there is no volatile form at temperatures found in ecosystems. Exchangeable K has a dominant role in the soil's potassium fertility. Potassium moves through soil *chromatographically,* as described in Chapter 5.

Supplying Potassium to Plants

Potassium fertilizers are usually very soluble substances. Surprisingly, the two major mineral sources of potassium in soils (micas and feldspars) have very low solubility. However, even the potassium dissolved in the soil solution is not very mobile in soils. As the water containing potassium flows through soil, the K^+ absorbs to cation exchange sites, replacing other cations; this cation exchange slows the movement of K^+ through soil. Some other important characteristics of potassium are the following:

- Potassium is not supplied from decomposing soil humus, except as an exchangeable ion on humus exchange sites. *Fresh plant residues* do contain the 1%–3% potassium common in plants, but that potassium is soluble and is immediately leachable from the dead plant tissue. Manures do have some potassium, but most available potassium is excreted in the urine.

Table 10-12 Examples of Potassium Gains, Losses, and Transformations (in kg/ha/yr) for Eight Different Cropping Systems*

Description of Change	*Grazed Bluegrass (N.C.)*	*Corn Grain (Ind.)*	*Soybean Seeds (Ark.)*	*Wheat (Kansas)*	*Irish Potatoes (Maine)*	*Cotton (Calif.)*	*Loblolly Pine (Miss.)*	*Douglas Fir (Wash.)*
Additions								
Added fertilizer	46	65	37	0	207	0	—	—
Irrigation, floodwater	—	4	—	—	—	50	—	—
Sediments added	4	—	4	2	3	1	4	4
Removals								
Harvested product	—	20	22	6	117	30	7	11
Leaching, loss	—	15	15	5	15	10	2	1
Erosion, runoff	6	10	10	5	tr	10	3	1
Recycling processes								
Uptake from soil	150	111	37	50	177	67	9	14
Manure from grazing	127	—	—	—	—	—	—	—
Plant residues left	—	91	15	44	60	37	2	4
Mineralization from humus	—	—	—	—	—	—	—	—

Source: Data from M. J. Frissel, ed., *Cycling of Mineral Nutrients in Agricultural Ecosystems,* Elsevier Science Publishing, New York, 1978, 202–243.

*A dash means that no measurement was made or the item does not apply to the system.

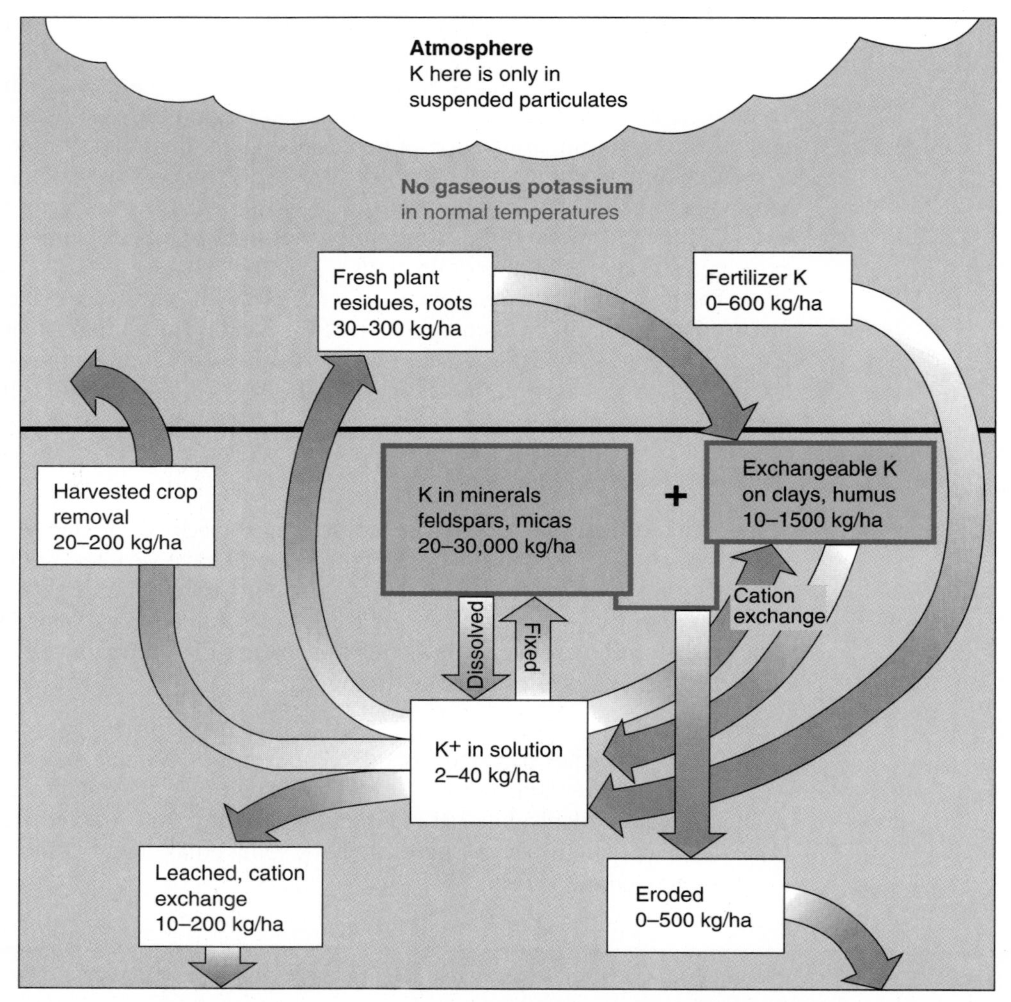

FIGURE 10-12 The potassium cycle, showing transformations, forms, and amounts of potassium in various soils. (Courtesy of Raymond W. Miller, Utah State University.)

- Potassium, because it is a positive ion, moves slowly through the soil. Potassium fertilizers should be placed in the soil where roots have good access to the potassium.
- The plants obtain *practically all of their potassium from soluble and exchangeable forms.* In acidic soils these forms are often in deficient amounts. In more arid regions potassium deficiencies in the soils are likely only in sands and a few other soils where heavy crop demands over the years have depleted exchangeable potassium.
- There is some competition for uptake among potassium, calcium, and magnesium ions. Because all of these ions are cations, the plant's internal charge balance is one cause for this competition.

Even with these guides, few scientists would have expected potassium deficiency in cotton when soil tests suggest there is adequate potassium. Yet potassium deficiency symptoms—bronzing of thick, leathery leaves, with no necrosis, that snap when bent—are found in cotton in the San Joaquin Valley of California. Also, the symptoms appear on *younger, not older,* leaves.

Cotton is a good example of the uniqueness of this condition: The carpel wall of the cotton bolls contains about 4 percent K and accounts for 60 percent of all the potassium accumulated by cotton plants. As the plants develop, the bolls become a *sink* for potassium during peak boll growth. The cotton bolls probably use translocated potassium at the expense of some potassium that the younger leaves would otherwise get.[19] Prune fruits also seem to be a potassium sink, which can cause potassium deficiency in a similar manner as prunes increase in size and mature.

Several factors should be kept in mind about nutrient requirements in general: first, plants may differ greatly in the *amounts* and in the *proportions* needed of various nutrients. Second, each plant usually has a growth stage during which the need for one or more nutrients is much higher than during the rest of the plant's growth cycle. If nutrients are limiting during this critical peak-use period, yields will be reduced. Third, seeming abnormalities in symptoms or response to a nutrient (the deficiency of potassium on younger leaves, for example) often have been logically explained when enough is known. One should expect to have some differences in nutrient needs (timing, amounts, proportions) for each plant species.

Additional examples of plant needs are shown in Table 10-1. Notice the heavy potassium needs of potatoes, sugarcane, and bananas. Leaching losses occur; potassium does move chromatographically through soil as large amounts of water containing other cations percolate. Notice the lack of any volatilization product and *no potassium* mineralized from humus.

Cost per unit of potassium is the dominant criterion for choosing a potassium fertilizer. The fertilizer KCl is usually the cheapest, but the more expensive sulfate or nitrate forms are sometimes chosen in preference to KCl. The sulfate, nitrate, or phosphate supplies that nutrient plus potassium; also, some plants, such as avocados and potatoes, may be injured by large amounts of chloride. Excess chloride in potatoes lowers starch and makes them poor for french fries. Damage can also occur if chlorides in foliage spray or irrigation water are left on the leaves. Tobacco burns better when the high potassium need is partly supplied as the sulfate rather than all as the chloride.

The cost of allowing potassium deficiency to develop in prunes in California was given in dollar values as follows:[20]

Extent of K Deficiency	*Value per Hectare*
Trees with no K deficiency	\$2470–\$6175
Trees with severe K deficiency	Less than \$617
Slight K deficiency symptoms	\$2643 average
No K deficiency symptoms	\$4383 average
Net value if fertilized with K*	\$1541

*Cost for K fertilizer/year is about \$200.

Generally, even hidden hunger (no visual symptoms) will reduce production and cost the producer in lost yields.

Managing Soil Potassium

Sugar-producing plants, potatoes, and tobacco are all heavy users of potassium (Figures 10-13 and 10-14). Most plants have the highest potassium requirements during vegetative

[19] Bill L. Weir, Thomas A. Kerby, Bruce A. Roberts, Duane S. Mikkelsen, and Richard H. Garber, "Potassium Deficiency Syndrome of Cotton," *California Agriculture* (Sept.–Oct. 1986), 13–14.

[20] William H. Olson, Kiyoto Uriu, Robert M. Carlson, William H. Krueger, and James Pearson, "Correcting Potassium Deficiency in Prune Trees Is Profitable," *California Agriculture* **41** (May–June 1987), 20–21.

FIGURE 10-13 Tobacco in the southeastern United States has fairly high potassium requirements because the whole plant top is harvested and high rainfall often leaches potassium from the soil. In some instances such large amounts of potassium are needed that some of it is added as potassium sulfate, rather than all as potassium chloride, the cheaper source. High chloride concentrations in the tobacco cause slow curing and poor burning qualities. Other potassium salts (K_2SO_4 and KNO_3) could be used for part of the potassium requirement to avoid this problem. (*Source:* USDA—Soil Conservation Service.)

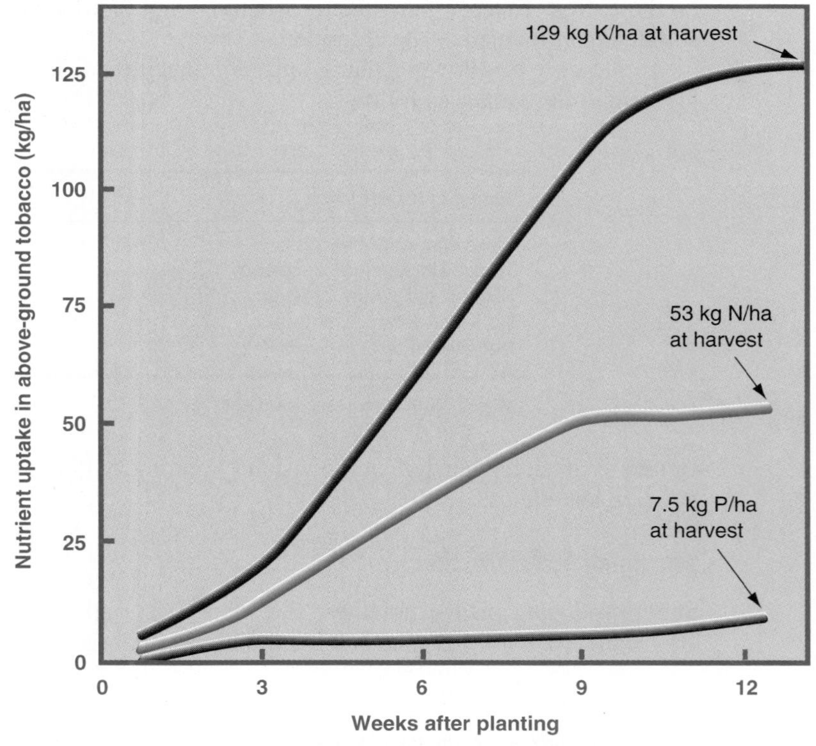

FIGURE 10-14 Comparison of the potassium, nitrogen, and phosphorus requirements of tobacco. (*Source:* Recalculated and redrawn from C. D. Raper, Jr., and C. B. McCants, *Tobacco Science* **10** [1966], 109.)

growth. As with phosphorus, it is important to have adequate potassium available near young seedlings, but not in amounts large enough to cause salt damage.

Little can be done to alter the available potassium in deficient soils other than adding potassium fertilizers. The soil test for potassium is simple and quite good; it should be used to predict deficiencies. Deficiencies should be expected in soils that are low in micas (the more soluble mineral source), soils that are low in clay (few exchange sites), and soils of pH 4 to 6 (leached by high rainfall, which is common to areas of acidic soils).

Management of soil potassium emphasizes (1) maximizing efficient use of added potassium, (2) minimizing luxury consumption, and (3) trying to make maximum use of natural potassium sources. The following practices are most useful:

- Avoid heavy applications of potassium fertilizers; use smaller split applications. Heavy additions allow unnecessary and expensive luxury consumption. High chloride levels from KCl, if used, may be toxic to some plants. Split applications are especially important on sandy soils where leaching losses are likely. However, low mobility of the potassium in soils is a problem in applying split applications on the soil surface.
- Maintain soil pH at about 6 to 6.5 with lime to reduce potassium losses from leaching (more calcium rather than potassium is leached).
- Returning crop residues and manures adds large amounts of potassium, some of which remains in the soil as exchangeable forms or, in less leached soils, as soluble potassium. Because of the large amounts of potassium in grasses, the removal of lawn clippings in marginal potassium soils can finally cause potassium deficiency unless fertilizer-K is added.

Materials Supplying Potassium

Potassium chloride (KCl), sometimes called *muriate of potash,* is the principal fertilizer material supplying potassium; second in importance is **potassium sulfate (K_2SO_4).** **Potassium–magnesium sulfate** is increasing in significance because it supplies the additional nutrient magnesium. **Potassium nitrate (KNO_3)** is also an excellent fertilizer. Muriate of potash is usually 95 percent KCl, equivalent to 60 percent K_2O. In Saskatchewan, Canada, the largest known potash reserves in the world are now being mined; potassium is also mined in New Mexico, Utah, and California.

Other materials less popular, but *not less suitable,* are **potassium nitrate** and **potassium polyphosphates.** Table 10-13 lists common potassium fertilizer materials.

Table 10-13 Common Potassium Fertilizers Given in Order of Percentage Use in the United States (Beginning with the Most Used)

Material	*K_2O (%)*	*Other Nutrients (%)*
Muriate of potash (KCl)	50–60	Unneeded Cl—may be toxic
Sulfate of potash (K_2SO_4)	45–50	18% sulfur (S), as sulfate
Potassium-magnesium sulfate	18–22	18%–23% sulfur (S), 11% magnesium (Mg)
Potassium nitrate	37–44	13% nitrogen (N), as nitrate
Potassium polyphosphate	20–44	25%–50% phosphorus (P_2O_5)

There are two ways of avoiding extreme fixation of the phosphorus in soluble solid fertilizers. Granulation provides the first method and banding the second.

—R. L. Cook and B. G. Ellis

Nature goes her own way and all that to us seems an exception is really according to order.

—Johann von Goethe

Questions

1. Draw a nitrogen cycle including these boxes: N_2-fixation, mineralization, nitrification, denitrification, and leaching.
2. About how much nitrogen is released during decomposition of soil humus each season?
3. (a) How fast is nitrification: completed in hours, or days, or weeks? (b) Explain why nitrification rates differ in different environments.
4. (a) Is leaching loss of nitrogen a large loss, typically? (b) Under which conditions will the losses be large? When will the losses be small?
5. Tabulate the necessary conditions for (a) ammonia volatilization and (b) denitrification losses of nitrogen.
6. Briefly tabulate some typical *N additions* and *N losses* to show how a soil's nitrogen balance changes. (Remember, N losses include removal in crops.)
7. (a) Why are nitrogen fertilizers only 40%–70% efficient? (b) What happens to that part not used?
8. (a) What are the purposes for using controlled-release fertilizers? (b) Indicate how several of these fertilizers work. (c) Why are these materials not commonly used?
9. Briefly discuss the problem of adequate phosphate uptake by plants as influenced by (a) solubility, (b) mass flow, (c) mobility in soil, and (d) cold temperatures.
10. Explain why plant efficiency of phosphorus fertilizers is only about 10%–30%, much lower than efficiency of nitrogen fertilizers.
11. (a) How much phosphorus is lost by leaching? (b) How much phosphorus is lost by erosion?
12. Draw the P cycle and emphasize some of the major differences between the P cycle and the N cycle.
13. (a) What sources of potassium do plants use during a growing season? (b) What plants have high potassium requirements? (c) Give the ionic form of potassium in solution.
14. Explain why humid areas are likely to have inadequate available potassium, whereas arid regions are likely to have adequate available potassium.
15. (a) Is there a volatile potassium form? (b) Is potassium released during humus decomposition? Explain. (c) What does *chromatographic* movement of K^+ during leaching mean?

Calcium, Magnesium, Sulfur, and Micronutrients

Plant nutrition is of unique importance in the realm of life on Earth and in the affairs of man.

—E. Epstein

He who sees things grow from the beginning will have the best view of them.

—Aristotle

11:1 Preview and Important Facts

PREVIEW

Calcium (Ca^{2+}) and magnesium (Mg^{2+}) have many chemical similarities in soils. Both have a 2+ charge, both are in the alkaline earth metal group of elements in the chemical periodic table, and both are found together in many minerals. Usually both elements occur in fairly large, soluble amounts in soils and are supplied to plants largely by **mass flow.** Calcium occupies more of the cation exchange sites than does any cation, except when $Al(OH)_2^{+}$ and Al^{3+} exceed it in strongly acidic soils. Plants contain about 0.4 percent to 2.5 percent calcium. A deficiency of calcium causes developing buds to deform or die (for example, corn leaves do not uncurl). In plants and in the soil, calcium is not very mobile.

Magnesium reacts similarly to calcium in soils. Plants contain only about 0.1 percent to 0.4 percent Mg, and about one-fifth of that becomes part of chlorophyll molecules. Although not very mobile in soils, in plants magnesium is quite mobile. One visual deficiency symptom of magnesium is chlorotic tissue between veins on older leaves as Mg^{2+} is moved to younger tissue in the plant. Magnesium deficiencies are not widespread but can be common on acidic sands and leached soils; deficiencies can be promoted by cool, wet, and cloudy weather.

Sulfur, in contrast to calcium and magnesium, is supplied from decomposing organic matter and from several moderately soluble minerals, but not generally from the anion exchange capacity. In many ways sulfur is similar to nitrogen. Sulfate (SO_4^{2-}) is the form used by plants; it is soluble in water and easily leached, much like nitrate. Sulfur is part of three amino acids that are essential to life; nitrogen is part of all amino acids. Deficiencies of sulfur slow the growth of plants and produce an overall yellowing of the younger leaves.

Seven micronutrients are essential to plants. Boron and molybdenum are anions and are *least available in strongly acidic soils* because their soluble forms are leached out. In contrast, copper, iron, manganese, and zinc are cations and are *least available in alkaline (basic) soils,* mostly because of the low solubility of the hydroxides and carbonates they form. Chlorine, the seventh micronutrient, is the anion chloride and is seldom deficient in soils.

IMPORTANT FACTS TO KNOW

1. The climatic areas most likely to have either adequate or deficient amounts of each of the following nutrients in soils: calcium, magnesium, sulfur, boron, iron, and zinc
2. The unique soil conditions and plant characteristics that result sometimes in calcium deficiency in peanuts
3. The relation of *grass tetany* and magnesium
4. The materials added to soils to supply calcium and/or magnesium if lime is not needed
5. The general soil chemistry (adsorption, mobility in soils, solubility) of the ion forms of calcium, magnesium, and sulfur
6. The micronutrient forms in soil solution, the method of applying them, and why application method may be more important than the material cost of the fertilizer material used
7. The origin, action, and value of siderophores and similar natural substances
8. The soil chemistry similarities between nitrogen and sulfur
9. Why chelates, both natural and manufactured, are widely involved in mobilizing micronutrient metals

11:2 Soil Calcium

Calcium occurs in soils in many minerals and is more plentiful in soils than are other plant nutrients. Plagioclase feldspar, several of the dark minerals, limestone, dolomite, and gypsum are all sources of calcium. Because of the wide variety of calcium minerals, calcium deficiency is rare, but deficiencies can occur in sands (mostly quartz minerals) and in acidic leached soils where lime has not been added.

Mobility of Calcium in Soils

Calcium is taken up by plants as $\mathbf{Ca^{2+}}$ from both soluble and exchangeable Ca^{2+} sources. Although calcium is not very mobile in soil because it is a strongly adsorbed cation, large amounts can be leached in wet climates simply because of the large *total amounts of calcium* in soils. The calcium will move by chromatographic flow down through soils in percolating water. Calcium dominates the cation exchange capacity because of the large amounts of calcium in the soil solution. Mass flow usually supplies adequate calcium to roots for uptake, even though Ca^{2+} is absorbed only through the root *tips.* Sometimes, for example, to supply the high calcium need for peanut growth, inadequate calcium is absorbed by roots unless the soluble calcium level is quite high (Detail 11-1).

Plant Needs for Calcium

Calcium is needed in cells that are dividing (buds, new growth tips) and forms calcium pectate (a cementing structure) in the middle lamella of the plate between dividing cells. Also, calcium is needed for physical integrity and normal functioning of cell membranes. A defi-

Detail 11-1 Peanuts Have Unusual Calcium Needs

The need for calcium (Ca) in peanuts is exceptionally high; calcium uptake by the peanut plant is a unique phenomenon. Calcium is often the essential element most commonly deficient for peanuts grown in noncalcareous soils. With inadequate Ca the peanut will have blackened plumules, high incidences of pod rot, and unfilled pods; seed peanuts will have much poorer germination.

The *uniqueness* of Ca uptake is that peanut fruit (the pod and nut) absorbs nutrients *directly from the soil solution around them* rather than getting the nutrients from those absorbed by other roots and translocated to the nut. So, if the Ca is not in the solution or exchangeable sites near the peanut fruit, the peanut will not get enough.

Peanuts are often grown in sandy acidic soil (low Ca) and in droughty soils (peanuts tolerate drought), which will decrease Ca solubility and mobility in soil solution. Soil-test Ca is the best way to estimate needs. Getting and maintaining adequate soluble Ca in the top 15 cm to 20 cm (6 in to 8 in) is particularly difficult in sandy soils, which are so easily leached. The 60-day period from bloom to maturity is especially critical. Small-seeded peanuts got enough Ca when the Mehlich-1 soil test is 200 parts per million; larger-seeded Virginia-type peanuts needed a Mehlich-1 test value of about 525 ppm Ca. (See Chapter 12 for Mehlich test information.)

Materials added to keep a high soluble Ca can include the following:

- **Calcium chloride ($CaCl_2$),** applied at bloom, is expensive and easily leached.
- **Gypsum ($CaSO_4 \cdot 2H_2O$),** applied at bloom, is less soluble and expensive but doesn't change soil pH and is usually adequate.
- **By-products** and **phosphogypsum,** where available, added at bloom
- **Pre-plant aglime** incorporated to shallow depths of 5 cm to 12 cm (may raise pH more than wanted)

Usually, unless lime recommended for pH adjustment is enough for the peanut's Ca demand, the material of choice is gypsum.

Source: Gary J. Gascho, "Aglime: A Low-Cost Alternative Source of Calcium for Peanuts," *Better Crops* **80** (1996), No. 1, 10–13.

ciency of calcium results in deformation of new leaves and causes necrotic tissue of new growth or even death of buds. Plants use less of calcium (1.0%–2.5% of plant dry weight) than of potassium (1.5%–4%) but more than of magnesium (0.08%–0.4%).

Because greenhouse media often are soilless and do not provide mineral sources of calcium, adequate calcium and magnesium must be added either as needed lime or as soluble fertilizer. Calcium deficiency is common in greenhouse-grown poinsettias, and soluble calcium nitrate is often added.[1] Gypsum is not soluble enough, so calcium nitrate is used even though it is relatively expensive.

Calcium Fertilizers

Calcium fertilizers are few because most soils with low calcium are acidic soils; when these soils are limed to adjust soil pH and/or to eliminate toxic aluminum and manganese, the lime is mostly calcium carbonate. If pH adjustment is not wanted or needed but calcium still needs to be added, gypsum ($CaSO_4 \cdot 2H_2O$) is most often selected. Gypsum is commonly added to

[1] Allen D. Owings, "Controlling Calcium," *Greenhouse Grower* **14** (1996), No. 8; 29, 32.

peanuts to supply enough calcium but yet to avoid raising the pH, which might cause other problems. In Georgia gypsum increased alfalfa yield 25 percent. However, because alfalfa is not very tolerant of toxic aluminum, addition of calcium carbonate (lime) increased yields nearly 50 percent.

11:3 Soil Magnesium

The magnesium minerals in soils are moderately soluble and commonly supply enough magnesium in soils for plant use. The **Mg^{2+}** ion reacts similarly to the calcium ion but is smaller and forms secondary minerals (sulfates and carbonates) that are more soluble than those of calcium.

Mobility of Magnesium in Soils

Sufficient Mg^{2+} is supplied to most plants from soluble and exchangeable forms. Its strength of adsorption to the cation exchange sites is strong, similar to that of Ca^{2+}. The magnesium ion migrates chromatographically through soils just as does calcium ion, but the lower total amounts in soil result in lower total leaching losses.

Plant Needs for Magnesium

Adequate magnesium is supplied to plant roots by **mass flow.** About one-fifth of the absorbed magnesium is used in the chlorophyll molecules of the plant. Magnesium is also essential to stabilize ribosome structure and is an activator for numerous critical enzymes. In plants Mg^{2+} reacts more like K^+ than like Ca^{2+}; it is quite mobile. A deficiency of magnesium causes interveinal chlorosis of older leaves because magnesium is mobilized from older leaves and transported to younger tissues. Pastures, corn, potatoes, oil palm, cotton, citrus, tobacco, and sugar beets respond to added magnesium when other crops might be growing well.

In some areas **grass tetany** (hypomagnesemia) is a serious disease of domestic animals that is caused by low levels of magnesium in their blood.[2] Heavy applications of potassium and/or ammonium fertilizers on pastures reduce plant uptake of magnesium and can cause grass tetany. The high concentration of potassium and ammonium in poultry litter can also cause grass tetany if large applications of the litter are put on pastures. Lower rates or split applications of these organic wastes are recommended in such situations.

Magnesium Fertilizers

Magnesium deficiency, as for calcium deficiency, is usually corrected by adding lime to the acidic soil; however, some limes have little magnesium, so it is critical to add **dolomitic lime** (Ca plus Mg) if a deficiency of magnesium exists. Several other moderately soluble magnesium salts (minerals) are available. The most commonly used Mg fertilizers are **potassium magnesium sulfate** and **magnesium sulfate** (epsom salts).

---

[2]S. R. Wilkinson and J. A. Stuedemann, "Tetany Hazard of Grass as Affected by Fertilization with Nitrogen, Potassium, or Poultry Litter and Methods of Grass Tetany Prevention," in *Grass Tetany,* American Society of Agronomy, Crop Science Society of America, and Soil Science Society of America, Madison, WI, 1979, 93–121.

Extended captions and credits to photographs are in Appendix B.

Plate 1 Lost natural resources, timber and soils; soils are more critical. (R. W. Miller)

Plate 2 Land is critical to some. Terraces in the Himalayas, Northern India. (R. W. Miller)

Plate 3 Few resources or many people keep living standards low. (R. W. Miller)

Plate 4 Poverty limits equipment and farm inputs available for use. (K. R. Allred)

Plate 5 Pegmatite; slow cooling forms large crystals; granite composition. (R. W. Miller)

Plate 6 Prismatic structure in a natric horizon in Montana. (A. R. Southard)

Extended captions and credits to photographs are in Appendix B.

Plate 7 Placic horizons (orange) cemented by iron and manganese, Ecuador. (S. W. Buol)

Plate 8 Orange mottles in deeper layers of an Oxisol in Australia. (A. R. Southard)

Plate 9 Variable ease of weathering makes colorful canyonlands, Utah. (R. W. Miller)

Plate 10 Tops of coarse columnar structure, in a Natrustalf, Australia. (A. R. Southard)

Plate 11 Erosion of sandstones formed Monument Valley, Utah. (A. R. Southard)

Plate 12 Chunks of carbonate-cemented hard-pan from canal construction. (R. W. Miller)

Extended captions and credits to photographs are in Appendix B.

Plate 13 A thick **E** horizon over a natric horizon in Montana. (A. R. Southard)

Plate 14 Soil parent material can change in just a few feet sometimes. (A. R. Southard)

Plate 15 Oxisolic soil in Australia (called a lateritic podzol) with a thick, leached **E** horizon. (A. R. Southard)

Plate 16 Molokai Oxisol growing sugarcane on Oahu, Hawaii. (A. R. Southard)

Extended captions and credits to photographs are in Appendix B.

Plate 17 Unstable slopes cause many kinds of damage. (Bureau of Land Management)

Plate 18 Cold in the Altiplano, Bolivia, makes food production a struggle. (R. W. Miller)

Plate 19 Montmorillonite clays swell and shrink forming cracks, India. (R. W. Miller)

Plate 20 Closeup of very coarse prisms in a soil in Montana. (A. R. Southard)

Plate 21 Hard basalt over volcanic cinders in Cape Verdes Islands. (A. R. Southard)

Extended captions and credits to photographs are in Appendix B.

Plate 22 Many years of liming needed for root development; acid Oxisol, Brazil. (S. W. Buol)

Plate 23 Ironstone in The Gambia, Africa, erosion along the coast. (R. W. Miller)

Plate 24 Manure patties made for fuel rather than as a fertilizer, India. (R. W. Miller)

Plate 25 Kaolinitic clayey soils (Guatemala) is used to make tiles and bricks. (R. W. Miller)

Plate 26 Burned sugarcane still has large amounts of plant residues. (R. W. Miller)

Extended captions and credits to photographs are in Appendix B.

Plate 27 Ganat maintenance keeps water available in arid Iran. (B. Anderson)

Plate 28 Cashew trees in droughty Africa do better with adequate water. (R. W. Miller)

Plate 29 Mollisol (calcareous sandstone) next to red Oxisol (sediments). (S. W. Buol)

Plate 30 Ironstone cap over folded sediments in New South Wales. (A. R. Southard)

Plate 31 Hardened titanium oxide crust on Lanai, HI. Few plants grow. (A. R. Southard)

Plate 32 Spodic Quartzipsamment near Perth, Australia. (A. R. Southard)

Extended captions and credits to photographs are in Appendix B.

Plate 33 Iron deficiency in peaches in Utah. **Top,** Sometimes only one tree of many around it will be affected. **Bottom,** Closeup of interveinal chlorosis. (R. L. Smith)

Plate 34 Iron deficiency in one tree of the cherry orchard, Utah. (D. W. James)

Plate 35 Potassium deficiency showing varied visual symptoms. **Top,** Clover in Pullman; **Center,** Alfalfa in Utah; **Bottom,** Potato in Columbia Basin, WA. (Top, R. W. Miller; Center and bottom, D. W. James)

Extended captions and credits to photographs are in Appendix B.

Plate 36 Phosphorus-deficient corn, Iowa. Purpling color. (R. L. Smith)

Plate 37 Manganese toxicity (*crinkle leaf*) in cotton. (Potash and Phosphate Institute, PPI)

Plate 38 Zn deficiency in corn, Utah, has interveinal chlorosis. (D. W. James)

Plate 39 Manganese deficiency in cotton. (Potash and Phosphate Institute, PPI)

Plate 40 Boron toxicity on potato. Note the cupped leaves and edge necrosis. (D. W. James)

Plate 41 Typical *greening* from nitrogen on grass just where applied. (K. R. Allred)

11:4 Soil Sulfur

Sulfur, as a constituent in three of the nearly two dozen amino acids, is an essential part of proteins. It is found in several vitamins and in oils of plants in the mustard and onion families. In protein materials sulfur content should be about 6 percent to 8 percent as high as nitrogen content, although it is 10 percent to 15 percent of nitrogen values in most soil humus.

For many years sulfur was neglected by researchers. With the common additions of sulfates as ammonium, potassium, and regular superphosphate fertilizers, it was unintentionally added. During those times few crops exhibited obvious sulfur deficiency symptoms.

Several factors have increased the need for sulfur fertilizers: (1) the lower amounts of sulfate added incidentally with other nutrients; (2) the lower pollution from sulfur oxides into air, later brought down in precipitation; and (3) higher plant yields, which put greater nutrient demands on the soils.

Sources of Sulfur

The availability of sulfur to plants is hard to predict because, like nitrogen, major portions may come from soil organic matter. Sulfur from soil organic matter depends on microbial action and its dependency on climate. Soils contain many sources of sulfur. The mineral *pyrite* (FeS_2, fool's gold) is common to most soils and oxidizes to sulfuric acid and ferric oxide. Moderately soluble gypsum is found in many arid soils; even quite soluble sodium sulfates are found in some arid soils.

Rainfall dissolves the sulfur oxides evolved during the burning of plant-derived fuels (wood, coal, and oil), from range and forest fires, and from the roasting of ores in smelters. Rainwater combined with sulfur oxides produces sulfuric acid. This acidic rainfall is corrosive enough to damage metals and other building materials. It can, through a sequence of reactions, also cause the death of fish in bodies of water where large amounts of the acid fall. The sulfate ion is relatively soluble and generally is leached downward by water. In arid soils sulfate often precipitates as moderately soluble gypsum ($CaSO_4 \cdot 2H_2O$). The sulfur cycle is shown in Figure 11-1.

Sulfur-bearing pesticides and fertilizers—such as ammonium sulfate, superphosphate, and potassium sulfate—are agricultural chemicals that add sulfur to soils. The incidental addition of sulfur to soil in agricultural chemicals, particularly in fertilizers, is the major reason why sulfur deficiencies are not more common.

Characteristics of Soil Sulfur

Sulfur and nitrogen share many features in common. First, in leached soils decomposition of organic matter can release major portions of sulfur. Mineralization supplies a large fraction of the plants' sulfur needs. Second, both nitrogen and sulfur have several chemical forms under different soil conditions. The following table itemizes some of these comparisons.

Item	*Sulfur*	*Nitrogen*
Oxidized negative ion	SO_4^{2-}	NO_3^-
Intermediate oxidations	S, SO_3^{2-}	N_2, NO_2^-, others
The most reduced ion form	S^{2-}	NH_4^+
Amino acid form	—C—SH	—C—NH_2
The most leachable form	SO_4^{2-}	NO_3^-
Gaseous forms	CH_3SCH_3, H_2S, SO_2, SO_3	NH_3, N_2O, NO, N_2, NO_2

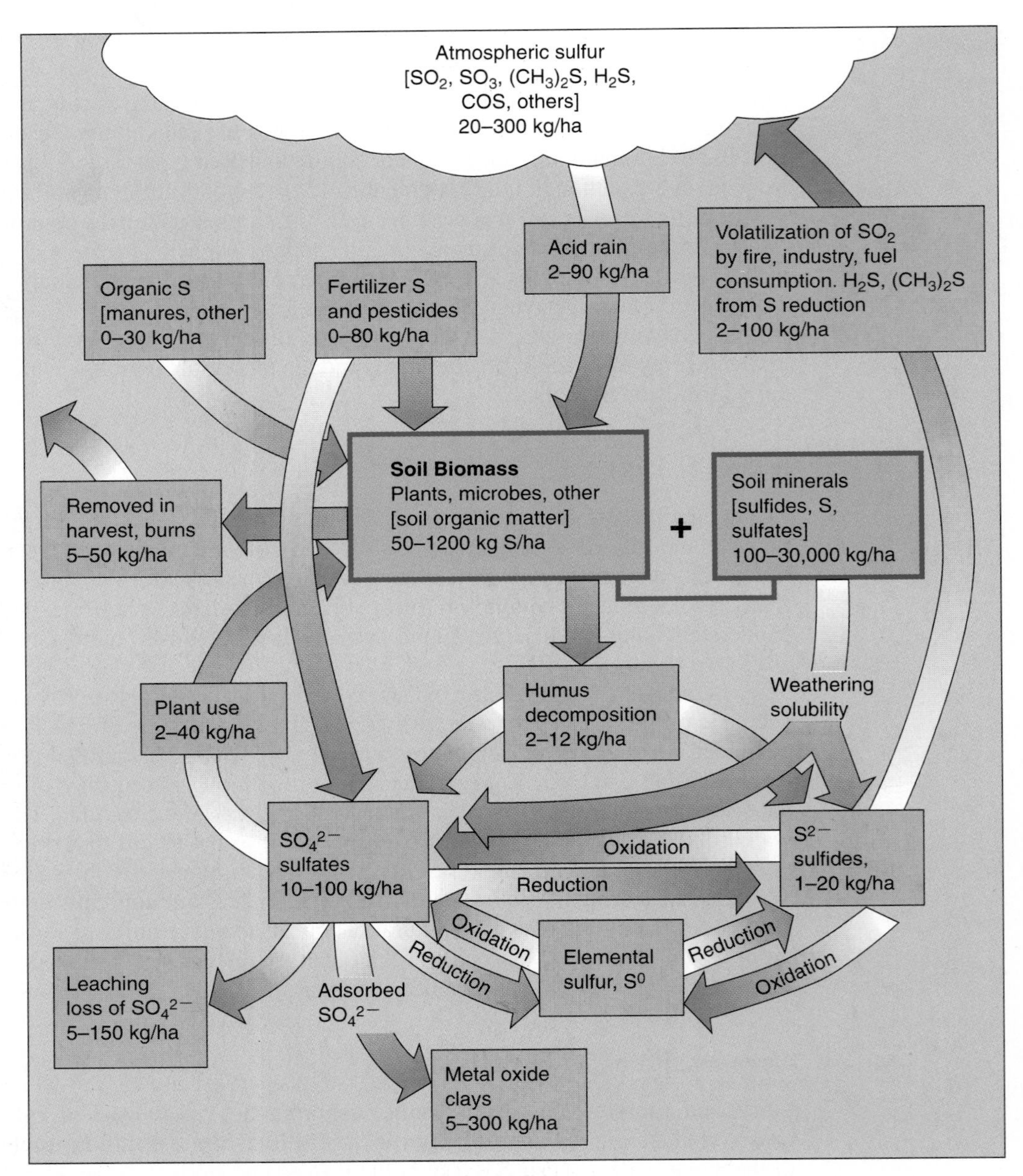

FIGURE 11-1 The sulfur cycle, showing transformations, forms, and amounts of sulfur in various soils. (Courtesy of Raymond W. Miller, Utah State University.)

Although sulfur can be supplied from decomposing soil humus, the various inorganic sulfur minerals may also supply considerable sulfur to plants. Because of the inorganic forms, available sulfur is less dependent than is nitrogen upon microbial action.

The C:S ratio operates much as the C:N ratio. If residues low in sulfur are added, some soil sulfate will be immobilized into microbes and humus residues. The estimated ratio thresholds are as follows:

C:S ratio in the residue added to soil		
<200:1 (sulfate released to soil solution)	200:1–400:1 (neither gain nor loss of sulfate to solution)	>400:1 (immobilization of soil sulfate into humus)

Sulfur-oxidizing bacteria, mostly the genera *Thiobacillus,* oxidize elemental sulfur (S^0) or the sulfide (S^{2-}) plus iron pyrite forms (FeS and FeS_2) to sulfuric acid (H_2SO_4). From pyrites the reactions can occur both by purely chemical and by several biological pathways to form sulfuric acid. In anaerobic conditions microbes can reduce sulfate to sulfide, where it precipitates or causes toxicity to plants as H_2S gas (*Akioki* disease in some paddy rice). When aeration occurs or when elemental sulfur is added, oxidation of the sulfides or sulfur to sulfate results in acidification of the soil. (Detail 11-2).

Managing Soil Sulfur

Increasing restrictions to reduce air pollution and greater use of higher-analysis (purer) fertilizers will reduce incidental sulfur additions. The result will be an increased incidence of sulfur deficiency.

The factors that affect nitrogen release from soil humus also affect sulfur release. Sulfate, as an anion, is easily leached. In arid climates where drainage water flows through high-sulfur soils, the irrigation water often carries adequate sulfate for plant needs.

Soils most likely to have sulfur deficiency are those that are sandy, low in organic matter, nonirrigated (no added sulfate in water), well leached, and far removed from highly industrialized areas (less airborne sulfur). Crops that have high sulfur requirements (corn, sorghum, peanuts, tobacco, cotton) and are fertilized with nonsulfur-containing materials might also have sulfur deficiency. Sulfur deficiencies are observed in several rice-producing areas where a shift from sulfur-containing fertilizers (e.g., ammonium sulfate) to high-analysis nitrogen (urea) has caused sulfur deficiency. Without well-established soil tests for sulfur deficiency, it would seem a good precaution to use some sulfate-containing fertilizers periodically on acid, sandy soils low in humus.

Sulfur Fertilizers or Amendments

Common ways to add sulfur are to deliberately select *ammonium sulfate* or *potassium sulfate* fertilizers if you expect sulfur to be low. *Ammonium thiosulfate* [$(NH_4)_2S_2O_3$] is a popular liquid fertilizer. It is noncorrosive and has good storage features. In soil it forms ammonium sulfate and elemental sulfur. *Ammonium polysulfide* (NH_4S_x, where *x* may take on various integers) is used as both a fertilizer and to reclaim high-pH soils in arid regions. **Elemental sulfur** suspensions (S^0) and gypsum ($CaSO_4 \cdot 2H_2O$) are also used as sulfur amendments. One source[3] lists twenty-seven different sulfur carriers that are used as fertilizers or amendments.

11:5 Soil Micronutrients

The micronutrients are boron, iron, manganese, zinc, copper, chlorine, and molybdenum. These elements are essential to plant growth, but they are used only in minute quantities in contrast to the macronutrients, which comprise a proportionally larger percentage of plant

[3] S. L. Tisdale, W. L. Nelson, J. D. Beaton, and John L. Havlin, *Soil Fertility and Fertilizers,* 5th ed., Macmillan, New York, 1993, 285.

Detail 11-2 Acidic Sulfate Soils

Acidic sulfate soils are the nastiest soils in the world.[a] They generate sulfuric acid and some lower their pH even to pH 2. Acid and aluminum can leak into streams and kill vegetation and aquatic life.

Soil with enough sulfides (FeS_2, etc.) to become strongly acidic when drained and aerated enough for cultivation are termed **acidic sulfate soils.** The Dutch refer to them as *Katteklei* (**cat clays**). When allowed to develop acidity, these soils are usually more acidic than pH 4. Before drainage such soils may have normal soil pHs. Lands inundated with waters that contain sulf*ates,* particularly salt (ocean) waters, accumulate sulfur compounds, which in poorly aerated soil are bacterially reduced to sulf*ides.* Such soils are not usually very acidic when first drained of water.

When the soil is drained and then aerated, the sulfide is oxidized to sulfate by a combination of chemical and bacterial actions, forming sulfuric acid. The extent of acid development depends on the amount of sulfide in the soil and the conditions and time of oxidation. If pyrite (iron sulfide, FeS_2) is present, the oxidized iron accentuates the acidity, but not as much as aluminum in normal acid soils because the iron oxides are less soluble than aluminum oxides.

The slow oxidation of mineral sulfides in soils is nonbiological until soil pH reaches an acidity of pH 4. Below pH 4 the bacteria *Thiobacillus ferrooxidans* are the most active oxidizers and the acidity builds up rapidly.[b] The chemical reactions are as follows:[c]

Nonbiological

$$\underset{\text{(pyrite)}}{2FeS_2} + 2H_2O + 7O_2 \longrightarrow \underset{\text{(ferrous sulfate)}}{2FeSO_4} + \underset{\text{(sulfuric acid)}}{2H_2SO_4}$$

Accelerated by bacteria (Thiobacillus ferrooxidans)

$$\underset{\text{(ferrous sulfate)}}{4FeSO_4} + O_2 + 2H_2SO_4 \longrightarrow \underset{\text{(ferric sulfate)}}{2Fe_2(SO_4)_3} + 2H_2O$$

Rapid in acid pH (nonbiological)

$$FeS_2 + 7Fe_2(SO_4)_3 + 8H_2O \longrightarrow \underset{\text{(ferrous sulfate)}}{15FeSO_4} + 8H_2SO_4$$

Acidic sulfate soils contain a **sulfuric horizon,** which has a pH of the 1:1 soil:water mixture lower than 3.5. Some other evidences of *sulfide* content (yellow color, mineralogy) are also needed. *Sulfaquents* great groups include all these acid sulfate soils. (See Chapter 7 for general soil classification nomenclature.)

Strong acidity can result in possible toxicities by aluminum and iron (if the solution is acid enough), soluble salts (unless leached), toxic manganese, and hydrogen sulfide (H_2S) gas. Hydrogen sulfide, often formed in paddy soils, causes the rice disease known by its Japanese name, *akiochi,* which prevents rice plant roots from absorbing nutrients.

Management techniques are extremely variable and depend on many specific facts—that is, the extent of acid formation, the thickness of the sulfide layer, leaching possibilities, and the value of the land area. The general approaches to reclamation are as follows:

1. *Keep the area flooded.* Flooded (anaerobic) soil inhibits acid development, which requires oxidation. This solution almost limits the use of the area to growing mostly paddy rice.
2. *Control the water table.* If a nonacidifying layer covers the sulfuric horizon, drainage to keep only the sulfuric layer under water (anaerobic) is possible.
3. *Lime and leach.* The primary way to reclaim these soils, as for any acid soil, is to lime them. This solution is possible but not always practical. Normal soils may require from 11 to 45 Mg/ha (5–20 t/a) of lime in a 20-year period, whereas acid sulfate soils may need from several metric tons per hectare per year up to even 224 Mg/ha (100 t/a) within a 10-year period or less.

If these soils are leached during early years of acidification, lime requirements are lowered. Leaching, however, is difficult because of the high water table common to these soils and low permeability of the clay. Because acid sulfate soils are often in reclaimed swamps and salt marshes, seawater is sometimes available for *preliminary* leaching.[d,e]

---

[a] D. L. Dent and L. J. Pons, "A World Perspective on Acid Sulfate Soils," *Geoderma* **67** (1995), 263–276.
[b] G. J. M. W. Arkesteyn, "Pyrite Oxidation in Acid Sulfate Soils: The Role of Microorganisms," *Plant and Soil* **54** (1980), 119–134.
[c] Darwin L. Sorenson, Walter A. Kneib, Donald B. Porcella, and Bland Z. Richardson, "Determining the Lime Requirement for the Blackbird Mine Spoil," *Journal of Environmental Quality* **9** (1980), No. 1; 162–166.
[d] Charles R. Lee, et al., "Restoration of Problem Soil Materials at Corps of Engineers Construction Sites," Instruction Report EL-85-2, U.S. Army Corps of Engineers, May 1985, H-1 to H-14.
[e] C. Ckharoenchamratcheep, C. J. Smith, S. Satawathananont, and W. H. Patrick, "Reduction and Oxidation of Acid Sulfate Soils of Thailand," *Soil Science Society of America Journal* **51** (1987), 630–634.

weight. Except for chlorine (chloride), the dominant role of micronutrients is as activators in numerous enzyme systems. **Chloride** affects root growth, but many of its functions are not clearly understood. **Boron** is another micronutrient whose function is not fully understood. A lack of boron reduces numbers of flowers and kills growing tips. Production is greatly reduced.

The origin of micronutrients in soils is the slowly weathering minerals; the micronutrients come from minerals mixed in small amounts with the common primary minerals. Many chloride salts and some boron salts are soluble; the metals zinc, copper, manganese, and iron are more soluble in acidic solution (hence, more available in acidic soils) and become less soluble as pH increases. Molybdenum is more soluble in basic soil (it reacts much as phosphate does). In strongly acid soils, manganese, zinc, and copper may dissolve to form toxic concentrations that actually hinder plant growth.

Boron is one of the most added micronutrients to correct a deficiency. It is commonly needed in humid areas because available boron can be leached. In contrast to the need for boron in humid areas, zinc and iron are the nutrients most often deficient in soils of arid regions, especially on calcareous soils. The deficiencies of manganese, copper, and molybdenum are less common. Although chloride is generally adequate for nutrition, higher addition levels of chloride do seem to benefit growth by reducing certain plant diseases.

11:6 Soil Boron

Boron is essential for the growth of new cells. It is not readily mobile in the plant, and a boron deficiency causes the terminal bud to cease growth, followed by the death of young leaves. Without adequate boron the number and retention of flowers is reduced and pollen germination and pollen tube growth are reduced. The result is that less fruit develops. Foliar boron sprays are needed on almonds, apples, and pears in California, Oregon, and Washington, even though visual symptoms of that need are not always evident.[4]

Soil Chemistry of Boron

Boron (B) forms a weak acid. In the soil solution at most soil pH values, it will occur as *nonionized* H_3BO_3; at pH values greater than 8.5, it will occur as $B(OH)_4^-$. Boron is a *non*metal. Deficiencies are most common in high-rainfall areas, particularly in the Atlantic coastal plain; the Southeastern states; northern Michigan, Wisconsin, and Minnesota; the Pacific coastal area; and the Pacific Northwest. The most prominent boron mineral in leached soils is *tourmaline,* a very slightly soluble mineral. In less leached soils, various soluble borates may exist.

In soils boron has four major sources: (1) its primary rocks and minerals; (2) combined in soil organic matter; (3) adsorbed on colloidal clay and hydrous oxide surfaces, much as is phosphorus; and (4) as the boric acid (H_3BO_3) or $B(OH)_4^-$ ion in solution. Freshly precipitated aluminum hydroxides adsorb large amounts of boron, so that liming acidic soils frequently causes a boron deficiency as soluble boron adsorbs on the new metal oxide precipitates.

Boron Deficiency and Added Amendments

Boron deficiency in grape vines in the San Joaquin Valley of California drastically reduces fruit set. The cost of adding enough boron is relatively inexpensive, and it would seem logical to add a little *insurance* boron, but adding *too much boron must be carefully avoided.* The concentration difference between adequate and toxic boron is not very large. Boron can

---

[4]Patrick Brown, Hening Hu, Agnes Nyomora, and Mark Freeman, "Foliar Boron Application Enhances Almond Yields," *Better Crops* **80** (1996), No. 1, 20–24.

Table 11-1 Minimum Needed Boron Concentrations for Optimum Yields

*0.2 mg B/ha-15 cm**	*0.2 to 1.0 mg B/ha-15 cm*	*1.0 to 2.0 mg B/ha-15 cm*
Small grains	Tobacco	Apples
Corn	Tomatoes	Alfalfa
Soybeans	Peaches, pears, cherries	Clovers
Peas, beans	Peanuts	Mustards
Potatoes	Carrots	Celery

*Boron extracted from soils by the *hot water* test.

Table 11-2 Threshold Concentrations for Boron Concentrations in Field Capacity Water

Sensitive Crops		***Semitolerant Crops***		***Tolerant Crops***	
0.6 mg B*	Citrus fruits		Sesame		Sorghum
			Peas		Alfalfa
1.0 mg B	Grapes	2.0 mg B	Carrots	8.0 mg B	Oats
	Pecans		Potatoes		Tomatoes
	Onions		Leafy vegetables		Sugar beets
	Wheat		Barley	20.1 mg B	Cotton
1.6 mg B	Strawberries	4.1 mg B	Corn	30 mg B	Asparagus
	Beans		Tobacco		

Source: Selected data from R. Karen and F. T. Bingham, "Boron in Water, Soils, and Plants," *Advances in Soil Science* **1** (1985), 229–265.

*Toxicity above these levels begins to decrease yields. Units are in milligrams of boron per hectare-15 cm deep, assuming that a ha-cm weighs 2 million kilograms.

accumulate to toxic concentrations. In some arid regions the accumulation of boron to toxic levels already exists. Tables 11-1 and 11-2 list minimum desired boron contents and toxicity thresholds for selected crops.

Some disorders attributed to boron deficiency include *canker* of beets, *hollow stem* of cauliflower, *cracked stem* of celery, *water core* of rutabagas, and *stem-end russet* of tomatoes.

The most common boron amendment is borax (**sodium tetraborate, $Na_2BO_4O_7 \cdot 5H_2O$**), which is 14 percent boron. It is sold under various names and is quite soluble. **Solubor** (20% boron) is a modified borate completely water soluble and used in many liquids and mixes. Low-solubility **frits** are made by mixing glass with boron and other nutrients, melting this mixture, and then cooling and shattering it. These frits have a low rate of nutrient release, which is just fast enough. They protect against excessive solubility (toxicity) and limit leaching losses in sands and in high-rainfall areas.

11:7 Iron in Soils

Iron is an important part of the plants' oxidation–reduction reactions. As much as 75 percent of the cell iron is associated with chloroplasts. Iron is a structural component of cytochromes, heme, and numerous other electron-transfer systems, including *nitrogenase* enzymes necessary for the fixation of dinitrogen gas.

In aerated soils iron oxides are one of the *least soluble* of soil minerals. When soil is limed, any soluble ferric iron (Fe^{3+}) readily forms one of many hydrous oxides, all of which have low solubility. In intensively weathered soils (e.g., Oxisols and some Ultisols) most of the primary minerals have been weathered and the more soluble materials leached away. The hydrous iron oxides remain because they are the *most resistant residues.* Most of the basic cations, some silica, and even considerable alumina have been leached away. Some Oxisols have 50 percent to 80 percent clay consisting primarily of iron and aluminum hydrous oxides. *The problem with soil iron is that iron has very low solubility in soil solutions and waters* (even added soluble-iron chemicals readily precipitate as low-solubility minerals), and the *challenge is how to keep iron sufficiently soluble for plants to absorb enough of it.* In strongly acidic solutions, below pH 5, iron becomes increasingly soluble and is rarely deficient.

The reduced ferrous iron (Fe^{2+}), formed in anaerobic conditions, is more soluble than the ferric ion, but the ferrous iron is easily and rapidly oxidized in aerated soils and its solubility becomes that of hydrous iron oxides. Alternating aerobic-anaerobic conditions cause some solubilization and precipitation cycles. Iron solubilization occurs in anaerobic (waterlogged) periods, then precipitation occurs when the soil becomes aerated (dried or drained) and allows iron oxide cementation to develop; soft *plinthite* eventually forms *ironstone.*

Given these solubility problems, some major questions about iron are as follows:

- How is iron made available?
- Why isn't iron deficient in almost all soils?
- Which materials and methods are used for supplying needed iron?

Iron in Soil Solution—Chelates and Availability

The solution pH has a dominant influence on iron solubility. At about pH 3 iron is soluble enough to supply plant needs, but plants are usually poisoned by toxic levels of aluminum. The solubility of Fe^{3+} decreases about a thousandfold per pH unit rise. In soils of pH suitable for most crops, simple solubility of iron will not supply enough iron for plants (Figure 11-2).

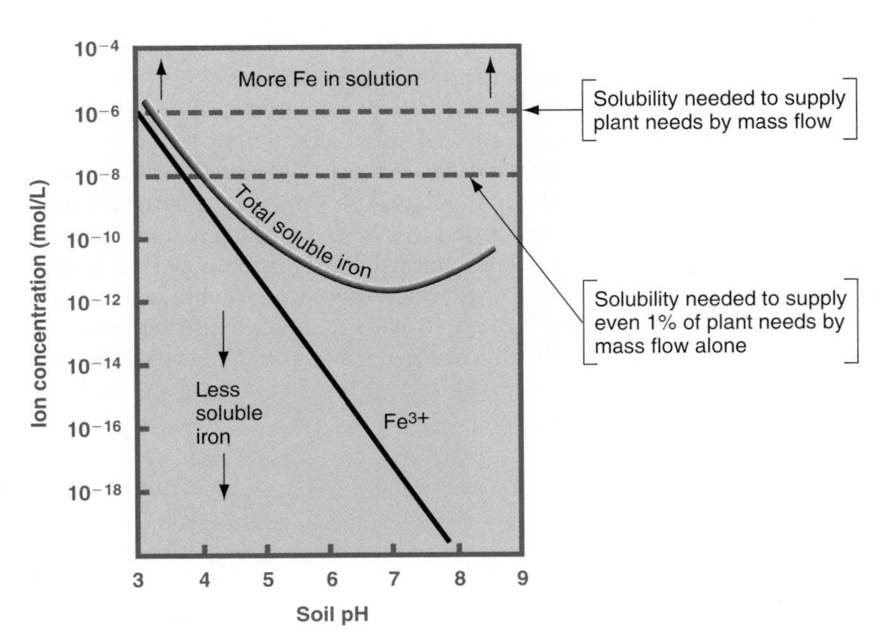

FIGURE 11-2 Illustration of iron solubility in aerobic soil solution. Only near pH 3 to 4 will most soils supply enough soluble iron from mineral sources to supply plant needs. (*Source:* Modified after Willard Lindsay, Chapter 17 in *The Plant Root and Its Environment,* E. W. Carson (ed.), University Press of Virginia, Charlottesville, 1974. Redrawn by Raymond W. Miller, Utah State University.)

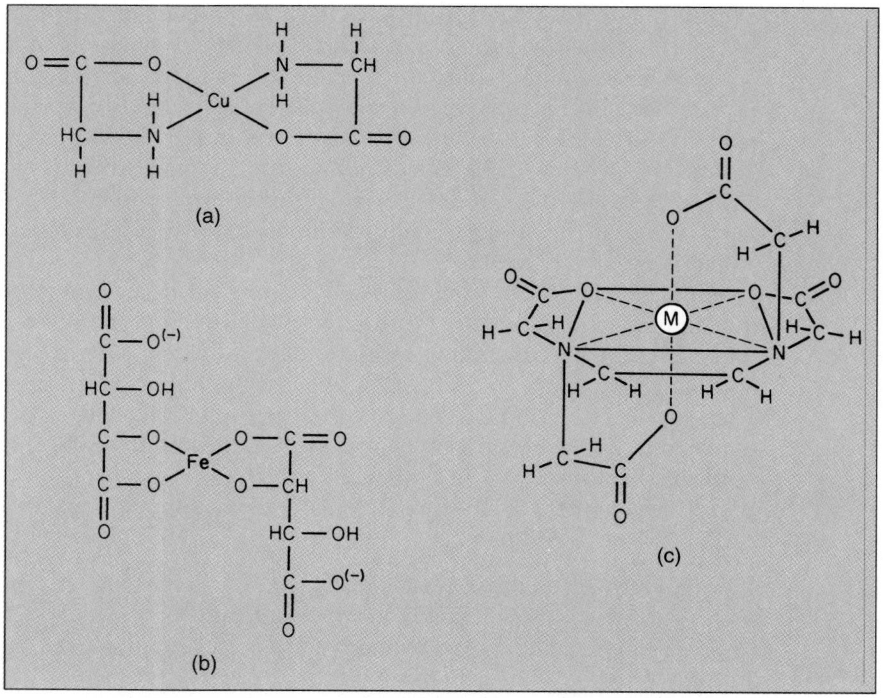

FIGURE 11-3 Example of ligands bonding to a metal. The resulting substances are called *chelates.* (a) Copper chelate involving the common amino acid glycine; (b) iron with tartaric acid; (c) the ligand ethylenediaminetetraacetic acid (EDTA) is shown bonding to any of several metals (*M*) forming chelates. Common metals bonding with ligands are Fe, Ca, Mn, Mg, Cu, Zn, and heavy metals. Other well-known natural ligands are citric acid, proteins, amino acids, polyphenols, and fulvic and humic acids. (Courtesy of Raymond W. Miller, Utah State University.)

Although the total soluble mineral iron is inadequate for plants, most plants are not deficient in iron. *What makes iron available?* Many soluble organic substances can react with iron to bond it into soluble forms that are quite mobile in the soil solution. These many substances come from root exudates, humus decomposition, microbial cell exudates, animal manures, and even polyphenols excreted from leaf surfaces and dissolved in rainfall. When an organic substance bonds to the metal by two or more contacts, the organic substance is called a chelating **ligand** (Figure 11-3). The ligand plus metal is called a **chelate**[5] (key′-late). Metals bonded to these ligands are carried in soil water. Many natural and synthetic chelates are available; some natural chelates are insoluble in water. Certain plants, when under iron stress (iron deficiency), can produce substances that solubilize or chelate soil iron. These substances have been called by various names, such as **siderophores** and **iron chelate reductases** (Detail 11-3).[6] Production of these substances by plants makes the plant more competitive to obtain iron; deficiencies of iron are less frequent for siderophore-producing plants.

[5]Chelate is derived from the Greek word *chele,* which means a *crab's claw.* The intent was to refer to the pincherlike manner in which the ligand bonds and holds the metal.

[6]D. A. Konstant, "Root Enzymes Are in Control," *Agricultural Research* **39** (1991), No. 9; 21; and M. Treeby, H. Marschner, and R. Romheld, "Mobilization of Iron and Other Micronutrient Cations from a Calcareous Soil by Plant-Borne, Microbial, and Synthetic Metal Chelators," *Plant and Soil* **114** (1989), 217–226.

Detail 11-3 Siderophores—Iron Scavengers in the Soil

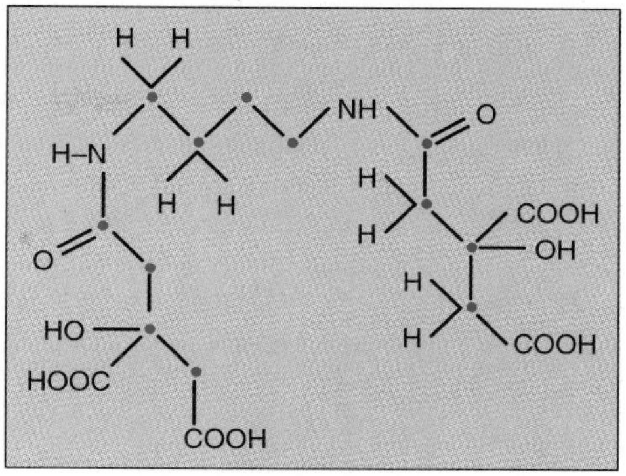

Iron (Fe) in soils is noted for its insolubility. Although plants need only relatively small amounts, iron is often deficient to some plants, particularly in neutral and alkaline soils. It has been observed that many microorganisms can secrete low molecular weight *ligands* that have a high affinity to bond to the oxidized ferric iron ion (becoming a *chelate* when bonded to a metal). These ligands have been named **siderophores.** Siderophores are capable of maintaining relatively high soluble Fe concentrations. The large number of microbes in the rhizosphere may help contribute iron to plant roots or compete with roots for the chelated iron. Recently, a fungus of the genus *Rhizopus* has been shown to produce siderophores, one of which is called **rhizoferrin.** This siderophore structure is shown at right. At concentrations of 5 μm the rhizoferrin siderophore alleviated Fe deficiency in Fe-stressed tomato in pH 7.3-buffered solution as effectively as 5 μm of FeEDDHA, one of the most effective commercial chelates. Control plants became highly chlorotic within 6 days.

Some plants may be gluttons for certain elements. A mutant pea plant was one that seemed unable to say no to iron, as well as to copper, magnesium, manganese, zinc, and potassium. Often the pea would absorb iron until it contained toxic amounts and died. The pea was discovered while looking for a way to increase iron in food plants for human buildup of iron. The pea seemed to have a known enzyme—ferric-chelate reductase—that is always turned on. The enzyme has long been known to help reduce Fe^{3+} to Fe^{2+}, a form plants can absorb. Speculation is that the reductase alters the chemical oxidation state of *sulfur* in a protein that makes up the openings to microscopic channels in plant root cells' outer membrane. If the presence of the metals turns on the reductase, which then alters the sulfur and its protein, the *channels* stay open and metals keep being absorbed into the cells. There are many theories, one of which involves speculation that **efflux pumps** move the excess of certain elements from inside cells to the exterior. These are known for calcium. One scientist suggests that perhaps plants screened to find those with high-reductase activity may be more suitable to grow on nutrient-deficient sites.

Sources: M. Shenker, R. Ghirlando, I. Oliver, M. Helmann, Y. Hadar, and Y. Chen, "Chemical Structure and Biological Activity of a Siderophore Produced by *Rhizopus arrhizus*," *Soil Science Society of America Journal* **59** (1995), 837–843; and Sandy Miller Hays, "Secrets of Soil Nutrient Uptake," *Agricultural Research* **41** (1993), No. 12, 18–19.

One ligand may supply four arms from one molecule; they form rings by bonding to the metal. Multiple bonds make the chelate very stable. The stability of a chelate is affected by the number of rings formed, the metal ion, and the solution pH. Usually, bonds are stronger with metals in approximately this order:

$$Fe^{3+} > Al^{3+} > Cu^{2+} > Co^{2+} > Zn^{2+} > Fe^{2+} > Mn^{2+} > Ca^{2+} = Mg^{2+}$$

Chelates keep large amounts of metals in a mobile form in which the metal can move to the roots (mass flow or diffusion). Roots are able to use the metal ion held in the chelate, although the mechanism is not yet clear. Soil organic matter is, therefore, usually important, but not necessarily sufficient, for increasing iron availability to plants. Synthetic chelates are often used in foliage sprays or as soil amendments to provide iron, zinc, manganese, and copper.

Problem Soils, Susceptible Plants, and Iron Amendments

Iron deficiencies are most common in calcareous soils, in arid soils, and in soils cropped to high-iron-demand plants (nursery ornamental trees and shrubs, fruits, corn, soybeans, sorghum, beans); see Table 11-3. High levels of bicarbonate and phosphates lower the iron availability to plants because of the formation of relatively insoluble iron salts. All iron salts are less soluble in basic media, which is the reason plants grown in arid soils are the most often deficient in iron.

Broadleaved plants that are iron deficient exhibit the typical interveinal chlorosis—light-colored areas between the darker veins on young leaves (see Figure 11-4).

Additions of soluble chelates usually correct iron deficiencies. Fast-acting, foliar, chelate sprays are usually not effective very long. They may need to be added two to four times a year for year-round growth. Applications to the soil last longer but act slower and require 5 to 15 times more of the expensive chelates ($15–$100 per kilogram). Some organic amendments (manures and sludges) may supply natural soluble chelates that help nutrition.

Table 11-3 Crops Sensitive to Low Iron Availability in Soils*

Sensitive Crops (Deficiency Is Most Likely)	*Moderately Tolerant Crops*	*Tolerant Crops (Infrequently Have Deficiency)*
Berries	Alfalfa	Alfalfa
Citrus	Small grains	Barley
Several fruits	Some fruit trees	Corn
Sorghum	Many grasses	Cotton
Grapes	Rice	Potatoes
Many ornamentals	Soybeans	Rice
Soybeans	Vegetables	Vegetables

Source: Selected data from J. J. Mortvedt, "Do You Really Know Your Fertilizers? Part 4: Iron, Manganese, and Molybdenum," *Farm Chemicals* **143** (1980), No. 12; 42.

*Some crops are listed in more than one category because of inadequate testing, varying conditions, or different plant varieties.

Iron	11	18	27	32	43	ppm
Chlorophyll	.3	.7	1.3	1.6	1.8	mg/g

FIGURE 11-4 A deficiency of iron causes chlorosis (whitening) of citrus leaves in California. The amount of iron (parts per million of dry weight) is associated with the amount of chlorophyll (in milligrams per gram of fresh weight)—the more iron, the more chlorophyll. (Courtesy of Ellis F. Wallihan, University of California.)

Table 11-4 Partial List of Common Carriers of Iron

Fertilizer Material	*Formula*	*Percent Iron*
Ferrous sulfate	$FeSO_4 \cdot 7H_2O$	19
Ferrous ammonium phosphate	$Fe(NH_4)PO_4 \cdot H_2O$	29
Ferrous ammonium sulfate	$(NH_4)_2SO_4 \cdot FeSO_4 \cdot 6H_2O$	14
Iron frits (crushed glass)	—	Varies
Iron chelates	Many kinds	5–14
Iron-sul	Mixture of $FeO(OH)$, $KFe_3(OH)_6(SO_4)_2$, FeS_2, and $CuFeS_2$	20

Several of the most common iron carriers are given in Table 11-4. The added *soluble, inorganic iron sources* revert rapidly to insoluble forms, unless they are sprayed on foliage or injected into tree trunks or limbs.

11:8 Zinc in Soils

Zinc is essential for numerous enzyme systems and is capable of forming many stable bonds with nitrogen and sulfur ligands. Unique among the quartet of metals—copper, iron, manganese, and zinc—only zinc does not exhibit multiple valences. Zinc is not subject to oxidation–reduction reactions in the soil–plant system.

Zinc in the Soil Solution

Zinc occurs in solution as Zn^{2+}. As a positive ion it is quite immobile in soil. Above pH 7.7 it becomes $Zn(OH)^+$, and at pH 9.1 it precipitates as $Zn(OH)_2$. Zinc will form soluble carbonates. Generally, zinc does not form particularly insoluble inorganic forms, but it does bond strongly with sulfide, forming ZnS (sphalerite). In rice paddy soils, zinc can become deficient because of its tendency to combine with sulfide produced from decomposing humus under anaerobic conditions. Zinc, in contrast to iron or manganese, *is less soluble in anaerobic conditions than in aerobic soil.*

Problem Soils, Susceptible Plants, and Zinc Amendments

Zinc deficiencies occur mostly in basic soils, in limed soils, and in soils cropped to high-zinc-demand plants (corn, onions, pecans, sorghum, deciduous fruits); (see Table 11-5). Calcareous soils have both a high pH and carbonates to which zinc adsorbs. Where topsoil (with its humus) has been eroded or cut off in land grading, zinc deficiency is more likely than if cut-and-fill had not occurred. Sands and anaerobic soils may also be likely to have lower soluble zinc than clays or aerobic soils.

Visual symptoms of zinc deficiency are sometimes distinctive. The low zinc mobility in plants causes some interveinal chlorosis, *but in both younger and some older leaves.* A zinc enzyme is involved in auxin production. Low auxin production reduces stem elongation, which results in shortened internodes and bunched leaves on the ends of branches (rosettes). Leaves are often smaller and thicker than normal, and early leaf fall may occur. Names of many zinc deficiencies indicate some of the visual symptoms: *white bud* of corn, *little leaf* of cotton, *mottle leaf* in citrus, and *fern leaf* in Russet Burbank potato (Figure 11-5).

Table 11-5 Crops Sensitive to Low Levels of Available Zinc

Sensitive Crops (Often Deficient)	*Intermediate Sensitivity*	*Insensitive (Seldom Deficient)*
Beans	Alfalfa	Carrots
Citrus	Barley	Forage Grasses
Corn	Clovers	Mustards
Deciduous fruits	Cotton	Oats
Grapes	Potatoes	Peas
Onions	Tomatoes	Rye
Pecans	Wheat	Safflower
Rice		
Soybeans		

Sources: Selected data from *Zinc in Crop Nutrition,* International Lead-Zinc Research Organization and the Zinc Institute, New York, 1974; J. J. Mortvedt, "Do You Really Know Your Fertilizers? Part 3: Zinc and Copper," *Farm Chemicals* **143** (1980), No. 11; 56.

(a)

(b)

FIGURE 11-5 Zinc deficiency in snap beans in Idaho. In (a) the rows in the foreground had no applied zinc and grew poorly. The taller plants behind them had been given 10 lb of Zn per acre as an application to the soil. In (b), note the chlorosis in the younger leaves. (Courtesy of Dale T. Westerman, USDA—ARS Soil and Water Management Research Station, Kimberly, Idaho.)

Table 11-6 Sources of Zinc

Source	Formula	Percent Zinc
Zinc sulfates	$ZnSO_4 \cdot xH_2O$	23–35
Zinc oxide	ZnO	78
Zinc carbonate	$ZnCO_3$	52
Zinc sulfide	ZnS	67
Zinc in glass frits	—	Varies
Zinc chelates	Various kinds	9–14
Zinc phosphate	$Zn_3(PO_4)_2$	51

Solubility of soil zinc is increased by stronger acidity. Solubility increases about a hundredfold for each unit that pH is *lowered,* although one study measured only changes of thirtyfold per pH unit between pH 5 and 7. Zinc deficiencies are most expected at high pH, particularly in calcareous soils.

Various zinc sources are available to correct zinc shortages (see Table 11-6). Most zinc salts are soluble enough to supply needed zinc if their total amount is increased in deficient soils, until the soluble zinc reverts to low solubility zinc oxide. Zinc sulfate is widely used.

The small amount of zinc needed permits use of foliar sprays of zinc chelates for rapid correction of deficiencies. These sprays are used mostly on trees and ornamentals. Rates are about 0.5 kg/ha to 2 kg/ha of Zn for foliar chelate sprays but may be 10 kg/ha to 20 kg/ha of Zn for inorganic amendments added to soil. Foliar sprays are not always satisfactory; fertilization of avocado trees with tracer zinc (radioactive ^{65}Zn) as the sulfate, the oxide, and the metalosate (chelate) was not adequate for seedlings in the greenhouse.[7] Adding Zn compounds to soil was more effective.

11:9 Manganese in Soils

Manganese is involved in many enzyme systems and in electron transport. In solution it occurs as the Mn^{2+} ion; in oxidized soils most of the manganese precipitates as insoluble MnO_2 (pyrolusite). As with zinc, manganese solubility increases about a hundredfold per pH unit more acidic. Organic-matter decomposition aids manganese solubility by furnishing electrons to reduce manganese (from +4 in MnO_2 to soluble Mn^{2+}) as the decomposition proceeds.

Toxicity, Problem Soils, and Deficiency Symptoms

Toxic concentrations of manganese are more common than are toxic levels of zinc, iron, or copper—partly because some soils have high total Mn contents and partly because strongly acidic soils can dissolve toxic concentrations of manganese. *Crinkle leaf* of cotton is caused by toxic levels of manganese. Toxic levels usually occur only in strongly acidic soils. Lime addition will control Mn toxicity.

Manganese deficiencies are most common in sands, organic soils, high-pH calcareous soils, and in soils growing fruits, small grains, and leafy vegetables. Manganese-deficient plants have interveinal chlorosis of younger leaves. The deficiency has been given such descriptive names as *marsh spot* of peas, *gray speck* of oats, and *speckled yellows* of sugar beets.

[7]David E. Crowley, Woody Smith, Ben Faber, and John A. Manthey, "Zinc Fertilization of Avocado Trees," *Soil Management, Fertilization, and Irrigation* **31** (1996), No. 2; 224–229.

Table 11-7 Crops Sensitive to Low Levels of Available Manganese*

Sensitive Crops (Often Deficient)	*Moderately Sensitive*	*Tolerant Crops (Seldom Deficient)*
Alfalfa	Corn	Corn
Citrus	Cotton	Cotton
Fruit trees	Potatoes	Field beans
Oats	Rice	Fruit trees
Onions	Vegetables	Rice
Potatoes		Vegetables
Sugar beets		

Source: Selected data from J. J. Mortvedt, "Do You Really Know Your Fertilizers? Part 3: Iron, Manganese, and Molybdenum," *Farm Chemicals* **143** (1980), No. 12; 42.

*Some crops are listed in more than one column because results show different conclusions, perhaps because of the plant variety and/or growing conditions.

Table 11-8 Sources of Manganese Amendments

Fertilizer Material	*Formula*	*Percent Manganese*
Manganese sulfate	$MnSO_4 \cdot 4H_2O$	26–28
Manganous oxide	MnO	41–68
Manganese dioxide	MnO_2	63
Glass frits	—	Variable
Manganese chelates	Numerous kinds	5–12

Plant Tolerance and Manganese Amendments

Table 11-7 lists plants tolerant of low levels of available manganese. As with iron, zinc, and copper, the most used inorganic manganese material is the sulfate salt. Several materials used are shown in Table 11-8. Rates of addition may range from 0.5 kg Mn/ha in some foliar sprays to 20 kg/ha to 25 kg/ha of Mn for additions to some soils.

11:10 Copper in Soils

Copper exists in soils mostly as **cupric (Cu^{2+})** and less as **cuprous (Cu^{+})** ions. Copper is essential in many plant enzymes (*oxidases,* for example) and is involved in many electron transfers. Plants absorb copper as the cupric ion, but the solution forms are Cu^{2+} in strongly acidic soil, $CuOH^{+}$ in mildly acidic soil, and a lot of $Cu(OH)_2$ at pH values near neutral and more alkaline. The most common copper mineral in soils is chalcopyrite ($CuFeS_2$) with copper in the Cu^{+} form. Other copper sulfides also exist. Most copper minerals are of very low solubility. Some available copper comes from exchangeable forms, some from less exchangeable forms, and some from soluble organic complexes or chelates. Copper is strongly adsorbed to many solids (clays, aluminum and iron hydrous oxides, manganese oxides). *Copper generally forms stronger metal-organic bonds than do the other metal ions.* Copper solubility is also pH-dependent, increasing about a hundredfold for each pH unit lowering (more acidic).

Problem Soils and Susceptible Plants

Copper deficiency exists for the following reasons:

- Copper bonds strongly to organic substances. Excessive straw additions may cause copper immobilization. Organic soils are often Cu-deficient. Newly cultivated organic soils have had copper shortages frequently enough to have the problem given the name *reclamation disease.*
- Sandy soils often have low total copper contents.
- Calcareous soils, with the high pH of 8.0 to 8.4, have low copper solubility. Seldom is the availability reduced enough to cause deficient levels.
- Copper competes with other metals (mostly with aluminum, zinc, iron, and phosphate) for uptake by the plant.

Copper deficiencies are fewer than for most other micronutrients, except perhaps for molybdenum and chloride. *Toxicities* occur mostly near copper ore deposits or where copper is smelted and volatile copper or solid wastes accumulate.

Visual deficiency symptoms vary. Yellowing of younger leaves, some off-colors (bluish greens), some small dead spots, and leaf curl are common symptoms. Plants sensitive to low available copper include alfalfa, barley, rice, carrots, citrus, onions, wheat, and oats. Plants that seldom show deficiencies include beans, asparagus, peas, potatoes, rye, and soybeans. Rye and triticale seem to be very tolerant of low available copper levels; in contrast, wheat may exhibit deficiency at relatively average available copper levels.

Copper Amendments and Their Use

Copper applications, like those of zinc, have been quite successful. Often only a few kilograms per hectare have been adequate to correct copper shortages for many years, and in a few instances for several decades. The most used carrier is copper sulfate, commonly known as *blue vitriol,* which has been used to control algal growth in water (even in swimming pools). This material and other copper sources are shown in Table 11-9. Of those materials in the table only the sulfates and chelates are considered soluble.

Typical application rates are from 0.2 kg Cu/ha in foliar chelate sprays up to as high as 20 kg Cu/ha for the mineral forms applied to the soil. In Western Australia one location with only 1.2 kg to 2.5 kg Cu/ha applied to the soil aided growth and plant content for up to 35 years.

Table 11-9 Sources of Copper Used for Fertilizers

Fertilizer Material	*Formula*	*Percent Copper*
Tenorite, copper oxide	CuO	75
Copper sulfate	$CuSO_4 \cdot 5H_2O$	25–35
Basic copper sulfates	$CuSO_4 \cdot x Cu(OH)_2$	12–50
Copper ammonium phosphate	$Cu(NH_4)PO_4 \cdot H_2O$	32
Glass frits	—	Varies
Copper chelates	Various kinds	8–14

11:11 Molybdenum in Soils

Molybdenum occurs in the soil solution as the **MoO_4^{2-}** (molybdate) ion. It exists in very low amounts in soil but is needed by plants in very small quantities. Most plant molybdenum exists as part of the enzyme *nitrate reductase.* In plants fixing N_2 molybdenum is also needed in the nitrogen-fixing enzyme *nitrogenase.*

The molybdate ion has many reactions similar to those of phosphate. It is strongly adsorbed to iron and aluminum hydrous oxides. Molybdenum is more soluble (available) as the pH rises to values of 7 or 8 as the hydroxyl ion competes with the molybdate ion for adsorption. Solubility of molybdenum increases about tenfold per unit rise in pH above pH 7. In more acidic soils the solubility may be nearer a hundredfold increase with each unit increase in pH. Lime addition increases available molybdenum. High concentrations of soluble manganese and/or copper reduce molybdate absorption by plants.

Problem Soils and Susceptible Plants

Molybdenum deficiencies will be most common in acidic sandy soils, where leaching losses, strong molybdate adsorption, and few molybdenum minerals exist. Soils high in metal oxides (sesquioxides) have low molybdenum availabilities. In Australia and New Zealand large soil areas are molybdenum deficient; acidic sandy soils of the U.S. Atlantic and Gulf coasts are likewise low in molybdenum.

Crops sensitive to low molybdate levels include legumes, crucifers (cauliflower, Brussels sprouts, broccoli), and citrus. Moderate sensitivity to low molybdate levels include cotton, leafy vegetables, corn, tomatoes, and sweet potatoes. *Whiptail* of cauliflower and *yellow leaf spot* of cashew are molybdate deficiencies. When legumes are grown, with their high molybdate demands, chances for deficiency increase.

Even molybdenum, with normally low concentrations in soils, can occasionally exist in toxic concentrations, but the toxicities are usually to *grazing animals,* not to plants. Such soils are typically high in organic matter and have a neutral to alkaline pH. The toxicity is really an imbalance between copper and molybdenum. The low copper causes stunted animal growth and bone deformation called *molybdenosis.* Feeding or injecting copper or adding copper fertilizer to the grazing area usually corrects the problem.

Molybdenum Amendments and Their Use

The low amounts of molybdenum needed in plants and the adequate solubility of most molybdate sources make correcting molybdenum deficiency relatively simple (see Table 11-10). Only 40–400 g (0.04–0.4 kg) per hectare are needed. The fertilizer may be applied as a foliar spray or *dusted or adsorbed to seed* before planting. Spraying cashews with 0.03 percent foliar spray (weight/volume) corrected symptoms, but not quickly. Up to three months

Table 11-10 Sources of Molybdenum

Source	*Formula*	*Percent Molybdenum*
Ammonium molybdate	$(NH_4)_6Mo_7O_{24} \cdot 2H_2O$	54
Molybdenum trioxide	MoO_3	66
Glass frits	—	1–30
Sodium molybdate	$Na_2MoO_4 \cdot 2H_2O$	39

was required before the symptoms were gone.[8] Liming the soil to about pH 5.3 also corrected the deficiency but took even longer than did the molybdenum spray.

11:12 Chloride in Soils

Chlorine exists in soils almost entirely as the **chloride ion (Cl^-),** a very soluble and mobile ion. Chloride has little tendency to react with anything in soil. Its role in plants is believed to be osmotic and in balancing cell cationic charges. Amounts of chloride in plants range in values similar to sulfur (0.2%). In some salt-tolerant plants up to 10 percent chloride has been measured. If plants sensitive to chloride have more than 1 percent to 2 percent chloride, yields are often reduced. Some of these sensitive crops include fruit trees, berry and vine crops, many woody ornamental plants, tobacco, and avocados.

Some Unique Features of Chloride

Chloride cycles easily in the environment. It is supplied to the air by volcanoes and sea spray, to water by water-softener wastes, industrial effluents, road deicing salt, food wastes, and sewage. It is also added to the soil in animal manures, KCl fertilizers, and rainfall or irrigation waters.

Chloride may accumulate in toxic amounts. Soluble salts, which hinder plant growth, usually include chloride as the most abundant anion. Irrigation water containing high chloride contents, when sprayed and left to dry on the foliage, may cause salt burn.

Some diseases, particularly *take-all root rot,* have been decreased by using chloride-containing fertilizers.[9] Some others (*stripe-rusts, leaf rust,* and *tan spot* of wheat) seem to be reduced by adequate chloride (above nutritional needs). For example, banding about 40 kg/ha of chloride has been recommended on winter wheat to reduce *take-all root rot.* Total additions of 100 kg/ha to 130 kg/ha of chloride have been used on winter wheat—part at planting time in the fall, but most of it in February or March. Some recent tests have not had such effective results.

Chloride Amendments and Their Use

Very little has been done with chloride amendments because no field plots exhibit deficiency and most fertilizers contain some chloride as a contaminant. With the additional benefits attributed to concentrations of chloride higher than needed as a nutrient, more study of large additions (30–50 kg/ha) is needed. Potassium chloride is the most used fertilizer containing large amounts of chloride. Other soluble chlorides are available, such as those of ammonium, calcium, magnesium, and sodium. Each has unique advantages and disadvantages.

11:13 Beneficial Elements

A number of elements are not yet proven to be needed by most plants but they are essential or beneficial to some plants. These elements include silicon (Si), sodium (Na), vanadium (V), cobalt (Co), and nickel (Ni). Nickel has been added to the list of 16 elements by some researchers and might be essential element number 17. The five elements are at least *beneficial* elements in which growth of a few kinds of plants or situations is benefitted. However, they do not seem to be needed by most plants for growth and reproduction.

---

[8]C. C. Subbaich, P. Mankandan, and Y. Jopshi, "Yellow Leaf Spot of Cashew: A Case of Molybdenum Deficiency," *Plant and Soil* **94** (1986), 35–42.

[9]R. J. Goos, "Chloride Fertilization," *Crops and Soils Magazine* **39** (1987), No. 6, 12–13.

Availability of Cu, Fe, Mn, and Zn . . . decreases as soil pH increases, so most deficiencies may occur in neutral and calcareous soils. Conversely, toxicities of these micronutrients, especially Mn, may occur in very acid soils.

—John J. Mortvedt

Truth comes home to the mind so naturally that when we learn it for the first time, it seems as though we do no more than recall it to our memory.

—Bernard Le Bovier de Fontenelle

Questions

1. How plentiful in soils are the nutrients *calcium, magnesium,* and *sulfur*?
2. (a) Explain why in the production of peanuts adequate *calcium* might be the most critical nutrient to keep available. (b) Explain how calcium availability is maintained.
3. How is *grass tetany* in cattle related to the nutrient *magnesium*?
4. What functions in the plant are performed by *calcium* and *magnesium*?
5. Tabulate four of the many similarities between soil *sulfur* and soil *nitrogen.*
6. To what extent is sulfur deficiency likely to be more common in the future? Explain.
7. (a) Which ion forms of sulfur are most common in the soil solution? (b) Which natural soil sources supply sulfur for plants?
8. How do the following soils differ from each other: acidic sulfate soils, cat clays, sulfuric horizons, and *potential* acidic sulfate soils?
9. How do the following management practices allow cultivation of acidic sulfate soils: (a) growing paddy rice and (b) controlling the depth of the water table?
10. Briefly discuss a combination management scheme that would involve crop selection, controlled water table depth, and liming. Rely on information derived from this chapter and from previous chapters.
11. (a) What is the form of *boron* in soil solution? (b) In which climatic region is deficiency most likely? Explain.
12. (a) Although *iron* occurs in large amounts in soils, iron is often deficient. Explain why. (b) If iron compounds have such low solubility, why are there not more crops exhibiting iron deficiency? (c) How is iron deficiency corrected?
13. Describe the visual deficiency symptoms for *iron, magnesium, zinc,* and *sulfur.*
14. (a) What is the ionic form of *zinc* in the soil solution? (b) Explain the mobility of zinc in soils.
15. Which materials are used to correct deficiencies in (a) zinc, (b) iron, (c) sulfur, and (d) boron?
16. List the common ion forms of (a) manganese, (b) zinc, (c) nickel, and (d) copper in soil solution.
17. Which crops are most likely to exhibit a molybdenum deficiency? Explain why.
18. If soluble boron amendments are easily applied to correct a boron deficiency, why is it still hazardous to apply enough for several years in one addition?

Diagnosis of Soils and Plants

As the 1990s progress, the world's farmers are being faced with the daunting task of feeding some 93 million more people every year, but with 24 billion fewer tons of topsoil than the year before.

—Lester Brown

12:1 Preview and Important Facts

PREVIEW

Fertilizers, lime, and water are the easiest growth factors to manipulate for increased crop growth. The costs of fertilizers may be 20 percent to 50 percent (about 30 percent on average) of the variable cash inputs into a crop. It is not only wise to add correctly the kinds and amounts of fertilizer needed, but now, it is also environmentally critical.

Soil and plant diagnosis is essential for predicting future crop fertilizer and lime needs and to explain past crop yields. The plant itself is the final proof of which nutrients were, or are, available to the plant, but plant analysis has its limitations. First, we only measure the total amount of nutrients in the plant. Factors other than soil availability—weed competition, diseases, or poor management (dryness, drainage, nutrient balance, pH control)—affect the amounts of nutrients absorbed, but the specific effects are not identified. A second, and more important, limitation is that by the time a plant can be tested, it may be too late to add fertilizer. Chemical analysis of the plants can provide helpful information for the overall evaluation of the soil conditions for plant growth over years of study; however, plant analysis is less satisfactory than soil analysis for making decisions about fertilizing the current crop. Some important uses are made of plant analyses for long-growing crops (orchards, turf, sugarcane, bananas). Only a few crops of high value (such as potatoes) have much baseline data permitting use of plant analysis for current crop fertilizer modifications.

Soils are tested for pH, soluble salts, available nutrients, element toxicities, and various other properties to aid in predicting nutrient and lime needs for optimum plant growth. It is possible to predict the amount of phosphorus or potassium fertilizer and of lime to add with considerable, but incomplete, accuracy.

Soil analysis is no panacea. It will *not* eliminate unsatisfactory plant growth when the cause is dry weather, compacted soils, critically low or high temperatures, inadequate soil drainage and low oxygen in the root zone, improper placement of fertilizer, salt accumulation, plant diseases, toxic elements, insect damage, competition from weeds or tree roots, or untimely operations.

Small additions of chemicals can sometimes cause spectacular increases in plant growth. Because of this, there will always be a number of magical growth additives offered for sale that are accompanied by claims of unusual effectiveness. Such products can be identified by a few key descriptors—*enzymatic, hormones, super, wonder, secret, miracle, biological, organic, stimulant,* and *conditioner.* Seek information from state agricultural universities or local extension representatives about such products. Although caution should be used when considering testimonial-promoted substances, there are legitimately useful microbial and hormonal additives. Examples of such materials are *urease inhibitors, Rhizobia inocula* (to ensure legume nodulation), *mycorrhizal associations,* and *fungus inocula* (to control sicklepod, a hard-to-control weed in soybeans).[1] One plant growth regulator, **ethephon** (Cerone®), used in amounts of about 2 L/ha (less than 1 qt/a), reduces heights of cereal grain plants and increases stem diameter, reducing lodging of the plant.[2]

IMPORTANT FACTS TO KNOW

1. The general guidelines for collecting a good-quality soil sample to be used for analysis and recommendations
2. The tests used to measure lime requirement and salt hazard in soils
3. The problems inherent in tests used for nitrogen recommendations and what to do about them
4. The reason that dilute acids are used to extract phosphates in acidic soils, but bicarbonates are used with alkaline soils, and why neither solution works satisfactorily with the other group of soils
5. The general status of soil tests to evaluate micronutrient needs
6. The advantages and disadvantages of relying on plant total analysis for making fertilizer recommendations
7. The special sampling needs and plant parts to sample when doing plant total analysis
8. The definition of *critical nutrient range* and why research continues to attempt to define these values
9. The general visual nutrient deficiency symptoms for nitrogen, phosphorus, potassium, and iron
10. The concept of *hidden hunger*
11. The advantages and limitations of using computers in formulating recommendations
12. Factors affecting yields or crop responses to fertilizers that soil tests do not evaluate
13. The meaning of *fertilizer grade* numbers
14. Important characteristics of fertilizers, such as their (a) salt effect, (b) present and potential acidity, and (c) solubility and mobility in soil
15. The calculation of elemental contents and fertilizer grade
16. The calculation of weights of nutrient carriers to prepare fertilizer mixtures

12:2 Soil Sampling

No amount of care in preparation and analysis can overcome the problems of careless or inappropriate sampling in the field. Good judgment in sampling soil is often better than any single set of rules. The overriding guide should be to *take a sample so that it represents what*

---

[1] Anonymous, "Fungus Saves Soybeans from Weeds," *Ag Consultant* (Jan. 1987), 9.

[2] Ann J. Rippy and Frank J. Wooding, "Use of a Plant Growth Regulator on Barley to Prevent Lodging," *Agroborealis* **18** (1986), 9–12.

it is intended to represent. If the sample is to represent the top 30 cm over the entire field, take many subsamples that have equal portions of soil from the 30-cm depth and composite them into one sample. Some soils of uneven relief, or that have been graded or leveled, may have different depths of topsoil over much different subsoils, such as dense, clayey **B** horizons. Some scientists recommend *not mixing* layers of soil of *obviously different color;* that is, they recommend using only the portion of each 30-cm soil core above such a different subsoil portion. Avoid abnormal areas and sample them separately.

With the need to reduce the application of fertilizer in order to prevent pollution, the interest in *variable-rate* fertilizer applications requires many more samples in the first few years. Soils with obvious differences in the horizons with depth would be sampled as separate composite samples. More samples are taken, often on a rigid grid system. In precision agriculture each composite sample may represent no more than one-third or one-half hectare.

Depth and Number of Samples

Which soil depth should be sampled? Most early workers sampled to *plow depth,* which was interpreted to be about 15 cm to 20 cm deep. At present many laboratories request samples to be the top 30 cm. *Sample the depth recommended by the laboratory that will analyze your sample.* It is common to sample different depths for different tests. For salt evaluation, depths may be 1 cm to 10 cm or more. For no-till fields and established sod crops, sampling depths might be shallow. For nitrate and sulfate measurements, at least two depths are needed, 0 cm to 30 cm and 30 cm to 60 cm. Some laboratories even request a third sample at depths of 60 cm to 90 cm, but these deep samples may be poorly taken because of the difficulty in sampling that deep.

How many subsamples should make up the single composite sample sent to the laboratory? Unfortunately cost and time, rather than sample quality, usually determine the number of subsamples taken for each composite sample. Statistical studies have led to the recommendation that 15 to 20 cores or subsamples be collected and combined into one good composite sample. In practice people often take few subsamples, sometimes only two to five. If only a few subsamples are used, there is risk that the sample is not representative of the whole area. For the deeper soil depths, perhaps only one-third as many subsamples are necessary for the same level of confidence and quality. The best recommendation is to *follow the suggestions given by the laboratory that will do the tests.* (See also the sampling guide in Table 12-1).

Table 12-1 Suggested Guide for Sampling Soil

Number of subsamples to make each composite:	
Phosphorus (P)	Take about 25 subsamples/field
Potassium (K)	Take about 5 subsamples/field
Nitrate	Take about 15 subsamples/field
When to sample:	
pH, P, and K	Every 3–5 years, at least 15 cm–20 cm (6 in–8 in) deep
Nitrate, sulfate	Annually at least to 60 cm (2 ft) deep in 30-cm increments
Changes for reduced tillage fields:	
pH	Sample shallow, 5 cm–10 cm (2 in–4 in) deep
Phosphorus, potassium	Sample at least 15 cm–20 cm (6 in–8 in) deep

Source: Wayne E. Sabbe, "Let's Rebuild Confidence in Soil and Plant Tissue Testing," *Ag Consultant* (Jan. 1987), 10–11.

Sampling Frequency, Time, and Location

When should sampling be done? Many recommendations are rather vague, such as "anytime at least one month prior to planting." New land, or land that is new to you, should be tested yearly for a few years. As the land's fertilizer needs for each area and for the plants grown become known, the more standard recommendation of sampling once every 2 or 3 years can be adopted. Concern about environmental pollution will probably necessitate sampling almost every year in the future. Sample prior to the crop responding most to the fertilizer to be added.

Exactly when during the year sampling should be done is not well defined. Soils are analyzed to see which reserves the soil can make available to the crop and which additional fertilizer might be needed. The supplies that the soil can make available are best taken into account when sampling is done after the soil has rested since the previous crop was harvested. *The best sample is one taken as close to planting as will permit its analysis and preparation for fertilizer purchase and addition.* Because this sampling time may be in cold, wet, early spring, samples are often taken in late fall or some other more convenient time, but some accuracy may be lost thusly. Fall sampling in areas where winters are cold and microbial activity is low may be acceptable. In warmer climates any prolonged time between sampling and planting the new crop may provide inaccurate data.

When sampling row crop fields, sample in the middles away from rows to *avoid sampling old fertilizer bands.* The intent is to represent the majority of the soil and to avoid including abnormal areas.

Uniformity of Sampling Areas[3]

Before sampling, a field should be examined for differences in soil characteristics and past treatment. Consider uniformity of productivity, topography, soil texture and structure, drainage, color and depth of topsoil, and past management. If these features are uniform throughout the field, each composite sample can represent areas as large as about 5 hectares (12.5 acres). If there is a great variation in any of these features, the field should be divided and a composite sample taken from each delineated area (Figure 12-1).

In the new *precision agriculture* (see Chapter 13), the spatial variability in the field is considered. In order to provide the detail to map out small areas, more soil sampling is required, at least initially. Usually the sampling is done on a grid layout, and each grid is sampled by compositing 5 to 8 subsamples around that grid point. A portable global positioning system (GPS) receiver (see Chapter 2) will help identify exact locations of the sample site (Figure 12-2). Lightweight GPS units of 1 kg to 2 kg are available for location accuracies within a meter or less. Taking and analyzing many soil samples, one composite from each grid point (each grid point represents about one-fourth of a hectare), is time consuming and expensive, but the data must be obtained. This will be about 5 to 25 times more samples than would be taken in the conventional manner described earlier. Once the different areas are delineated, fewer samples will be needed.

To decide how to sample a field, it is necessary to decide which soil areas are uniform. Nonuniform areas are sometimes obvious—small knolls, wet depressions, cut-and-fill sites. Other nonuniform areas are not visually obvious—depth variations in glacial till, outwash covering, and alluvium or loess covers. One clue to soil variation is the variability of the plant cover. It is presently possible to *see* photographically even small differences in plant cover from location to location within a field.

---

[3]J. C. Shickluna, "Sampling Soils for Fertilizer and Lime Recommendation," *MSU Ag Facts, Extension Bulletin E-498,* Cooperative Extension Service, Michigan State University, East Lansing, 1981.

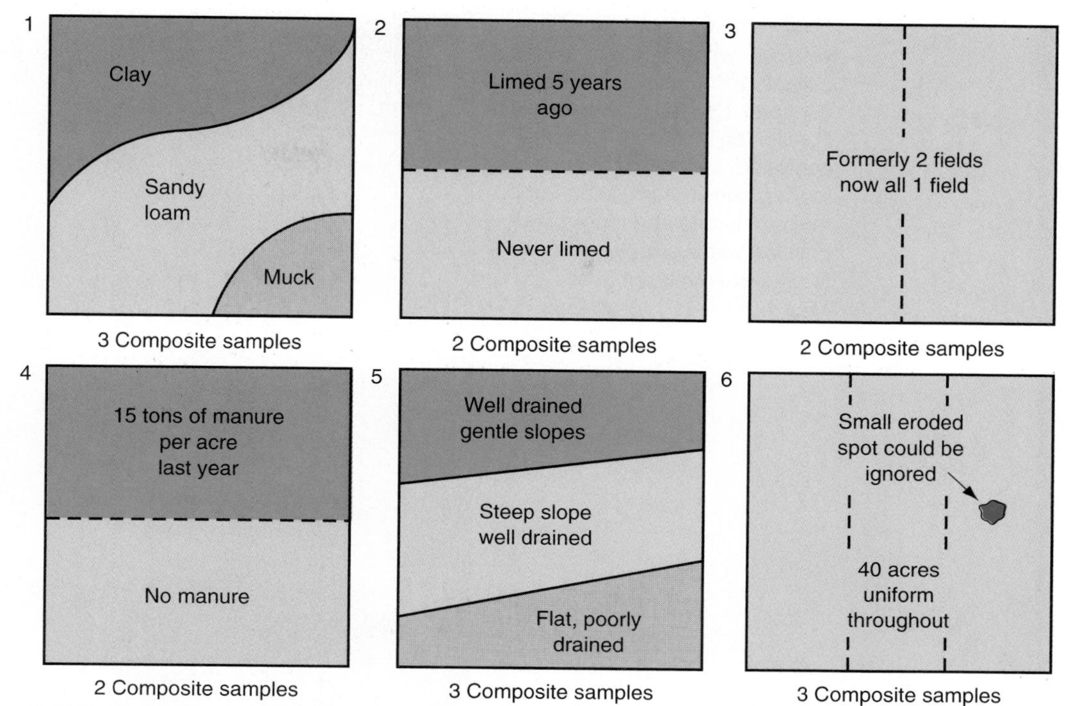

FIGURE 12-1 Conditions should be studied before sampling a field. Variations in soils, productivity, drainage, and past management will all determine the number of samples that should be taken to obtain a representative soil sample. The two textural classes and muck in Field 1 require three composite samples. Due to differences in management of Fields 2, 3, and 4, two composite samples should be taken in each. Differences in topography and drainage represented in Field 5 necessitate three composite samples. Field 6 is a uniform 16-hectare (40-acre) field; but because of its size, it should have three composite samples to ensure a representative soil analysis of the entire field. The examples of sample numbers to be considered are only approximate; the actual sampling area is determined by soil uniformity at the test site.

Providing Detailed Soil and Cropping Background

The more complete the information provided on a field's soil and cropping background, the more appropriate will be the fertilizer recommendation interpreted from the soil analysis. The following information should accompany the sample when it is sent to the soil testing laboratory:

- Previous crop grown
- Crop or crops to be grown
- Realistic yield goal
- When the field was last limed and fertilized and rates of application
- Whether or not the field will be manured (kind of material and rate of application)
- Depth of plowing
- Soil series or management group, if known
- Whether drainage is good, intermediate, or poor
- Whether or not irrigation is to be used
- Other special problems, conditions, or management that may affect plant growth (such as temperature, geographic location, elevation, hardpans, types of mulching, ridge tillage, and covering with plastic sheeting)

FIGURE 12-2 Knowing where you are when you sample is critical in precision farming (Chapter 13). A portable Global Positioning System receiver equipped with a bar-code reader, shown here, is one unit that can locate and record your position within a meter or less. (Courtesy of *Agriculture Research Magazine,* "Fine-Tuning Agricultural Inputs," January 1993, 17.)

12:3 Soil Tests and Recommendations

To obtain the maximum growth response from an added nutrient element, all other essential elements must be in adequate, but not injurious, amounts. Justus von Liebig (1803–1873) postulated this concept in the **law of the minimum,** which states that *the growth of any plant is limited most by the essential plant nutrient* [or growth factor] *present in the least relative amount.*

No quick laboratory test can duplicate the actual uptake action by roots, and laboratory tests are arbitrary and empirical. Good correlations of soil tests with crop response in the field require hundreds of tests on each crop of interest, each different soil, each climatic variation, and even each different management system. Some of the better established tests throughout the United States, which can be expected to provide correct and useful data in most instances, include those for soil pH, soluble salts, lime requirement, available phosphorus, and available potassium.

Soil Acidity Evaluation

The **soil reaction (soil pH)** is usually measured with a pH meter on a 1:1 or a 1:2 soil–water suspension or a 1:2 soil–0.01 $molar_c$ $CaCl_2$ suspension. In all instances the laboratories should know how to evaluate the pH data, which varies slightly according to the method used. One might consider lime additions if pH is below 6.

The **lime requirement** of a soil is the amount of lime required to bring about the desired change in soil pH. The pH is determined in a soil–buffer suspension. Table 12-2 provides an abbreviated example of a table used to estimate the lime needed from pH values obtained in the soil–buffer test.

Soluble Salts Evaluation

The reference method used for evaluating soluble salt hazard is the conductivity of the **saturated soil paste extract;** however, 1:1 and 1:2 soil–water suspensions are frequently used instead of a saturated paste extract because the saturation paste extract requires more work and more time. Salt values from these diluted samples should be multiplied by some factor to estimate what the salt value would be if measured on the saturated paste extract. An example of this adjustment is shown in Table 12-3. The multiplication factors used are not exact because the amounts and kinds of clays alter the saturation water content on which the estimation factors are based.

For example, assume that a loam soil has a *field capacity* of 20 percent water. Its *saturation percentage* is estimated (using a general rule of thumb) to be about *double* the field capacity value, thus 40 percent. The 1:1 soil–water suspension is soil at 100 percent water content (= 20 g of soil mixed in 20 g of water). Thus, the salt in this loam's 1:1 suspension is

Table 12-2 Amounts of Lime Required to Bring Mineral and Organic Soils to the Indicated pH According to Soil Buffer pH of the SMP Buffer

	Agricultural Ground Limestone[a] *(t/a)*[b]			
	Mineral Soils			*Organic Soils*[c]
Soil Buffer pH	*7.0*[d]	*6.5*[d]	*6.0*[d]	*5.2*[d]
6.8	1.4	1.2	1.0	0.7
6.6	3.4	2.9	2.4	1.8
6.4	5.5	4.7	3.8	2,9
6.2	7.5	6.4	5.2	4.0
6.0	9.6	8.1	6.6	5.1
5.8	11.7	9.8	8.0	6.2
5.6	13.7	11.6	9.4	7.3
5.4	15.8	13.4	10.9	8.4
5.2	17.9	15.1	12.3	9.4
5.0	20.0	16.9	13.7	10.5
4.8	22.1	18.6	15.1	11.6

Source: Modified from "Recommended Chemical Soil Test Procedures for the North Central Region," Bulletin 499, North Dakota Agricultural Experiment Station, North Dakota State University, Fargo, 1975.

[a]Agricultural ground lime of 90% total neutralizing power (TNP) or $CaCO_3$ equivalent, and fineness of 40% <100 mesh, 50% <60 mesh, 70% <20 mesh, and 95% <8 mesh.

[b]To convert tons per acre to metric tons per hectare, multiply by 2.24.

[c]Because of lower mineral contents, organic soils are often suitable when limed only to pH 5.0 to 5.5.

[d]Desired pH level for the soil.

Table 12-3 Estimation of Saturation-Soil-Paste Values from Measured Values Using 1:1 or 1:2 Soil–Water Suspensions

Suspension Whose Solution Conductivity Is Measured in Laboratory Method	*Water Content of Sample* Paste (≈ 2x FC) (%)	Suspension (%)	*Multiplication Factor to Estimate Paste Extract Conductivity*
1:1 soil–water suspension			
Loamy sand texture	24	100	3.5–4
Loam texture	34	100	2.5–3
Clay loam texture	44	100	2–2.5
1:2 soil–water suspension			
Loamy sand texture	24	200	6–8
Loam texture	34	200	5–6
Clay loam texture	44	200	4–5

dissolved in 2.5 times more water (100% water divided by 40% water) than would be in the saturated paste. Soluble salt concentrations in the 1:1 suspension will be 2.5 times more dilute. To approximate the conductivity that would be measured in the saturation paste extract, *multiply the 1:1 conductivity value by 2.5.* Note that this estimation has two major errors: (1) It is the water percentage at saturation that is being estimated; (2) some salts, such as gypsum, dissolve more as more water is added, so the salt estimate from a diluted sample would be too high if such salts are prevalent.

Soil-Test Nitrogen

Good predictive soil tests for nitrogen are often not available, although some tests for residual nitrate in the top 60 cm to 90 cm (2 ft–3 ft) of soil are extensively used in some localities. The complex nitrogen transformations (biological denitrification, humus decomposition, immobilization, leaching, ammonia volatilization) are rapid and sensitive to temperature and water, which makes finding a satisfactory *quick soil test* for nitrogen difficult. When nitrogen tests are used, a measurement of total soil nitrate to 60 cm (2 ft) is the most commonly used. This test is essentially a test of nitrate released by *field incubation of soil,* but the incubation time involved is not really known. An example involving North Dakota soils is shown in Table 12-4.

Most states provide nitrogen recommendations based, at least partly, on data from many years of field plot trials involving various crops, different soils, various management practices, and different fertilizers and fertilizer rates. A nitrogen fertilizer recommendation considers previous crops, estimates of carryover nitrogen from the past season, nitrogen needed to decompose residues, the kind of crop, the projected yield, and an evaluation of the effect of the farm location (climate); see Table 12-5.

Soil-Test Phosphorus

Phosphorus soil tests have been quite reliable despite the fact that they extract little phosphorus from *organic* sources. Generally, at least two different kinds of extractants are needed. Dilute acid extractants have proven best for extracting phosphorus from the less soluble phosphate minerals that are the phosphorus carriers in acidic soils. Sodium bicarbonate extractants are most used on alkaline soils because acids will dissolve soil carbonates and become neutralized and ineffective. The phosphorus carriers in alkaline soils are dominantly calcium

Table 12-4 Nitrate–Nitrogen (NO_3–N) Soil Fertility Values (lb/a-2 ft) and N Recommendations for Different Yield Goals for Oil and Confectionary Sunflowers (North Dakota)

	Nitrogen Fertility Ratings					
Yield Goal: Oil or Confectionary (lb/a)[a]	*Very Low (VL)*	*Low (L)*	*Medium (M)*	*High (H)*	*Very High (VH)*	*Soil Plus Fertilizer N Needed*
1000	0–12	13–24	25–36	37–49	50+	50
1200	0–15	16–30	31–45	46–59	60+	60
1400	0–17	18–34	35–52	53–69	70+	70
1600	0–20	21–40	41–60	61–79	80+	80
1800	0–22	23–45	46–70	71–89	90+	90
2000	0–25	26–50	51–80	81–99	100+	100
2200	0–27	28–55	56–85	86–109	110+	110
2400	0–30	31–60	61–90	91–119	120+	120
2600	0–33	34–66	67–101	102–134	135+	135
2800	0–36	37–72	73–112	113–149	150+	150
3000	0–41	42–81	82–123	124–169	170+	170

Source: Fertilizing Sunflowers, Cooperative Extension Service, North Dakota State University, Fargo.
[a]To convert pounds per acre to kilograms per hectare, multiply by 1.12.

Table 12-5 Guide for Estimating Nitrogen Needs (lb/a)[a] for Corn and Grain Sorghum Following Certain Crops

	Yield Levels (bu/a)[b]				
Previous Crop	*100–110*	*111–125*	*126–150*	*151–175*	*176–200*
	Lb of N to Add per Acre				
Good legume stand (alfalfa, red clover, sweet clover, 5 plants per square foot or more)	40	70	100	120	150
Average legume stand (alfalfa, red clover, sweet clover, 2–4 plants per square foot)	60	100	140	160	180
Soybeans, legume seeding of alfalfa, red clover, sweet clover	100	120	160	190	220
Corn, small grain, or grass crops	120	140	170	200	230

Source: R. K. Stivers, et al., mimeographed report, Department of Agronomy, Purdue University, Lafayette, IN, 1979.
[a]To convert pounds of N per acre to kilograms per hectare, multiply by 1.12.
[b]To convert bushels per acre of corn and sorghum to kilograms per hectare, multiply by 56 × 1.12.

phosphates and adsorbed to carbonates. Solutions widely used in the United States include the following:

- **Bray-1:** 0.025 molar HCl + 0.03 molar NH_4F (for acidic soils)
- **Mehlich-1:** 0.05 molar HCl + 0.025 $molar_c$ H_2SO_4 (for acidic soils)
- **Olsen's bicarbonate:** 0.5 molar $NaHCO_3$ at pH 8.5 (for neutral and alkaline soils)
- **Mehlich-3:** 0.2 molar acetic acid + 0.25 molar ammonium nitrate + 0.015 molar NH_4F + 0.013 molar HNO_3 + 0.001 molar EDTA

Table 12-6 Phosphate and Potash Fertilizer Recommendations (lb/a)[a] for Corn and Grain Sorghum (Indiana)

Soil Test Level[b] (lb/a)			For Yield Goals (bu/a)[a]									
			100–110		111–125		126–150		151–175		176–200	
P		*K*	P_2O_5	K_2O	P_2O_5	K_2O	P_2O_5	K_2O	P_2O_5	K_2O	P_2O_5	K_2O
			Add These lb per Acre of P_2O_5 or K_2O									
0–10	Very low	0–80	100	100	110	120	120	150	130	180	150	200
11–20	Low	81–150	70	70	80	90	90	120	100	140	120	160
21–30	Medium	151–210	50	50	60	60	60	70	70	90	80	120
31–70	High	211–300	30	30	30	30	40	40	50	60	50	80
71+	Very high	301+	0	0	0	0	10	0	10	0	10	0

Source: R. K. Stivers, et al., mimeographed report, Department of Agronomy, Purdue University, Lafayette, IN, 1979.
[a]To convert bushels per acre of corn and sorghum to kilograms per hectare, multiply by 56 × 1.12. To convert pounds per acre to kilograms per hectare, multiply by 1.12.
[b]Soil phosphorus and potassium are extracted with Bray P_1 solution and neutral normal ammonium acetate, respectively.

The Bray 1 method has been used to measure both available phosphorus and potassium in acidic soils. Because most available potassium is from soluble + exchangeable forms, the H^+ and NH_4^+ ions are adequate extractors of K^+. Table 12-6 illustrates some soil test extraction values and the recommendations for both phosphorus and potassium for corn and grain sorghum in Indiana. The NH_4F in the Bray-1 and Mehlich-3 extracts is a good extractant of phosphorus from aluminum phosphate sources. The Mehlich-3 is used as a single extractant for K, P, and micronutrient metals.

A popular procedure, using a **1 molar ammonium bicarbonate** extraction and containing dilute DTPA chelate (ligand), is widely used in neutral and alkaline soils as a *single extractant for K, P, NO_3^-, Zn, Cu, Fe, and Mn.*[4]

Soil-Test Potassium

Most soils supply available potassium for a season from the small *soluble* and the larger *exchangeable* reservoirs. Many test extractants, most of them fairly successful, have been tried. Table 12-6 illustrates the recommendations based on the potassium extracted by neutral 1 molar ammonium acetate. The *critical level* of soluble + exchangeable potassium varies with the test and crop, but it is about 180 kg to 280 kg of K/ha. Above these extracted values plant response to added K is not usually expected, except for high-K-requiring crops such as tobacco, bananas, and potatoes.

Some laboratories involve the *cation exchange capacity* (CEC) of a soil with the concept that as the CEC is higher the potassium is less available; thus, the critical level of potassium needed is higher. The optimum value for corn in Ohio is determined by the following formula:

$$\text{optimum K soil test in lb K/acre} = 220 + (5 \times \text{CEC})$$

On a sandy soil of CEC = 9 cmoles_c/kg, the needed K value is 265 lb/a. On a clayey soil of CEC = 23 cmoles_c/kg, the needed K value is 335 lb/a.

[4]P. N. Soltanpour and A. P. Schwab, "A New Soil Test for Simultaneous Extraction of Macro- and Micro-nutrients in Alkaline Soils," *Communications in Soil Science and Plant Analysis* **8** (1977), 195–207.

Table 12-7 Approximate Nutrient Levels in Soils as Estimated by Extraction of Soil with 1 Molar Ammonium Bicarbonate + 0.05 M DTPA Chelate

Concentration of Extracted Element (mg/kg soil)							
P	*K*	*Fe*	*Zn*	*Cu*	*Mn*	*Estimated Level of Availability in Soil*[a]	*Response to Added Fertilizer*
0–3	0–60	0–2.0	0–0.5	0–0.2	0–1.8	Low	Expected
4–7	61–120	2.1–4.0	0.6–1.0	(0.2)[b]	(1.8)[b]	Medium	Possible
8–11	120+	4.0+	1.0+	0.2+	1.8+	High	Unlikely
>11	—	—	—	—	—	Very high	Rarely

Source: P. N. Soltanpour and P. Schwab, "A New Soil Test for Simultaneous Extraction of Macro- and Micro-Nutrients in Alkaline Soils," *Communications in Soil Science and Plant Analysis* **8** (1977), 195–207.
[a]The plant is expected to respond to fertilizer if the test shows *low.* Response at higher concentrations is less sure.
[b]These estimates are for the critical level. Too few data are available for further separation into groups.

The 1 molar ammonium bicarbonate + DTPA chelate extract is receiving a lot of interest for estimating the soil fertility of P, K, Zn, Fe, Mn, and Cu (and even NO_3^-) all in one simple extract. The values correlate well to values with sodium bicarbonate for available phosphorus (although only half as much phosphorus is extracted). Correlation for potassium also seems to be good. An example of recommendations is given in Table 12-7.

Soil-Test Calcium and Magnesium

Most calcium and magnesium tests are related to the need for lime. Well-limed soils and neutral and alkaline soils usually contain sufficient calcium for good growth of most plants. If liming (to minimize diseases) is not done, **gypsum** may be used to supply needed calcium.

Magnesium deficiency is more common than that of calcium, particularly in plants growing in coarse-textured acidic soils. Magnesium deficiency can also occur in soils limed over the years with calcic limestone or marl, because the added lime contains little or no magnesium.

The following soil test criteria are used to interpret magnesium levels in the soil and to make recommendations for magnesium additions:

- Exchangeable soil magnesium
- Percentage magnesium saturation of soil colloids
- Ratio of potassium to magnesium

Interpretive values for magnesium levels in Indiana soils are as follows:[5]

Humus and Texture of Soil	*Inadequate Levels of Magnesium*
Light colored (low humus), coarse textured	0–75 lb of Mg per acre (0–84 kg/ha)
Light colored (low humus), medium to fine textured	0–100 lb of Mg per acre (0–112 kg/ha)
All other Indiana soils	0–200 lb of Mg per acre (0–224 kg/ha)

Suggested magnesium saturation percentages (of the CEC) range from the suggested minimum value of 3 percent in Michigan to 10 percent in Missouri and Pennsylvania.

[5]R. K. Stivers, et al., mimeographed report, Department of Agronomy, Purdue University, Lafayette, Indiana, 1979.

Researchers in Ohio suggest that soil used for the growth of animal forage have a minimum of 15 percent basic cation saturation by magnesium.

Adequate levels of available soil magnesium are important not only for normal plant growth and maximum yields but also for inhibiting nutritional and metabolic disorders referred to as *grass tetany* (*staggers*, or hypomagnesemia) in grazing cattle. Potassium-induced magnesium deficiency may occur in plants if potassium exceeds magnesium as a percentage of the total available basic cations.[6]

Magnesium deficiency can be corrected using dolomitic limestone ($MgCO_3 \cdot CaCO_3$) or soluble sources of magnesium sulfate (epsom salts) and sulfate of potash magnesia.

Soil-Test Sulfur

Sulfur tests are less well developed than those for phosphorus, potassium, and lime. Fewer soils exhibit sulfur deficiency, yet the same requirements must be met to develop a good test: a large number of experimental sites and plots for field correlation. The result is that fewer soil sites are available. Developing a test for sulfur meets with some of the same problems as developing a test for nitrogen: There are many ionic forms, it is intimately associated with changes in soil humus, and the most common available form to plants is the soluble, mobile SO_4^{2-} anion.

Soil-Test Boron

Hot-water-extractable boron is generally considered the best measure of the available form of boron in soils. The following interpretations for boron levels have been established:

- Less than 1.0 ppm: *too low* for normal plant growth
- 1.0–5.0 ppm: *adequate* for normal plant growth
- More than 5.0 ppm: *may be excessive or toxic* to certain plants

Crops exhibit considerable variability in their response to boron. Under Michigan conditions, application rates of 1.5 lb to 3 lb of elemental boron per acre (1.7–3.4 kg/ha) are recommended for highly responsive crops and 0.5 lb/a to 1.0 lb/a for low- to medium-responsive crops.[7]

Soil-Test Zinc, Iron, Manganese, and Copper

Considerable effort has been made to develop good predictive tests for the micronutrient metals, but with limited success. The relatively fewer soil locations having these nutrients deficient limits the ease and extent of field correlation study. Because the metals have increased solubility in acids, it was logical to use an acidic extractant (0.1 or 1 molar HCl is common). Some laboratories make predictions of these nutrient needs from data using such extractants. For example, the Mehlich-1 extract has been used for zinc and manganese evaluation. Even the use of Coca-Cola as an extractant is reported to be about as good as the DTPA extractants;[8] Coca-Cola® is a strong solution of carbonic acid and sugar.

---

[6]C. B. Elkins, C. S. Hoveland, R. L. Haaland, and W. A. Griffey, "Wet Soils Increase Risk of Grass Tetany Deaths," *Crops and Soils Magazine* **30** (1977), No. 2, 20–21.

[7]M. L. Vittosh, D. D. Warncke, B. D. Knezek, and R. E. Lucas, "Secondary and Micronutrients for Vegetables and Field Crops," *Extension Bulletin E-486,* Cooperative Extension Service, Michigan State University, East Lansing, 1981.

[8]Ewald Schnug, Juergen Fleckenstein, and Silvia Haneklaus, "Coca Cola® Is It! The Ubiquitous Extractant For Micronutrients in Soil," *Communications in Soil and Plant Analysis* **27** (1996), Nos. 5–8; 1721–1730.

Table 12-8 Zinc Application (Recommended lb/a) for Corn, Sorghum, Soybeans, and Pinto Beans (Kansas)[a]

Management and Crop	Area of State	Zinc Soil Test (ppm Zn)[b] Low (0–0.5)	Medium (0.51–1.0)	High (Above 1.0)
		Lb/a of Zn to Add		
Irrigated corn, sorghum, soybeans, and pinto beans	Entire	8–10	2–5	None
Nonirrigated corn	Eastern (humid)	8–10	2–5	None
Sorghum, soybeans, and pinto beans	Entire	2–5	None	None

Source: D. A. Whitney, R. Ellis, Jr., L. Murphy, and G. Herron, *Identifying and Correcting Zinc and Iron Deficiency in Field Crops,* Cooperative Extension Service, Kansas State University, Manhattan, 1975.
[a]Based on the use of zinc sulfate as source of zinc.
[b]DTPA-extractable zinc.

The ligand DTPA has been extensively studied in Colorado and in a few other states.[9] Its use has spread to many laboratories (see Table 12-8). The accuracy of DTPA-extractable metals to predict nutrient adequacy or deficiency is not proven by local correlation in many areas and on many crops where it is used.

Soil-Test Molybdenum and Nickel

Soil tests for molybdenum and nickel are not done routinely because such deficiencies are not common. The total molybdenum content of soils is low, ranging from traces to 24 ppm and averaging about 1 ppm to 2 ppm. Oxalates, water, and anion exchange resins have been used to extract available molybdenum from soils. Soils containing 0.1 ppm to 0.2 ppm of oxalate-extractable molybdenum generally are not deficient in this element. Nickel is more often toxic than deficient.

12:4 Soil Testing: How Good Is It?

Good fertilizer recommendations result from accuracy of numerous details such as the following: First, the farmer must send in an accurate cropping history, an accurate projection of his expected yield, and an accurately taken soil sample for analysis. Second, the laboratory must analyze the soil sample correctly and have good correlation to field plot data on which to evaluate the soil's test data. Third, the test evaluator must be trained in the job and know the area and plants for which he or she makes predictions (Figure 12-3). Fourth, the farmer must correctly select good seed, plant on time and at the correct density, manage the crop adequately, and follow the fertilizer recommendation. Finally, the recommendation cannot account for adverse weather (heavy rains, hail, drought, cold periods), pest damage, or poor management. It is easy to imagine from this partial list that some producers will not be satisfied with the results they obtain for a given field in a given year. But what is the cause of the problem that brings dissatisfaction? Was the problem caused by an error in fertilizer addition,

[9]W. L. Lindsay and W. A. Norvell, "Development of a DTPA Soil Test for Zinc, Iron, Manganese, and Copper," *Soil Science Society of America Journal* **42** (1978), 421–428.

FIGURE 12-3 Dryland bananas in wet Puerto Rico. To make fertilizer recommendations for this crop requires knowing the nutrient needs of bananas, the growing period, the rainfall pattern, and the phosphorus fixation of the tropical soils. Experience is a great help. (*Source:* USDA—Soil Conservation Service.)

by the soil sampler, by an incorrect yield prediction, by management error, or by the laboratory's recommendation?

Soil tests, as now used, are generally quite good for phosphorus, potassium, soluble salts, pH, and lime requirements. Tests in use for nitrate nitrogen are useful, and correlations are improving, for specific crops and areas. Tests for the other nutrients are improving and locally may be quite good and useful. Varying degrees of success are obtained using tests for nitrogen, magnesium, sulfur, boron, zinc, and iron. Because soil N tests are still not widely established, *most nitrogen recommendations are still made from the data of many years of field plot trials.*

Recommendations Are Good but Not Perfect

There are many opportunities for error in making fertilizer recommendations. To eliminate most errors the correlations would require thousands of field plots, for every crop and variety, in all soils and climates, and in all management levels, which would entail horrendous costs. Accurate field plot testing may cost $20,000 per field test; these are difficult to fund today. The extent to which correlations of soil tests to field response are updated and done accurately determines the quality of the recommendations. All these limitations are part of the *experimental error* that exists in making fertilizer predictions.

Fertilizer predictions are not exact, and they are done better by some persons and laboratories than by others. Most N, P, and K fertilizers and lime recommendations are valuable and quite accurate. Laboratories and their scientists usually suggest that the recommendations should be modified by the individual users (1) according to their own knowledge of their fields, (2) according to their management level, (3) on the basis of responses they have observed in past years, and (4) by the prediction of the weather pattern for this crop season. Being accurate is not easy, but soil testing helps.

Why Recommendations May Be in Error

Why aren't the predictive tests more accurate? The laboratory analyses are easy to do well, so the greatest likelihood for errors arise from neglecting the following principles:

1. **The soil sample must be taken accurately.** All of the analyses and interpretations are based on a field's soil sample. Each laboratory lists directions for taking the sample (depth, time, composite of subsamples).
2. **An accurate field history is needed.** Give an accurate history of each field's use and fertilization in recent years.
3. **Laboratory tests need an up-to-date correlation to field plots.** This is critical and is the responsibility of the testing laboratory. *Laboratory data have no meaning until the chemical test values are carefully correlated to crop responses to different fertilizer levels in field plot trials.* Old correlation data, data from obsolete plant varieties, different management, changed rotations, or plant densities change recommendations. Up-to-date field correlations with current varieties, current fertilizers, and current management systems are essential.
4. **Good computer integration programs are essential.** Many growth factors are involved in making predictions. A human mind cannot continually and consistently do the integration of the many factors possible that a computer can do. A vast computer memory plus a wise, trained scientist make a formidable combination. However, all data and interrelationships in a computer were measured, interpreted, and installed in the computer by people, so the output can only be as good as the quality of the input.
5. **A wise analyst and predictor helps.** Even with extreme care in all work with natural systems, exceptions and special cases will continually be encountered. Data for evaluating some factors of growth is lacking; nevertheless, the predictor still must make a recommendation. An experienced scientist has his or her own accumulated knowledge and feel of when and how an adjustment should be made for a sampled area. Many of these uniquely qualified scientists have extensive field experience themselves.

12:5 Analysis of Plants

The only way to be sure that nutrients are available to plants is to measure the nutrients that plants have taken up. Many of the following conditions may alter nutrient uptake, even though the nutrient is adequately soluble:

- Low soil temperature
- Rapid plant growth
- Low soil water contents
- Poor aeration
- Antagonistic or synergistic interactions among nutrients
- Root damage

Plant Analysis vs. Soil Testing

Should we change from using soil tests to using plant analysis for predicting fertilizer needs? Because the plant is the final arbiter of whether a nutrient is truly available, this might at first seem to be a logical change; however, the plant cannot be analyzed until it is growing. Using the plant analysis means there is no information for preplanting or during-planting fertilization. Further, it cannot be known if a plant is deficient in an element until the plant has been seen or measured. By that time the plant's growth is reduced for lack of nutrients and it may be too late to adequately correct the damage done to yields. Only with a few long-growing crops (sugarcane, turf, citrus, forests, bananas) might plant analysis alone be suitable for predicting fertilizer needs.

Quick Tests in the Field

For many years field tests were used for quick estimates of nitrogen, phosphorus, and potassium amounts in young growing plants. They were done on chopped plant material to measure the amounts in the plant juices (sap). Such tests can be useful, but the field tests were not precise enough. Better portable test kits are now available, and quick field measurements are again being used for evaluating nitrate and potassium status on crops that are fertilized several times during growth (as with drip and sprinkler systems). The petioles (or other suggested plant part) are collected from about twenty plants. The petioles are chopped and mixed. The sap is squeezed from the chopped material and the nutrient content is measured. The time needed for analysis is only a few minutes. Test kits cost between $100 and $300. Sampling can be done as often as every week. Some recommended test values (for Florida) are in Table 12-9.

Total Plant Analysis

Total plant analysis is a laboratory measurement of the plant contents of each element analyzed. The increasing ability to do total plant analysis quickly and easily has increased the use of the testing. Quantitative values of any of the plant nutrients are possible within hours, but more commonly it can take a few days. This fast turnaround time from sampling to content values makes it possible to institute fertility adjustments within a few days by applying fertilizers in irrigation water, as foliar sprays, in drip systems, or even as top dressings.

Sampling for Total Plant Analyses

The tissue sampled for analysis should be selected on the basis of *physiological age* (development stage) rather than chronological (calendar) age. Ideally, normal healthy plants that represent most of the field should be sampled. If most plants are not looking well, a sample from a normal healthy plant should be obtained at the same time of sampling any problem plant areas so that the diagnostician can base conclusions on comparative tests.

The sensitive part of the plant is often close to the developing fruiting body. Orange tree leaves from *nonfruiting* terminal branches contained more than 50 percent more nitrogen and phosphorus, 220 percent more potassium, and 35 percent less magnesium than leaves behind young fruits. Micronutrient contents vary within individual corn leaves. Boron, iron, and manganese accumulate in leaf margins, so even *the portion of a leaf* used for analysis can influence test results. Sampling should be done according to the instructions of the laboratory doing the testing. If the grower collects *old petioles* as the sample and the laboratory interprets results from a database of *young, whole leaves,* interpretations will be meaningless.

Usually the element composition in a leaf varies during growth. Nitrogen, phosphorus, potassium, and zinc levels in leaves usually *decrease* as the growing season progresses; contents of calcium, magnesium, boron, manganese, and iron *increase* during the growing season.

Table 12-9 Nitrogen and Potassium Levels Suggested As Being Satisfactory for Several Vegetable Crops As Measured in Quick Field Tests. Tests are for Florida Conditions But May Be Applicable in Other Areas.

		Fresh Petiole Sap Concentration in Parts per Million (ppm)	
Crop	*Crop Development Stage*	*Nitrate-N*	*K*
Broccoli and collard	Six-leaf stage	800–1000	No information
	1 week prior to first harvest	500–800	
	First harvest	300–500	
Cucumber	First blossom	800–1000	No information
	Fruits 8 cm long	600–800	
	First harvest	400–600	
Eggplant	First fruit 5 cm long	1200–1600	4500–5000
	First harvest	1000–1200	4000–4500
	Mid harvest	800–1000	3500–4000
Muskmelon	First blossom	1000–1200	No information
	Fruits 5 cm long	800–1000	
	First harvest	700–800	
Pepper	First flower buds	1400–1600	3200–3500
	First open flowers	1400–1600	3000–3200
	Fruits half-grown	1200–1400	3000–3200
	First harvest	800–1000	2400–3000
	Second harvest	600–800	2000–2400
Potato	Plants 20 cm tall	1200–1400	4500–5000
	First open flowers	1000–1400	4500–5000
	50% of flowers open	1000–1200	4000–4500
	100% of flowers open	900–1200	3500–4000
	Tops falling over	600–900	2500–3000
Squash	First blossom	900–1000	No information
	First harvest	800–900	
Tomato (field)	First buds	1000–1200	3500–4000
	First open flowers	600–800	3500–4000
	Fruits 2–3 cm diameter	400–600	3000–3500
	Fruits 5 cm diameter	400–600	3000–3500
	First harvest	300–400	2500–3000
	Second harvest	200–400	2000–2500
Watermelon	Vines 15 cm long	1200–1500	4000–5000
	Fruits 5 cm long	1000–1200	4000–5000
	Fruits half mature	800–1000	3500–4000
	At first harvest	600–800	3000–3500

Source: Modified from Jean D. Aylsworth, "Zap the Sap for Precise Fertigation," *American Vegetable Grower* **44** (1996), No. 2; 32–35.

Some suggested sampling instructions for vegetable crops, fruit and nut crops, field crops, and ornamentals and flowers are presented in Tables 12-10 and 12-11. Even concentrations of nutrients in *bark* may be useful in determining nutrient needs for some crops, such as cassava. In young seedlings of annual crops (corn, grains, beans), the *whole plant* is sampled. As the plant becomes more mature, the *youngest mature leaf* seems to be the preferred plant part. When this leaf is near the fruit body (corn ear, fruit terminal, grape cluster), nutrients it

Table 12-10 Suggested Sampling Instructions

Stage of Growth	Plant Part to Sample	Number of Plants to Sample
Selected Vegetable Crops		
Potato, prior to or in bloom	3rd to 6th leaf from growing tip	20–30
Head crops (cabbage, etc.), prior to heading	1st mature leaves from center whorl	10–20
Tomato, prior to or during bloom	3rd or 4th leaf from growing tip	20–25
Root crops (carrots, onions, beets), prior to root or bulb enlargement	Center mature leaves	20–30
Celery, midgrowth (30–70 cm tall)	Petiole of youngest mature leaf	15–30
Leaf crops (lettuce, spinach, etc.), midgrowth	Youngest mature leaf	35–55
Sweet corn, at tasseling	Entire leaf at the ear node	20–30
Melons (watermelon, cucumber, etc.), early stages of growth prior to fruit set	Mature leaves near the base portion of plant on main stem	20–30
Selected Field Crops		
Corn, tasseling to silking	Entire leaf at ear node	20–30
Beans, soybeans, initial flower	2 or 3 fully developed leaves	20–30
Small grains, rice, prior to heading	4 uppermost leaves	50–100
Hay, pasture, forage grasses; optimum forage stage	4 uppermost leaf blades	40–50
Alfalfa, clovers, prior to 1/10 bloom	Mature leaf blades, top 1/3 of plant	40–50
Sugar beets, midseason	Fully mature leaves midway between younger center and oldest leaves	30–40
Tobacco, before bloom	Uppermost fully developed leaf	8–12
Grain sorghum, heading or prior	2nd leaf from top of plant	15–25
Sugar cane, up to 4 months old	3rd or 4th mature leaf from top	15–25
Peanuts, bloom stage or prior	Mature leaves	40–50
Cotton, prior to or when first squares appear	Youngest mature leaves on main stem	30–40

Source: Modified from John E. Bowen, "Plant Parts to Sample," *Crops and Soils Magazine* **31,** No. 3, Dec. 1978.

might have contained are often robbed to help feed the growing fruit body. For ornamental trees only the developed leaves on the current year's growth are recommended for sampling.

Preparation of Plant Samples

Dry the plant material samples for at least 24 hours to remove moisture and prevent spoilage during mailing. Samples should not be mailed in air-tight containers (plastic bags) because the sample will spoil if the material contains appreciable water.

Samples contaminated by soils, sprays, or other materials must be cleaned by a quick (one minute or less) washing of the *fresh green tissue* because nutrients such as potassium, sodium, nitrate, and chloride may be readily leached from the plant material. The plant tissues should be washed *quickly* in 0.1 percent to 0.3 percent detergent solution using a *phosphate-free* material to avoid phosphorus contamination. The tissue should be rinsed thoroughly but quickly in distilled water as soon as possible.

Table 12-11 Suggested Sampling Instructions

Stage of Growth	Plant Part to Sample	Number of Plants to Sample
Selected Fruit and Nut Crops		
Apple, apricot, almond, prune, peach, pear, cherry, midseason	Leaves near base of current year's growth	50–100
Strawberry, midseason	Youngest mature leaves	50–75
Pecan, 6–8 weeks after bloom	Leaves from terminal shoots, pairs from middle of compound leaf	30–45
Walnut, 6–8 weeks after bloom	Middle leaflet pairs of mature shoots	30–35
Lemon, lime, midseason	Latest mature leaves on nonfruiting terminals	20–30
Orange, midseason	Spring cycle leaves, 4–7 months old	20–30
Grapes, end of bloom period	Petioles adjacent to fruit clusters	60–100
Raspberry, midseason	Youngest mature leaves on 1st canes	20–40
Selected Ornamentals and Flowers		
Ornamental trees, current year's growth	Fully developed leaves	30–100
Ornamental shrubs, current year's growth	Fully developed leaves	30–100
Turf, during normal growing season	Leaf blades (hand harvest)	1/2 pint
Roses, during flower production	Upper leaves on flower stems	20–30
Chrysanthemums, flowering or prior	Upper leaves on flower stems	20–30
Carnations, unpinched plants	4th or 5th leaf pair from base	20–30
Carnations, pinched plants	5th and 6th leaf pairs from top of primary laterals	20–30
Poinsettias, flowering or prior	Newest fully mature leaves	15–20

Source: Modified from John E. Bowen, "Plant Parts to Sample," *Crops and Soils Magazine* **31**, No. 3, Dec. 1978.

Interpreting Plant Analyses

The diagnostician must have a complete history (plant variety, planting density, geographic location for climate, date planted) of the management factors to make an accurate and practical interpretation of test results. An essential element is classified as deficient when the element is below the **threshold** (critical) **nutrient levels.** These levels change with plant variety, plant part, time of season, climate, other nutrient levels, soil pH, root health, disease, and almost any other plant growth factor. This interpretation, not the laboratory analysis, is the most difficult problem in using plant analysis.

Nutrient sufficiency levels for various crops are listed in Tables 12-12 and 12-13. The assumption is that plant contents below the lower values indicate a deficiency of that nutrient in that plant. There is no general agreement on these threshold values for crops.

Critical Nutrient Range or Threshold Values

The **critical nutrient range (CNR)** has been referred to by various names. In the preceding section, the *threshold* or *critical* nutrient level was used to refer to that CNR range. Figure 12-4 illustrates the relationships among several factors, including CNR.

Table 12-12 Plant Nutrient Sufficiency Levels (Threshold or CNR Values) for Selected Crops

Nutrients	*Units*	*Alfalfa: Top 6 in (15 cm) Sampled Prior to Initial Flowering*	*Corn: Ear Leaf Sampled at Initial Silk*	*Potatoes: Petioles from Most Recently Matured Leaf Sampled in Midseason*	*Soybeans: Upper Fully Developed Leaf Sampled Prior to Initial Flowering*	*Sugar Beets: Center Fully Developed Leaf Sampled in Midseason*	*Vegetables: Top Fully Developed Leaves*	*Wheat: Upper Leaves Sampled Prior to Initial Bloom*
Nitrogen	%	3.76–5.50	2.76–3.50	2.50–4.00	4.26–5.50	3.01–4.50	2.50–4.00	2.59–3.00
Phosphorus	%	0.26–0.70	0.25–0.40	0.18–0.22	0.26–0.50	0.26–0.50	0.25–0.80	0.21–0.50
Potassium	%	2.01–3.50	1.17–2.50	6.0–9.0	1.71–2.50	2.01–6.00	2.00–9.00	1.51–3.00
Calcium	%	1.76–3.00	0.21–1.00	0.36–0.50	0.36–2.00	0.36–1.20	0.35–2.00	0.21–1.00
Magnesium	%	0.31–1.00	0.16–0.60	0.17–0.22	0.26–1.00	0.36–1.00	0.25–1.00	0.16–1.00
Sulfur	%	0.31–0.50	0.16–0.50	0.21–0.50	0.21–0.40	0.21–0.50	0.16–0.50	0.20–0.40
Manganese	ppm*	30–100	20–150	30–200	21–100	21–150	30–200	16–200
Iron	ppm	30–250	21–250	30–300	51–350	51–200	50–250	11–300
Boron	ppm	31–80	4–25	15–40	21–55	26–80	30–60	6–40
Copper	ppm	11–30	6–20	7–30	10–30	11–40	8–20	6–50
Zinc	ppm	21–70	20–70	30–100	21–50	19–60	30–100	21–70
Molybdenum	ppm	1.0–5.0	0.1–2.0	0.5–4.0	1.0–5.0	0.15–5.0	0.5–5.0	0.03–5.0

Source: M. L. Vitosh, D. D. Warncke, B. D. Knezek, and R. E. Lucas, "Secondary and Micronutrients for Vegetables and Field Crops," Extension Bulletin E-486, Cooperative Extension Service, Michigan State University, East Lansing, 1981.

**ppm* means *parts per million,* or the micrograms of element per gram of dry plant material.

Table 12-13 Plant Tissue Nutrient Levels for Various Floricultural Crops (Critical Levels below Which the Plant Should Respond to Added Nutrient)

	Dry Weight (%)					*Micrograms per Gram Dry*				
Plant	*N*	*P*	*K*	*Ca*	*Mg*	*Mn*	*Fe*	*Cu*	*B*	*Zn*
Roses	3.0	0.2	1.8	1.0	0.25	30	50	5	30	15
Carnations	3.0	0.05	2.0	0.6	0.15	30	30	5	25	15
Chrysanthemums	4.5	0.18	2.2	0.5	0.14	200	125	5	25	7
Poinsettias	3.5	0.2	1.0	0.5	0.2	—	—	—	—	—
Geraniums	2.4	0.28	0.6	0.8	0.14	9	60	5.5	18	6

Source: Selected data from R. A. Criley and W. H. Carlson, "Tissue Analysis Standards for Various Floricultural Crops," *Florists' Review* **146** (1970), No. 3771; 19–20.

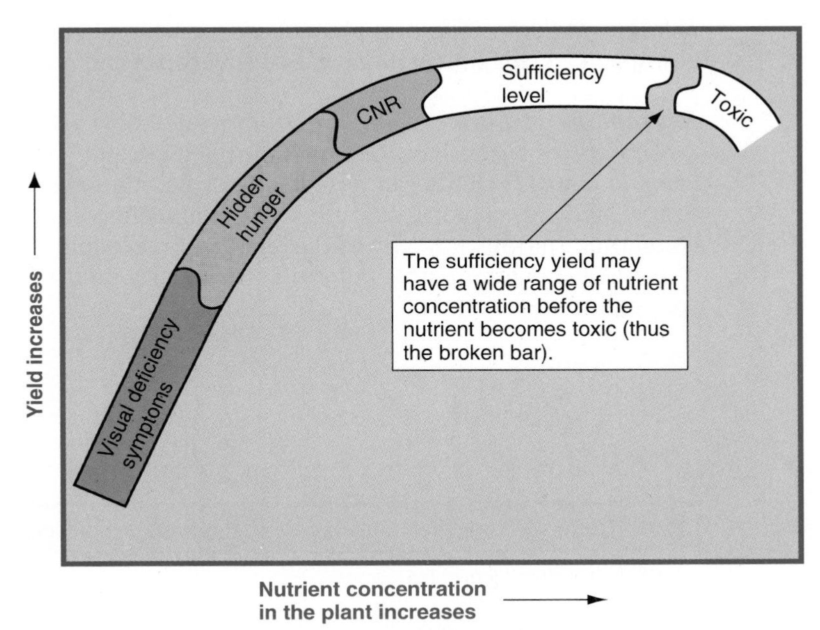

FIGURE 12-4 Relationships of plant growth and visual deficiency symptoms, hidden hunger, critical nutrient range (CNR), sufficiency levels, and toxic concentrations. (Courtesy of Raymond W. Miller, Utah State University.)

The difficulty in pinning down a single critical value for each nutrient in each plant variety has made it necessary to use a range of values for CNR and nutrient *ratios.* One approach to using ratios is the **DRIS** (**d**iagnosis and **r**ecommendation **i**ntegrated **s**ystem), which is based on the concept that the *balance of nutrients* is more critical than the *total amounts.* Certain interactions and relationships occur that can be indicated by various ratios. For example, N:P, N:K, and K:P ratios are used. Both *deficient* and *toxic* levels of elements are determined. The approach sounds simple, but it is complex and will require extensive testing to prove its accuracy and value.[10,11]

[10]C. A. Jones and J. E. Bowen, "Comparative DRIS and Crop Log Diagnosis of Sugarcane Tissue Analysis," *Agronomy Journal* **73** (1981), 941–945.

[11]Milton B. Jones, D. Michael Center, Charles E. Vaughn, and Fremont L. Bell, "Using DRIS to Assay Nutrients in Subclover," *California Agriculture* (Sept.–Oct. 1986), 19–21.

Visual Nutrient Deficiency Symptoms

Plants exhibit external symptoms of starvation as a result of nutrient deficiency or imbalance. Certain nutrient deficiencies cause a reduction in the formation of the green pigment, chlorophyll, giving deficient plants a distinctive yellowish to whitish appearance (**chlorosis**). Unfortunately, chlorosis can also be caused by diseases, insect damage, salt accumulation, and other growth stresses, as well as by several nutrient deficiencies. Some visual deficiency symptoms in plants are easily diagnosed; other visual changes can be symptomatic of many deficiencies. Abnormal growth (color, size) is a clue that something is wrong in growth, but it may not be easy to determine the cause. General patterns for some nutrient deficiencies are shown in Table 12-14.

Nutrient Deficiency Symptoms in Specific Crops

Nitrogen in Corn The corn plant may become light green or yellow (chlorotic), and the oldest leaf by this time will have died (**necrotic**) and turned light-to-dark brown.

Phosphorus in Corn Young phosphorus-deficient corn plants are stunted and *dark green* in color because of their high nitrogen content (stunted growth). Sugars accumulate and increase anthocyanin (reddish-purple) pigment. With a decrease in chlorophyll (green pigment) content, the anthocyanin pigment predominates and the leaf has purple coloration, starting at the tip of the leaf and proceeding along the edges. Some varieties of corn have natural purple coloring, particularly on the stems. In legumes, white spots are random in the leaves.

Table 12-14 Abbreviated Outline of Visual Symptoms of Nutrient Deficiencies

Symptom	Nutrient
I. Chlorotic foliage dominates	
A. Entire leaf chlorotic	
1. Older leaves chlorotic, then necrotic, then leaf drop	Nitrogen
2. Leaves on all parts affected	Sulfur
B. Interveinal chlorosis	
1. Only older and recently mature leaves show symptoms	Magnesium
2. Only younger leaves show interveinal chlorosis	Iron
a. Also tan or gray necrotic spots in chlorotic areas	Manganese
b. Tips and lobes remain green, then rapid necrosis	Copper
c. Young leaves small, internodes short; appear as rosette	Zinc
II. Chlorosis not dominant feature	
A. Symptoms appear at base of plant	
1. All leaves dark green at first, then possible yellowing, growth stunted, purple coloring in older leaves	Phosphorus
2. Margins of older leaves chlorotic, then burn; small, whitish spots scattered over old leaves; spots on some new leaves on cotton	Potassium
B. Symptoms on new growth of plant	
1. Terminal buds die; young leaves thick, leathery, and chlorotic	Boron
2. Margins of young leaves fail to form; growing points cease to develop; bud leaves don't uncurl; roots are short and thick	Calcium

Potassium in Corn The plant is stunted and internodes are shortened when potassium is inadequate. Yellowing of the leaves starts at the tip of *older* leaves and proceeds along the edges to the base of the leaf. Eventually, the edges become brown and die (are necrotic). It has often been termed *leaf scorch.* An excess of soluble salt produces a somewhat similar yellowing at the tips and may be confused with potassium deficiency.

Nitrogen in Cotton A light green or yellow chlorosis becomes general over the entire leaf. The nitrogen-deficient leaf is smaller than leaves from normal, healthy plants. Yellowing starts with the older, basal leaves and progresses up the plant. In advanced stages of nitrogen deficiency, the leaves turn brown and eventually fall from the plant.

Potassium in Cotton The leaf is chlorotic with yellow spots between veins and along tips and margins of the leaf. The leaf tends to curl downward at the margins. At the time of boll formation, reddish spots may appear on leaves near developing bolls.

Manganese Deficiency in Beans On navy beans, soybeans, and garden beans, a manganese deficiency appears first as a mottled (spotty) effect on new leaves (Figure 12-5) and is generally uniform over the leaf while the veins remain green. If symptoms develop early in the season, when the bean plants are small, a side-dress application of manganese sulfate or a spray application of a soluble manganese salt can generally correct the deficiency.

FIGURE 12-5 On soil with pH of 7.5 and deficient in available manganese, a normal green color of navy bean leaves was exhibited only after 134 kg/ha (120 lb/a) of manganese (as manganese sulfate) was applied (right); 45 kg/ha (40 lb/a) of manganese did not supply sufficient manganese for normal color (center); and where no manganese was applied to the soil (left), the leaves were nearly white, except for the veins. (Courtesy of J. Rumpel, B. G. Ellis, and J. F. Davis, Michigan State University.)

12:6 Making Fertilizer Recommendations

The objective in making fertilizer recommendations is to predict accurately the amount of each nutrient needed for growth of a particular plant, both in quality and quantity of plant material wanted. Recommendations for nutrients can be based on different plant production schemes. Different laboratories analyzing the same soil might use a different scheme and suggest that a different amount of fertilizer be applied. The following three schemes are commonly in use:

Prescription Method: Recommend only the nutrients needed for that year's optimum crop yield. This scheme takes into account other nutrient additions (manure, residues, etc.) that will be added to the crop.

Prescription plus Buildup: The nutrients needed for the crop are recommended *plus* more of the nutrient that is considered important to help build up the soil's nutrient level in general to improve growth or plant quality.

Rotation Need: The total fertilizer needed during a full rotation (at least two crops) may be recommended to be added before a certain crop. As an example, in a corn–soybean rotation, enough nitrogen, phosphorus, and potassium may be added to the corn to carry over to the soybeans that follow. The decision must be made whether or not the cost of two applications (or the environmental hazard) is cheaper than the lowered efficiency of fertilizer added so far ahead of the soybean crop—or whether two applications are even permitted.

Developing a Fertilizer Recommendation

To develop a correlation between soil test values on a soil and yield response to added fertilizer on that soil in field plots, a large number of field plot trials are essential. For example, different rates of the nutrients (N, P, K, etc.) are applied to small plots in the field and a crop is grown for yield data. The soil sample is tested with the selected laboratory extractant. The values are obtained for many soil samples from many different field locations. The many soil test results are grouped according to the soil test values obtained and are compared to yields from the various field sites, each site having many plots with different fertilizer rates added.

Suppose a phosphorus test was being established, the large number of soil sites were tested, and the soil test values determined. The soil test values and crop yields might be as those given in Table 12-15. From this data it would be obvious that if the soil tested 2.2 mg/kg of P, yields would be increased by adding at least 90 kg/ha of P_2O_5: there might be a yield increase if even more P is added. For soils that have test values of 11 to 16, there would be little yield increase with an addition of more than 60 kg/ha of P_2O_5. For soils with test values

Table 12-15 Example of a Correlation Tabulation of Soil-Test P and Alfalfa Yield Response to Fertilizer P in a Variety of Soils

	Alfalfa Yields at Different P_2O_5 Rates (kg/ha)			
Soil-Test P Values (mg/kg)	*0*	*30 kg/ha*	*60 kg/ha*	*90 kg/ha*
0–3	1880	2800	3660	4500
4–6	2480	3410	4240	4750
7–10	3140	4050	4610	4900
11–16	3860	4530	4800	4880
17–24	4210	4820	4910	4950

in the 17 to 24 mg/kg of phosphorus range, only 30 or 40 kg/ha of P_2O_5 additions would increase yields significantly.

Tables (such as Table 12-14) are needed for each crop (or even variety), for a selection of soils, for different climatic areas, and for each nutrient. Usually, there is considerable grouping of the test values, similar crops, and kinds of soils. To develop good correlation tables, an extraordinary number of field plots are needed and the data must be compiled and grouped to obtain *average* values of fertilizer needed for yield increases.

Once the general amount of needed fertilizer is known, there are various details to be considered in preparing the recommendation, including the following:

1. *Subtract* a certain amount of recommended fertilizer for each ton of manure to be added (such as 10 kg P_2O_5 per ton added).
2. *Subtract* a certain amount of recommended fertilizer for the residual P or N left over from previous crops that will be usable by this crop. Tables of these amounts are prepared from numerous experimental data and are probably programmed into the computer.
3. *Add* or *subtract* from the recommended fertilizer base the amount of fertilizer needed to meet yield goals that are larger or smaller than the reference value.
4. *Add* or *subtract* fertilizer depending on the area climate (crops selected to fit short season, extra cool seasons, or expected heavy leaching).
5. *Add* or *subtract* recommended nitrogen, phosphorus, or sulfur because of unusually high or low humus or crop residue levels in the soil. These nutrients will be supplied from (or used up by) decomposing organic materials. Soil tests do not measure the effects of changes in soil humus.

Computer Use in Recommendations

Computers are now widely used in recommendations. They allow for integration of many interactions, cautions, and associations of properties. For example, a computer could be coded to print out a warning of possible sodic soil conditions if the pH entered is more than 8.5. Several factors could be tied together so that a warning occurs whenever that *combination of factors* all occur. In the past, this integration of data was done by people who did it more slowly than computers and often forgot to do it altogether. The computer does not forget the combinations put into it, but a word of caution: *A computer readout is only as good as the data put into it.* Setting up the computer with poor or incorrect data can only result in poor or incorrect interpretations.

Soil Test Example

A good soil test laboratory provides sample containers and instructions for sampling and filling out information about the sample. Most laboratories require the following information:

- The client's name and address
- Sample identification
- The crop to be grown
- The **last crop grown**
- The last **lime addition**
- Any **second crop** to follow the next crop

To help get uniformity and to keep it simple, code numbers are often given for a list of crops, such as 084 for sweet corn, 026 for home lawn, or 099 for sweet potato. One laboratory information sheet lists 152 plants by code number. Figure 12-6 illustrates a soil test report. All test values are converted (proportioned by mathematical factor) to what is called a **soil test**

NCDA Agronomic Division 4300 Reedy Creek Road Raleigh, NC 27607-6465 (919) 733-2655 **Report No: 02269**

Soil Test Report

Grower: **Public, John**
PO Box 000
Raleigh, NC 27607

Copies to: County Extension Director
Farm Supply

Farm: Taylor

10/7/96 **SERVING N.C. CITIZENS FOR OVER 50 YEARS** **Wake County**

Agronomist Comments: A -- 3, 12, $

This is an example report for predictive and diagnostic samples. Predictive reports are generally mailed without review. Diagnostic samples are reviewed by an agronomist before mailing. Comments and recommendations from the agronomist regarding fertility problems and corrective action required will appear in this area. Diagnostic samples include soluble salts in addition to the components of a predictive soil test. Soil test reports can be accessed on the World Wide Web when the test results have been completed. For questions regarding the test results call M. Ray Tucker or J. Kent Messick @ (919) 733-2656

REMEMBER: THE SOIL TEST IS NO BETTER THAN THE SAMPLE

Field Information / Applied Lime / Recommendations

Sample No.	*Last Crop*	*Mo*	*Yr*	*T/A*	*Crop or Year*	*Lime*	*N*	*P_2O_5*	*K_2O*	*Mg*	*Cu*	*Zn*	*B*	*Mn*	*See Note*
TF001	Corn, Grain				1st Crop: Corn, Grain	.3T	120-160	0	10-30	0	0	0		0	3
					2nd Crop:										

Test Results

Soil Class	*HM%*	*W/V*	*CEC*	*BS%*	*Ac*	*pH*	*P-I*	*K-I*	*Ca%*	*Mg%*	*Mn-I*	*Mn-AI (1)*	*Mn-AI (2)*	*Zn-I*	*Zn-AI*	*Cu-I*	*S-I*	*SS-I*	*NO_3-N*	*NH_4-N*	*Na*
MIN	0.71	1.33	4.3	77.0	1.0	5.7	232	67	59.0	10.0	60	53		172	172	133	57				0.1

Field Information / Applied Lime / Recommendations

Sample No.	*Last Crop*	*Mo*	*Yr*	*T/A*	*Crop or Year*	*Lime*	*N*	*P_2O_5*	*K_2O*	*Mg*	*Cu*	*Zn*	*B*	*Mn*	*See Note*
TF100	Small Grains	12	1995	0.5	1st Crop: Soybeans	.5T	0	0	10-30	0	0	0		0	3
					2nd Crop:										

Test Results

Soil Class	*HM%*	*W/V*	*CEC*	*BS%*	*Ac*	*pH*	*P-I*	*K-I*	*Ca%*	*Mg%*	*Mn-I*	*Mn-AI (1)*	*Mn-AI (2)*	*Zn-I*	*Zn-AI*	*Cu-I*	*S-I*	*SS-I*	*NO_3-N*	*NH_4-N*	*Na*
MIN	1.49	1.36	3.4	59.0	1.4	5.6	351	63	40.0	9.0	70	52		228	228	200	43				0.1

HM% = humus, W/V = weight/volume of soil, Ac= acidity, P-I= phosphorus index (same for all other -I)

FIGURE 12-6 Example soil test report with recommendations for (1) grain corn following last year's grain corn and (2) soybeans following last year's small grains. (See Table 12-16 for nutrient soil test index values.) Notice no nitrogen recommended for N-fixing soybeans. *Notes* are not shown but are additional specific information. (Data provided by M. Ray Tucker and the Agronomic Division, North Carolina Department of Agriculture, 4300 Reedy Creek Road, Raleigh, NC 27607.)

index so that all tests are interpreted by the producer from the same numerical reference values. The soil test index used for Figure 12-6 is given in Table 12-16. The laboratory providing Figure 12-6 suggests sampling soils 0 cm to 10 cm deep for sod and 0 cm to 20 cm deep for cultivated fields.

Plant Analysis Example

Chemical analysis of plant material is extensively done. The laboratory providing the sample report in Figure 12-8 converts all numbers to an **interpretive index scale,** which is illustrated in Figure 12-7. In this scheme both the actual content values and the interpretive index values are given in the report. Ratios of elements (N:S, N:K, K:Ca, and Ca:Mg) are given for those wishing to use additional evaluation systems, such as the DRIS analysis. In Figure 12-7 the strawberry plant samples include a *good* sample and one from an area of poor growth, labelled a *bad* sample.

Table 12-16 Soil Test Index Values for Predicting Crop Response to Nutrient Application

Soil Test Index		*Crop Response to Nutrient Application*				
Range	*Rating*	*Phosphorus*	*Potassium*	*Manganese*	*Zinc*	*Copper*
0–10	Very low	Very high	Very high	Very high	Very high	Very high
11–25	Low	High	High	High	High	High
26–50	Medium	Medium*	Medium*	None	None	None
51–100	High	None	Low–none	None	None	None
100+	Very high	None	None	None	None	None

Data provided by M. Ray Tucker and the Agronomic Division, North Carolina Department of Agriculture, 4300 Reedy Creek Road, Raleigh, NC 27607.
*Response decreases as soil test index increases.

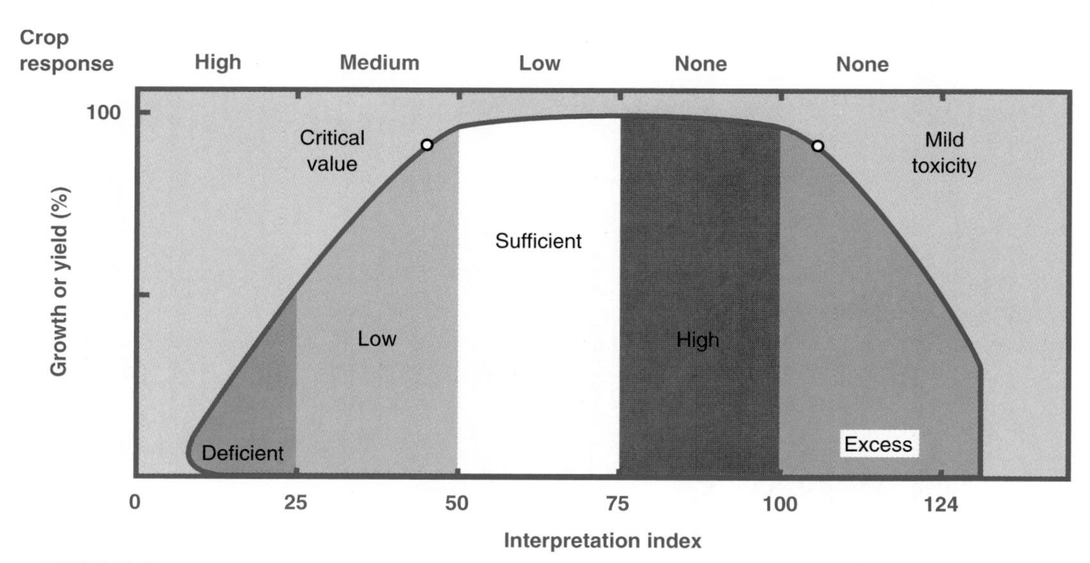

FIGURE 12-7 The numerical *interpretive index* used for nutrients in plant analysis reports. A value of 50 or higher is adequate. Higher than 100 may cause yield decrease from toxicity or some imbalance of nutrients. (Data provided by M. Ray Tucker and the Agronomic Division, North Carolina Department of Agriculture, 4300 Reedy Creek Road, Raleigh, NC 27607.)

NCDA Agronomic Division 4300 Reedy Creek Rd Raleigh, NC 27607-6465 (919) 733-2655 **Report No: P00574**

Plant Analysis Report

Grower: **Education Sample**
4300 Reedy Creek Road
Raleigh, NC 27607-6465

Copies to: County Extension Director

Farm: Strawberry

Wake County

Sample Information

Sample ID: Good
Crop: Strawberry
Plant Appearance: Normal

Laboratory Results

N%	P%	K%	Ca%	Mg%	S%	Fe-ppm	Mn-ppm	Zn-ppm	Cu-ppm	B-ppm	Mo-ppm	Cl%	Na%	Ni-ppm	Cd-ppm	Pb-ppm
4.52	0.65	2.15	0.85	0.40	0.25	75.0	35.0	25.0	6.0	18.0			0.02			

Interpretation Indexes

N	P	K	Ca	Mg	S	Fe	Mn	Zn	Cu	B	Mo
87	95	68	58	68	60	52	50	55	56	32-L	

DRIS Interpretation: *Most limiting— —Least limiting*

Ratios

N:S	N:K	K:Ca	K:Mg	Ca:Mg	Fe:Mn
18.1	2.10	2.53	5.38	2.13	2.14

Nitrate N (ppm): 5500

Recommendations:
Boron is low in the tissue. Nitrogen and other elements look very good. Petiole nitrate nitrogen is within the desired range for early stages of growth. To correct the boron problem, I would inject 0.125 lb of boron per acre. One application should be sufficient. Be careful not to over apply boron since it can become toxic.

Sample Information

Sample ID: Bad
Crop: Strawberry
Plant Appearance: Yellowing

Laboratory Results

N%	P%	K%	Ca%	Mg%	S%	Fe-ppm	Mn-ppm	Zn-ppm	Cu-ppm	B-ppm	Mo-ppm	Cl%	Na%	Ni-ppm	Cd-ppm	Pb-ppm
2.75	0.28	1.90	0.65	0.34	0.15	75.0	55.0	28.0	8.0	34.0			0.02			

Interpretation Indexes

N	P	K	Ca	Mg	S	Fe	Mn	Zn	Cu	B	Mo
45-L	60	64	54	61	50	52	52	57	60	59	

DRIS Interpretation: *Most limiting— —Least limiting*

Ratios

N:S	N:K	K:Ca	K:Mg	Ca:Mg	Fe:Mn
18.3	1.45	2.92	5.59	1.91	1.36

Nitrate N (ppm): 1500

Recommendations:
Leaf and petiole nitrogen values are low for this stage of growth. I would feed 1.5 lbs of nitrogen per acre per day until nitrogen increases to the desired range. At that time, you may want to decrease nitrogen feeding rate to 0.75 lb N per acre per day. You can feed the nitrogen once or twice weekly. Sulfur is also at the lower limit of the desired range. The N:S ratio is at the upper limit of the desired range. Once weekly, I would inject 10 lbs Epsom Salt per acre to supply sulfur. Concentrations of all other elements are within the desired ranges.

FIGURE 12-8 Example plant analysis report showing results and recommendations for strawberries, in which a portion were not growing well (Sample ID: bad). Both absolute values (top line) and interpretation indexes (second line) are given. See Figure 12-8 for Interpretation indexes. (Data provided by M. Ray Tucker and the Agronomic Division, North Carolina Department of Agriculture, 4300 Reedy Creek Road, Raleigh, NC 27607.)

Waste Analysis Example

The major wastes are animal manures. The N, P, and K in these materials are valued at about \$32 to \$45 per Mg or hectare-cm (\$35–\$50/t or acre-in). It is suggested that North Carolina livestock and poultry farms generate enough waste to meet 20 percent of the annual nitrogen, 60 percent of phosphate, and 40 percent of potassium requirements for all crops except pastures. Both solid and solution (wastes or regular water) samples are analyzed in Figure 12-9. The material may even have value as lime. It is estimated that 50 percent to 75 percent of the nutrient quantities will become available for the crop within the first month. Potentially harmful elements (Ni, Cd, Pb, Na) are measured.

Nematode Diagnostic Assay Example

In selected areas particular problems may justify special emphasis. In North Carolina, for example, nematode problems are significant enough that routine assays are done on submitted samples. A **hazard index** is used to indicate the potential for damage:

Hazard Index	*Potential for Damage*	*Action Codes*
0–19	Very low	**A:** Nematodes are not a threat to current crop
20–39	Low	**B:** Nematode population is low but may cause some plant stress
40–59	Moderate	**C:** Population is high enough to cause considerable damage
60–79	High	**D:** Use a nematode-resistant variety
80–100	Very high	**E:** Use crop rotation with nonhost crop

Sampling techniques and storage for samplings for nematode diagnosis are somewhat unique, and instructions from the laboratory need to be followed. An example report is given in Figure 12-10.

Assumptions in Fertilizer Recommendations

Some assumptions are made in interpreting soil tests. If assumptions made are *not* valid for the tested soil, the recommendations may be inaccurate for the given soil. Some of the assumptions used are the following:

Assumption 1: The soil is as deep as the plant normally roots. This is from about 1 m (39 in) for shallow-rooted crops to nearly 2 m (79 in) for corn, alfalfa, and other deep-rooted crops. If your soil is only 50 cm to 80 cm deep, there is only a partial root zone available to supply growth needs.

Assumption 2: No conditions inhibit root growth or root penetration. If inhibiting layers (plowpans, genetic hardpans, compacted and poorly aerated soils) exist, they may effectively limit the root zone as in Assumption 1 above.

Assumption 3: Other growth factors are assumed to be adequate. It is assumed that periods of drought, excess wetness, unusually cold or hot periods, and toxic levels of salts or other elements do not occur. It is usual in recommending nitrogen fertilizer for dryland crops to give a range of rates for fertilizer addition—high rates for high-rainfall years, medium rates for average years, and low rates for drought years—which allows a grower to make his or her own prediction of what the year's rainfall will be.

Assumption 4: Management is good. Suitable control of pH is assumed. A grower may plant too late, not control weeds well, select a poor variety or quality of seed, have poor timing in planting or adding split fertilizer applications and pesticides, irrigate too infrequently or too little, and get too thin or too dense a stand. All of these actions can reduce yields and make a farmer doubt the value of the fertilizer recommendation.

NCDA Agronomic Division 4300 Reedy Creek Road Raleigh, NC 27607-6465 (919) 733-2655 **Report No: W00585 W**

Waste Analysis Report

Grower: **Educational Sample**
4300 Reedy Creek Rd.
Raleigh, NC 27607-6465

Copies to: County Extension Director

Farm: Swine and Broiler

Wake County

Sample Info.	Laboratory Results (parts per million unless otherwise noted).																	
Sample ID: 1	*DM%*	*N*	*P*	*K*	*Ca*	*Mg*	*S*	*Fe*	*Mn*	*Zn*	*Cu*	*B*	*Mo*	*Cl*	*Na*	*Ni*	*Cd*	*Pb*
		343	58.6	351	137	52.7	31.6	2.70	0.26	0.74	0.30	0.35			124			

Waste Code: ALS ***Description:*** Swine Lagoon Liq.	Nutrients Available for First Crop lbs/1000 gallons													Other Elements *lbs/1000 gallons*			
Application Method	*N*	P_2O_5	K_2O	*Ca*	*Mg*	*S*	*Fe*	*Mn*	*Zn*	*Cu*	*B*	*Mo*	*Cl*	*Na*	*Ni*	*Cd*	*Pb*
Irrigation	1.3	0.78	2.8	0.80	0.31	0.18	0.02	T	T	T	T			1.0			
Soil Incorp	2.3	0.90	3.2	0.91	0.35	0.21	0.02	T	0.01	T	T						

Recommendations:
Nutrients available for the first crop are based on estimates of mineralization rate and projected loss for the application method listed. Concentrations of zinc and other metals are not excessive. The waste product should not cause production or environmental problems if utilized according to recommended practices. Monitor nutrient buildup and soil pH with a soil test no less than every two years. I would apply the waste at rates needed to supply nitrogen for crop production.

Sample Info.	Laboratory Results (parts per million unless otherwise noted).																	
Sample ID: 2	*DM%*	*N*	*P*	*K*	*Ca*	*Mg*	*S*	*Fe*	*Mn*	*Zn*	*Cu*	*B*	*Mo*	*Cl*	*Na*	*Ni*	*Cd*	*Pb*
	73.98	40519	15898	34890	28923	6827	8021	4269	761	641	663	50.9			10636			

Waste Code: SLB ***Description:*** Broiler Stockpiled	Nutrients Available for First Crop lbs/ton (wet basis)													Other Elements *lbs/ton (wet basis)*			
Application Method	*N*	P_2O_5	K_2O	*Ca*	*Mg*	*S*	*Fe*	*Mn*	*Zn*	*Cu*	*B*	*Mo*	*Cl*	*Na*	*Ni*	*Cd*	*Pb*
Broadcast	27.6	32.3	49.6	25.7	6.1	7.1	3.8	0.68	0.76	0.82	.05			15.7			
Soil Incrop	36.0	40.4	55.8	32.1	7.6	8.9	4.7	0.84	0.85	0.93	.06						

FIGURE 12-9 Example waste analysis report for (1) swine lagoon liquid waste and (2) poultry broiler stockpiled solid waste. Most agricultural test laboratories still report to farmers in pounds per gallon or pounds per ton to avoid confusion. Nutrients available the first crop vary with the application method (see report) and the climate of the area. Broiler litter can be higher in nutrients than cattle manures. (Data provided by C. Ray Campbell and the Agronomic Division, North Carolina Department of Agriculture, 4300 Reedy Creek Road, Raleigh, NC 27607.)

NCDA Agronomic Division 4300 Reedy Creek Rd. Raleigh, NC 27607-6465 (919) 733-2655

Nematode Diagnostic Assay Report

Report No: 00233

Grower: **Educational Sample**
4300 Reedy Creek Road
Raleigh, NC 27607-6465

Copies To: County Ext. Dir. - Sampson

10/ 8/96 *Farm:* **Sampson County** 006

Nematologist's Comments

Root-knot nematodes are causing the damage seen in the garden. Read the enclosed information on methods for managing root-knot in the home garden. The ring nematode population is high enough to be stressing the lawn, but grass is usually not damaged greatly by ring nematodes unless other problems are also present. Maintaining optimum growing conditions should help the grass tolerate the nematodes.

Field vege: Recommendations

Last Crop	Crop	Action Codes	Nema Notes
garden	garden	C	

Nematodes/500 cc Soil

Nematode	#	Nematode	#	Nematode	#
Root Knot	9850**				
Ring	30				

Nematodes/Gram Root

Nematode	#	Nematode	#	Nematode	#
Root Knot	1000**				

Field yard: Recommendations

Last Crop	Crop	Action Codes	Nema Notes
grass	grass	A	006

Nematodes/500 cc Soil

Nematode	#	Nematode	#	Nematode	#
Root Knot	40				
Lesion	10				
Dagger	80				
Ring	1480*				

Nematodes/Gram Root

Nematode	#	Nematode	#	Nematode	#
Ring	40				

Nematologist's Comments

(none)

Field 001: Recommendations

Last Crop	Crop	Action Codes	Nema Notes
corn	small grain	A	
	soybean	D	2-4

Hazard Index

Crop	Nematode: Root Knot	Lesion	Dagger
small grain			
soybean	30-90	0-5	0-5

Nematodes/500 cc Soil

Nematode	#	Nematode	#	Nematode	#
Root Knot	1100@	Stubby Root	50		
Lesion	140@	Dagger	120@		
Stunt	9200				
Spiral	5600				

Field 004: Recommendations

Last Crop	Crop	Action Codes	Nema Notes
alfalfa	alfalfa	D and E	

Hazard Index

Crop	Nematode: Root Knot	Lesion
alfalfa	20-50	55-80

Nematodes/500 cc Soil

Nematode	#	Nematode	#	Nematode	#
Root Knot	3200@				
Lesion	440@				

FIGURE 12-10 Example nematode diagnostic assay report for nematode hazard in (1) garden crops, (2) lawn grass, (3) corn, and (4) alfalfa. Nematode counts are made and recommendations for five possible kinds of action (*A-B-C-D-E Action Codes,* given in the text) are given. Notice that both soybean and alfalfa have serious nematode damage (Action codes D and E). (Data provided by Jack Imbriani and the Agronomic Division, North Carolina Department of Agriculture, 4300 Reedy Creek Road, Raleigh, NC 27607.)

With all of the challenges involved in getting a good representative soil sample, and all the factors necessary to make accurate fertilizer predictions, it is amazing how often and how well the recommendations work. Soil tests predict many situations well; their use will continue to get better.

12:7 Balancing Soil Nutrients

Crops are fertilized to produce better or larger yields, often at a decreased cost per unit of production. To accomplish these goals it is essential to know how much of the critical nutrients the crop will need. Nutrients removed from soil by plants vary with the variety of plant, its stage of growth, and its yield (see Chapter 13, Table 13-1). Crop plants usually contain more nitrogen than any other fertilizer nutrient. The second-highest concentration of fertilizer nutrient is potassium, followed by calcium, phosphorus, magnesium, and sulfur. Bananas, corn, potatoes, peanuts, and sorghum have high nitrogen needs. Alfalfa fixes its own. Bananas, sorghum, pineapples, and corn have high phosphorus requirements. Bananas, pineapples, potatoes, and alfalfa have high potassium needs.

If the different plant requirements are coupled with the in-field soil variations, the fertilizer needs within different areas of the field will also vary, and these different areas should be fertilized differently. Precision agriculture (discussed in Chapter 13) is one way this is done. The effectiveness of such variable treatment can now be measured by grain sensors (Figure 12-11).

Nutrients Supplied by Soils

Soils vary in the nutrients they can supply to various plants. Unfortunately, a widely satisfactory soil test for nitrogen is not available. Most soil nitrogen comes from decomposing organic matter, and quick soil tests for nitrogen or phosphorus do not measure the potential release from the organic source because this measurement usually requires an incubation time of a couple of

FIGURE 12-11 A grain yield measuring system automatically plots the location with the global positioning system (GPS) and maps the yield by location on the field. It measures yield differences as it harvests and can be reviewed later to locate field areas of low and high yields. (Courtesy of USDA—Agricultural Research Service.)

weeks. The nitrate test is an attempt to measure nitrogen released by soil organic matter in the days or weeks prior to sampling, but the incubation times and leaching losses are often unknown.

Soil tests do not tell us exactly how many kilograms of a nutrient are actually available from an area of each soil. The test values allow us to group the soils into categories of soils that have *very deficient, deficient, marginal, adequate,* or *excess* amounts of available nutrients. As an example, consider the soil test numbers obtained by a laboratory, as illustrated in Figure 12-12. The values are illustrated as a gauge. Soils with high soil test values may not require any additional fertilizer; some may even have toxic levels of nutrients. The lower the soil test values, the more fertilizer will need to be added to optimize yields.

12:8 Fertilizer Guarantee (Grade)

All states require that fertilizers offered for sale be accurately labeled with the **grade** (minimum guaranteed percentage of N, P, and K), the weight of the material, the manufacturer, and the manufacturer's address. Some states require additional information, such as the composition of the **filler** (materials such as sand or lime sometimes added to make the fertilizer to convenient bulk weights) and the amount of acidity the fertilizer forms when it reacts in soil.

The most useful information is the grade. The grade gives, in order, (1) the **percentage total nitrogen,** measured as elemental N, (2) the **percentage available phosphorus,** listed as that phosphorus soluble in ammonium citrate solution and *calculated as phosphorus pentoxide,* P_2O_5, and (3) the **percentage water-soluble potassium,** *calculated as potassium oxide,* K_2O. These quantities are often listed on fertilizer labels as 10-16-12—meaning that this fertilizer contains 10 percent by weight total nitrogen, 16 percent available phosphorus (calculated as P_2O_5), and 12 percent water-soluble potassium (calculated as K_2O). A slow but definite effort is being made to change these conventions and to report only N, P, and K *element percentages.* One reason for the resistance to change is that P or K reported as element percentages gives lower numbers and makes the fertilizer appear to be a lower grade. For example, a 0-45-0 triple superphosphate is only a 0-19-0 material if reported as percentage of P rather than percentage of P_2O_5. It is convenient to remember that the order of listing is alphabetical by element: nitrogen, phosphorus, and potassium. Figure 12-13 shows several examples of fertilizer labels. State chemists periodically sample and analyze commercial fertilizers to verify the grades listed. Fertilizer companies can be prosecuted if the grade displayed on their bags is higher than the laboratory analysis shows it to be.

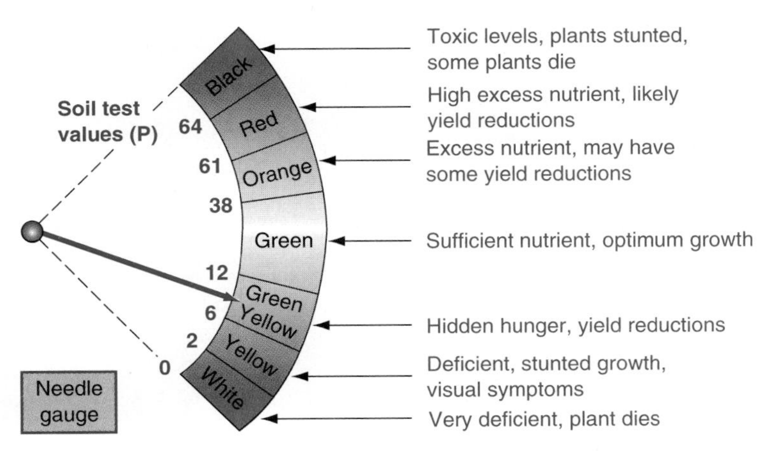

FIGURE 12-12 Representation of a soil test and its meaning. The phosphorus soil test values are hypothetical and are shown to illustrate that the numerical values are only used for purposes of grouping the soils into categories, such as deficient, marginal, or toxic levels of available nutrients. (Courtesy of Raymond W. Miller, Utah State University.)

FIGURE 12-13 Examples of fertilizer labels showing grades. In (a) the grade is 23-23-0, and additional information, the form of the nitrogen, is given in the box. In (b) a slow-release nitrogen fertilizer has 38 percent total nitrogen (grade is 38-0-0), but a small unlisted part of it is readily soluble. A chelate material (c) is not added to supply N, P, or K, and its grade is often 0-0-0. Grade information is not descriptive of such a material. A mixed (formulated) fertilizer containing many nutrients (d) is characterized by both grade and additional tabulated materials. (Courtesy of Raymond W. Miller, Utah State University.)

(a)

(b)

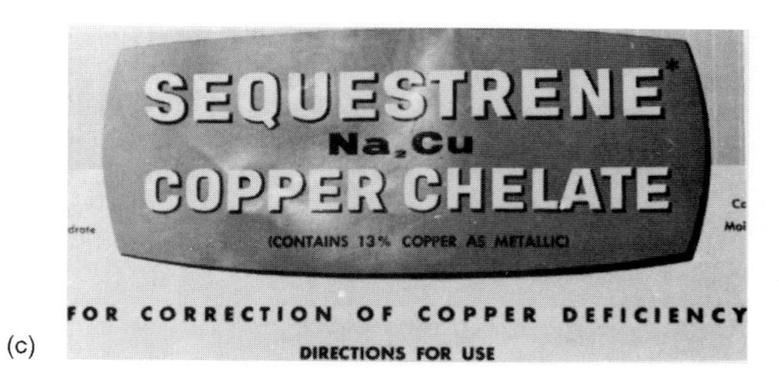

(c)

(d)

Notice that the fertilizer grade only gives information about nitrogen, phosphorus, and potassium. Often sulfur is also included, calculated as elemental sulfur. In a value such as 10-20-5-4S, the last number—4S—means 4 percent sulfur. Agricultural minerals such as chelates and micronutrients usually have no grade. They might be 0-0-0 but be a zinc chelate. Thus, *fertilizer grade gives information only about the nitrogen, phosphorus, and potassium contents,* unless the additional numbers are given with the element's symbol.

12:9 Some Characteristics of Fertilizers

Soluble fertilizers are usually soluble salts. Phosphates are generally low salt; chlorides, sulfates, and nitrates are high in their immediate salt effects. Use of the chlorides, sulfates and nitrates of ammonium and potassium should be managed as if soluble salts were being added—*they are.*

Acidity and Basicity of Fertilizers

Fertilizers may affect soil acidity. All potassium fertilizers are neutral, except potassium nitrate, which is residually slightly base forming. Superphosphate and triple superphosphate are neutral, but both monoammonium phosphate and diammonium phosphate are acid forming. Most nitrogen fertilizers, and all ammonium materials and many organic nitrogen fertilizers, are acid forming. Acidity is developed by microbial oxidation of ammonium cations to nitrate anions:

$$\underset{\text{(ammonium)}}{NH_4^+} + \underset{\text{(oxygen)}}{2O_2} \xrightarrow{\text{Bacteria}} \underset{\text{(nitrate)}}{NO_3^-} + \underset{\text{(acidity)}}{2H^+} + \underset{\text{(water)}}{H_2O}$$

Although many fertilizers acidify soil over time, many are alkaline (basic) initially. Urea, when hydrolyzed by urease, may cause a pH of 9.5 to 10 in the area of the pellet or band. This is why NH_3 volatilization can occur rapidly from surface applications of urea. Diammonium phosphate also has a high pH initially, as does anhydrous ammonia and some liquid nitrogen.

Solubility and Nutrient Mobility

Nutrient mobility is a function of ion charge, the ion's tendency to form insoluble compounds (precipitate), adsorption to surfaces of soil solids, soil texture, water movement, and concentration of other ions. Phosphate ions precipitate, or are adsorbed to, soil solids; phosphates are the *least mobile anions,* usually moving only a centimeter or two from where they were placed in the soil. The K^+ and NH_4^+ ions also move slowly because of their attraction to the cation exchange sites. Nitrate and sulfate will be quite mobile, moving mostly wherever water moves. Thus, soluble nitrogen in soil eventually is oxidized to nitrate, a *mobile ion.* In contrast to nitrate movement, neither phosphate nor potassium move very far or very fast in soils. These nutrients need to be placed in the root zone where the grower wants them to be; they stay mostly where they are placed.

12:10 Fertilizer Calculations

The usual calculations on fertilizer materials are (1) nutrient percentage, (2) weight of bulk fertilizer to use for an area, and (3) weights of materials to use in making mixed bulk fertilizer for adding several nutrients. Such a mixture might be based on a fertilizer recommendation from a tested soil sample. The more complex these calculations are, the more suitably they can be adapted to computer programs for calculation.

Calculating Nutrient Percentage

All nutrients should be expressed as a percentage of an element (zinc, nitrogen, phosphorus, etc.). Unfortunately, early chemists reported elements as *burned solids* that produced oxides; thus, we are still using the convention of reporting phosphorus as P_2O_5 equivalent and potassium as K_2O equivalent—which is unfortunate because *fertilizers do not contain* P_2O_5 *or* K_2O.

A close approximation of the percentage of a nutrient in a fertilizer can be obtained by using the fertilizer formula and chemical atomic weights. Urea has a formula of $(NH_2)_2CO$. The molecular weight (sum of all atomic weights) is 60.056 (see the periodic table in Appendix C). The *percentage nitrogen* is calculated as follows:

$$\frac{2\text{N (or N}_2)}{\text{H}_4\text{N}_2\text{CO}}(100) = \frac{28.014\text{ g}}{60.056\text{ g}}(100) = 46.6\%$$

Because fertilizers are seldom pure, the calculations should be rounded downward to the nearest whole integer:

$$(28\text{ g}/60\text{ g})\,(100) = 46\%$$

Rounded-off values may still be high. For another example, see Calculation 12-1.

Calculating Simple Fertilizer Mixtures

To combine various single-nutrient sources into a mixed fertilizer, do it one of two ways: (1) Mix the two or three materials to get a mixture with the proportions wanted, e.g., high N, low P, and low K, or (2) mix the materials to an exact grade, adding a filler material to get a final weight previously selected.

For example, calculate the amount of ammonium nitrate (34-0-0) and treble superphosphate (TSP, 0-45-0) to make 1000 kg of a 15-10-0 mixture. The final mixture must have 15 percent of 1000 kg, which equals 150 kg of N. It also must have 10 percent of 1000 kg, which equals 100 kg of P_2O_5. To get 150 kg of N, each 100 kg of ammonium nitrate has 34 kg of N. Thus,

$$\left(\frac{150\text{ kg of N}}{1000\text{ kg of mix}}\right)\left(\frac{100\%\text{ of 34-0-0}}{34\%\text{ N}}\right) = \frac{15{,}000\text{ kg}}{34}$$
$$= 441\text{ kg of ammonium nitrate/1000 kg}$$

To get 100 kg of P_2O_5, each 100 kg of TSP has 45 kg of P_2O_5. So,

$$\left(\frac{100\text{ kg of P}_2\text{O}_5}{1000\text{ kg of mix}}\right)\left(\frac{100\%\text{ of 0-45-0}}{45\%\text{ P}_2\text{O}_5}\right) = \frac{10{,}000\text{ kg}}{45}$$
$$= 222\text{ kg of TSP needed/1000 kg}$$

Adding the ammonium nitrate (441 kg) and the treble superphosphate (222 kg) is 663 kg of material needed. The total mixture was to be 1000 kg, so the additional material is 337 kg ($1000 - 663 = 337$) and is made up by filler or lime or whatever additive is wanted.

Calculation 12-1 Calculation of Percentage of Nutrients

Problem Calculate the percentage of N, P, and P_2O_5 in diammonium phosphate as if it were pure. Formula is $(NH_4)_2HPO_4$.

Solution

1. To calculate percentage N, put the atomic weight of two nitrogens over the molecular weight of the $(NH_4)2HPO_4$ and multiply by 100 to get percentage. Use values to the closest tenth.

$$\% \text{ N} = \left(\frac{2\text{N}}{(\text{NH}_4)_2\text{HPO}_4}\right)\left(\frac{100}{}\right) = \left(\frac{28 \text{ g}}{132 \text{ g}}\right)\left(\frac{100}{}\right) = 21\%$$

Most commercial $(NH_4)_2HPO_4$ fertilizer is less pure and less fully ammoniated, so it will have percentages of N about 16%–18%.

2. To calculate percentage P, use the same procedure as for N, but remember that the formula has only one P atom, whereas there are two N in step 1 above.

$$\% \text{ P} = \left(\frac{\text{P}}{(\text{NH}_4)_2\text{HPO}_4}\right)\left(\frac{100}{}\right) = \left(\frac{31 \text{ g}}{132 \text{ g}}\right)\left(\frac{100}{}\right) = 23.5\%$$

3. To calculate the percentage of P_2O_5, use the relation of P_2O_5 to two P (fertilizer grades are calculated to this formula). Using the atomic and molecular weights and the calculated 23.5% P from step 2 gives

$$\left(\frac{23.5\%}{}\right)\left(\frac{\text{P}_2\text{O}_5}{2\text{P}}\right) = \left(\frac{23.5\%}{}\right)\left(\frac{142 \text{ g}}{62 \text{ g}}\right) = (23.5)\ (2.29) = 53.8\%\ \text{P}_2\text{O}_5$$

Impure diammonium phosphate fertilizers usually have about 46%–48% P_2O_5. Some may reach nearly 50%–53%.

To mix a fertilizer with N, P, and K all in it, use the same pattern. However, if the sum of weights of the three carriers (the N source, the P source, and the K source) add up to more weight than you calculated it for (e.g., 1000 kg), it cannot be done. The sources used do not have a high enough percentage of nutrient to be mixed as you wished. If this happens, you have two choices: (1) Lower the mixture's grade (and thus add more mix per hectare) or (2) select one or more different carriers with a higher nutrient content. For example, you could select urea with 46 percent N rather than ammonium sulfate with its 21 percent N.

Weights of Fertilizer to Add

If it is desired to add 80 kg of N and 40 kg of P_2O_5 per hectare, how much ammonium nitrate (34% N) and treble superphosphate (45% P_2O_5) should be added? For 80 kg of N,

$$\left(\frac{80 \text{ kg of N}}{1 \text{ ha}}\right)\left(\frac{100\% \text{ of 34-0-0}}{34\% \text{ N}}\right) = \frac{8000 \text{ kg of 34-0-0}}{34 \text{ ha}}$$

$$= \frac{235 \text{ kg of 34-0-0}}{1 \text{ ha}}$$

For 40 kg of P_2O_5, use the same procedure:

$$\left(\frac{40 \text{ kg of } P_2O_5}{1 \text{ ha}}\right)\left(\frac{100\% \text{ of 0-45-0}}{45\% \; P_2O_5}\right) = \frac{4000 \text{ kg of 0-45-0}}{45 \text{ ha}}$$

$$= \frac{88.9 \text{ kg of 0-45-0}}{1 \text{ ha}}$$

Mix 235 kg of ammonium nitrate plus 89 kg of treble superphosphate times the number of hectares to be fertilized, and set the drill to spread the 324 kg of mix per hectare. Keep in mind that kg/ha times 0.89 equals lb/a. Field applicators may not be adjustable to closer than about 10 kg/ha.

It is also easy to calculate the amount of fertilizer to apply to a small area. Consider adding the rate above to a garden plot that is 100 ft × 100 ft = 10,000 ft^2. The 10,000 ft^2 is

$$\left(\frac{10{,}000 \text{ ft}^2}{1 \text{ plot}}\right)\left(\frac{1 \text{ acre}}{43{,}560 \text{ ft}^2}\right) = \frac{0.23 \text{ acre}}{1 \text{ plot}}$$

Adding the mixture calculated earlier, 324 kg of mix per hectare, the amount to add is

$$\left(\frac{324 \text{ kg}}{1 \text{ ha}}\right)\left(\frac{1 \text{ ha}}{2.47 \text{ acres}}\right)\left(\frac{2.2 \text{ lb}}{1 \text{ kg}}\right)\left(\frac{0.23 \text{ acre}}{\text{plot}}\right) = \frac{66 \text{ lb}}{\text{plot}}$$

Calculating Mixed Fertilizers to Predetermined Weights

Often you want to make a particular amount of a mixture. Suppose you want to make a material that contains N and P in the ratio of 2:1 (2 kg of N per 1 kg of P_2O_5). Calculate for 100 kg of the mixture and then make as many 100-kg multiples as you wish to have. To do this, use **simultaneous equations.** Make as many equations that explain the information as you have variables. For only N and P (two variables), do as follows:

$$X + Y = 100 \text{ kg of mix} \qquad \text{(Eq. 12-1a)}$$

where X = kg of N carrier and Y = kg of P carrier to use per 100 kg of mixture. If we use ammonium nitrate (34% N) and treble superphosphate (TSP, 45% P_2O_5),

$$X \text{ kg of AN} + Y \text{ kg of TSP} = 100 \text{ kg of mixture} \qquad \text{(Eq. 12-1b)}$$

We need to make a mixture with 2 kg of N per 1 kg of P_2O_5, so

$$\left(\frac{X \text{ kg of AN}}{}\right)\left(\frac{34\% \text{ N}}{100\% \text{ AN}}\right) = \left(\frac{X}{}\right)\left(\frac{34}{100}\right) \text{kg of N} \qquad \text{(Eq. 12-2)}$$

$$\left(\frac{Y \text{ kg of TSP}}{}\right)\left(\frac{45\% \; P_2O_5}{100\% \text{ TSP}}\right) = \left(\frac{Y}{}\right)\left(\frac{45}{100}\right) \text{kg of } P_2O_5 \qquad \text{(Eq. 12-3)}$$

Because we want 2 kg of N for each 1 kg of P_2O_5, we must have twice as much of P_2O_5 to equal the amount of N wanted. Thus,

$$X \text{ kg of N} = 2(Y \text{ kg of } P_2O_5) \qquad \text{(Eq. 12-4)}$$

This is the key step in this problem. Substitute into Equation 12-4 the values from Equations 12-2 and 12-3 to give the following:

$$\left(\frac{X}{}\right)\left(\frac{34}{100}\right) = \left(\frac{2Y}{}\right)\left(\frac{45}{100}\right) \qquad \text{(Eq. 12-5)}$$

Multiplying each side by 100 and simplifying yields

$$34X = 90Y \qquad \text{or} \qquad 34X - 90Y = 0 \qquad \text{(Eq. 12-6)}$$

Recall Equation 12-1; $X + Y = 100$. So Equations 12-1 and 12-6 will be the two **simultaneous equations** used to solve the problem. Eliminate one of the unknowns; we will choose to eliminate X. To do so, multiply Equation 12-1 by 34 (so that X will cancel out), rearrange Equation 12-6, and subtract the second equation from the first (multiply the second equation by -1 and add the two equations):

$$\begin{aligned} 34X - 90Y &= 0 \\ -34X - 34Y &= -3400 \\ -0X - 124Y &= -3400 \end{aligned} \qquad Y = 27.42 \text{ kg of TSP needed/100 kg mix}$$

Then

$$\text{kg of } P_2O_5 = \left(\frac{27.42 \text{ kg of TSP}}{100 \text{ kg of mix}}\right)\left(\frac{45\% \ P_2O_5}{100\% \text{ TSP}}\right) = 12.34 \text{ kg of } P_2O_5\text{/100 kg mix}$$

Substitute the value of Y into Equation 12-1:

$$\begin{aligned} X + 27.42 \text{ kg} &= 100 \text{ kg of mixture} \\ X &= 100 \text{ kg} - 27.42 = 72.58 \text{ kg of AN needed} \end{aligned}$$

Then

$$\text{kg of N} = \left(\frac{72.58 \text{ kg of AN}}{100 \text{ kg of mix}}\right)\left(\frac{34\% \text{ N}}{100\% \text{ AN}}\right) = 24.68 \text{ kg of N/100 kg mix}$$

To check the work, see if the ratios of N to P_2O_5 are 2:1. Remember, we rounded numbers off to the closest hundredth.

$$24.68 \text{ kg of N} = 2(12.34) \text{ kg of } P_2O_5$$

The ratios are correct.

Calculation 12-2 gives an example of the more complicated method of determining a mixture of all three nutrients. This requires three equations (always as many equations as there are unknowns). Use pairs of equations and solve for one unknown at a time.

Calculation 12-2 Calculating an N-P-K Mixture

Problem Calculate the weight of urea (46% N), treble superphosphate (TSP = 45% P_2O_5), and KCl (60% K_2O) to make up 100 kg of a mixture having a ratio of N-P_2O_5-K_2O of 4-2-1. When the mixture is made, calculate its grade.

Solution 1A Two methods for solving this will be shown. The first technique seeks to make a 4-2-1 ratio using simple mathematics and logic. Calculate the mixture needed to provide 4 kg of N, 2 kg of P_2O_5, and 1 kg of K_2O (a 4-2-1 ratio).

$$\left(\frac{4 \text{ kg of N}}{}\right)\left(\frac{100\% \text{ urea}}{46\% \text{ N}}\right) = 8.696 \text{ kg of urea} \qquad \text{(Eq. S12-1)}$$

$$\left(\frac{2 \text{ kg of } P_2O_5}{}\right)\left(\frac{100\% \text{ TSP}}{45\% \ P_2O_5}\right) = 4.444 \text{ kg of TSP} \qquad \text{(Eq. S12-2)}$$

$$\left(\frac{1 \text{ kg of } K_2O}{}\right)\left(\frac{100\% \text{ KCl}}{60\% \ K_2O}\right) = 1.667 \text{ kg of KCl} \qquad \text{(Eq. S12-3)}$$

Add the three materials together:

$$8.696 \text{ kg urea} + 4.444 \text{ kg TSP} + 1.667 \text{ kg KCl} = 14.807 \text{ kg}$$

Now the mixture of 14.81 kg has the 4-2-1 ratio wanted. To make 100 kg of a mixture with the same ratio,

$$\frac{100 \text{ kg}}{14.81 \text{ kg}} = 6.752 \text{ multiples}$$

The required 100 kg of mix is 6.752 times more than the 14.81 kg. So multiply the amount of each ingredient used by 6.752. Thus,

$$8.696 \text{ kg of urea} \times 6.752 = 58.715 \text{ kg of urea needed}$$

$$4.444 \text{ kg of TSP} \times 6.752 = 30.006 \text{ kg of TSP needed}$$

$$1.668 \text{ kg of KCl} \times 6.752 = 11.256 \text{ kg of KCl}$$

$$\text{Total weight} = 99.977 \text{ kg}$$

Solution 1B This problem can also be calculated, as in the text, using simultaneous equations. With three unknowns (N, P, and K), three equations are needed. The following three equations can be used (there is at least one other equation possible):

$$X \text{ kg} + Y \text{ kg} + Z \text{ kg} = 100 \text{ kg} \qquad \text{(Eq. S12-4)}$$

$$\left(\frac{46\%}{100\%}\right)\left(\frac{X}{}\right) = \left(\frac{2}{}\right)\left(\frac{45\%}{100\%}\right)\left(\frac{Y}{}\right) \qquad [1 \text{ kg N} = 2 \text{ (kg } P_2O_5)] \qquad \text{(Eq. S12-5)}$$

Clearing fractions and transposing X and Y to the same side, we get $4600X - 9000Y = 0$.

$$\left(\frac{45\%}{100\%}\right)\left(\frac{Y}{}\right) = \left(\frac{2}{}\right)\left(\frac{60\%}{100\%}\right)\left(\frac{Z}{}\right) \qquad \text{(Eq. S12-6)}$$

Calculation 12-2, Cont'd Calculating an N-P-K Mixture

Clear and transpose: $4500Y - 12{,}000Z = 0$.

Equations S12-5 and S12-6 could be simplified by dividing each by 100. The three equations, then, are

$$X \text{ kg} + Y \text{ kg} + Z \text{ kg} = 100 \text{ kg} \qquad \text{(Eq. S12-7)}$$

$$46X - 90Y = 0 \qquad \text{(Eq. S12-8)}$$

$$45Y - 120Z = 0 \qquad \text{(Eq. S12-9)}$$

Eliminate, successively, the X (Equations S12-7 and S12-8), then Y, and obtain the first value for Z (= kg KCl). Your answers will be the same as calculated by Solution 1A.

Solution 2 The grade of the material is obtained by multiplying the quantity of each material by its nutrient content:

$$\text{kg of N} = (58.715 \text{ kg of urea}) (46\%/100\%) = 27.0 \text{ kg}$$

$$\text{kg of } P_2O_5 = (30.006 \text{ kg of TSP}) (45\%/100\%) = 13.5 \text{ kg}$$

$$\text{kg of } K_2O = (11.256 \text{ kg of KCl}) (60\%/100\%) = 6.75 \text{ kg}$$

The grade usually must be in *whole numbers, never rounded upward,* and is 27-13-6 (grade).

We used to use one fertilizer rate for the country, then the farm, then the field, and now . . . [we can] vary application rates within the field.

—Dick Stiltz (manager, fertilizer plant)

I know of no pursuit in which more real and important services can be rendered to any country than by improving its agriculture, its breed of useful animals, and other branches of a husbandman's cares.

—George Washington

Questions

1. List six of the laboratory tests made on soil samples, and indicate what each test indicates.
2. If two different extractants for phosphate remove different amounts of phosphate from the *same soils,* how can both of these tests possibly be used to make good fertilizer recommendations?
3. In detail, tell what to do regarding each of the following items related to taking a soil sample: (a) depth to sample, (b) number of subsamples per composite sample, (c) areas to avoid, and (d) when to take the sample.
4. Changes in laboratory procedures, particularly using a different soil–water ratio for salt, are very critical in interpreting soluble-salt data. Explain.
5. More accepted soil tests are available for potassium and phosphorus than for nitrogen. Explain why it is so difficult to get a good diagnostic test for nitrogen.

6. What important source of phosphorus in soils is little measured by the usual soil test, and how might this affect a farmer's use of that recommendation?
7. Many different salt and dilute acid extracts have worked well to extract available soil potassium and predict needs for potassium. Explain why this would be expected.
8. As soils are limed, any calcium deficiency will probably be solved, but a magnesium deficiency may not be solved. Explain.
9. How good are soil tests for predicting soils deficient in micronutrient metals? Explain the likely reasons.
10. Plants are the final evaluator of whether or not nutrients are available to the plant. (a) Why aren't plant tests rather than soil tests used more? (b) On which plants can plant tests be used?
11. Briefly discuss the *critical nutrient range* (CNR) and indicate its relationship to hidden hunger.
12. Describe briefly the general visual deficiency symptoms for nitrogen, phosphorus, potassium, and iron.
13. What is meant by *prescription plus buildup*?
14. What are some of the assumptions by those making fertilizer recommendations that may be untrue about an actual field soil?
15. How have environmental concerns altered our view and practice of adding fertilizers?
16. If a fertilizer grade is given as 16-20-10-5S-1Zn, indicate exactly what each number means.
17. Briefly discuss the *saltiness, potential acidity,* and *mobility* of most N, P, and K fertilizers.
18. Calculate the amount of urea (46% N) and of triple superphosphate (0-45-0) to make 100 kg of a 10-20-0 mixed fertilizer.
19. Calculate the percentage P (element, not P_2O_5) in pure phosphoric acid (H_3PO_4).
20. Calculate the amount of urea (46% N), ammonium phosphate (11-48-0), and potassium sulfate (0-0-50) to make 1000 kg of mixed 20-10-5 fertilizer.
21. Calculate the amount of urea (46% N) and treble superphosphate (0-45-0) to add to a plot of 280 m^2 so that the rate is the same as 150 kg N/ha and 60 kg P_2O_5/ha.

13 Fertilizer Management and Precision Agriculture

The farmer is covetous of his dollar, and with reason . . . He knows how many strokes of labor it represents. His bones ache with the day's work that earned it.

—Ralph Waldo Emerson

When land productivity decreases, the city resident is the first to tighten his belt.

—M. Graham Netting

13:1 Preview and Important Facts

PREVIEW

There are numerous methods of adding fertilizers. Broadcasting on the soil surface, followed by plowing or disking it in, is common. Inserting a strip of fertilizer within a few centimeters of planted seed of a row crop is also widely done. Application as surface strips on the soil near the crop after the row crop is growing, as a foliage spray, or in irrigation water are other common methods.

As more people looked at ways to improve agriculture, they recognized that soil areas, even within a single field, can vary and react to growing plants differently. While scientists were wondering how best to solve such problems, soil variability models and the personal computer emerged as fantastic aids. With the introduction of GIS mapping from databases and the GPS to locate an area on a field precisely, site-specific farming became a real possibility. Application rates across a field could be changed as the fertilization was being done. This capability did require prepared maps of the delineated soil areas that required differential treatment.

Although many new techniques are still being explored, many have already been implemented. Measuring grain yields on-the-go with an in-cab computer system is now done; being aware of where the yields change allows one to know where in the field to look for solutions to low yields. Tools to assure precise planting depths and the health of a crop (remote sensing) are in use. Some of the marvels of today are likely to become the standard practices of tomorrow.

IMPORTANT FACTS TO KNOW

1. The various ways fertilizers are added: *starter, broadcast, deep banding, split applications, side* or *top dressing, fertigation,* and *foliar*
2. The value of split applications and the nutrients for which the method is most suitable
3. The advantages and disadvantages of *fertigation* and *foliar application* as methods of adding fertilizers
4. The unique fertilization problems in paddy rice and possible solutions
5. The meaning of *site-specific farming* and *variable-rate technology* (VRT)
6. The concept, information needed, and equipment needed for site-specific farming
7. The use of plant *health* to evaluate where problem areas exist, and how it is done
8. The use of the GIS and GPS in precision agriculture

13:2 Goals and Concerns in Using Fertilizers

Crop producers have several goals in using fertilizers:

- To increase yields
- To reduce costs per unit of production
- To increase plant quality
- To reduce certain diseases
- To prevent environmental pollution

The first two of these goals are the most common reasons for adding fertilizers.

Efficient land managers spend perhaps 20 percent of all production costs on chemical fertilizers and expect yield increases of up to 50 percent. But fertilization is *not profitable* when (1) water is the first limiting factor, (2) other growth hindrances—such as insects, diseases, strongly acidic soils, or cold temperatures—control the growth, and (3) the increased yield has less market value than the cost of buying and applying the fertilizer.

Environmental controls and penalties are management concerns. Careless and polluting fertilizer applications can become costly errors as environmental concerns, regulations, and pressures increase. The pressure to be sure to control nonpoint sources of pollutants will increase. Research to justify fertilizer needs and to verify the extent of pollution caused by various practices may become a requirement rather than a voluntary action of the agricultural business community. Site-specific management may become necessary in many situations as we consider the possible impact of our management on pollution of the environment.

Nutrients Removed by Plants

Nutrients removed from soil by plant growth vary with the variety of plant, its stage of growth, and its yield (Table 13-1). Crop plants usually contain more nitrogen than any other fertilizer nutrient. The second-highest concentration of a fertilizer nutrient is potassium, followed by calcium, phosphorus, magnesium, and sulfur.

The values in Table 13-1 emphasize the different nutrient requirements of various crops. Bananas, corn, potatoes, peanuts, and sorghum have high nitrogen needs (alfalfa fixes its own N). Bananas, sorghum, pineapple, and corn have high phosphorus requirements. Bananas, pineapple, potatoes, and alfalfa have high potassium needs.

Table 13-1 Major and Secondary Essential Nutrients Contained in the Entire Plant, with Yield Indicated

		Nutrients Contained in Crop (lb/a per Crop)[a]					
Crop	Crop Yield per Acre[a]	Nitrogen (N)	Phosphorus (P_2O_5)	Potassium (K_2O)	Calcium (Ca)	Magnesium (Mg)	Sulfur (S)
Alfalfa	5 tons	250	60	225	160	25	23
Corn	150 bu	220	80	195	58	50	33
Cotton	1.5 bales	95	50	60	28	8	4
Coastal Bermuda grass	6 tons	150	60	180	33	22	40
Soybeans	40 bu	145	40	75	7	9	7
Rice	2500 lb	185	51	18	20	15	18
Tobacco	2800 lb	95	25	190	105	24	21
Wheat	60 bu	125	50	110	16	18	16
Oats	100 bu	100	40	120	14	20	20
Potatoes	400 bu	200	55	310	50	15	18
Peanuts	3000 lb	220	45	120	105	28	25
Sorghum grain	8000 lb	260	110	220	445	36	38
Banana	1200 plants	400	400	1500	300	156	[b]
Coffee	1784 lb	27	4	43	46	61	16
Oil palm	13,382 lb	80	18	120	64	18	[b]
Pineapple	15,000 plants	134	107	535	102	53	[b]

Source: Magnesium-Sulfur, Essential Plant Nutrients, International Minerals and Chemical Corp., Libertyville, IL, undated.

[a]To convert lb/a to kg/ha, multiply lb/a by 1.12. Weight of a bushel of wheat, 60 lb; corn, 56; soybeans, 60; oats, 32; sorghum, 56; potatoes, 60; rice, 45 lb.

[b]No information.

13:3 Techniques of Fertilizer Applications

What is the best way to add fertilizer? There is no single best way. Some crops spread over the field hinder surface traffic a few weeks after germination (melons, cucumbers, squash, pumpkins). Environmental concerns preclude putting on large applications before planting if much of it might be leached or lost before the plant can use it. Plants on sandy soils need frequent, small applications. There are many concerns when selecting the manner in which to add fertilizers. The most common ways used for fertilizer application in the fields are the following techniques; their definitions are given in the Glossary.

Starter (pop-up)
Broadcast
Deep banding
Split application
Side dressing, strip placement
Top dressing
Point injector
Fertigation
Foliar sprays

Starter (Pop-Up) Fertilizers

Starter fertilizer, for our purposes, includes any addition of fertilizer *with the seed* during planting, dribbled into the soil *in a strip near the seed* or placed *in a band* anywhere within about 5 cm (2 in) of the seed (Figure 13-1). The key nutrients added are phosphates and

FIGURE 13-1 Side placement of fertilizer in two bands placed 5 cm (2 in) below the level of lima bean seed. This is called banded *starter* fertilizer.
(*Source:* USDA.)

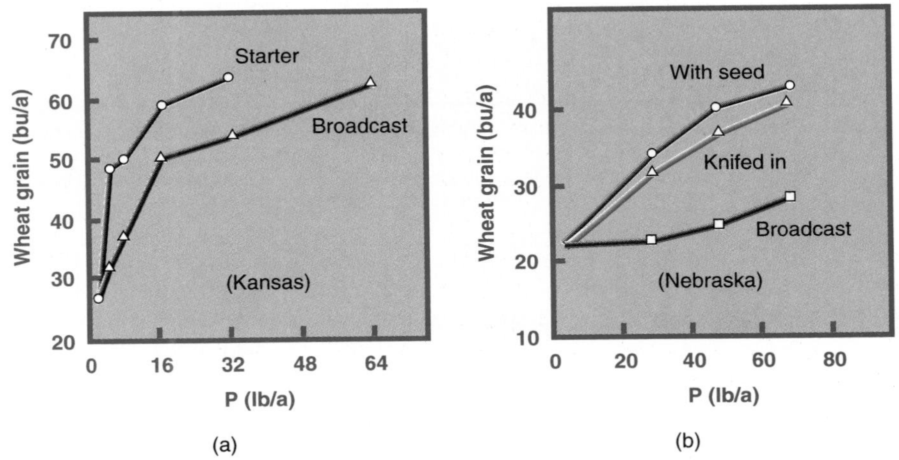

FIGURE 13-2 Effects of starter fertilizer on wheat yields compared to broadcast and to knifed-in applications: (a) 9-18-0 fertilizer, (b) 10-34-0 fertilizer. (bu/a × 67.4 = kg/ha; lb/a × 1.12 = kg/ha.)
(*Source:* Modified from John L. Haolin, "Optimizing Wheat Inputs for Profit," *Solutions* **30** [1986], 50–60.)

potassium because of their low mobility in soils. Nitrogen will usually be added, too, but it is mobile enough to be side-dressed later during peak crop need.

Three conditions favor obtaining a good response from adding starter fertilizer: (1) a soil that is too cold for good nutrient absorption by roots, (2) a soil test value that indicates a low nutrient level, and (3) a fast-growing plant. Crops on soils having adequate soil test values may still respond to starter fertilizer in cool periods.[1] Figure 13-2 and Table 13-2 illustrate examples of growth response to applied starter fertilizer.

Some applications require the avoidance of adding fertilizer in contact with the seed because *salt damage* is possible, unless care is used. *Not more than about 10 kg/ha to 12 kg/ha*

[1]P. A. Costigan, "The Effect of Soil Temperature on the Response of Lettuce Seedlings to Starter Fertilizer," *Plant and Soil* **93** (1986), 183–193.

Table 13-2 Yield Increases (kg/ha) of Corn Grain from Use of Starter Fertilizer on High-Soil-Test Land in Iowa

*Starter Used**	*1983*	*1984*	*1985*	*1986*	*Average*
Check (no fertilizer)	155	152	165	207	170
5-17-0 fertilizer	157	157	173	213	175
6-18-6 fertilizer	163	158	168	213	177
7-18-6-3S-0.5Zn	165	166	182	220	183

*Starter was put on as a liquid at 7 gal/a (= 65 L/ha). Note that the last material in the table has 3% sulfur and 0.5% zinc.

of nitrogen plus potassium should be put in contact with the seed. Phosphates have much less salt effect and have less stringent restrictions for seed contact (see Figure 13-2). As the application rate of material is increased, it is increasingly important to have some soil between the seed and the fertilizer. With low fertilizer rates, a separation soil thickness of about 2 cm will usually protect the seed from salt effects; with high fertilizer rates, the distance (soil thickness) between the seed and the fertilizer should be a minimum of 4 cm to 5 cm.

Starter fertilizer (1) is usually more efficient than other placements, (2) is only one part of many fertilizer programs, (3) is effective for helping early growth, (4) is particularly helpful on cold soils (Figure 13-3), and (5) is probably more important for phosphate and potassium than for nitrogen.

Broadcast Application

Broadcast applications are uniform surface applications of fertilizers. The fertilizer may be left on the surface or mixed into the soil to different degrees by plowing, disking, or having minimal covering as seeders splash soil over the fertilizer (as by cereal grain drills). Irrigation water or rainfall may move soluble materials (urea, sulfates, nitrates, etc.) down into the soil. Usually, *the lowest fertilizer efficiency is obtained by broadcasting.*

If broadcasting is often so inefficient, why is it used? The many reasons for continued use include the following:

- Often broadcast is the only practical method for putting fertilizer on established pastures, on forests, on turf, on paddy rice during growth, or on partly grown crops. Some injection techniques on established crops are being studied, but root damage will occur.
- On low-fertility soils it is often necessary to add large amounts of fertilizer or to build up available levels of the nutrient throughout the soil. This is best done by broadcast followed by incorporation tillage.
- Broadcast is an easy and cheap method that is often chosen because of a need for speed, a lack of injection equipment, or a personal preference. A grower may know that the added phosphate will be less efficient but may decide that it will not be lost and will, over many years, gradually build up soil reserves. In some developing countries limited equipment is available. Surface application, with or without incorporation by hand work, is the only choice available.
- Broadcast or a modification of it (*top dressing, strip dressing, split application*) is the most feasible method, except for fertigation, of adding fertilizer after the crop has begun growth. Any attempt to insert the fertilizer into the soil close to the plant will tear some roots. However, some shallow knifing of fertilizers can be done carefully. On much of the early no-till acreage, fertilizers were applied broadcast. The problem of how to fertilize in no-till crops and in reduced-till farming is still being studied.

(a)

(b)

FIGURE 13-3 Grapes, like many other crops, have trouble absorbing phosphorus when the soil is cold. Phosphorus deficiency—(a) the lower leaves yellow and (b) increasing red dot masses—occurs in grapes as growers expand plantings onto higher-elevation soils that are also shallow and acidic. The symptoms shown above are typical of those found in grapes in the hill areas of the Napa Valley and Sierra Nevada in California. Petiole phosphorus in deficient plants was about 0.04 percent to 0.14 percent compared to normal values of 0.3 percent to 0.6 percent. As little as 0.4 lb of phosphorus per vine caused much improved growth, but the best practices are yet to be determined. (*Source:* By permission of *California Agriculture;* James A. Cook, William R. Ward, and Alan S. Wicks, "Phosphorus Deficiency in California Vineyards," *California Agriculture* [May–June 1983], 16–18.)

Incorporation with narrow knives, fluted coulters, or dribbling liquid fertilizers on the soil has helped improve the fertilizer efficiency by getting some phosphorus and potassium fertilizer down into the soil.

The efficiency of broadcast fertilizer usually increases under the following conditions:

- If the soil is shaded by vegetation, the soil surface will be shaded and stay moist longer; thus, some roots will explore the soil at the surface, as happens in pastures, and use more of the immobile nutrients there.
- When the broadcast fertilizer is incorporated into the soil, more fertilizer is deep in the soil where roots stay active (because the soil stays moist) most of the time.
- If cool spring air temperatures are warming, the surface soil may be warmer and allow better nutrient uptake in the shallow surface soil.
- If sprinkler irrigation or rainfall comes at desirable times, the water will move nitrates down into the soil where roots have better access to it.

Deep Banding

Deep banding is the application of strips of fertilizer into the soil either (1) without regard to where the seed is planted exactly or (2) to the side and below the seed *at the same time when planting.* Most often, application at the time of planting puts the fertilizer 10 cm to 25 cm deep

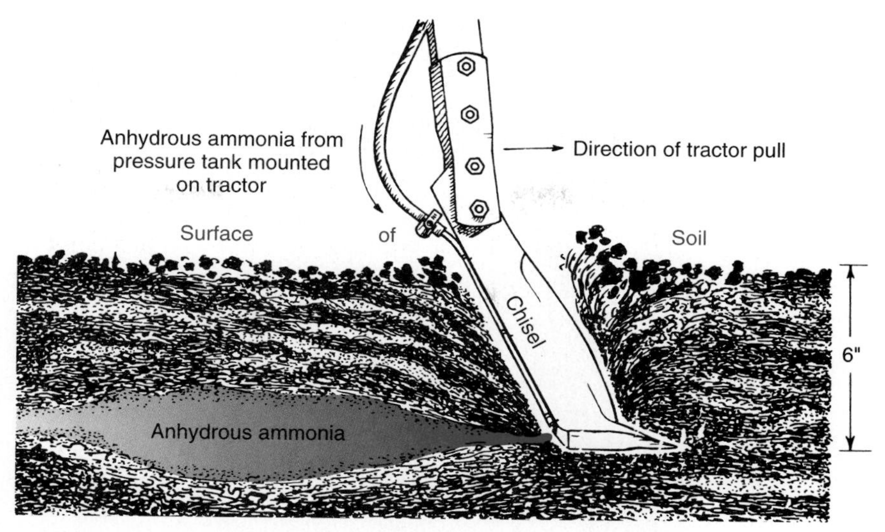

FIGURE 13-4 Under pressure and behind a chisel at a depth of approximately 15 cm (6 in), anhydrous ammonia is released as a gas and is adsorbed on the surface of clay and humus particles for use by plants. (*Source: Crops and Soils* magazine.)

and about 5 cm to 8 cm to the side of the seeds. Anhydrous ammonia fertilizer may be applied in either of these two methods (Figure 13-4).

Ammonia is transported under moderate pressure in tanks, and it changes to gas at normal atmospheric pressure. Normal knifed-in injections are placed about 8 cm to 20 cm deep, thus allowing the gas to be absorbed into the soil water (it is very soluble in water) before much escapes into the atmosphere. Applications into quite dry soil, especially sandy soil, should be avoided. Gaseous losses of ammonia can be excessive in sandy and dry soil. The ammonia gas dissolves readily in water to form ammonium hydroxide, which may immediately cause a high pH, above 9 to 9.5, and some ammonia can volatilize. The gaseous ammonia is toxic to both plant roots and to crop foliage, so it is seldom added except in preplant or during planting.

Deep banding has the disadvantages of any deep tillage: the need for stronger equipment and higher energy costs to pull the equipment through the soil. Before deep banding is used, it should be determined that this method will increase yields appreciably above those obtained with simpler and less costly methods.

It has been suggested that deep banding in *dryland* crops might be beneficial. The concept is that the nonirrigated soil dries in the surface soil layers for long time periods, and deeply placed fertilizer might be in moister soil longer and thus available to roots longer. In some dryland conditions deep banding could be detrimental if the deeply placed fertilizer keeps the plant in a vegetative stage too long. If the plant runs out of profile water before it sets seed, grain yields will be low.

Banding of fertilizer is putting the most fertilizer where the most roots have access to it.[2] This is a most important concept for phosphorus. This guideline suggests that applications close to and below the seed (starter banding) should be best. If the fertilizer was mixed into a little larger volume of soil, more of it would be available to more roots.

Keeping fertilizer phosphorus in bands lessens contact of the phosphate with soil (Figure 13-5). This minimum contact with soil is thought to reduce the rate and amount of soluble phosphate precipitated by calcium, iron, and aluminum and to reduce adsorption to

---

[2]D. M. Sleight, D. H. Sander, and G. A. Peterson, "Effect of Fertilizer Phosphorus Placement on Availability of Phosphorus," *Soil Science Society of America Journal* **48** (1984), 336–340.

FIGURE 13-5 No-till system of tillage in Ohio, where corn is being planted into 3000 lb (1364 kg) of soybean residues per acre and fertilizer is banded to one side of the corn row at a depth of about 4 in to 5 in (10 cm–13 cm). (Courtesy of USDA—Soil Conservation Service.)

mineral surfaces. The benefit of this application pattern is sometimes questioned. For example, the ideal tight band with minimal mixing of fertilizer with soil would have low soil contact but also have a low volume of phosphate accessible to roots and may have toxic salt levels in the band. Which factor is most important, having more accessible volume to roots or less contact with soil? Most banded fertilizers are not tight bands at all and perhaps already have extensive contact in the soil for precipitation and adsorption to soil particles. Generally, *banded phosphates have been more efficient than broadcast applications.*

Split Application

When the total applied fertilizer is split into several portions that are applied at two or more different times, the system is called a **split application.** Any method, or combination of methods, of application can be used. Split applications are usually done for the following reasons:

- If large applications are needed, it may be more efficient to hold back some of the fertilizer and add it after the crop is growing well and has high nutrient demands.
- In some conditions, as in sands, the soil may have low retention or high losses of nitrogen. Multiple but small additions are necessary to reduce losses and keep adequate amounts available to the plant.
- Control of vegetative growth in early stages by *not adding fertilizer early* may be desirable, such as with cotton and tobacco. Higher fertilizer demands during later stages of growth may require supplemental fertilizer in midgrowth stages.

- The most efficient fertilizer use (to reduce costs or minimize pollution of runoff waters) may require several well-timed but small applications. The choice the grower makes is between ease of application and efficiency. These choices become more varied as the crop season (sugarcane, turf, rice) becomes longer.
- The number of applications may be related to the method of application or damages. If fertilizer is added in irrigation water, several smaller applications may be the safest and most efficient method.

Usually for nitrogen, efficiency is increased by split applications. For phosphorus and potassium, split applications may give lower efficiency because they are immobile nutrients and, when surface-applied, they do not move down to where most roots are growing.

Side or Top Dressing

Side dressing is usually a surface or shallow-band application put on after the crop is growing (another split application). These applications may be broadcast, stripped on the soil surface, inserted into the soil at shallow depths, or put on in liquid fertilizers (dribbled on the soil). The following principles should be kept in mind when considering side and top dressing:

- Losses of nitrogen from the soil surface by ammonia volatilization will be a major concern when urea or ammonium fertilizers are applied on calcareous and other alkaline soils. The fertilizer should be irrigated into the soil as soon as possible after application.
- Fertilizer should be added to surface-irrigated crops in the furrows rather than to ridges so that irrigation water will move it into the root zone. If sprinkler irrigation or rainwater is used to move the fertilizer, location of the fertilizer applied is less critical because practically all of the soil surface will be leached under young crops.
- Side or top dressing is not very effective for phosphorus and potassium unless the crop *shades* the soil and keeps the soil surface moist so roots can explore it.

Point Injector Application

A **point injection** allows putting potassium and phosphate fertilizers down into the soil where roots grow, while preventing great damage to roots. The concept has been used in small plots and gardens for a long time. A pointed stick or rod is used to poke a hole in the soil and fertilizer is poured into the hole before covering it over. This is done around fruit trees, vines, shrubs, large garden plants (tomatoes), and isolated berry, melon, and squash vines. For field use, mechanized equipment is needed. Rolling point injectors (Figure 13-6) work fairly well, but they are still somewhat experimental and costly. Large units with 1.5-m-diameter wheels, 10 to 20 injectors per wheel, and 8 to 10 wheels are available.

Fertigation

Fertigation is the application of fertilizers by injecting them into irrigation water. The fertilizer is applied in large quantities of water, *not as a foliar spray* (a foliar spray barely wets the leaves with the solution). Fertigation has been very successful for nitrogen applied through center pivots, other large sprinkler systems, in drip irrigation, and as metered applications of liquid fertilizers to greenhouse plants through spaghetti-tubing irrigation lines.

If the irrigation water already contains appreciable salts, ammonium polyphosphates and anhydrous ammonia can cause salt precipitations, which plug lines and nozzles. The circular center-pivot sprinkler systems are easily adapted to fertilization in irrigation water. High fertilizer efficiencies reported by various application methods are shown in Table 13-3.

FIGURE 13-6 Rolling-point injector fertilizer application. The equipment can inject either liquid or gas into the soil, even during early stages of crop growth. It does not prune the roots nor require a very high tractive energy, as is needed by a knifing-in application. Also, surface residues are little affected. The injector shown is one of the early models using 1-cm-diameter spokes that inject the liquid about 10 cm (4 in) deep. The liquid must be kept at sufficient pressure to prevent plugging of the outlet holes (about 3 mm in the side of the spoke near the solid end of each spoke). Some problems are getting uniform applications, keeping the wheel seals from leaking (because of the corrosive fertilizer solutions), and volatilization of some nitrogen from the holes, which do not always close. (Photo and information courtesy of James L. Baker, Department of Agricultural Engineering, Iowa State University, Ames.)

Reports on the potential of applying fertilizers in irrigation water are enthusiastic. Nebraska has reported that adding nitrogen in irrigation water is 30 percent to 50 percent more efficient than any preplant application. Some Texas growers are putting 50 percent less nitrogen in various irrigation systems and are getting equal or better yields, compared to yields from conventional methods of adding nitrogen fertilizers to soil.

The use of irrigation water to apply nutrients allows the addition of fertilizers when the crop can most benefit from it. If the need and conditions indicate that more fertilizer can be beneficial, it can be applied immediately and conveniently.

Fertigation is most effective on crops where nutrient retention is low (sands and sandy soils with low humus) and for mobile nutrients, such as nitrate and sulfate. In sands potassium, magnesium, and boron are also quite mobile.

Table 13-3 Fertilizer Efficiencies by Various Application Methods

Nutrient and Method of Application	*Efficiency*
Nitrogen by fertigation	95% possible
Nitrogen, in surface flow irrigation water	50%–70%
Solid nitrogen, applied to soil surface	30%–50%
Potassium, by fertigation	80% possible
Phosphorus, by fertigation	45% possible
Nitrogen, foliar spray (on citrus)	11%–75%

Source: Modified from Tom Milligan, "Tooling Up for Fertilization," *Irrigation Age* **6** (No. 3), Nov. 1972; 6–8.

Fertilizers are not the only chemicals that are added in water. **Chemigation** is the application of *any chemical* in irrigation water. Pesticides of various kinds can be applied. These additions are also called **herbigation, insectigation, fungigation,** and **nemagation.**[3]

Even with all the current interest in fertigation and chemigation, they are not simple substitutes for old techniques. They are good tools to use *for certain situations.* Some of the precautions and limitations to chemigation are as follows:

- Chemigation (including fertigation) is a surface application. For phosphorus and potassium the surface applications usually are less effective than if placed *in the soil* before planting.
- The technique is only economical if *irrigation must be done anyway,* or if the system is already set up.
- Application of fertilizer is only as uniform as the water application. If the system is poorly designed (nozzles too small, too large, or too far apart), application of water will not be uniform. If it is windy, applications will not be uniform; much of the chemical may even end up in adjacent fields. If the applications include pesticides, problems could be severe.
- If pesticides are used, a number of concerns need to be considered.[4] Pesticides are dangerous and need to be applied at carefully controlled rates, only on the area specified, and only during calm or low-wind periods. Currently large sprinkler systems do not generally have the precision, nor is the care taken, to avoid problems in windy periods. Often application is continuous all day and all night without adequate attention to *changes* in wind. For fertilizer application this hazard is less serious, but the nonuniformity of application is still a problem. For certain pesticide applications the lack of supervision, stoppages, and the wind hazard make chemigation with pesticides a serious hazard.

Legal restrictions may limit the application of some pesticides and fertilizers by fertigation or chemigation. Also, an irrigation system must be designed to ensure proper safety precautions, such as *backflow* prevention devices, which prevent possible contamination of groundwater with chemicals and fertilizers injected into the well water.

Foliar Application

When only the foliage is wetted to allow maximum absorption through the leaf, the spray is referred to as **foliar application.** Nutrients from foliar sprays move into the plant both through the leaf stomata (openings in leaves for gas exchange) and through parts of the epidermis (outer layer of cells). Sometimes translocation of absorbed nutrients within the plant may be slow. The relative mobility of nutrients indicates the variability of translocation (see Table 13-4).

All the nitrogen needed for a growing crop is seldom applied only in spray form, but supplemental amounts are often included in sprays, supplying pesticides, micronutrients, or other materials. Florida citrus trees are sprayed with three or four applications per year of about 1.1 kg each of nitrogen and of potassium (K_2O) per tree per application.

---

[3]A. W. Johnson, J. R. Young, E. D. Threadgill, C. C. Dowler, and D. R. Sumner, "Chemigation's Strong Future," *Agrichemical Age* (Feb. 1987), 8–9.
[4]D. M. Clark, "Letters: Costing Out Chemigation," *Agrichemical Age* (June 1987), 4.

Table 13-4 The Approximate Mobilities of Elements in Plants, with Decreasing Mobility As One Goes Down the Column. Symptoms of Immobile Elements Show Up in the Younger Leaves and Parts of Plants.

Mobile Elements	*Partially Mobile*	*Immobile Elements*
Potassium	Zinc	Calcium
Phosphorus	Copper	Boron
Chlorine	Manganese	Iron
Sulfur	Molybdenum	
Magnesium		

In summary, the following are some facts and suggestions for the use of foliar fertilizer applications:

- Foliar sprays are best suited for applications of small amounts of nutrients (1–10 kg/ha) for quick plant uptake and response. This method is mostly used for micronutrients (Figure 13-7), but it can give a boost in plant growth by adding a few kilograms of nitrogen, potassium, or phosphorus per hectare. Pineapples have been fed up to 75% of their nitrogen requirement as urea and 50% of their phosphorus and potassium in foliar spray. This requires frequent applications.
- Iron chelates, because of immobilization in soil, are frequently applied as foliar sprays to solve iron deficiencies (Figure 13-8). Soil applications of zinc, manganese, copper, and iron require many times more material than when added in sprays.
- Most sprays on foliage require a *wetting agent* to aid the liquid in spreading over the leaf and entering into the plant openings. Commercial wetting agents for this purpose are available. A *sticking agent* may help retain the liquid on the plant leaves.
- Foliar sprays have been helpful when other conditions reduce root uptake (cold or diseases such as root rot or nematode damage).
- Foliar sprays can yield a quick response—in a few days, in many situations. The effects of foliar spray are also often short lived because of the small amounts of nutrients absorbed. Several applications may be needed.

FIGURE 13-7 Boron deficiency in Thompson seedless grapes (right) drastically reduces fruit set. The limited areas of deficiency in California occur on soils of granite origin in the San Joaquin Valley. Treatment cost is relatively low, so whole blocks of vineyards are fertilized where boron is known to be low. About 3 lb of B each 2 to 3 years is sprayed onto the soil and lesser amounts are sprayed on the foliage, applied by aircraft, or put on with a sulfur-dusting machine. (*Source:* By permission of *California Agriculture;* Peter Christensen, "Boron Application in Vineyards," *California Agriculture* [March–Apr. 1986], 17–18.)

(a)

(b)

FIGURE 13-8 Iron deficiency on peaches in a calcareous Utah soil. (a) Close-up of leaves showing no deficiency (right) to increasing deficiency (left). Note the darker veins. (b) View of peach trees; light foliage is chlorotic leaves from iron deficiency. (Courtesy of Raymond W. Miller, Utah State University.)

- The spray left on the leaf can cause salt burn if it is too concentrated. Concentrations recommended seldom exceed 1%–2%. Many phosphates are damaging, and the maximum orthophosphate ($H_2PO_4^-$) concentration tolerated in spray without plant damage was 0.5% for corn and 0.4% for soybeans.

Fertilizing in Paddy and Other Waterlogged (Anaerobic) Soils

Paddy rice is the growing of rice in anaerobic soil, kept anaerobic by maintaining a layer of water on the soil (Figure 13-9). The depth of the water layer may vary, but it is usually between about 7 cm to 15 cm deep. Although certain varieties of rice can be grown in normal, aerated soils, most rice is grown in water-covered, anaerobic soils. In areas of heavy rains it is often one of only a few crops that tolerate the excessive wetness, sometimes growing in water 30 cm to 120 cm (1 ft to 4 ft) deep.

The flooded condition favors a number of growth factors:

- The pH becomes less acidic or less basic when flooded.
- Phosphorus and iron are more soluble and available.
- Fewer weeds can grow in anaerobic soils to compete with the rice.
- N_2-fixation is increased, including more free-living algae that fix nitrogen.
- Fewer soil-borne diseases occur.
- The water supplies some nutrients.
- The terracing to hold a water layer greatly reduces soil erosion.
- Yields of rice grain are higher.

Soils that may have a pH between 4 and 5 when drained will often develop a pH higher than 6 when flooded for paddy. Soils whose tests for available phosphorus suggest a low phosphorus availability are often adequate in phosphorus when used for paddy.

Anaerobic soils do have some disadvantages. Soil organic materials can produce toxic decomposition products, especially in the first few weeks of inundation. Nitrogen is lost by denitrification. Optimum conditions for denitrification are mild acidity, high amounts of easily decomposable soil organic matter, a lack of free gaseous oxygen in the soil solution, and warm temperatures. Nitrogen added to paddy as nitrates may have more than 50 percent loss

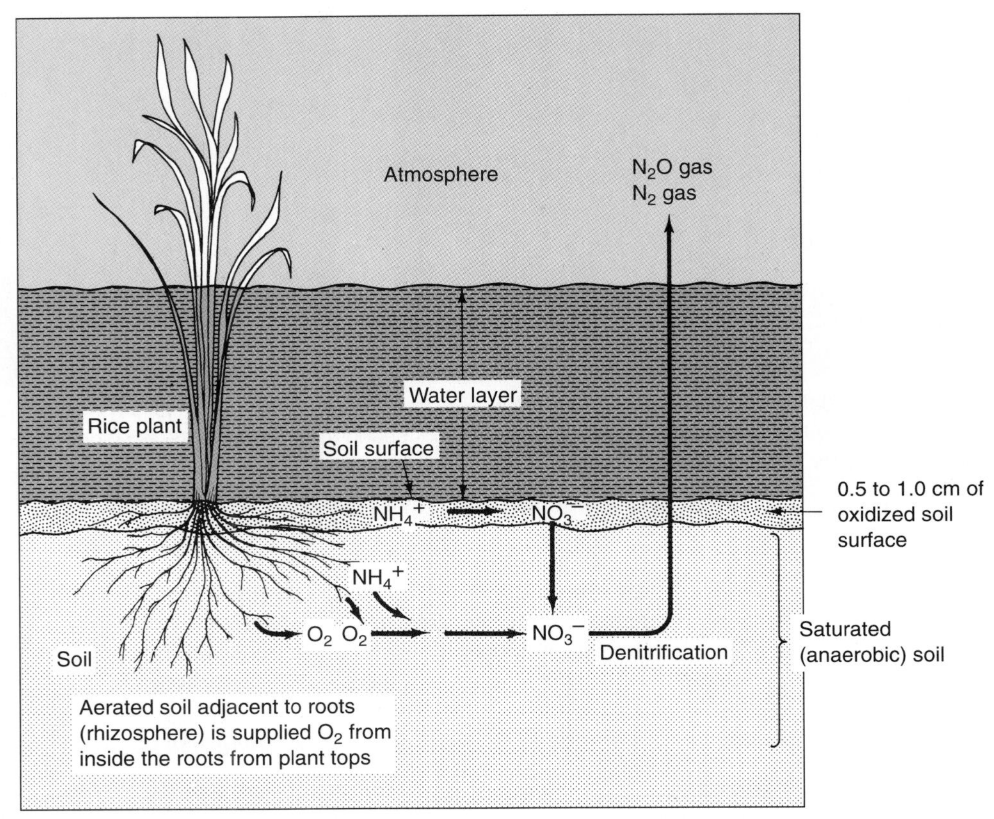

FIGURE 13-9 Diagrammatic cross-section of a rice paddy, emphasizing the hazard of losing nitrogen by denitrification. Ammonium ion forms are oxidized to nitrate in the thin oxidized soil surface or near roots by oxygen brought down inside the plant to the roots. (Courtesy of Raymond W. Miller, Utah State University.)

by denitrification. In the presence of rapidly decomposing organic material, anaerobic conditions *can be developed in less than a day* in warm soils. In paddy with slow water replacement and stagnant conditions, most of the root zone and deeper water is anaerobic within a few days of flooding (Figure 13-10). It may require 3 or 4 weeks to reach its most anaerobic level.

The extent to which soil around roots is made aerobic by the oxygen brought down internally in the plant is not well documented. It is believed that the major oxidation of NH_4^+ to NO_3^- occurs in the thin, partially oxidized 0.5 cm of surface soil in the rhizosphere or in the water layer itself. In one laboratory study using tracer nitrogen ($^{15}NH_4Cl$) and a heavy nitrogen application (more than 400 kg N/ha; 356 lb N/ha), the root zone caused a loss of 18 percent of the nitrogen during 40 days.[5]

The practical management of nitrogen in paddy is a problem. There is increasing use of urea and less use of ammonium sulfate (because the sulfate is more costly). Much of the urea is lost by ammonia volatilization from high-pH water caused by algae growth. The best efficiency of added nitrogen was with ammonium sulfate. The best urea effi-

[5]K. R. Reddy and W. H. Patrick, Jr., "Fate of Fertilizer Nitrogen in the Rice Root Zone," *Soil Science Society of America Journal* **50** (1986), 649–651.

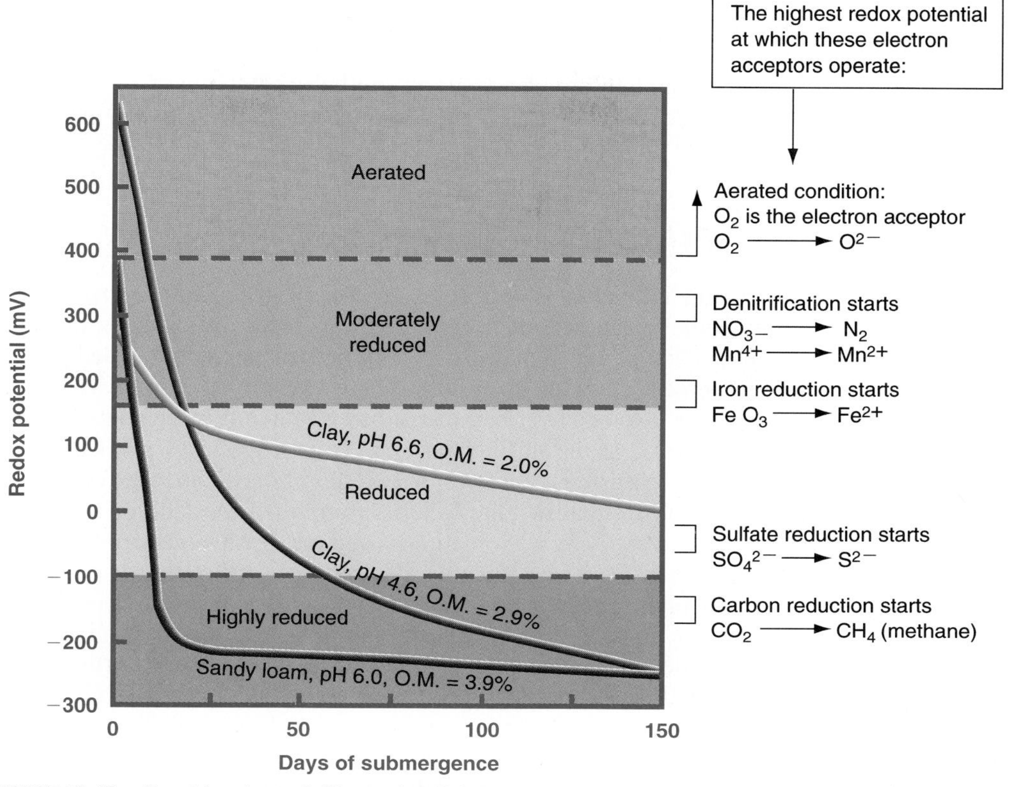

FIGURE 13-10 Combined graph illustrating (1) the rate at which three soils, following flooding, decreased in redox potential and became increasingly anaerobic, (2) the approximate extent of reduction of the different redox potentials, and (3) the approximate redox values at which NO_3^-, Mn^{4+}, Fe^{3+}, SO_4^{2-}, and carbon become electron acceptors in place of O_2, which is absent. (Courtesy of Raymond W. Miller; data selected and combined from F. N. Ponnamperuma, "Some Aspects of the Physical Chemistry of Paddy Soils," and W. H. Patrick, Jr., "The Role of Inorganic Redox Systems in Controlling Reduction in Paddy Soils," both in *Proceedings of Symposium on Paddy Soil,* Institute of Soil Science, Academia Sinica, Science Press, Beijing, and Springer-Verlag, New York, 1981.)

ciency occurred when preplant urea was incorporated into soil *and* when later nitrogen additions were made after the rice canopy was mostly closed (the closed canopy shaded the water and reduced algal growth).[6] The use of large urea granules about 10 mm (3/8 in) in diameter reduced the nitrogen loss to less than half as much as was lost from using prilled urea.[7]

Flooded rice paddies may also require sulfur or zinc. Correcting zinc deficiency in rice in California is a good example. Zinc deficiency (alkali disease) is common on basic soils, sodic soils, and soils with calcium carbonate, especially in the Sacramento and San Joaquin valleys. Soils extracted with a chelate (DTPA) and testing lower than 0.50 ppm zinc usually

---

[6]A. C. B. M. Van der Kruijs, J. C. P. M. Jacobs, P. D. J. Van der Vorm, and A. Van Diest, "Recovery of Fertilizer-Nitrogen by Rice Grown in a Greenhouse Under Varying Soil and Climatic Conditions," in Institute of Soil Science and Academia Sinica (eds.), *Proceedings of Symposium on Paddy Soil and Springer-Verlag,* New York, 1981, 678–688.

[7]K. Sudhakara and R. Prasad, "Ammonia Volatilization Losses from Prilled Urea, Urea Supergranules (USG), and Coated USG in Rice Fields," *Plant and Soil* **94** (1986), 293–295.

respond to zinc additions. Because zinc moves very little in soils, it must be placed where it is needed. In flooded rice, zinc is most available in the top 2.5 cm of soil. Satisfactory correction of zinc deficiency is made with 9.0 kg/ha of zinc sulfate or zinc lignosulfonate (chelate) applied broadcast on the soil before flooding. Coating the water-sown seed with 0.9 kg calculated as actual zinc (zinc sulfate or zinc oxide) per 45.5 kg of seed has also solved the problem.

13:4 Fertilizer Efficiency

Fertilizer efficiency is defined as *the percentage of added fertilizer that is actually used by the plants.* In general fertilizer usage the expected efficiencies are approximately 30 percent to 70 percent of added nitrogen, 5 percent to 30 percent of added phosphorus, and 50 percent to 80 percent of added potassium. These values can be improved by using special care; the values can also be lower because of bad weather or carelessness.

Looking at the fertility problem from a grower's point of view, the addition of any amount of fertilizer is of interest only if it *profitably* enhances yields, resulting in either larger yields or better quality—the benefits from fertilizer addition must exceed the costs of purchasing and applying it. Sometimes even the inconvenience of fertilizing at a busy time can be a major concern in evaluating the proposed fertilization.

In Figure 13-11 an illustration of yields and profits from added fertilizer is shown. *Maximum profits are rarely at maximum yields* because the last increments of fertilizer to produce a little more yield cost more than the yield increase is worth. The diagram also illustrates that a good producer can afford to use higher amounts of fertilizer (and still make a profit) than could a producer who is not managing a farm well. Too often research is reported with *significantly increased yields* from some experimental practice without any concern given to whether the increase is a *profitable increase.* The downward curve of Figure 13-11, as excessive fertilization increases, is due to reduced crop yields, which may be caused by salt problems (too much soluble fertilizer), imbalanced plant nutrition, or increased susceptibility to disease. Before attempting to increase yields on farms, the economics of such increases must be assessed carefully.

A number of factors affect fertilizer efficiency and crop responses to added fertilizer. The plant itself will vary because of the nutrients it needs and its kind of root system. Water availability, temperatures, pests (weeds, insects, other organisms), and management are all important. To evaluate the many factors, detailed crop-budget-analysis software is available. One of these is the **MEY (Maximum Economic Yield)** analysis package. This program attempts to consider the economic effects of each input in a crop's production.[8] Best management practices (BMPs; see Chapter 19) lead to MEY, and maximum economic yields lead to sustainable agriculture. These three concepts are compatible and should be part of every producer's agriculturalist vocabulary and activity.

Slow-release fertilizers may increase efficiency. Environmental concerns require minimal leaching of fertilizer materials. In one example the use of slow-release materials on citrus, which has year-round growth, reduced the frequency of addition (saving time) and reduced leaching losses (being environmentally responsible). The higher cost of the slow-release material is partly offset by the benefits and lower application costs.[9]

---

[8]Harold F. Rutz, Jr., "MEY Analysis Software Looks at Cost Per Unit," *Solutions* **32** (1988), No. 3; 38–40.
[9]Maria D. Raigon, E. Primo Yufera, A. Maquieira, and Rosa Puchades, "The Use of Slow-Release Fertilizers in Citrus," *Journal of Horticultural Science* **71** (1996), 349–359.

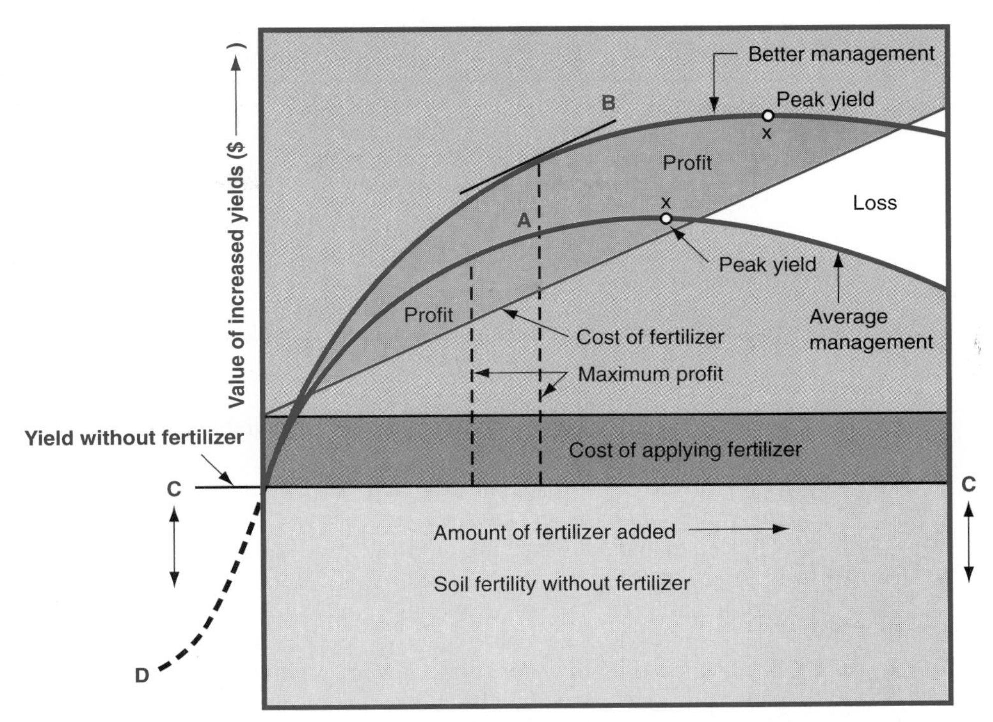

FIGURE 13-11 Representation of the increased profit resulting from added fertilizers. Notice that profit per dollar cost of additional fertilizer decreases as high fertility levels are reached. The initial soil fertility level (line C–C) varies with the soil and may be higher or lower than shown along the growth curve D–B or D–A. The maximum profits from fertilizer (where the line tangent to the curve is parallel to the fertilizer cost line) are higher and at a higher fertilizer addition for well-managed crops than for poorer management. Maximum yields (*x*) are seldom maximum profits because of slight growth increments per unit of fertilizer (or water or other treatment) added near maximum growth. (Courtesy of Raymond W. Miller, Utah State University.)

Plant Root Systems

Since early studies of why certain plants grew well on a given soil while other varieties or plant species did not, it has been known that some plants are better nutrient scavengers than others. For example, dicotyledons (especially legumes) feed strongly on divalent cations such as Ca^{2+}, whereas grasses feed better on monovalent cations such as K^{+}.

How well roots absorb fertilizers depends upon their distribution in the soil. The smaller the root system (shallow, few in number) and the shorter the growing season, the more dependent the plant is on fertilization. Plant growth rates and size must also be considered because small, slow-growing plants (short grasses) have low growth rates and low total demands. When not restricted by compacted soils, rocks, or toxic salts, the root systems of some common plants are grouped as shown in Table 13-5.

Fertilizer-Water-Application Interactions

Nutrients in soils move in water to root surfaces, where they can be absorbed. *The availability of nutrients to plants is directly proportional to soil water content.* Surface soils are usually highest in nutrients; when it dries, those nutrients in the dry soil become less available. A wet surface soil may increase the hazard of diseases to crops such as cucumbers and

Table 13-5 General Rooting Patterns for Various Common Plants

Deep and Numerous Roots, About 1.8 m	*Intermediate Root System, About 1.2 m*	*Shallow or Few in Number, About 0.6 m*
Grapes	Corn	Potatoes
Sorghum	Peas	Beans
Alfalfa	Soybeans	Turf, short grasses
Tomatoes	Tall grasses	Lettuce
Orchards	Cotton	Peppers
Sweet clover	Small grains	Onions

strawberries, but a dry surface soil reduces available nutrients in the soil layer or maximum nutrients and root proliferation.

Surface flow irrigation can alter nitrogen and sulfate levels in soils because of the excessive leaching that occurs at the tops of most fields. In contrast, sprinkler irrigation could be adjusted to avoid all leaching of nitrates and sulfates; the wetting depth can be controlled and is somewhat uniform. In the sandhills of Nebraska, the most efficient system involved fertigation that wetted the soil to about 45 cm (18 in) deep in 12 hours when irrigation was needed.

Some intriguing management systems involve fertilizer added in *drip* irrigation systems. Efficiencies of both water and fertilizers are high. If the drip is surface-applied and is adding phosphate, the constant high soil water content (near field capacity) helps movement of some phosphate further into the soil. In one study on acidic sandy loam, placing phosphate fertilizer near a drip outlet increased available phosphate to a depth of at least 22 cm deep. Drip lines buried 20 cm to 25 cm deep, with water filtered and adjusted to pH 6.0 to 6.5 to minimize plugging emitters, allowed fertilization with N, P, and K for years. Tillage was shallow to avoid damage to lines. Some lines are buried as deep as 45 cm (almost 18 in), and irrigations can be as frequent as several times *daily* (each irrigation with 1 mm of water containing fertilizer).

Frequent irrigation with a drip system may cause periods of partially *anaerobic* conditions. Also, to minimize plugging problems, water must be carefully filtered. More than 25 percent of Israeli irrigation is using drip, and use of drip irrigation in other parts of the world is increasing (Figure 13-12).

Another concern when using drip systems is the higher costs of water-soluble fertilizers, particularly the phosphates. Additionally, a *fumigant* is usually used to kill microbes that could produce emitter-plugging slime.

Timing and Climate

The effectiveness of fertilizer depends on having it *where* the plant can get it and *when* the plant needs it. To avoid excess losses by leaching or volatilization and reduce adsorption and precipitation losses of phosphorus, fertilizers are often applied in split applications. Heavy rainfall following a top dress of urea may leach much of the urea away. A lack of rainfall may allow volatilization losses. A cold or extremely hot period following fertilization may reduce the plant growth more than expected. Applications too early or too late will reduce their effectiveness, even if they were late because another crop needed attention.

13:5 Fertilizer Management Techniques

Prior to the regulations on polluting the environment, some lower efficiencies of fertilizer were often accepted to allow a saving in labor or to permit application in a more convenient manner or time. Maximum fertilizer efficiency was often thought not worth the changes required to get

FIGURE 13-12 Drip irrigation on vegetable crops in Senegal, west Africa, to attempt to get into the winter market in Europe. With an 8-month dry season, drip irrigation is practical. This is one of the simplest drip approaches—straight lines and one line per row on top of the ground. (Courtesy of Raymond W. Miller, Utah State University.)

it. The value of the lost fertilizer caused by management changes was traded for gains in time or savings in equipment. Sometimes the value of a crop justified the addition of insurance fertilizer, small excesses that guaranteed that there would be enough for maximum production.

Fertilization for High Efficiency

Environmental-pollution concerns and the need to follow established (even regulatory) *best management practices (BMP)* require new evaluations of these concepts (see Chapter 19). In some instances certain choices could cause large or extensive amounts of pollution.

One goal in fertilization is to achieve *optimum fertilizer efficiency,* and the result is usually to cause minimal pollution. Some of the guides to approach *optimum fertilization* follow.

- Avoid single, large fertilizer additions of N or K (more than 23 kg/ha of element; 50 lb/a) on sandy soils when rainfall or overirrigation can cause leaching. Use split applications when the crop has grown enough to get maximum uptake quickly.
- Reduce ammonia volatilization losses by avoiding surface-broadcast applications of urea and ammonia solutions on moist calcareous soil during the warm growing period, unless the fertilizers are immediately incorporated by tillage or watered into the soil.
- Reduce denitrification losses by avoiding heavy nitrate fertilizer additions on poorly drained soils, such as rice paddies and poorly drained clayey soils in wet climates. Use the ammonium forms and attempt to prevent cycles of aerated and nonaerated soil, which cause oxidation to nitrate, then loss by denitrification.
- Band water-soluble phosphorus fertilizers; broadcast and incorporate low-solubility materials.On soils of very low available phosphorus, banding may be less productive than broadcast mixing. On high phosphorus-fixing soils (those high in iron and aluminum [acidic soils] or high in calcium [calcareous soils]) banding is most efficient.

- Use a small amount (up to 11–22 kg/ha of element; 10–20 lb/a) of starter fertilizer with or near the seed at planting. Avoid larger salt-damaging amounts of nitrogen and potassium fertilizers near seed or roots.
- Nitrogen and potassium fertilizers are soluble salts. Avoid large additions of fertilizers at distances less than 3 cm, and preferably at least 5 cm to 7 cm, from the seed, to avoid salt burn.
- Urea and nitrate are very soluble in water. They can be moved, even leached from the root zone, by rain or excess irrigation.
- Combining ammonium and phosphates together in bands helps increase the efficient use of phosphates.
- Foliar applications of nutrients (in concentrations of less than 1% solutions) provide rapid uptake and use of small quantities of nutrients (a few kilograms per hectare). Foliar applications are very effective for micronutrients, particularly for iron, zinc, and manganese.
- Know the nutrient demands of the crop. Large amounts of nitrogen are needed for corn, sorghum, bananas, peanuts, and potatoes. Many legumes require large amounts of phosphorus and potassium but fix some or all of their nitrogen needs. Bananas, corn, and potatoes have high phosphorus needs. Tobacco and crops producing sugars and starches need high levels of potassium.
- Be willing to try *test strips* on your fields, which are modifications of the recommended fertilizer to add. For example, add more fertilizer, add a micronutrient, reduce rates, add another nutrient, and so on. Good fertilizer use requires integrating knowledge of your field and management over many years of different crops.
- Fertilize and irrigate at the correct times. Timing may be as important as the variation in amount of fertilizer.
- Remember Liebig's *law of the minimum.* The growth factor in the least relative amount will limit the growth. Fertilizer is only effective when other growth factors are adequate. Some of these other factors include (1) using responsive crop varieties, (2) having correct plant densities, (3) employing correct water control, (4) timing properly the management operations, (5) being free of disease, (6) having no physical soil problems, (7) having adjusted the soil pH to an appropriate level for that crop, (8) having other nutrients in adequate amounts, and (9) having optimum weather.
- Large concentrations of fertilizers in concentrated bands may hinder quick plant use because of (1) too high a salt content for nitrification and (2) hindrance to root entry into the band because of salt or ammonia toxicities in and near the bands.
- Test soils frequently (some fields every year) and follow the recommendations, generally. During some years phosphorus, potassium, or other nutrients may be adequate in some fields for a year or more, especially if the crop is a low-demand type.

There are other considerations when planning fertilization. For instance, high soil levels of ammonium (NH_4^+) and nitrate (NO_3^-) minimize atmospheric nitrogen (N_2) fixation by microorganisms. The cultivar selected may have particular specific nutrient requirements that should be taken into account. Some rice cultivars, for example, produce satisfactory yields under *low* levels of soil nitrogen, phosphorus, potassium, sulfur, and zinc. At least three rice cultivars produce high yields on low-phosphorus soils, whereas others have low yields at low soil phosphorus levels.[10]

---

[10]M. Mahadevappa, H. Ikehashi, and F. N. Ponnamperuma, "Research on Varietal Tolerance for Phosphorus-Deficient Rice Soils," *International Rice Research Institute Newsletter* **4** (Feb. 1979), No. 1, 9–10.

Variable-Rate Fertilization

For decades farmers have felt comfortable treating different fields differently. They fertilized, rotated crops, sprayed, and irrigated each field as separate units; then, they combined some fields into larger single fields or separated large fields into several smaller fields. One field might be a challenge to manage because it was nothing but sticky gumbo clay. Another field might have had a soil variation through the middle where an old drainage canal was filled up with topsoil. Another field might have been graded (cut and filled) for irrigation and had a bad spot in the upper corner, where a knoll had been cut off down into the clayey subsoil. These are the kinds of situations most farmers have experienced.

As they walked through their fields, looking at the growing crops, it was obvious that each field had different plant growth in different spots. One area had poor growth because of compaction by field traffic. Another area had been planted when the poorly drained spot had been too wet and it crusted over, reducing the number of emerging seedlings, reducing the plant density (stand). Another area had inadequate drainage and many plants had drowned in puddled water.

Those observations of yield differences across a field have led to the question of whether those different areas should be treated differently. Recall that in Chapter 12 the advice for taking a soil sample was to avoid unusual areas and not include subsamples from those abnormal areas into the composite sample for the field. Maybe it is time to consider changing that approach somewhat.

Portable computers can be used to change many things on-the-go, such as depth of planting, rate of planting, rate of fertilization, and rate of spraying pesticides. Two slight problems exist: when to change applications and which changes to make. To determine differences in areas and then map the field so these different areas can be identified requires *spatial variability measurements.* **Spatial variability** means just what the name suggests: *changes in the property measured with distance across the field* (see Chapter 2). These measurements of the property at the different locations are called **site-specific measurements.** Thus, **site-specific management** is changing the treatment done to different portions of the field (fertilization, spraying, planting) as those portions of the field are traversed.

13:6 Understanding Precision Agriculture

Many fields have several soil areas that react differently in their ability to grow plants under a particular management. Those fields are not uniform, homogeneous fields. When fields are not homogeneous, applying a single fertilizer rate over the whole field means that more fertilizer was added to fertile parts and too little was added to low-fertility portions. The latter areas did not produce as much as they could, and the fertile areas probably had fertilizer washed into waters, contaminating them.

Site-Specific Management

Is it possible or economical to try to treat the unlike areas of a field differently? Yes. The ability to treat areas of a field differently is the emerging **site-specific management,** which is also called **spatial variability management, precision agriculture,** and **high-tech agriculture.** It is partly the product of adaptable computers to control changes rapidly and automatically—changes in fertilizer or pesticide added. Whether or not it is economical is still in question.

Precision agriculture is an attempt to treat increasingly smaller portions of land differentially with fertilizer application, spraying, and other growth-promoting activities. Basically it involves the process of using precise devices to collect precise data and turn it into precise decisions, but buying technology won't necessarily make you a precise farmer.

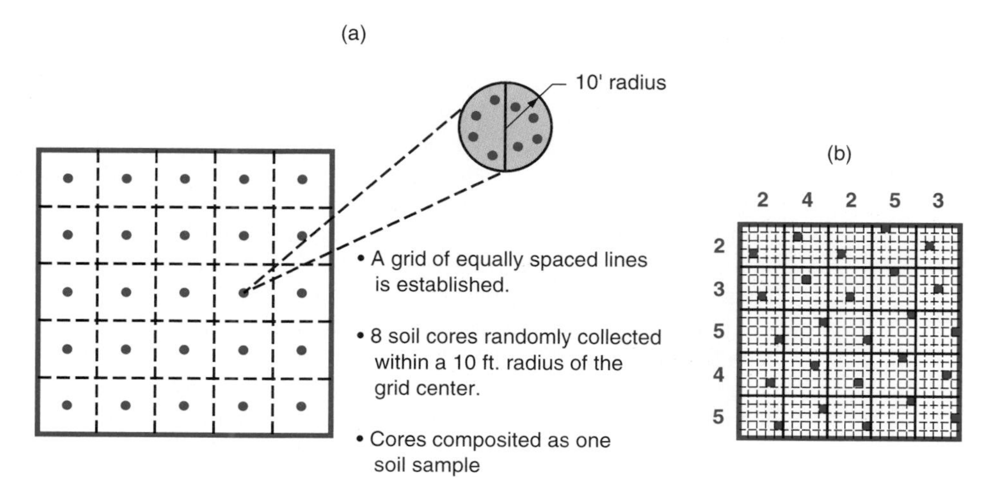
(a)
10' radius
• A grid of equally spaced lines is established.
• 8 soil cores randomly collected within a 10 ft. radius of the grid center.
• Cores composited as one soil sample
Systematic grid square sampling pattern
(b)
2 4 2 5 3
2
3
5
4
5
Systematic unaligned grid

FIGURE 13-13 Schematic layout for (a) square grid sampling and (b) systematic unaligned grid. Although the distances between sample sites and numbers of subsamples per composite differ, the general concepts above suggest acceptable approaches for detailed sampling. The random sample sites (blackened squares) in (b) are selected to eliminate bias in sample location. (Courtesy of N. C. Wollenhaupt and R. P. Wolkowski, "Grid Soil Sampling," *Better Crops* **78** [1994], No. 4; 6, 7, 9.)

The proper starting place involves **record-keeping** of cropping practices and conditions, weather, and the results (yields, etc.). If you didn't do it before, you must do it now to become a precision producer.

Delineating Different Areas for Different Treatment

To treat different portions of a field differently requires (1) the identification and delineation of each portion of the field from other portions of the field, and (2) knowing the treatment changes that are needed for each portion. To identify and separate the field portions is *delineating the soil's spatial variation* into soil units that have a number of measurable characteristics: yields, organic matter contents, compaction, fertility, or plant growth variations.

Once the areas of a field are identified and their differences in characteristics are known, the next question is, *What and how should the different areas be treated?* Unfortunately, field portions that produce different yields or plant growth can be identified, but *why* those differences exist is still the more difficult problem. Farming knowledge and experience must be applied to each location.

Spatial Variability Mapping Using Soil or Crop Data[11]

The known spatial variability in yields and nutrient levels in many fields coupled with environmental issues related to *excessive* fertilization now provide a strong incentive to adapt variable-rate application technology. The first step is to make a map showing the variable levels of fertility, which is done by taking soil samples on a grid pattern. Distances might be from 10 m to 80 m between grid points, where a local composite sample is taken (Figure 13-13).

[11]D. C. Penney, S. C. Nolan, R. C. McKenzie, T. W. Goddard, and L. Kryzanowski, "Yield and Nutrient Mapping for Site Specific Fertilizer Management," *Communications in Soil Science and Plant Analysis* 27 (1996), Nos. 5–8, 1265–1279.

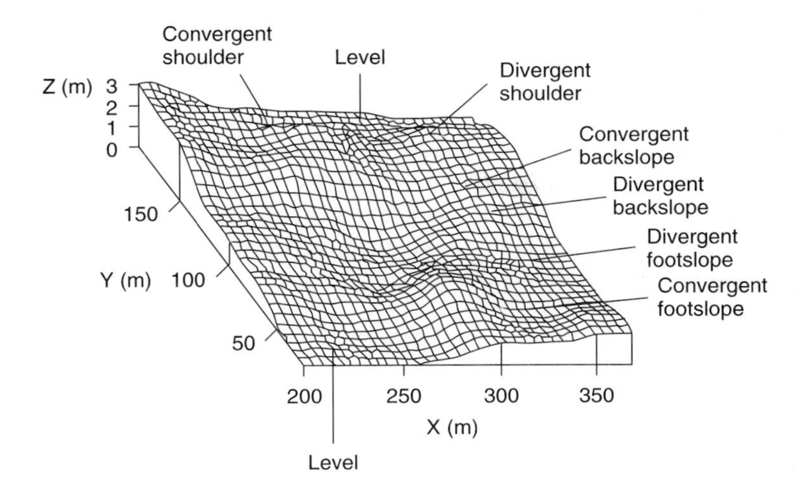

FIGURE 13-14 Wetness, erosion, and sediment fill can all be related to land relief. The GIS system permits involvement of landform elements (slopes, etc.) as shown here, as they might affect yield. This figure shows seven position elements that might be critical in evaluating a nonlevel or graded field for precision farming. There may be more. (Courtesy of D. C. Penney, S. C. Nolan, R. C. McKenzie, T. W. Goddard, and L. Krysznowski, "Yield and Nutrient Mapping for Site-Specific Fertilizer Management," *Communications in Soil Science and Plant Analysis* **27** [1996], Nos. 5–8; 1265–1279.)

The delineation of soil areas is done by inspection or by computer programs that *smooth* the data for all the grid point samples. To do this each year would be very costly, so it must be assumed that the data collected to delineate areas are going to be generally true for several years; at least, the spot boundaries should not greatly change. A few locations might be checked yearly for an update on each property (available P or K, salt, etc.) within its boundaries on the map. Yield differences can be measured by computer systems set up in the harvester. The yield is measured as the combine moves across the field and cuts the grain. As much as 50 percent differences in yield within 200 m (221 yd) have been observed. If a Differential Global Positioning System (**DGPS;** see Chapter 2) is added, the yield differences of each portion of the field, measured within plus or minus 1 m to 5 m, are claimed possible.[12] The spatial variation in yield can then be made into a map.

One writer cautions users to *control their expectations.* The yield monitor may raise as many questions as it answers, but it will provide information to evaluate management practices so you can work toward answers. A yield monitor is not a magic bullet; they need to be understood. As one producer stated, "A red dot on a yield map is just that—a red dot—until you understand what made it red and if it should have been red in the first place."[13] It is quite likely that crop yield maps will not necessarily match soil test maps. Yield variability is most likely to be caused by differences in *compaction, water management, tillage,* and *pest problems* as much as by fertility levels (Figure 13-14). Equipment costs for a yield monitor are \$3000 to \$5000; an added DGPS system is another \$1500 to \$5000.

Spatial-Variability Mapping Using Soil Inspection

A few new analytical techniques are adaptable to detect spatial differences in the soil profile—electromagnetic conductivity meters (**EMs**) and ground-penetrating radar (**GPR**). Until now profile observations meant the use of a shovel, a backhoe, an auger, or some other probe to physically look at the soil and take samples for analysis. For some information this is still necessary. For other useful information nondestructive methods can be used.

The **electromagnetic conductivity (EM)** system sends an electromagnetic signal into the soil and measures changes in the signal that bounces back to the meter. This is done on-the-go. The method is most useful when two *contrasting soil properties* (layers) are involved.

---

[12]Tim Aughenbaugh, "Yield Monitors: New Tools for On-Farm Testing," *Successful Farmer* **94** (1996), No. 7; 40.
[13]Tim Aughenbaugh, *ibid.*

The information obtained are the differences in conductivity of the waves. The problem then is to find out why the difference occurs and try to quantify the observed changes in conductivity. The EM system has been used to sense different depths to a claypan in Missouri (shallow claypans greatly reduce yields).[14]

The **ground-penetrating radar (GPR)** uses a similar concept: sending and receiving radar waves. The radar waves bouncing back identify the differences that exist in layer densities. It has been used to detect the presence of plowpans and other compaction layers. It is used to assess depths and continuity of bedrock beneath building sites and to gauge subgrade conditions beneath highway beds.[15] The GPR was developed by the U. S. Military to locate networks of tunnels in the jungles of Vietnam.

Ground-penetrating radar measures differences; the user needs then to identify the cause of the differences. It has been used as a noninvasive technology to measure the entry of salt water and the distribution of salty subsoil areas.[16] The radar penetrates up to about 8 m deep and reflects a measurable *wave change* from voids, compacted layers, solid metal, wood or plastic surfaces, and water tables. It has been used to locate buried tanks, drums, landfill debris, geologic structure, water levels, buried pipes, tunnels, land mines, deteriorating cement masses, buried bodies, large buried bones (dinosaur), archaeological sites and buried objects, structures within thick ice (buried WW II aircraft in the Greenland icecap), and mining (Is the vein changing? Are there dangerous rock flaws?).[17] Thermal infrared (IR) is often used with ground-penetrating radar, because it is possible to do a larger area faster. Thermal IR senses a difference in heat absorption of buried materials or layered substances.

Spatial-Variability Mapping Using Plant Health[18,19,20,21]

Measuring the health of the crop will locate (delineate) problem and nonproblem areas but not usually indicate what the problems are. The differences in crop growth in a field are best done by remote sensing (aerial photography). In fact, it has been suggested that rather than sampling soils on a grid and then making the map of soil variations, the aerial photo could be used to determine *where to sample* the soil for the most information (not on a grid).

The major changes occurring in recent remote sensing are the use of video cameras and the use of several color wavelengths. Numerical data identifying the different areas are obtained, often by using several video cameras on a plane (or planned satellite), each camera with a different light filter—one for near-infrared, one for red, and one for yellow-green. Scientists have been working on establishing spectral *signatures* for dozens of plants, soil and water conditions, weeds, excess salts, and infestations of plant diseases and insect pests.

For example, in a sorghum field the healthy plants will show up as *magenta* and iron chlorotic plants as *pink*. The *normalized difference vegetation index* (NDVI) used shows good-health plants in white grading to the poorest, black, through lavender, purple, green, yellow, orange, and brown to gray.

---

[14]Greg D. Horstmeier, "Electronic Shovels," *Farm Journal* **120** (1996), No. 4; 16–17.
[15]Greg D. Horstmeier, *ibid.*
[16]Sun F. Shih and Don L. Myhre, "Ground-Penetrating Radar for Salt-Affected Soil Assessment," *Journal of Irrigation and Drainage Engineering* **120** (1994), No. 2; 222–233.
[17]Leon Peters, Jr., Jeffrey J. Daniels, and Jonathan D. Young, "Ground-Penetrating Radar as a Subsurface Environmental Sensing Tool," *Proceedings of the Institute of Electrical and Electronic Engineers* **82** (1994), No. 12; 1802–1822.
[18]R. F. Denison, D. Bryant, A. Abshahi, R. O. Miller, and W. E. Wildman, "Image Processing Extracts More Information from Color Infrared Aerial Photos," *California Agriculture* **50** (1996), No. 3; 9–13.
[19]Jim De Quattro, "Orbiting Eye Will See Where Crops Need Help," *Agricultural Research* **44** (1996), No. 4; 12–14.
[20]Janet White, "Crop Management Goes High-Tech," *California Agriculture* **50** (1996), No. 3; 8.
[21]Reuben B. Beverly, "Video Image Analysis as a Nondestructive Measure of Plant Vigor For Precision Agriculture," *Communications in Soil Science and Plant Analysis* **27** (1996), Nos. 3–4; 607–614.

In 1995 planes were flying for remote sensing over the states of California, Illinois, Indiana, Iowa, Missouri, and Nebraska. Customers get picture images before planting and at weekly intervals. The weekly map shows four views: (1) before planting, as a reference; (2) and (3) vegetation maps, one with 16 gradients in vegetation quality; and (4) the *change* that has occurred *since the previous image.* Remote sensing images cannot replace field monitoring, but they can provide other information and pinpoint where problems exist in the field.

Four commercial satellites are planned to be launched in 1999. The joint partnership, called *RESOURCES21,* plans to launch the 1200-lb satellites 450 miles up to circle the Earth, collect data, and beam the information back about twice a week on both the East coast and the West coast.[22] The images will be sent to subscribers with 1-day turnaround to be used for solving field problems and following the plant status.

13:7 Getting Started in Precision Farming

The first step in precision farming is to determine if it is needed. Do the fields have enough spatial variability to have areas of different plant growth? This is most easily determined with measurements of yield with a variable-yield harvester. If the answer is yes, then consider some degree of involvement in precision farming, as discussed in the rest of this chapter.

Starting the Changeover to Precision Farming[23]

First, when getting into precision farming, *start slowly.* It is a toll road, not an easy street. Consider it a learning experience that may pay in the long run, even if not immediately. Second, *start keeping records;* they will be a part of the yearly input to your system. You will initially accumulate much more data than you are used to collecting. Third, begin to *accumulate the necessary equipment and capability* to (1) store the data collected, (2) do data analysis for making decisions, and (3) apply the **variable-rate technology (VRT).** This accumulation of data will continue from one year to the next.

Data Collection

All data must be tied to locations in the field. Field boundaries, slope and aspect, water relations, drainage, soil texture, fertility variables, and rooting depth are some things that are needed. Conventional soil survey maps are not likely to be accurate enough for more than general information on kinds of soils and approximate boundaries.

Set up **ground control points (GCPs)** using the differential global positioning system (DGPS). Set up no fewer than four points, which is possible to do within ≤ 0.1 m (about 4 in). These points provide a precise setting for all maps and data that year and for years to come. The GCPs should appear in all aerial photographs and video images taken thereafter; they are the starting point for units using on-the-go DGPS referencing later. All sampling will be done with reference to the GCPs and entered into a common geophysical information system (GIS) that is to be used. The data are processed and decisions made as where to change planting rates, how deep to plant, whether anything needs to be done to certain areas (such as add lime), which kinds of plants and varieties will be used, and which changes in rates of application should be made and, if so, where?

It is becoming necessary to improve overall yields and provide needed nutrients without unnecessary environmental pollution from overapplication. Often BMPs are required

---

[22]Agricultural Research Service, "Ag Satellites to Launch in '99," *Ag Consultant* **52** (1996), No. 4; 7.
[23]E. Lynn Usery, Stuart Pocknee, and Broughton Boydell, "Precision Farming Data Management Using Geographic Information Systems," *Photogrammetric Engineering and Remote Sensing* **61** (1995), No. 11; 1383–1391.

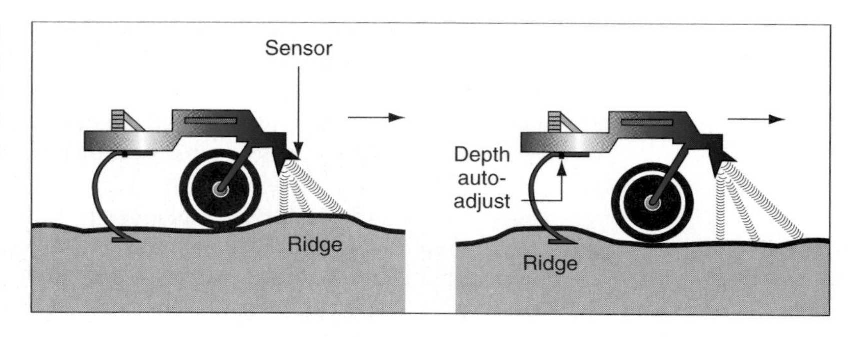

FIGURE 13-15 Diagram of a planter with an ultrasonic sensor that maintains the depth of planting, relative to the soil surface, set by the operator even though the ground level may have some ridges and depressions in it.

farming procedures, and the keeping of more and better records will be necessary to avoid loss of general permits. It may be necessary to show that the BMPs are being followed in order to retain a simple but required *general permit* (see Chapter 19).

13:8 Examples of Site-Specific Farming

Site-specific farming is new and still has many obstacles that must be overcome. The equipment is expensive and requires specialists to set it up and repair it. Habits (based on conventional agriculture) and attitudes are often slow to change. Convincing a producer to invest money and management time to change to new, expensive, high-tech methods will require some clear evidence and examples of benefits.

Precision Planting[24,25]

Quality planting *is* possible. Skips and overlaps can be eliminated and planting depths can be precise. A wheat planter in Saskatchewan, Canada, planting nearly 30,000 ha per year uses a differential global positioning system (DGPS), which he claims is capable of 10 cm (4 in) accuracy. This accuracy reduces planting skips and overlaps and even permits planting at night.

Automatic control of planting depth is a useful modification. Seeding too deep is energy-expensive, and planting either too deep or too shallow can greatly reduce stand density. Seeding depth can be controlled by use of an **ultrasonic sensor** (Figure 13-15). Planting small seeds 5 cm (2 in) deep when optimum depth is 2.5 cm can reduce emergence by 30 percent. This planter can also run at night since depth of planting is automatically adjusted.

Variable-Rate Fertilization (VRF)[26]

Fertilizing at different rates across a field requires a prepared map of spatial nutrient levels and a computer-controlled variable fertilizer applicator. An example of the overall picture of precision farming is given in Figure 13-16.

In one Canadian study, areas of high nitrate were associated with high salinity and *low* yield. If a uniform nitrogen rate had been used, these high-nitrate areas would have been overfertilized. The grid sampling resulted in changes in application rates that varied from 0 to

---

[24]Anonymous, "GPS Adds to Accuracy," *Successful Farming* **94** (1996), No. 4; 17.
[25]Anonymous, "New Ultrasonic Depth Control," *Successful Farming* **94** (1996), No. 6; 20.
[26]D. C. Penney, S. C. Nolar, R. C. McKenzie, T. W. Goddard, and L. Kryzanowski, "Yield and Nutrient Mapping for Site Specific Fertilizer Management," *Communications in Soil Science and Plant Analysis* **27** (1996), Nos. 5–8; 1265–1279.

FIGURE 13-16 An overall view of site-specific nutrient management indicating the various high-tech tools that would be involved in most instances. Crop scouting also helps to locate problems that might need solving. (Courtesy of H. F. Reetz, Jr., "Site-Specific Nutrient Management Systems for the 1990s," *Better Crops* **78** [Fall 1994], No. 4; 14–19.)

more than 160 kg N/ha. Estimated phosphorus needs ranged from 10 kg to 45 kg P_2O_5/ha. Potassium would not have been applied if the average value for the field had been used. The spatial variability from the grid-point soil samples showed that 20 of the 40 grid samples were below the critical K level.

Growing sugar beets with variable-rate fertilization in Minnesota is a good example of the overall process.[27] On 25 ha and 28 ha fields, soil samples were collected in the conventional manner and also on a grid (a little more than 1 ha per grid point). Nitrate contents were included in the analyses. The grid samples clearly indicated that some areas of the fields were underfertilized and others were overfertilized if a blanket single application rate was used over the entire field (Table 13-6). The variable-rate treatments had higher yields and sugar contents. Additional costs in the variable-rate system were $62/ha ($25/a) for soil samples and extra fertilizer. The gross difference in gross return was about $210/ha ($85/a) for the variable-rate treatment; the gross difference in return was about $148/ha for the conventional approach.

Site-Specific Herbicide Application[28]

Some herbicides have rates dependent upon soil components. High soil organic-matter contents often require that more herbicide be added to be effective. The herbicides adsorb to the organic matter and become inactive or less effective in controlling weeds.

---

[27]Larry J. Smith and Doug Rains, "Grid Soil Testing and Variable-Rate Fertilization for Profitable Sugarbeet Production," *Better Crops* **80** (1996), No. 3; 30–31.

[28]Mike Holmberg, "Variable Rates with Consistent Control," *Successful Farming* **94** (1996), No. 2; 34–35.

Table 13-6 Summary of Soil Nitrate-N and Fertilizer Recommendations for Two Years Crops of Sugar Beets in Wisconsin, Comparing Conventional and Variable-Rate Systems

*Factor**		*Year 1*	*Year 2*
Available soil nitrate-N from conventional sampling		95	83
Average available soil nitrate-N from grid sampling		81	76
Range in available soil nitrate-N from grid sampling		21–180	39–102
Nitrogen recommendation based on conventional samples		25	37
Nitrogen recommendation, variable-rate samples:	Average	63	78
	Range	0–100	26–100
Percentage of field under fertilized based on conventional samples		65	79

Source: Modified after Larry J. Smith and Doug Rains, "Grid Soil Testing and Variable-Rate Fertilization for Profitable Sugarbeet Production," *Better Crops* **80** (1996), No. 3; 30–31.
*Nitrate-N was the sum of the 0–60 cm depth plus 80% of that in the 60 cm–120 cm depth which exceeds 34 kg/ha.

Pesticide application rates must be precise. If the soil organic matter is variable (it varied from 3.3% to 8.9% in one Minnesota field), poor grass control is risked if an average rate is used over the field. Using a lower rate on high organic-matter areas will not control the weeds; using a higher rate over such a field is too high on the low-organic-matter areas and is environmentally irresponsible. A lot of careful work and monitoring of variable spray rates is still needed.

Important Factors Affecting Yields

Interestingly, a current ranking from the Director of Brookside Laboratories may be a surprise.[29] As you look at areas of fields that do better or worse than other areas, think of some of the factors in the following list as possible reasons for the growth quality differences. The Director stresses that the list is dynamic; the order can change if certain factors such as crop rotation or insect/weed problems are not a factor in a given year in a given field. The conditions are ordered from most important to least important:

1. Drainage (water, and stresses)
2. Crop variety (resistance, root systems, ability to adapt to stresses)
3. Insect and weed problems (nematodes, etc.)
4. Crop rotation (synergistic effect)
5. Tillage (type, timing, wet or dry soil)
6. Compaction
7. pH (liming, extreme pH variation)
8. Herbicides (misapplication, drift)
9. Subsoil condition (acid or alkaline, clay layer, fragipan, etc.)
10. Fertility placement (ridge-till, no-till)
11. **SOIL FERTILITY**
12. Plant population (most fields have a narrow optimum plant population)

Whether or not you agree with the general listing, it emphasizes an important concept: *There are many possible causes for differences in growth that should be considered when evaluating an area.* Notice that this list is not even comprehensive. It lacks *toxicities* of some kinds, *aeration* (clayey soils), *topsoil depths,* soluble salts, drought, and soil *texture variations,* to name a few.

[29]Mark Flock, "Agronomist Tallies Top 12 Yield Factors—Precision Mapping or Precision Farming?" *@griculture on Line, @g/Innovator,* Internet, http://www.agriculture.com/contents/aginn/ai029609.html, August 5, 1996; 1.

Precision agriculture is not the acquisition of high-tech equipment so much as it is to collect and wisely use the *information* obtained by using that technology. Technology does not replace people; it places higher demands on the knowledge and training of those people. The increasing concern about the environment and controls to protect the environment will add additional restrictions, costs, and needs.

Numerous growth models involve fertilization, irrigation, and tillage activities, many of which are listed in **Appendix A.**

There is nothing more difficult to take in hand, more perilous to conduct, or more uncertain in its success, than to take the lead in the introduction of a new order of things.

—*Nicolo Machiavelli,* The Prince

Questions

1. (a) Because many fertilizers are broadcast but not incorporated and are of low efficiency, why is broadcasting used? (b) Is broadcast plus incorporation better? Discuss.
2. State one major reason (two, if you can) for using each of these fertilizing techniques in preference to broadcasting: (a) deep banding, (b) split applications, (c) fertigation, (d) foliar application, and (e) starter.
3. State at least one disadvantage for each fertilizing technique in question 2.
4. (a) Discuss the extensive use of starter-banding applications. (b) Is it the best method for most crops? Explain.
5. Evaluate fertigation and chemigation as (a) economic and (b) safe techniques.
6. How does *fertigation* differ from *foliar application*?
7. (a) Describe the aeration conditions in paddy rice. (b) How do these conditions influence fertilization techniques?
8. Discuss some of the factors, other than fertilizer application techniques, that affect fertilizer efficiency.
9. How does Liebig's *law of the minimum* enter into the expected response from added fertilizer?
10. Are individual farmers encouraged to set up some of their own simple and small fertilizer test strips that are different from the fertilizer recommendation given them? Discuss.
11. What is meant by the term *spatial variation* in a field?
12. Why is *spatial variation* important to a producer, and what can he or she do about it?
13. What are some of the general requirements for doing *precision farming*?
14. How are portions of fields that have different properties delineated to allow site-specific farming?
15. What is the purpose of remote sensing to determine plant *health*? How is it helpful?
16. Discuss how the following activities help in site-specific farming: (a) monitoring yield, (b) grid soil sampling, (c) remote sensing, and (d) soil depth inspection.

Tillage Systems and Alternatives

The earth rewards richly the knowing and diligent but punishes inexorably the ignorant and slothful.

—W. C. Loudermilk

If there is a better way to do it, find it.

—Thomas Edison

14:1 Preview and Important Facts

PREVIEW

For thousands of years people have tilled the soil to grow crops. Early plows were forked or crooked sticks pulled behind people or oxen. The ancient Greeks and Romans used a scratch plow, a metal-tipped implement, pulled behind oxen. About 2000 years ago a primitive moldboard plow was used in China. The moldboard was reinvented in northern Europe about the eleventh century in response to the difficulty experienced in working the heavy soils there. Tillage methods in much of the world are virtually unchanged from what they were centuries ago (see Figures 14-1 and 14-2). Sometimes ancient tillage techniques are used because of custom. More often the tillage systems are used because of limitations in equipment, money, knowledge, or environment.

In developed nations tillage has changed radically over the past 200 years. The opening of the American Midwest presented new challenges because of the heavy native sod. The cast-iron plow used in the East and in Europe broke too easily, scoured poorly, and rusted. A blacksmith named John Deere revolutionized agriculture. He introduced the steel moldboard plow in 1837, reportedly making the first plowshares from saw blades. The popularity of this invention, and others to follow, catapulted him to fame. When the 1900s began, crop residue was thought to be harmful to the subsequent crop or difficult to plant in, so the stubble was commonly burned. Tillage was mostly animal-powered, and crop rows were about 1 m (3 ft) wide to accommodate horses. Now tillage has become a mechanized and sophisticated process.

The zenith in *extensive* tillage in the United States was reached in the 1970s, when heavy equipment prepared large, clean fields, devoid of even fence-row vegetation. All

FIGURE 14-1 A farmer in southern Senegal using a *caillande* (a special hoe) to scalp the sod and turn it over to form a ridge from two sides. The weeds in the sod will gradually rot, and the bare strip will be used to plant peanuts, yams, and other crops. (Courtesy of C. K. Kline, Michigan State University.)

FIGURE 14-2 *Yuntas* (oxen plows) incorporating wheat and fertilizer in Carrasco Province, near Cochabamba, Bolivia. These parents of students are donating their time to plant school land to wheat to raise funds for the school. Annual precipitation is about 61 cm (24 in) and the mean annual temperature is 61°F (16°C). (Courtesy of Walter Carrera M., Ing. Agron., Bolivia.)

residues were mixed into the soil. Farmers would plow, disk, and smooth in one or more passes; then they would fertilize, plant, and spray preemergent herbicides in another pass. Technical abilities often exceeded good reasoning. This pinnacle of success was expensive in energy, harmful because of compacted and eroded soil, and damaging to the environment. Topsoil in a field subjected to intensive tillage for a few generations may have about half the organic carbon of a virgin grass pasture. Increased energy costs, concern about soil erosion losses, and environmental concerns have focused new attention on our tillage practices. New farming systems involving *reduced tillage, conservation tillage,* and *no-till* are receiving considerable attention from scientists and farmers.

Reduced-tillage and no-tillage practices are increasingly popular (see Figure 14-3). Less tillage requires less fuel and less time; however, less tillage also intensifies the challenges of weed control, insect control, and seedbed preparation. Heavier, more rugged equipment is needed to plant in no-till soil. Fields may appear to be trashy and poorly managed. The reduced-tillage farmer must be a *better* farmer than the conventional-till farmer to produce the same yields.

Reduced tillage is not for all crops or all soils. To decide whether to attempt reduced tillage, the land user must understand the tillage options, their advantages, and their disadvantages.

IMPORTANT FACTS TO KNOW

1. The reasons for tillage
2. How conventional tillage differs from reduced tillage
3. The potential damage from tillage
4. The advantages of reduced tillage
5. The definition of *conservation tillage*
6. Problems of no-till, including (a) seedbed preparation, (b) planting, (c) fertilizer placement, (d) weed control, (e) insect control, and (f) liming
7. Recommended practices in reduced tillage related to (a) seedbed preparation, (b) planting, (c) weed control, (d) insect control, (e) erosion control, (f) fertilizer and lime application, and (g) irrigation techniques
8. Impact of various tillage practices on environmental quality

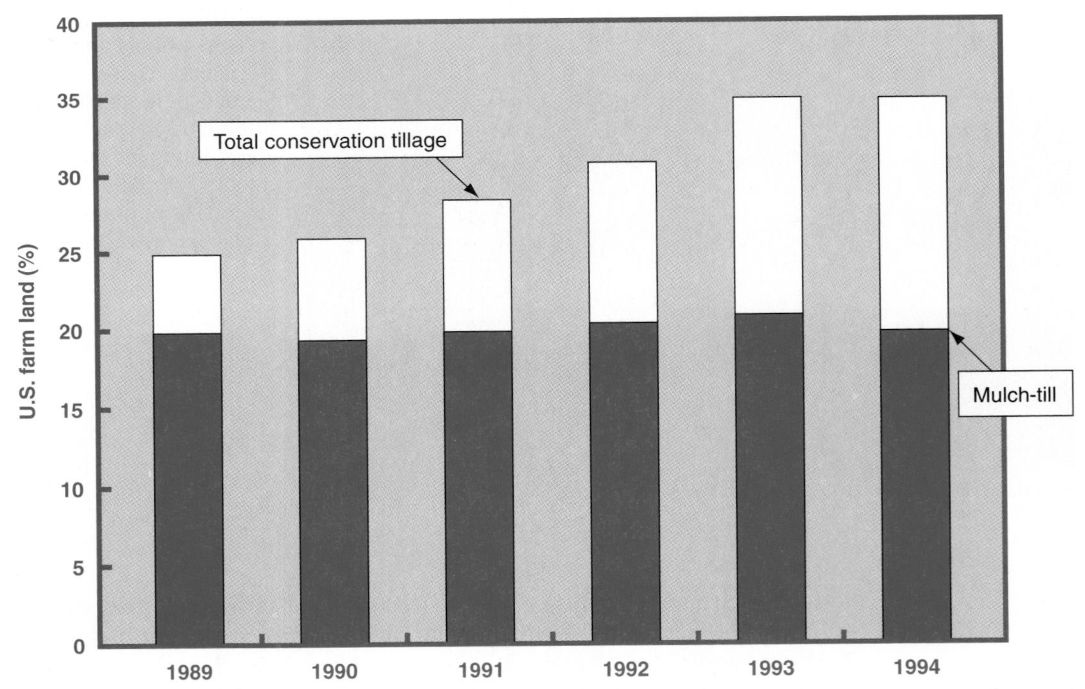

FIGURE 14-3 Yearly trends in acres farmed using conservation tillage. Mulch-till accounts for most conservation tillage acreage in the United States but its use is not increasing. The overall increase in conservation tillage acreage is due to increases in no-till and ridge-till. In 1995 30 percent of soybean acres were in no-till, as were 18 percent of corn and 3.5 percent of cotton. (*Source:* Conservation Technology Information Center, West Lafayette, IN.)

14:2 Purposes for Tillage

Tillage is hard, expensive work. When people first began to cultivate crops, they did only what tillage was necessary to plant and to control weeds. As cultivation became more sophisticated, tillage operations and equipment were altered and specialized. Powered machinery and cheap fossil fuels brought in the age of maximum tillage—a loose, fine seedbed; weedless fields; and extensive mixing of soils, lime, plant residues, and fertilizers.

Now we recognize that this extensive tillage was often destructive to the long-term sustainability of agriculture. Compaction of soils, the need for soil conservation, and increased costs of energy and labor have led to a reappraisal of our concepts about tillage.

The major purposes for tillage are to prepare an adequate seedbed and to control weeds. Tillage is also done to improve aeration, increase water infiltration, make furrows for irrigation, and incorporate crop residues or farm chemicals. The importance of the various reasons for tillage varies with geographic location, soil differences, crops grown, and climate.

Seeded Preparation

Tillage to prepare the seedbed is most needed when planting small and high-cost seed (lettuce, tomatoes, clovers). It is less critical for vigorous large-seeded plants (corn, wheat, soybeans, dry beans, sorghum). The preferred soil-covering depth for a large-seeded crop such as wheat is about four times the seed diameter—or even more if greater depths are necessary to place the seed into moist soil. The challenge of preparing a suitable seedbed is greatly increased for small-seeded crops such as alfalfa, where only three times the small seed diameter would be an ideal seeding depth. The soil layer covering such seed may be only 3 mm (0.1 in) deep and highly susceptible to desiccation unless firmly packed. At the same time the soil should allow good water infiltration and root penetration and be free of hard crusts that hinder seedling emergence.

Anyone who has attempted to operate a planter in soil covered by bulky plant residues has experienced the frustration of needing to constantly remove accumulated trash from the equipment and trying to achieve uniform planting depths. Tillage to incorporate crop residue greatly facilitates planting. For example, if a no-till planter goes over a pile of residue 5 cm to 10 cm (2 in–4 in) thick, the seed may be planted up in the residue (not the soil), where the seed dries out and dies, resulting in patches with low plant populations.

Weed Control

Once crop seeds germinate, weed control is essential. Weeds compete with the crop for water, nutrients, and light. Until herbicides became available in the 1940s, tillage for weed control was an integral part of crop production. Weeds were plowed under in land preparation and killed by cultivation once or several times during the early stages of crop growth. But cultivation was only partially effective; many weeds were in the plant row, and they were only removed by hand; also, the cultivator always injured some shallow crop roots. For weed control, herbicides are generally more effective than tillage and they eliminate extra trips over the field, but tillage for weed control is still effective and is still used. Certain crop systems require weed control by means other than herbicides. Crop rotation is one solution. Throughout the world, particularly in developing countries, weed control by tillage and by hand is still common; often the amount of area cultivated is limited by the farmer's ability to control weeds.

Besides the real need for weed control, there is a common desire among farmers for an aesthetically pleasing field. The clean field has been the ideal for many land users; however, a clean field offers more than the intangible benefit of a pleasing appearance. Burial of plant and weed residues has the practical effect of reduced infestations of pests.

Loosening the Soil

Farming equipment compacts soil and develops soil pans which may reduce or inhibit root and water penetration. The heavier the machinery and the more fine and moist the soil, the greater will be the compaction problem. A hardpan forming at, or just below, the normal depth of tillage is called a **plowpan.** In one soil in the southern Great Plains, 35 years of cropping to grain sorghum reduced infiltration rates from 23 cm/h to 0.5 cm/h—an enormous change in the quality of the soil.

Corrective tillage may be useful on several types of natural or man-made soil pans. These **pans** (hard, compacted, or impermeable layers) include not only the man-made plowpans but also **claypans, fragipans,** and cemented horizons (see Chapter 2). These natural pans formed in the geologic past by soluble cements that moved downward and precipitated in layers of soil. These pans can be ripped and opened by tillage, allowing water and root penetration. Deep chiseling is an energy-expensive operation but may be justified by the greater production that results (Figure 14-4).

In Clovis, California, a 2.1 m (7 ft) shank chisel was used to break deep silica-cemented hardpans common in the San Joaquin soil series. The 2.1-m ripper requires four D-9 caterpillar tractors to pull and push it along (Figure 14-5). An unusual but less rare tillage is called *rock plowing.* Productive for fruits and vegetables, the Chekika *very gravelly loam* soil is

FIGURE 14-4 Tillage pans can be broken or shattered by some type of chisel or sweep set at a depth of about 30 cm (1 ft). Note that the greatest compaction occurs in this soil at, and just below, the plow depth. (Courtesy of Caterpillar Tractor Co.)

FIGURE 14-5 A 2.1-m (7-ft)-tall ripper chisel used near Clovis, California, to rip deep hardpans in the San Joaquin soil series. The large ripper requires enormous power (four D-9 Caterpillar tractors) to pull it through such deep and hardened soil masses. Note the man standing to the right of the chisel. Smaller rippers are used for shallower depths to increase root and water penetration. (Courtesy of USDA—Soil Conservation Service; photo by G. Kennedy.)

profitable in Dade County, Florida, in spite of the unusual yearly soil loosening, which requires a unique chisel bullet with small wings (Figure 14-6).

Shaping the Soil

Surface flow irrigation, or ridges, to allow drainage of excess water are often essential. In the United States about 50 percent of irrigated acreage is irrigated by gravity flow systems such as furrow or flood systems. Tillage is needed to form the ridges and furrows and to smooth the land surface, to avoid hills and valleys in the field, over which surface flow irrigation would otherwise be difficult or impossible.

Incorporation of Lime and Fertilizer

Lime, to be most efficient, should be mixed with the soil. If it is added on the soil surface, it will move down slowly, and the soil acidity at deeper depths will be corrected slowly. Leaving excessive lime at the soil surface favors some fungal diseases. Also, many management schemes involve broadcasting fertilizers and then incorporating them into the soil by tillage.

Control of Insects and Disease

The timely plowing under of crop residues is an effective means of controlling certain insects. The Hessian fly, a destructive pest in many wheat fields, can be controlled by plowing under infested wheat stubble and volunteer wheat. The wheat jointworm is held in check by destroying all volunteer grain. Plowing under corn stalks reduces the next year's crop of European corn borers. Timely cultivation aids in reducing grasshopper infestations by drying out their eggs. The cotton boll weevil and the pink boll worm in cotton are held in check by early destruction and plowing under of cotton stalks. Sometimes, depending upon the time and place, legal sanctions can be brought against farmers who do not plow under residues that harbor certain pests.

Improving Water Relations

Tillage breaks up soil crusts, loosens soil, and generally increases water intake when water is applied immediately after tillage. Unfortunately, a clean-tilled field is likely to crust again after a heavy rain. Crop residues help keep cultivated soil porous. In Minnesota, when residues were partially incorporated to 15 cm (6 in) with a chisel cultivator in the fall, they provided

FIGURE 14-6 Rock plowing. In Dade County, Florida, this Chekika very gravelly loam requires the loosening of gravel yearly by rock-plowing. The soil requires special management. It has a shallow water table, 35 percent to 60 percent gravel and limestone bedrock, often starting at 12 cm to 20 cm deep; however, the excellent climate and market allow a wide variety of fruits and vegetables to be grown. (Courtesy of USDA—NRCS and University of Florida, *Soil Survey of Dade County Area, Florida.*)

eight times more infiltration before runoff started and four times more infiltration during runoff than occurred on spring-tilled land.

The higher water storage in the root zone resulting from deep tillage permits irrigation before planting and fewer irrigations during crop growth. Tillage on a clay loam to a depth of 41 cm (16 in) was still beneficial after 3 years.

The practice of growing no crop for a year for the purposes of increasing the total soil-stored water is called **fallow.** Weeds and volunteer growth are controlled by two to four weed-control tillage passes (cutting blades slide along below the soil surface). Water storage is often small, but the storage of even a few centimeters of water can greatly affect crop yields in areas of limited rainfall. In the northern Great Plains of the United States, 2.5 cm of water produces on average an extra 161 kg/ha (2.4 bu/a) of wheat. Except in low-rainfall areas of less than 200 mm to 300 mm (8 in to 12 in), where a profitable yield is impossible without fallow, the value of allowing land to lie idle a full crop year is questionable. When fallow-year rainfall greatly exceeds the soil's water-holding capacity, water tables rise, creating wet soils and saline seeps. This is a common problem in the dryland wheat region of Montana and the Dakotas. Tillage systems that would improve over-winter water storage without fallow could be highly beneficial in the wheat belt. A study in eastern Washington demonstrated that using a paratiller (an implement that shatters compacted subsoil layers with little effect on the soil surface) improved over-winter water storage by 8 cm (3 in).[1]

---

[1]L. F. Elliott, K. E. Saxton, and R. I. Papendick, "The Effect of Residue Management and Paratillage on Soil Water Conservation and Spring Barley Yields," *Journal of Soil and Water Conservation* **50** (1995), 656–658.

Aerating and Warming

Compacted or crusted soil may have inadequate air exchange. This poor aeration can limit plant growth in some clayey soils, but it is unlikely to be a problem in sandy soils. Tillage may be necessary to relieve the problem of poor surface aeration. Also, soil cover reduces the amount of heat radiation reaching the soil and therefore tends to keep the soil cooler and wetter. Tillage that exposes lower, wetter soil layers to the sun will speed warming and drying of the soil.

Reasons for Minimizing Tillage

The emerging goal for agriculture is to develop systems that are *permanent* or *sustainable.* Agriculture must supply food, fiber, and chemicals to an increasing world population. This production should be as economical as possible for both producers and consumers. Meanwhile, the soil must not be permitted to be further degraded or destroyed. To accomplish this goal improved farming systems must be adopted to reduce soil erosion and use few resources, especially fuel.

The reasons for minimizing tillage are primarily to reduce erosion of soil and to save time and fuel. A conventional farming system that uses the moldboard plow, disking, herbicides, planting, and cultivation consumes about 72 liters of diesel fuel per hectare (7.7 gal/a). In comparison, no-till farming consumes about 38 liters per hectare (4.1 gal/a).

Erosion of soil has finally been recognized by decision makers as the bankrupting policy of soil management that it is. Topsoil must be kept in place. Pressure from national and state soil conservation groups and controls by the U.S. Environmental Protection Agency to prevent erosion of soil sediments into waters have encouraged the adoption of reduced tillage. Reducing tillage reduces erosion.

Benefits growers can realize from reducing tillage intensity include the following:

- Soil erosion decreases.
- More area can be planted in less time.
- Topsoil drying decreases and subsoil water storage improves.
- Double-cropping (two crops per growing year) is more practical because of reduced time needed for seedbed preparation.
- Steep or otherwise marginal land may be farmable without excessive erosion.
- Fuel costs decrease.
- Pesticides can be substituted for tillage, often with lower cost and greater effectiveness.
- Subsoil compaction decreases.

14:3 Tillage Terminology

With the change from a simple moldboard-disk-harrow sequence to disk plows, rotary plows, chisel plows, cultivators, subsoilers, and other equipment, many terms have been added to the grower's vocabulary. Traditionally, the term **primary tillage** referred to the first and deepest operation. Primary tillage is done to loosen the soil, often by use of a moldboard plow, large disk plow, or chisel plow. **Secondary tillage** follows primary tillage with implements such as disks, cultivators, and harrows. Secondary tillage is done to kill weeds, incorporate pesticides and fertilizers, and prepare a well-pulverized seedbed. With modern tillage systems designed to minimize soil manipulation, the terms *primary tillage* and *secondary tillage* often do not apply. Some of the terms used to describe tillage are following (See Glossary for full definitions). Also, some examples of common equipment are shown in Figure 14-7.

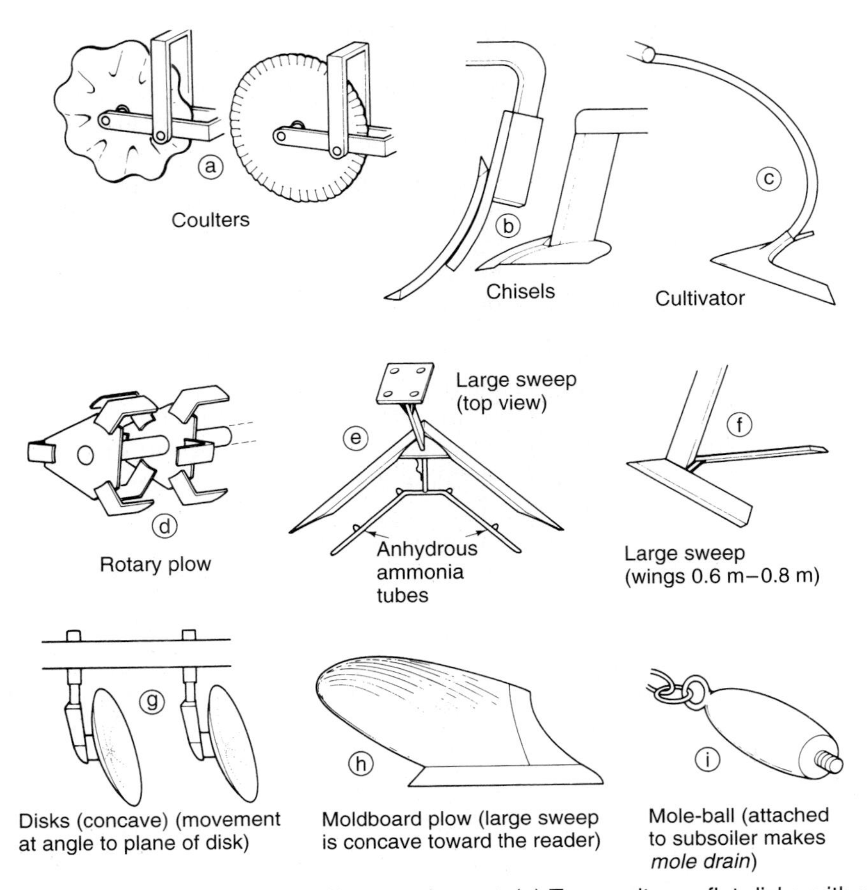

FIGURE 14-7 Examples of common tillage equipment. (a) Two coulters—flat disks with wavy cutting edges to open a slit for a planter chisel or to cut surface litter to avoid plugging the planter. Chisels in (b) are used to loosen the soil without removing surface litter. Cultivators (c) are small sweeps to cut weeds 3 cm–6 cm below the soil surface. Rotary plows (d) are *rototillers,* which allow vigorous shallow mixing 10 cm–25 cm deep. Large sweeps (e and f) are used to do shallow minimal tillage. Preplant anhydrous nitrogen could be added as shown in (e). Disks (g) are concave and pulled at slight angles to the plane of the disk, thus forcing the disk to dig into the soil. The moldboard plow (h) was the implement of the tillage revolution in the 1920s and 1930s. It inverts soil and buries all residues, producing a clean-field look. Various modifications, such as pulling the mole-ball (i), allow additional beneficial actions all in one pass over the field. (Courtesy of Raymond W. Miller, Utah State University.)

Bedder: A sweep like a small moldboard plow with curved sides on both sides (Figure 14-7).

Chisel: A narrow shank pulled through the soil to rip it.

Chisel plow: Large chisels pulled through the soil to rip it 15 cm to 30 cm (6 in–12 in) deep.

Conservation tillage: Any tillage system that leaves at least 30% of the surface covered by plant residues for control of erosion by water; for controlling erosion by wind, at least 1120 kg/ha (1000 lb/a) of small-grain straw equivalent during the critical wind erosion period.

Conventional tillage: Traditionally a moldboard plowing, two disking operations, then harrowing and seeding. Any tillage that leaves less than 15% crop residue cover.

FIGURE 14-8 Establishing the ridge tillage system using bedders. This system allows some drainage in soils with drainage problems. It also allows faster soil warming where seed will be planted and is used for some furrow irrigation, although ridges are narrow. The method involves extra tillage over nonridge systems. (Courtesy of the Buffalo Line, Fleischer Manufacturers.)

Coulter: A sharpened disk, with a straight or fluted edge to cut residue or open a slit in soil.
Cultivator: Several small sweeps to pull between crop rows to cut and kill weeds.
Disk: A combination of concave-shaped disks for shallow tillage.
Disk harrow: Closely spaced disks plus harrow to break large clods to smaller units.
Disk plow: Large concave-shaped disks to turn soil to depths of 15 cm to 40 cm (6 in–16 in).
Cross-slot drill: This planter opens a slot (furrow), drops the seed, and then covers the slot and seed with soil to retain soil moisture and to firm the soil.
Fallow: Land left a year without a crop to conserve water and control weeds.
Harrow: A set of vertical tines that break the larger clods into smaller pieces.
Moldboard plow: A curved-face implement for deep (15 cm–40 cm or 6 in–16 in), primary tillage which inverts the soil almost upside down and buries most of the surface plant residues.
Mulch-till: A conservation tillage system in which the entire soil surface is tilled.
No-till: No tillage except to insert seed, and sometimes fertilizer, in a slit (see Figure 14-9).
Paraplow: A series of chisels that go down and then bend to the side at a 45° angle. Soil flows over each bent wing (leg) and falls back, shattering but not mixing the soil.
Reduced tillage: Any combination of tillage operations less than conventional tillage. Different degrees of reduced tillage include *minimum tillage, conservation tillage, and no-till.*
Ridge furrow: The formation of alternate ridges and furrows.
Ridge-till: Ridges formed by sweeps or other implements.
Rod weeding: Pulling a sweep or blade at a shallow depth to cut and kill weeds.
Roller harrow: Toothed disks (10 cm, or 4 in apart) on a wide-diameter drum to pulverize and smooth the seedbed.
Rotary plow: Blades that rotate into soil to mix it (rototill).
Shovels: A term often used for narrow-winged sweeps.
Slot mulcher: Cuts a slot into the soil and fills it with grain straw mulch to absorb water.
Strip tillage: Till only strips of soil into a loose seedbed between untilled strips.

FIGURE 14-9 (a) No-till seeding of corn in Indiana into a wheat cover crop grown to protect the soil and retain leachable nutrients. A fluted coulter cuts a narrow strip of soil, which is the total seedbed preparation. Fertilizer is banded in the same operation. (b) Soybeans growing in grain stubble from no-till planting. (*Source:* USDA—Soil Conservation Service; photos by Bob Steele and J. B. McDonald.)

Stubble mulch: Only part of the crop residue is partially incorporated into the soil. Most of the residue is left anchored in the soil but exposed at the soil surface.

Subsoiler bedder: A bedder built onto a subsoil chisel.

Sweep plow: A sweep, up to 2 m wide, with shanks about 15 cm long underneath, which break up the soil to a depth of 20 cm–25 cm.

Sweeps: Any chisel tip that has *wings,* long or short (cultivator, sweep plow, etc.).

14:4 Understanding Reduced Tillage

Nature has no provision to till the soil, except slowly by insects and other macroorganisms or by the slow churning of Vertisols. The natural vegetation of the dense forests and grasslands do not require tillage. People initiated tillage because they wanted to grow large areas of a single crop, to produce blemish-free foods, and to produce maximum yields. People also wanted to irrigate dry lands and control weeds. Tillage is one way people have tried to improve on nature's methods. Although tillage can be beneficial, it can also be devastating, especially in the long term. How much tillage is too much? Can practices other than tillage compensate for omitted tillage operations? The following sections provide some insights into the questions surrounding tillage.

Physical Properties of Soil

Tillage profoundly affects the physical properties of soil. Any traffic over the field tends to compact the soil. Tillage loosens soil to the depth of tillage but leaves the subsoil more compacted with each pass. The no-till soil often has a greater number of small pores; also, soil protection and aggregation in no-till soils are better at the surface because of the crop residues there. Compared to conventional-till soils, reduced-tillage soils at the time of spring planting tend to be wetter and cooler. This condition often delays planting and reduces ger-

(a)

(b)

FIGURE 14-10 Compaction of soils into pans hinders water and root penetration. Tillage is needed to open paths through the pan. (a) Increased growth in corn as a result of ribbon tillage (narrow but deep tillage, right) compared to no-till (left). Ribbon tillage is narrow but deep strip tillage. The plow pans and other shallow pans can be broken by chisels or deep fluted coulters. (b) Soybeans. (Courtesy of Albert C. Trouse, Jr., Tilth International, Auburn, AL.)

mination and early seedling growth. One management option is to rotate tillage (such as deep plowing every few years; less tillage other years) to optimize the pros and cons.[2]

Less tillage favors more earthworm activities and burrowing-mixing actions by other soil macroanimals; these activities improve the soil by increasing its water infiltration rate. Freeze-thaw cycles over the winter ameliorate some surface compaction. Compaction may gradually develop even under reduced tillage. Periodic chisel plowing or ripping in the top 30 cm to 40 cm (12 in–16 in) may be beneficial to the soil (Figure 14-10). The inclusion of deep-rooted crops in a rotation will help soil permeability as old roots decompose, leaving vertical channels.

Much of the improvement in soil physical properties from conservation tillage arises from the increase in organic matter. A study in South Carolina demonstrated that after several years of conservation tillage, organic matter in the upper 5 cm (2 in) of soil was increasing to about double that found in plots with conventional tillage. Crop yields were also greater under conservation tillage. Below 15 cm (6 in) organic matter contents were unaffected by conservation tillage (Figure 14-11).

Seedbed Preparation

Techniques that prepare a *minimum* seedbed are well adapted to large-seeded crops such as corn, wheat, and soybeans. Fine-seeded crops, however, are difficult to germinate without a firm, uniform, and pulverized seedbed. Forage crops and many vegetable crops often require extensive tillage for an acceptable seedbed. Where no-till is used, strong and heavy planters are required to rip through trash and stubble (Figure 14-12). With reduced tillage, seedling

[2]Steve McGill, "Why Some Farmers Are Rotating Tillage Practices," *The Furrow* **101** (1996), No. 4; 7–8.

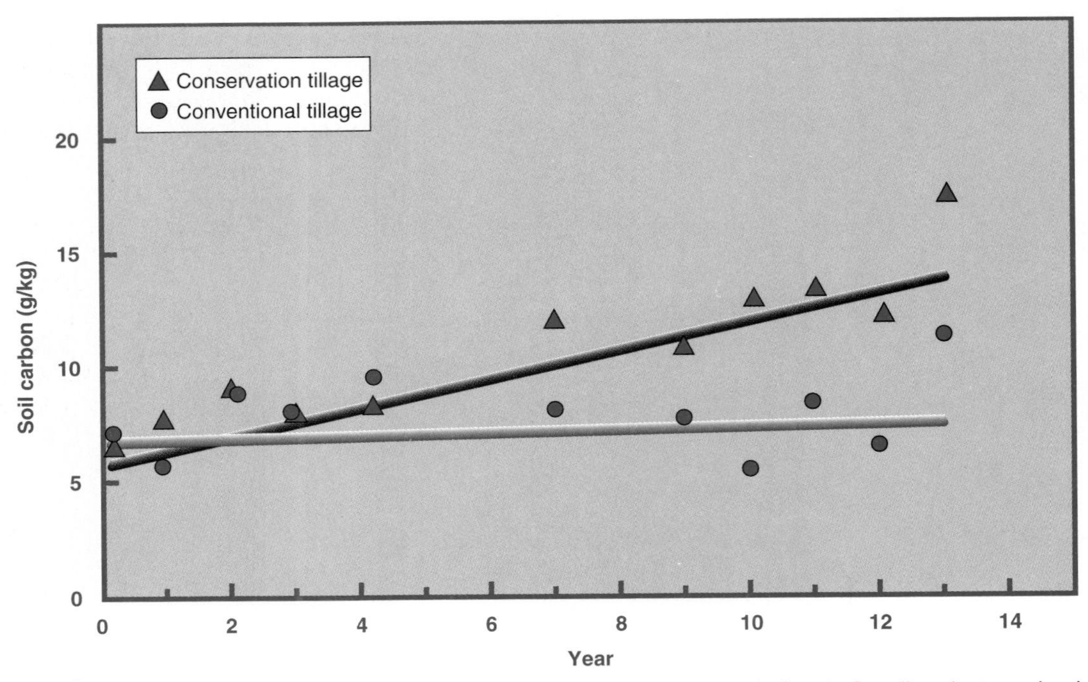

FIGURE 14-11 Carbon content of upper 5 cm (2 in) of loamy sand in South Carolina that received either conventional or conservation tillage for 14 years. Carbon is measured to indicate concentrations of organic matter (g/kg carbon $\times$ 0.172 $\approx$ % organic matter). Conventional tillage had little effect on carbon content, whereas conservation tillage steadily increased carbon contents. (Modified from P. G. Hunt, D. L. Karlen, T. A. Matheny, and V. L. Quisenberry, "Changes in Carbon Content of a Norfolk Loamy Sand After 14 Years of Conservation or Conventional Tillage," *Journal of Soil and Water Conservation* **51** [1996], 225–258.)

FIGURE 14-12 This heavy no-till drill is seeding dryland wheat on the Clarkston watershed in northern Utah. This Yielder-type research model drill was built by Yielder Drill, Inc., of Spokane, Washington. It weighs about 8864 kg (19,500 lb), nearly 10 times more than conventional drills of similar width. It has large dual disks to allow deep placement of fertilizer up to 20 cm (8 in) deep. It can also surface band pesticides. (Courtesy of V. P. Rasmussen, Utah State University.)

emergence is usually reduced and stand density is decreased. Typically about 10 percent to 15 percent more seed is used to compensate for this effect.

Weed Control

Weed control is a major problem. Weeds include any unwanted plants, including those arising from lost seeds from previous crops. If farmers do not control weeds, their crops fail. Reduced tillage puts the burden of weed control on herbicides.

A serious concern to growers is that the increased costs of herbicides may eat up any savings on fuel and labor from no-till. One study on corn in 17 states reported that herbicides in no-till cost 34 percent more than in conventional tillage. Also, the high organic matter left on the surface absorbs some herbicide and requires that more be added to get to the weeds (Figure 14-13).

With reduced tillage, control of many increased populations of perennial weeds will require increased herbicide use. Large populations of rhizomatous grasses (quackgrass, johnsongrass, bermudagrass), nutsedge, and hemp dogbane may develop in reduced-tillage systems. In contrast, large-seeded broadleaf weeds (cocklebur, velvetleaf) usually are less of a problem and require less herbicide. In response to the need for better herbicide-tillage systems, genetically improved crops are entering the market (see Detail 14-1). A crop that is immune to the effects of certain nonselective herbicides could make the use of that herbicide much more effective and may further revolutionize farming systems.

FIGURE 14-13 No-tillage (zero tillage) is shown with soybeans being planted in last year's corn field with no prior tillage. The large amount of residues would absorb large amounts of any preemergence herbicide, increasing the amounts needed to accomplish good weed control. Often, postemergence, rather than preemergence, herbicides are applied as a more economical method of weed control. (*Source:* USDA—Soil Conservation Service.)

Detail 14-1 Biotechnology and Farming Systems

Recent advances in biotechnology may dramatically change farming practices, improve the environment, and save the farmer money.

One such advance is the introduction of **Roundup Ready Soybeans.** These soybeans are immune to the effects of the Monsanto glyphosate herbicide, Roundup. This development allows farmers to spray their soybean crop with a nonselective, contact herbicide to kill all plants in the field except soybeans. No-till soybean farmers could be the big winners as weed control becomes easier and more successful than in the past. Farmers buying the seed relinquish the right to save seed and to use sources of glyphosate other than Roundup.

Another notable advance is the introduction of **Bt cotton.** This cotton contains the *Bacillus thuringensis* toxin and is lethal to bollworms. One of the problems with spraying insecticides is that beneficial insects, such as those that feed on harmful insects, are killed along with the pests. Many farmers observe that the more they spray the more they *have to* spray. Cotton is a pampered crop, receiving many expensive inputs and many applications of pesticide. Now farmers using Bt cotton will require less insecticide, which means that more beneficial insects will survive, which means that still less insecticide is needed. Adoption of this new technology may help clean up the environment and save money that otherwise would have been spent on chemicals.

Insect Control

Insect problems, both in the soil and above ground, will increase in reduced-tillage systems. Army ants, armyworms, corn borers, and many other insects will increase rapidly when not disturbed by tillage. Fortunately, better pesticides and alternative treatments (Detail 14-1) are being developed for some of these. Plant residues keep the soil surface moist and make conditions favorable for slugs, deer-mice, weevils, and many other pests. Some of these are only minor problems under conventional tillage, but birds, rodents, and plant diseases may become more severe problems because of seeds and residues left on the soil surface. Rotation crops must be carefully selected. Many insects that damage corn, for example, are commonly associated with many grass weeds or ryegrass crops. Some of these pests are the armyworm, lined stalk borer, hop vine borer, and potato stem borer.

Soil insects are more difficult to control than above-ground insects. In one survey in Ohio in no-till corn, black cutworms had attacked 15 percent of the plants in no-till but only 1 percent of the plants in conventional tillage.

Fertilizer and Lime Applications

Phosphorus and potassium are best placed into the soil (either band or broadcast and plow in) because they are not very mobile in soil (Figure 14-14). Likewise, lime is mixed as intimately with the soil as is practical because it is not very soluble or mobile. Reducing tillage limits the preferred methods of addition that can be selected. Nitrogen in the nitrate form is quite mobile in the soil, but volatilization losses of both ammonia from surface-applied urea or gases from denitrification in wet soil may be greater than in conventional tillage.

In soils where reduced tillage is used, most of the lime and fertilizers can be concentrated near the surface. Recently greater effort has been made to band fertilizer a few centimeters into the soil. The shaded soil surface and higher soil-water content in reduced tillage causes higher root densities to develop in the soil's top 5 cm; as much as 10 times as many roots were measured in no-till soil surfaces in climates where conventional tillage allowed extensive drying. Nitrification in this shallow soil layer can strongly acidify the soil surface.

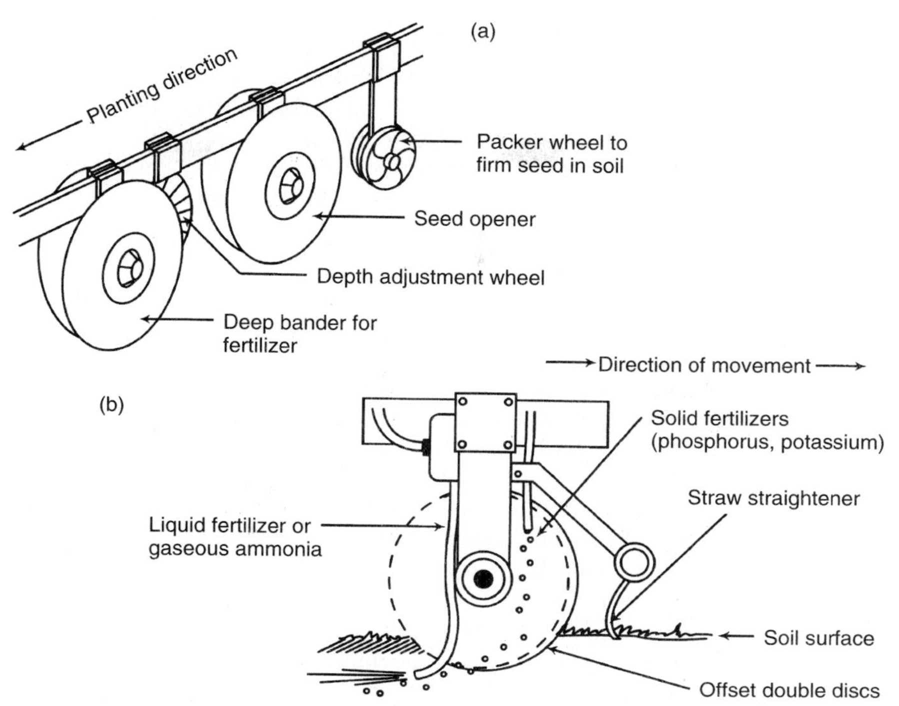

FIGURE 14-14 Diagram (a) of a combined fertilizer–planter using double discs to cut through the surface mulch, band fertilizer, plant seed, and firm the soil onto the seed. Diagram (b) shows how liquid or solid fertilizers, or both, can be added with a double-disc assembly. Notice the straw straightener to minimize plugging by residues. A narrow rotary plow could be attached in front of these discs to *strip till.* A planter double-disc could also be added (a). (*Source:* Modified by Raymond W. Miller from "Go for Yield," a publication of Yielder™ (manufacturer of a no-till drill), S. 4305 University Road, Spokane, WA 99206.)

All of these factors suggest that yields with no-till could be reduced if short droughts occurred. Dryness would hinder the mass of roots in the shallow soil from absorbing nutrients.

Reduced tillage may affect phosphorus uptake in opposing ways. First, the surface residues keep the soil cooler and wetter. If the temperature is lower than about 10°C (50°F) for corn, phosphorus uptake will be reduced. Second, the higher water content in reduced-tillage soils will *increase* phosphorus uptake by increasing the amount of diffusion, the main mechanism for the movement of phosphorus to roots. Third, fertilizer placement in bands in more conventional tillage prolongs phosphorus availability.

Organic Toxins

Reduced tillage, especially no-till, provides conditions favoring the production of organic toxins. This production by one plant of substances toxic to another plant is called *allelopathy* (see Chapter 6). Layers of residues on the soil remain wet, sometimes resulting in production of allelopathic materials. Wheat residues produce some toxins that reduce growth of the following wheat crop during early growth stages. The cause of toxin production is the wetness of heavy residues and anaerobic conditions in some thick accumulations. Corn residues can be allelopathic to the following crop of corn, too. Removing most residue from the immediate seed row eliminates the major allelopathic effects. Most problems of allelopathy are in early growth stages.

Special Situations

Certain crops pose special problems for reduced-tillage systems. Small sugar beet seeds must be planted at precise depths. Potato planters need special adaptation to insert the large seed potato. Harvesting root crops requires some tillage as underground parts are dug. Also, the soil compaction or aeration can affect the shape and quality of the root or tuber. In no-till, poor aeration causes sugar beets to increase branching of roots. When potato tubers are not *hilled-over* (covered with soil), the parts that are exposed to the sun turn green and are toxic to humans. In conventional tillage these tubers would be covered by tillage (hilling). With both cotton and tobacco, weed control with reduced tillage has been marginal and difficult. Many other crops can have unique difficulties, such as fungal diseases on the wet plant residues when growing cucumbers and strawberries. Each of these problems must be viewed as unique, and compromises must be made accordingly.

Rotations can help control pests. The more drastic the rotation, the greater will be the impact on weeds. This concept also applies somewhat to other pests: insects, diseases, and macrofauna. The selection of plants in the rotation is a very important decision.

Reduced tillage is difficult in arid climates where surface irrigation is planned (Figure 14-15). Although sprinkler and drip irrigation are possible with reduced tillage, furrow irrigation without clean furrows is problematic. The surface residues plug the furrows so that the water erodes the ridges. Also, furrows must be remade or rebuilt for each crop. Border irrigation could be used, but plant residues would still hinder flow rates down the field, causing excess deep percolation at the upper ends of the fields.

FIGURE 14-15 Furrow irrigation of carrots with siphon tubes (spiles) in the Lompoc Valley, California. The need to make ridges and furrows encourages clean tillage as a convenience. Some row crops (sugar beets, onions, carrots, celery, tomatoes, etc.) are difficult to plant, and furrows are hard to make in trash-covered soils. Fortunately, many areas needing furrow irrigation have little soil erosion hazard from rain during the cropping season. Alternatives to ridge furrow needs are to use drip or sprinkler systems, both of which are more expensive. (*Source:* USDA—Soil Conservation Service.)

(a)

(b)

FIGURE 14-16 These sand dunes (a) near Pasco, Washington, appear to be anything but inviting as profitable farmland. In (b) potatoes are uncovered by wind erosion of the sandy soil. However, 5 years after enough leveling (some 3-m [10-ft] cuts) to permit the use of center-pivot sprinklers, potatoes were producing 60,480 kg/ha (54,000 lb/a), wheat yielded 5376 kg/ha (80 bu/a), and alfalfa produced 17.9 Mg/ha (8 t/a). (*Source:* Jerry Schleicher, "Sand Dune Potatoes," *Irrigation Age* **9** [Sept. 1975], No. 9; 25–26.)

Erosion Control

Soils subject to serious wind erosion need careful management to maintain a protective residue cover. Only about 38 percent of highly erodible land in the United States is adequately treated by conservation practices.[3] In areas with windy periods, sandy and organic soils have the most severe erosion hazard (Figure 14-16). Erosion by wind is increased by clean tillage because the protective organic mulch is removed by tillage. Although no more organic matter is produced with reduced tillage, that which exists lasts longer; therefore, soils may have fewer of the benefits of humus and active decomposition but more of the beneficial cover to protect against erosion.

The registration of good postemergence herbicides for use on vegetables has made it possible to grow crops such as rye and barley as (protective) winter cover crops. These crops are killed by sprays just before the tillage and planting of vegetables.

---

[3]B. Gomez, "Assessing the Impact of the 1985 Farm Bill on Sediment-Related Nonpoint Source Pollution," *Journal of Soil and Water Conservation* **50** (1995), 374–377.

Vegetative cover is the best safeguard against erosion, especially wind erosion. Tillage implements differ greatly in the amount of residue they leave on the soil surface (Table 14-1). Growers using *conservation tillage* should use a system that will leave at least 30 percent residue after planting.

14:5 Tillage and Sustainability

Reduced tillage will reduce soil sediments that move into the air and water, thus reduced tillage seems desirable. However, reduced tillage also increases the amounts of pesticides applied. The tendency is to use the more persistent types for long-term weed control. These are more likely to stay potent long enough to contaminate surface and ground waters. Typically, 15 percent to 40 percent more pesticide is used in reduced tillage than in conventional tillage. Plus, with more pesticides used and more nutrient accumulation in the uppermost layer of soil, those soil particles that do erode are more threatening to the environment than they would be without chemicals and nutrients affixed. Beyond the scope of this book is another set of issues: the economic sustainability of farming systems. Considerable research is underway to determine

Table 14-1 Percent of Crop Residue Remaining on Soil Surface Following Each Tillage Operation

	Type of Residue * (%)	
Operation	*Nonfragile*	*Fragile*
Over-winter weathering, no tillage	70–95	65–85
Plows		
Moldboard plow	0–10	0–5
Disk plow	10–20	5–15
Chisel plow with sweeps	70–85	50–60
Paraplow	80–90	75–85
Disks		
Offset or tandem, heavy, 25 cm (10 in) spacing	25–50	10–25
Offset or tandem, finishing, 18 cm–23 cm (7–9 in) spacing	40–70	25–40
Undercutters & Cultivators		
Stubble mulch sweeps 50–76 cm (23–30 in) wide	80–90	65–75
Primary field cultivator 30–50 cm (12–20 in) sweeps	60–80	55–75
Plain rotary rodweeder	80–90	50–60
Single sweep row cultivator 76 cm (30 in) wide rows	75–90	55–70
Harrows		
Spike tooth	70–90	60–80
Cultipacker	60–80	50–70
Springtooth	60–80	50–70
Other		
Rotary hoe	85–90	80–90
Anhydrous ammonia applicator	75–85	45–70
Bedder or lister	15–30	5–20
Ridge-till planter	40–60	20–40
No-till planter, fluted coulters	65–85	55–80

Example: Amount of a fragile residue (sunflower) remaining after overwintering, disk plowing, and spike tooth harrowing: $0.75 \times 0.10 \times 0.70 = 0.0525$, or about 5%.

Source: "Soil Management," Deere & Company, Moline, IL, 1993.

*Nonfragile residues include alfalfa, barley, corn, rice, sorghum, tobacco, and wheat. Fragile residues include cotton, peanuts, potatoes, soybeans, sunflowers, and most vegetables.

best management practices under various conditions and climates. In one ten-year study in Montana, the economics of a conventional fallow–wheat rotation was compared to a wheat–barley rotation and continuous no-till wheat. In this climate (cool and 356 mm rain) continuous no-till wheat was most profitable.[4] Erosion losses without fallow would certainly be smaller. However, this conclusion is strictly site-specific; the issues surrounding farming systems are many-faceted.

The Pesticide or Tillage Dilemma

With our present stage of technology, many growers feel trapped in a winless dilemma with only three undesirable options:

- Use tillage sparingly and compensate with pesticides, thereby incurring the wrath of citizens, organizations, or agencies for releasing chemicals into the environment.
- Use chemicals sparingly and compensate with tillage, thereby incurring the wrath of citizens, organizations, or agencies for leaving the soil vulnerable to erosion.
- Use both chemicals and tillage sparingly, with the likelihood of lower yields and lower quality products from more intensive management, at least in the short run.

One might well ask which is worse for the environment, chemicals or tillage? Using chemicals may lower the quality of soil or water by leaving residues, rendering them less than completely safe. Tillage seems more natural but consumes lots of air-polluting fossil fuel and leaves the soil vulnerable to erosion. Implied in the policies of the USDA and others is that soil *quantity* retained in place is of greater importance than soil *quality* and, therefore, keeping the soil in place by reducing tillage and increasing chemicals is favored. Indeed, worrying about the *quality* of soil is a luxury only those *with soil* can afford.

Which Tillage System Is Best?

To suggest that there is a *best* tillage is to ignore that there are special needs pertaining to each climate, each soil, each crop, each rotation, each social and economic setting, and each farmer's preferences. The following items are a partial summary of the considerations to ponder when planning a tillage system for a field or a farm:

1. In the rare case where soil erosion is not a problem but pests are, conventional tillage may be the best system because pests are easier to control using conventional tillage. Substantial amounts of pesticides may still be needed.
2. Use conservation tillage or no-till if soil erosion control is a critical requirement. Conservation tillage is much more adaptable to various farming systems than is no-till. Two reduced-tillage suggestions to consider are the following:
   - Use the paraplow to loosen soil, but avoid inverting the soil.
   - Use ridge-till to avoid waterlogging and to warm the seeding zone earlier (Figure 14-17).
   - Rotate tillage practices to optimize pest control and profitability.

[4]J. K. Aase and G. M. Schaefer, "Economics of Tillage Practices and Spring Wheat and Barley Crop Sequence in the Northern Great Plains," *Journal of Soil and Water Conservation* **51** (1996), 167–170.

FIGURE 14-17 An example of ridge-till tillage illustrating the sequential operations from (a) early-spring preplant conditions through (b) planting, (c) precultivation, and finally (d) midsummer postcultivation. (*Source:* Modified and redrawn by Raymond W. Miller from "Conservation Systems for Row Crops," University of Nebraska publication EC76-714, as published in "Ridge-Till Row Crop Production," Cooperative Extension Service Publication C-662, Kansas State University [1985].)

(a) **Early spring** after cornstalks are chopped (optional).

(b) **Late spring** after ridge is scalped off by 25- to 35-cm-wide sweeps to lay bare the ridge top that is now about one-third of the row spacing width. Plant the seed, band the fertilizer, and apply preemergence herbicide.

(c) **Summer** after considerable residue decomposition and before cultivation.

(d) **Summer** after one or two cultivations to control weeds and to rebuild ridges.

3. Reduced tillage requires better management skills and planning than does conventional tillage. Reduced tillage will reduce soil preparation and planting time, *but it may not reduce total costs and probably will not increase yields.*
4. Seeding practices are different with reduced tillage. Special planters capable of cutting through residues and placing seed uniformly, precisely, and firmly are required. Seed varieties used should have quick emergence, tolerance to cold, disease resistance, and the ability to perform in a high plant population.
5. The least adaptable soils to reduced tillage are clayey soils in cold climates where crop residues cause slow drying and warming. In such areas the least amount of residue capable of controlling erosion should be used. Sandy, highly erodible soils used for small grains are most adaptable to no-till or low-intensity conservation tillage.
6. Preparation of a clean, uniform seedbed is needed for small-seeded crops, such as tomatoes, lettuce, carrots, alfalfa, and clovers.
7. Fertilizer and lime additions as strips on the soil surface can be reasonably effective. On basic soils ammonium and urea placed on the surface are subject to large volatilization losses.
8. Crop rotations, rather than pesticides or tillage as a primary recourse against pests, may be as good or better financially in the long run. Where feasible, rotations reduce weed and insect problems, distribute the work load, and allow more options in tillage, fertility, and crop residue management.

9. Cover crops grown in the winter may add fertility to the soil (if a nitrogen fixer is planted) and also protect against erosion, thereby allowing cleaner tillage and fewer chemicals. In many locations the scarcity of water precludes using a cover crop.
10. Where extensive use of herbicides seems to be the best alternative, growers should beware of herbicide buildup that can harm sensitive crops. Also, growers can often use lower rates of herbicides on young weeds rather than higher, more expensive rates of preemergence herbicides.

But we're at a point where we have to ask ourselves if we are the beneficiaries of our progress—or its victims.

—Robert Redford

Questions

1. Describe the following tillage implements: (a) bedder, (b) chisel, (c) coulter, (d) cultivator, (e) disk, (f) harrow, and (g) rotary plow.
2. Define (a) *conservation tillage,* (b) *reduced tillage,* (c) *conventional tillage,* and (d) *minimum tillage.*
3. Discuss the purposes of tillage.
4. Discuss the advantages of reduced tillage.
5. Discuss the disadvantages of reduced tillage.
6. What advantages does conservation tillage have over no-till systems?
7. How does reduced tillage affect physical properties of soils?
8. List some recommendations and precautions for weed control in reduced tillage?
9. How effective are fertilizer and lime in no-till systems? Explain.
10. When considering reduced tillage, which particular problems occur if the land is to be planted with small-seeded crops?
11. How does tillage or the lack of it influence pollution of the environment?
12. To what extent can one define *the best tillage system?* Explain.
13. Write a paragraph explaining why reduced tillage should be used on *most* farm land.

Soil Erosion

Liberty exists in proportion to wholesome restraint.

—Daniel Webster

What is the use of a house if you haven't got a tolerable planet to put it on?

—Henry David Thoreau

15:1 Preview and Important Facts

PREVIEW

We treat our soils like dirt. Soil is not widely recognized as the treasure and essential, *nonrenewable resource* that it is. The lure of quick money and the need for wood products has caused us to abuse forests, whose soils could then erode. Highways and homes require ever more clearing of vegetation from the land, concentrating runoff to wear more gullies and erode steep slopes. Growing annual row crops where nature grew perennial grasses has left farm soil unnaturally exposed to the elements.

Soil erosion is the removal of soil by water and wind. Erosion is slight from soils that are covered by dense grass or forest, but erosion is enormous from steep or poorly covered soils, especially those exposed to heavy rainfall or strong winds. Well-aggregated soils resist erosion, but pulverized silts and fine sands are easily eroded.

The tragedy of erosion is twofold. First, when the topsoil erodes, growers must practice farming or forestry on the exposed **B** horizon, which is less fertile and harder to till than the topsoil. In a 10-year Indiana study, severely eroded soil produced 9 percent to 18 percent lower corn yields and 17 percent to 24 percent lower soybean yields than slightly eroded sites.[1] Second, the topsoil carried by wind and water eventually comes to rest as sediment. Erosion is the major cause of nonpoint source pollution. Not only are the eroding materials rich in pesticides and fertilizers that upset ecosystems at the point of sedimentation, but the sediments themselves are also prob-

[1]G. A. Weesies, S. J. Livingston, W. D. Hosteter, and D. L. Schertz, "Effect of Soil Erosion on Crop Yield in Indiana: Results of a 10-Year Study," *Journal of Soil and Water Conservation* **49** (1994), 597–600.

FIGURE 15-1 Erosion in past centuries has left some scenic treasures, such as this Torriorthents-Badland-Rock outcrop in Coal Mine Canyon, Arizona. As farmland it is still spectacular but not as welcome a sight. It is very dry (150 mm–250 mm; 6 in–10 in) and supports only very poor grazing. (Courtesy of USDA—NRSC, the Bureau of Indian Affairs and the Arizona Agricultural Experiment Station, *Soil Survey of Hopi Area, Arizona, Parts of Coconino and Navajo Counties.* Photo by Fred Kootswatewa, Hopi Tribe.)

lematic. In 1988 flooding of the Ganges and other rivers devastated Bangladesh; thousands died as most of the country was inundated. Solutions to these problems are beyond national realms. The sediments that caused the river bed to rise and overflow the banks originated in heavily cultivated lands upstream in neighboring India. Dust from wind erosion can be likewise devastating as it fouls machinery and lungs. In California, where wind-borne particulates are especially severe, the poor visibility caused by a dust storm in 1991 resulted in a 104-car pileup and at least 15 deaths.

Natural or **geologic soil erosion** preceded farming and will continue regardless of farming practices (Figure 15-1). The problem addressed in this chapter is not geologic erosion, but the accelerated erosion related to human activities. Most cropland in the United States is losing soil and producing sediments at an unacceptable rate. In a few areas, such as Kentucky, catastrophic erosion continues (Figure 15-2). Construction (Figure 15-3) and extending clean cultivation to sloping lands also accelerate erosion losses. In the last 100 years more than 40 percent of the original topsoil has been lost from the rolling hills of the Palouse of eastern Washington and adjacent areas in Idaho and Oregon. This is an erosion rate of 2.5 cm (1 in) every 15 years—nature required about 800 years to form the amount of topsoil lost in only 15 years. Few consumers realize the high cost in soil for the amount of food harvested:

- An average of 545 kg of topsoil was lost in the rolling hills of the Palouse of eastern Washington (and adjacent areas of Idaho and Oregon) for every 27 kg (1 bu) of wheat produced. This is a ratio of 20 kg of soil lost per kilogram of wheat grown.
- In the cornbelt only 0.67 kg corn is produced for each kilogram of soil eroded.[2]

Such high soil losses reflect myopic policies of soil management.

Stopping erosion and sedimentation is impossible, but both of these problems can be greatly reduced. It is not always easy. Only when the dust clouds of the 1930s settled in on

[2]Nani G. Bhowmik, et al., "Conceptual Models of Erosion and Sedimentation in Illinois: Vol. 1. Project Summary," Illinois Scientific Surveys Joint Report 1, Illinois State Geological Survey and Illinois Natural History Survey, Champaign, 1984.

FIGURE 15-2 Erosion by water in 1961 in this Kentucky land was ignored until it became severe. It was probably further ignored because of the high cost to control the chaotic condition. Government financial aid plus regulatory laws to control erosion are needed to minimize this kind of waste of our soil resources. (Courtesy of USDA—Soil Conservation Service.)

FIGURE 15-3 Construction areas such as this highway often have limited vegetative cover. Steep slopes easily slip or erode. One storm saturated this West Virginia clayey soil and caused the slip. The repair cost in 1970 was $418,000. Prevention might have been possible by (1) a better routing of the highway, (2) better water disposal, or (3) more deeply rooted vegetation. (Courtesy of USDA—Soil Conservation Service.)

Washington and New York City did policy makers take seriously the erosion tragedy unfolding in the Southern plains. Equally tragic was the loss of denuded soil from the once-forested regions of the Mississippi basin. Hugh Bennett and Walter Clay Lowdermilk from the USDA surveyed global erosion in the 1930s. Testifying before a Congressional committee, they placed a thick towel on the committee's polished table, then poured a cup of water onto it. The towel soaked up the water. Then, without a word, they removed the towel and poured another cup of water on the bare table, allowing water to splash onto congressmen and trickle off the table. This demonstration convinced the committee of the dire consequences of removing soil from hillslopes.[3] Since 1935 over $30 billion has been spent on erosion control in the United States, but erosion is still a major agronomic and economic problem.

In 1981 a renewed emphasis to control erosion began. Associated with this renewed effort were increasingly strict environmental laws directed against pollution of waters. Between

[3]Daniel J. Hillel, "Out of the Earth," The Free Press, New York, 1991, 159.

1982 and 1992 progress was made in reducing erosion (Figure 15-4) by taking marginal lands out of cultivation and by promoting conservation tillage. Curiously, however, even though the average annual erosion loss per acre in the United States is above the maximum allowable level of 5 ton/acre (11 kg/ha), only rarely is a farmer penalized for violating the standard. The control of soil erosion is already past due, but is better done now than later.

IMPORTANT FACTS TO KNOW

1. Relative rate of erosion under different climates
2. Tolerable rates of erosion
3. The value of lost topsoil
4. Factors that affect erosion by water
5. The importance of soil cover in reducing erosion
6. The purpose of the Universal Soil Loss Equation (USLE)
7. How the Revised Universal Soil Loss Equation (RUSLE) differs from the USLE
8. Factors represented in the wind erosion equation
9. Various ways to control soil erosion by wind

15:2 Nature of Water Erosion

Soil erosion destroys man-made structures; fills reservoirs, lakes, and rivers with wasted soil sediment; and badly damages the land. Whether it is called *mud, silt,* or *sediment,* it is soil material that might have been kept in place on top of the land, where it can support plant growth, and where plants can in turn stabilize the landscape. Eroded sediment usually is the richest part of the soil, the nutritive topsoil containing most of the organic matter. The cost of dredging the several billion tons of sediment from rivers and harbors *each year* is about fifteen times more than the cost of holding the soil on the land from which it eroded. More than 12,300 hectare-meters (1 million acre-feet) of sediment settles annually in reservoirs, reducing water storage by an equal volume (Figure 15-5). The water that cannot be stored, because

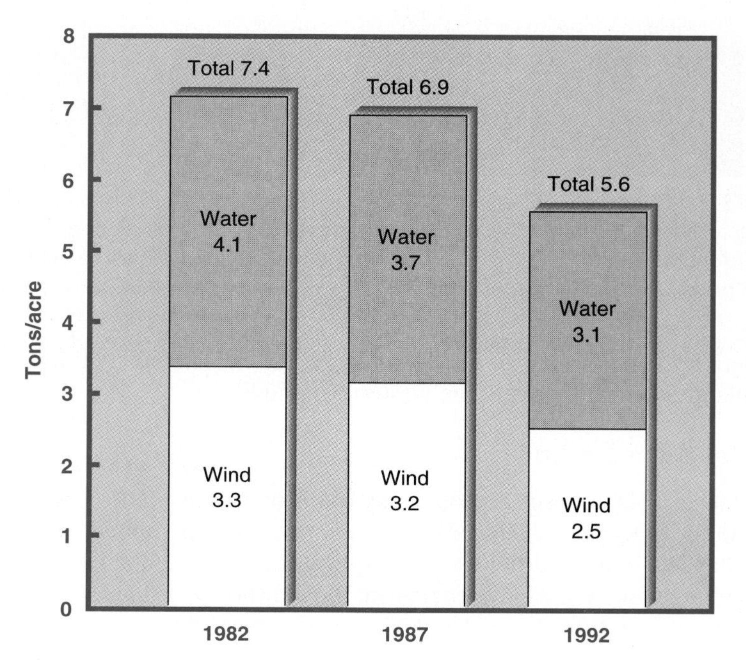

FIGURE 15-4 Average annual soil erosion from cropland in the United States by wind and water. Erosion has declined since 1982 but still exceeds the tolerable limit of 5 t/a (11.2 Mg/ha). (*Source:* Adapted from Natural Resources Conservation Service, "Summary Report, 1992 National Resources Inventory," USDA, Washington, DC, 1994, revised 1995.)

(a)

(b)

FIGURE 15-5 (a) The city of Ballinger, Texas, used the water stored behind this dam as the city water supply from 1920 to 1952. (b) By the early 1970s soil erosion sediments filled the lake to a height of more than 35 ft (10.7 m), destroying the ability of the dam to hold water. Soil sediments are the principal cause of water pollution. (*Source:* Environmental Protection Agency.)

of lowered reservoir capacity due to silting, could irrigate 100,000 hectares (250,000 acres) of cropland in the dry areas of the western United States.

Causes of Water Erosion

Aggregated soils (soil peds or clods) may be disintegrated by the direct impact of falling raindrops. In the process of disintegration, sand, silt, clay, and humus in the aggregate are separated. Clay and humus particles may be splashed as far as 150 cm (5 ft) from the point of impact; the larger silt and sand particles are moved shorter distances. Less dense humus floats away with surface flow; the water also suspends some clay and silt in the muddy runoff. This is **soil detachment** and surface erosion; the cause is raindrop impact on exposed soil and runoff flow. The soil productivity is lessened, and the sediments pollute surface waters.

Water erosion of soil starts when raindrops strike bare soil peds and clods, separating particles and causing the finer particles to move with the flowing water as suspended sediments. This muddy water moves downhill, scouring channels along the way. Each subsequent rain erodes additional amounts of soil until erosion has transformed an area into gullies, rills, and eroded land with reduced productivity.

People bare the soil when they use irresponsible agricultural practices, removing protective plant cover by excessive plowing, burning crop residues, overgrazing ranges and pastures, and overcutting forests. Also, drastic soil disturbance results from using heavy machinery in road and building construction and surface mining and from using off-road vehicles in easily erodible areas. Soil disturbance is especially disastrous in arid and Arctic regions, where water deficit and cold weather slow the reestablishment of protective vegetation.

Most soils in permafrost regions have 15 cm to 30 cm of organic matter on the surface, which insulates against warming. Many of these soils have ice lenses. Any fire, land clearing, or construction activity that exposes bare soil to the sun causes the ice in the soil to melt (Figure 15-6). Following the melting, slopes fail and soil caves in, channeling water that causes erosion and often produces pits or gullies (Figure 15-7).

FIGURE 15-6 Ice lense in permafrost in Alaska (between arrows). Such ice lenses are common in soils of northern and central Alaska but recede with continuous field crop production. (Courtesy of Mark P. Kinney, USDA—Soil Conservation Service, Fairbanks, Alaska.)

FIGURE 15-7 This cave-in pit in a farmer's field in the permafrost region of Alaska was caused by melting of ice lenses and subsequent vertical erosion. Note the large size of the pit by comparing it with the man indicated by the arrow. (*Source:* USDA—Soil Conservation Service.)

Erosion by water is classified as **raindrop splash** erosion, **sheet** erosion, and **channelized-flow** (or *rill and gully*) erosion.

Raindrop Splash Erosion

Raindrops fall with an approximate speed of 900 cm/s (30 ft/s). When raindrops strike bare soil, they may beat it into flowing mud, which splashes as far as 60 cm high and 150 cm away (Figure 15-8).

(a)

(b)

FIGURE 15-8 The impact of falling raindrops on bare soil (a) beats the soil into flowing mud, and soil erosion has begun degradation of the soil. (b) This farm in Ripley, Mississippi, should have been left in trees. (*Source:* USDA—Soil Conservation Service.)

The soils most readily detached by raindrop splash are fine sands and silts. Coarser particles are not shifted about as much because of their greater weight. Most soils of finer texture, such as clays and clay loams, are not readily detached because of the strong forces of cohesion that keep them aggregated.

Clays can be dispersed by repeated freezing–thawing actions, by high exchangeable sodium, by excessive tillage, and by allowing destruction of the soil organic matter.

During heavy rain bare soil aggregates are disrupted, splashed, shifted about, and packed together more closely. As muddy water flows down through natural openings of the soil into the profile below, the soil sediment plugs the pore openings. The result is a nonaggregated soil surface, which dries and forms a crust that is only slowly permeable to air and water. Water that should move downward into the soil profile during subsequent storms or irrigation now flows over the wet, clogged surface, carrying soil particles with it to pollute surface waters.

Sheet and Channelized-Flow Erosion

The more-or-less uniform depletion of the soil surface by nonturbulent (laminar) water is called **sheet** erosion. As water moves over the surface of the soil, some of it concentrates in low places to cut deeper depressions or channels. Continued flow develops minor channels called **rills.** Later, major rills and large **gullies** may be formed by the scouring action of increasing volumes of channeled muddy water. For practical purposes the difference between a rill and a gully is that a rill can be erased by tillage, and a gully is too wide or deep to be so easily removed. Volume, speed, and turbulence of flow make channelized flow much more erosive than laminar flow. Although gully formation can be catastrophic, expanding gullies are no longer common on farmland (Figure 15-9). Sheet and rill erosion, however, is still rampant.

15:3 Factors Affecting Erosion by Water

In 1965 the **Universal Soil Loss Equation (USLE)** was developed to estimate sheet and rill erosion losses from cultivated fields in the United States east of the Rocky Mountains. The equation was later adapted for use in other cultivated areas of the United States, in Europe,

FIGURE 15-9 Control of erosion can be costly. In today's world, cost may be a problem but not an excuse. This waterway in Ohio on 3 percent to 15 percent slopes is rock-lined. Someone did what was necessary with a lot of sweat and toil. (Courtesy of USDA—NRSC and Ohio Department of Natural Resources, *Soil Survey of Jefferson County, Ohio.*)

and in tropical western Africa. It can also be used for predicting erosion on rangelands and forestlands in the United States and for most situations in the tropics.

The USLE is an empirical formula that is used by obtaining numerical values from tables and multiplying the values together. In 1993 a revised USLE (RUSLE) was released. The RUSLE employs the same basic features of the USLE, but it employs a computer program rather than a series of tables.

The USLE is a useful tool, easy to understand, and is still in use in some Natural Resource Conservation Service (NRCS) field offices. Because of the conceptual simplicity of the USLE, and because of its fundamental similarity to the RUSLE, the USLE is readily explained below. Using the USLE, the expected soil loss is determined from five factors: the rainfall, the erodibility of the selected soil, the length and steepness of the ground slope, the crop grown in the soil, and the land practices used. The Universal Soil Loss Equation[4,5] is as follows:

$$A = R \cdot K \cdot LS \cdot C \cdot P$$

where

A = *erosion soil loss* in tons per acre per year
R = *rainfall* factor
K = *soil erodibility* factor
LS = *field length and slope* factor
C = *vegetative cover and management* factor
P = *practices* used for erosion control

A practicing conservationist using either the USLE or the newer RUSLE normally works in a small area, maybe a county, and often will need only one or two rainfall factors (R), values for only a few soils (K), and only a few plant cover systems (C). The remaining data can be tabulated quite easily for the small area and few, if any, extensive computations need to be done in the field. Additional information on the RUSLE is given later in Detail 15-1.

Rainfall Factor (R)

The **rainfall factor (*R*)** is a product of the kinetic energy (falling force) of rainfall times its *maximum 30-minute intensity* of fall (Figure 15-10). Obviously, fluctuations in duration and intensities of storms from year to year complicate obtaining average R values for any location. The R values shown in Figure 15-10 are for 22-year averages. The most erosive rainfall patterns in the United States are in the southern states, particularly in coastal regions. Compared to the Gulf Coast, the regions of the Pacific Northwest with equivalent rainfall *amounts* have much lower R values. Rain near the Gulf Coast comes in violent storms with large raindrops, in contrast to the gentle, misty rains of the Northwest. Arid lands have lower R values but may still have extensive erosion because of poor plant cover or steep slopes.

Soil Erodibility Factor (K)

Soil erodibility (*K*) is the ease with which a soil can be eroded. Values have been determined for most soils in the United States (Table 15-1), and range from 1.0 (most easily eroded) to 0.01 (least easily eroded). Most soils have a K value between 0.2 and 0.5. Soils high in silt and very fine sand are more easily eroded than other soils. Organic matter, larger structural aggregates, and rapid soil permeability all reduce the K factor.

---

[4] W. H. Weschmeier and D. D. Smith, "Predicting Rainfall-Erosion Losses: A Guide to Conservation Planning," USDA Agriculture Handbook 537, 1978.
[5] F. R. Troeh, J. A. Hobbs, and R. L. Donahue, "Soil and Water Conservation," 2nd ed., Prentice-Hall, Englewood Cliffs, NJ, 1991, 530.

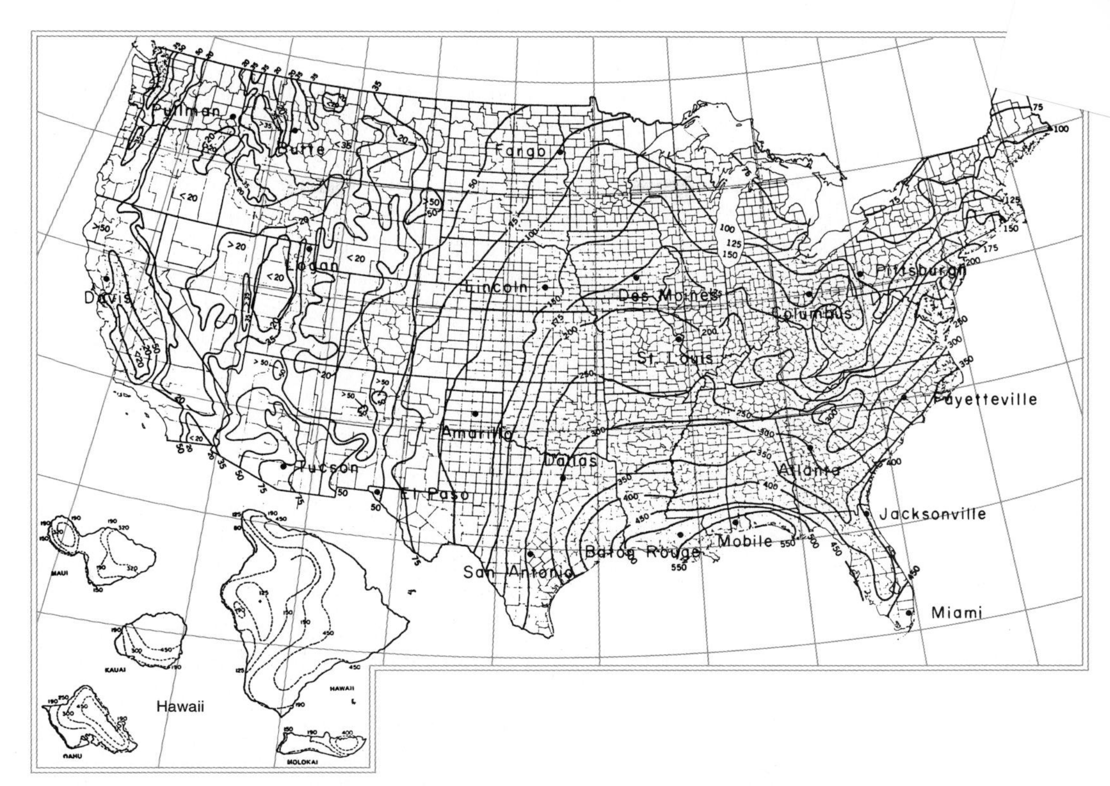

FIGURE 15-10 Average annual values for *R*, the rainfall erosion index. Individual values for small areas may differ considerably from these average values because elevation or mountains alter the amount and intensity of rainfall. Elevation differences can occur away from mountains. Death Valley—below sea level. Valleys in Utah—4000 ft. Altaplano of Bolivia—10,000 ft–12,000 ft. Mountains cause air lift-cooling-rain. So elevated plains can be different in effect from mountains themselves. (*Source:* W. H. Wischmeier and D. D. Smith, *Predicting Rainfall Erosion Losses: A Guide to Conservation Planning,* Agriculture Handbook 537, USDA, Washington, DC, 1978.)

Table 15-1 Selected Values for Soil Erosion Hazard *K* for Selected Phases of Soil Series

Phases of Soil Series and Location	*K*	*Phases of Soil Series and Location*	*K*
Albia gravelley loam (NY)	0.03	Austin clay (TX)	0.29
Tifton loamy sand (GA)	0.10	Mansic clay loam (KS)	0.32
Hector loam (OK)	0.17	Marshall silt loam (IA)	0.33
Molokai clay (HI)	0.20	Shelby loam (IN)	0.41
Boswell fine sandy loam (MS)	0.25	Keene silt loam (OH)	0.48
Cecil sandy loam (NC)	0.28	Dunkirk silt loam (NY)	0.69

Source: USDA—Soil Conservation Service, *Engineering Field Manual,* 1975, 2-3 to 2-29.

Length-Slope Factor (LS)

The **length-slope value** (***LS***) is the ratio of soil loss from the slope in question to the soil loss from the reference, which is a slope 22.1 m (72.5 ft) long and of a uniform 9 percent grade (Table 15-2). The reference slope has an *LS* factor of 1.00.

Longer slopes increase erosion because water accumulates and increases in speed, collecting more cutting sediment and doing proportionally more damage. Doubling slope steepness (percentage grade) usually more than doubles the erosion; on long slopes it may triple the erosion. Doubling slope length (with the same grade) increases erosion about 20 percent to 40 percent. *Convex* (dome-shaped) slopes have more erosion than indicated by calculations using the average slope; *concave* slopes have less erosion than indicated by calculations using average slopes.

Cover and Management Factor (C)

The **cover and management factor** (***C***) considers the type and density of vegetative cover on the soil and all related management practices, such as time between operations, tillage, watering, and fertilization. This *C* factor is complicated because of the wide range of possibilities in cover materials, management, and the manner in which crop residues can be left on soil.

Table 15-3 lists a few selected values from a wide range of research-verified factors. Often values are determined on a regional basis to accommodate the farming practices in the region. Each value is the ratio of erosion under specified cover and management to the amount of erosion from continuous fallow. If the value is 0.33, for example, the cover and management factor reduces erosion to one-third that which would occur on soil under continuous fallow. Lower *C* values mean greater protection and less erosion.

Practice Factor (P)

The **practice factor** (***P***) recognizes the influence of various combinations of contour planting, strip cropping, and terracing relative to the erosion potential of simple, up-and-down-slope cultivation. Table 15-4 lists several of these practices and their effect on erosion. The total benefit of terracing is not evident in the table because each terrace is treated as a separate slope, which is less steep than the original slope, thereby improving the *LS* factor.

Table 15-2 Combined *LS* Factor (Slope Length and Steepness) for the Soil Erosion Equation

	Slope Length (ft)[a]							
Slope Percentage	*25*	*50*	*100*	*200*	*400*	*600*	*800*	*1000*
0.5	0.073	0.083	0.096	0.110	0.126	0.137	0.145	0.152
2	0.13	0.16	0.20	0.25	0.31	0.35	0.38	0.40
4	0.23	0.30	0.40	0.53	0.70	0.82	0.92	1.01
8	0.50	0.70	0.99	1.41	1.98	2.43	2.81	3.14
12	0.90	1.28	1.80	2.65	3.61	4.42	5.11	5.71
16	1.42	2.01	2.84	4.01	5.68	6.95	8.03	8.98
20	2.04	2.88	4.08	5.77	8.16	10.0	11.5	12.9

[a]Feet times 0.3048 equals meters.

Table 15-3 Examples of *C* Factors for Use in Ohio, as Compiled by the USDA and Ohio State University

Vegetation	Tillage			
	Autumn Conventional	*Spring Conventional*	*Spring Conservation*	*No-Till*
Rotations				
Corn grown continuously	0.40	0.36	0.27	0.10
Corn, soybeans	0.42	0.37	0.24	0.12
Corn, oats, meadow	0.072	0.065	0.042	0.040
Corn, corn, oats, meadow	0.13	0.12	0.10	0.064
Corn, oats, meadow, meadow	0.055	0.050	0.033	0.033
Wheat and soybeans (double crop)	—	0.20	—	0.11
Permanent Pasture poor condition				0.04
Good condition, grass				0.003
Good condition, broadleaf				0.01
Woodland poor condition				0.004
Good condition				0.001

Source: Modified from G. O. Schwab, D. D. Fangmeirer, and W. J. Elliot, *Soil and Water Management Systems,* 4th edition, John Wiley & Sons, New York, 1996, 101.

Table 15-4 The *P* Factor, Which Is the Ratio of the Erosion Resultant from the Practice Described to that Which Would Occur with Up-and-Down-Slope Cultivation

Vegetative Condition	*P Value*
A 1%–2% slope, contoured, in 400-ft lengths	0.60
A 3%–5% slope, contoured, in 300-ft lengths	0.50
A 6%–8% slope, contoured, in 200-ft lengths	0.50
A 9%–12% slope, contoured, in 120-ft lengths	0.60
A 21%–25% slope, contoured, in 50-ft lengths	0.90
A 1%–2% slope, contour strip cropping (row crop—grain), 130-ft strip width, 800-ft slope length (along direction of strip)	0.60
A 6%–8% slope, contour strip cropping (row crop—grain), 100-ft strip width, 400-ft strip length	0.50
A 13%–16% slope, contour strip cropping (row crop—grain), 80-ft strip width, 160-ft strip length	0.70
A 21%–25% slope, contour strip cropping (row crop—grain), 50-ft strip width, 100-ft strip length	0.90
A 6%–8% slope, contour strip cropping with a 4-year rotation of 1 row crop, 1 small grain, 2 meadow	0.25

Source: Selected data from W. H. Wischmeier and D. D. Smith, *Predicting Rainfall Erosion Losses: A Guide to Conservation Planning,* Agriculture Handbook 537, USDA, Washington, DC, 1978.

Calculation of Erosion by Water

To calculate erosion, determine the value for each of the five factors in the USLE and multiply all factors together to obtain the product (Calculation 15-1). The soil loss calculated by the USLE is for soil that leaves the field or slope, not for in-field movement of sediment. Notice that the values given in the tables in this chapter are in United States units and estimate annual erosion loss in *tons per acre.* One can multiply tons per acre by 2.24 to obtain Mg/ha. For other conversion factors see Table 15-5. The calculation of erosion is simple (see Calculation 15-1); validating the numerical values to be used for each factor and situation is the difficult part.

Table 15-5 Conversion Factors for Universal Soil Loss Equation (USLE) Factors

To Convert:	*From U.S. Customary Units:*	*Multiply By:*	*To Obtain:*	*SI Units*
Rainfall intensity, *i* or *I*	$\frac{\text{inch}}{\text{hour}}$	25.4	$\frac{\text{millimeter}}{\text{hour}}$	$\frac{\text{mm}}{\text{h}^{a}}$
Rainfall energy per unit of rainfall, *e*	$\frac{\text{foot-ton}}{\text{acre-inch}}$	2.638×10^{-4}	$\frac{\text{megajoule}}{\text{hectare}\cdot\text{millimeter}}$	$\frac{\text{MJ}^{b}}{\text{ha}\cdot\text{mm}}$
Storm energy, *E*	$\frac{\text{foot-ton}}{\text{acre}}$	0.006701	$\frac{\text{megajoule}}{\text{hectare}}$	$\frac{\text{MJ}^{c}}{\text{ha}}$
Storm erosivity, E_I	$\frac{\text{foot-ton}\cdot\text{inch}}{\text{acre}\cdot\text{hour}}$	0.1702	$\frac{\text{megajoule}\cdot\text{millimeter}}{\text{hectare}\cdot\text{hour}}$	$\frac{\text{MJ}\cdot\text{mm}}{\text{ha}\cdot\text{h}}$
Storm erosivity, E_I	$\frac{\text{hundreds of foot-ton}\cdot\text{inch}^{d}}{\text{acre}\cdot\text{hour}}$	17.02	$\frac{\text{megajoule}\cdot\text{millimeter}}{\text{hectare}\cdot\text{hour}}$	$\frac{\text{MJ}\cdot\text{mm}}{\text{ha}\cdot\text{h}}$
Annual erosivity, R^{e}	$\frac{\text{hundreds of foot-ton}\cdot\text{inch}}{\text{acre}\cdot\text{hour}\cdot\text{year}}$	17.02	$\frac{\text{megajoule}\cdot\text{millimeter}}{\text{hectare}\cdot\text{hour}\cdot\text{year}}$	$\frac{\text{MJ}\cdot\text{mm}}{\text{ha}\cdot\text{h}\cdot\text{y}}$
Soil erodibility, K^{f}	$\frac{\text{ton}\cdot\text{acre}\cdot\text{hour}}{\text{hundreds of acre}\cdot\text{foot-ton}\cdot\text{inch}}$	0.1317	$\frac{\text{metric ton}\cdot\text{hectare}\cdot\text{hour}}{\text{hectare}\cdot\text{megajoule}\cdot\text{millimeter}}$	$\frac{\text{t}\cdot\text{ha}\cdot\text{h}}{\text{ha}\cdot\text{MJ}\cdot\text{mm}}$
Soil loss, *A*	$\frac{\text{ton}}{\text{acre}}$	2.242	$\frac{\text{metric ton}}{\text{hectare}}$	$\frac{\text{t}}{\text{ha}}$
Soil loss, *A*	$\frac{\text{ton}}{\text{acre}}$	0.2242	$\frac{\text{kilogram}}{\text{meter}^{2}}$	$\frac{\text{kg}}{\text{m}^{2}}$

Source: G. R. Foster, D. K. McCool, K. G. Renard, and W. C. Moldenhauer, "Conversion of the Universal Soil Loss Equation to SI Metric Units," *Journal of Soil and Water Conservation* **36** (1981), No. 6; 355–359.

[a] Hour and year are written in U.S. customary units as *hr* and *yr* and in SI units as *h* and *y*. The difference is helpful for distinguishing between U.S. customary and SI units.

[b] The prefix *mega* (*M*) has a multiplication factor of 1×10^{6}.

[c] To convert ft-ton to megajoule, multiply by 2.712×10^{-3}. To convert acre to hectare, multiply by 0.4071.

[d] The notation, "hundreds of" means numerical values should be multiplied by 100 to obtain true numerical values in given units. For example, $R = 125$ (hundreds of ft-ton·in/acre·hr) = 12,500 ft-tonf·in/acre·hr. The converse is true for "hundreds of" in the denominator of a fraction.

[e] Erosivity, E_I or *R*, can be converted from a value in U.S. customary units to a value in units of newton/hour (N/h) by multiplying by 1.702.

[f] Soil erodibility, *K*, can be converted from a value in U.S. customary units to a value in units of metric ton·hectare/newton·hour (t·h/ha·N) by multiplying by 1.317.

Calculation 15-1 Sample Calculation of Erosion by Water

Problem What erosion loss might be expected on a Keene silt loam in Ohio, on a 1000 ft-long 2 percent slope, using conventional spring tillage for a corn–soybean rotation?

Solution

The factors needed are as follows:

Rainfall (R)	150 (Figure 15-10)
Erodibility (K)	0.48 (Table 15-1)
Slope (LS)	0.40 (Table 15-2)
Crop (C)	0.37 (Table 15-3)
Erosion-control practice (P)	1 (none)

Using the Universal Soil Loss Equation,

$$A = R \cdot K \cdot LS \cdot C \cdot P$$

and substituting the values above,

$$A = 150 \times 0.48 \times 0.40 \times 0.37 \times 1 = 10.7 \text{ t/a/yr}$$

The loss of 10.7 t/a/yr from water erosion is unacceptable. The farmer could reduce erosion substantially by including small grains or meadow in the rotation, by using conservation tillage, and by using an erosion-control practice such as contour farming (Table 15-4).

The RUSLE offers greater precision and adaptability (see Detail 15-1) than the USLE. The RUSLE has been generally well received. Meanwhile, the USDA has been working on yet another software system intended to take the place of the RUSLE. The new system, called *WEPP* (Water Erosion Prediction Project) is not empirical, but rather is based on scientific modelling of the actual processes of erosion (see also **Appendix A**). It is more powerful and sophisticated than the RUSLE but requires more input data and a knowledgeable conservationist for accurate predictions. An early version of the WEPP is available from the Soil and Water Conservation Society; it can also be downloaded from the Internet. How readily farmers and conservationists will accept the WEPP is unknown. Other erosion models are CREAMS (chemical movement, runoff, erosion), EUROSEM (an erosion and sediment transport model), KINEROS, and KINEROS2 (erosion and sediment transport models).

15:4 Erosion Tolerance (*T*)

To receive federal benefits from agriculture programs, farmers in the United States must have in operation a **Conservation Compliance Plan.** Control of erosion is one requirement in a conservation compliance plan. **Highly erodible land** must have erosion reduced to a predetermined tolerance level. The **erosion tolerance level (*T*)** is usually 5 t/a/year. Some believe that the value of T for some soils should be as low as 1 to 2 t/a. One definition of T is the *maximum rate of annual soil loss that will permit crop productivity to be maintained indefinitely.* The tolerance value is supposed to approach the rate of natural replacement of the soil. Some soils, such as those forming from hard bedrock, may produce much less than 5 t/a of new soil annually.

Relating the T value to the USLE, the actual erosion must not be greater than the tolerance level; in other words, $\boldsymbol{A}$ should be $\boldsymbol{\leq T}$.

The **erodibility index (*EI*) is**

$$EI = \frac{R \cdot K \cdot LS}{T}$$

Detail 15-1 Revised Universal Soil Loss Equation

The Universal Soil Loss Equation (USLE) has been criticized for being less than accurate. Because the equation is used to determine eligibility for payments to farmers, an inaccuracy could have enormous impact on a person's livelihood. To address this concern a Revised Universal Soil Loss Equation (RUSLE) was developed, in the form of a computer model. The computer software, distributed by the Soil and Water Conservation Society, has been updated and improved several times. The USLE and the RUSLE are conceptually similar and use the same factors. However, some of the values of factors used in the two equations differ substantially. The primary difference is that the USLE involves tables of data collected from very few studies, whereas the RUSLE involves computerized computations of data collected from very large databases. RUSLE factors are probably more accurate. Specific differences in the factors used in the equations are explained below.

***R* Factor:** For the USLE, values were computed rigorously for the eastern United States but from very few weather stations in the western United States. For the RUSLE, values are slightly changed for the eastern United States, but values for the western United States were obtained using data from more than 1000 weather stations, allowing much greater precision. RUSLE also computes a corrected *R* value for the effects of rain falling on a flat, ponded surface.

***K* Factor:** For the USLE, values are based on texture, organic matter, structure, and permeability. Values for the RUSLE are very similar but recognize the seasonal differences in structure and permeability due to freezing–thawing, soil moisture content, and soil compaction. The two *K* values can differ by more than 20%.

***LS* Factor:** For the USLE, values are calculated based on length and steepness of slope, regardless of land use, and complexity of slope. The RUSLE uses different functions to compute the *LS* value. It can accommodate complex slopes and slopes too short to be considered by the USLE. Values for the RUSLE differ from USLE values by as much as 57%.

***C* Factor:** For the USLE, values are based on cropping sequences, surface residue and roughness, and canopy cover, all weighted over crop growth stages. These factors are condensed into a series of tables. For the RUSLE the *C* factor is divided into subfactors (prior land use, crop canopy, surface roughness, surface cover) to improve flexibility. A new subfactor is calculated every time a tillage operation changes soil conditions. For conventional farming *C* factors for the USLE and RUSLE are similar. For conservation tillage and no-till, the RUSLE factor is substantially lower. For no-till the USLE may overestimate erosion by almost 300%.

***P* Factor:** For the USLE, values are based on practices that slow runoff. For the RUSLE, routines are more sophisticated and take into account hydrologic soil groups, slope, row grade, ridge height, and storm intensity. The RUSLE computes the effects of stripcropping based on transport of sediments both *to* the strips and *through* the strips.

Source: Modified from K. G. Renard, G. R. Foster, D. C. Yoder, and D. K. McCool, "RUSLE Revisited: Status, Questions, Answers, and the Future," *Journal of Soil and Water Conservation* **49** (1994), 213–220.

When $EI = 8$ or greater, the land is considered to be **highly erodible land.** To be in conservation compliance the plan for soil erosion control must be activated to bring all erosion eventually down to T or less.

The values T, R, and K are constant for a given field. The T value is set by governing authorities, R is a function of climate, and K is an inherent property of the soil. Therefore, a landowner's alternatives for controlling erosion are combinations of slope features (LS), cropping (C), and erosion-control practices (P). A determination of cropping options for a soil is given in Calculation 15-2.

Calculation 15-2 Determining an Acceptable C Value Based on Soil Loss Tolerance (T)

Problem The soil in Calculation 15-1 has a T value of 5 t/a/yr. If the grower is unwilling to use erosion-control practices or to alter the length or slope of the field, which cropping options would keep the soil erosion within the tolerance?

Solution Erosion (A) should be $\leq T$. If $T \geq A$, the USLE can be rewritten:

$$T \leq R \cdot K \cdot LS \cdot C \cdot P$$

and solving for C,

$$C \leq \frac{T}{(R \cdot K \cdot LS \cdot P)}$$

Using the values from Calculation 15-1,

$$C \leq \frac{5}{[(150)(0.48)(0.40)(1)]}$$

or,

$$C \leq 0.17$$

From Table 15-3, the farmer could achieve a tolerable rate of erosion with various cropping systems. The corn, corn oats, meadow rotation has a C value of 0.13 with conventional autumn tillage and, therefore, meets the criterion but with little margin for error. By switching to no-till, the farmer could meet the tolerable erosion level with any crop rotation. This calculation is based on the simplified assumption that erosion by wind is negligible. Actually, erosion from water and wind combined cannot exceed the tolerable rate.

15:5 Water Erosion Control

Soil erosion is reduced by controlling either soil detachment or soil sediment transport or both. The various methods of reducing detachment are much the same, whether protecting against erosion caused by wind or water. Methods for controlling transport are vastly different.

Controlling Soil Detachment

Soil detachment can be controlled by cropping or other vegetative cover practices that keep the soil covered as much as possible. The effectiveness of soil cover to control detachment is reflected in the C factor of the USLE. When plants, either living or dead, cover the soil surface, the energy of falling raindrops is dissipated by the springy vegetation. As raindrops fall the vegetation absorbs the energy, then the water gently slides off to be adsorbed into the soil.

Maintaining protective cover requires deliberate action. During winter months, when cool-area lands might be left barren, planting a cover crop of cool-season grasses, legumes, or small grains protects the soil surface. Dryland grain farms are often left in fallow. Fallow soils are particularly susceptible to erosion. To protect them the grain stubble should not be plowed into the soil, but rather left standing or only partly incorporated.

Stubble mulches help to control soil erosion. Crop residues on or near the surface reduce the impact of falling raindrops, help hold winter snows, protect topsoil from winds, improve soil structure, and increase the infiltration of rainwater.

Leaving a surface mulch of crop residues is an effective way to slow water runoff and lessen raindrop destruction of soil aggregates. **Conservation tillage** maintains at least 30 percent of the soil surface covered by residues (see Figure 15-11 and Calculation 15-3).

The need to control erosion on problem areas has accelerated the search for especially suitable plants. A Cape Cod beach grass is now a widely used dune stabilizer; emerald crown vetch in the corn belt stabilizes road banks where soil is too shallow for grass. Pink Lady winterberry, a shrub used for windbreaks and for wildlife food, came from a stone wall in Beijing, China. Tegmar intermediate wheat grass, brought from Turkey, is used in diversions and waterways in the West, and Nortran tufted hairgrass developed from breeding materials is used in Iceland and Alaska.

FIGURE 15-11 Visual comparison of various amounts of soil cover by corn or soybean crop residues. Estimates should be made looking straight down at the ground. Scanning the field from the edge might result in an overestimated residue amount. (*Source:* USDA—Soil Conservation Service.)

Calculation 15-3 Measuring Residue Percentage

When residues are left on the soil, how is the percentage coverage determined? To be in *conservation tillage,* the soil must have at least 30 percent cover.

Problem What is the percentage cover of a field covered by plant residues?

Solution Use a 100-ft tape or cord marked at each 1-ft interval. Then follow these steps:

1. Select an area representative of the whole field.
2. Anchor one end of the tape or cord and stretch it diagonally across several rows, then fix the other end of tape or cord.
3. Count the number of residues that are *under* the tape or cord at each of the 1-ft interval marks. For corn residues count only those with at least a 1/4-in diameter. Read the contact points only on one side of the tape or cord. Repeat three times at different locations in the field and average the counts (see illustration below).
4. The number of contact points equals the percentage cover. If you have 31 contact points at the 1-ft-interval marks, the coverage is 31%.

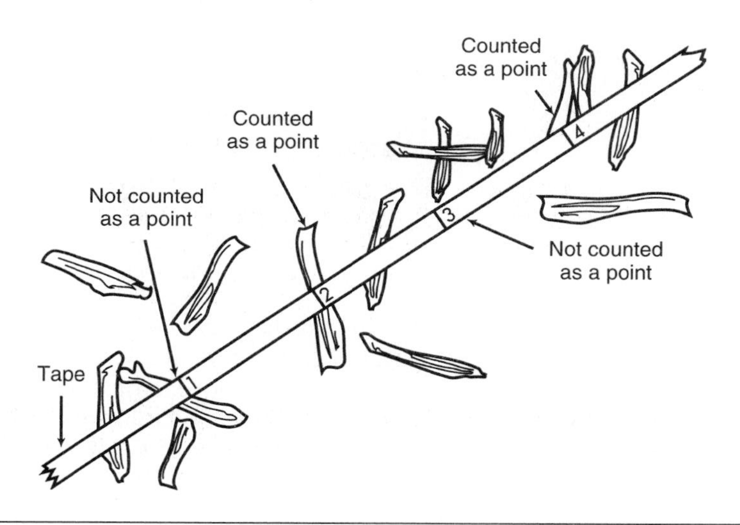

Source: Kelly O'Brien-Wray, "Helping Producers with Conservation Compliance," *Solutions* **36** (1992), No. 3; 18–22.

Controlling Soil Transport

Soil transport is hindered by slowing the eroding water, reducing the steepness of the slope, constructing barriers or terraces, cultivating on the contour, or cropping with contour strips (Figures 15-12 to 15-14). The effectiveness of methods for controlling soil transport is reflected in the P factor of the USLE. Success in achieving adequate solutions depends on the cost, whether equipment can be used, and the manager's desire to make long-range plans for use of the land.

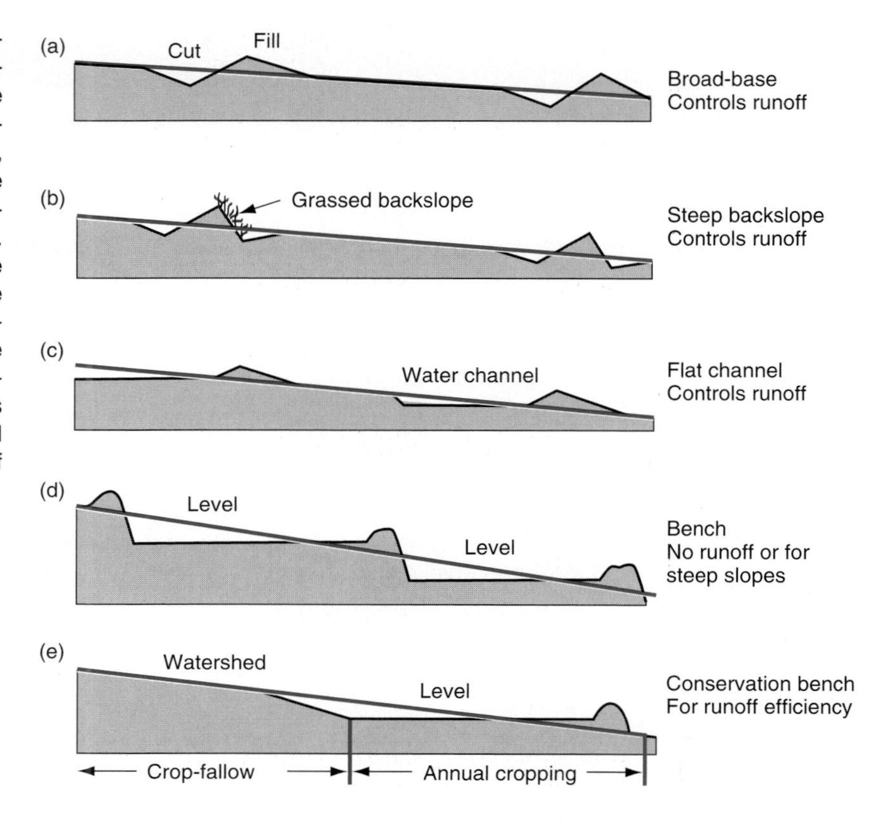

FIGURE 15-12 Cross-sections of several kinds of terraces. In the broad-base terraces (a) all of the surface area is planted (steepness is accentuated in all drawings). Thus, (a), (b), and (c) are quite similar in use. The bench terrace (d) is much more expensive if it is on steep slopes. However, gentle slopes, as in the plains of Texas, can easily be made into terraces leveled in both directions to hinder all runoff losses. The conservation bench terrace (e) attempts to take advantage of areas with runoff to produce good annual yields on nearly level portions of sloping soil areas. (Courtesy of Raymond W. Miller, Utah State University.)

Terraces can be used to reshape the land in areas of high-intensity rainfall. Terraces reduce the slope steepness and, therefore, the velocity of runoff water. Figure 15-12b diagrams a cross-section of a steep back-slope terrace. Terracing is usually recommended only for intensively used, eroding cropland. Land that has grass or other dense perennial cover seldom needs terracing for water runoff erosion control. Terraces are costly to construct and require annual maintenance (Figure 15-15). They are feasible where arable land is in short supply or valuable crops can be grown. In some areas terraces on extremely steep hillsides have been created.

Although well-designed and adequately maintained terraces are very effective in reducing soil transport, they are not included in estimating the *P* factor in the Universal Soil Loss Equation. Their effectiveness results from reducing the effective slope length and gradient. Terraces are considered as permanent changes and new *LS* values are determined for terraced fields.

Contour cultivation is tilling and planting at right angles to the natural slope of the land. On terraced fields contour tillage should be parallel to the terraces. Contour tillage successfully controls erosion during low-intensity rainfall on moderate slopes of 2 percent to 8 percent. Ridges formed during contour tillage are effective in reducing erosion. Contour tillage combined with terracing or contour strip cropping is more effective than contour tillage alone.

Contour strip cropping is the practice of planting, on the contour, strips of intensively cultivated crops alternating with strips of sod-forming crops. Erosion sediments from the clean-

(a)

(b)

FIGURE 15-13 Terraces are usually intended to reduce erosion and to control water runoff. Terraces can be essential, but they are also expensive. (a) Thai farmers learning to construct bench terraces in China; (b) Ethiopians making bench terraces in rocky hills;—(continued on next page) (By permission, Food and Agricultural Organization, United Nations. Photo a by F. Botts; photo b by D. Craig.)

tilled strips are filtered out and retained on the sod strips. The greater the proportion of sod-crop strips to cultivated strips, the less the erosion. If the sod strips occupied one-half of the field, the P value would be half that of contour cultivation alone. The width of each strip should be a multiple of the width of the machinery used to plant and harvest the crops. Soil erodibility and the crops to be grown influence the grower's choice of strip width. For a 5 percent slope, typical strip widths are about 10 m (33 ft) for a sod crop and 25 m (82 ft) for a cultivated crop.

Filter strips are relatively new technology designed to control transport of nutrient- and pesticide-enriched sediments. Usually filter strips are planted on the lower end of a field to prevent sediments from entering adjacent ditches or streams. In an Iowa study, bromegrass strips 3 m (10 ft) wide reduced the sediment load in runoff by 70 percent, and strips 9.1 m (30 ft) wide removed 85 percent. Strips wider than 9.1 m were found to be unnecessary.[6]

---

[6]C. A. Robinson, M. Ghaffarzadeh, and R. M. Cruse, "Vegetative Filter Strip Effects on Sediment Concentration in Cropland Runoff," *Journal of Soil and Water Conservation* **51** (1996), 227–230.

FIGURE 15-13, cont'd. (c) bench terraces in East Java being prepared for rice paddy; (d) bench terraces in the deep loess hills in northern Shaanxi Province, China, to prevent excessive erosion into the Yellow River. (By permission, Food and Agricultural Organization, United Nations, Photos c and d by F. Botts.)

(c)

(d)

Cover crops are crops grown during the off-season. Certain cash crops (peanuts, cotton, soybeans, corn silage, etc.) do not produce enough residue to provide adequate ground cover. Cover crops, usually legumes or cool-season grasses, are used to provide the needed protection against erosion. In addition, legumes such as subterranean clover, hairy vetch, or crimson clover can add nitrogen to the soil, reducing the farmer's fertilizer expenses for the main cash crop. Any good cover crop will increase soil organic matter, improve water infiltration, and reduce runoff. Some farmers harvest the cover crop by grazing or haying the forage. Other farmers kill the crop with herbicides before spring planting but leave all plant materials in the field for soil enrichment. In semiarid farming systems where water is the limiting factor, growers are hesitant to use cover crops that may deplete the soil water storage.

FIGURE 15-14 Contour strip cropping of hay and corn near North Springville, Wisconsin. The grassed diversion ditch (downslope at extreme right center) carries away runoff water. True contour strips would be of unequal widths, making cultivation awkward. These strips of equal width are a compromise between true contours and cultivation conveniences. (*Source:* USDA—Soil Conservation Service.)

FIGURE 15-15 A steep back-slope terrace in Missouri; the steep backside slope is grassed for protection from erosion. This Ida silt loam has 14 percent to 25 percent slopes and can erode easily without controls. (Courtesy of USDA—NRSC and the Missouri Agricultural Experiment Station, *Soil Survey of Atchison County, Missouri.*)

FIGURE 15-16 Example of a tropical desert in central Asia that is threatening productive farmland. Such conditions require immediate and often costly measures. Dune encroachment has threatened Cairo, Egypt, for decades. (*Source:* By permission, from *Food, Fiber and the Arid Lands,* by William G. McGinnies, Bram J. Goldman, and Patricia Paylore, eds., University of Arizona Press, Tucson, © 1971.)

15:6 Nature of Wind Erosion

Wind erosion is accelerated when winds are strong and when soil is dry, weakly aggregated, and bare. Winds segregate dry humus, clay, silt, and sands; the least dense are carried the farthest. Even moderate wind velocities can keep small particles of humus, clay, and silt (particles <0.05 mm in diameter) in **suspension.** Intermediate-sized particles, such as fine sands 0.05 mm to 0.5 mm in diameter, are moved by wind in a succession of bounces known as **saltation.** Coarser sands (0.5 mm–1.0 mm in diameter) are not usually airborne but rather are rolled along the soil surface (Figure 15-16). This kind of erosion is called **surface creep.** Very coarse sand (1 mm–2 mm in diameter), gravels, peds, and clods are too large to be rolled by the wind, so wind-eroded soils have surfaces covered with coarse fragments larger than 1 mm in diameter. This very coarse desert soil surface is known as **desert pavement.** Even under semiarid conditions soils having a broad range of particle sizes become more coarse in texture. The texture of medium- and fine-textured soils may not be altered by wind erosion.

An estimated 1.9 billion Mg (2.1 billion t) of soil erodes from U.S. croplands yearly, 44 percent of which is stripped off by wind erosion. Annual wind-erosion losses exceed 11 Mg/ha (5 t/a) on 3.2 million hectares (8 million acres) of cropland, 40,500 hectares (100,000 acres) of pastureland, and 12.1 million hectares (30 million acres) of rangeland. On most lands east of the Mississippi River, wind erosion is negligible. The

USDA reports 0 Mg/ha wind erosion from cultivated land in 22 states.[7] In contrast, cropland in Nevada averages 50 Mg/ha (22 t/a) and Wyoming 47 Mg/ha (21 t/a) annual loss from wind erosion.

Wind erosion is most severe in areas of arid and semiarid climates, which make up one-third of the land surface of the world, excluding polar deserts. Warm, **tropical** deserts such as the Sahara are found in high-pressure regions between 20° and 30° latitude north or south of the equator. Warm or cool **topographic deserts** are found on the leeward (eastern) side of mountain ranges because air masses become warmer and drier as they descend. The hot Mohave desert in southern California and the cool deserts of eastern Washington and Oregon exemplify this topographic effect. Arid and semiarid lands are especially susceptible to wind erosion because these lands have the following:

- Less vegetation, therefore less cover
- Less clay, therefore less aggregation
- Less soil moisture, therefore lighter-weight soil

Dry soils that are poorly aggregated and produce little vegetation may erode to produce soils that are even more coarse and less able to contain water. This downward spiral in the quality of arid lands is called **desertification.**

Although wind erosion in desert areas is especially severe, the social impact of wind erosion is greater in semiarid regions because more people live and farm in these regions. Very serious wind erosion has occurred, and continues to occur, during dry seasons in northern Africa, the former Soviet Union, the Middle East, China, India, Pakistan, Australia, Argentina, Peru, western Canada, and in the Great Plains of the United States. Expanding cultivation into marginal climates and soils seems economically sensible during periods of favorable rainfall; during dry years, however, crops fail and strong winds cause serious erosion damage.

Wind erosion can be severe in some humid regions. Strong winds may develop shifting dunes from humid-region beach sands along oceans throughout the world. This hazard is especially serious in the United States along the Atlantic and Gulf Coasts, and the Great Lakes. Drained and bare Histosols (peats and mucks) are subject to wind erosion.

15:7 Factors Affecting Erosion by Wind

Erosion by wind increases where soil is less cohesive, loose particles are smaller, land cover is lighter, and wind speeds are higher. Soils low in clay but high in fine sands and coarse silts are usually weakly structured. These soil particles are easily detached and transported if wind speeds exceed 20 km/hr (13 mi/hr). Wet soils, because they are denser, are less easily detached and transported by winds.

An equation was developed in 1965 to assess erosion by wind. That equation, called the **Wind Erosion Equation (WEQ),** is less straight-forward than the USLE and is generally solved by computer. The soil and climate data requirements for accurate use of the Wind Erosion Equation are complex and site-specific. The basic concepts of the WEQ, without the exhaustive mathematical treatments, are explained on the following pages.

---

[7]Natural Resource Conservation Service, "Summary Report, 1992 National Resources Inventory," USDA, Washington, DC, 1994, revised 1995.

The WEQ[8] estimates annual erosion by wind (E) as follows:

$$E = f(I,K,C,L,V)$$

where

E = *estimated annual erosion* loss in tons per acre (t/a $\times$ 2.24 = Mg/ha)
f = indicates that erosion is a *function* of the various factors
I = *soil erodibility index*
K = *surface roughness factor,* sometimes called the *Rf* factor
C = *climate* factor
L = unsheltered *length of field*
V = *vegetative cover* factor

Factors for the WEQ

The **Erodibility Index (I)** is based on texture and aggregation. The values of I range from 0 (stony) to more than 300 (very fine, nonaggregated sand). The formula, $I = 234\ e^{(F)(-0.04)}$ determines I based on the percentage of nonerodible aggregates larger than 0.83 mm in diameter (F). In addition to an I value, soils are assigned to a wind erodibility group (Table 15-6).

Ridge Roughness Factor (K) values vary from 1.0 for a smooth soil to 0.5 if the surface has ridges optimally placed about 3 inches high. The actual value is computed from a fourth-order polynomial function of ridge height.

Climate Factor (C) is an index of climate erosivity, relative to the climate at Garden City, Kansas, which was arbitrarily set equal to 1.0, or 100 percent (see Table 15-7). The C factor is based on wind speed, rainfall, and temperature.

Table 15-6 Soil Erodibility Index Values (I) for Wind Erosion of Various Soil Textures. Wind Erodibility Group (WEG) Designations are Also Given.

Properties of Soil Surface	*WEG*	*Erodibility Index (I)*
Sand	1	310–160 (avg. = 220)
Loamy sands and sapric organic materials	2	134
Sandy loams	3	86
Clay, silty clay, noncalcareous clay loam, noncalcareous silty clay loam with more than 35% clay; calcareous loams, silt loam, clay loam, and silty clay loam	4	86
Noncalcareous loams and silt loams with less than 20% clay, sandy clay loam, sandy clay, and hemic organic materials	5	56
Noncalcareous loams and silt loams with more than 20% clay, noncalcareous clay loam with less than 35% clay	6	48
Silt, noncalcareous silty clay loam with less than 35% clay, fibric organic material	7	38
Wet or rocky soils not susceptible to erosion	8	0

Source: Soil Conservation Service, *National Soil Survey Handbook,* USDA, Washington, DC, 1993, Exhibit 617–618.

[8]E. L. Skidmore, P. S. Fisher, and N. P. Woodruff, "Wind Erosion Equation: Computer Solution and Application," *Soil Science Society of America Proceedings* **34** (1970), 931–935.

Table 15-7 Approximate Climate Factor (*C*) Values for Land Near Various U.S. Cities

Location	*C Value*	*Location*	*C Value*
Portland, OR	0.03	Kingsville, TX	0.50
Fort Worth, TX	0.14	Cheyenne, WY	0.79
Sioux Falls, SD	0.24	Farmington, NM	0.84
Reno, NV	0.29	Garden City, KS	1.00
Denver, CO	0.37	Tucson, AZ	1.43
Miles City, MT	0.44	Las Vegas, NV	3.25

Length of Field Factor (*L*) is a distance (in feet) of unsheltered area in the direction of the prevailing wind; however, there is no simple arithmetic relation between field length and erosion. When using the WEQ, the *I, K,* and *C* factors are multiplied together to form a composite factor. The use of the *L* factor is complex and beyond the scope of this book. In actual applications conservationists read the effects of the *L* factor from computer-generated printouts. For a large, open field (10,000 ft) the *L* factor has no effect and can be ignored. For small fields ignoring the *L* factor produces an erroneously high erosion estimate. For example, consider a hypothetical field near Garden City, Kansas, that is more than 10,000 feet long, has no vegetative cover, and suffers an annual wind-erosion loss of 5 t/a. If that field were divided into quarter-sections 2600 feet long, the *L* factor indicates that wind erosion would decrease from 5 to 3 t/a.

Vegetative Cover Factor (*V*) is calculated from tables according to the kind and amount of residue, stubble, or growing plants. Some plant materials are more valuable than their weight would indicate. The equivalent cover for various residues is as follows:

Grain sorghum, flat	1.5 times actual weight
Grain sorghum, 12 in tall	2.3 times actual weight
Grain sorghum, 20 in tall	3 times actual weight
Wheat stubble, standing	6 times actual weight
Desert range vegetation	7–8 times actual weight

As with the *L* factor, the effects of vegetation are complex and cannot be simply multiplied to obtain other factors.

Calculating Erosion by Wind

The simplified calculation is done in two steps as follows:

Step 1: Partial estimate $E_A = I \times K \times C$; if necessary, the product would then be slightly modified by the *L* factor.

Step 2: Final estimated $E = E_A$ as modified by the *V* factor. The graph in Figure 15-17 is used to determine *E* from E_A and the *V* factor (Calculation 15-4).

The complexity of the tables and the determination of values for *L* and *V* in the complete procedure have prompted the development of new computer software. The Wind Erosion Protection System (WEPS; see **Appendix A**) is a computer model that takes into account (1) wind speed and direction, (2) soil characteristics, (3) crop growth, (4) types of tillage, and (5) erosion mechanics.[9]

[9]W. Doral Kemper, "Farmers Look to Science for Anti-Erosion Plan," *Agricultural Research* (April 1988), 7–10.

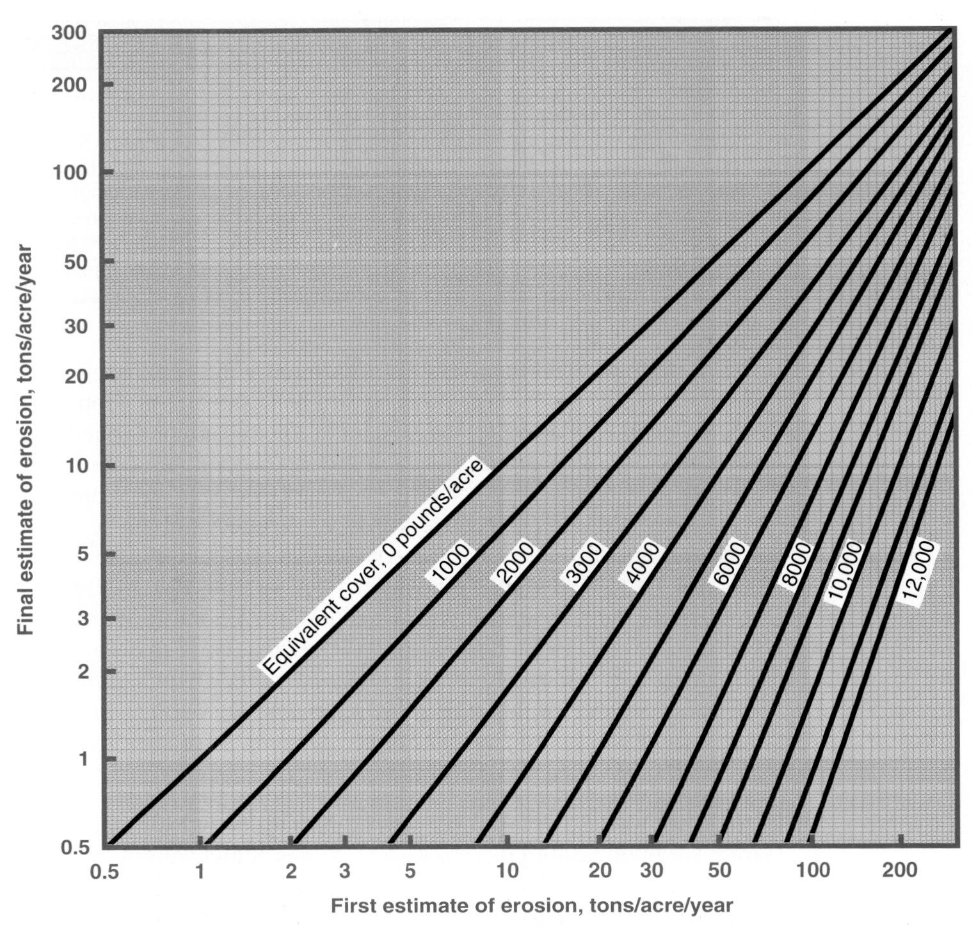

FIGURE 15-17 Completion of the calculation for soil erosion by wind. Find E_A, the *first estimate of erosion,* along the horizontal scale. Read upward until the line for *equivalent cover* is intersected. Follow that point horizontally left to the vertical scale. That intersect value is the *final value, E,* calculated for erosion. (*Note:* t/a/yr × 2.242 = Mg/ha/yr.)

15:8 Wind Erosion Control Techniques

Some factors that influence wind erosion cannot be controlled by the land manager. The soil erodibility index (I) and the climate (C factor) are not readily controllable; however, the soil's erodibility index can be altered slowly by changes in soil humus levels and tillage practices. The C factor is set on a geographical or county basis by the National Resource Conservation Service. Soil surface roughness (K), field length (L), and vegetative cover (V) can be controlled by the land manager.

When soils are fine or medium textured (not sandy), soil surface roughness can be increased by establishing ridges in a field by tillage implements. Research has demonstrated that soil ridges 5 cm to 10 cm (2 in–4 in) in height are most effective. Lower ridges were not effective either in reducing wind velocity or in trapping soil. Higher ridges increased wind velocity over the ridge tops and increased erosion.

Field length (L) can be shortened to reduce wind erosion by establishing porous-to-wind mechanical or vegetative windbreaks at suitable intervals at right angles to the most ero-

Calculation 15-4 Sample Wind-Erosion Calculation

Problem For a sorghum field near Kingsville, Texas, calculate the predicted wind erosion under the following conditions: Stubble is flat and has an actual weight of about 2400 lb/a; the soil is a noncalcareous loam with more than 20 percent clay; the field is large, nearly smooth, and without windbreaks.

Solution The solution illustrated is an estimate. In actual practice either a computer or graphs and tables would be used to obtain more precise values.

1. Estimate factors for the WEQ: $E = f(I, K, C, L, V)$

   $I = 48$ (Table 15-6)

   $K = 0.9$ (estimated for nearly smooth surface)

   $C = 0.50$ (Table 15-7)

   $L =$ (can be ignored for large field without windbreaks)

   $V =$ (see step 3)

2. Determine E_A:

$$E_A = I \cdot K \cdot C$$
$$= (48)\ (0.9)\ (0.50)$$
$$= 21.6 \text{ t/a/yr}$$

3. Calculate the effective cover as 1.5 times the actual weight of 2400 lb/acre because it is sorghum rather than wheat. This equals 3600 lb/a. Then, using the graph of Figure 15-17, find 21.6 along the First Estimate (E_A) axis and follow it up to where it intersects the extrapolated 3600 line. At the point of intersection, follow a line to the left to obtain the final estimate of erosion (E) of about 3.2 t/a/yr.

sive winds. Porosity of about 40 percent to 50 percent seems to be ideal; and the higher the barrier, the more effective it is.

Vegetative cover (V) is the most important factor in wind erosion under the control of the land manager (Figure 15-18). The erosion-resisting value of vegetation depends on the amount of plant material, its coarseness or fineness, its height when standing, and whether it is living or dead, standing or flattened. Because the original research was conducted on flattened wheat straw, the relative value of V in the wind erodibility equation is based on this condition. A system has been developed to convert the principal kinds and conditions of plant materials to their equivalent of flattened wheat straw to obtain their V value for use in the equation.[10]

As indicated by the USLE and WEQ, and as confirmed by field studies, factors relating to vegetation (C in the USLE, V in the WEQ) have the greatest remedial effect on erosion. For controlling erosion by either water or wind, the more vegetation and the greater the

[10] F. R. Troeh, J. A. Hobbs, and R. L. Donahue, *Soil and Water Conservation,* 1991, 147–149.

FIGURE 15-18 Practical wind erosion control in South Dakota. Multiple rows of trees called *shelterbelts* protect farm buildings from prairie winds. Forage crops and small grains are often planted in blocks, but row crops like corn are planted on the contour. (Courtesy of USDA.)

duration of vegetative cover, the better. For this reason, the primary weapons against erosion are residue management through conservation tillage, the use of cover crops, and the adoption of cropping systems that avoid bare strips of soil during row-cropping and avoid bare fields during the off-season.

Although windbreaks are primarily for erosion control and comfort (around homes), windbreaks in other areas may have other values. Corn sheltered by trees grew taller and tasseled earlier than other corn in the field (Figure 15-19).[11] A study in Germany demonstrated a 19.7 percent increase in sugar in sheltered sugar beets. In Nebraska windbreaks increased yields of cantaloupes, snap beans, and asparagus; protected melons matured a week earlier. An east-west windbreak collects warmth from the south in spring and protects against hot winds from the south in summer. Other benefits include the following:

- The trees will shelter natural predators, flickers, and woodpeckers that consume cutworms, grasshoppers, and beetles.
- The wood can be valuable as fuel.
- A wildlife habitat is produced.

Studies indicate that deep-rooted trees don't seem to compete very much with the crops.

[11]John J. Regan, "Windbreaks for Corn?" *The Furrow* **101,** Issue 5 (1996), 7–8.

FIGURE 15-19 Do windbreaks help crops, or do they compete with crops and shade them? Studies in Nebraska, Minnesota, and Germany suggest that in windy areas shelterbreaks may increase yield, speed maturity, provide homes for enemies of crop pests (such as birds), and provide wildlife habitat. (Courtesy of The Furrow, Deere & Company, Moline, IL.)

15:9 Decision Case

The concept of decision case studies was introduced in Chapter 6. The following is an additional decision case study on considering the impact of conservation compliance: Larry J. Grabau and Mark V. Kane, "Stratton Farm: A Case of Conservation Compliance," *Journal of Natural Resources and Life Sciences Education* **21** (1992), No. 1; 20–26.

> *. . . one cannot very well assess in monetary terms the future value of the soil that is saved today. . .*
>
> —**Daniel Hillel**

How dangerous it is to reason from insufficient data.

—Sherlock Holmes

Questions

1. What is the typical tolerable erosion rate (*T* value) in t/a/yr?
2. Explain the damages caused by allowing rainfall to hit bare soil.
3. List practices that decrease detachment of soil by raindrops.
4. Discuss the damages of erosion from two points of view: (a) quality of the soil lost and (b) rates of loss versus the rates of soil formation.
5. Define (a) *sheet erosion,* (b) *rill erosion,* and (c) *gully erosion.*
6. Define each term in the Universal Soil Loss Equation (USLE).
7. List the condition for each factor in the USLE that will allow (a) very large amounts of erosion and (b) very low amounts of erosion.
8. In general terms, how does the RUSLE differ from the USLE?
9. Define and briefly discuss these terms: (a) *highly erodible land;* (b) *erosivity index;* and (c) *conservation compliance.*
10. Which causes the greatest increase in soil erosion by water: doubling the slope percentage or doubling the slope length?
11. Describe characteristics of vegetative cover that would generally allow the least soil erosion.
12. When relating erosion tolerance to the USLE, explain why *T/RK* is considered a constant.
13. Define each term in the Wind Erosion Equation (WEQ).
14. (a) Which soil factors influence the soil's wind erodibility index (*I*)? (b) Describe a soil with a high *I* value.
15. Discuss the importance of soil cover for controlling wind erosion.
16. Briefly discuss methods to control wind erosion.

Water Resources and Irrigation

I have made water flow in dry channels and have given an unfailing supply to the people. I have changed desert plains into well-watered land.

—Hammurabi

In the war of sun and dryness against living things, life has its secrets of survival. Life, no matter on what level, must be moist or it will disappear.

—John Steinbeck

16:1 Preview and Important Facts

PREVIEW

Societies experience water stress when they approach yearly water supplies of 2000 m^3 (530,000 gal) of water per person (nationwide need, not just household use). By the year 2000 six east African and all five north African countries will be in this category. Six of these countries will have less than 1000 m^3 (265,000 gal) of water per capita.[1] As a comparison, the United States has about 10,000 m^3 (2.65 million gal) per person, Jordan has only 160 m^3 (42,400 gal), Mexico has 4000 m^3 (1,060,000 gal), Canada has 109,000 m^3 (28.8 million gal), and Israel has 370 m^3 (98,050 gal).[2]

In a sobering review of the fresh-water status of the world, the message is clearly and unquestionably a wake-up call:[3] Water in many countries is in short—even stressful—supply. As is frequently stated when comparing the relative importance of shortages of petroleum and shortages of water, "There are other sources of energy; for water there is no substitute." Even people with sufficient water may not have adequate supplies of *clean* water; by the year 2000 a third of the world's population will not have access to clean water.

---

[1] M. Falkenmark, "The Massive Water Scarcity Now Threatening Africa—Why Isn't It Being Addressed?" *Ambio* **18** (1989), 112–118.

[2] World Resources Institute, *World Resources 1992–93,* Oxford University Press, New York, 1992, 130.

[3] Sandra Postel, "Water in Agriculture," Chapter 5 in Peter H. Gleick, (ed.), *Water In Crisis,* Oxford University Press, New York, 1993, 55–66.

The world's voracious demand for water is causing stress in many areas. We are attempting to meet insatiable demands for water by historical supply methods—dig more wells and dam and divert more streams—without recognizing that the supply has limits, both ecological and economical. People are generally quick to assume rights to use water, but they are slow to shoulder the responsibility and obligation to preserve and protect it.

The value of water has long been known, and its importance has many monuments and evidences around the world (Figures 16-1 and 16-2). Raised waterways are found throughout the lands conquered by ancient Rome. In many arid Middle Eastern countries, underground springs (*ganats*) and canals were laboriously dug. Underground canals eliminated the need to lift the water to the soil surface and reduced evaporation losses during transport (see Figure 16-3).

Surface waters make up about 75 percent of the U.S. water being used, although surface waters are only a small fraction of the total fresh-water supply. Subterranean **groundwaters** are extensive, but they are expensive to use because of pumping costs. The porous strata called **aquifers** that contain these waters occur at various depths underground and may have very slow to very fast water flow and recharge rates. Surface waters and groundwaters also vary in composition.

During a typical growing season plants use about 20 cm to 100 cm (8 in–40 in) of water, and 40 cm to 70 cm (16 in–30 in) is average for most crops. Irrigation is water-expensive. This water use is larger where crops are dense and grow longer in hot, dry climates. Plants vary in their rooting depths, in their susceptibility to damage when water is available slowly, and in the growth stage, during which water is most critical.

Irrigation waters vary in quality. The primary concerns are the salt content and the proportion of cations that is sodium. High salt or high relative sodium ion contents are undesirable.

Irrigation water can be added in many ways. Various gravity-flow techniques are the cheapest and most frequently used. Sprinklers are expensive. Drip techniques are most efficient in water use but are also expensive. The scientific approach to irrigation (when to irrigate and

FIGURE 16-1 Roman aqueduct (by the painter Zemo Diemer, the original now in the German Museum in Munich). Many remnants of the aqueducts remain to furnish information on the shape, size, and extent of the Roman systems. (*Source:* By permission, from *Food, Fiber and the Arid Lands,* by William G. McGinnies, Bram J. Goldman, and Patricia Paylore, eds., University of Arizona Press, Tucson, © 1971; and from *Arid Lands in Perspective,* William G. McGinnies and Bram J. Goldman, eds., University of Arizona Press, Tucson, © 1969.)

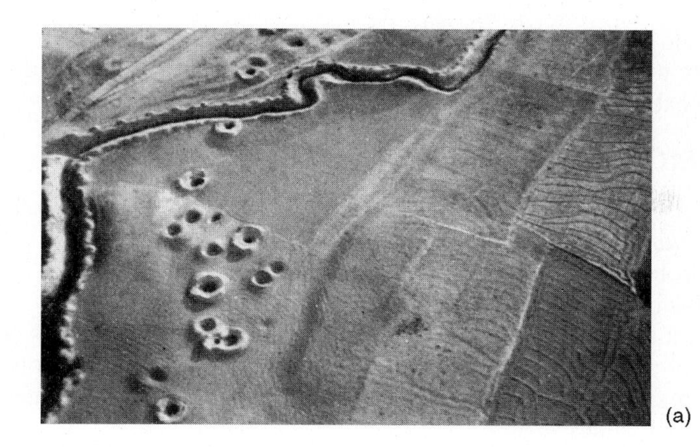

(a)

(b)

FIGURE 16-2 Ancient civilizations developed unique ways to get water from below ground. Underground tunnels (*ganats*) tapped the water-carrying strata. Access holes to the soil surface needed to be frequent (a), and sometimes the tunnels needed reinforcement (b). Near Shiraz, Iran. (Courtesy of Bruce Anderson, Utah State University.)

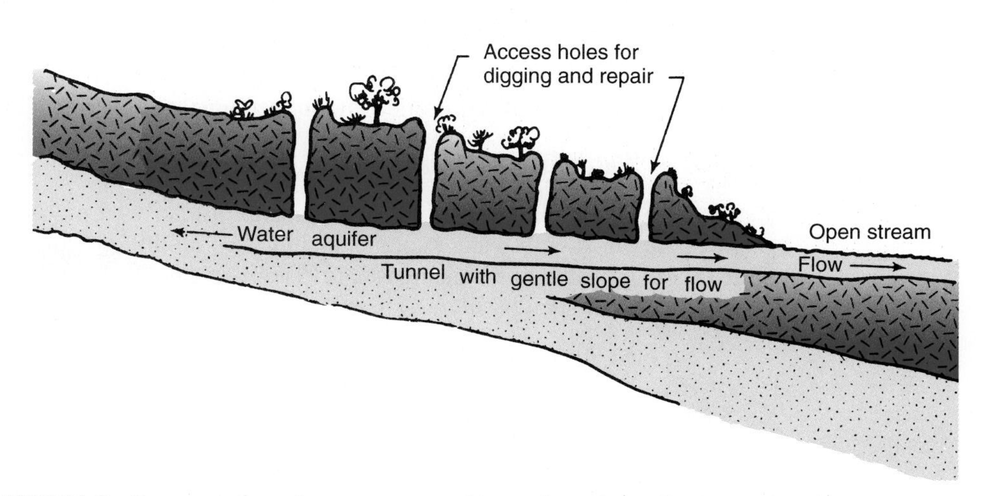

FIGURE 16-3 Cross-section of a *ganat,* an ancient well made by digging slightly sloping tunnels into aquifers to obtain irrigation water. These are still in use. (Courtesy of Raymond W. Miller, Utah State University.)

how much to add) has been extensively studied. With the demands of an increasing population on finite yearly water supplies, many changes in techniques, supplies, and water laws are expected in the next few decades.

IMPORTANT FACTS TO KNOW

1. The definitions of *aquifer, spring, saltwater intrusion boundary, recharge area,* and *artesian wells*
2. The hydrologic cycle and composition of aquifers
3. The status of world fresh-water supplies
4. The extent to which overuse of surface waters and groundwaters is occurring
5. The major parameters determining the quality of water for irrigation use
6. The approximate water requirements of most crops and how climate affects these requirement values
7. The soil depth from which the plant extracts most of its water
8. How to determine the time to irrigate and the amount of water to add
9. The advantages and disadvantages of irrigating with surface flow in furrows, basin and border strips, sprinkler systems, and drip or trickle systems
10. The techniques and potentials for reducing the irrigation water required for crops

16:2 World and U.S. Water Resources

The Earth is estimated to have about 1359 million km^3 (327 million mi^3) of water; 97.22 percent of it is contained in oceans, 2.15 percent is in glaciers and icebergs, and about 0.03 percent of it is circulated annually by precipitation, transpiration, and evaporation. This 0.03 percent is the critical amount that includes snow and rain, flowing surface water, underground recharge water, and atmospheric evaporative water vapor—most of the agricultural, industrial, and culinary fresh water used daily (Figure 16-4). This natural water circulation system is called the **hydrologic cycle.**

Water is not evenly distributed over the Earth. Most population centers developed near good sources of water, and people became accustomed to having adequate water. Some areas outgrew their water supplies and looked elsewhere for new supplies. The massive California Central Valley project has water flowing hundreds of miles in large aqueducts. The Bureau of Reclamation constructed large dams in various areas and distributed expensive water to farmlands at low costs. Similar developments grew in India, China, and the Soviet Union. As populations increased, so did their demands for water. The population of nearly 6 billion people in 1996 is projected to be 10 billion by 2050 and 12 billion by the year 2100. More and more water will need to be controlled and moved unless something changes.

Examples of Water Stressed Areas

Life requires water. Nowhere is this more apparent than where water supplies are meager. Looming problems with water supplies include (1) rapidly growing populations, (2) uncertainties about global climatic changes, and (3) possible conflicts over shared water resources. Some examples of such problems follow:[4, 5]

- More than 90% of future population growth will be in the developing countries, where clean water is often not available. More than one-third of all deaths in these countries are children under 5 years old; the rates of infant mortality in Africa are

[4] Peter H. Gleick, "Water in the 21st Century," Chapter 9 in Peter H. Gleick, (ed.), *Water in Crisis,* Oxford University Press, New York, 1993, 105–126.
[5] Sandra Postel, "Water in Agriculture," *ibid.,* 63.

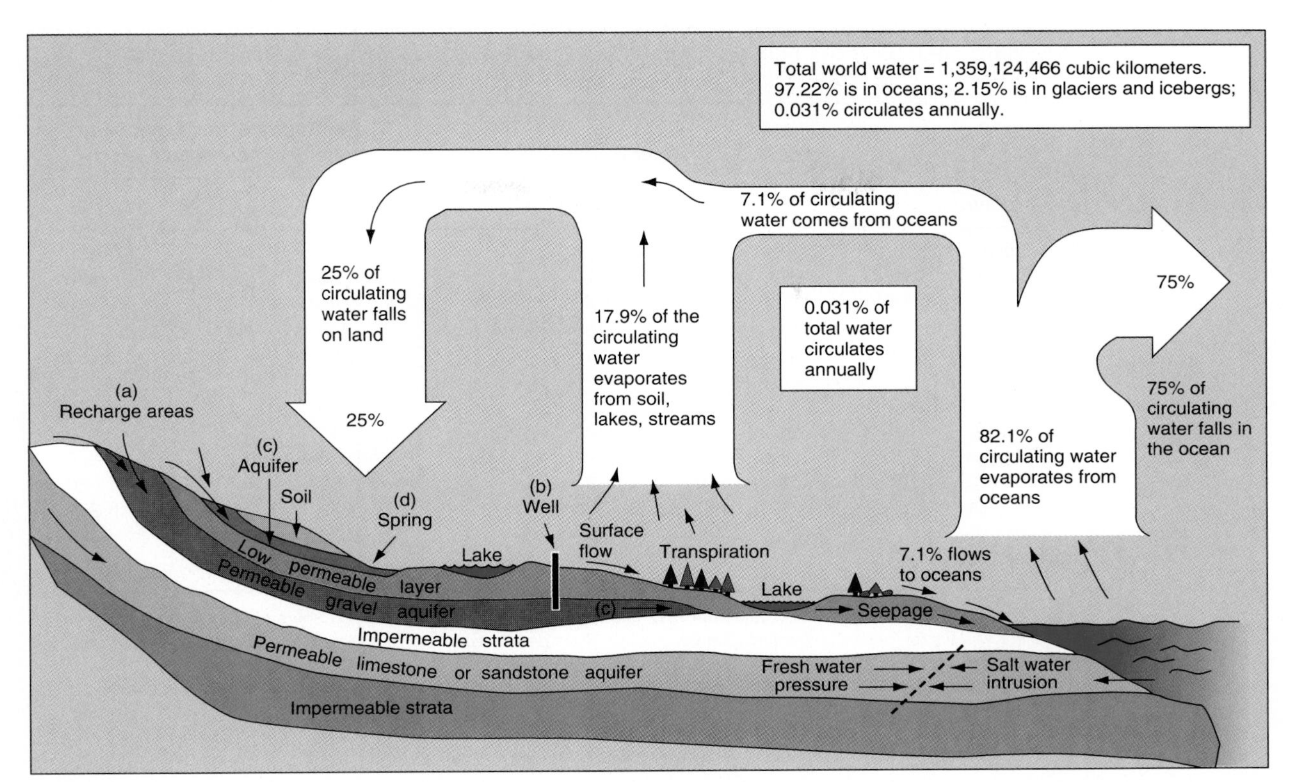

FIGURE 16-4 Hydrologic cycle, showing relative water distribution. Only about 0.03% of the total Earth water recycles each year. (Percentage numbers are the portion of recycled water moving in that form.) Groundwaters are recharged from this circulated water. Salt-water intrusion is caused by pumping out so much fresh water from underground sources that nearby ocean pressure fills the aquifer layer with salt water. (a) Recharge areas are land surfaces where porous strata are open to the soil surface. (b) Wells may flow or require pumping. If the level of recharge area water is higher than the well top, water can flow by gravity. This is called *artesian well flow.* (c) Aquifers often are coarse sediment laid down from former geologic erosion. These tend to become thinner or pinch out farther from the deposition front. (d) A spring is an exposed opening in an aquifer. (Courtesy of Raymond W. Miller, Utah State University.)

nearly 15 times greater than in North America. One-third of the population (more than 2 billion people) lack access to clean water.

- Nine countries of the Middle East are using more than 100% of their *renewable fresh water.* This is only possible by pumping groundwater faster than recharge rates and by expensive desalinization of seawater (Table 16-1).
- By the year 2000 chronic water shortages are likely in northern China, all of northern Africa, parts of India and Mexico, much of the Middle East, and portions of the western United States. China decreased irrigated acreage by 930,000 ha (2.297 million a) (almost 2%) since 1980. The Soviet Union ceased irrigation on 2.9 million ha (7.16 million a) between 1971 and 1985, which was equal to new areas brought into cultivation.
- Egypt depends entirely on the Nile River water, which originates in Sudan and Ethiopia to the south. Virtually all of Egypt's cropland must be irrigated. Ethiopia, which controls 80% of the Nile's flow, plans to divert and use more of that water, which could be catastrophic for Egypt.
- Jordan and Israel get most of their water from the Jordan River basin. Already Israel exceeds renewable supplies by 15%. Jordan exceeds its sustainable supply, and its population is doubling every 17 years (4.1% annual growth).

Table 16-1 Use of Country's Fresh-Water Supply and the Available Water Per Capita in 1990 and Projected to 2025 for Middle-Eastern Countries*

Country	Water Withdrawals[a] (%) (sustainable supply = 100%)	Per Capita Water Availability[b] (m^3 per person per year)	
		1990	2025
Libya	374	160	60
Qatar	174	50	20
United Arab Emirates	140	190	110
Yemen	135	240	80
Jordan	110	260	80
Israel	110	470	310
Saudi Arabia	106	160	50
Kuwait	>100	<10	<10
Egypt	97	1070	620
Malta	92	80	80
Tunisia	53	530	330
Iran	39	2080	960
Morocco	37	1200	680

**Source:* Modified and selected data from P. H. Gleick, "Effects of Climate Change on Shared Fresh Water Resources," in I. M. Mintzer, (ed.), *Confronting Climate Change: Risks, Implications and Responses,* Cambridge University Press, Cambridge, 1992, 106.

[a]More than 100% means more than yearly renewal is used by pumping more water than the recharge rate and by desalinization of sea water.

[b]500 m^3/yr (132,500 gal) might be enough for a semiarid society that has sophisticated water management, but 2000 m^3 (530,000 gal) is considered a suitable amount to reduce stress from scarce water.

- Water tables beneath Beijing, China, have been dropping 1 m to 2 m yearly, and one-third of its wells have gone dry. One study suggests that local farmers could lose 30% to 40% of their current water supply within 10 years. When reservoir levels in 1985 dropped to record lows, all farmers except vegetable farmers had irrigation water cut off.
- Three-fourths of Saudi Arabia's water comes from *nonrenewable groundwaters.* Some farms have already been abandoned because of high pumping costs and low water yields from wells.
- The Ogallala aquifer (from southern South Dakota to northwest Texas) has diminished supplies by 24% because of overpumping (pumping more than the recharge rate) especially in the Texas High Plains. Irrigated acreage using Ogallala water decreased 34% between the peak year of 1974 and 1989.
- Russia's Aral Sea, once the world's fourth largest, has lost 40% of its surface area since 1960, its volume dropped by two-thirds, its salinity level increased threefold, and virtually all native fish (and the fishing industry) are gone. All of these changes were caused by diverting water from the two major rivers that supply the Aral Sea to use in irrigation. Saving the sea must involve major reductions in irrigation.

Irrigated Areas and Available Fresh Water

Irrigated lands are only about 16 percent of the world's cropland, but they produce about one-third of the total harvest. Decreases in irrigation will have major repercussions on food production. The countries irrigating the most area are shown in Table 16-2. The top 20 and selected other states in the United States are shown in Table 16-3.

Table 16-2 Net Irrigated Area of the Top 10 Countries and Selected Others, 1989

Country	*Net Irrigated Area (thousands of hectares)*	*Share of Total Cropland Irrigated (%)*
China	45,349	47
India	45,039	25
Soviet Union	21,064	9
United States	20,162	11
Pakistan	16,222	78
Indonesia	7,550	36
Iran	5,750	39
Mexico	5,150	21
Thailand	4,230	19
Romania	3,420	33
Italy	3,100	26
Japan	2,868	62
Brazil	2,700	3
Egypt	2,585	100
World	235,299	16

Source: Modified and selected portions from S. Postel, *Last Oasis: Facing Water Scarcity,* W. W. Norton & Company, New York, 1992.

Table 16-3 Net Irrigated Area, Top 20 States, 1992, and Number of Farms, 1994

State	*Irrigated Area (1000 acres)*	*Number of Farms (1000 farms)*	*State*	*Irrigated Area (1000 acres)*	*Number of Farms (1000 farms)*
California	7245*	48.3	Florida	1416	7.2
Nebraska	5980	16.0	Wyoming	1374	4.5
Texas	5101	13.6	Utah	1085	9.0
Idaho	3184	13.4	Louisiana	821	2.5
Colorado	2999	12.3	Arizona	752	3.0
Arkansas	2853	5.6	Missouri	702	1.6
Kansas	2502	5.1	New Mexico	685	7.1
Montana	1936	7.9	Mississippi	647	1.1
Oregon	1587	9.5	Georgia	620	2.5
Washington	1587	9.7	Nevada	520	1.5

Source: Arranged and calculated by Duane T. Gardiner from "Farm and Ranch Irrigation Survey," 1994, and 1992 *Census of Agriculture.* U.S. Department of Commerce, Bureau of the Census, Washington, DC, 1994.
*Acres × 0.405 = hectares.

The quantities of fresh water are finite. Once the limit of available fresh water (river flow from rains and rechargeable groundwaters) is reached, the water per capita depends on how much can actually be used and what the population numbers are. Estimates of renewable water and populations are continually being assessed. Table 16-4 lists data for the United States and selected other countries. Truly, the United States has good fresh water supplies, even for double the 1997 population, assuming an overall need of 2,000 m^3/yr (530,000 gal) per capita of fresh water.[6] However, the arid West and Great Plains are struggling for adequate water in some areas even now.

[6] Sandra Postel, *Last Oasis: Facing Water Scarcity,* W. W. Norton & Company, New York, 1992, 185.

Table 16-4 Population, Total Renewable Fresh Water and Renewable Fresh Water Per Capita for the United States and Other Selected Countries

Country	1990 Population (millions of people)	Annual Renewable Water Resources	
		Total (km^3/yr)	1990 per capita (thousand m^3/yr)
United States	249.2	2,478	9.94
Canada	26.5	2,901	109.37
Mexico	88.6	357	4.03
China	1,139.1	2,800	2.47
India	853.1	1,850	2.17
USSR	288.6	4,384	15.22
Australia	16.9	343	20.48
Brazil	150.4	5,190	34.52
Israel	4.6	1.7	0.37
Japan	123.5	547	4.43
Jordan	4.0	0.7	0.16
Nigeria	108.5	261	2.31
Poland	38.4	49.4	1.29
Spain	39.2	110	2.80

Source: Selected data and modified from World Resources Institute, *World Resources 1992–93,* Oxford University Press, New York, 1992, 129–133.

U.S. Fresh Surface Waters

For irrigation 50 percent of applied water comes from wells. Surface water is usually cheaper and easier to use than groundwater. The composition of surface waters varies with the terrain through which the waters flow, because the dissolving minerals that enter the water solution vary. The kinds and amounts of materials that are dissolved in surface waters exhibit extreme variations.

Diversion of surface flow water has been cheaper than the pumping of groundwater. However, as greater demands are put on fresh water supplies, we will have to look at alternatives for replacement, such as more pumping of groundwater, treatment and recycling of wastewaters, reverse osmosis, and distillation. Table 16-5 lists the costs of providing some of these sources in comparison with new surface diversion projects. All will be costly.

U.S. Groundwaters

Much of the recent increase in irrigated land area has been possible because of a greater use of **groundwaters,** those waters in underground reservoirs in the deeper soil and substrata. These underground reservoirs are usually found in porous rock formations called **aquifers** (see Figure 16-3). Most aquifers extend from a few kilometers to 20 km to 30 km (12 mi–18 mi) long but may connect for hundreds of kilometers. These porous strata can be sands, gravels, porous sandstones, and channels in partly dissolved limestones.

The part of the Earth's crust between the land surface and the water table (aquifer surface) is called the **vadose zone.** It is normally unsaturated, but it includes the capillary fringe above the water table. At times, such as after prolonged rain, portions will be partly saturated. The upper part of the vadose zone is the *root zone*—that zone occupied by roots. That depth between the root zone and water table is called the *intermediate vadose zone.*

Table 16-5 Costs of Providing Fresh Water by Various Means

Water Source and Technology	*Per Acre-Foot (U.S. dollars)*	*Per Thousand m^3 (U.S. dollars)*
Interbasin diversion projects (surface flow)	100–200	123–246
Groundwater development	72	88
Groundwater recharge	95–112	118–138
Recycling wastewater (secondary treatment)	62–104	77–128
Recycling wastewater (secondary treatment plus)*	162–393	200–485
Reverse osmosis (of brackish water)	97–322	120–397
Distillation for pure water	530–880	654–1085

Source: Selected data and modified from P. Rogers, "Assessment of Water Resources: Technology For Supply," in D. J. McLaren and B. J. Skinner (eds.), *Resources and World Development,* Dahlem Konferenzen, Berlin, John Wiley and Sons, Chichester, UK, 1987, 415.

*Secondary treatment plus N, plus phosphate reduction, plus filtration and carbon adsorption for volatile and other organics.

The vadose zone is a favorable environment for attenuation of contaminants. Trace elements have time and space for precipitation, adsorption, and cation exchange. Organic substances can be adsorbed or undergo bioremediation (breakdown by microorganisms). Pathogenic bacteria and viruses will be intercepted, adsorbed, or partially decomposed. The vadose zone has a key role as the source of recycled nutrients and pollution of groundwater. Its ability to *filter* percolating contaminated water is far from complete, however.

How Much Groundwater?

How important are groundwaters in comparison to surface waters? Total groundwater stored is about 25 times more than the amount of water in all the world's lakes, rivers, and streams combined.[7, 8] Groundwater is a vast and important resource. In the United States approximately 15 quadrillion gallons (56 quadrillion liters) of water are stored within 0.5 mi (0.8 km) of the land surface. Groundwater supplies about 25% of all fresh water used. Fifty percent of U.S. citizens obtain all or part of their drinking water from groundwater; and 95% of rural households depend on it totally. Commercially, groundwater is employed extensively in agricultural practices, particularly for irrigation and for various industries.

Groundwaters generally have a *more constant temperature, less sediment, less dissolved material,* and are *more omnipresent* than surface waters. With these groundwater advantages, why are surface waters so much more extensively used? Cost is the primary reason. Drilling wells and pumping to lift water are much more expensive than are the surface reservoirs and simple gravity-flow diversion canals needed for surface waters. Both ancient and modern users have found unique ways to tap groundwater to avoid the problem of lifting it. *Horizontal wells* called *ganats* (or *khanats,* or *kharezes*) have been used in Iraq, Iran, Afghanistan, and other countries for thousands of years (see Figures 16-2 and 16-3).

---

[7]Robert A. Weimer, "Prevent Groundwater Contamination Before It's Too Late," *Water and Wastes Engineering* **17** (1980), No. 2; 30–33, 63.

[8]Brana Label, *EPA Groundwater Research Programs,* EPA/600/S8-004, Environmental Protection Agency, Washington, DC, 1986.

Wise Use of Groundwaters

A serious problem in using groundwaters is the rate of recharge of the aquifer. Some aquifers are quite porous, and percolating recharge water can readily move hundreds of meters in a few days to refill the underground reservoir. Retaining adequate water in these aquifers depends upon the rainfall needed to recharge these reservoirs and the rate of water use from each aquifer.

In contrast to rapid-recharge aquifers, some aquifers allow very slow water intake and flow; water may flow only a few hundred meters per year. The volume decrease in the Ogallala aquifer was mentioned previously. Another aquifer, the Carrizo–Wilcox sandstone aquifer in Texas, permits water movement at the rate of only 2 m to 16 m (7 ft–53 ft) per year.[9] If recharge areas were 160 km (96 mi) away, natural recharge would take 3000 to 4000 years. If an aquifer layer is tapped by too many wells, the water is removed faster than recharge can refill it—a condition called *mining.*[10] The water table drops and the flow from wells decreases or stops entirely. As aquifer water (which helps to support the land above) is removed and not replaced, the surface land may sink (*subside*). For example, Mexico City, the world's most populous city, removes water from a large aquifer. However, the water quality lessens with depth and the aquifer has been dropping about 1 m per year. The clay cap is drying and cracking, making the aquifer more susceptible to contamination. The land area has been slowly subsiding. One well casing, once a few centimeters above the ground, is now exposed 7 m above the ground level.[11]

As groundwater use increases, a balance of extraction and recharge is essential (Detail 16-1). For example, groundwater supplies about 40 percent of Arizona's water, but it is being pumped out almost five times faster than it is being recharged.[12] In south-central Kansas farmers above the Equus Beds aquifer north of Wichita were limited by 1979 state regulations to 450 mm (17.6 in) of groundwater per year, about two-thirds of optimum needs.[13] In western Kansas water resources are in even worse condition. The western 60 percent of the United States, except for the general area of Colorado, is using water faster than it is recharging. Overuse ranges from 8.5 percent to 77.2 percent annually. The Arkansas–Oklahoma area and the Texas–Gulf regions have more than 60 percent overdraft.

Do We Need a Water Ethic?

With the increasing pressures for more water and our historical approach of exploiting more of the water resources, perhaps there is a need to change our perspective about water. The United States has good water supplies, but in the Western states, many areas are having difficulty obtaining adequate water for their present lifestyle. In those areas water is valuable. To secure our water future, everyone should treat water as the treasure that it is. For too many of us, water simply flows from the faucet, the well, or the canal; we are insensitive to water in its natural setting of the wild river or the intricate wetland as the support of all life. It has been said that we need a *water ethic,* which means in essence to ". . . make the protection of water ecosystems a central goal in all that we do."[14] Such a water ethic would persuade peo-

---

[9]Keith Young, *Geology: The Paradox of Earth and Man,* Houghton Mifflin, Boston, 1975, 120–121.

[10]Sometimes the term *mining* is used to mean *one time extraction* of fossil groundwater that can be extracted but *will not recharge* in any reasonable time period.

[11]Dani Shannon, "Mexico City Aquifer Depletion Degrading Water Quality," *Environmental Science & Technology* **29** (1995), No. 6; 254A.

[12]Anonymous, "Like Having Your Dad Die," *Time,* Mar. 7, 1977, 80.

[13]Ron Larsen, "Water Control Imperative in Kansas," *Irrigation Age* **13** (1979), No. 5; 38–41.

[14]Sandra Postel, *Last Oasis: Facing Water Scarcity,* W. W. Norton & Company, New York, 1992, 185.

Detail 16-1 Natural Canals and Reservoirs

To solve the problem of inadequate water distribution, water is often transported great distances in expensive canals. Israeli scientists in their water-short country discovered that some of the aquifers were continuous from north to south. Pumping water in the north from Lake Tiberius (Galilee) into depleted wells replenished the aquifer's water, which flowed as far as 242 km (150 mi) south, where it could be pumped out for use.*

Many other areas are now using deep wells as water disposal sinks. In Orange County, California, treated sewage effluent pumped into deep wells helps keep salt water intrusion from moving farther inland and also recharges the aquifer for use by others.

*Michael Overman, *Water,* Doubleday, New York, 1979, 11–14.

ple to consider and evaluate two contrasting views on an equal footing. The contrasting views are a *reduction in water use* versus conventional projects aimed at *expanding access to more and more of the dwindling fresh-water supplies.* Further, one writer states that, "No quick fix is going to solve agriculture's water problems any time soon. . . . The struggle for a secure water future will not end until societies recognize water's natural limits and begin to bring human numbers and demands into line with them."[15] This concept suggests that just as we are now looking seriously at sustainable agriculture, we must also do that with all parts of the system. This concept is *sustainable irrigation,* although many more details must be probed; it emphasizes a need to search to improve water use efficiency energetically and to try to conserve water in other ways.

16:3 Irrigation Water Quality

The highest quality water is required for drinking. Progressively lesser quality water is essential for swimming, industry, and irrigation. If carefully used, water for irrigation can have considerable amounts of many substances in it without causing problems. Critical, however, are the amounts of soluble salts, the proportion of sodium to calcium and magnesium, and the absence of certain metals or toxic materials such as selenium, cadmium, chloride, sodium, and boron.

Many criteria important in assessing water quality for other uses (taste, color, odor, turbidity, temperature, hardness, pH, BOD, or COD, nutrient content—N and P—and pathogenic organisms) are sometimes, but not usually, important for irrigation water.

Turbidity is water opaqueness caused by the presence of suspended solids of clays, silts, sands, and organic materials. These materials may fill up irrigation canals or reservoirs, seal up the surface pores of soils, grind away turbine blades of electrical generators, and clog sprinkler and trickle irrigation systems. Sediment problems can be reduced by the use of settling basins and filters, but this is expensive.

Water temperature is of limited concern in irrigation, except where it may be cold enough to reduce growth, such as for flooded rice production. Usually the soil heat and incoming solar radiation modify the water's initial temperature enough that it does not greatly affect plant growth.

[15]Sandra Postel, "Water in Agriculture," *ibid.,* 63.

BOD (biological oxygen demand) or **COD** (chemical oxygen demand) are measures of how much of the oxygen dissolved in water will be used as the organic material and certain chemicals in the water are decomposed (oxidized). The BOD or COD is higher as organic materials such as algae, plant residues, or manures increase in the water (Figure 16-5). These same materials are usually high in nitrogen and phosphorus, which are nutrients for growth of algae.

Sludge and manure effluents have high BODs and CODs and are undesirable for aquatic life (fish, algae, protozoa). Such waters would also be more rapidly depleted of oxygen when added to poorly drained soils. High BOD waters when used on soil can cause poor aeration (inadequate oxygen) conditions faster than would low-BOD waters.

Pathogenic organisms are disease organisms. It is becoming more common to find pathogenic organisms in natural freshwater sources in the United States; they are almost certain to be present in inadequately treated sewage effluents. If sewage effluents are to be used for irrigation (or disposal) on agricultural land, they should first be certified by the Public Health Service to be free of viable pathogenic organisms. Usually these effluents are not permitted to be used on foods, particularly any that would be eaten fresh.

Salinity

Salinity, or concentration of *total soluble salts* (TSS), is one of the most critical criteria for irrigation water quality. Salts affect plants by increasing the osmotic pressure of water, making the plant exert more energy to absorb soil water. Salt concentration of even a few tenths of a percent by weight can hinder plant growth. Salt contents are measured by electrical conductivity (EC) in siemens meter^{-1} (S m^{-1}), or decisiemens meter^{-1} (dS m^{-1}) for soil solutions, or millisiemens meter^{-1} (mS m^{-1}) for waters. Previously these measurements were reported as mmhos/cm and micromhos/cm, respectively. A typical water classification is given in Figure 16-6. The arbitrarily selected boundaries are only approximate. A second system of classifying irrigation water is given in Table 16-6. Most irrigation waters do not exceed about

FIGURE 16-5 A small stream ponded near areas of animal habitation where manure washed into the stream results in the growth of algae (the light-colored floating material). This algal-growth condition indicates an undesirably high level of nutrients (especially of nitrogen and phosphorus) called *eutrophication.* Large amounts of decomposable organic wastes, including the dead algae, result in high BOD and depleted oxygen levels in the water. (Courtesy of Raymond W. Miller, Utah State University.)

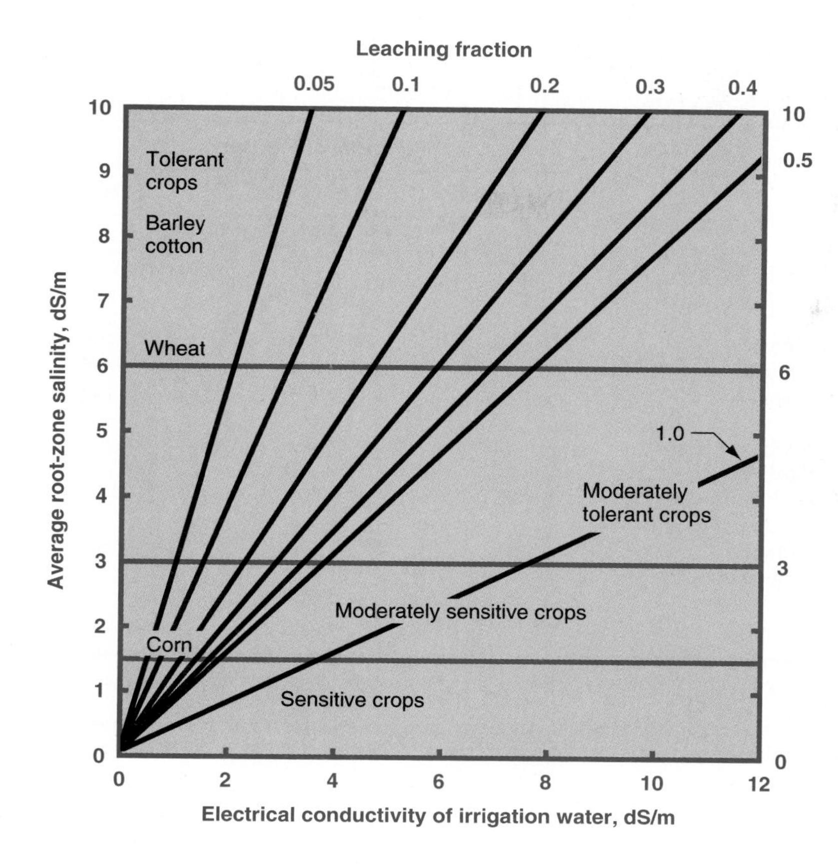

FIGURE 16-6 Chart for salt content in water for irrigation. The four categories vertically group water into *root-zone* water for *sensitive plants* (0–1.5 dS/m), *moderately sensitive* (1.5 dS/m–3.0 dS/m), *moderately tolerant* (3.0 dS/m–6.0 dS/m), and *tolerant plants* (6 dS/m–10 dS/m). The *leaching fraction* (*L*) lines separate the suitable water (to the left of each *L* line) from more hazardous water (to the right of each *L* line). The *L* line to use depends on the amount of leaching done each irrigation to replace root-zone water. If leaching was greater than the *L* = 0.5 shown, water higher in salt could be used. This chart emphasizes that management, frequency of irrigation, and the crop chosen determine the suitable salt contents in irrigation water. (Modified and redrawn from J. D. Rhoades, "Reclamation and Management of Salt-Affected Soils After Drainage," *Proceedings of the First Annual Western Provincial Conference Rationalization of Water and Soil Research and Management,* Lethbridge, Alberta, Canada, November 27–December 2, 1982; 123–197; and Parker F. Pratt and Donald L. Suarez, "Irrigation Water Quality Assessments," Chapter 11, in K. Tanji (ed.), *Agricultural Salinity Assessment and Management,* American Society of Civil Engineers, Publishers, 345 East 47th Street, New York, NY 10017, 1990, 504–529.)

2 dS/m in EC. Many drainage and groundwaters are in the range of 2 to 10 dS/m, but few plants can grow using irrigation waters of more than 10 dS/m EC.

In Figure 16-6 water quality is based on the idea that salt can be higher and the water still be usable if more leaching is done each irrigation. In the figure, salt contents to the left of the leaching fraction lines are EC values in irrigation water that should cause no yield reduction. Notice that as the leaching fraction increases, more salty water is suitable for a given crop without yield reduction. The technique to minimize a plant's damage from soluble salts is to keep the salt as dilute as possible and use the most salt-tolerant plants.

The salts in waters are comprised mostly of six ions: calcium, magnesium and sodium cations, and the anions of chloride, sulfate, and the carbonate–bicarbonate couplet. The proportions vary greatly, depending on the rocks and soils over and through which the water has flowed. The composition of several river waters is given in Table 16-7. Notice that some of these waters are quite salty, more than 2 dS/m. Also, most of them are in relatively dry areas, which makes it likely that they are saltier than humid climate rivers.

Whether a given water is usable or unsuitable for plants depends on the plant grown, the amount of leaching permitted during each irrigation, and how dry the soil is allowed to get before the next irrigation. If the soil is allowed to dry, salts will concentrate as water is lost or move to the surface with evaporating water instead of moving deeper into the soil by leaching. The more leaching that occurs and the wetter the soil is kept, the higher is the salt content in soil that can be tolerated by plants. One cannot say this water is okay just because its salt

Table 16-6 Guidelines for Irrigation Water Quality Established by the World Food and Agriculture Organization (FAO)

Water Constituent	Intensity of Problem[a]		
	No Problem	Moderate	Severe
Salinity (decisiemens meter^{-1})	<0.70	0.70–3.0	>3.0
Permeability (rate of infiltration affected)			
Salinity (decisiemens meter^{-1})	>0.5	0.5–0.2	<0.2
Adjusted SAR; soils are:			
Dominantly montmorillonite	<6	6–9	>9
Dominantly illite–vermiculite	<8	8–16	>16
Dominantly kaolinite–sesquioxides	<16	16–24	>24
Specific ion toxicity			
Sodium (as adjusted SAR) (sprinkler)	<3	3–9	>9
Chloride (mmol/L) (sprinkler)	<3	>3	>10
Boron (mmol/L)[b] as B	<0.70	0.70–3.0	>3.0
Miscellaneous			
NO_3^-—N or NH_4^+—N (mmol/L)	<5	5–30	>30
HCO_3^- (mmol/L) as damage by overhead sprinkler	<1.5	1.51–8.5	>8.5
pH	6.5–8.4		0–5, 9.5+

Source: Modified from R. S. Ayres and D. W. Westcott, "Water Quality for Agriculture," Irrigation and Drainage Paper 29, FAO, Rome, 1976; rev. 1986.

[a]Based on the assumptions that the soils are sandy loam to clay loams, have good drainage, are in arid to semiarid climates, that irrigation is sprinkler or surface, that root depths are normal for deep soil, and that the guidelines are only approximate.

[b]Assumes molecular weight = mole$_c$ weight (one charge) because it is slightly ionized or nonionized.

Table 16-7 Composition of Various River Waters

Property (ions in mmol/L)	Feather, Nicolaus CA	Grand, Wakpala SD	Missouri, Nebraska City, NE	Salt, Near Stewart Mtn., AZ	Sevier, Lynndyl UT	Pecos, Artesia NM
EC, dS/m	0.10	0.94	0.91	1.56	2.03	3.26
Ca^{++}	0.45	2.00	4.06	3.15	3.71	16.98
Mg^{++}	0.36	0.79	1.92	1.35	6.05	9.07
Na^+	0.20	7.08	3.02	9.62	10.62	11.38
K^+	0.04	0.19	0.10	0.17	0.15	0.08
HCO^-	0.86	6.29	3.24	3.21	5.21	3.11
Cl^-	0.08	0.19	1.78	10.12	9.31	12.13
SO^{2-}	0.16	3.43	4.05	0.89	5.96	22.39
SAR*	0.30	6.00	1.80	6.40	4.80	3.20

Source: Modified after J. D. Rhoades, A. Kandiah, and A. M. Mashali, "The Use of Saline Waters for Crop Production," *FAO Irrigation and Drainage Paper* **48,** Food and Agriculture Organization of the United Nations, Rome, 1992, 78.

*SAR = $Na/\sqrt{(Ca + Mg)}$, where all concentrations are in mmol/L.

content is relatively low. Unless adequate leaching occurs, salt will increase and salt problems will progressively get worse.

Sodium Hazard (Sodicity)

High concentrations of sodium are undesirable in water because sodium adsorbs onto the soil cation exchange sites, causing soil aggregates to break down (disperse), sealing the pores of the soil and making it less permeable to water flow. The tendency for sodium to increase its proportion on the cation exchange sites at the expense of other types of cations is estimated by the ratio of sodium content to the content of the square root of calcium plus magnesium in the water. This is called the **sodium adsorption ratio (SAR).** A small SAR value indicates a desirable low-sodium content.

The Food and Agricultural Organization (FAO) guideline (Table 16-6) refers to the sodicity problem as *permeability.* The guide includes salt concentrations because, at very low salt levels, the soil particle flocculation (which occurs with any high salt concentration) is lost and permeability decreases. Waters that are very low in salt (<0.2 dS m^{-1}) accentuate poor permeability.

The FAO guidelines recognize also that the problem of sodium is most severe with montmorillonitic soils and least with kaolinitic and sesquioxide (metal oxide) clays that have slight swelling. An adjusted SAR can be used to correct for the precipitation of calcium and magnesium with bicarbonate and carbonate ions in the water added. This calculation produces higher values for adjusted SAR than for the unadjusted SAR and a truer picture of the sodicity of the soil. Most adjusted SAR values of average waters are about 10 percent to 15 percent higher than the unadjusted SAR. Waters low in salt and high in SAR favor poor soil permeability eventually. Waters high or low in salt but with low SAR values favor good soil permeability.

The calculation of adjusted SAR has been changed from earlier procedures a decade ago. Although the calculation is still undergoing modification, it is complex and is now done by a computer model or by use of a number of tables. (Calculation of the adjusted SAR is not provided in this book.)[16]

Toxicities

Boron is the most commonly encountered element found in toxic concentrations in water, with threshold levels (maximum nontoxic levels) about 0.5 g/m^3 to 0.75 g/m^3 in water. Because it is quite soluble, boron is found in water where drainage and geologic strata supply boron source minerals. The problem of boron levels for plants is accentuated because the range between nutritionally deficient and toxic levels of boron is relatively narrow. Boron cannot be precipitated or otherwise easily removed from water. The only known remedy is to dilute high-boron water with low-boron water or to grow boron-tolerant crops.

Chloride and **bicarbonate** may cause toxicity. Many plants (avocado, tobacco, berries) are sensitive to high chloride concentrations—and sometimes to high **sodium** levels—in their leaves. Bicarbonates and carbonates promote precipitation of calcium as calcium carbonate (lime) during drying periods, resulting in a higher SAR in the water (higher sodium hazard) because of the lowered calcium content.

Long-term applications of toxic elements may become an additional restriction, eventually. One list recommends maximums that should be permitted with the idea that the metals tend to accumulate in soils and can become a growth hindrance to the plant or to the animal that consumes the plant.

---

[16] J. D. Rhoades, A. Kandiah, and A. M. Mashali, "The Use of Saline Waters for Crop Production," *FAO Irrigation and Drainage Paper* **48,** Food and Agriculture Organization of the United Nations, Rome, Italy, 1992, 52–67. The booklet states that a floppy disk of the model for adjusted SAR is available from FAO or from J. D. Rhoades (U.S. Salinity Laboratory, USDA, Riverside, CA).

In localized areas other elements toxic to plants or animals, such as lithium (California) and selenium (Wyoming), may contaminate water and require analysis for waters of those areas or kinds of uses.

16:4 Water Needs of Plants

Most commercial crops cannot store water to carry them through a dry period; plants need a continuous water supply. Any serious reduction in available water reduces plant growth to some extent. The growth reduction becomes more pronounced as the time of dryness increases, the rate of transpiration increases, and the rate of water movement from drying soil areas to root surfaces decreases (because thinner water films in soils are held with greater force). To irrigate efficiently it is necessary to know the amount of water needed by the chosen crop and the method of applying it that will provide the best results.

Consumptive use is increased by conditions that increase evaporation: warm days, dry air, wind if the atmosphere is dry, and maximum plant-available water in the soil. Some typical consumptive use (evapotranspiration, *ET*) values are given in Table 16-8; note that total water needs vary from year to year as climatic conditions vary. **Daily consumptive use** ranges from low values of about 2 mm (0.1 in) to a maximum of about 1 cm to 1.5 cm (0.4 in–0.6 in). For example, peak water use in midsummer in California is generalized as follows:[17]

Area	*Daily Water Use by Plants*
Coastal fog belt	0.25–0.38 cm (0.10–0.15 in)
Coastal valley	0.50–0.64 cm (0.20–0.25 in)
Interior valleys	0.64–0.76 cm (0.25–0.30 in)
Desert areas	0.64–1.02 cm (0.25–0.40 in)

Some additional ET values and conditions are given in Chapter 4.

16:5 Amount and Frequency of Irrigation

The objective of irrigation is to add the amount of water a plant needs when the plant needs it. The increased demands for water by cities and industry will pressure growers to use less water and to use that water as efficiently as possible; however, a water shortage reduces vegetative growth, forces premature seed production, or both. Some factors to be considered when planning the amount and frequency of irrigation are as follows:

- The depth and distribution of plant roots
- The amount of water retained within the rooting depth
- The minimum water potential to be maintained in the root zone
- The rate of water use by the plant (consumptive use)
- Whether or not adequate irrigation water is available to add when it is needed

Plant Root Systems

Rooting depths for any given plant are altered by shallow hardpans, dense soil, and clayey soil texture, as well as by the watering and tillage methods. The discussion that follows assumes deep, medium-textured soil without physical limitations for roots—almost ideal conditions.

---

[17] L. N. Brown and L. J. Booher, "Irrigation on Steep Land," California Agricultural Experiment Station—Extension Service Circular 561, 1972.

Table 16-8 Selected Total Evapotranspiration Values for Crop Season or Crop Year Growth When the Crop Is Well Watered*

Crop	*Location*	*Crop Duration*	***Evapotranspiration*** *in*	*mm*
Alfalfa	North Dakota	143 days (summer)	23.4	594
	Nevada	124 days (summer)	39.9	1013
Grass	Canada	—	22.8	579
	Davis, Calif.	12 months	51.8	1316
Barley	Wyoming	May–Aug.	15.2	386
	Mesa, Ariz.	Dec.–May	25.3	643
Beans	South Dakota	105 days	16.4	417
	Davis, Calif.	92 days	15.9	404
Corn	Ohio	124 days	18.5	470
	Bushland, Tex.	122 days	24.3	617
Potatoes	Alberta, Can.	—	19.9	505
	Phoenix, Ariz.	Feb.–June	24.3	617
Rice, flooded	Davis, Calif.	150 days	36.2	919
Sorghum	Kansas	—	21.7	551
	Mesa, Ariz.	July–Nov.	25.4	645
Wheat, hard	South Dakota	—	16.3	414
	Bushland, Tex.	Oct.–June	28.3	719
Sugar beets	Montana	Apr.–Sept.	22.5	571
	Kansas	Apr.–Nov.	36.5	927
Safflower	Southern Idaho	Apr.–Sept.	25.0	635
Soybeans	South Dakota	—	15.7	399
Cotton	Arvin, Calif.	12 months	35.9	912
	Mesa, Ariz.	Apr.–Nov.	41.2	1046
Cabbage, late	Mesa, Ariz.	Sept.–Mar.	24.9	632
Lettuce	Mesa, Ariz.	Sept.–Dec.	8.5	216
Peas, green	Alberta, Can.	—	13.4	340
Tomatoes	Alberta, Can.	—	14.4	366
	Davis, Calif.	May–Oct.	26.8	689
Apples	Wenatchee, Wash.	Apr.–Nov.	41.7	1059
Oranges	Phoenix, Ariz.	12 months	39.1	993
Turf	Reno, Nevada	112 days	21.8	554

Source: Selected and modified from a tabulation by Marvin E. Jensen, ed., Consumptive Use of Water and Irrigation Water Requirements, American Society of Civil Engineers, New York, 1973.

*Notice that values for a given crop usually increase from cooler Northern areas to warmer Southern regions. Notice also that in the table the crop durations and times of the year vary considerably.

The effective rooting depth for numerous plants is given by showing the suggested irrigation depth for some crops (Figure 16-7). To understand the importance of keeping the upper soil layers moist, a rule of thumb for water uptake by plants is:

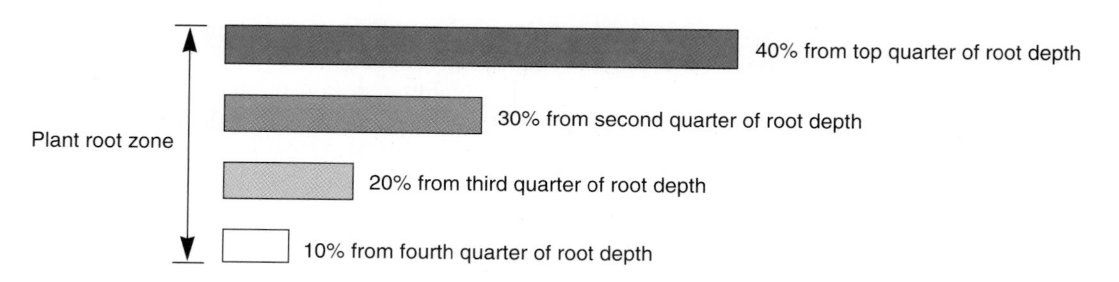

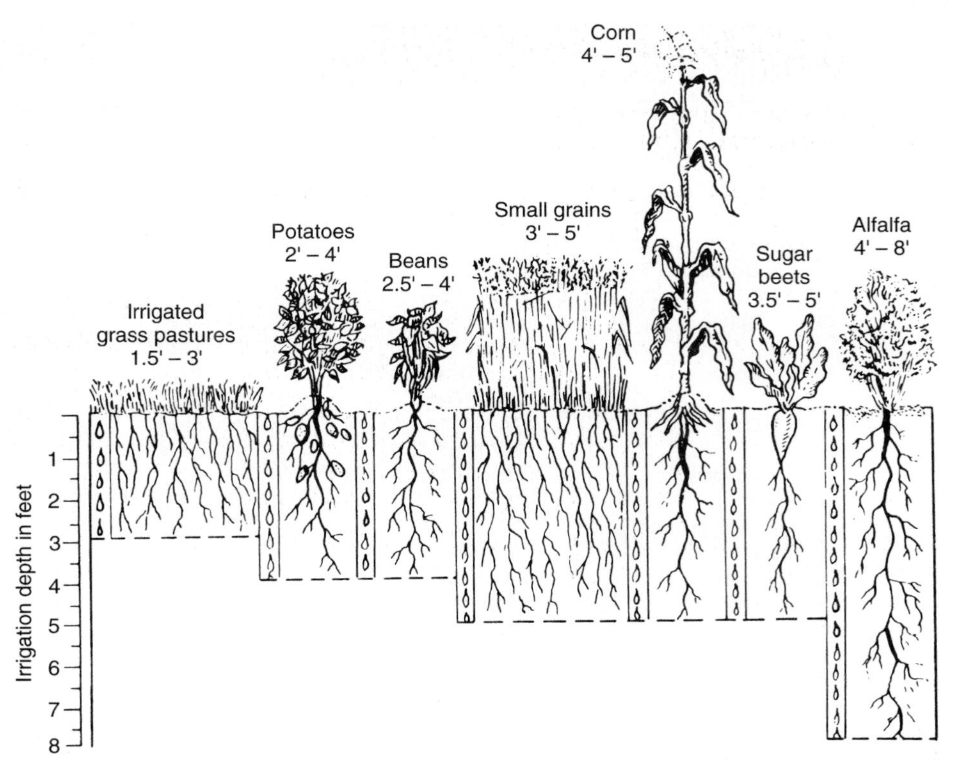

FIGURE 16-7 Normal irrigation depth for common farm crops varies from 1½ ft for some grass pastures to as much as 8 ft for alfalfa. Some not shown are carrots and peas (2–3 ft), cantaloupe (3–5 ft), grapes (5 ft), and tomatoes (3–5 ft). Feet × 30.5 = cm. (*Source:* Irrigation on Western Farms, USDA Agricultural Information Bulletin 199, 1959.)

Obviously, most roots and the uptake of most water exist in shallow soil depths. Allowing the surface 30 cm to 60 cm (1 ft–2 ft) of soil to dry will greatly limit the plant uptake of water and nutrients.

Minimum Water Potential

Although plants can readily use soil water held with water potentials approaching −1500 kPa (permanent wilting point), most of the plant-available water is held in high potentials, from −33 kPa to −100 kPa (or −10 kPa to −100 kPa for sands). Irrigation is recommended when about 50 percent of plant-available water has been used in the zone of maximum root activity, probably within the 15 cm to 60 cm (6 in–24 in) depth (Figure 16-8). Except in ripening seed crops or sugar crops, the moisture content in soil should never approach the permanent wilting point unless rapid maturing is needed.

In Arizona—one of the high-yielding, cotton-producing states—cotton yields were increased 25 percent by irrigating four times in July rather than the usual two times.[18] The increase in yield is thought to be due to less stress to the plant during its critical boll-producing

[18] Dennis Senft, "Extra Irrigations Protect Cotton," *Agricultural Research* **40** (1992), No. 2; 9.

Extended captions and credits to photographs are in Appendix B.

Plate 42 Applying solution potassium as a variable rate for precision. (USDA—ARS)

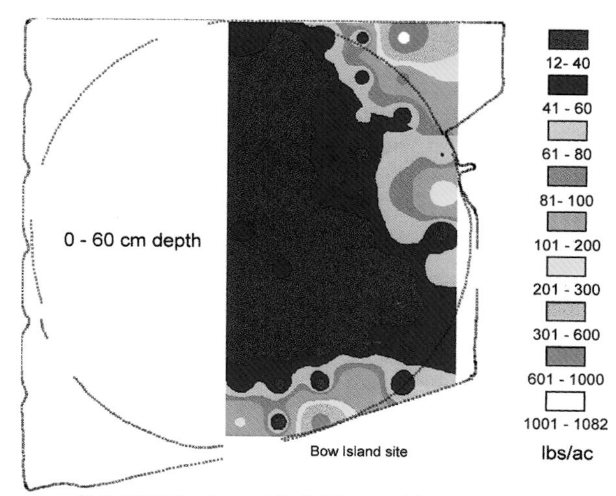

Plate 43 Mapped soil nitrogen variation in Canada. (C. McKenzie and PPI)

Plate 44 How do you locate yourself on land without good landmarks? (B. E. Frazier)

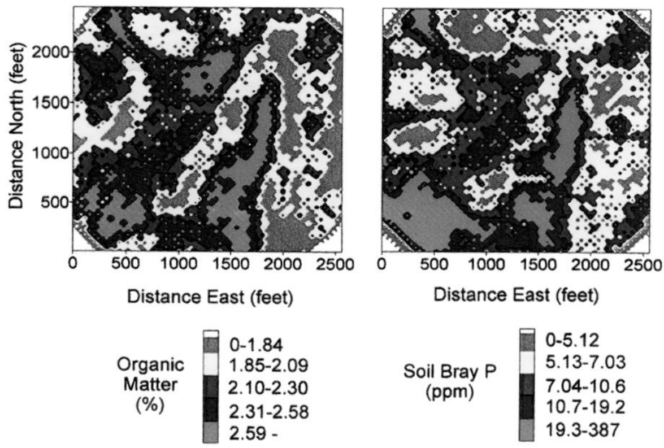

Plate 45 Variation in organic matter and Bray-P, Nebraska. (T. Blackmer, et al. and PPI)

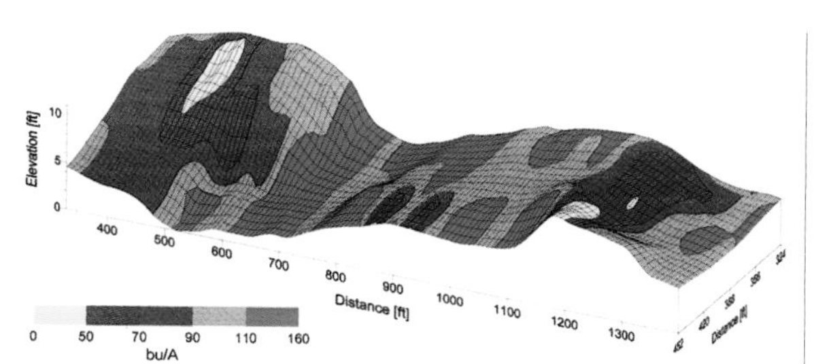

Plate 46 Mapped corn yield on land in Canada (G. Kachanoski, et al. and PPI)

Plate 47 Aerial infrared shot over Texas. In IR, green shows red. (Brantwood Publ.)

Extended captions and credits to photographs are in Appendix B.

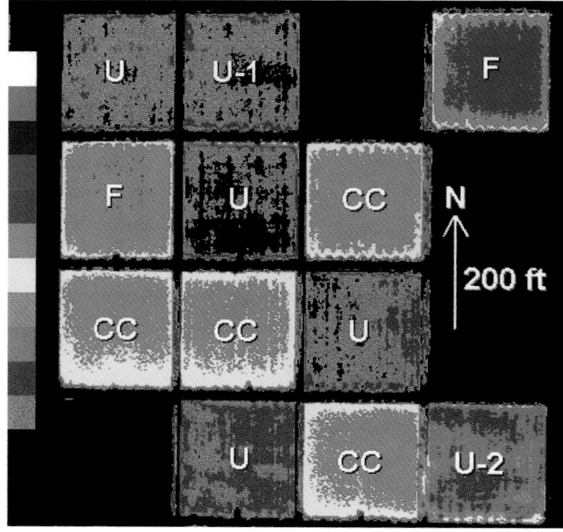

Plate 48 Infrared aerial photos show field areas with good and poor plant growth which help to know where to look for differences. **Left,** uniformly treated sudangrass; **right,** upper right one-fourth of left photo with various fertilizer treatments. Growth is visibly different. (R. F. Denison)

Plate 49 Pedestals: surface rocks protect the soil beneath them, Australia. (A. R. Southard)

Plate 50 *Hard-setting surfaces* slow erosion of soft material beneath. (A. R. Southard)

Plate 51 Extensive erosion on a Natrustalf, a high-Na soil, Australia. (A. R. Southard)

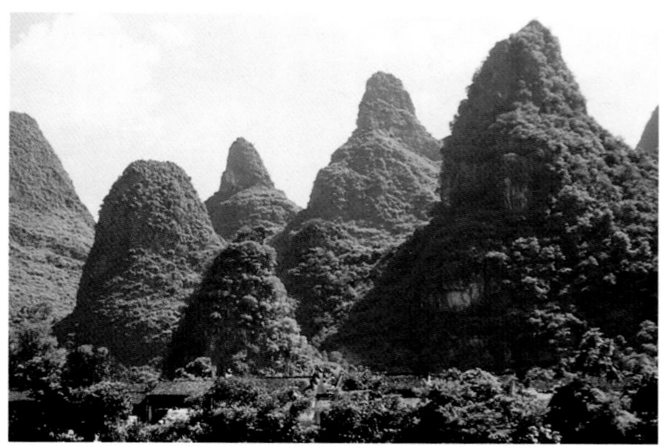

Plate 52 Limestone pinnacles from erosion along Li River, China. (A. R. Southard)

Extended captions and credits to photographs are in Appendix B.

Plate 53 Severe gulley erosion in New Mexico. See columnar structure. (A. R. Southard)

Plate 54 Machu Picchu, Peru, Inca terracing to control water and erosion. (R. W. Miller)

Plate 55 Greasewood and salt (summer snow) cover this Nevada soil. (A. R. Southard)

Plate 56 Controlling erosion with grass covered steep-back terraces. A lot of cost but a lot of control; Missouri. (USDA—NRSC)

Plate 57 Mulch on the soil helps control soil erosion and raindrop splash. (Yetter Mftg Co)

Extended captions and credits to photographs are in Appendix B.

Plate 58 Even tractors can be powered by electricity. **Left,** tractor pulling 7000 lb trailer up slope. **Right,** solar cells on barn roof recharge batteries overnight. (S. Heckeroth)

Plate 59 Center pivot sprinklers in Oregon. Over 100 circles are computer-operated automatically. (USDA—ARS)

Plate 60 Erosion cannot be controlled when cultivating slopes like these in the Andes of Venezuela. (R. W. Miller)

Extended captions and credits to photographs are in Appendix B.

Plate 61 Peat in Vancouver, water table 1 m; vacuum dry top for peat moss. (R. W. Miller)

Plate 62 Dredged and filled boater home sites in Florida, near sea level. (EPA—Documerica)

Plate 63 Greenhouse-style waste water treatment system called Solar Aquatics™. (S. Peterson)

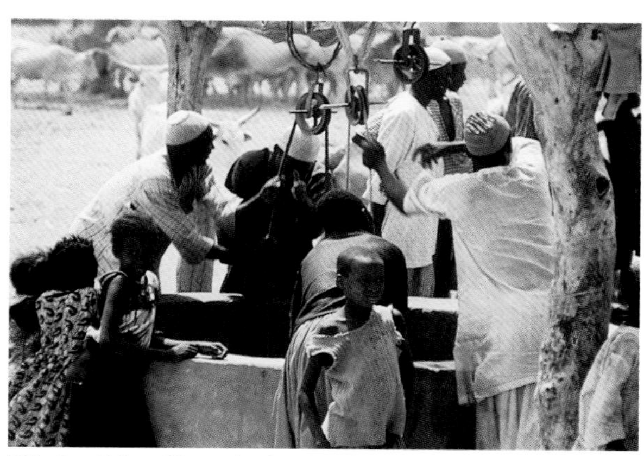

Plate 64 One help for better water—a new well in The Gambia, Africa. (R. W. Miller)

Plate 65 Kennecott's copper mine waste is all in tailings ponds now. (R. W. Miller)

Plate 66 Dairy farm on poorly drained land in Wisconsin. (USDA—NRSC)

Extended captions and credits to photographs are in Appendix B.

Plate 67 Thin volcanic ash over white lime, Montana. (A. R. Southard)

Plate 68 In Pawnee, Colorado, an undisturbed soil core for rangeland studies. (Courtesy of Colorado State University)

Plate 69 Ice chunks in a city park in New Brunswick, New Jersey. (Courtesy of R. W. Miller)

Plate 70 Windbreaks protect against fierce winds on the plateaus in Nebraska. (USDA—NRCS)

Plate 71 Beauty is found in the flowers grown in soil or in soilless culture. (E. Israelson)

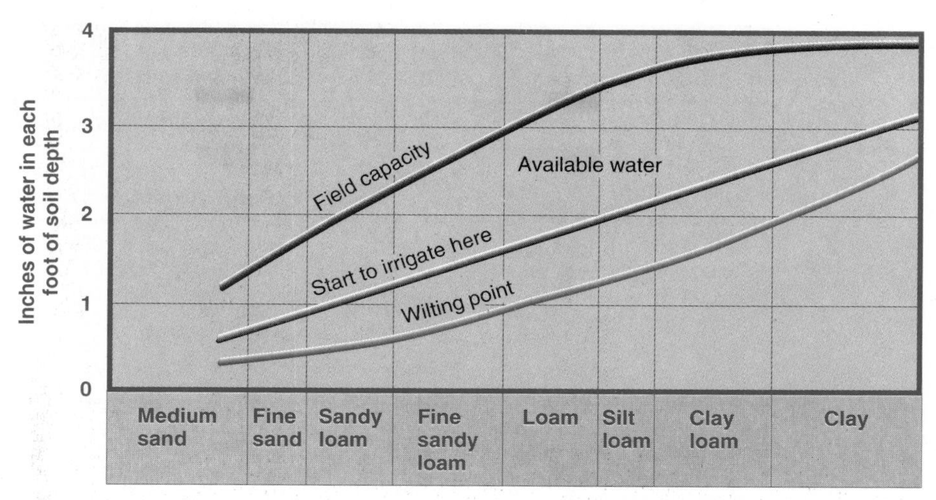

FIGURE 16-8 The amount and frequency of irrigation are determined partly by soil texture. A silt loam soil has the largest available water capacity and, therefore, should require irrigation water less frequently than a sand or a clay soil. Feet × 30.5 = cm. (*Source:* "Water," *Yearbook of Agriculture,* USDA, Washington, DC, 1955.)

stage of growth. Approximately the same amount of total water for July was used, but only one-fourth was added per irrigation in each of the four irrigations. The frequent irrigations kept wet more of the time the surface soil, where most roots and nutrients are located. Cotton plants under water stress shed some fruits. Water stress may also cause some root die-back during boll loading, again reducing water uptake.

Calculating When to Irrigate

The objective of irrigation is to keep adequate water available to crops or, if water is in short supply, to use what is available most effectively. The most critical time to keep water available for fruits, nuts, grains, or cotton is the several weeks following flowering. If the desired yield is vegetative growth rather than fruit, the critical period may be the latter half of the growth period prior to flowering, when most *size* growth occurs.

The method of calculating CU or ET from climatic data is too complex for a short discussion, but Table 16-9 indicates how the calculated data can vary if the equation used is not calibrated with measured data for the area. For example, the Thornthwaite equation in Table 16-9 gives values that overestimate ET in Copenhagen, Denmark, by 36 percent and underestimates it in Kimberly, Idaho, by 58 percent.

In comparison to estimating CU or ET from climatic-data formulas, the use of the weather station class-A pan evaporation to estimate ET is quite accurate and simple.[19] The assumption is that all effects of climate on ET (temperature, wind, relative humidity, day length, etc.) act similarly on water in a pan and on water in crops. After measuring loss from a pan, the next requirement is to correlate the pan loss to actual previously measured losses for that crop, then use the pan water loss relationship thereafter to figure ET values for the

[19] G. W. Bloemen, "A High-Accuracy Recording Pan-Evaporimeter and Some of Its Possibilities," *Journal of Hydrology* **39** (1978), 159–173.

Table 16-9 Estimated Potential *CU* or *ET* Values as a Percentage of Actual Measured *ET* for Various Locations by Several Widely Used Equations[a]

Location	Equation for Estimating CU or ET[b]				
	Th	*Pen*	*J-H*	*B-C*	*C-H*
	(calculated ET as % of measured ET)				
Aspendale, Austria	64	133	69	75	91
Copenhagen, Denmark	136	130	66	135	113
Ruzizi, Zaire	66	87	108	82	69
Brawley, California	60	106	102	85	82
Kimberly, Idaho	42	86	76	56	66
Coshocton, Ohio	65	89	80	81	81

Source: Selected data from M. E. Jensen, ed., *Consumptive Use of Water and Irrigation Water Requirements,* American Society of Civil Engineers, New York, 1974.

[a]The table illustrates that such equations should not be used unless it is established that they are suitable to the area where data are wanted.

[b]Abbreviations for equation names are: *Th* = Thornthwaite, *Pen* = Penman, *J-H* = Jensen-Haise, *B-C* = Blaney-Criddle, *C-H* = Christiansen-Hargreaves.

crop. Examples are given in Calculations 16-1 and 16-2. Many models are available for computer calculations and determination of irrigation timing.

Irrigation timing by most commercial agriculturalists is planned based on (1) an art learned over the years, (2) the use of simple measurements, or (3) careful measurements estimating consumptive use.

The first method, the art of irrigation learned by experience, can be effective in determining watering timing, but extra water beyond minimum need is usually applied. The irrigator looks at early signs of dryness: darkened foliage color, slight temporary wilts in the hot afternoon, and the wetness of the soil down to 30 cm to 40 cm (12 in–16 in).

The second method, the use of simple measurements in the field, determines irrigation need and timing by the measurement of soil water with tensiometer or gravimetric soil water measurements. When the moisture readings of instruments reach a predetermined value (e.g., −50 kPa or other water-potential value), irrigation is needed. The gauge of a moisture sensor can be wired to trip an automatic sprinkler or other device for automatic watering systems. Tensiometers have been successfully used for many situations, but they are most useful in crops where water content is kept relatively high.

The third method, called the **water budget method,** requires consumptive use data; it is a scientific approach to detailed scheduling. Both the time for irrigation and the estimated amount of water to add are calculated by measuring both the amount of water the soil holds and how much of that water has been used. When the soil reaches a predetermined dryness, irrigation is recommended. The major difficulty is in determining the consumptive use (CU) or evapotranspiration (ET). (These are almost the same amount.) The following two procedures are used:

- **Measuring the water lost from a free water surface:** Such as loss from a weather station class-A pan.
- **Calculating CU or ET values from climate:** Solar radiation, average temperatures, and other factors such as relative humidity and wind. Many equations are used. No one formula is suitable for all climatic areas[20] (see Table 16-9).

[20]R. J. Hanks and G. L. Ashcroft, *Applied Soil Physics,* Unit 4: "Soil-Plant-Atmosphere Relations," Springer-Verlag, New York, 1980, 99–124.

Calculation 16-1 Calculation of Irrigation Frequency*

Problem If a deciduous orchard is to be irrigated when half of the available water in the top 76 cm (30 in) is used up, how often must the orchard be irrigated during August when the class-A pan is evaporating 5.6 cm (2.2 in) of water per week? The clay loam soil holds 5.0 cm (2.0 in) of plant-available water in the top foot of soil and 4.3 cm (1.7 in) of plant-available water in each additional foot of subsoil.

Solution

1. First, the total plant-available water in the top 76 cm (30 in), when the soil is wetted, is 5.0 cm (in the top 30 cm) plus 4.32 cm (in each 30 cm of soil). Thus, 5.0 cm for the top 30 cm plus 6.62 cm in the next 46 cm equals 11.6 cm (4.52 in).
2. Only half of the plant-available water can be used before irrigating again. Half of 11.6 cm is 5.8 cm (2.26 in).
3. The last problem to solve is how fast the water is used. The class-A pan loses 5.6 cm (2.18 in) per week or 0.80 cm (0.312 in) per day.

 Referring to Table 6-4, deciduous orchards in August are seen to have a consumptive use estimated to be 65% as high as the class-A pan evaporation. So the daily use is as follows:

 (pan loss)(percentage by plant)

 = 0.08 cm (0.65)

 = 0.52 cm used by the orchard per day (= 0.20 in)

4. The final step is to see how long the 5.8 cm (2.26 in) in the soil that can be used (see step 2) will last if 0.52 cm (0.20 in) per day is used.

$$\frac{5.8 \text{ cm total water}}{0.52 \text{ cm water per day}} = 11.2 \text{ days}$$

 Irrigation should be about every 11 or 12 days.

The methods actually used by laboratories and consulting specialists are more complex—some are computerized, and all take into account any added rainfall when consumptive use (CU) is calculated.

*Further reading: E. C. Stegman, "Microcomputer Applications to Irrigation System Management," *North Dakota Farm Research* **43** (Nov.–Dec. 1985).

16:6 Methods of Applying Water

Once the seasonal and daily water use is determined, the next question is how to apply the required amount of water most effectively. Irrigation water can be applied by methods ranging from haphazard flooding to enormous center-pivot mobile sprinklers that cover 53 hectares (130 acres) in one circular sweep (Figure 16-9). The use of sprinklers and, more recently, trickle (drip) irrigation has made it possible to irrigate almost all arable soils, even those of rolling hills, sands, and steep slopes. However, making irrigation possible does not necessarily make it economical, practical, or even desirable.

Border-Strip and Check-Basin Irrigation

The **border-strip** method of irrigation is illustrated in Figures 16-10 and 16-11. In **border irrigation** soil ridges keep the water flowing down a strip of land. Border irrigation works well if the soil surface is level in the direction perpendicular to water flow and the slope in the direction of water flow is gentle. Percolation losses will occur in the intake end of the

Calculation 16-2 Simplified Scientific Irrigation Scheduling

The use by most farmers of scientific methods to determine timing and quantity of irrigation requires that these methods be convenient. Complex equipment (expensive computers, weather stations, etc.) or formulas that require frequent recalculation to avoid errors are not useful for fieldwork. However, local areas may be able to establish and use the simple systems to be described here.

Having no summer rainfall, the San Joaquin Valley of California has a yearly summer climate that seldom deviates more than about 10 percent from the average summer climate. For each crop and planting date a certain soil can be programmed for irrigation in the manner indicated in the graph.* The program can be set up from a master plan created months ahead. These data are required for the initial plan: (1) class-A pan evaporation data for the area, (2) the crop coefficients (see Table 6-4), (3) the crop *ET* value, (4) the soil's depth and water-holding capacity, (5) the crop sensitivity to water stress, which determines allowable depletions during each irrigation cycle, and (6) any modifications based on the cultural practices used. The result is a graph (such as that shown) for each crop, each soil, and each planting date, all available in this convenient form before planting is ever started.

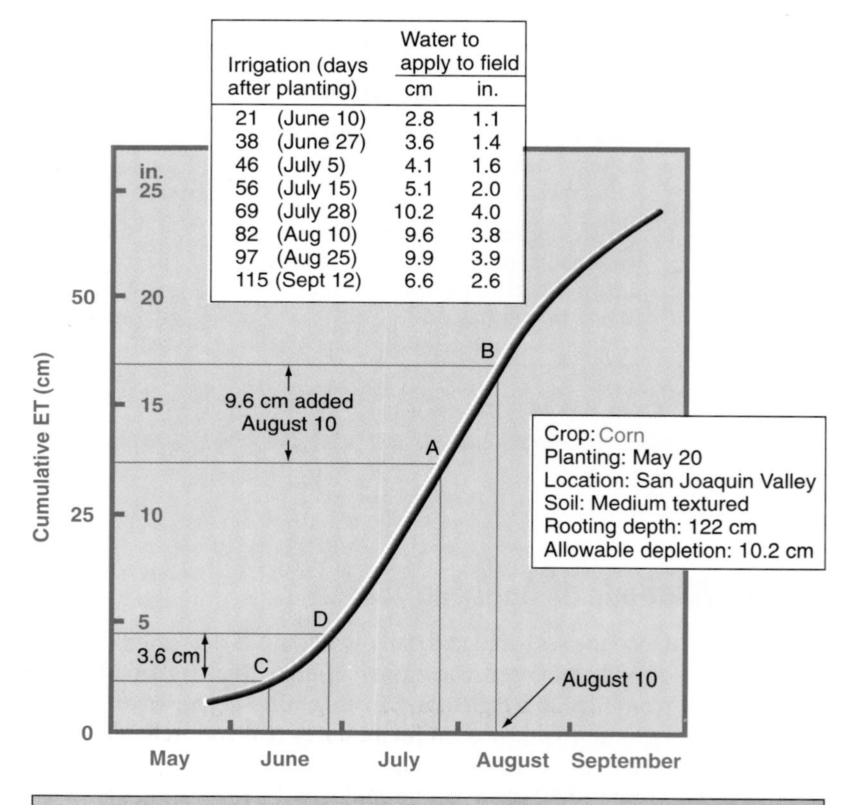

Irrigation (days after planting)	Water to apply to field	
	cm	in.
21 (June 10)	2.8	1.1
38 (June 27)	3.6	1.4
46 (July 5)	4.1	1.6
56 (July 15)	5.1	2.0
69 (July 28)	10.2	4.0
82 (Aug 10)	9.6	3.8
97 (Aug 25)	9.9	3.9
115 (Sept 12)	6.6	2.6

Notice that irrigation is scheduled during early growth although only 3 centimeters of water is needed to wet the soil. This is because of the shallow roots of the corn in the seedling stage.

*Redrawn by Raymond Miller from Elias Fereres, Patricia M. Kitlas, Richard E. Goldfien, William O. Pruit, and Robert M. Hagan, "Simplified but Scientific Irrigation Scheduling," *California Agriculture* **35** (1981), Nos. 5–6; 19–21.

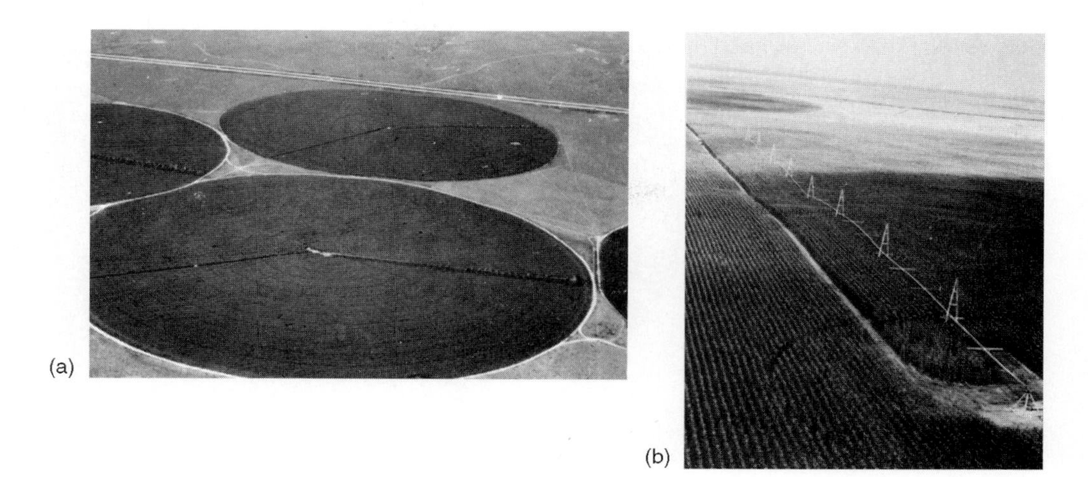

FIGURE 16-9 A typical center-pivot sprinkler system is shown in these photos, ranging from a high aerial view (a), to a nearer view (b), to a close-up of the mobile unit (c). One unit covers 53 ha (131 a), nearly ¼ mi^2. Water is supplied by a well at the center pivot. An automatic fertilizer injection system is sometimes located at the center-pivot area. (*Source:* USDA—Agricultural Research Service, Ft. Collins, CO; photos by Dale F. Heermann.)

field, and runoff is usually appreciable. Loss of 20 percent to 45 percent of applied water is common in border-irrigated fields when runoff water is not reused.

Basin irrigation (creating soil-ridged basins to hold water) can be quite efficient and has been widely used for pastures and for orchards where each tree is within its own check or basin (Figure 16-12). Basin irrigation is not suitable for highly permeable soils (sands, organic soils), for irregularly sloped land, for crops harmed by temporary flooding (tomatoes, beans, corn in early stages, lettuce), or for very slowly permeable soil (clays).

Furrow Irrigation

Furrow irrigation, including small, close furrows called **corrugations,** is the oldest form of irrigation. In this method water flows by gravity from a main ditch and down each furrow. The crop is usually planted atop the ridges before water is applied. About 40 percent of all irrigated land is furrow irrigated.

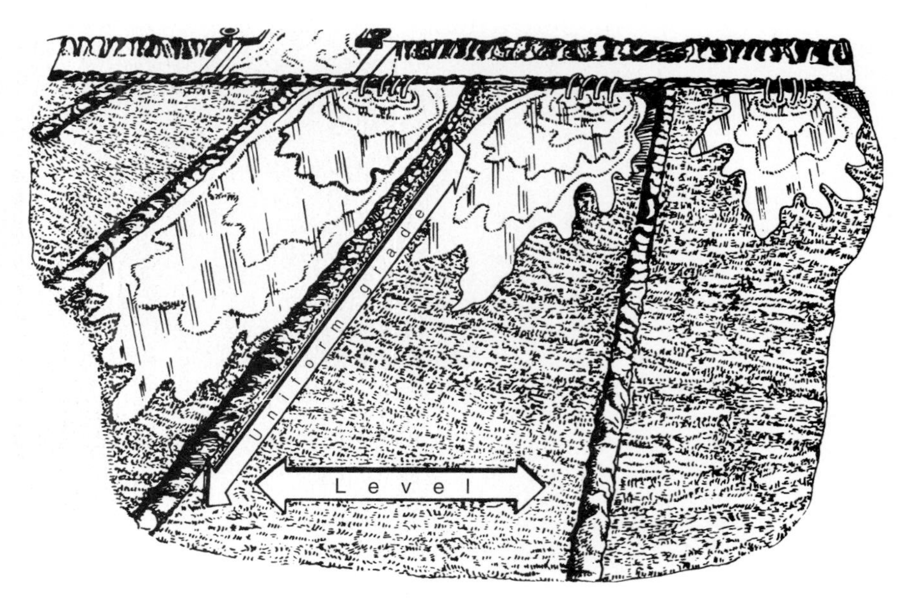

FIGURE 16-10 The field has been prepared for border irrigation by building small levees around each leveled area; then the areas are flooded to irrigate them. (*Source:* USDA.)

FIGURE 16-11 Border irrigation of pecan trees near Fabens, Texas. This type of irrigation requires large streams of water, a soil that is not permeable too rapidly, and a nearly level land surface perpendicular to the direction of water flow. (*Source:* U.S. Department of the Interior—Bureau of Reclamation; photo by H. L. Personius.)

Field crops such as corn and cotton have a furrow to carry water between all planted rows. Crops that are planted in double rows or beds (wide enough for two or more crop rows between furrows) are irrigated by directing the water between the beds. Crops planted in a wide spacing (berries, grapes, orchards) usually have two furrows for irrigation between adjacent rows of plants.

One problem associated with all surface flow methods of irrigation is *deep percolation and runoff water loss.* Deep percolation occurs at the head of the field. To reduce the deep percolation, the full length of the furrow needs to be wet quickly so that soaking will be nearly the same duration over the full length of the field. Water runs down the furrow faster if the

FIGURE 16-12 Where a field slopes in more than one direction, basin flooding can still be done satisfactorily by making the ridges of the check basins (dikes) on the land contour. Each tree or several trees are enclosed by a small dike. As each dike is filled with water, it is cut to allow water to flow into the next lower dike. Stanislaus County, California. (*Source:* USDA—Soil Conservation Service.)

furrow is smooth rather than rough. Some irrigators in California have dragged torpedoes (25-cm-diameter steel cylinders, coned at the front end, filled with cement, and about a meter long) down nonwheel-track rows to smooth the furrow.[21] The most furrow smoothing was in the cloddy, clayey soil. If the soil is moist, torpedoes slicken the furrow more than if the soil is dry; the slicking might even partially seal the soil. Water advance rates increased about 15 percent to 30 percent in the studied fields.

In general, soil erosion is excessive when the furrow method of irrigation is used on rows that have a slope of more than 2 percent; ideally, the slope of the furrows should be less than 0.25 percent. However, on erosive slopes in Idaho, Kentucky bluegrass was established in each furrow. This reduced erosion to a minimum (Figure 16-13).

The goal of the irrigator should be to obtain the maximum flow of water down each furrow without causing excessive erosion. In this way water will soak into the soil at a fairly uniform rate all along the furrow instead of wetting the soil at the upper ends of the rows deeply and wetting the lower ends to shallow depths. Water should soak two to three times as long as the time required to initially wet the entire row length. For example, if it takes 1 hour to wet the lower end of a row, the water should be run another hour for sandy loam, or up to 3 more hours for soil as fine as clay loam. To reduce water loss by runoff, the amount of water running down the furrows should be reduced when the water reaches the lower end of the furrow.

Following are several common methods of controlling the distribution of water in surface irrigation:

- Large lateral ditches across the field with smaller equalizing ditches leading directly to each furrow or border
- Large lateral ditches with siphon tubes leading to each furrow (Figure 16-14)
- Field lateral ditch with spiles (small straight pipes) leading directly through the bank to each row or border

[21]Lawrence J. Schwanki, Blaine R. Hanson, and Anthanosios Panoras, "Furrow Torpedoes Improve Water Irrigation Advance," *California Agriculture* **46** (1992), No. 6; 15–17.

FIGURE 16-13 Erosion in furrows during furrow irrigation was controlled by planting and maintaining Kentucky bluegrass in them. Beans are on the ridges, but corn, wheat, and barley were also grown satisfactorily. However, sugar beet yields with grass in the furrows were not satisfactory (Idaho). (Courtesy of John Cary, USDA—Agricultural Research Service, Kimberly, ID; used with permission.)

FIGURE 16-14 Furrow irrigation of cotton, showing plastic siphon tubes in use for moving water from the main ditch to each furrow. Notice the temporary dam in the lower right to keep the water level where siphons are set to a nearly uniform elevation for more equal water flows in each furrow. (Courtesy of Drue W. Dunn, Oklahoma Extension Service.)

FIGURE 16-15 Irrigation water is transported by pipe to the field where large openings in the pipe (gates) occur at each furrow. Note the canvas (or plastic) sleeves that lead the water into the furrows (at arrows) without causing excessive erosion. (Courtesy of Atto C. Wilke, Texas Agricultural Experiment Station, Lubbock.)

- Irrigation pipe with large openings (gates) emptying into each furrow (Figure 16-15)
- Buried pipe to carry the water to the field, with risers (vertical pipe outlets) emptying into each furrow or series of furrows

Surge Flow Surface Irrigation[22]

The low water efficiency of surface flow irrigation has encouraged more use of sprinkler and drip systems. Now a technique of automated surface flow, called **surge flow,** may help those farmers who prefer surface furrow irrigation. Surge flow delivers water intermittently. Larger flows for short on–off periods have long been known to wet a longer furrow distance more quickly than the same amount of water added continuously at a smaller flow rate. For example, applying 100 gallons of water during 100 minutes produced the following results when applied in different patterns:

Flow Rate (gal/min)	*Time Flowing*	*Furrow Distance Wetted (ft)*
10	Continuously	240
15	⅔ of time	350
20	½ of time	490
30	⅓ of time	600

Surge intervals may be from 1 minute to 10 minutes or more (1 minute on, 1 minute off).

The reason surge flow works is that the first short burst of water smoothes the furrow and breaks down clods. Second, the surface soil particles reorient and partially seal the surface, reducing intake. Third, in swelling clay soils the cracks may close during the *off* period, allowing greater water delivery further down the furrow. Thus, water runs over the wetted portion faster the second time and allows the entire furrow to wet before large amounts infiltrate at the upper part of the field. Programmable controllers can be modified at any time to adjust the frequency or flow rate to reduce tailwater runoff. The surge valve is used on gated pipe (level and sloping), concrete-lined ditches, and border strips. Costs in 1995 were about $30 to $125 per hectare.[23]

[22]A. Alvin Bishop, "Surge Flow," *Crops and Soils Magazine* **33** (1980), No. 2; 13–16.
[23]Jay A. Belt, "Surge Offers Hope for Surface Irrigation Efficiency," *Irrigation Journal* **43** (1993), No. 2; 16, 18, 20, 22–23.

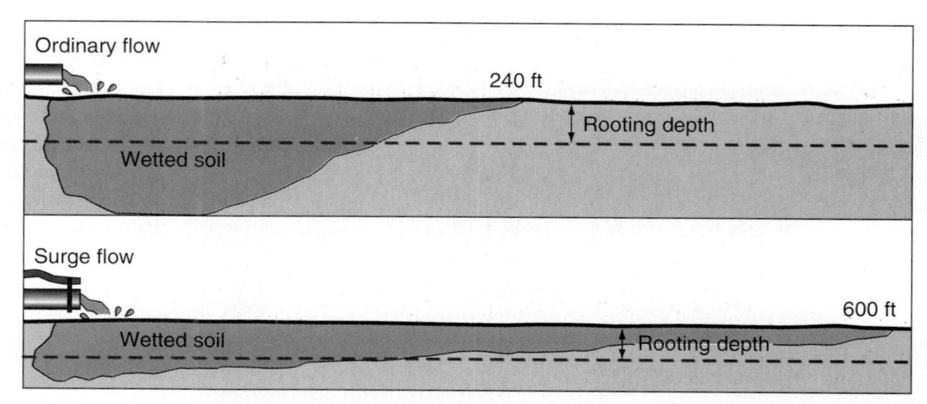

FIGURE 16-16 The same quantity of water applied intermittently, but in larger flow rates, wets farther down the irrigation row but lessens deep percolation loss at the upper end compared with continuous flow of an equal amount of water for the same time period. On susceptible soils, the larger heads of water may increase on-field erosion. This method of intermittently adding water is called *surge flow.* (Courtesy of Raymond W. Miller, Utah State University.)

The advantage of surge flow is in minimizing the deep percolation of water at the head of the furrow in order to wet the soil at the end (Figure 16-16). Although it is impossible to get uniform wetting from top to bottom, the surge flow method greatly increases efficient water use. The automated system involves gated pipe valves mechanically adjustable for flow rate by opening and closing pneumatically (or by other mechanisms) for surge flow. Air pressure to run the pneumatic valves is produced with a small compressor.

Cablegation is a surface-flow-controlled pipe system that is placed on grade and delivers water out plastic distribution tubes. Timing the opening and closing of outlets as the pipe plug moves down the pipe is predetermined. The operation is done with a fishing reel-like cable that slowly opens new outlets at one end as it slowly slides down the pipe reducing and closing on the longest-flowing uppermost outlets on the end as irrigation is completed. The number of furrows irrigated and the flow times and rates are adjustable (Figure 16-17). Cablegation is more efficient than siphon tubes and is usable in some situations not suitable to surge flow.

Sprinkler Methods

Sprinkler systems are the overhead application of water in simulated rainfall. Sprinklers are able to irrigate both normal lands and the unlevel and sandy lands—neither of which is readily irrigated otherwise except by drip irrigation—(Figures 16-18 and 16-19).[24]

Sprinkler systems have decided advantages over surface-flow methods. More uniform application of water is usually possible than with surface irrigation (except in high winds). In California some crops are given a light sprinkler irrigation just after planting to germinate the seed; then surface irrigation is used the rest of the season. With sprinklers the amount of water added can be calculated and applied with great precision and usually at the rate desired; watering ditches are eliminated, the physical labor is usually less, and fertilizer can be applied simultaneously through the system. For example, on the sandhills

[24] J. W. Cary, "Irrigating Row Crops from Sod Furrows to Reduce Erosion," *Soil Science Society of America Journal* **50** (1986), 1299–1302.

FIGURE 16-17 Cablegation is a surface-flow irrigation scheme that uses a traveling plug, which allows upslope furrows to reduce flow as it allows water to flow out new downslope gates. The plug is attached by cable to a manual or automated control that lets the cable out at a predetermined rate. (Courtesy of Harry L. Manges, "Automating Surface Irrigation Today, and Tomorrow," *Irrigation Journal* **44** (1994), No. 1; 18–20, 22–23.)

FIGURE 16-18 Portable or solid-set systems, similar to the one shown, are the least expensive sprinkler systems. After each area is wetted, portable systems must be moved by hand to a new location. This system near Morganfield, Kentucky, applies 2.5 cm (1 in) of water per hour, pumped from a 3.0-ha (7.5-a) pond containing water accumulated from a 21.9-ha (54-a) watershed. The pond, at upper left, is also used for recreation. Grassed waterways reduce the amount of sediment carried into the pond. (*Source:* USDA—Soil Conservation Service.)

(a)

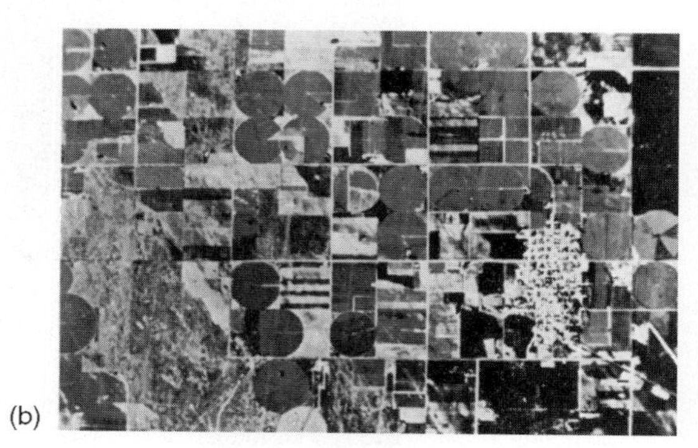

(b)

FIGURE 16-19 (a) Many automatic-drive sprinkler systems are available. This one, photographed from the outer end, is a center-pivot system near Klamath Falls, Oregon, irrigating wheat. The entire circle of 51.8 ha (128 a) can be irrigated in either 11 or 22 hours. The system has 14 tower stations, each with its own drive system. The spray is a heavy mist. This system now costs over $70,000. (b) The extent of center-pivot systems is indicated in this space satellite photo over Texas. Each circle is a 56.7-ha (140-a) area. Such scenes are also common over Nebraska and neighboring states. (*Sources:* (a) USDA—Soil Conservation Service; (b) Brantwood Publications, cover of *Irrigation Journal* [July–Aug. 1974]. Price updated.)

of Nebraska, fertilizer is applied in small regular increments as the crop needs it to prevent large leaching losses. Yields are around 8467 kg/ha (135 bu/a) of corn on these soils by this method.

There are concerns, other than costs, in using sprinkler systems. The nonuniform application of water and the chemicals in the water can cause problems (Detail 16-2). Wetting foliage and wetting the soil surface increase the hazard from many fungi and bacterial diseases. For example, significant reduction of the fungus *Botryosphaeria dothidea* (panicle and shoot blight of pistachio) was observed when sprinkler irrigation times were reduced from 24 to 12 hours in Sacramento Valley and from 48 to 24 hours in the San Joaquin Valley of California.[25] The disease has optimum development at about 80°C to 85°C. Spores germinate within 2 hours after a period of wetness but need about 12 hours to penetrate the petioles and leaves. The orchards had previous losses of more than 76 percent using 24- to 46-hour-long irrigations. Shorter irrigations reduced fruit losses to as much as one-third the loss in longer irrigation sites.

Modifications of sprinkler systems include changes in the height of sprinklers from high overhead ones of early center pivots to tubes suspended down to just above the crop canopy. The low-energy precision applicators (LEPA sprinklers) can actually deliver water close to the crop through drop tubes extending down from the sprinkler arm. One producer in Lubbock, Texas, during two years dropped water use 47 percent and electricity 32 percent while crop yield rose about one-third.

[25] Themis J. Michailides, David P. Morgan, Joseph A. Grant, and William H. Olson, "Shorter Sprinkler Irrigations Reduce Botryosphaeria Blight of Pistachio," *California Agriculture* **46** (1992), No. 6; 28–32.

Detail 16-2 Sprinkler Water Application Uniformity

There are good and not-so-good ways of applying water by sprinkler systems. Recently **low-energy precision applicators (LEPA)** have become one way to increase application uniformity. In Texas studies of center pivots using nozzles that shoot water with pressures near 60 psi (pounds per square inch) had as high as 15%–20% of the water that *didn't reach weighing lysimeters* (3 ft × 3 ft) in the field. With nozzles at 30 psi, the loss was 10%–15%. With LEPA systems (6 psi) the loss was only about 4%.[a]

Generally, **emission uniformities** over 90% are considered excellent, 80%–90% are good, 70%–80% are fair, and less than 70% are poor. A study of 112 systems in the San Joaquin Valley of California showed 62% were good to excellent, with a mean value of all systems of 80.3%. A very large portion of distribution uniformity problems were caused by *poor pressure regulation* and *plugged nozzles*.[b]

Where fertilizer or pesticide is applied in the water, drift (movement away from the target area) can be reduced by equipment adjustments. Some of these alterations are as follows:[c]

1. Select nozzle type to produce large drops. Large drops drift less.
2. Use lowest pressure feasible. High pressures form small droplets.
3. Lower the boom height. Wind speed, thus drift, increases with height above ground.
4. Spray when wind speeds are low (less than 10 mph). Wind blows the spray off target.
5. Increase nozzle size. Larger nozzles form larger droplets at a given pressure.
6. Do not spray when air is completely calm or an inversion exists. Spray can slowly move downwind before it falls.
7. Use a drift control additive, if needed. These increase the droplet size produced.

[a]Don Comis, "Lower Water Pressure, Less Water Loss," *Agricultural Research* **40** (1992), No. 5; 23.
[b]Dale Handley, Henry J. Vaux, Jr., and Nigel Pickering, "Evaluating Low-Volume Irrigation Systems for Emission Uniformity," *California Agriculture* **37** (1983), Nos. 1–2; 10–12.
[c]Anonymous, "Seven Ways to Reduce Drift," *Solutions* **37** (1993), No. 1; 36–37.

To avoid flowing water on the soil surface, fabric socks are placed on the bubble tubes hanging down just 30 cm to 40 cm above ground so the sock actually drags over the ground surface. LEPA systems should include basin tillage (creates pits or small dikes in furrows to hold water in place).[26]

Sprinkler systems come in all shapes, sizes, and kinds. Small, portable, rotating heads, such as those used in home gardening, are often employed agriculturally. *Solid-set* stationary systems and *mobile* sprinkler lines come in many sizes. **Center-pivot** or **lateral-move line** systems are usually installed on large acreages. Another innovation is large **sprinkler guns** that deliver up to 4.54 m^3/min (1200 gal/min), sprinkling a circle of 139 m (620 ft) diameter.[27] These guns have large nozzle openings that do not easily plug with debris in the water; sewage effluents can be spread easily in this way (Figure 16-20).

Sprinkler systems have two major disadvantages: their high cost and plugging of the nozzles by debris in the water. Costs for solid-set systems range from about $1000 upward per hectare (about $400 per acre) for installation.

[26]Dale F. Heermann, Harold R. Duke, and Gerald W. Buchleiter, "Irrigation Systems in Transition—Center Pivots and Linear Moves," *Irrigation Journal* **44** (1994), No. 4; 16–23.
[27]Bob Rupar, "The Big Gun Boom," *Irrigation Journal* **25** (1975), No. 1; 16–18.

FIGURE 16-20 Large sprinkler guns, such as this single unit in western Washington, have large nozzle openings and do not easily become plugged. They are used to spread manure slurries, as shown here, containing 5%–8% solids, to apply other effluents, or to irrigate normally. Because the water falls with considerable force, sprinkler guns are best used on established crop cover, such as sugarcane and pastures, rather than on barren or newly planted land. (Courtesy of Darrell Turner, Washington State University.)

Precision farming uses variable-rate treatments (VRT) of lime, fertilizer, and pesticides. It can also involve variable irrigation rates. The system now under development uses a miniature computer to read the position of the pivot in the field (Figure 16-21). Then, using a digitized map of the field with delineated areas, the computer sets a specific flow rate for each nozzle. Irrigation rates are varied for the water each area needs.

Drip (Trickle) Irrigation[28]

The most efficient method of watering is **drip (trickle) irrigation.** As the name suggests, drip irrigation is the frequent, slow application of dripping water to soil through small outlets (**emitters**) located along small plastic delivery lines, 1.3 cm to 2.5 cm (0.5 in–1.0 in) in diameter. The application rate is so slow (often less than 3.7 L [1 gal] per hour per emitter) that surface water flow is almost nil. Water movement is by saturated and unsaturated (capillary) flow, and seldom is all the surface of the field wetted. The small emitters can be placed at any frequency wanted. For example, a large tree in an orchard may have four to eight emitters spaced around it; a grapevine may have one or two emitters.

Why drip irrigation? Costs are only 50 percent to 75 percent of some sprinkler systems; the system uses smaller lines (slower water flow) at low water pressures; and once the system is installed, the labor cost is low. Drip irrigation is adaptable to very steep hills, although differential hydraulic pressure in lines at the top and bottom of the slope must be considered.

[28] "Drip/Trickle Irrigation in Action," *Proceedings of the Third International Drip/Trickle Irrigation Congress,* Fresno, CA, Nov. 18–21, 1985, American Society of Agricultural Engineers, St. Joseph, MI, 1986.

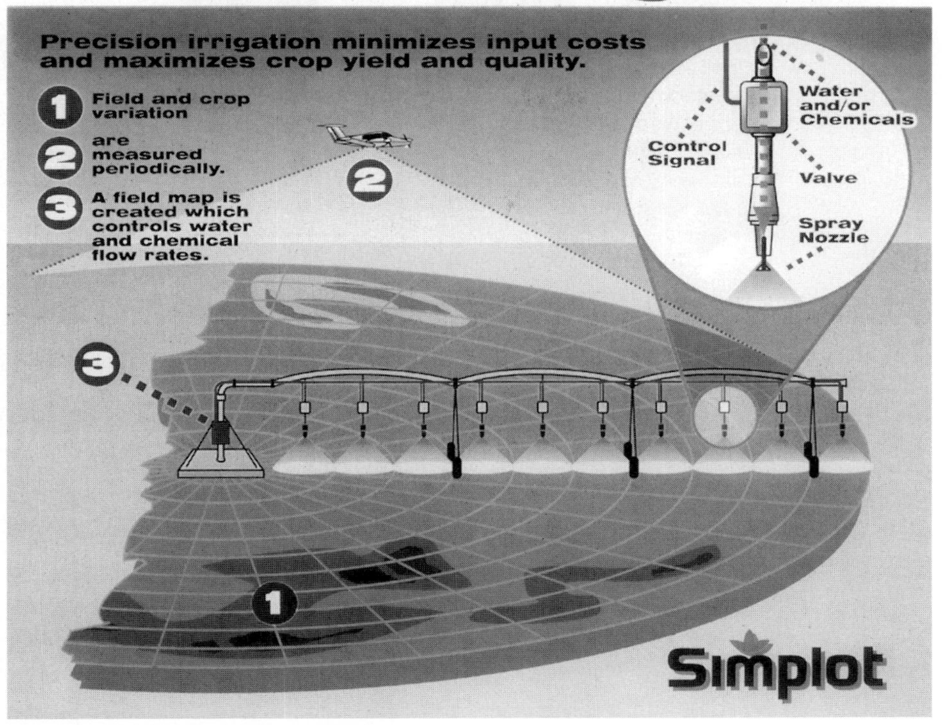

FIGURE 16-21 Proposed precision irrigation control is used on a field that (a) has the field delineated into variable production areas, (b) is monitored periodically by aerial photos for crop growth and "health" and (c) a newly altered field map automatically adjusts flow rates. (Courtesy of Precision Irrigation Systems, J. R. Simplot Co. and *Ag Consultant Magazine* **52** (1996), No. 2; 4.)

Of increasing importance, drip systems conserve water because of lower distribution and evaporation losses; savings of 20 percent to 50 percent are expected. Water of higher salt content can also be used because of the nearly constant, high soil moisture maintained in part of the root zone, and salt is constantly being moved away from the plant roots to the outer part of the wetted soil volume. Drip systems are currently used on orchards, in vineyards, on sugarcane in Hawaii, and for numerous other crops. Drip irrigation is popular in water-short Israel and is even used for corn.

Drip irrigation is not without its problems. The small emitters plug easily, so careful filtering at the water source is essential. Without water added to promote leaching, salts can accumulate, although this occurs at the periphery of the wetting area. Also, the plant root zone is often less deep and extensive with drip irrigation, and a water failure can quickly cause water-stress problems to the plant. Drip water lines may require special cultivation and harvesting arrangements. In sugarcane harvesting (18–24 months after planting), the plastic lines are considered expendable and are wasted (destroyed during harvest).

Drip irrigation is usually a water application to the soil surface, but subsurface application lines and emitters are also used. Drip irrigation allows wetting only the root zone desired. For example, if plants are widely spaced (such as grapes or orchards), only the soil that needs to be wet is drip irrigated (Figure 16-22). For many crops (tomatoes, melons,

FIGURE 16-22 Replanting young grapes in Sonoma County, California. Drip lines for these nursery starts wet only the small areas where each young rootstock is rooting. This helps control weeds and greatly minimizes water use. (Courtesy of *California Agriculture Magazine;* photo by Jack Kelly Clark.)

cucumbers, strawberries), buried lines allow wetting the root zone soil without wetting all the surface soil. Wet surface soil causes rot and spotting of fruit lying on the wet ground. Underground lines and computer-controlled linkage for irrigation scheduling allow a yield of 100 tons of ripe tomatoes per acre (224 Mg/ha) compared to the area average of about 26 t/a (58 Mg/ha).[29]

A simple modification of drip irrigation on gently sloping land is **bubbler irrigation,** a method using low water pressure and open, standing outlet tubes. The rate of water flow is determined by line size, water pressure, and the height of each standing outlet tube (Figure 16-23). The large flow opening (which may be a fully open tube end) reduces the problem of clogged emitters. Setting flow by elevation adjustment can be tedious and frustrating if the area is unlevel and the line system has many outlets.

---

[29]Marcia Wood, "Underground Drip Irrigation Yields Record Tomato Harvest," *Agricultural Research* **36** (1988), No. 7; 14.

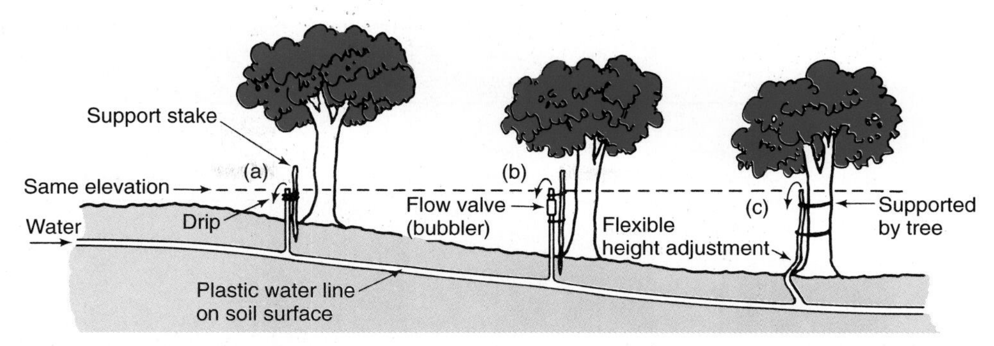

FIGURE 16-23 Diagram of a bubbler irrigation system, a variation of drip irrigation. The open outlet system controls waterflow by (a) its pipe elevation, (b) with a valve, or (c) by both. The riser lines can be flexible and attached to the grape trellis or tree (as in c) to simplify connections. A flexible line with extra length allows vertical adjustment. (Courtesy of Raymond W. Miller, Utah State University.)

The special problems of drip irrigation have prompted proposals of criteria for water quality in such systems. One proposed quality scheme is given in abridged form in Table 16-10. It classifies water based on three categories of constituents:

- *Physical, suspended solid material* such as sands, silts, clays, humus, other suspended organic materials, and small organisms.
- *Chemical precipitation,* such as carbonates, gypsum, hydroxides of iron and manganese, fertilizer phosphates, dissolved salts, and substances that might precipitate, thereby clogging emitters.
- *Biological,* the microorganisms whose actions produce slimes, filaments, and chemical deposits that plug emitters.

Rating these three categories (averaging the two columns for chemical precipitation) in the order *physical-chemical-biological,* a 0-0-0 water should be excellent for use in a drip system, whereas a 10-10-10 water would be almost unusable. If the rating is between 0 and 10 after adding the three numbers, few problems are anticipated. A rating of 10-20 indicates some problems will occur, and 20-30 indicates severe problems. Above 10, filtration or other remedial measures would be required.

Subirrigation

Some areas have unique soil properties that permit irrigation by raising the localized water table. In this method, called **subirrigation,** water is applied to an area via open ditches or by buried pipes and flows into the water table to raise it to the root-zone level. Subirrigation is effective on land with sandy, peat, or muck soils from a root depth to 1.5 m to 3 m (5 ft–10 ft), a level soil surface, and an impermeable layer beneath the root zone to hold the water. A typical subirrigation area in northern Utah has river terraces of loamy sands 2.1 m to 2.4 m (7 ft–8 ft) deep over impermeable clay deposits from ancient Lake Bonneville (Great Salt Lake). Tile drains with controllable gates (closures) allow early spring drainage. As the water table becomes sufficiently low, the gates in the drains are closed to restrict further water drainage loss and, later in the season, to allow subirrigation of the crop.

Table 16-10 Proposed Scheme for Classifying Waters as to Their Suitability for Use in Drip (Trickle) Irrigation

Numerical Rating[a]	*Suspended Solids*[b] *(max. mg/L)*	*Dissolved Solids*[c] *(max. mg/L)*	*Iron and/or Manganese*[c] *(max. mg/L)*	*Bacteria Population*[d] *(no./mL)*
(Best)				
0	<10	<100	<0.1	<100
1	20	200	0.2	1,000
2	30	300	0.3	2,000
4	50	500	0.5	4,000
6	90	800	0.7	10,000
8	120	1,200	0.9	30,000
10	>160	>1,600	>1.1	>50,000
(Poorest)				

Source: These data have been selected and modified from D. A. Bucks, F. S. Nakayama, and R. G. Gilbert, "Trickle Irrigation Water Quality and Preventive Maintenance," *Agricultural Water Management* **2** (1979), 149–162.
[a]The complete proposal has 11 rating classes from 0 to 10; only 7 are listed here.
[b]These are sands, silts, clays, organic substances of many kinds, bacteria, phytoplankton, and zooplankton.
[c]Dissolved solids are mostly soluble salts. Columns 3 and 4 make up the chemical portion of water quality and affect the plugging of emitters by precipitation. If water pH is 7.5 or higher, increase suspended solids rating by 2.
[d]Action of microorganisms increases potential to plug emitters. If snails are common, increase the rating by 4.

Irrigation by controlling the water table drainage line is being tried in North Carolina.[30] A quality system requires careful attention to arrange the depth of drain lines and the water capacity of the open ditch drain so that the water put back into the drain to raise the water table will work without requiring excessive water.

Subirrigation is commonly practiced in tile-drained or tube-drained peat or muck soils by closing the outlets and raising the water table to the bottom of the rhizosphere.

16:7 Special Irrigation Techniques

Modifications of irrigation techniques are employed in special circumstances. Recently, innovations in furrow irrigation, automated water delivery, and sprinkler system improvements have been subjects of investigation. Special techniques may not decrease consumptive water use for a given yield, but they may improve efficiency by reducing the leaching of nutrients, water losses by drainage, and so on.

Alternate-Row Irrigation

Grain sorghum in the Great Plains of Texas, corn of Nebraska, and cotton in California and Texas have been irrigated by applying water to **alternate rows** during a given irrigation and then watering the missed rows in the next irrigation period. Water savings are large here (nearly 50%), and yields are normal or only slightly reduced. Alternate-row watering allows more rapid coverage of a field during an irrigation period, saves water, and requires less labor. An alternate-row system also leaves some soil dry enough to absorb any rainfall that comes, which reduces erosion and runoff loss and maximizes use of rainfall. Time intervals between irrigations must be shorter because only half as much water is added to the area at a given irrigation.

[30]Robert Evans, "Switch-Hitting with Subsurface Drainage/Irrigation Systems," *Irrigation Journal* **44** (1994), No. 4; 8, 12, 14, 16, 17.

FIGURE 16-24 Pomegranate trees in Israel grown in *negarins* (mini watersheds). A small watershed area for each tree supplies runoff water to a small catchment basin in the center of which the tree is planted. In this area of Israel, with an annual rainfall of 150 mm–200 mm (6 in–8 in), orchards could not grow without some method of increasing the water available to them. The distance between trees indicates the watershed area for each tree. (By permission, from *Food, Fiber and the Arid Lands,* by William G. McGinnies, Bram J. Goldman, and Patricia Paylore, eds., University of Arizona Press, Tucson, 1969.)

Mini Watersheds or Catchment Basins

Where water is deficient but wells or surface water is too expensive or is unavailable, wide spacing of crops (hills of corn, widely spaced fruit trees or vines), each fed by runoff water, is sometimes possible (Figure 16-24). Normally, the plants must be quite tolerant of drought but able to produce better yields when rainfall is more plentiful than during drier periods. Numerous modifications of these **mini watershed** patterns are possible. **Contour furrows** (ditches or furrows dug along the contour at various distances apart) are used on arid rangelands to concentrate enough water in and near furrows to reestablish grasses. These increase yields of better grasses on parts of a range.

All of these processes *concentrate sparse water onto parts of the area.* This concentrated water allows production of desired plants that generally grow unsatisfactorily, or not at all, without water concentration.

Irrigating Clay Soils

Soils with high clay percentages (more than 40%) have special problems. Clayey soils in the United States usually contain montmorillonite, illite, or kaolinite. Of these, montmorillonitic clay soils are the greatest problem because of large volume changes that occur when the clays are wetted and dried, their great stickiness and plasticity, and their exceptionally high water-retention capacities.

The three major problems of clay soils are (1) inadequate aeration when wet, (2) slow water infiltration, and (3) a limited moisture range that is suitable for tilling. The Sharkey clay, an alluvial soil of the Mississippi River delta area, exemplifies two of these problems. Its pores are all very small and drain no water when fully wetted—all pore space is filled with water having water potentials lower than -33kPa.[31] When dried and cracked, this clay readily absorbs several inches of rain, but after it becomes wetted, additional rain runs off the surface. Only open-ditch drains are helpful in draining these clays.

Treat clayey soils as shallow soils. Because poor aeration reduces rooting depths, *frequent but shallow irrigations* are a practical approach to irrigation of many clayey soils. When cracks in dry clayey soils are not wide, wetting and quick swelling soon close those cracks and reduce water intake. One clay studied decreased from a very rapid infiltration rate

[31] W. M. Broadfoot, "The Fame of Sharkey Clay," *Forests and People* **12** (1962), No. 1; 30, 40.

of 25 cm/hr (9.8 in/hr) to a slow rate of 0.6 cm/hr (0.23 in/hr) while absorbing only 2 cm (0.8 in) of water, a reduction to less than one-fortieth of the maximum rate. Because of the greater soil surfaces exposed to the wind and sun, cracking of clayey soils increases evaporation losses 12 percent to 30 percent more than if there were no cracks.

The rate of water infiltration is a major problem when irrigating clay soils. Where deep and extensive cracks occur, wetting to capacity (up to 15%–20% of the soil volume) is relatively easy. But if the clay does not crack extensively or is not allowed to dry enough, water intake is slow. A surface mulch of organic residues and soil tillage help some, but usually not enough. Methods for holding water on the soil longer include lengthening the rows (longer fields or serpentine rows), using check furrows (small dams across the furrow), slow rate of sprinkling, and drip irrigation.

Slow infiltration rates in the range of 0.76 cm/hr (0.3 in/hr) or less after the initial few minutes require 10 to 20 hours to wet clay soils to normal rooting depth. Maximum row lengths for nearly level clays are about 244 m to 366 m (800 ft–1200 ft). Where field lengths are rigidly established, irrigation furrows can be lengthened by a serpentine scheme, as shown in Figure 16-25. **Serpentine** schemes are made by cutting passageways across the furrow direction to connect several channels so that the water flows back and forth like a snake's (serpent's) trail,

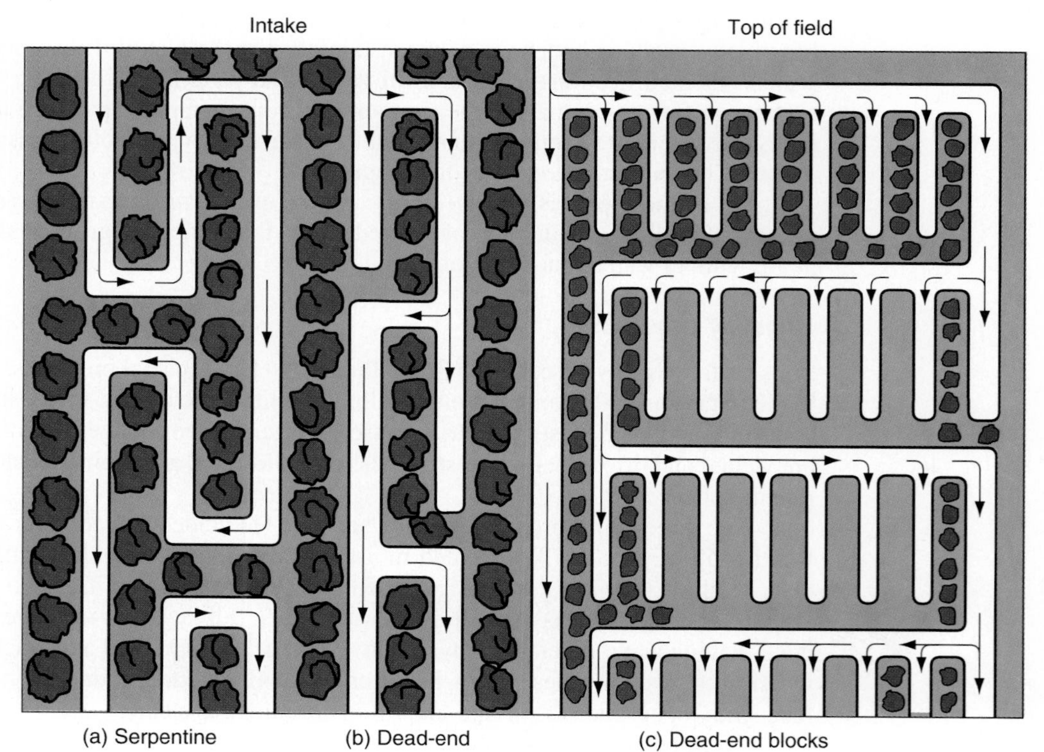

FIGURE 16-25 Serpentine channels increase infiltration on clayey soils by maximizing stream flow, minimizing erosion and water runoff, and increasing water-soil contact time by means of longer furrows. (a) A typical serpentine involves three meander furrows. (b) The dead-end serpentine keeps a large soil area in contact with water as flow occurs in only half of it. (c) Modifications are possible for special uses, such as onion sets or other intensive crops. Serpentines require more work than simple irrigation furrows and should be used only where the benefits exceed disadvantages. (Courtesy of Raymond W. Miller, Utah State University.)

greatly increasing the effective length of the furrow. The many patterns have been given names such as *dead-end serpentine* or *block serpentine* and *meander furrows.* Longer furrows allow larger volumes of water to be added per furrow, which increases the soil-water contact surface and, thus, total water intake.

Irrigating with Salty Water

In some world areas water shortages are relieved by using brackish (slightly to moderately salty) water for irrigation, although most salty waters have the problems of high sodium (undesirable SAR), high chloride (toxic on foliage of many plants, especially citrus and avocados), and the salt itself (osmotic effects). Countries such as Israel have developed extensive canal and reservoir systems where both low-salt and salty waters are mixed to obtain usable water. These irrigation techniques are selected: first, to hold the soil near field capacity to keep salt as dilute as possible; second, to avoid application techniques that wet the foliage damaging the leaves; and third, to leach accumulated salts periodically. To accomplish these objectives, three general rules follow:

- Apply water at or below soil surface. Sprinklers should be used only if they avoid wilting the foliage (such as sprinkling before plant emergence or below-canopy to avoid salt-burn damage).
- Keep water additions almost continuous, but at or below field capacity so that most flow is unsaturated. This maintains adequate aeration.
- Enough water should be added to keep salts moving downward, thus avoiding salt buildup in the root zone.

These requirements are difficult to meet and are best satisfied by some form of drip irrigation. Due to the need for high moisture levels and because of the high sodium (SAR) problem, sands seem to be most adaptable to irrigation with salty water. Where the soil includes considerable clay and silt, the use of salty water is less likely to be suitable.

16:8 Irrigating Efficiently

Efficiency is the watchword. The competition for water will force reductions in irrigation water. To continue to produce, growers will have to do with less water. The definition for **water use efficiency (WUE)** is the *portion of the added water that is used by the plant.* It can also be compared to the *amount of growth produced divided by the amount of water used by the plant.* This relationship can be given in several ways:

- Dry weight of produce per volume of water used, such as kg of grain per hectare-cm of water added (or in lb/a-in).
- **Transpiration ratio (TR),** which is the kilograms of water used per kilogram of dry yield produced. This is the inverse of the previous system. *TR* values of 400–1000 are common.

Increasing water demands or scant water supplies usually encourage more efficient water use. In Israel, where water requirements are carefully measured, a maximum quantity of water is allotted on the basis of soil properties and the crop to be grown.[32] It has 48.7 percent of its cropland under **microirrigation,** a term for drip plus other low-pressure, small outlet systems. Cypress, with limited fresh water, has 71.4 percent of its citrus, grapes, olives, vegetables, and nuts under microirrigation.[33]

---

[32] Ron Ross, "Israel . . . Where Irrigation Is Art," *Irrigation Age* **13** (1978), No. 1; 6–9.
[33] Sandra Postel, *Last Oasis: Facing Water Scarcity, ibid.,* 104–106.

Table 16-11 Comparison of Furrow Irrigation and Drip-Under-Plastic for Melons in Texas

Item	*Drip-Under-Plastic*	*Furrow*
Rainfall	65 mm	65 mm
Irrigation	11 cm	32.8 cm
Number of irrigations	8	7
Nitrogen added	28 kg	72 kg
Yield (boxes of melons/ha)	1,235	741
Water efficiency (boxes/cm)	112	22.6
N efficiency (boxes/kg)	44	10

Source: Modified from Guy Fipps, "Melons Demonstrate Drip Under Plastic Efficiency," *Irrigation Journal* **43** (1993), No. 7; 8, 11–13.

Buried drip lines are becoming more used. Burying lines 20 cm to 30 cm deep, below shallow tillage, allows wetting the root zone with minimal wetting of the soil surface. This is vital because evapotranspiration is the greatest water loss in irrigation. In the Rio Grande Valley of Texas, a plastic cover with a drip line underneath is very effective in saving water; the vegetables and melons are ready for market two weeks sooner (Table 16-11).

An estimation of some water use efficiencies by the various irrigation techniques and modifications is tabulated in the following:[34, 35]

- Worldwide, irrigation efficiency is estimated to be about 37%. Much of the water lost is from surface runoff and can be used again. Surface flow efficiency is only about 50%–65% where it is well done.
- Surge flow irrigation can reduce water loss over regular furrow irrigation by 10%–40%, averaging about 25% reduction.
- Sprinkler systems average 60%–70% efficiency; well designed systems should be 85% efficient. A spray system on drops from the sprinkler arm is about 91%, and LEPA systems have reached 96% efficiency with low pressure.

Although these systems can operate with high water efficiency, many of them do not, and there is room for improvement. The cost of major projects is becoming so great that even if the water were there to be used, the costs to develop such large projects cause too high a cost for water to the grower. The hope is for an ever increasing effort to use less water to do what we used to do with more.

16:9 A Future for Irrigated Agriculture?

Irrigated agriculture will always have drainage waters, either surface flow or internal flow to the vadose zone and groundwaters. Irrigation drainage waters will always carry soluble salts. These salts will cause a degradation of most environments or waters into which they are carried. A permanent irrigation agriculture does require the sacrifice of

[34] Dale F. Heermann, Harold R. Duke, and Gerald W. Buchleiter, *ibid.*
[35] Sandra Postel, "Water and Agriculture," *ibid.*

some value elsewhere. For example, estimates of damages to agriculture, households, water utilities, and industry from increased salt in the Colorado River (from damming, reentry of salty irrigation drainage, other human-caused salt increases) are $311 million to $831 million, depending on the natural assumed salt level of the river before people altered it.

Successful irrigation depends on adequate drainage; this implies some off-site damages from carried salts. Demands on water will require that irrigation water charges be real charges, not kept low by subsidies. This will increase costs and require more efficient irrigation, which, in turn, will reduce leaching of salts and require more careful attention to the leaching requirement. What do we do with the salts? All questions cannot be satisfactorily answered at this time. One thing is sure: People will still want to eat and some means will be provided to assure that food is available, if at all possible. Whether irrigation increases or decreases will depend on how well we manage the water, the salt, and the land. Will we be able to compete in the future price for available water?

16:10 A Decision Case Study

The concept of decision case studies was introduced in Chapter 6. The following is an additional decision case study on considering the impact of a water quality study: J. W. Bauder, "Assessing Extension Program Impact: Case Study of a Water Quality Program," *Journal of Natural Resources and Life Sciences Education* **22** (1993), No. 2; 138–144.

As our case is new so must we think and act anew.

—Abraham Lincoln

There is no medicine like hope, no incentive so great, and no tonic so powerful as expectation of something better tomorrow.

—O. S. Marden

Questions

1. (a) What is an *aquifer?* (b) What are the materials from which aquifers form?
2. (a) What causes an artesian well to flow? (b) Why might a flowing well later cease to flow naturally but can be pumped?
3. (a) Why would some aquifers produce lower flow from wells and recharge more slowly than other aquifers? (b) Define *overuse* of groundwater.
4. (a) What causes saltwater intrusion? (b) What can be done to minimize it?
5. What are the advantages and disadvantages of groundwater compared to surface waters as sources of irrigation water?
6. Which areas of the United States are overusing groundwaters?
7. How inadequate are the world water supplies? Give examples.
8. What is a water ethic? Do you think we need one? Discuss.
9. In water budget irrigation scheduling: (a) How is the rate and amount of water loss measured? (b) How is the amount of water to be added determined?
10. Discuss briefly how dry a soil should be allowed to get before irrigating it. Consider (a) rooting depth, (b) the different soil depths, and (c) the nature of the crop grown.

11. Define and describe each of the following irrigation techniques: (a) *furrow,* (b) *basin,* (c) *border,* (d) *drip,* and (e) *sprinkler.*
12. What are the disadvantages of furrow irrigation, and how is the extent of these disadvantages reduced?
13. List the disadvantages and advantages of using (a) drip systems and (b) sprinkler systems.
14. Discuss the filtering requirement of drip irrigation systems.
15. Discuss the techniques for irrigating sloping lands, sands, or shallow rolling-land soils.
16. What are some surface flow techniques used to increase wetting depth of poorly permeable (clayey) soils?
17. (a) What is *surge flow?* (b) What is *cablegation?* (c) Why are they of interest?

Wetlands and Land Drainage

Sometimes it's more important to discover what one cannot do than what one can do.

—**Lin Yutang**

17:1 Preview and Important Facts

PREVIEW

Wetlands are land areas with periods of wetness. In the past many of us have viewed wetlands as soils to be drained so that they can be used more effectively. Today wetlands are considered to have great value as they exist and we know that to change them would cause serious damage to the environment in the following ways:

- Destroying habitat for many animals
- Altering groundwater recharge sources
- Destroying water storage
- Eliminating sediment reduction
- Inhibiting water purification
- Reducing biodiversity
- Limiting outdoor recreation
- Reducing the capacity of a flood modifying system

Even for municipalities and other governmental agencies to drain or destroy wetlands requires intensive review. Land trades or establishing *new* wetlands are sometimes done to replace the lands to be drained.

Many lands that have been cultivated for decades, even centuries, have required drainage to be productive. Some of these soils are drained wetlands but others have developed drainage problems because of changes in use, irrigation, or deterioration of the soil.

Artificial drainage is used on more than 10 percent of the world's cropland and on about one-third of the cropland in Canada and the United States. The extent of worldwide drainage utilization is listed in Table 17-1. In humid areas drainage enhances plant growth by increasing the oxygen supply for plant roots and permits timely planting and harvesting of crops. In

Table 17-1 Estimates of Croplands Artificially Drained*

Area	*Total Cropland (10^3 acres)*	*Cropland Artificially Drained (10^3 acres)*	*Cropland Drained (%)*
World	3,627,497	383,492	10.6
Asia	1,178,020	78,756	6.7
Africa	522,090	5,925	1.1
North and Central America	674,531	167,195	24.8
South America	220,858	19,276	8.7
Europe	353,988	87,844	24.8
Oceana	116,599	2,234	1.9
USSR	561,411	22,335	4.0
Canada	108,148	36,741	34.0
Mexico	63,692	3,385	5.3
United States	467,718	147,766	31.6

Source: Calculated from *Food and Agriculture Production Yearbook, 1974* and P. P. Nosenko and I. S. Zonn, *Land Drainage in the World,* ICID Biennial Bulletin, International Commission of Irrigation and Drainage, 1976.
*Acres × 0.405 = hectares.

arid regions many soils have or accumulate soluble salts that need to be removed. Moving water through the profile to dissolve the salts sometimes requires that drains be installed when the soils are not very permeable.

Drainage is the removal of excess gravitational water from soils by natural or artificial means. Unless otherwise specified, *drainage* refers to artificial systems for removal of water in soils.

IMPORTANT FACTS TO KNOW

1. The value of wetlands to people and the environment
2. The definition of *wetlands* and restrictions on conversion (drainage) of present wetlands
3. The soil and vegetation clues that indicate poor drainage exists
4. The numerous benefits of drainage for agricultural crop production
5. Some of the undesirable results from drainage
6. The drainage characteristics and nature of paddy rice culture
7. How to improve soil drainage by *smoothing, open-ditch drains,* and *mole drains*
8. The techniques of *ridge tillage* and *beds* to help use poorly drained soils
9. The need and details of drainage in salty soils for purposes of reclamation
10. How water enters drain lines from the soil and how to avoid plugging

17:2 Wetlands Definition and Nature

Many poorly drained soils—wetlands—have long been considered as potential cropland, if the excess water could be removed. Although most wetlands were considered less valuable than adjacent land with good drainage, they were still considered worth draining. Since the 1780s, 22 states have lost 50% or more of their wetlands and 10 states have lost 70% or more.[1]

[1]Tina Adler, "Two Views of a Swamp: Scientists Dispute Legislator's Take on Wetlands," *Science News* **148** (1995), July 22, 56–57.

Definition of Wetlands

A **wetland** is defined as *an area of predominantly hydric soil that can support a prevalence of water-loving plants.*[2] Wetlands are inundated or saturated by surface or groundwater often enough and long enough for growth of mostly vegetation adapted to saturated soil conditions, such as cattails, willow trees, sedges, rushes, or other water-loving plants.[3] The jurisdictional agency, the U.S. Army Corps of Engineers, offers a longer definition:

> Areas that are inundated or saturated by surface or ground water at a frequency and duration sufficient to support, and that under normal circumstances do support, a prevalence of vegetation typically adapted for life in saturated soil condition. Wetlands generally include swamps, marshes, bogs and similar areas.[4]

Clearly from this definition a **wetland** must exhibit three criteria:

- Have soil that is subject to waterlogging, which is called **hydric soil**
- Have, in its natural state, a predominance of wetland plants
- Have indicators of wetland hydrology (ponding, high water table)

Is a wetland by any other name just a wetland? Not quite. There are **converted wetlands.** These are lands farmed or converted which have drainage to improve crop production done after 23 December 1985. These soils have considerable restrictions or federal penalties if used for crops. **Farmed wetlands** are wetlands only partially drained or altered before 23 December 1985 to improve crop production. These can continue to be used without additional drainage or alteration. Finally, a category called **prior converted wetlands** are those converted to cropland use before 23 December 1985 and were fully drained or altered so they no longer fit the criteria of saturated soil or water-loving plants. These latter soils are considered cultivated soils and have no restrictions in use relative to regulations in the wetland legislation. Thus, if conversion to cropland by additional drainage occurred after December 1985, there may be federal penalties or restrictions on its cropping use.

Importance of Wetlands

Why the concern about draining wetlands? It is clear that large portions of wetlands have been drained in the last 200 years. The U.S. Government Land Acts of 1849, 1850, and 1860 encouraged drainage of wetlands on the assumption that they were threats to public health or were public nuisances. In the past three decades the public's attitudes have changed. It is now widely known that wetlands provide many benefits. Some of these benefits are crucial. Wetlands aid in the following ways:

- They preserve biodiversity. About 40% of the threatened and endangered species rely on wetlands.
- They are a water supply, directly used by people or by helping to recharge the groundwater.

---

[2]Doug Snyder (compiler), "What Farmers Should Know About Wetlands," *Journal of Soil and Water Conservation* **50** (1995), No. 6; 630–632.

[3]**HYDRIC soils,** produced by the University of Florida, is a CD-ROM for teaching about wetland communities. $75 [Contact Ami Neiberger, University of Florida, 116 Mowry Road, P.O. Box 110810, Gainesville, FL 32611-0810; ph (904)392-2411, fax (904)392-7902.]

[4]John Lyon, "Wetlands: How to Avoid Getting Soaked," *Professional Surveyor,* Jan.–Feb. 1995, 16–18.

- They reduce the energy of flood waters (flood control) and help regulate surface flow by storing some of the water and providing a flat place for the flow to dissipate.
- They provide pressure to prevent saline water intrusion into aquifers.
- They act as a buffer to protect against natural forces (hurricanes, strong winds, wave break).
- They catch sediment to keep it from moving further downstream.
- They retain nutrients.
- They remove toxicants from the collected water.
- They are a source of natural products (peat, timber, fruit, wildlife meats, cane, reeds, resins, medicinal products).
- They produce energy (peat, hydroelectric power, wood).
- They are a means for transporting water.
- They retain and maintain a gene bank of many organisms.
- They are a source of food and habitat for many resident and migrating animals and birds.
- They improve water quality by retaining water long enough for nature's processes to act.
- They can be specialized cropland for crops such as rice and cranberries.

Surely, this list of benefits is sufficient to raise an alarm to maintain and reserve extensive areas of wetlands. Wetlands are valuable, although they can be nuisances to some people and havens for some diseases and pests.

There was enough concern over the continued drainage of more and more wetlands that President Bush in 1990 gave a directive to Congress saying that there should be *no net loss of wetlands in the future.* He based his authority on Section 404 of the 1977 Clean Water Act (and more recent additional controls).[5] To drain new land the owner must obtain a permit from the U.S. Corps of Engineers. Often now to get a permit to drain one area, another area must be allowed to revert to wetlands or new wetlands must be constructed to meet the *no net loss* requirement (Detail 17-1).

Hydric Soils and Hydrophytic Plants

A **hydric soil** is one that is saturated, flooded, or ponded long enough during the growing season to develop anaerobic conditions in the upper part. The U.S. Corps of Engineers states that this inundation must occur for the consecutive days equal to or greater than one-eighth (12.5%) of the growing season.[6] Table 17-2 lists some soil indicators and some primary indicators of wetlands. The details are purposely abridged for space considerations.

Hydrophytic plants have adaptations that allow them to survive, even thrive, in saturated or inundated soils. Some of these adaptations are *adventitious (prop) roots, stooling, and buttressing of the plant base.* Some of these plants are cattails, button bush, speckled alder, soft rush, and red maple. According to one author, one list of wetland plants has itemized 7000 species.[7]

---

[5]Taylor A. DeLaney, "Benefits to Downstream Flood Attenuation and Water Quality As a Result of Constructed Wetlands in Agricultural Landscapes," *Journal of Soil and Water Conservation* **50** (1995), 620–626.

[6]Christopher L. Lant, Steven E. Kraft, and Keith R. Gillman, "The 1990 Farm Bill and Water Quality in Corn Belt Watersheds: Conserving Remaining Wetlands and Restoring Farmed Wetlands," *Journal of Soil and Water Conservation* **50** (Mar.–Apr.), 1995, 201–205.

[7]Carl E. Tammi, "On-Site Identification and Delineation of Wetlands," Chapter 3 in Donald M. Kent (ed.), *Applied Wetlands Science and Technology,* Lewis Publishers, Ann Arbor, MI, 1994, 35–59.

Detail 17-1 Building Wetlands in the Desert?

It may sound strange, but it is true: Wetlands are being built in the desert. Where water is short, wetlands still exist. In some low areas with poorly permeable soils and seepage or drainage coming down off upslope areas, soils, even in deserts, can be wetlands. However, to *build* wetlands in arid sites is new.

Scientists in Tucson and Phoenix, Arizona, are doing just that in rainfall areas of 250 mm to 300 mm annually (10 in–12 in). The intent in Tucson is to improve clean-up of the city's reclaimed water. Because all of the city's water is from groundwater, the wetland will help cleanse the water as it goes to recharge the aquifer. The wetland will be built along the Santa Cruz River. There will be some planted habitat in the 7-hectare site.

In Phoenix three experimental wetlands are designed to help in removing pesticides and metals from secondary-treated sewage effluent; the effluent is currently discharged into the Salt River.

Source: David R. Rosenbaum, "Wetlands Bloom in Desert," *Engineering News-Record* **235** (1995), No. 24; 19–20.

Table 17-2 Some Indicators of Hydric (Wetland) Soils and Some Primary Indicators*

Soil indicators (not all are required in each soil)

- Organic soils, except folists (leaf mats).
- Histic epipedon. A root mat top horizon (see Chapter 2).
- Sulfidic material. Sulfur is reduced to sulfide within the top 30 cm.
- Gleyed horizon. Chroma is less than 2 (gray); some mottles of high chroma (bright) of yellow-orange color.
- Aquic moisture regime. Waterlogged 7 to 15 days during growing season (see Chapter 7).
- All *poorly drained* and *very poorly drained* soils.

Primary indicators of wetlands (not all are required)

- Hydrophytic plants dominate. Cattails, sedges, many other trees, shrubs, and reeds.
- One abundant plant species has one or more of these water-plants adaptations: pneumatophores (knees), prop roots, buttressed trunks, and floating leaves.
- Significant patches of peat mosses.

*Modified from Christopher L. Lant, Steven E. Kraft, and Keith R. Gillman, "The 1990 Farm Bill and Water Quality in Corn Belt Watersheds: Conserving Remaining Wetlands and Restoring Farmed Wetlands," *Journal of Soil and Water Conservation* **50** (1995), Mar.–Apr., 201–205.

17:3 Wetlands Identification and Delineation

The identification of a wetland and delineating its boundaries are not easy. The litigation of a lumber company and the U.S. Government concerning 6800 ha (16,796 a) of timbered land (that it can not be harvested if it is wetlands) indicates the uncertainty in what is wetland (Figure 17-1). One EPA scientist is quoted as saying, "I go out to a site and for legal purposes I have to draw a line in the sand and say, 'this is and this isn't a wetland.' It's kind of an artificial call from an ecological standpoint."[8]

[8]Donald M. Kent, "Introduction," Chapter 1 in Donald M. Kent (ed.), *Applied Wetlands Science and Technology,* Lewis Publishers, Ann Arbor, MI, 1994, 1–11.

FIGURE 17-1 Is this a wetland? With such deep water, even on cropland, one might say, yes. But, it is not wetland. The water stays only a few days a year, the water is aerated, and the soil does not become anaerobic, at least not for long. What are the boundary conditions? (*Source:* Reprinted from *Ceres*, the FAO Review on Agriculture and Development.)

The intent of identifying and controlling drainage or destruction of wetlands is to conserve valuable natural areas from extermination. However, the process of identifying where a wetland begins and ends, if it is marginal, and then delineating its boundaries is more difficult. As discussed in Chapters 2 and 13, soil is a continuum, often changing only gradually from one location to another. Wetlands and their boundaries, too, have spatial variation. At a wetland boundary, one side is defined as *not wetland* and the other side is said to *be wetland.* One observer calls finding those boundaries a "multi-million dollar question."

Some soils are wet and others are very wet. To use them for crop plants usually requires that they have some drainage. For soils that will be used for crops and drained for that or some other reason, it is important to understand the soil, why it is poorly drained, what drainage is needed, and what the soil will be like after drainage.

17:4 Soils That Need Drainage

Much can be learned about the internal drainage of a soil by digging into it. The light-gray colors of gley indicate long periods of continuous saturation. Periodic aeration usually produces orange-red mottles caused by oxidation of iron to various iron oxides (such as rust). This coloring develops in the most easily oxidized parts (cracks and other large channels, such as old root channels). The rest of the soil may be gray.

Except for rice, almost all crop plants of economic importance grow best when soil pores contain air that is easily exchangeable, because some space is required for the release of CO_2 from plant root respiration and bacterial respiration and for the entry of O_2 from the atmosphere. The purpose of artificial drainage is to increase aeration to the growing plant roots.

Causes of Poor Drainage

A soil may need artificial drainage because of a high water table that should be lowered or because of excess surface water that cannot move off the surface or downward into the soil fast enough to keep from suffocating plant roots. If the condition persists, it causes an oxygen (O_2) deficiency. The results are (1) a reduction in root respiration and hence in growth, (2) increased resistance to water and nutrient movement inside the root, and (3) formation of substances (such as manganese cations) of a kind or in concentrations that are toxic to plants (Detail 17-2).

Usually, poor internal drainage is caused by shallow depths to bedrock or by low-permeability clayey layers (Figure 17-2). Depressions with clayey bottoms will pond water, eventually forming peat and muck soils after thousands of years. Most organic soils need drainage; they formed because of poor drainage. Clayey soils in flat relief in humid climates are usually poorly drained. Soils may also have poor drainage conditions simply because they accumulate more water (high rainfall or collection of runoff water) than they can dissipate by their slow natural drainage. After heavy rains poorly drained soils take longer to lose surface water (Figures 17-1 and 17-2). Such soils may have poor-drainage *indicator plants,* such as marsh grasses, sedges, cattails, and willows. Standing water most of the year is an obvious clue (Figure 17-3). Surface salt accumulations during dry periods and light-gray gley soil are other visual clues to poor drainage.

Redox Potentials and Poor Drainage

There are various degrees of poor drainage. Poorly drained soil conditions are usually indicated by the **redox potential,** the relative electron concentration in the soil. In well-aerated soils the free gaseous oxygen accepts the electrons produced during decomposition of organic matter (Figure 17-4). If the free oxygen is used up, *nitrate* becomes the first electron acceptor and denitrification occurs, producing N_2, N_2O, and NO gases. Manganese as MnO_2 can also be reduced, releasing soluble Mn^{2+}. As the nitrate and manganese dioxide are used up, ferric iron can be reduced to ferrous (Fe^{2+}). Eventually, in very poorly aerated conditions, sulfate or sulfur will be reduced to sulfide. Even carbon dioxide can be reduced to methane gas (reduced carbon form) and be given off in very poorly aerated conditions. Thus, the development of maximum anaerobic conditions occurs with (1) exclusion of air (waterlogging), (2) warm temperatures for organic-matter decomposition, and (3) large amounts of easily oxidized organic material.

17:5 Paddy Rice

Rice is the major food grain for about 60 percent of the world's population. It is grown in the tropics and subtropics. Only about 1 percent is grown in the United States.

Rice is an aquatic plant and the only major food crop that can germinate and grow throughout its life under continuous shallow water. It can also be grown as an upland crop, much like wheat or corn. However, *paddy rice* is usually grown in soil covered with a layer of water 5 cm to 10 cm (2 in–4 in) deep because yields are much greater than when grown on drained soils (see Figure 13-9 in Chapter 13).

All rice in the United States is grown with flood irrigation, and about 80 percent of the rice in Asia is also so grown. Thirteen percent of Asian rice is grown on dry land (rain-fed,

Detail 17-2 Waterlogging Effects on Plants

Some symptoms of waterlogging include drooping leaves (apparent wilting and epinasty—leaves curve downward at margins), decreased stem growth rate, leaf abscission (prepares to drop), leaf chlorosis (pale coloring), adventitious (secondary) root formation, decreased root growth, death of smaller roots, absence of fruits, and reduced yields. The overriding effect of soil flooding is the limited diffusion of oxygen to roots. The plants that are tolerant of waterlogging have a good flow of oxygen from shoots to roots inside the plant, but the majority of plants require most of their oxygen from soil air around the roots. This oxygen flow internally in waterlogged-tolerant plants apparently occurs through larger air spaces within stems and roots than in those parts in plants not tolerant to waterlogging. Plants that adapt to waterlogged conditions during growth do so by forming larger internal air spaces, even at the expense of some cell destruction and the dissolution of some cellulose in cell walls.

The exact effects of waterlogging damage are rapid but still not clearly understood. The following changes and effects are those generally proposed:

1. A lack of adequate O_2 to roots initiates changes of the amino acid methionine to S-adenosylmethionine (SAM). SAM is converted by an enzyme to 1-amino-cyclopropane-1-carboxylic acid (ACC). The ACC can then be converted to ethylene *but only in the presence of* O_2. Ethylene ($H_2C = CH_2$) in high concentrations is known to alter plant growth.

   In anaerobic (inadequate free oxygen) conditions growing plants that are not tolerant of waterlogging (poor O_2 transfer to roots through stems) have the ACC produced in roots translocated to stems and petioles where O_2 is available. The ACC is then quickly changed to ethylene. The presence of high ethylene concentrations in petioles causes rapid expansion of cells on the upward side, causing the edges of leaves to droop (**epinasty**). This feature can be seen in less than a day. Other stimulators converting SAM to ACC are increased amounts of the plant hormone indoleacetic acid (IAA), wounds to the plant, and senescence (old age).

2. Apparent wilting of leaves, although not necessarily a general loss of leaf turgor, may be a combination of reduced water permeability to roots (low oxygen lowers roots' abilities to absorb water) and a water loss from leaves (which is unchanged) that exceeds intake. Rapid closure of stomata allows leaves to regain turgor but hinders growth by hindering CO_2 uptake into the leaf.

3. Contents of certain growth hormones, such as gibberellic acid (GA) and abscisic acid (ABA), are reduced in the transport parts of the plant, perhaps another chemical signal forcing the stomata to close. Reduced transport of IAA from leaves may cause the epinastic response by petioles. Increases in concentrations of growth substances (auxins) because of reduced transport from roots to the stem may be responsible for adventitious (new secondary) root formation.

4. The plant system can also transport upward some toxins produced in roots by anaerobic conditions.

5. Anaerobic energy transformations are poor; thus, growth rates are slowed because of limited energy.

6. Some toxic substances are produced, including hydrogen sulfide, butyric acid, and volatile fatty acid components of carbohydrate decomposition.

Sources: (1) K. A. Smith and P. D. Robertson, "Effect of Ethylene on Root Extension of Cereals," *Nature* **234** (1971), No. 5325; 148–149. (2) Makota Kawase, "Anatomical and Morphological Adaptation of Plants to Waterlogging," *HortScience* **16** (1981), No. 1; 30–34. (3) Kent J. Bradford and Shang Fa Yang, "Physiological Responses of Plants to Waterlogging," *HortScience* **16** (1981), No. 1; 25–30.

FIGURE 17-2 This Coker clay soil (Vertisol) near Vilas, Oregon, has poor internal and surface drainage. All Vertisols have very slow infiltration when saturated. During periods of high rainfall or surface flooding, this home site is a sticky mess. (*Source:* USDA—Soil Conservation Service.)

FIGURE 17-3 A wetland (swamp). This Cypress mucky clay soil is poorly drained, is in lakebeds, oxbows, and along stream channels. It is ponded most of the time and frequently flooded. It is mostly useful for habitat and grows bald cypress with an understory of buttonbush, palmetto, maidencane, lizardtail, and duckweed. (Courtesy of USDA—Natural Resources Conservation Service and the Louisiana Agricultural Experiment Station.)

no levees) and about 7 percent is rain-fed, but with levees around each field to retain rainwater. One reason for paddy culture is that it provides good weed control in naturally wet soils. Rotating rice with nonpaddy crops helps in weed control of both crops.

Most rice soils throughout the world have low hydraulic conductivity either naturally or artificially. To reduce the hydraulic conductivity, the soil is *puddled.* **Puddling** consists of repeatedly tilling the saturated soil to achieve an artificial tillage pan with soupy mud above it. The objectives of puddling are to reduce percolation losses of standing water and dissolved

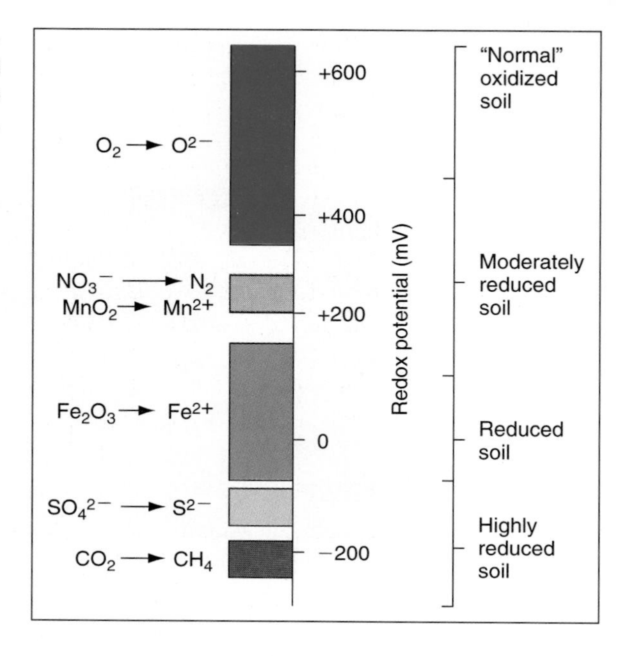

FIGURE 17-4 Example of the range in redox potentials in waterlogged soils and the location in the redox range where the various electron acceptors are active. (Courtesy of Raymond W. Miller, from data of W. H. Patrick, Jr., "The Role of Inorganic Redox Systems in Controlling Reduction in Paddy Soils," *Proceedings of the Symposium on Paddy Soil,* Institute of Soil Science, Academia Sinica, Science Press, Beijing, and Springer-Verlag, New York, 1981.)

plant nutrients, to control weeds, and to facilitate hand transplanting of rice seedlings where this practice is followed.

Rice can transport oxygen from leaves to roots internally in contrast to most crops that depend on oxygen transport through the soil pores to roots.

In the United States rice is grown on about 1.2 million hectares (3 million acres), having an average yield of more than 5000 kg/ha (about 4500 lb/a). The six rice-producing states' rank in acreage is Arkansas > Louisiana > California > Texas > Mississippi > Missouri. Throughout the tropical and subtropical world, rice is grown in 89 countries.[9]

Rice paddies are laid out in large beds surrounded by a levee called a *bund.* Seed may be planted with a small-grain drill and each bed then wetted enough to germinate them. When the plants grow up to a good stand (five-leaf stage), more water is added to form a depth of 5 cm to 10 cm. This depth is maintained until near maturity. Fields are then drained 2 to 3 weeks prior to harvest to facilitate combine harvesting. In some places barely sprouted rice seeds are sown from an aircraft. Rice seeds have the ability to sprout and grow through shallow planting in soil or through shallow water on the soil surface but *not through soil and water.* Most paddy rice outside the United States uses seedling transplants.

Rice tolerates a soil pH of 4.5 to 7.0; above about pH 7.0, Zn deficiency may occur. When flooded the pH of such soils shifts toward neutrality. Even somewhat saline soils can be used for wetland rice because water ponding dilutes the salt and thus the plasmolytic effect (death of plant cells by desiccation).[10, 11, 12] Reclamation of polders from the sea often uses paddy rice as the first reclamation crop, if the climate is suitable.

Fertilizer N is necessary to maximize yields of rice. Under lowland rice culture ammonium-nitrogen is more efficient than nitrate sources because of the hazard of denitrifi-

---

[9] S. K. DeDatta and S. S. Hundal, "Effects of Organic Matter Management on Land Preparation and Structural Regeneration in Rice-Based Cropping Systems," in *Organic Matter and Rice,* International Rice Research Institute, Los Banos, The Philippines, 1984, 399–416.
[10] Frans R. Moorman and Nico van Breemen, *Rice: Soil, Water, Land,* International Rice Research Institute, Los Banos, The Philippines, 1978, 185.
[11] International Rice Research Institute, *Field Problems of Tropical Rice,* IRRI, Los Banos, The Philippines, 1983, 172.
[12] D. S. Mikkelson and D. M. Brandon, "Zinc Deficiency in California Rice," *California Agriculture* **38** (1984).

cation of the nitrate. When urea, the cheapest source of solid N fertilizer, is topdressed into flood waters, the urea increases the pH of the water. This can cause volatilization loss as gaseous NH_3. Ammonia loss was measured by N isotope (^{15}N) techniques in the Philippines. During the 30 days following application of 80 kg N/ha (71 lb/a) of urea, 55 percent was lost by ammonia volatilization.[13] Techniques to release urea fertilizers more slowly into this hostile environment include using supergranules, sulfur-coated urea, urea mixed into large mud balls, and the use of nitrification inhibitors.

When flooded rice is followed by wheat or soybeans, phosphorus deficiency is common, for the following reasons:[14] (1) During flooding, phosphorus and iron become reduced and are more soluble and more available to rice than in aerobic soil; and (2) during the months after drainage of paddy, iron becomes reoxidized and the larger amounts of solubilized iron combine with soil solution phosphorus to make phosphorus and iron less available to the following upland crop.

17:6 Drainage Benefits

Some of the more obvious benefits of artificial drainage are as follows:

- Wet soils are usually the most fertile soils (high clay and high organic matter). Drainage permits them to be used for more productive purposes.
- Drained soils warm earlier in the spring. In temperate climates wet soils warm more slowly in the spring than drained soils because water requires four to five times more heat to raise a unit weight of it 1° than is needed for the same weight of dry soil minerals. Plant growth and other chemical reactions are slowed approximately 25% for each 4.7° C (10° F) decrease in soil temperature. Plant uptake of phosphorus, potassium, and most other nutrients is increased by drainage (aeration).
- Proper drainage makes the entire field more uniform in soil moisture (elimination of wet spots) and thus results in earlier, more predictable, and more efficient tractor tillage, planting, and harvesting operations.
- Removal of surplus water increases aerobic (well-aerated) microbial activity by permitting air (with its 20% oxygen) to replace water in more soil pore spaces. The resulting larger population of aerobic microorganisms produces more organic matter decomposition and, thus, more potential plant nutrients are made available.
- Drainage decreases the potential losses of nitrogen from the soil by microbial denitrification (which occurs under anaerobic conditions).
- Drainage reduces the buildup of toxic substances in the soil, such as soluble salts, ethylene gas, methane gas, butyric acid, sulfides, and excessive ferrous and manganous ions.
- Drained land is adapted to a wider variety of more valuable crops. Wet soils can produce well only with the limited crop types that can tolerate wet roots. Water ponded in small surface depressions, even when the profile is not too wet, can kill many crops, such as seedling corn.
- Drainage permits a deeper penetration by plant roots, thereby increasing the amounts of nutrients available to growing plants and resulting in greater crop yields. Deeper roots also make plants more drought resistant.
- Drained land is more capable of supporting buildings and roadbeds.

---

[13] S. K. DeDatta, et al., "Comparison of Total N Loss and Ammonia Volatilization in Lowland Rice Using Simple Techniques," Agronomy Abstracts, (1986), 197.

[14] R. N. Sah, D. S. Mikkelsen, and A. A. Hafez, "Phosphorus Behavior in Flooded-Drained Soils: Iron Transformation and Phosphorus Sorption," *Soil Science Society of America Journal* **53** (1989), 1723–1729.

FIGURE 17-5 Controlled drainage and contour planting in contour blocks help conserve soil planted to pineapple. Notice the diversion terraces and sodded waterways (upper center). The high rainfall in this Maui Pineapple Co. field in Hawaii requires controlled surface drainage to avoid severe erosion. (Courtesy of USDA—Soil Conservation Service; photo by Arnold Nowotny.)

- Septic systems perform best in well-drained soils; so do sanitary landfills.
- Controlled drainage allows use of lands for high value crops but with lessened erosion (Figure 17-5).
- Mosquito and fly populations are reduced by adequate drainage of wetlands.
- The market value of drained land is greater than that of comparable ponded ground. Drainage also preserves the land for arable use.
- Adequate drainage is essential for reclaiming saline, saline-sodic, and sodic soils.
- Drainage systems reduce heaving (pushing out of the ground) of plants by ice crystals forming around the plant crowns.

17:7 Drainage: Yes or No?

Drainage is expensive, so it must be given careful economic consideration. Currently only land previously drained can continue to be drained without special permission or restrictions. **Swampbuster** is a term given to a part of the Food Security Act of 1985, which denies federal subsidies to persons draining and farming previously undrained wetlands outside of strict regulations. This law is an attempt to control and reduce conversion of wetlands to farmlands, thereby helping to retain wetland habitats.[15] This law concerns conversion of wetlands to new croplands, not the drainage of land already cropped. The law deals only with land on which drainage was initiated after the law was enacted.

Even where legally permissible, some wet soils should *not* be drained, as explained in the following cases:

- Land where drainage would adversely affect the environment (e.g., by reducing the number and variety of aquatic animals, such as beaver, muskrat, and waterfowl). Income from leasing, hunting, and trapping rights before establishing drainage systems may even exceed probable crop income after drainage. Environmental protection of some areas may also override the economic benefits of draining the land for crop production.

[15]*Federal Register,* Sept. 17, 1987.

- Land where drainage would lower the ambient (surrounding) water table and thereby decrease the level and volume of water flowing into nearby streams, ponds, springs, and shallow wells.
- Land where excessive drainage would adversely affect capillary water rise in deep sandy soils (Psamments) and deep organic soils (Histosols). Such soils may lack sufficient water after drainage because of a slow and low capillary rise from the water table. This hazard is less on medium- and fine-textured soils, because the soil pores are smaller and more continuous, which aids capillary action.
- Wet soils that contain excess amounts of iron disulfide (FeS_2, pyrite, fool's gold) should not be drained. Upon being drained, the iron disulfide oxidizes to ferrous sulfate and *sulfuric acid,* thereby lowering the pH, sometimes even to a toxic pH 2.0.
- Land where drainage, particularly for soils low in fertility, may cost more than the increased value of crops grown on the soil.

17:8 Drainage System Selection

The two chief types of drainage systems are *surface* and *subsurface.* **Surface drainage** is achieved primarily by constructing gently sloping open ditches for water collection and by smoothing the soil surface and creating enough slope to facilitate runoff toward the drainage ditches. The most prevalent kind of **subsurface drainage** consists of burying conduits (sections of tile or porous plastic tubing) on specified grades (slopes).

There are no technical restrictions on the use of surface drains; they can be used on all kinds of soils. Subsurface drains, however, require a soil profile that is sufficiently porous to water to allow percolation through it to the buried drain tile or porous tubing. Many of the finest-textured clay soils have such small pores between the particles that they are too slowly permeable to be drained adequately by any *subsurface* drainage system. Such soils must be drained by a *surface* drainage technique or an open-ditch system.

Subsoil Drainage: Will It Work?

It is often difficult to determine whether a soil will drain rapidly enough to permit the use of some form of subsurface drainage. With such a large potential investment, it is usually worthwhile to make some field and laboratory determinations to find which drainage system is best suited to any particular soil area. Pore spaces that are nearly full of water at field capacity do not transmit much water down through the pores to drain lines. The soil pore air space through which gravitational water moves is the total pore space minus the pore space occupied by water held at field capacity (−33 kPa matric potential).[16] This value is known as the **drainage capacity.** As a percentage, drainage capacity is determined as follows:

percent drainage capacity = (% soil volume that is pore space) − (% water-filled pore space at field capacity, on a volume basis)

$$DC_p = E_p - P_v$$

In Mississippi, for example, the surface soil of Memphis silt loam (a Typic Hapludalf developed from loess) has a total pore space of 59 percent and a field capacity of 20 percent (Table 17-3). The drainage capacity is, therefore,

$$59\% - 20\% = 39\%$$

which is a readily drainable soil.

[16]Percentage moisture times the ratio of bulk soil density to water density equals volume percentage of water. This gives the percentage of the soil *volume* containing water at that moisture content.

Table 17-3 Total Pore Space, Pore Volume at −⅓ Matric Potential and Drainage Capacity of Three Mississippi Soils

Soil Type	*Depth (cm)**	*Total Pore Space (%)*	*Pore Volume at −⅓ bar of Matric Potential (%)*	*Drainage Capacity (%)*
Memphis silt loam	0–15	59	20	39
Bosket sandy loam	0–20	51	39	12
Sharkey clay	0–15	51	81	0

Source: W. M. Broadfoot and W. A. Raney, "Properties Affecting Water Relations and Management of 14 Mississippi Soils," Mississippi Agricultural Experiment Station Bulletin 521, 1954.
*Centimeters × 0.4 = inches.

Soils with drainage capacities greater than about 10 percent can be drained with subsurface systems. From Table 17-3, the Bosket sandy loam (a Mollic Hapludalf), for instance, with a drainage capacity of 12 percent, will drain fairly readily. On the other hand, Sharkey clay, a Vertic Haplaquept—because its moisture retention at −33 kPa is greater than the total dry soil pore space—*will not drain through tile drains.* The Sharkey clay expands greatly when wet, so the total pore space wet is greater than when it is dry. For this reason the total *dry* pore space percentage is exceeded by the water-filled pore space percentage. This soil can be drained only by surface ditches because water will not flow through the soil into tile or tube drains at a rate sufficient to remove the surplus water.

Drainage Systems: Which One?

Some of the more important relationships to be considered when choosing a drainage system are given in Table 17-4. The major factors are soil impermeability (the cause of the poor drainage to begin with) and the topographic features of the area.

17:9 Surface Drainage Systems

Surface drainage systems are best adapted to drain flat or nearly flat soils that are (1) slowly permeable, (2) shallow over rock or fine clay, (3) have surface depressions that trap water, (4) receive runoff or seepage from upslope areas, (5) require the removal of excess irrigation water, or (6) require lowering of the water table.[17] Four principal surface drainage systems are *open ditch, smoothing, ridge-till,* and *bedding.*

Open Ditches

Open ditches must be designed to conform to topography, natural drainways, land use, and soil characteristics. Such ditches are usually parallel and designed as a grid or in a herringbone pattern. Spacing between the parallel ditches is determined by the amount and intensity of the rainfall in relation to the infiltration capacity of the soil. The depth and width of the ditches, as well as the spacing, must be determined by using all available local data. Data on rainfall intensity are readily available from the U.S. Weather Service. Farm field drains are normally de-

[17]USDA—Soil Conservation Service, *National Handbook of Conservation Practices,* USDA, Washington, DC, 1977, 607-A-1.

Table 17-4 Soil and Topographic Parameters Important in Selecting the Type of Drainage System to Install on a Poorly Drained Area

Parameter	*Suggested Drainage System*
• Sloping areas	Interceptor lines in low areas or seepage spots
• Closed basins, level areas	Outlet problems, may need to pump
• Deep, permeable sands	Any system is adequate.
• Deep, impermeable clays	Careful irrigation management; mole drains, surface and open-ditch drains
• Shallow, permeable soil over impermeable layers	Tube or tile drains just above impermeable layer; careful irrigation management
• Deep (9–12 ft, 2.8–3.7 m thick), impermeable soils over coarse sands, gravels	Sump or well* drainage; surface and open-ditch drains in humid areas
• Water table fluctuates with irrigation	Tube drain system on a grid and careful irrigation control
• Water table fluctuates with rainfall	Surface drainage better; also, consider tube or tile drains
• Ponded water in fields	Surface grading (sloping) and surface drains

Source: Selected, modified, and tabulated from W. W. Donnan and G. O. Schwab, "Current Drainage Methods in the USA," in *Drainage for Agriculture,* J. Van Schilfgaarde, ed., No. 17 in the Agronomy Series, American Society of Agronomy, Madison, Wis., 1974, pp. 93–114.
*Shallow wells or sumps with the water collected are pumped to surface drainage lines.

signed to carry a maximum volume of water from the maximum expected 1-hour rainfall once in 5 years. When larger areas are drained, they may be designed to direct the surface flow adequately from a maximum 1-hour storm that might be expected only once in 20 years.

Field ditches should be at least 30 cm deep and, if machinery is to cross them, have side slopes of 8:1 (8 linear units horizontal to 1 unit vertical). If the drain isn't to be crossed with machinery, it is cheaper to build ditches at the angle of repose (natural stability) of the soil. A stable slope for loam, for example, is about 18° for clay, 27° for silt, and 45° for peat, muck, or sand. Seeding an adapted grass on the bottom and sides of the drain ditches helps to stabilize them but can slow water flow. Timothy and reed canarygrass are widely recommended. Do not use bermudagrass or quackgrass because these grasses will spread to the fields.

There is an alternative technique to simple ditches. It is known as the **open-W drainage system** and establishes side slopes of about 8:1 and moves the excavated soil toward the middles between each two drains (making a slight hump that results in a W-shaped soil cross-section, the ditches forming the low points). When dry enough, crops can be planted *in the ditches* at the same time as planting between them. Rows should run parallel to the ditches to avoid crossing the ditches with tillage implements, damaging the ditches. This system is similar to the drainage-by-beds system (see Bedding, on following page).

Open ditches do not satisfactorily drain any field if the soil surface is not uniformly smooth and sloping toward the ditches. The field should be plowed or disked and then leveled by a landplane or leveler. Laser-beam-guided equipment can greatly aid land leveling. Greater leveling efficiency is achieved if the technique is followed as shown in Figure 17-6.

FIGURE 17-6 Surface drainage system showing (a) a system of field ditches with the land smoothed to facilitate surface flow toward the ditches and (b) the recommended pattern for operation of a leveler for smoothing or grading. (*Sources:* (a) U.S. Department of Agriculture; (b) U.S. Department of the Interior.)

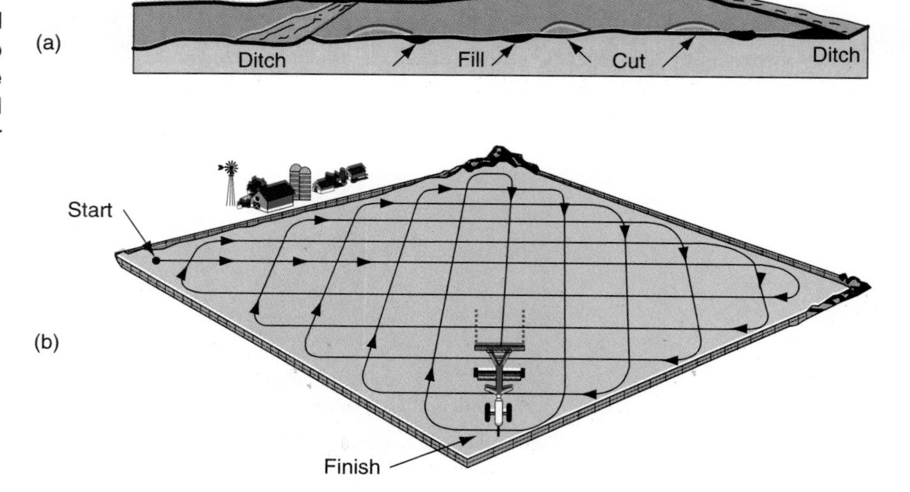

Drainage by Smoothing

Smoothing (eliminating minor ridges and depressions of the field without altering the general topography) and **grading** (making a uniform slope) usually need 2 years to complete. After the initial work the first year, a second smoothing is done the second year, once the soil has settled. The quality of the smoothing is best viewed immediately after a storm. If puddles persist—even small ones, a half meter across—the smoothing is not adequate. Yearly repairs are desirable. Rapid ditchers are available for making nonpermanent shallow-surface field ditches.

Maintenance of open ditches and drainage ways should be done at least once a year after crop harvest or more often if the need arises. Grasses and other weeds often grow rampant. Sediment must be removed, willow sprouts and water weeds removed, and, in some areas, tumbleweeds disposed of. Often a suitable herbicide may be used on growing plants, followed by burning to dispose of the plant residues.

Bedding

Fine-textured, fairly level fields in areas of high rainfall are usually drained most advantageously by the construction of **beds.** The system of **bedding** is similar to the open-W drainage system. One difference is that the drainage-by-beds system usually has a higher crown and is narrower between drains, as depicted in Figure 17-7. The more water to be disposed of, the narrower must be the beds and the higher the crown. The bed widths should be in multiples of the width of the tillage and planting implements to be used. Maintenance of bedding drainage includes plowing a back furrow at the crest of the ridge and plowing uphill, leaving the final furrows (dead furrows) at the drains.

Ridge-Till

In some soils the drainage needed is slight and mostly during the growing season to remove excess water from current storms. Instead of planting on level land, as is done where rainfall adds sufficient water, planting on the ridges of furrowed land is done. This is called **ridge-till** (Figure 17-8). In the ridge-till system the furrow is to drain excess water and the ridge is to keep a portion of the root zone well drained to aid seedling growth and even later growth.

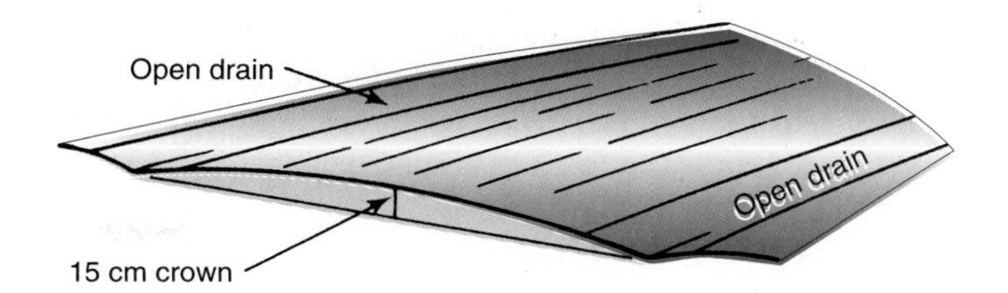

Open-W drainage system

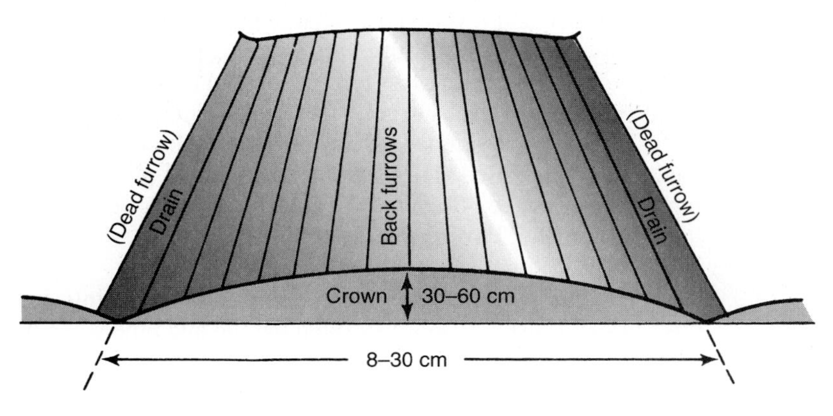

Drainage-by-beds system

FIGURE 17-7 Comparison of the open-W drainage system of open-ditch technique (above) and the drainage-by-beds system (below). (*Source:* Adapted from USDA—Soil Conservation Service.)

FIGURE 17-8 Ridge-tillage is like furrowing for furrow irrigation. Ridges are made and seeds are planted in them so that seedling plants have good drainage. The furrows allow drainage of excess water in high rainfall areas. As with hand-made mounds in Africa and other places, the size and height of the ridges could be varied for the crop or climate. (Courtesy of *Agricultural Research Magazine* **41** (1993), No. 9; 7.)

This concept has been used for many years in modified ways. The drainage-by-beds method uses widened ridges that hold more than one row of crops. In areas where machinery is too expensive but the rain is excessive, yams or other crops may be planted in the tops of mounds, which allow drainage all around them.

17:10 Subsurface Drainage Systems

The principal subsurface drainage systems include tile drainage, tube drainage, mole drainage, sump-and-pump drainage, and special vertical techniques, such as relief wells, pumped wells, and inverted or recharge wells. Subsurface drainage systems have become increasingly popular in recent years because prices of productive cropland make installing drainage systems on wet, less expensive land more feasible than acquiring better land. Crops can be planted over subsurface drainage systems but not over most surface drainage ditches.

Tile and tube drainage have the same technical layout parameters. They differ only in the kinds of pipe used. **Tiles** are short sections of pipe made from fired clay or concrete. The pipe pieces may be 30 cm to 60 cm (1 ft to 2 ft) long and 10 cm to 30 cm (4 in to 12 in) or more in diameter. **Porous plastic tubes** are made in similar diameters but one section may be 100 m (328 ft) long.

To inhibit the entrance of soil particles into openings in the sides of tile or plastic drain pipe, gravel is usually poured around the tile and pipe when it is laid or before backfilling. In some locations gravel or chat (ground stone) is scarce. In these areas Drainguard®, a nylon substitute for gravel, has been tried and proved satisfactory. A discussion of filters tried outside the United States is given in Detail 17-3.

Tile Drainage Systems

Satisfactory layout of a tile drainage system requires considerable planning and a great deal of experience. Technical assistance in layout and installation of tile drainage systems can be obtained from local Natural Resource Conservation Service (**NRCS**) offices (formerly Soil Conservation Service).

A tile drainage system can operate satisfactorily for a century or more if properly planned, adequately constructed, and carefully maintained. Concrete and fired-clay tile are both satisfactory drain materials, but in strongly acidic soils the concrete is dissolved by the acid. The depth and spacing of the lines of tile vary with the crops grown and the kind of soil. Soils with slow downward movement of water should have shallower placings of the lines of tile, and the lines should be laid closer together. Alfalfa and orchards need drainage with a depth of tile of about 1.2 m. Corn needs intermediate depths, and grasses and small grains can get along adequately with tile lines placed about 60 cm deep. Horizontal spacings may vary from 12 m to more than 90 m (40 ft–300 ft) between tile lines, depending on soil drainage capacity, which is related to soil texture and soil structure. Clayey soils need close spacings.

In low-permeability clay and clay loam soils, tile depth should not exceed 1 m and horizontal spacing no more than 21 m (about 70 ft). Tile lines in silt loam soil can be placed 1.2 m deep and 30 m (about 100 ft) apart. The respective maximum spacing and depth suggested for sandy and organic soils are 92 m (300 ft) and 1.4 m, respectively. In irrigated, arid soils it is common for tile to be laid 1.8 m or deeper.

Outlets for tile lines should be screened to prevent rodents and insects from plugging them. They should also be encased in cement, with a suitable apron to prevent undercutting by flowing water. The last 3 m of tile back from the outlet should be cemented at the joints. Other tile in the lines is placed end to end to permit water to seep in between the sections of tile (Figure 17-9).

Detail 17-3 Filters Proved in Other Countries

Although gravel is the customary material used in the United States as an envelope filter around newly laid subsurface drains, cheaper materials have been effective in other countries. These materials include peat, lime, gypsum, and small-grain straws.

Peat has proven satisfactory as a filter material in Belgium, the Netherlands, and Great Britain. Small-grain straws have also been used as drain filters in Great Britain. Calcium oxide (CaO) has been used successfully in Hungary for three different soil conditions. On acid soils CaO precipitates soluble iron, which reduces its buildup inside the drain pipe. Calcium oxide flocculates clay soils, making them more permeable to water and also less likely to wash into the drain tubes. On sodic soils CaO is mixed with gypsum ($CaSO_4$) to replace soil sodium ions (Na^+) with calcium ions (Ca^{2+}), which increases the movement of water through the soil particles (better hydraulic conductivity).

Source: Food and Agriculture Organization of the United Nations, "Drainage Materials," Irrigation and Drainage Paper 9, FAO, Rome, 1972, 122 pp.

Trees and shrubs near tile lines should be removed so that the roots cannot grow into cracks between the joints of tile. A small sink hole in the soil above a tile line indicates that one of the sections has been broken or displaced. It should be repaired before the whole tile system is ruined by being plugged with soil. Sometimes an outlet will erode and render the entire drainage system useless. Timely maintenance pays big dividends in extending the useful life of a tile drainage system.

Plastic Tube Drainage Systems

A major disadvantage of tile drainage is its cost, particularly in labor. Plastic tube drains are less expensive than are tile drains. The major advantages of plastic tube lines are savings in installation labor, near-foolproof alignment, and lower materials cost.

Corrugated plastic drain tubing ranges from 10 cm to more than 50 cm in diameter. All diameter sizes perform satisfactorily when installed in sand, loam, clay, and muck soils according to manufacturer's directions. The tubing comes in 76-m (250-ft) lengths weighing about 32 kg (70 lb) , and it has easy-to-connect couplers to form an endless tube (Figure 17-10). Laser-guided machines with a single operator can easily lay more than 1.6 km (1 mi) of tube line daily with a vertical-alignment accuracy to within a few *millimeters*. This eliminates some of the problems of poor drainage flow resulting from humps or dips in the tubing line. Although installation is rapid, the laser-guided laying equipment is expensive. Thus, plastic tubing drainage is excellent for installation in large fields needing extensive drains over long distances. Clay or concrete tile is a more rigid material than plastic and can still be used economically on short lines and special small drainage systems. Rodents sometimes cause failure of plastic drainage tubing by chewing holes in it.

Mole Drainage Systems

Mole drains are most useful when temporary drainage is needed, such as for salty land that needs rapid reclamation but that may not require permanent improved drainage once the excess salt is removed. Generally, mole drains can be cheaply and rapidly installed or even redone when needed every few years.

FIGURE 17-9 A tile drainage system will be satisfactory for a century or more if properly planned, adequately constructed, and carefully maintained. (a) A tile ditching machine seen in operation. West Virginia. (b) A tile line that has been laid but not yet covered. Oregon. (c) A line of tile that has not yet been completed but is already draining the field into an open ditch. The outlet will be screened to keep out rodents. Washington. (*Source:* USDA.)

(a)

(b)

(c)

(a)

(b)

(c)

FIGURE 17-10 (a) A laser-guided plastic drainage tube layer ensures great accuracy in the control of the line elevation and rapidity of installation. This installation near Florence, South Carolina, can lay drains 6 ft (1.8 m) deep at rates of 3000 ft (915 m) per hour, and can work at a 40-acre (16-hectare) field with only one setting of the laser plane command post shown in the left foreground. (b) A trench-cutting, laser-guided layer with following disks, which partially cover the tube to fix it in place. Careful covering is essential to minimize crushing the tube. Installation cost in 1989 was about $2.50–$3.00 per foot plus gravel and pipe. The tube often used has a nylon web welded around the plastic tubing that eliminates the need for gravel. (c) A less rapid trench-cutting tube layer, but with a less costly machine, in operation near Lewiston, Utah. The trailing box screens soil or gravel added to cover the tube first so large rocks or clods do not jar tile out of alignment. (*Sources:* (a) USDA—Agricultural Research Service, from J. L. Fouss; (b) United Enterprises, Orem, UT; (c) USDA—Soil Conservation Service.)

Mole drains are shallow, short-lived drainage channels created by pulling a torpedo-shaped object (the *mole*) through fine-textured soil about 50 cm to 60 cm (20 in to 24 in) deep (Figure 17-11). Even in the first year some channels will partly fill with soil. To compensate for this loss of drainage capacity, the channels can be dug only a few feet apart, more closely than is needed originally. Thin, perforated, plastic liners are available for an additional cost, but they extend the mole drain channel life considerably.

Sump-and-Pump Drainage Systems

Sump-and-pump drainage is used for areas in which water removal is not possible by gravity flow. **Sump-and-pump** systems collect water in a low-lying area and then remove it by pumping. The sump is a low area where water will flow from surface or subsurface runoff or drainage. To avoid collapse of the sides of the sump depression, it should be lined with sand, then gravel, and on top a layer of coarse gravel or broken rock. The sump motor and pump should be mounted rigidly above the sump depression and a float valve installed to start and stop the motor and pump.

FIGURE 17-11 Organic soils usually require drainage because of high water tables. A mole drainage machine, in operation in Florida, is making a cut from the main drainage ditch and continuing the mole drainage line across the field. The mole drain *plow* has a knife-edged shank that eases the pull by cutting through the soil, and the torpedo-shaped mole tool has a trailing chain and ball to smooth the tunnel a second time. (*Source:* USDA—Soil Conservation Service.)

With increasing energy costs, every practical effort should be made to avoid sump-and-pump drainage if gravity drainage is practical (shape the fields to permit gravity flow of drainage water). The *polder* lands reclaimed from the sea have the surface level of the land below sea level. Pumping is necessary to remove drainage water. *Windmills* often furnished the power to lift the water to elevated drainageways.

Vertical Drainage Systems

The principal kinds of vertical drainage systems are vertical subsurface drainage inlets, inverted wells, pumped wells, and relief wells.

Vertical Subsurface Drainage Inlets An increasing use of herbicides to kill weeds on terraced cropland causes runoff of sediments with attendant herbicides that often kill grasses used in stabilizing grassed waterways. Furthermore, land now used for grassed waterways could be producing crops under a different drainage system. **Vertical subsurface drainage,** an alternative technique for moving runoff water from higher-level to lower-level terraces, consists of installing vertical pipes with screened tops into which surplus water enters. The water then moves through a subsurface drainage system to a lower terrace or a protected outlet.

Inverted Wells In this vertical well drainage system, *excess water is directed into a hole in the ground* (the reverse of the usual water flow direction in a well; hence the name *inverted wells*). Inverted shallow wells near Fresno, California, are being used in soil underlaid with hardpan, which ponds the surface water. The wet fields become hatching grounds for mosquitoes when water stands for as long as a week. A screw auger drills holes about 23 cm in diameter and 3 m to 3.5 m deep where needed to drain the wet spots. The bore holes, filled with coarse gravel and several centimeters of soil on top, drain the soil in less than 24 hours, causing mosquito eggs to dry and die. An added benefit of the drainage is that the soil can be cropped earlier.

The first law of ecology: In nature you can never do just one thing, so always expect the unexpected; or there are numerous effects, often unpredictable, to everything we do.

—Anonymous

The danger is not what nature will do with man, but what man will do with nature.

—Evan Esar

Questions

1. Define a *wetland* and discuss the problem of delineating its boundaries.
2. What is *hydric soil* and how does it differ from a soil that is not hydric?
3. What are six or seven benefits to people of retaining wetlands?
4. What are the points you would argue in a debate to save wetlands if given pressures for food production, and draining more wetlands is proposed to produce food?
5. What is the law now about draining wetlands on your property and anywhere else?
6. What evidence is visible that might indicate that a soil has very poor drainage? (Consider vegetation, appearance in rainy weather, and soil colors.)
7. (a) What is *gley*? (b) What are *mottles*? (c) What are the colors of gleyed and of mottled soils?
8. State how drainage affects these items: (a) rate of warming, (b) the variety of adaptable crops, (c) toxic organic substances produced, (d) depth of the root zone, and (e) salt accumulation or removal.
9. Recent *swampbuster* laws restrict indiscriminant drainage of wetlands or swamps. What is the justification given?
10. Name some kinds of toxins produced in poorly drained soils.
11. Define *paddy rice* and describe the changes in aeration and soil pH as they are flooded.
12. Because all plants need oxygen for root respiration, where do rice roots in an anaerobic paddy get oxygen?
13. List advantages and disadvantages of (a) open-ditch drains, (b) tile drains, and (c) mole drains.
14. (a) What is the composition of the *tile* in tile drains? (b) How does water seep into the tile lines? (c) What problems can occur to tile lines?
15. Explain the use of lasers in installing drain lines.

Pollution of Water, Air, and Soils

The far more difficult problems [in controlling affronts to law and decency] in a complex society are those where there is not evil intent, no deliberate intrusion on the security of others, not even outrageous stupidity, just normal, law-abiding, enlightened folks going about their daily business.

—Lawrence W. Libby

It is not alone what we do, but also what we do not do, for which we are accountable.

—Moliere (1622–1673)

18:1 Preview and Important Facts

PREVIEW

The principal cause of soil and water degradation (pollution) is people. Devising policies to reduce or eliminate the adverse cumulative effects of human actions is difficult politically; effective policies restrict actions that many people regard as personal and beneficial. Early anxieties about pollutants mainly concerned cancer. Now we are concerned with pollutant effects on many kinds of diseases, physiological changes, and susceptibility to diseases.

Pollution is *adding something to air, water, or soil that makes it less desirable for human use or less able to maintain nature's balance.* Most activities have the potential to pollute with **noise, wastes, toxins, tastes, odors, carcinogens,** and other undesirable products.

Pollution happens as people use resources. Nature is able to handle reasonable amounts of most natural substances, but we have become more consumptive and more numerous, discarding larger and larger amounts of wastes. We have manufactured hard-to-decompose materials—metal tools, homes, and machines; resistant synthetics and plastics; and countless varieties of glass, ceramics, and heavy metals. We have invented poisons and toxins to control unwanted pests. We are overdue in facing the havoc we have wrought. The question now is whether we can afford the reclamation without major changes in the lifestyles to which most people in developed countries have become accustomed.

IMPORTANT FACTS TO KNOW

1. The definitions of *bioaccumulation, pollution, eutrophication, half-life, heavy metals, PANs, pesticides, carcinogenic, teratogenic,* and *radioactive*
2. The seriousness of nutrient pollution of waters and which alternatives are available
3. The meaning of *nonpoint-source pollution*
4. The general problem of heavy-metal contamination
5. The reason we will continue to use pesticides for at least several decades more and the results if pesticides were to be banned from use
6. The health hazards from infection and from heavy-metal bioaccumulation when applying sewage sludges and various other solid wastes to cropped soils
7. The extent to which the soluble-salt hazard is increasing worldwide and whether the problem is likely to lessen or worsen
8. The extent of radioactive fallout on soils and the alternatives available to clean up polluted soils
9. The extent and kinds of damages caused by soil erosional sediments

18:2 Threats to the Environment

When attempting to make changes, we need to know what the problems are and what they threaten. Is pollution a threat to important things, or is being environmentally conscious just a current fad? It is *not* a fad. *There are real problems that need fixing.* What are the cumulative threats of all our polluting actions? Among others, these threats include *loss of biodiversity, landscape modification, overexploitation, introduction of alien species, changes in biogeochemical cycles, pollution of groundwaters,* and *soil erosion.*

The **loss of biodiversity** is the extinction of one or more species. An extinct species, whatever its apparent or unknown value, is gone forever.

Landscape modification is replacing nature's complex ecosystems with large areas of monoculture crops, drained wetlands, housing units, and timber for construction. Smaller areas of habitats exist. Tigers, elephants, buffalo, and rhinoceros that need large areas on which to live are now in small isolated areas, and that limitation threatens their existence. Deer, with houses on their winter range, use home gardens and ornamentals for winter food. Innumerable smaller organisms have less habitat available in cities, urban housing areas, and even farmlands.

Overexploitation threatens loss of species. Excessive ocean fishing, trapping for furs, trading animal parts (ivory, rhinoceros horns, bear claws, tiger paws), excessive hunting, and lumbering all cause changes in habitats and the unnecessary killing of animals and destruction of plants. Fossil fuels are being wastefully used and will soon be in short supply.

Introduction of alien species may have good or bad effects. The water hyacinth, brought to the New Orleans Cotton Exposition in 1884 as a bottled gift in one display, now is a weed that clogs many drainage waterways (Figure 18-1). The African honey bee, a ferocious and dangerous bee, was accidentally helped in its move to South America. It has since found its way into the United States. Many enemies of local pests are brought into the United States to control the pests. This is called **biological pest control,** one technique of many in **integrated pest management (IPM),** which also includes *cultural control,* use of *pesticides, genetic resistance,* and other techniques. How these introduced species might alter other systems is not yet fully known.

Cumulative changes in biogeochemical cycles include changes to climate and the atmosphere. Depletion of the stratospheric ozone layer by chlorofluorocarbons (CFCs) can allow greater ultraviolet radiation (UV, which causes skin cancer) to reach the Earth. The increase of carbon dioxide and methane in the air cause the well-known greenhouse effect, which is the warming of the Earth. The results of these changes may be catastrophic; the possibilities are being debated, and whether the changes are already at a point of no return is an ongoing disagreement.

FIGURE 18-1 Unexpected problems from the water hyacinth cause major water flow problems in canals and other waterways. (Courtesy of USDA—Agricultural Research Service.)

Pollution of fresh water is recognized by most people and has received considerable attention. Extensive pollution from city sewage and industrial wastes has done great damage to many lakes. Lake Washington near Seattle and Lake Erie near Buffalo, New York, are examples of those problem lakes in the 1960s, which are now recovering (Figure 18-2). The problems came from large populations, excessive untreated or partially treated sewage, and industrial wastes that were dumped into relatively small bodies of water—too much material into too little water.

Salination is the accumulation of soluble salts. All irrigation waters have soluble salts. The more efficient irrigation is, the less water that is added to leach out salts. Less leaching of soils in arid climates causes salt accumulation as the added water is evaporated and transpired.

Truly, the pollutants and their consequences to the world are at critical stages. It will take some time to repair the damages already wrought. Some damages may yet be hard to resolve. These facts of pollution could fit the statement attributed to President Harry S. Truman, "*I never give them hell. I just tell the truth and they think it's hell.*" The truth about the extent and nature of some pollution is like that.

18:3 Pollution Terminology

In considering sensitive topics, such as pollution, the meanings for terms used must be clear and exact. Some terms listed have more complete definitions given in the Glossary.

Acceptable daily intake (ADI): Amount of material, plus a "safety margin," safe to ingest.
Acute toxicity: Symptoms are exhibited by the organism in a short period of time.
Antagonism: Substance A reduces the detrimental effects of a second substance B.

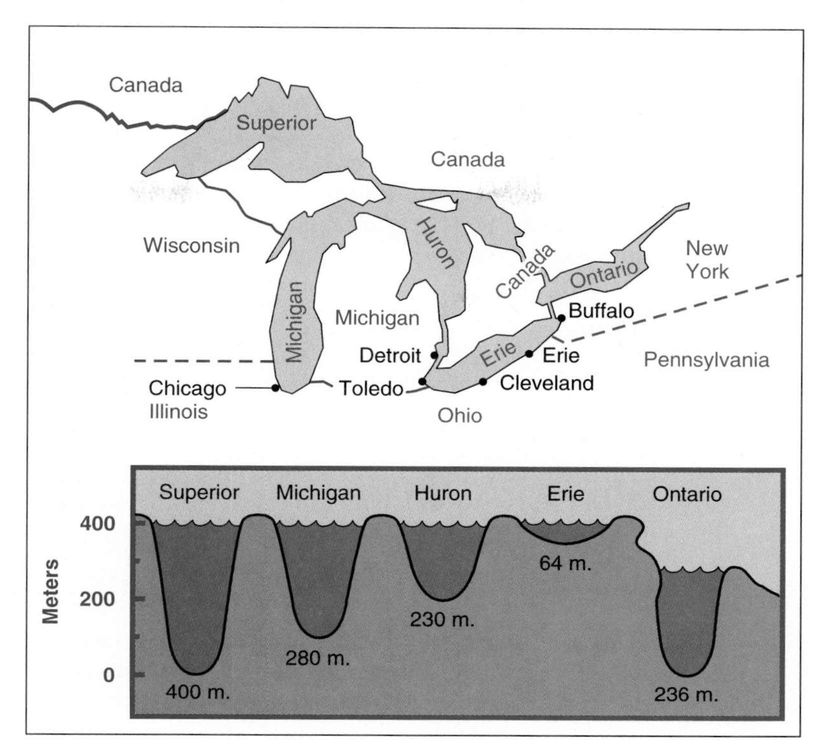

FIGURE 18-2 Lake Erie, one of the Great Lakes in the northeastern United States, was devastated by eutrophication in the late 1950s and 1960s from the partially treated sewage dumped in it from the five large cities and smaller population centers around the lake. This cross-section shows how shallow the lake is and one reason why the other four lakes fared better than Erie. The lake recovered well when dumping was reduced. (Drawn by Raymond W. Miller.)

Bioaccumulate (biomagnify): Increased concentrations accumulated by an organism over time. *Biomagnify* usually means increasing the concentration upward in the food chain.

Cancer potency factor: Causing no more than *one additional cancer per million people.*

Carcinogenic: The ability to cause cancer in animals and, by implication, in humans.

Chlorinated hydrocarbon: Organic pesticides with chloride groups on carbon atom.

Chronic toxicity: Effects or symptoms show up slowly, months or even years after exposure.

Eutrophic (eutrophication): Water overly enriched with nutrients; forms algal bloom.

Food chain (food web): Energy and material sequentially transferred through many organisms, one being food for the next. The food web is intermeshing of many food chains.

Half-life: The time for half of a substance to be destroyed, made inactive, or to be lost.

Heavy metals: High-atomic-weight metals such as cadmium, lead, mercury, nickel, and chromium, which are toxic to animal life and bioaccumulate.

Mutation: A random but inheritable gene change, which can occur naturally.

PANs: A group of chemicals (**p**eroxy**a**cetyl**n**itrate**s**) present in photochemical smog and extremely toxic to plants and irritating to eyes and membranes of the nose and throat.

Reference dose (RFD): Level of daily exposure that, over 70 years, has no negative effect.

Risk-benefit analysis: Evaluation of pesticides when considering registration versus permitted residue levels.
Synergism: Two factors interact causing a greater effect than the sum of the two substances separately.
Teratogenic: Causes tissue deformations resulting in birth defects.
Threshold level: Maximum level of a substance tolerated without ill effects.
Xenobiotic: Substance foreign to natural systems and decomposes slowly, usually.

18:4 Plant Nutrients

Nitrogen and phosphorus, which are added in agricultural operations, are the major nutrients of concern in water pollution. Some of the expected effects of various forms of nitrogen are shown in Table 18-1. Phosphorus exerts its greatest influence by increasing eutrophication. Concentrations of 0.1 parts per million (ppm = mg/kg) of phosphorus usually cause algal growth. Nitrate-N above 10 ppm can cause health problems in infants.

Eutrophication

Fertilizers increase algal growth in surface waters (and mosquitos in shallow water) into which they are washed. **Eutrophication** is defined as an *overabundance of nutrients in water,* which causes accelerated algae and water-plant growth. Often phosphorus (which is usually scarce in water because it has low solubility) is the major cause of eutrophication, but nitrogen and other nutrients also contribute to the problem. The considerable masses of dead algae and other organic material undergo decomposition continually. Because decomposition uses up oxygen from the water, these large masses of decomposing tissue deplete the gaseous oxygen (O_2) in water, causing other aquatic life to die and near-anaerobic (low O_2) conditions of the water body.

It is likely that most surface waters already have adequate nitrates in them to encourage growth of water plants and N_2-fixation by algae. The major mechanism to limit algal

Table 18-1 Some of the Potentially Adverse Environmental and Health Effects Caused by Forms of Nitrogen

Effect	*Causative Agent*
Environmental quality	
Eutrophication	Nitrogen sources in surface waters
Corrosive damage	HNO_3 in rainfall (acid rain)
Ozone layer depletion	Nitrous oxides from fuels, denitrification, and industrial stack emissions
Human health	
Methemoglobinemia in infants and elderly; also livestock	Excess NO_3^- and NO_2^- in water and food
Respiratory illness	PANS and other nitrogen oxides
Cancer	Nitrosamines from NO_2^- and secondary amines in food

Source: Modified and abbreviated from D. R. Keeney, "Nitrogen Management for Maximum Efficiency and Minimum Pollution," in *Nitrogen in Agricultural Soils,* F. J. Stevenson, ed., American Society of Agronomy, Madison, WI, 1982, 605–649.

growth is to minimize *phosphate* additions. Phosphorus is not a mobile nutrient in soils; it too easily forms insoluble substances with many common cations. Because of its insoluble nature, phosphorus is often the limiting nutrient for algae and other plants growing in water. The usual ways phosphate enters waters are (1) from municipal sewage with its high phosphate detergents, (2) from direct dumping of wastes (sewage, animal wastes, industrial wastes), and (3) piggy back as phosphate adsorbed onto soil particles eroded into waters.

In Minnesota more than 95 percent of measured phosphorus loss from soils was carried away in eroded sediment. In Canada, of the phosphates applied in late fall on top of frozen ground—a nearly optimum condition for maximum spring runoff losses as soluble phosphorus—only 10 percent (4.7 kg/ha, or 4.2 lb/a) was dissolved and carried off the field.[1] Canadian water quality criteria suggest a maximum of 0.15 mg/L of phosphorus as acceptable. Interestingly, one study from 6 years of data showed that natural phosphorus losses from *unfertilized* fields *exceeded* the quality criteria guidelines, and this was from semiarid dryland wheat soils (too little water to run off, not a highly fertile soil).

Unnecessary pollution should be avoided by wise use of fertilizers, but *zero* tolerance is impossible, unrealistic, and catastrophic to the production of food supplies. In another example, a concentration as low as 0.01 mg/L of inorganic phosphorus enhanced algae growth in waters.

Nitrogen in Groundwater

Nearly all nitrogen fertilizers (except organic and other slow-release fertilizers) are very soluble in water, and the final oxidized form, nitrate, moves readily in the water. Heavy rain or irrigation following nitrogen application can wash the fertilizer off the soil into surface flow and carry urea and nitrates deep into the soil to drainage waters; sandy soils are particularly susceptible to this.

Fertilizers are not the only source of soluble nitrogen in soil water. Nitrogen also comes from decomposing organic matter and ashes from burns (Figure 18-3). Approximately 37 percent of the U.S. total soluble nitrogen comes from soil humus, 22 percent from human and animal manures, 18 percent originally from fixation by soil bacteria and algae, and 9 percent from rainfall. *About 13 percent comes from fertilizers* added to soils. Nitrogen in rain and snow comes from dissolved nitrogen oxides. Sometimes the source of soluble nitrogen moving into waters is unusual. For example, in a deep geologic substratum (27 m; 89 ft) in Nebraska, nitrate concentrations were high (25 ppm–85 ppm), and it seemed the nitrates would in time be washed into groundwaters. Investigation proved that the source was not fertilizer but deeply buried ammonium and decomposing organic matter deposited 35,000 years ago, before it was covered by loess.[2]

The most important factors that influence the amount of nitrate movement to groundwater or surface waters are the following:

- The amount of nitrate dissolved in the soil solution
- The rate of its use by plants
- The rate of immobilization into soil microorganisms or newly synthesized soil organic matter
- The amount of water available for runoff and leaching through the soil
- Soil permeability

---

[1] W. Nicholaichuk and D. W. L. Read, "Nutrient Runoff from Fertilized and Unfertilized Fields in Western Canada," *Journal of Environmental Quality* **7** (1978), No. 4; 542–544.

[2] J. S. Boyce, J. Muir, E. D. Seim, and R. A. Olson, "Scientists Trace Ancient Nitrogen in Deep Nebraska Soils," *Farm, Ranch, and Home Quarterly* **22** (1976), No. 4; 2–4.

FIGURE 18-3 In Venezuela thousands of acres of brush and trees were pushed down by these land clearers, the dead vegetation pushed into piled rows and burned. The ashes were readily washed into groundwater, streams, and lakes. Ashes from natural wildfire burns would have the same fate.
(Courtesy of Raymond W. Miller, Utah State University.)

Soils heavily fertilized above recommended levels and soils naturally high in fertility are potential sources of nitrate contamination in groundwaters or runoff waters. If much of the soluble nitrate in soil solution is quickly absorbed by a rapidly growing plant or by multiplying populations of soil microorganisms, less is available to become leached. Nitrates are soluble ions but are not leached unless enough water exists to leach through the soil and the soil is permeable enough to allow leaching. Some examples of losses of nitrogen and concentrations of nitrate that could be leached are shown in Table 18-2. Any soil with a large amount of readily decomposable organic material can, in favorable climates and with adequate leaching, supply considerable nitrate to waters. These can be rangeland, forest, urban, or cultivated soils.

Methemoglobinemia

Nitrate becomes toxic to any animal with a disrupted digestive tract in which the conditions cause microbes to reduce nitr*ate* ion to nitr*ite* ion in large amounts; this effect is called **methemoglobinemia.** The nitrite is absorbed into the bloodstream, where it oxidizes *oxyhemoglobin* (the oxygen carrier) to *methemoglobin* (which cannot carry oxygen)—a young animal suffocates. The blue coloring (*blue baby*) is called *cyanosis.* When 70 percent of the hemoglobin is changed, death may occur. In mammals 0 to 3 months old, the stomach's lower acidity and the particular bacteria it contains (later killed by stomach acids) causes a high rate of conversion of nitrate to nitrite. In healthy, older mammals and adults the stomach acids and rapid absorption and excretion of unused nitrates make this nitrate poisoning unlikely,[3] but it can be a problem in sick and elderly adult mammals.

The nitrate content permitted by public health regulations for drinking water is 45 ppm nitrate (or 10 ppm nitrate-nitrogen). Many wells have been tested that approach or exceed this value; some are near the 100-ppm nitrate level. Obviously there are some justifiable concerns with nitrate concentrations.[4, 5]

Although deaths of infants from methemoglobinemia are rare, other possible effects of reduced oxygen transport may be extensive. Death is not the only possible damage.

Nitrate contents may be hazardous to grown animals. The bacteria that reduce nitrate to nitrite are found in adult ruminants (cows and sheep), as well as in infant pigs and chickens,

---

[3]Gary W. Hegert, "Consequences of Nitrate in Groundwater," *Solutions* **30** (1986), No. 5; 24–31.
[4]Nebraska Cooperative Extension Service, "Living with Nitrate," Paper EC 81-2400, 1981.
[5]W. L. Magette, R. A. Weismuller, J. S. Angle, and R. B. Brinsfield, "A Nitrate Groundwater Standard for the 1990 Farm Bill," *Journal of Soil and Water Conservation* **44** (1989), 491–494.

Table 18-2 Loss of Nitrates from Various Sites by Leaching or Runoff

Site Description and Time Interval	Annual Nitrogen Loss (kg N/ha)
• Ontario, Canada, tile-drained Humaquepts,[a] in drainage water	
Clay, corn-soybeans-vegetables, 110 kg N/ha added, 85 kg N recommended	64
Clay, soybeans-wheat-barley-corn, 35 kg N added, 50 kg N recommended/ha	16
Sand, corn, 150 kg N added, 150 kg N recommended/ha	4
Sand, corn, 200 kg N added, 150 kg N recommended/ha, high humus	49
• Ontario, Canada, tile-drained organic Borosaprists,[b] in drainage water	
Well decomposed, onions, 70 kg N added yearly, site 1/ha	96
Well decomposed, onions, 70 kg N added yearly, site 2/ha	188
Well decomposed, sandy loam at 50 cm, onions-carrots, 50 kg N added/ha	202
• New Hampshire, hardwood forest	
Clearcut area, no vegetation removed, herbicide-inhibited regrowth	97
Control area not cut or sprayed	2
• Alsea River, Oregon, coniferous forest	
Clearcut area with slash burned	15
Control area, no vegetation removed	4
	Nitrate Concentrations (mg/L)
• Intact forest sites,[c] on soil water	
Indiana, maple and beech, no root uptake of N	91.0
Indiana, maple and beech, control (normal forest cover)	0.7
New Mexico, ponderosa pine, no root uptake of N	2.7
New Mexico, ponderosa pine, control (normal forest cover)	0.04
Oregon, western hemlock, no root uptake of N	33.0
Oregon, western hemlock, control (normal forest cover)	1.1
Washington, red alder, no root uptake of N	70.0
Washington, red alder, control (normal forest cover)	17.0

Sources: (1) P. M. Vitousek, J. R. Gosz, C. C. Grier, J. M. Melillo, W. A. Reiners, and R. L. Todd, "Nitrate Losses from Disturbed Ecosystems," *Science* **204** (May 4, 1979), 469–474. (2) M. H. Miller, "Contribution of Nitrogen and Phosphorus to Subsurface Drainage Water from Intensively Cropped Mineral and Organic Soils in Ontario," *Journal of Environmental Quality* **8** (1979), No. 1; 42–48.

[a]Normally poorly drained soil high in humus.

[b]An organic, well-decomposed, cold-area soil.

[c]Control plots did have normal plants with root uptake; other plots were isolated portions where tree roots, etc., were cut by a trench to avoid major root activity.

and in the secum and colon of the horse. Thus, the following conditions could cause damage to animals from methemoglobinemia, to some degree, and should be given attention:

- The mammal infant about 0–3 months old is susceptible. Concentrations as low as 65 ppm nitrate have produced symptoms.
- Cattle and sheep (rumen) and horse (secum and colon) can develop the condition. It is especially common in cattle.
- High-nitrate vegetables or other foods (particularly spinach) stored where microbial growth is expected may magnify the problem to those eating the food.
- Damp forage materials containing high nitrate contents add to the hazard to those eating the forage.
- Grazing of newly greened pastures in spring or after heavy frost.

The *acute* dosages of nitrate for cattle are about 50 mg of nitrate-nitrogen taken in per kilogram of animal weight (25 g, or 1.5 oz, per large cow).[6] *Chronic* (lower level) intakes of 0.5 ppm *nitrite* or 100 ppm to 150 ppm *nitrate* may cause degeneration of vascular tissues of the brain, lungs, heart, liver, and kidneys.

Comparisons of analytical values of nitrate found in plants at the turn of the century with those of recent years give no evidence that plants today have higher nitrate contents as a result of using chemical fertilizers. However, total nitrate in the environment has increased.

A number of N-nitroso compounds are formed by reduction of nitrate to nitrite forming nitrous acid (HNO_2), which reacts with a variety of nitrogenous compounds. Some of these compounds are highly toxic, and some are animal carcinogens. The amount of nitrite formed in the *saliva* of the mouth is five times greater than the amount in most dietary sources and 12 times greater than nitrites in meat *curing* actions. The small hazard reduction by eliminating nitrite in meat curing would be overwhelmed by the increased hazard of the bacterium *Clostridium botulinum* (controlled by nitrite) that produces the toxin which causes botulism.[7]

18:5 Oxygen-Demanding Wastes

Chemical oxygen demand (COD) is a measure of decomposable material and other oxygen consumers in the wastewater or slurry. **Biological oxygen demand (BOD)** is a measure of the amount of oxygen required from water for biological decomposition of organic material. It is an indication of the oxygen stress that organic pollutants will have on living aquatic organisms. BODs are determined on a more restrictive basis than CODs and so are of smaller values than CODs for the same water samples. Organic materials (high CODs) cause microbes decomposing them to use the water's oxygen and make the system anaerobic, an undesirable condition.

Wastewaters Added to Soils

Wastewaters that must be cleaned up include domestic-sewage effluents as well as an almost unlimited variety of liquid industrial wastes. Many industrial wastewaters have been dumped into municipal sewage lines routinely. Tighter Environmental Protection Agency (EPA) restrictions now require that such liquids be treated extensively prior to disposal into municipal treatment plants or back into groundwater. Although irrigating with wastewaters partially cleans water by percolation through the soil, not all contaminants in the water are removed; soluble salts and organic chemicals may continue to flow with the water to groundwater or surface waters.

In general, the disposal of wastewaters on land or into waters is permitted only if it does not cause the following:

- Extensive groundwater pollution
- A direct public health hazard
- An accumulation in the soil or water of hazardous substances that can get into the food chain
- An accumulation of pollutants such as odors into the atmosphere
- Other aesthetic losses, within limits (Figure 18-4)

---

[6]G. B. Garner, W. H. Pfauder, G. E. Smith, and A. A. Case, "Nature and History of the Nitrate Problem," in *Science and Technology Guide,* Columbia Extension Division, University of Missouri, Columbia, 1979, 9800.

[7]Charles A. Black, "Reducing American Exposure to Nitrate, Nitrite, and Nitroso Compounds: The National Network to Prevent Birth Defects Proposal," Council for Agricultural Science and Technology, Ames, IA, 1989, 13.

(a)

(b)

FIGURE 18-4 Sewage wastewater being applied through a sprinkler irrigation system to a forest in summer (a) and in winter (b). A forest seems to be an ideal ecosystem for discharge of sewage effluent. To avoid damage, a scientific assessment must be made of each site as to the water capacity of the soil and the tolerance of the vegetation to chemicals in the effluent. Selection of spray sites must be done carefully to avoid runoff into surface water if freezing seals the soil, causing ponding or runoff. (Courtesy of William E. Sopper, Pennsylvania State University.)

Wash waters are one of the most common wastewaters. Wash waters in large quantities are used in vegetable and fruit processing plants. Dade County, Florida, is an example of a sensitive area.[8] More than 1.8 million people there obtain drinking water from a shallow, porous, limestone aquifer that is easily polluted (according to past experience). The county

[8] Dennis F. Howard, Sue M. Alspack, and Nancy D. Stevens, "Wastewater Disposal at Fruit and Vegetable Packing Facilities in Dade County, Florida," *Journal of Soil and Water Conservation* **45** (1990), 274–275.

FIGURE 18-5 Chisel injectors are used effectively to apply slurries of sludge and animal manure. This method conserves ammonium nitrogen and removes much of the undesirable appearance and smell associated with surface applications. (Courtesy of Big Wheels, Inc., Paxton, IL.)

also produces vegetables, fruits, and ornamentals worth more than $350 million a year on more than 32,000 hectares (80,000 acres). Composition of the effluent of 18 of 38 packing operations—particularly those of limes, mangoes, tomatoes, and potatoes—exceeded water quality standards for BOD and some other contaminants. The obvious result is to alter processing and treatment of effluents to improve water quality.

Sewage Sludge

For centuries people in Asia have recycled human waste through the soil to utilize the nutrients and organic materials it contains. In Western developed countries such use has not been popular, largely due to concern about spreading human diseases and the aesthetics of the practice. Animal manures are beneficial to crops; human waste could be just as useful. Municipal sewage systems have frequently buried residual sewage solid wastes (called **sewage sludge**) in garbage dumps or sanitary landfills. This is expensive and wasteful because the sludge, which contains organic material, nitrogen, phosphorus, micronutrients, and other substances, can be applied to land to improve plant growth (Figure 18-5).

Table 18-3 lists some of the components typically contained in sewage sludges. Although only about 2 percent of all U.S. cropland is needed to accommodate all sewage sludge currently produced, high-population areas such as New Jersey may require as much as 50 percent of its cropland on which to dispose of its sludge, if this is the only method of disposal.[9]

Although variable in chemical composition, sludges contain about 4.0 percent nitrogen (N), 2.0 percent phosphorus (P) (4.6% P_2O_5), and 0.4 percent K (0.5% K_2O) on a dry-weight basis. They also contain toxic and heavy metals, such as boron, cadmium, copper, mercury, nickel, lead, selenium, and zinc. There is real concern that toxic quantities of these metals will be detrimental to plants, animals, and people. Cadmium is of special concern because it is relatively soluble, mobile in plants, and toxic.

---

[9]Council for Agricultural Science and Technology, "Application of Sewage Sludge to Cropland: Appraisal of Potential Hazards of the Heavy Metals to Plants and Animals," CAST Report 6A, EPA 430/9-76-013, 1976, 1.

Table 18-3 Average Composition of Sewage Sludges from More Than 200 Municipalities in Eight States

Component	Concentration on a Dry-Weight Basis		
	Minimum	Maximum	Median[a]
		(percent)	
Organic carbon[b]	6.5	48.0	30.4
Total nitrogen	0.1	17.6	3.3
Total phosphorus	0.1	14.3	2.3
Total sulfur	0.6	1.5	1.1
Calcium	0.1	25.0	3.9
Sodium	0.01	3.1	0.2
Potassium	0.02	2.6	0.3
		(parts per million)	
Zinc	101	27,800	1,740
Copper	84	10,400	850
Nickel	2	3,515	82
Chromium	10	99,000	890
Cadmium	3	3,410	260
Lead	13	19,730	500
Mercury	1	10,600	5
Arsenic	6	230	10

Source: Selected data from L. E. Sommers, "Chemical Composition of Sewage Sludges and Analysis of Their Potential Use as Fertilizers," *Journal of Environmental Quality* **6** (1977), 225–232.

[a]Median is the middle number of all numbers ranked sequentially.

[b]Organic carbon times 1.5–1.8 estimates the organic-matter content. Most sewage sludges are between 40% and 60% organic matter, although some will be even higher.

Pathogenic (disease-producing) organisms present in some sewage wastewaters and sludges may cause cholera, diarrhea (amoebic and bacterial), hepatitis, pinworms, poliomyelitis, and tapeworms, so natural organic systems of agriculture may not be totally advantageous.

Using sludges can be hazardous. They may contain enough live viruses and viable intestinal worm eggs to require careful handling for several months. They should not be used on crops that are grazed (grasses, clovers) or those eaten fresh by humans (such as lettuce, carrots, and radishes). High concentrations of soluble salts may be troublesome also. Heavy metals—such as cadmium, zinc, chromium, copper, lead, cobalt, nickel, and mercury—accumulate by adsorption in the soil to which sludge containing them (Table 18-3) is applied and remain for centuries. These metals may be adsorbed by plants grown in contaminated soil and then be accumulated in animals eating those plants, perhaps reaching chronic toxic levels. Most heavy metals become quite insoluble in soil of about pH 6 or higher. Cadmium, being more highly soluble than other heavy metals, is a frequently found contaminant, as are nickel, copper, molybdenum, and zinc to a lesser extent. Federal regulations for sludge use require that the soil pH be kept at 6.5 or higher to reduce the solubility of heavy metals. *Annual sludge applications must not exceed 0.5 kg Cd/ha annually.*[10] *Accumulation totals* on a soil must not exceed 5 kg Cd/ha for soil with less than 5 $mmol_c$/kg cation exchange capacity (CEC, sandy soils), or 20 kg Cd/ha for soils with a CEC above 15 $mmol_c$/kg (clayey or high-humus soils). PCBs (polychlorinated biphenyls, which have long half-lives, similar to DDT) must not exceed 10 mg/kg in sludge.

---

[10]Environmental Protection Agency, *Federal Register* **44** (Sept. 13, 1979), No. 179; 53461–53462.

FIGURE 18-6 This beef cattle feedlot in Nebraska has 6000 cattle, which produce more than 300,000 pounds (136,079 kg), wet weight, of manure each day, excluding bedding, that must be disposed of without polluting the water and air environments. Who can supply an acceptable answer to this problem? (*Source:* USDA.)

The crops most likely to be high in heavy metals when grown on sludge-treated soil are leafy vegetables such as Swiss chard and spinach. The least likely foods are grains, fruits, and other seed or fruit products. As pollution controls reduce indiscriminant dumping into sewer lines, sewage sludge should become a product with lower amounts of potential hazards. If serious contaminants are avoided, sludge could be used more extensively on agricultural lands and with greater safety.

With the ban on ocean dumping (which went into effect 1 January 1992), solid wastes must be recycled, burned, or buried. Recycling allows sludge to be used as would manures. The city of Holyoke, Massachusetts, had to pay $87.50 per ton to dispose of wastes into distant landfills.[11] A private operator is now *composting* the sludge with wood chips for $72.50 a wet ton. The city of about 45,000 people produces more than 60 tons of sludge per day (32% solids). The state of Washington in 1992 passed the nation's first *biosolids law,* establishing a policy for the beneficial use of treated sludge.

Animal Manures

Today's animal manures are not the same as yesterday's. The push to increase the weight of beef cattle and to encourage appetite in many animals has promoted the practice of increasing salt in animal diets and confining them to small areas (Figure 18-6). Manures may have from several percent to more than 10 percent soluble salts by dry weight. Heavy application of manures to soils without periodic leaching could cause a salt hazard to plants in a few

---

[11]Robert Spencer, "Sludge Composting Takes Town out of Landfill," *Biocycle* **33** (1992), No. 1; 52–54.

years. Leaching those salts into groundwaters to free the root zone of salts may pollute groundwaters. Eutrophication may also result from simultaneously leaching nitrates or eroding portions of the manure into surface waters.

Some poultry and swine are fed enough disease-control medicines to leave significant amounts of antibiotics, copper, and some other metals in their manures. These chemicals accumulate in soils. Animal disease organisms in manures are also of concern—along with odors and aesthetics—in dealing with manure disposal.

Cadmium is added to soils in sewage sludge, manures, and even in phosphate fertilizers. The average content in 91 percent of rock phosphate (RP) mined sources of phosphate had 25 mg Cd/kg of RP, 185 mg Cr/kg of RP, 11 mg As/kg of RP, and 10 mg Pb/kg of RP. To reach the *maximum loading* permitted in German soils would require 1300 years if 20 kg/ha was added yearly. The heavy metal concentration limits for sewage biosolids in the United States are given in Table 18-4. It is not legally permissible to add much cadmium to soils.

Municipal Garbage, Composts, and Sanitary Landfills

The major technical problems of garbage disposal (other than the sheer volume of waste) are the toxic chemical cleaners, pesticides, solvents, and medicines contained therein, the leaching by water of garbage solubles, the volatilization of solvents, and the gases formed by anaerobic decomposition of organic wastes.

Most municipal wastes are disposed of by burial in **sanitary landfills** (soil-covered trenches or holes filled with garbage), although some cities are composting it or using it as fuel in power plants. The danger of pollution of groundwater by leaching and of the air as volatile gases escape has caused a tightening of regulations for land suitable for sanitary landfills (Figure 18-7).

Food-Processing Wastes

Food-processing wastes are as varied as the foods processed: pea pods; tomato and potato peels; soybean, peanut, and cottonseed pulp after extracting oils; sugarcane pulp; waste from cheese making; and any chemicals used in food processing.

These products (except some chemical processing solutions) are organic and can be composted, added to soil, burned, used in animal feeds, or buried. Food wastes contain considerable

Table 18-4 Heavy Metal Concentration Limits and Annual Load Limits of Sewage Biosolids in the United States, Effective February 1994

Element	*Maximum Limit in the Added Material (mg/kg of dry weight)*	*Maximum Annual Load Rate (kg/ha/yr)*
As	75	2.0
Cd	85	1.9
Cr	3000	60.0
Pb	4300	75.0
Hg	840	15.0
Mo	75	3.8
Ni	420	21.0
Se	100	4.9
Zn	7500	140.0

Source: Modified from U.S. Environmental Protection Agency, *Clean Water Act 40 CFR 503, Sludge Rule,* U.S. Government Printing Office, Washington, DC, 1993.

FIGURE 18-7 Disposal of solid wastes is most often done by burial in soil. These photos of a Salt Lake City sanitary landfill serving about 1 million people show wide trenches to be filled with garbage and covered by soil; the trenches are 30.5 m (100 ft) wide, 2.1 m (7 ft) deep, and 1.6 km (1 mi) long. No-burning laws increase the volume of garbage buried. Twelve trenches were filled in 8 years in Salt Lake City. (*Source:* USDA—Soil Conservation Service; photo by D. C. Schuhart.)

amounts of nitrogen and phosphorus; their disposal can result in large nitrate concentrations in the soil (similar to the disposal of fresh animal manure).

The most serious environmental threat posed by food-processing waste disposal is that of water pollution by nitrogen. If such materials are dumped or eroded into surface waters, they also reduce oxygen in the water because of high chemical oxygen demand (COD) values and cause eutrophication from nitrogen and phosphorus added by the wastes.

Treatment chemicals—such as sodium hydroxide (lye) in peeling potatoes, waste syrups, salts, and cleanup detergents—can become disposal problems under rigid disposal regulations. In one study of potato wastes in wastewater, the sodium hydroxide used to peel potatoes was beneficial as a feed. The waste solids were collected by filtration and used as 20 percent to 25 percent of the total feed ration for cattle. In studying other disposal alternatives, the potato waste slurry, when added to soil at rates as high as 435 kg N/ha (388 lb/a), was expected to cause excessive percolation of nitrate into the soil substratum and groundwaters. The temporary anaerobic condition formed by the ponded slurry caused large nitrate losses, but by denitrification; insignificant losses occurred in percolation water (1 ppm nitrate) because of the lowered nitrate content after denitrification.

18:6 Toxic Chemicals

Even innocuous-seeming chemicals can be toxic or hazardous under extreme conditions of use. There are about 4 million known chemicals and 60,000 in common use. Approximately 1000 new chemicals are introduced each year. All of these chemicals can be monitored, but the cost would be prohibitive. The EPA assembles data for regulation and control of about 129 **priority pollutants,** not all of which are necessarily toxic.

Toxic chemicals include many substances—pesticides, polychlorinated biphenyls (PCBs), heavy metals, oils and gasoline, solvents, and many others. In agricultural use mostly pesticides and heavy metals are of concern, although small amounts of oil, gasoline, and solvents are spilled or dumped on farmlands.

Pesticides

Pests are any noxious, destructive, or troublesome organisms; **pesticides** are chemicals to kill pest organisms. The *-cide* ending means *to kill.* Various kinds of pesticides have been used for a long time. The Greek poet Homer, for example, wrote about pesticides 1000 years before Christ. He referred to "pest-averting sulfur with its properties of divine and purifying fumigation." In the 1930s sulfur was still powdered onto the human body to clean up some fungus skin effects. It was also volatilized in closed greenhouses for fumigation even in the middle 1950s.

DDT, a Magic Substance[12, 13] The story of **DDT** is a classic example of a magic substance capable of unbelievably good pest control. Eventually it was also shown to be a problem. DDT was developed by a chemist in the 1880s, but not as a pesticide. In 1938 Paul Muller tried it as a pesticide (among many other chemicals). It killed almost all insects easily and quickly. The chemical had *low toxicity to people and animals,* was *inexpensive,* and was *long-lasting.* Its discovery brought Muller the Nobel prize in 1948. DDT was used during World War II to delouse people, to control mosquitoes for malaria control, on thatched-roof homes to control the chagas beetle, and for long-term general insect control. Millions of people were saved from typhus and malarial deaths by DDT.

DDT has two major disadvantages: (1) Its half-life in the environment is too long (10–25 years), and (2) it bioaccumulates in the fat of animals. Thus in 1962 Rachel Carson published a book called *Silent Spring* that emphasized the long-term, nonselective killing of insects by DDT in the environment. Birds ate the dead insects and *biomagnified* the DDT into their bodies. Eventually the accumulated DDT would in some manner cause a bird population decrease, maybe to zero—thus a silent spring. DDT was banned from most uses in the United States in the early 1970s.

Pesticides Today Hundreds of pesticides have been synthesized, but far fewer are being used extensively as suitable pesticides. There are now about 600 commercially important pesticides and more than 1500 registered for sale. Newer ones will be developed more slowly because of the high cost of testing required before they can be licensed for sale and use. The cost is about $6 to $10 million plus more than 4 years of testing and review for each pesticide brought to the market.

---

[12] Orie L. Loucks, "The Trial of DDT in Wisconsin," in *Patient Earth,* J. Harte and R. H. Socolaw (eds.), Holt, Rinehart and Winston, New York, 1971, 88–107.
[13] Bernard J. Nebel, *Environmental Science,* 3rd ed., Prentice-Hall, Englewood Cliffs, NJ, 1990, 404–414.

What criteria define a *good pesticide?* Many of the highly regarded pesticides are quite toxic to animals, so it is not low toxicity. Acceptable pesticides must have at least the following characteristics:

- *The pesticide must be short-lived in the environment.* It must not exist long enough to bioaccumulate and biomagnify extensively in the food chains. The residues must be gone from foodstuffs. Usually it must be dissipated within a week or two of application time. To allow workers into fields, it is desirable for many pesticides to be gone within a few days.
- *The pesticide must not be carcinogenic, teratogenic, or mutagenic.* There is no easy or clear-cut way to determine these characteristics with certainty. Much disagreement is voiced about many decisions related to these properties for pesticides.
- *The pesticide must be effective yet able to be safely handled.* The chemical must be retained on the sprayed or treated area, must not be overly volatile so that it moves extensively into the air, and must permit safe handling by the applicator when reasonable care is used.

To these characteristics could be added others. The chemicals are preferably low-cost, easily washed from the body and equipment, of low caustic nature to skin and eyes, have low corrosiveness to equipment, and exist as liquids at normal working temperatures.

Pesticides currently used range from those of low toxicity to animals (many herbicides) to those that are very toxic to animals (some of the organic phosphates). Most pesticides are extremely toxic but short-lived, although some herbicides (triazines) may remain active in the soil for more than a year.

The negligible-risk level for *noncarcinogenic substances* is usually set as shown in Figure 18-8. The no-observable-effect level (NOEL) is reduced from the 0.001 point for a carcinogenic substance to only 0.01 of the value of the NOEL. The safety factor is based on the concept that people are ten times more sensitive than are animals and that the most sensitive persons are ten times more sensitive than the average person. California's Proposition 65 states that a cancer-causing substance posing a negligible risk of exposure at regulatory levels causes not more than *one additional cancer per million people.* The **Delaney clause** (passed in 1958 in the Federal Food, Drug, and Cosmetic Act) *inhibits addition of any amount of any known carcinogen to foods.* A new reaffirmation of *zero addition,* by court ruling, may trigger new pesticide cancellations.[14] The EPA has added that substances to be included now on the carcinogen list must have information on *how* the substance causes cancer. The rule could eliminate some substances that cause cancer in animals but through mechanisms that don't apply to humans.[15]

Pesticide Problems and the Extent of Pollution Toxicity to animals is the major concern about pesticides. Safe levels of several of these materials are given in drinking-water standards and food standards. Unreasonable standard limits are sometimes established, such as *no measurable amount.* Analysts continue to perfect analytical methods that can detect smaller and smaller quantities. Also, some materials may be present naturally (e.g., formed during fires).

---

[14]Melnicoe Stinmann, "Delaney Clause Ruling May Trigger Pesticide Cancellations," *California Agriculture* **48** (1994), No. 1; 30–36.

[15]Allen Newman, "New Criteria to Be Used in Identifying Human Carcinogens," *Environmental Science & Technology* **29** (1995), No. 10; 448A–449A.

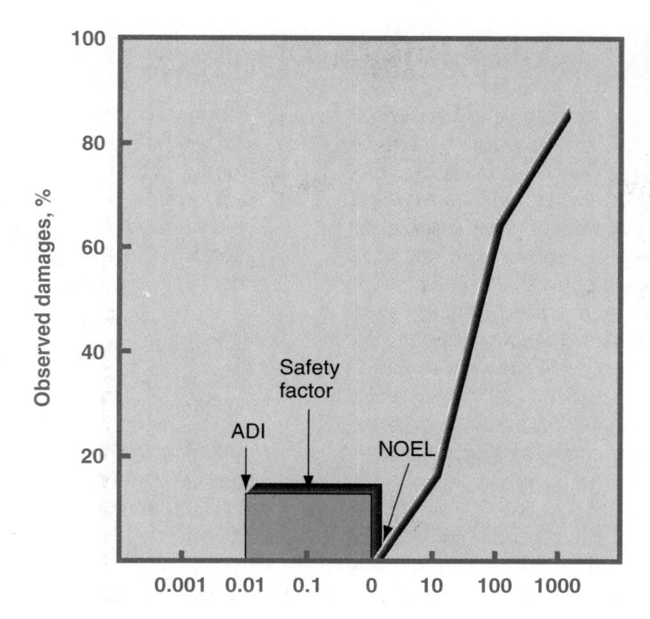

FIGURE 18-8 Approximate concept for setting safe dosage levels (*acceptable daily intake,* ADI) of a noncarcinogenic pollutant. The *no-observable-effect level* (NOEL) was observed in animal studies. (*Source:* Modified from Gary A. Beall, Christine M. Bruhn, Arthur L. Craigmill, and Carl K. Winter, "Pesticides and Your Food: How Safe is 'Safe,' " *California Agriculture* **45** [1991], No. 4; 4–8.)

Growers face an increasingly serious problem: *resistance.* Some organisms have built up resistance to certain frequently used pesticides by chemical changes or by natural selection of the more resistant organisms over time. Development of resistance forces growers to use larger pesticide applications. One example is the use of triazine herbicide, which keeps most weeds out of cornfields. Unfortunately, many weeds (pigweed, ragweed, lambs quarters) developed resistance.[16] The major way to fight resistance, currently, is to change pesticides for a while. For the reason of resistance, new pesticides must continually be sought.

Agriculture accounts for more than 67 percent of all pesticides used in the United States.[17] Numerous spills and data from monitoring of surface waters and groundwaters point unequivocally to agriculture as an important polluter. Part of today's pollution was put in the leaching cycle three decades ago; some is from present overuse, careless use, spills, and waste and wash waters.

Although *less than 5 percent of pesticides added are lost into groundwater,* the desired limit is zero contamination. The uncertainty about the hazards of many pesticides in such low concentrations has kept the permissible levels low. As a result many wells in various states have been closed, some for over 2 decades.[18] Groundwaters clean themselves, but often very slowly (Detail 18-1). *Many pollutants are already en route to groundwaters from past applications and disposal of pesticides.*

Industrial Sludge and Solid Waste

Industrial sludges are even more difficult than industrial solid wastes to dispose of tidily. Compositions of industrial sludges vary enormously; two common ones are boiler scale (calcium carbonates) and flue gas sludge. *Flue gas desulfurization sludge* (FGDS) is generated when lime

[16] "The Trouble with Triazines Is Resistance," *Agricultural Age,* Aug. 1986, p. 32.

[17] U.S. Department of Agriculture, *1983 Handbook of Agricultural Charts,* Agricultural Handbook 619, USDA, Washington, DC, 1983.

[18] G. R. Hallberg, "From Hoes to Herbicides: Agriculture and Groundwater Quality," *Journal of Soil and Water Conservation* **41** (1986), 357–364.

Detail 18-1 Pandora's Box: Groundwater Contamination

Everything that leaches into subsoils with waters has a good chance to end up in groundwater (aquifers). Half of all Americans and 95% of rural Americans get their household water from groundwater. We are finding pollutants from waste disposal that occurred decades ago now showing up in some groundwaters. Although we should be concerned about any hazard, pesticides have received the most attention. Twenty-two municipal wells were closed in 1989 in Fresno, California, because of contamination with the nematocide *dibromochlorpropane* (DBCP)—which had not been used in the area since 1977!

Pesticides are not the only, nor even the major, contaminants in groundwaters; they have just had high visibility by testing. Imagine the variety of substances from the following general sources of groundwater contamination:

1. *Septic tanks, cesspools, and privies.* Nearly one-fourth of U.S. homes use septic systems that dump relatively untreated wastes into the ground: human wastes, nitrates, detergents, household chemicals, and viruses.
2. *Surface impoundments* used by industries, municipalities, businesses, and farms. Supposedly, sealing liners cover the bottoms, but some are punctured and leak. Some states now require double liners, forcing the costly removal and safe disposal of the sediment already in such impoundments. The University of California expects their cost to remove and replace one-liner impoundments on nine experiment stations to be greater than $1.5 million.
3. *Agricultural activities* dispense pesticides, fertilizers, salts, fuels, and solvents from both agricultural lands and from urban homes and gardens.
4. *Landfills* may have almost anything in them. About 500 hazardous-waste facilities and 16,000 other landfills exist nationwide.
5. *Underground storage tanks,* 5 to 6 million of them, store gasoline, fuel oil, solvents, and numerous other chemicals.
6. *Abandoned wells,* if not properly sealed, can be conduits from surface contamination to groundwater.
7. *Accidents and illegal dumping* can cause major pollution in localized areas. A cleanup cost in Massachusetts—after 2000 gallons of gasoline had leaked from an underground tank near a municipal well—cost more than $3 million. Small spills and accidents happen daily and can gradually move into groundwaters.
8. *Highway deicing salts.*

Sources: Brenda Simmonds and Dennis Brosten, "DBCP Haunts California Wells," *Agrichemical Age* **34** (1990), No. 1; 11–12; Anonymous, "Citizens Guide to Ground-Water Protection," U.S. Environmental Protection Agency, *Social Issues Resources Series* **5** (April 1990), Article 9; 1–10; Dennis Brosten, "University and Applicators Grappling with Rinsate Morass," *Agrichemical Age* **31** (1987), No. 7; 8, 15.

$[Ca(OH)_2]$ or limestones slurries are used to trap sulfur oxides from escaping gases in coal-fired power plants. The waste contains fly ash (burned coal ash), calcium salts, and volatile elements such as mercury, arsenic, selenium, lead, and cadmium. Disposal restrictions to avoid heavy-metal contamination are similar to those discussed previously for sewage sludges.

Phosphogypsum, a by-product of phosphoric acid production, has been used as a non-lime calcium source for peanuts in the southeastern United States. In 1989 the EPA banned the material because it was claimed to have unsafe levels of *radionuclides* that could produce radioactive radon gas.[19] The material was cheap but was low in calcium (16%) and it caked,

[19]Anonymous, "Farewell Phosphogypsum," *Agricultural Age* **34** (1990), No. 4; 17, 18, 30.

making spreading it physically hard and even hazardous. It was also difficult to know exactly how much calcium was being applied. The more expensive landplaster (21%–25% calcium) is easier to spread and is of more constant composition than the phosphogypsum.

Oil sludges at oil terminals are difficult wastes to dispose of. Large quantities are produced as empty oil tankers are filled with ballast water on the return trips. The oil left in the tankers mixes with this ballast water and is accumulated at oil terminals when it is removed to fill tankers again with oil. At the Valdez oil terminal in Alaska, the oil sludge has been mixed with soil (7.0%–7.5% sludge); limed; fertilized with nitrogen, phosphorus, and potassium to aid breakdown; and planted to various grasses.[20] Annual grasses are most used to allow yearly liming, fertilization, and cultivation of the mixture.

Heavy metals (mercury, lead, cadmium, chromium, nickel, zinc, selenium, and many others) are elements that are very immobile in soils and last forever. Contamination in soils comes from additions in sludges, phosphate fertilizers, atmospheric fallout (from ore smelting), electroplating wastes, paint wastes, fumes from burning gasoline, natural gas, coal, and many other sources. Almost all heavy-metal tailings ponds have potential problems of (1) developing strong acidity as their pyrites oxidize and (2) accumulating soluble heavy metals.

In mid-Wales many abandoned metal mines are sources of pollution. Oxidizing pyrites form sulfuric acids that produce drainage waters with pH 2.6 that are heavily contaminated with aluminum, zinc, cadmium, and nickel.[21] The large quantities of aluminum (200–411 micrograms/L) were in the area drainage. The EPA has suggested that a 4-day aluminum concentration average should not exceed 87 μg/L to avoid damage to fish. Zinc levels in river waters near drainage entryways were three times higher than the recommended limits. Even drainage from acidic peaty soils have high aluminum contents.

Cadmium, because it is readily absorbed by plants, is a major concern in foods, as was discussed in the section on sewage sludges.

Lead also is a hazard when direct ingestion of lead-containing materials is common (children eating old paint chips; contaminated soil; contaminated dust on fruits, berries, vegetables). Metal smelters were common distributors of volatile metals (cadmium and lead) before emission stacks were scrubbed to remove volatiles. Many smelters are now closed, but their effects are still problems.

After comparing lead contents in 500-year-old frozen mummies from Greenland with the current contents in people in Denmark (1979) and in people in U.S. metropolitan areas, the following conclusions were reached.[22]

- U.S. metropolitan people had about seven times more lead than the Danish people.
- The 500-year-old mummies had about 1/30 the lead content of present-day Danish people.
- Present-day exposures are about 10 to 1000 times larger than those centuries ago.
- Some medieval samples had high lead levels, perhaps due to use of lead ceramic glaze, pewterware, lead water pipes, lead therapeutic agents, and lead to preserve certain beverages.

Water fowl and other animals have been found with lead poisoning caused by ingesting lead shot (from shotgun shells). It is estimated that 1 to 3 million North American ducks

[20]William W. Mitchell and G. Allen Mitchell, "Land Farming of Oil Sludge at Valdez Oil Terminal," *Agroborealis* **22** (1990), No. 1; 18–21.

[21]Ronald Fuge, Isan M. S. Laidlaw, William T. Perkins, and Kerry P. Rogers, "The Influence of Acidic Mine and Spoil Drainage on Water Quality in the Mid-Wales Area," *Environmental Geochemical Health* **13** (1991), No. 2; 70–75.

[22]Philippe Grandjean and Poul J. Jorgensen, "Retention of Lead and Cadmium in Prehistoric and Modern Human Teeth," *Environmental Research* **53** (1990), 6–15.

FIGURE 18-9 People damage land in many ways, including mixing material that is best left unmixed. These long piles of overburden to get at coal seams are destructive to many soils and hydraulic patterns. The size of this operation can be scaled by looking at the pickup truck (lower right, at the arrow). (Courtesy of USDA—Agricultural Research Service.)

and geese die yearly from lead poisoning after ingesting lead shot.[23] Such results prompted the banning of the use of lead shot in Canada and in the United States, but the lead shot already in the wetlands will be there a long time. Through bioaccumulation, poisoning occurred in hawks, mice (Pb elevated 5 to 64 times), a single mole (Pb elevated 35 to 1038 times), and green frogs (Pb elevated about 1000 times).[24]

Mining Spoils and Wastes

Mining in Thailand has left wastes rich in arseno-pyrite (arsenic pyrite). Surface and groundwaters exceed the World Health Organization (WHO) potable water guidelines for arsenic (of 10 μg/L) by as much as 500 times.[25] Skin lesions, an arsenic-related health symptom, are common in the area. Because the shallow aquifers have higher arsenic (As) than the deeper ones, use of deep wells may be essential. Can this water pollution really be remedied? Probably not without great cost. The mining waste is widespread and the arsenic contaminant is a product of slow weathering of the rock waste.

Many kinds of mine spoils and wastes occur around the world. Figure 18-9 depicts many of these areas that are poorly vegetated and often eroded as a result. Some uranium tailings have become superfund clean-up sites costing millions of dollars.

Toxic Elements Natural in Soil [26]

In 1856 a doctor at Fort Randall (now central South Dakota) observed a fatal disease that he linked with pastures in the area. The same problem (horses losing portions of their hooves and hair, listlessness, and even death) was noted in 1275 by Marco Polo. Although early reports are not wholly reliable or specific, we now know various plants that, in certain kinds of soils, are toxic to animals.

---

[23] Anton M. Scheuhammer and Kathy M. Dickson, "Patterns of Environmental Lead Exposure in Waterfowl in Eastern Canada," *Ambio* **25** (1996), No. 1; 14–20.

[24] W. Stansley and D. E. Roscoe, "The Uptake and Effects of Lead in Small Mammals and Frogs at a Trap and Skeet Range," *Archives of Environmental Contamination and Toxicology* **30** (1996), 220–226.

[25] M. Williams, F. Fordyce, A. Paijitprapapon, and P. Charoenchaisri, "Arsenic Contamination in Surface Drainage and Groundwater in Part of the Southeast Asian Tin Belt, Nakhon Si Thammarat Province, Southern Thailand," *Environmental Geology* **27** (1996), 16–33.

[26] Gary S. Banuelos, "Selenium-Loving Plants Cleanse the Soil," *Agricultural Research* **37** (1989), No. 5; 8–9.

(a)

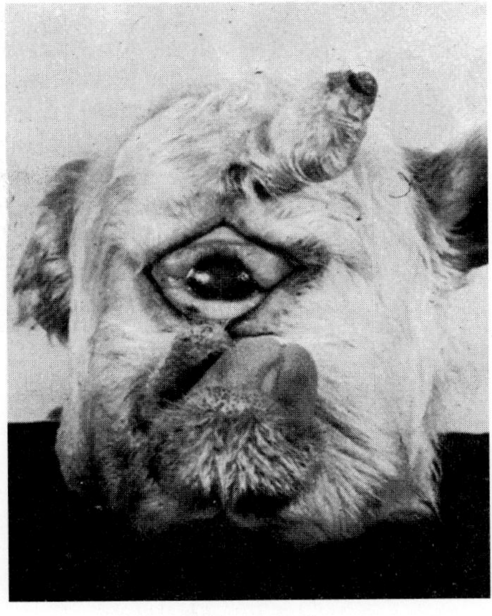
(b)

FIGURE 18-10 Deformation caused by eating poisonous range plants during pregnancy. (a) Deformed lamb from ewe fed locoweed (*Astragalus pubentissimus*). (b) Malformed head of a lamb from ewe fed false hellebore (*Veratrum californicum*) on day 14 of gestation. (Courtesy of USDA Poisonous Plant Research Laboratory, Logan, UT; photo by Lynn F. James.)

The economic loss from animal consumption of poisonous rangeland plants in the 17 western states of the United States is estimated at about $340 million annually.[27] These damages vary from reduced production to animal death. Following are some examples of individual losses:

- Sheep kills from grazing halogeton (*Halogeton glomeratus*) totaled 3900 in one year in northern Utah grazing areas.
- Losses from locoweed (*Astragalus* spp. and *Oxytropis* spp.) poisoning included 6000 sheep killed in eastern Utah and losses of $125,000 by one rancher in 1964.
- Grazing larkspur (*Delphinum* spp.) on mountain ranges caused the death of 103 mature cattle in one Forest Service allotment. U.S. ranchers in the intermountain West lose more cattle to larkspur than to any other poisonous plant.

The more than 200 range plants containing toxic substances usually cause problems because of the innate toxic nature of the plant or management of livestock; only a few plants are toxic because of excess or deficient absorption of soil elements. The more common toxic range plants are locoweed, halogeton, saltbush (*Atriplex nutallii*), goldenweed (*Oonipsis* spp.), larkspur, lupine, prince's plume (*Stanleya pinnata*), and woody aster (*Xylorrhiza* spp.) (Figure 18-10).

Selenium is a soil-supplied essential nutritive element whose concentration is critically important. Too little selenium in forage foods can cause *white-muscle disease* in animals; too much selenium causes *blind staggers, alkali disease,* or even death. Plants that accumulate

[27]Julie Corliss, "Toxic Encounters with Range Plants," *Agricultural Research* **39** (1991), No. 12; 4–7.

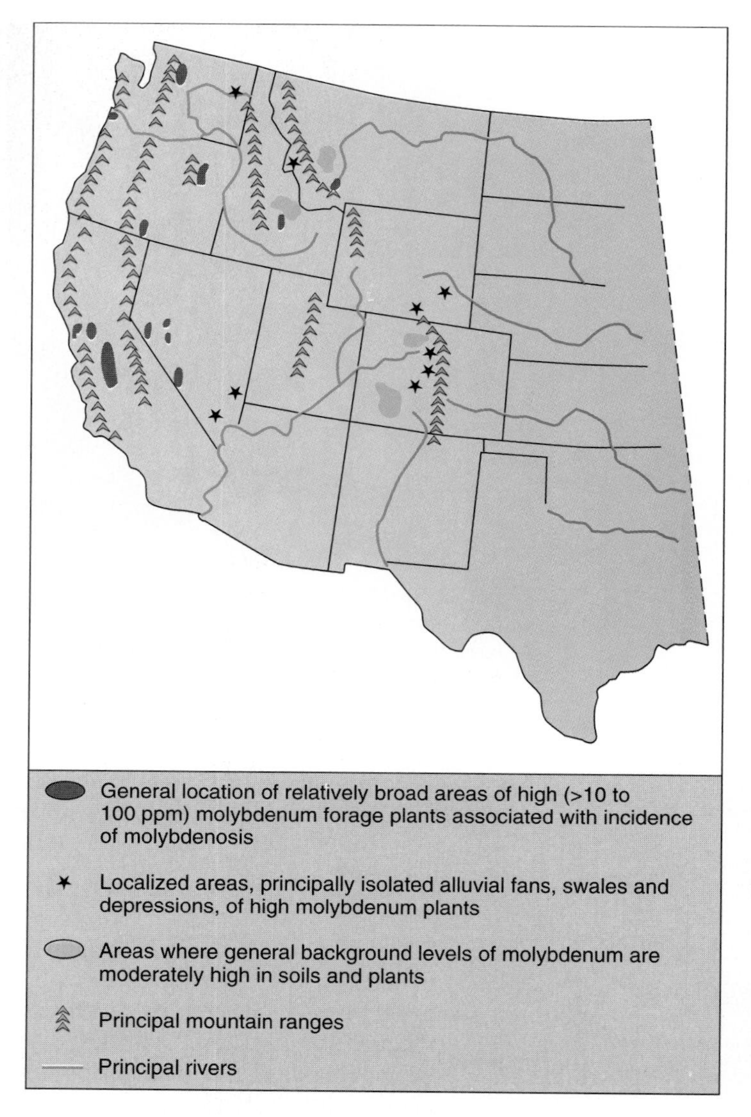

FIGURE 18-11 Areas of molybdenosis and other potential problem areas having high molybdenum in soils and plants. (*Source:* Modified from Joe Kubota, "The Poisoned Cattle of Willow Creek," *Soil Conservation* **40** [1975], No. 9; 18–21.)

very high selenium contents of 1000 mg/kg to 10,000 mg/kg include milkvetch (*Astragalus*), woody aster (*Machaeranthera*), mustard (*Brassica*), and prince's plume (*Stanleya*).

Molybdenosis, caused by excess molybdenum, is an animal disease that hinders utilization of copper and so results in poor growth. A common treatment is to feed supplemental copper. Areas that may cause molybdenosis in the West are *wet* alluvial soils formed from granite. Problem areas are shown in Figure 18-11. The Florida Everglades is the only area east of the Rockies where molybdenosis is known to be a problem.

For years ranchers thought that hungry cattle grazing pine needles may have abortions, premature delivery, or post-pregnancy complications. It is now known that isocupressic acid from Ponderosa pine needles can cause cows to abort.[28]

[28]Lynn F. James, "Mysterious Pine Toxin Is Identified," *Agricultural Research* **43** (1995), No. 1; 23.

FIGURE 18-12 Preharvest burning of sugarcane converts large amounts of trash leaves and weeds to ashes and partially burned waste. It also kills or chases out snakes! Burning releases particulates and gases into the air. Burning of straw and range grasses are common in many countries. (Courtesy of Raymond W. Miller, Utah State University.)

18:7 Particulates and Gases

Sugarcane Field Trash

Restrictions on burning (which have been recommended to reduce air pollution) seriously concern sugarcane growers, who customarily burn tons of cane leaf in the field before harvest. Such burning reduces weeds, removes hazardous snakes in tropical areas, reduces insect pests, and helps clear the field for harvesting (Figure 18-12).

Burning vegetation can produce carbon particulates, CO_2, CO, heat, and fiber-like silica particles from burning rice straw.[29] It is the silica particles that are of concern; it is feared that they may act in the same harmful way as asbestos fibers, but little is known about them yet. Rice is known to need considerable silica in growth (probably for straw strength), and much of this silica appears to become particulate fibers of silica when the straw burns.

Acidic Rain and Fog

Acidic rain, defined as rainfall (or melting snow) with enough dissolved acids to have a pH lower than 5.6, is found in many areas. The most severe conditions are near large industrial valleys. The major contributing acid is sulfuric acid (H_2SO_4), but nitric acid (HNO_3) adds to that acidity and is an oxidizer. Because these acids are in the air, they can be carried hundreds of miles from their origin.

The effect of acid is to dissolve carbonates (such as limestone and marble buildings), corrode metals, and kill much aquatic life. Some lakes have become almost devoid of fish and of most other large organisms. The acidic water increases the solubility of aluminum. Soluble aluminum is toxic to gill-breathing animals (fish) by causing loss of plasma- and hemolymph ions, leading to osmoregulatory failure.[30] On oxygenated fish gills the soluble soil aluminum may form a gelatinous aluminum hydroxide coating, which suffocates fish. In Norway and Sweden fish have died in more than 6500 lakes and in 7 Atlantic salmon-containing rivers. Similar fates have affected 1200 lakes in Ontario, Canada, and more than 200 lakes in the Adirondack Mountains of the northeastern United States and in the Alps.

---

[29]Bryan M. Jenkins, Scott Q. Turn, and Robert B. Williams, "Survey Documents Open Burning in the San Joaquin Valley," *California Agriculture* **45** (1991), No. 4; 12–16.

[30]B. O. Rosseland, T. D. Eldhuset, and M. Staurnes, "Environmental Effects of Aluminum," *Environmental and Geochemical Health* **12** (1990), Nos. 1, 2; 17–27.

Acid fog in southern California has been more damaging to plants than acidic rain.[31] The lower amount of water in the fog than in rains allows acidity to reach values of pH 2 to 3. A single 2-hour exposure to fog at pH 2.8 causes low injury to most plants. Marketability of cauliflower, spinach, and lettuce is reduced by pH 2.4 to 2.6. Reduced crop yields require pH values more acidic than pH 2. Injuries to plants from acidic fog are still less than the damages from ozone.

High acidity of water also increases soluble aluminum (Al), even to levels toxic to plants. Contents of Al in drinking water increase with acidity; an association of Al to Alzheimer's disease–Parkinsonism-dementia–is suspected by some medical researchers but is not yet proven.[32] Aluminum reduces the activities of gill enzymes important in the active uptake of ions.

The nitric and sulfuric acids in the atmosphere come from various oxides of nitrogen and sulfur accumulated in the atmosphere from denitrification, volatilized ammonia that is later oxidized in air, burning of fossil fuels, and the burning of vegetation at hot temperatures (especially in forest fires). Any high-temperature burning or heating of organic materials of plant origin, which is 1 percent to 4 percent nitrogen on a dry-weight basis, may evolve some of the nitrogen and sulfur as oxides, which then oxidize to nitric acid and sulfuric acid. The dominance of sulfuric acid in acidic rain is the reason power plants and other users attempt to buy *low-sulfur* coal. The EPA has regulations on the amount of these oxides that can be exhausted to air.

Agriculture produces pollutants both to the air and to water. Particulates (dust) are added by wind erosion of unprotected soil, by tillage, and by burning of many fuels and residues. Fuels release oxides of sulfur and nitrogen. Volatile pesticides are carried great distances in wind. Persons affected by asthma testify to the many pollens and spores of various types suspended in the atmosphere. Anaerobic decomposition (wetlands, ponds, silage, manure piles) provides methane, sulfur oxides, ammonia, and numerous odoriferous organic gases. Large amounts of heat and water vapor join these other pollutants. Even noise may be of concern in very limited sites.

Some fertilizer nitrogen produces nitrous oxide by denitrification. The EPA estimates global nitrous oxide emissions are in the range of 11 million to 17 million tons (about 10–15.5 million metric tons).[33] Only about 3 percent to 5 percent of these emissions come from nitrogen fertilizers. Major nitrous oxide sources are burning fossil fuels and biomass and release of nitrous oxide from tropical and subtropical forest soils.

Ozone Layer Depletion

The **ozone shield (O_3)** in the stratosphere, at 10 km to 15 km above the Earth, screens out as much as 99 percent of the ultraviolet (**UV**) rays that could otherwise be absorbed preferentially in proteins and nucleic acids. These would cause tissue damage and possible genetic mutations. The UV that does get through the ozone layer may cause sunburn and skin cancer. Actually, the UV rays form the ozone by splitting O_2 molecules into very reactive O* atoms. These reactive atoms recombine with other O_2 molecules to form ozone (O_3). This process is an equilibrium between ozone destruction and formation. Chlorine gas atoms are effective catalysts for ozone breakdown. Frequently the chlorine has been supplied by leakage of Freon gas (a chlorofluorocarbon) from refrigeration units and by various chlorofluorocarbons used

---

[31]Robert C. Musselman, Patrick M. McCool, and Jerry L. Sterrett, "Acid Fog Injures California Crops," *California Agriculture* **42** (1988), No. 4; 6–8.

[32]Trond Peder Flaten, "Geographical Associations Between Aluminum in Drinking Water and Death Rates with Dementia (Including Alzheimer's Disease), Parkinson's Disease, and Myotrophic Lateral Sclerosis in Norway," *Environmental and Geochemical Health* **12** (1990), Nos. 1, 2; 17–27.

[33]Dennis Brosten and Brenda Simmonds, "Do Fertilizers Affect the Atmosphere?" *Agricultural Age* **38** (1990), No. 2; 6–7.

FIGURE 18-13 Large areas of paddy rice produce large amounts of anaerobic gases, including methane, carbon monoxide, nitrogen oxides plus dinitrogen, and dimethyl sulfide or hydrogen sulfide. (By permission of Food and Agriculture Organization, Rome.)

in aerosol containers and as solvents for cleaning special electronic equipment. In 1974 the United States alone sprayed 230 million kg (506 million lb) of these gases into the atmosphere. These chemicals are now banned in aerosols in the United States.[34]

Other chemicals also destroy ozone. Carbon tetrachloride (a common grease solvent) and nitric oxide (NO) destroy ozone. The effect of NO supplies on ozone destruction is about 10 percent as important as those of chlorofluorocarbons. It is estimated that by the year 2100 about 1.5 percent to 3.5 percent of the ozone layer will be destroyed by NO. Some scientists say this estimate is too high. Whatever the facts really are, the NO produced in various agricultural activities adds to this hazard of undetermined seriousness.

At the Earth's surface ozone is not desirable. Almost 90 percent of crops lost to pollutants are lost because of ozone and sulfur dioxide; ozone is about ten times more toxic than sulfur dioxide.[35] It is estimated that reducing ozone by 25 percent would boost farmers' incomes $1.9 billion annually. Ozone destroys chloroplasts, weakens cell walls, and allows leaching of nutrients from plants. A stressed plant has higher sugar content, which increases damage from insects it attracts.

Methane

Methane (CH_4) is given off from rice paddies and swamp areas. About one-fourth of the 500 million tons of methane released yearly into the atmosphere comes from flooded rice fields (Figure 18-13; see also Detail 19-3).[36] A methane molecule traps heat about *30 times more*

---

[34]B. J. Nebel, *Environmental Science,* 2nd ed., Prentice-Hall, Englewood Cliffs, NJ, 1987, 388.
[35]Sandy Miller Hayes, "Wanted: Breathing Room for Crops," *Agricultural Research* **37** (1989), No. 7; 4–6.
[36]Anonymous, "IRRI Studies Role of Ricefield Methane in Global Climate Change," *The IRRI Reporter,* Dec. 1991, 1–2.

effectively than a carbon dioxide molecule (methane may cause 15 percent as much global warming as does carbon dioxide). Because about 80 percent of the methane from a rice paddy escapes from the root area *up through* plants, plant selection (breeding) is one approach to reducing methane loss from paddy.

Carbon Monoxide

Carbon monoxide (CO) occurs from incompletely oxidized carbon; burning or decomposition of any carbon source with limited access to oxygen can produce carbon monoxide. Landfills produce considerable CO. Carbon monoxide has about 200 times the affinity for hemoglobin as does oxygen, so CO causes loss of oxygen transport to body cells. Small amounts cause people to become nondiscriminatory, resulting in more accidents. High doses cause headaches, dizziness, and eventually death. Fortunately, soil bacteria (autotrophs) convert CO to CO_2. This is another good reason for planting parks and lawns in cities.

18:8 Radionuclides

Radioactive elements differ from nonradioactive *isotopes* (elements with similar properties) by the emission at some time of a high-energy particle. These emissions are **radiation** and include, among others, gamma rays, beta rays, alpha rays, and neutrons. Bombardment of a living body by enough radiation changes atoms of compounds into different elements, altering their action, even killing cells or portions of cells. Exposure to high radiation levels causes nausea, diarrhea, vomiting, hemorrhages, leukemia, sterilization, or death. At lower dosage levels cell membranes are damaged or destroyed, and leukemia (a low count of white blood cells) is common.

Radioactive elements have specific rates at which they emit radiation (**decay**). This rate is measured as the time needed for a mass of the element to reach a radiation rate that is half of what it was at time zero, when it began to radiate; this is called the **half-life** of the element. At the end of the two half-life periods, a mass of radioactive material would have one-fourth (one-half of one-half) as much radiation as it had at the initial starting time of measurement (Figure 18-14). The half-lives of some common radioactive elements are as follows:

^{238}Uranium	4,510,000,000 years
^{14}Carbon	5730 years
^{137}Cesium	30.2 years
^{90}Strontium	28.1 years
^{131}Iodine	8.0 days

Radioactivity is of great concern because of its invisibility, its insidious damages, and the fact that there is no known way to reduce or stop the radiation process. Shielding the source with energy-absorbing heavy metals (lead) and/or increasing the distance between source and victim (a thickness of solid materials and even air) are the only known means of protection. Atomic bomb testing has covered all areas of the Earth with radioactive strontium, cesium, and iodine. The radioactive element ^{90}strontium, taken up in pastures, becomes substituted into bones and milk in place of calcium; ^{137}cesium accumulates in muscle tissue. Claims in Arizona, Utah, and Nevada of increased leukemia deaths (2.4 times more than normal) during

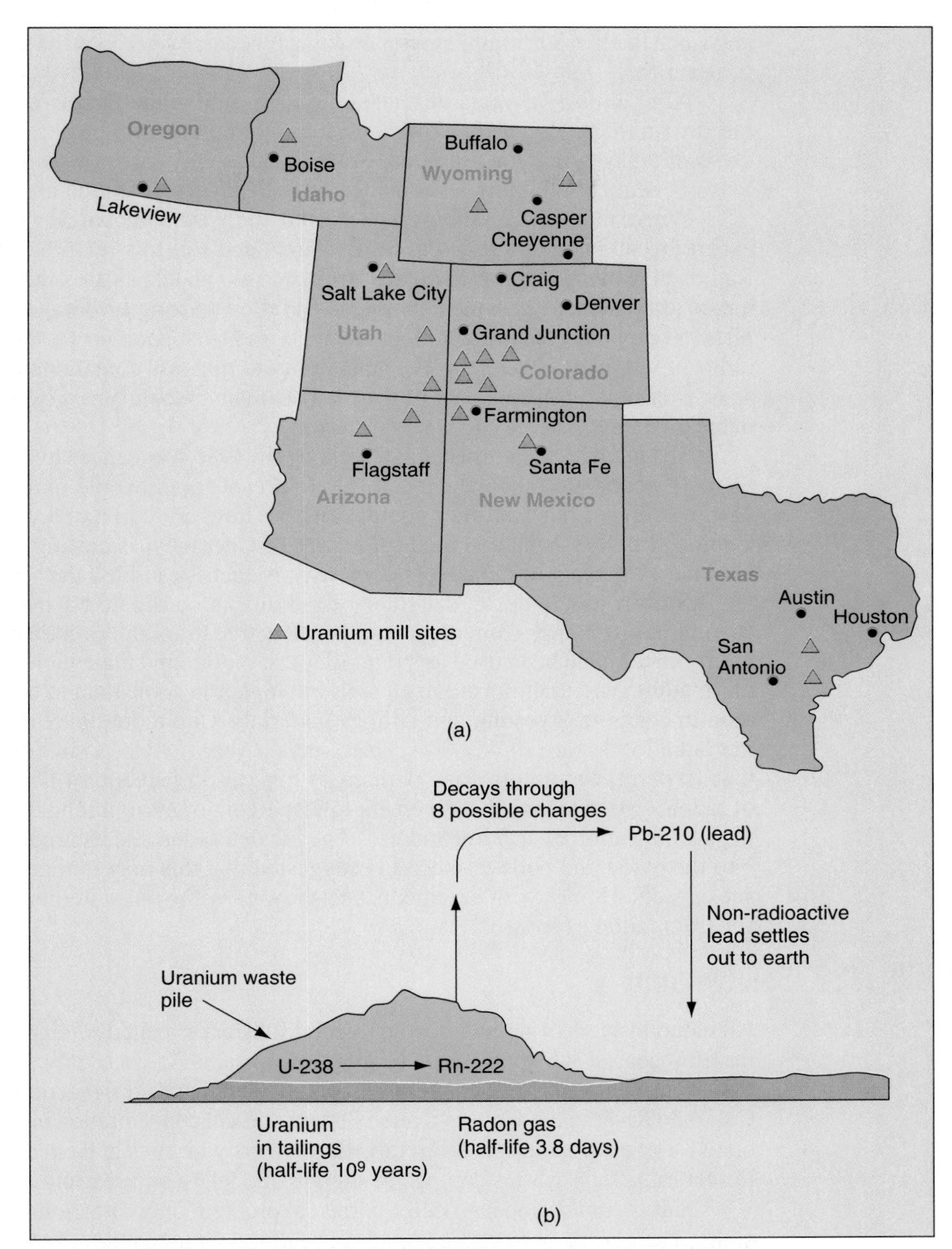

FIGURE 18-14 Uranium mine tailings, some of which are shown in map (a), contain unextracted radioactive uranium-238. Left as spoil piles these tailings now cause concern. Some of the sandy waste has found its way into cement, brick mortar, and as fill under sidewalks and buildings. In (b) the general decay of uranium (4.5-billion-year half-life) to radon gas and several other elements shows the radiation hazard. In some areas large amounts of sandy uranium tailings were used as filler under buildings and sidewalks or even in the cement. Major, expensive excavations have been undertaken to remove these tailings, where it is possible. This action is necessary to reduce radon gas pollution in the area and in the buildings. (Courtesy of Raymond W. Miller, Utah State University.)

and since the Nevada atomic bomb tests[37] are generally accepted as valid claims by the U.S. government.

After almost 40 years of nuclear bomb building, the Department of Energy (DOE) is faced with massive cleanup. Toxic residues from plutonium processing were stored in hundreds of tanks buried underground. Some of these tanks are now leaking. Thousands of gallons of chemicals and low-level nuclear wastes have been buried underground.

Where radioactive fallout has occurred, there is no known way to eliminate the hazard, except by scraping and carrying off contaminated soil to which the cations have absorbed. Radioactive elements in the soil are absorbed into plants, as are other elements and become a part of whatever consumes the plants. Wildlife grazing on forages on radioactive tailing piles or on plants containing fallout elements and later eaten for food is one way that isolated radioactivity may affect people. The inability to mitigate contamination by radioactive substances except by the passage of time makes many people strongly oppose development of nuclear energy systems and weapons testing.

The nuclear plant disaster at Chernobyl in 1986 was caused by human error. In a revisit there 10 years later, the incidence of thyroid cancers in people in and near Chernobyl was 18 times higher than in other populations.[38] These are correlated with ^{131}I (iodine) doses. Coping with the social and medical legacy of Chernobyl is costing Belarus 20 percent and Ukraine 12 percent of their gross domestic products to address the accident victims.

Radon gas, a decay daughter of radium, is considered a major indoor health risk, found in most homes from natural soil radioactive uranium. Some building stones and uranium wastes have been used as sand, fill, or in mortar and have increased the hazard to people. Radium and uranium are in all soils but higher in some than in others. It is usually common in concrete, gypsum, and tailings materials. One radon level reported was equated to the radiation hazard of 455,000 chest x-rays a year.[39] Also, a survey in 30 states indicated that 10 percent minimum, and perhaps as high as 35 percent, of homes had harmful levels of radon. Estimates are that 8 percent to 25 percent of current lung cancer deaths are related to past inhalation of airborne radon.[40] The inhaled radon and its progeny pass from the lungs into the blood and body tissues. It is suggested that this may initiate many types of soft tissue cancers. Homes with basements and showers with well water may increase risks of cancer from radon exposure.

18:9 Soluble Salts

All natural waters contain dissolved mineral substances called *soluble salts.* Some rainwaters, far from coastal salt sprays, may be very low in salts. As water flows over and through soils, it picks up salt loads. If the water has rapid evaporation as it flows on the surface (such as the Colorado River in the western United States), the salt concentration increases as water is evaporated. The erosion of salts and return-flow waters with salts in them all add to the load of salt. Deicer salts, salty wastes dumped in streams and lakes, and sea spray are all sources of salts.

Salt accumulation has been a perpetual problem of civilizations in arid and semiarid regions. The United Nations Food and Agricultural Organization (FAO) states that half of the irrigated farms in the world are damaged by salt (Figure 18-15). Scientists, in their attempts to increase water efficiency (less water per unit of crop yield), have sometimes increased salt problems when leaching is too little.

---

[37] "A Fallout of Nuclear Fear," *Time,* Mar. 12, 1979, 84.
[38] Christoph Hohenemser, "Chernobyl, 10 Years Later," *Environment* (1996), No. 3; 3–5.
[39] G. Tyler Miller, *Living in the Environment,* Wadsworth Publishing Company, Belmont, CA, 1990, 489.
[40] Douglas G. Mose, George W. Mushrush, and J. Eric Slone, "Environmental Factors Governing Indoor Radon," *Journal of Environmental Science and Health* **A31** (1996), No. 3; 553–577.

FIGURE 18-15 Salt accumulation (white crusts) in Isphahan, Iran. The salt in the surface is caused by irrigating a soil in an arid climate where most water loss was by evapotranspiration and few salts were removed by leaching. (Courtesy of Bruce Anderson, Utah State University.)

The environmental concern with soluble salts is the cost damages (crop losses, metal corrosion, and costly cleansing activities needed by industry). Salt washed from one field ends up in groundwaters or rivers to be used by someone else. The major question related to soluble salts is how to best manage the salt with the least cost and damage to everyone.

An example of costs from salts is given by the complicated United States–Mexico agreement, signed in 1973, concerning the Colorado River water flowing into Mexico.[41, 42]

---

[41]M. B. Holburt, "The 1973 Agreement on Colorado River Salinity Between the United States and Mexico," in *Irrigation Return Flow Quality Management,* J. P. Law, Jr., and G. V. Skogerboe, eds., Proceedings of National Conference, U.S. EPA and Colorado State University, Fort Collins, 1977, 325–333.

[42]Fred Pearce, "Banishing the Salt of the Earth," *New Scientist* **114** (1987), No. 1564; 53–56.

The agreement states that the United States will keep the salt content in the water behind Imperial Dam (the last U.S. dam on the Colorado before it enters Mexico) to 879 ppm salt. At places along the river, this value is sometimes exceeded. From the Imperial Dam, water goes either to Mexico (12% of the Colorado River flow) or to the Welton–Mohawk Irrigation District in southwestern Arizona. The drainage from the irrigation district is high in salt and must be cleaned.

To clean up the salty water of the lower Colorado River, an enormous desalting plant was planned. Some of its details are given in the following items:

- About 350 million liters (about 95 million gallons) of water per day need to be desalinized.
- For 20 years the saline drainage of the Welton–Mohawk District has been channeled to drain into the sea in the Gulf of California.
- The desalinization process used in the plant is *reverse osmosis,* which will cost about $264 per thousand cubic meters. This cost is estimated as about ten times more than the water is worth to farmers as irrigation water. Most of this cost will be borne by the taxpayer.

Some of the individual elements in soluble salts can also be problems. As total salts accumulate, hazardous concentrations of boron, selenium, molybdenum, or arsenic may increase. Often these hazards occur where regional soils and rocks contain high levels of these metals, and they are dissolved into drainage water. Soluble contents in solution phase in ponds in California are as follows:

Selenium	1.0 mg/L
Arsenic	5.0 mg/L
Boron	70.0 mg/L
Molybdenum	350.0 mg/L

Selenium has become a severe problem in some evaporation ponds and other salt accumulation areas (Detail 18-2).[43]

18:10 Soil Sediments as Pollutants

Eroded soil becomes a serious pollutant, both as a physical problem and because of chemicals it carries adsorbed to the particles' surfaces. The control of soil erosion will be beneficial to solving several pollution problems.

Sediment Problems

Soil itself becomes a spectacular pollutant when large amounts slide down to cover homes or roadways, when receded floodwaters leave behind mucky, muddy messes, or reservoirs and harbors fill with silt. Pollution by sediment modifies the environment in the following ways:

---

[43]Kenneth K. Tanji, Colin G. H. Ong, Randy A. Dahlgren, and Mitchell J. Herbel, "Salt Deposits in Evaporation Ponds: An Environmental Hazard?" *California Agriculture* **46** (1992), No. 6; 18–21.

Detail 18-2 Toxicity of Selenium

Selenium (Se) is an element not needed by plants but required by animals in small amounts to avoid white-muscle disease, which weakens the heart. Too much selenium causes severe liver and kidney damage, odoriferous body and breath, and eventually death.

Selenium is chemically similar to sulfur and has four soil forms: (1) selenide (Se^{2-}), (2) elemental Se^{0}, (3) selenites (SeO_3^{2-}), and (4) selenates (SeO_4^{2-}). Selenates are quite soluble, just as are sulfates, so they move and accumulate with other soluble salts.

The San Joaquin Valley in California has poor drainage and accumulates salts from the irrigation and other drainage waters. The Salton Sea, flooded long ago by ocean waters, continues to accumulate salts from added waters. In 1992 about 150,000 eared grebes died at the Salton Sea; selenium is suspected as part of the problem. In the San Joaquin Valley health problems with nesting birds were noted in 1983. In 1984 about 16,000 birds died of what is believed by some to be Selenium poisoning, although avian cholera was the cause listed by local wildlife specialists. Selenium is blamed for deformation and death of ruddy ducks, mallards, grebes, killdeer, coots, and other birds.

Studies on the selenium hazard in these areas are currently in progress to seek ways to reduce soluble selenium in the high-selenium soils of the area. High-uptake plants (mustard, milk vetch, prince's plume) are possible extractors to lower soil selenium.

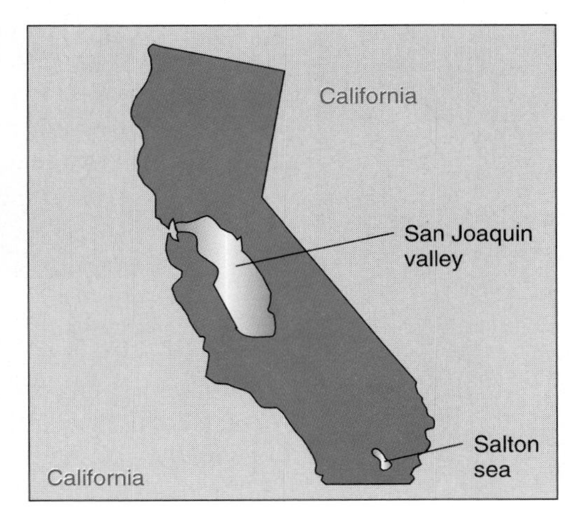

Sources: Gary S. Banuelos, "Selenium-Loving Plants Cleanse the Soil," *Agricultural Research* **37** (1989), No. 5; 8–9; Gary Banuelos and Gerrit Schrale, "Plants That Remove Selenium from Soils," *California Agriculture* **43** (1989), No. 3; 19–20; Robert H. Boyle, "The Killing Fields," *Sports Illustrated,* March 22, 1993, 62–69.

- Suspended sediment is usually eroded topsoil, the most fertile portion of soil (Figure 18-16). The eroded soil is deteriorated and the carried topsoil could deposit in places where the fertility is a liability, such as in bodies of water where eutrophication would be increased by the nutrients, especially phosphorus. In contrast the great deltas of major rivers are formed from deposited sediments and can be beneficial as productive cropland.
- Water reservoirs can be filled by sediments, decreasing their storage capacities. Tarbela Dam reservoir in Pakistan, one of the world's largest, has a silt load about 16 times larger than predicted by the dam's engineers. The life expectancy of the reservoir is now only 50 years. In India the Kosi Canal was so heavily silted in its 20-year existence from 1958 to 1978 because of overgrazing its watershed that it provided water for only one-seventh its original irrigation area of 570,000 hectares (1,407,900 acres) down to 81,000 hectares (200,070 acres).[44]

[44]Sid Gautam, "Dam Building No Longer Means 'Instant Progress,' " *Water and Sewage Works* **125** (Aug. 1978), 30–32.

FIGURE 18-16 All of this California orchard's lost soil ended up somewhere downslope or in the river. In areas with intensive cropping, this kind of erosion can and should be better controlled. (*Source:* USDA—Soil Conservation Service.)

- Suspended solids reduce sunlight penetration into water, thereby reducing production of microscopic-sized organisms, which begin the aquatic food chain. Suspended soils also cover lake bottom plants and fish eggs, to the detriment of both, and interfere with the gill action of fish, suffocating them.
- Deposition on land can cover good soil with poorer or even rocky sediments. Removing unwanted sediment—from agricultural land, residences, roadways, harbors, and streams—is toilsome and expensive.

It is estimated that only a small fraction of the cropland has excessive erosion of more than 33.6 Mg/ha (about 15 t/a). Other associated estimates are the following:[45]

- 3.5% of cropland, which is losing more than 56 Mg/ha (25 t/a), accounts for 32% of the total soil loss.
- 7% of cropland, which is losing more than 33.6 Mg/ha (15 t/a), accounts for 44% of the total soil loss (includes the estimate above).

In 1979 a national erosion estimate anticipated that at least 3.6 billion metric tons of soil would be lost annually by erosion, enough to cover 850,000 hectares (more than 2 million acres) with sediment nearly 30 cm (1 ft) deep. The major part of this erosion is natural and nearly impossible to remedy. Rangelands, which are enormous contributors to sediment pollution, are usually in low rainfall areas and are too arid to maintain good plant cover, so wind carries away the barren soil, and occasional flash floods wash away more (Figure 18-17).

It is not just erosion by croplands that is of concern. Wind erosion causes many kinds of damage. Dune stabilization has been worked at for decades to protect cities (Cairo, Egypt, for

[45]Clayton W. Ogg and Harry B. Pionke, "Water Quality and the New Farm Policy Initiative," *Journal of Soil and Water Conservation* **41** (1986), 85–88.

FIGURE 18-17 Wind erosion can be catastrophic. The great dust storms in the Dust Bowl (Oklahoma, Texas, Nebraska, etc.) in the 1930s resulted from cultivation of unstable soil in a windy and dry area. This 1965 view in Oklahoma makes clear that more permanent protection is needed on such areas. *(Courtesy of* USDA—Soil Conservation Service; photo by James E. Smith.)

example), and facilities in many areas are forced to close due to blowing sands. In China, storms originating in inner Mongolia sweep into Beijing (old Peking) and across 8 million hectares (nearly 20 million acres) of farmland and pastures. To protect against this wind, an enormous shelter belt 7000 km (4375 mi) long was begun in 1978. By 1985 the first phase of 6 million hectares (nearly 15 million acres) of barren land was planted from Heilongjiang province to Xinjiang Uyger region. About 51 percent of the seedlings survived in the area having 400 mm (15.6 in) of rainfall yearly. Peasants who plant will own the trees to bequeath to their children. The shelter belts have reduced water transpiration on croplands and yields are up by one-fifth. Chinese agronomists set a net benefit on the Green Wall of China at $630 million a year already. It is expected that the total forest wall will take several generations to plant.

Adsorbed Chemicals

Eroded sediment can carry appreciable amounts of phosphorus and pesticides and large quantities of nitrogen and organic matter to surface waters. One report suggests that 80 percent of phosphorus and 73 percent of nitrogen loadings of surface waters nationally are from eroded soil. Cropland erosion accounts for about one-third of these (27% of phosphorus and 24% of nitrogen) and costs of losses are $2.2 billion per year.[46] These authors state, however, that their most notable finding was that "cropland erosion control achieves acceptable phosphorus concentrations in only a few regions." There are sources other than croplands providing phosphorus. If the phosphorus pollutant has many sources, cropland controls alone seem inadequate to bring about major improvements in water quality. The quality is already adversely affected by other nonpoint-source pollution.

In 1977 the Great-Lakes-Pollution-from-Land-Use-Activities Reference Group (PLUARG), composed of U.S. and Canadian specialists, believed the Great Lakes were being appreciably polluted by runoff of phosphorus from agricultural and *urban* areas. They estimated that 41 percent of the total phosphorus load came from fine-textured sediments washed into the lakes.

---

[46]Leonard P. Glanessi, Henry M. Peskin, Pierre Crosson, and Cyndi Puffer, "Nonpoint-Source Pollution: Are Cropland Controls the Answer?" *Journal of Soil and Water Conservation* **41** (1986), 215–218.

18:11 Decision Case Studies

A brief introduction and discussion suggesting the use of *decision cases* was given in Chapter 6. Below are three other decision cases for use. Case 3 is more one of finances, rotational grazing, and milk production strategies.

1. Robert L. Mikkelsen, "Swine Waste Disposal Dilemma: A Case Study," *Journal of Natural Resources and Life Sciences Education* **24** (1995), No. 2; 169–172.
2. M. G. Allen, J. LR. McKenna, A. O. Abaye, and W. G. Camp, "Exterminators: The Politics of Chemical Fumigation—A Case Study," *Journal of Natural Resources and Life Sciences Education* **25** (1996), No. 2; 156–160.
3. Craig C. Sheaffer, Melvin J. Stanford, Charlene Chan-Muehlbauer, and Douglas Gunnick,"The Future of Walnut Creek Farm: A Decision Case Study," *Journal of Natural Resources and Life Sciences Education* **25** (1996), No. 1; 53–58.

We would be happy if we studied nature more in natural things and acted according to nature, whose rules are few, plain and most reasonable.

—William Penn

Where there is an inquisitive mind, there will always be a frontier.

—Anonymous

Questions

1. (a) What is meant by *pollution?* (b) Must a pollutant be toxic or carcinogenic? Explain.
2. (a) What is meant by *eutrophication?* (b) Explain why phosphates have greater influence than nitrates on eutrophication.
3. Probably the major pollution by nitrogen is in eutrophication and methemoglobinemia. Explain each of these and indicate the relative hazard of each compared to other hazards.
4. (a) To what extent was DDT more poisonous to animals than other currently accepted pesticides? (b) What are the criteria a pesticide must meet to be acceptable for use today?
5. To what extent is agriculture responsible for pollution with pesticides?
6. Discuss the problem of selenium in soils and wastewater impoundments.
7. (a) What are some of the common heavy metals? (b) Why are they of concern?
8. (a) Which undesirable by-products are added to soils in manures and sewage sludges that may make their use hazardous? (b) Explain the hazardous nature of each pollutant mentioned.
9. (a) If adding soluble salts in natural waters is usually a natural process, why are soluble salts treated as pollutants? (b) How serious is the salt problem?
10. Which hazardous gas is produced when organic substances are buried in landfills?
11. How is BOD (or COD) related to undesirable water conditions and waste disposal?
12. (a) What are the kinds and sources of radioactive pollutants most likely to occur on crops and in soils? (b) What is the nature of the hazard?
13. (a) How do soil sediments fit the definition of a pollutant? (b) What are the damages from soil sediments?
14. Although air pollutants have been given less attention than water pollutants, agriculture does produce some air pollutants. List and briefly discuss three of these.

Toward Environmental Integrity

Before we can begin 'saving the Earth,' we must accept the fact that we, the rich inhabitants of the industrialized nations, are the greatest danger to the future of the World.

—Sicco Mansholt (Dutch economist)

It is often the last key on the ring that opens the door.

—Anonymous

19:1 Preview and Important Facts

PREVIEW

What are your reference points for a baseline *healthy* ecosystem? A fascinating issue in each remediation attempt is determining what should be done and when the process has reached *an acceptable end point.* How clean is *clean?*

Because many people are not inclined to undertake clean-up activities unless it will benefit themselves, pollution control has been a *command-and-control* process. The numbers of regulations have mushroomed. They have worked with some success so far; much has been accomplished. However, if 3 aspirin are good for a headache, should 10 or 20 be even better? There is concern that the command-and-control approach can go beyond the beneficial point and that the law of diminishing returns will be active, as the increased regulatory needs begin to siphon off resources (talent, time, funds) and, by doing this, reduce our competitive environment.

Soils are nature's dispose-all, its sewage treatment plant, its water purifier, and, at times, also a pollutant. Soils are valuable as cleansers of the Earth's environments. The soil is a **physical filter** (sieving action), a **chemical filter** (absorption, precipitation), and a **biological filter** (decomposition of organic materials), as well as the receptacle for all things buried and disposed of beneath and on the surface.

Although the soil is the most universal and extensive substance that cleans waters and recycles wastes, it is not infinite in capacity. High land costs near most large cities increase the expense of waste disposal. Many toxins added to soils can build up to concentrations that become serious threats to plant and animal health. Some toxic substances become residual in the soils; centuries, even millennia, may pass before their lowered levels again permit normal use of the soil. Even harmful organic substances that will decompose eventually to nonharmful recycled elements of carbon, oxygen, hydrogen, phosphorus, nitrogen, and sulfur are dangerous until that decomposition is well along. Materials accumulate when they are added in larger amounts than can be accommodated by their decomposition rates. Materials that are toxic to soil microorganisms further slow recycling.

Solutions to the numerous pollution problems listed in Chapter 18, and many others not mentioned, are vigorously sought. Some partial remedies are in place; but additional remedies, closely related to agricultural activity and involving soils, are the topics for this chapter. The seemingly simple problem of **eutrophication** still commands a major interest. **Toxic substances** will always have our attention, but on which group of them do we concentrate our efforts in the next decade—there are so many of them. **Groundwater pollution,** already in unstoppable progress in many places, is difficult to reduce or clean up (Figure 19-1). We can slow down new pollution, but some contaminants, already en route to waters, may take years or decades to finally dissipate.

The wastes in soils cause new challenges. **Oil** and **chemical spills** onto soils are slow to change. **Bioremediation** (decomposition by microorganisms) is new, slow, and largely as yet untried. What can be done with **sewage sludges** that contain **heavy metals?** Heavy metals find a nearly permanent home in the soils. **Landfills** continue to be a menace to pollute groundwaters and evolve methane gas. We will need both natural and creative solutions, but to resolve the **soluble salts** problem will take unusual creativity (Figure 19-2). Some of the needed remedies we will discover. We will, in the process, come to realize that *avoiding future environmental problems* is as important as learning to remediate current contaminations.

FIGURE 19-1 Sanitary landfills! Out of sight—no problem. Many of these now are beginning to haunt us with the chemicals gradually being leached to groundwaters. Is there a way to stop them? Will it be necessary to dig them up and rebury them in water-tight units? No solution seems simple or cheap. This pit in Ada, Oklahoma, was begun before contamination concerns were widespread.
(*Source:* USDA—Soil Conservation Service, Ada, OK.)

FIGURE 19-2 What do we do with soluble salts? Until now, if we could wash them away, that was good enough. Now no one wants the salt in groundwater or in surface waters. This cotton field in Texas must either be reclaimed or abandoned. The country needs its land to be productive. What is best to be done? (*Source:* USDA—Natural Resources Conservation Service.)

IMPORTANT FACTS TO KNOW

1. The meaning of and the need for BMPs
2. Approaches to reducing contamination from nitrates
3. The remedial measures used to reduce *eutrophication, high BOD waters, heavy metal loadings from applied organic wastes,* and *radioactive materials*
4. The nature and longevity of the methane explosion hazard from organics buried in landfills
5. The legal basis on which the EPA controls many activities on lands and waters
6. The nature, cause, and remedial measures for acidic rain
7. The potential and techniques to (a) reduce additional accumulation of soluble salts and (b) reclaim land already salty
8. Techniques to reduce *particulate matter* going into air and *erosion sediments* in water
9. Approaches in biodegradation of contaminated soils
10. Some unique suggestions for remediation of some contaminants
11. Techniques to protect land from unwise use by people
12. The advantages and disadvantages of bioremediation

19:2 Legal Basis for Control

National water and air quality control is embedded in numerous laws, beginning with the basis for our zoning laws in 1926. An extensive Clean Water Act was passed in 1963, 1965,

1970, 1972, 1977, and 1987. Several environmental laws passed in 1987 include the Federal Water Pollution Control Act, the Safe Drinking Water Act, the Solid Waste Disposal Act, the Comprehensive Environmental Response Act, the Compensation and Liability (Superfund) Act, the Resource Conservation and Recovery Act, and the Federal Insecticide, Fungicide, and Rodenticide Act. These many acts, and almost yearly additions and alterations, *empower* and *require* the **Environmental Protection Agency (EPA)** to regulate water and air quality, and thereby many soil activities. The EPA must control management of land to the extent necessary to limit pollution of water and air.

The essential purpose of water and air quality control policies is to change certain of people's behaviors. Commonly people *do* change if there is evidence that a failure to change could be costly. Establishing certain actions as illegal and imposing penalties for failure to curb illegal actions are effective, even though they are unpleasant and restrictive of our freedoms.

Liability for damages done to an individual, group, city, state, or federal area is quite straightforward when the action causing the problem is a *deliberate act of defiance* or is done with the knowledge that it is illegal. The most common predicament for liability is likely to be because of *negligence,* in which the defendant caused a problem because he or she was not sufficiently informed or careful. Proving *causation* (that the action was the actual cause of the claimed damage) may be the most difficult aspect of liability. *Blameless contamination* denotes limited situations in which contamination occurs even *when normal and approved actions are followed.* Blameless contamination does not include such things as *spills, accidents, use at wrong time of year, application of chemicals by a person untrained in their use, and actions contrary to labels or BMPs.*

One of the frightening aspects of pollution regulation for those of us who consider ourselves environmentally-friendly and law-abiding people is that future legislation may impose cleanup and other liability for contamination that occurred from activities *that were entirely legal at the time they occurred.* Care, common sense, and a concern for others are needed as we protect our own environment.

For the past two decades considerable control has been exerted on **point sources** of pollution (pipelines, smokestacks, hauled wastes, and disposal canals). In Section 319 of the 1987 Water Quality Act, Congress extended the EPA's obligation to set up methods to control **nonpoint sources** of pollution (underground leaching, general erosion, or evolution of gases from large areas).

The problem in attempting to control nonpoint sources of pollution is knowing whom to blame and how much. Sources of pollution are widespread and may come from many land users in small amounts. In urban and city areas the pollution can come from everyone's yards, houses, or roadways. This problem has forced the EPA to require land users to practice what we might label *reasonable responsibility.* Freedom of action in the United States is difficult to stifle. The most direct way to deter actions that cause contamination or pollution is to declare those actions illegal. Consequently, state and federal agencies establish standards and then declare as illegal any actions that violate those standards. Sources of nonpoint pollution are reduced by making certain actions illegal. The list of permissible actions is termed *best management practices (BMPs).*

19:3 Best Management Practices

A **best management practice (BMP)** is a set of directions for an activity, such as fertilizing cotton, disposing of waste oil in lubrication shops, or disposing of spent solvents in dry cleaners, and replanting forests after logging. It is a guideline of whichever method seems to be the best currently known way (least polluting) to accomplish a necessary activity. Guidelines indicate methods of doing things, the quantities permitted to add, relationships to timing, and many other details.

An example of a far-reaching program involving BMPs was the program of the East Bay Municipal Utilities District in Oakland, California.[1] Using a combination of *discharge prevention* (or *discharge minimization* or *discharge estimation*) *permits* and *BMPs* as guides to meet the requirements, the District has decreased pollution into San Francisco Bay. The concentration in the influent to the wastewater treatment plant decreased the tetrachloroethylene to 42 percent, toluene to 16 percent, and xylene to 12 percent of contents prior to 1993. Since the start of the program in 1974, *heavy metals* discharged into the Bay have been reduced by 97 percent. More than 691 commercial discharge prevention permits and 281 discharge prevention permits to dealerships and repair shops have been issued. About 1500 accounts will be reviewed yearly for the next 8 years. To obtain a permit the business's facilities are checked, modification may be necessary (such as sealing drains), records must be kept, and a verification shown that hazardous materials are handled appropriately. During the anticipated lengthening of the permit from 1 to 5 years, *random unannounced inspections* will be conducted.

For rules to be effective, they must be enforceable. The Arizona Department of Environmental Quality (ADEQ) is an example of how this can be done.[2] A number of BMPs are established and producers can farm under a *general permit* as long as they follow the BMPs. If the BMPs are not followed, producers are shut down until each offender obtains an *individual permit,* which is expensive in time and paper work and takes 6 months to a year if things progress smoothly. The message to producers is that they should *implement BMPs to maintain a general permit at all costs.* Producers can be asked to verify that they are using BMPs, which is only easily done by keeping records (soil tests, recommendations, meetings with county agents, etc.) to show appropriate knowledge of the BMPs and adequate guidance.

In Pennsylvania, in 1993, a Nutrient Management Act was enacted. If a rancher with a cattle operation is following practices in the Department of Environmental Resources publication *Manure Management for Environmental Protection,* no special permit is required for manure utilization. If the law is violated, he must obtain a nutrient management plan to continue use of the manure.

Reducing Nutrient Pollution

Nitrate and phosphorus are the major nutrients of concern. Nitrate is leached into waters by rainfall and irrigation—any time water flushes through soil. Large amounts of nitrate for flushing can be in soils where nitrogen fertilizers have been applied but plants have not had time to absorb it. Decomposition of organic materials can also build up nitrate reservoirs if it is released faster than plants use it. Controls are usually to add less, time it better, and don't water excessively.

An example of some BMPs in Arizona is shown in Detail 19-1. Some of the **guidance practices (GP)** to meet BMP 1 in Detail 19-1 (limit N fertilizer added to that necessary for projected yields) are the following:

G.P. 1.1 Sample and analyze soils for nitrate contents

G.P. 1.2 Test irrigation water for nitrate and for compatibility with ammonia if fertigation is used

[1] Dan H. Kimm, Thomas C. Paulson, and Joe Damas, Jr., "An Ounce of Prevention," *Water Environment & Technology* (1995), No. 1; 38–41.

[2] J. Watson, E. Hassinger, K. Reffruschinni, M. Sheedy, and B. Anthony, "Best Management Practices Meeting Water Quality Goals," *Nutrient Management Special Supplement* to *Journal of Soil and Water Conservation* **49** (1994), No. 2; 39–43.

Detail 19-1 Best Management Practices (BMPs)

Statewide and nationally best management practices (BMPs) are being devised for all imaginable activities that might contribute to pollution of air and water. The two schemes listed below are from Arizona and apply generally to the entire state. The BMPs are to be followed to reduce pollution by (a) nitrogen from fertilizer applications and (b) various pollutants from animal-feeding operations (manures, washings).

1. BMPs for application of nitrogen fertilizers to crops
   a. Application shall be limited to the amount necessary to meet projected crop needs (eliminates freedom to add extra—insurance—fertilizer; requires knowledge of crop and soil needs).
   b. Applications shall be timed to be as efficient as possible.
   c. Method of addition shall be to add nitrogen to the area of maximum crop uptake.
   d. Irrigation shall be managed to minimize nitrogen loss by leaching and runoff.
   e. Tillage practices that maximize water and nitrogen uptake shall be used.
2. BMPs for animal feeding operations
   a. Minimize leaching and runoff losses of nitrogen by harvest, stockpiling, and disposing of manures as economically as is feasible.
   b. Have water control facilities (ponds, etc.) to control and dispose of nitrogen-contaminated water in case of a 25-year, 24-hour storm event equivalent as economically as is feasible.
   c. Close facilities, as necessary, in an economically feasible manner, to minimize nitrogen pollutant discharges.

Source: Modified from Brian E. Munson and Carrol Russell, "Environmental Regulation of Agriculture in Arizona," *Journal of Soil and Water Conservation* **45** (1990), No. 2, 249–253.

G.P. 1.4 Use application equipment properly calibrated
G.P. 1.7 Use slow-release nitrogen [if it is economically appropriate]
G.P. 1.8 Use appropriate plant analysis to guide nitrogen fertilizer applications

Nitrate is so soluble that restricting its movement is usually limited to keeping applications down to low concentrations and avoid allowing the soil to be flushed. The mechanisms to reduce nitrate leaching are the following:

- Keep additions to levels that plants require; do not add excess.
- Time the additions to avoid leaching by rain or irrigation.
- Avoid heavy (leaching) irrigations.
- Use slow-release nitrogen where appropriate.
- Use ponding, as for holding ponds for manures, to allow denitrification losses if the manures added are too high in nitrate.
- Flashboard partitions can be installed for adjustable flow restriction or other controls in drainage lines to regulate water table levels and to permit more denitrification of excess nitrate before it moves deeply into subsoils or into drainage ways.

Deeply rooted perennials may be able to retrieve some nitrates that are deeper than most roots penetrate. The major concept is that disposal of manure may load the soil with nitrates unless crops that can intercept the nitrate before it gets too deep are grown. A low-N_2-fixing alfalfa removed 393 kg/ha (350 lb/a), reed canarygrass 247 kg/ha (220 lb/a), and switchgrass

Lt. Ives's report, April 18, 1857: "Our reconnoitering parties have now been out in all directions, and everywhere have been headed off by impassable obstacles . . . The region last explored is, of course, altogether valueless. It can be approached only from the south, and after entering it there is nothing to do but to leave. Ours has been the first, and will doubtless be the last, party of whites to visit this profitless locality. It seems intended by nature that the Colorado river, along the greater portion of its lonely and majestic way, shall be forever unvisited and undistributed."

FIGURE 19-3 What BMPs could fit the Grand Canyon National Park, which ranges from 4 to 18 miles wide and is 280 miles long? We look at it now as useful—not worthless, as did Lt. Ives in 1857. We hope to control, to the degree feasible, further erosion and degradation, mostly to protect the Colorado River water users downstream. (Courtesy of Raymond W. Miller, Utah State University.)

152 kg/ha (135 lb/a). Some alfalfa roots will grow to 4.5 m to 5.4 m (15 ft–18 ft) if the soil is open.[3]

Reduction in **phosphate** is mostly controlled by (1) reducing municipal wastewater contents, (2) restricting phosphates used in detergents and other chemicals, and (3) reducing erosion, which has phosphorus riding piggy-back adsorbed to the soil particles eroded.

Best management practices are an approach to provide the best solutions known to avoid unnecessary pollution (see Detail 19-1). Which BMPs can be installed to control erosion or other damage in Grand Canyon National Park (Figure 19-3)?

19:4 Managing Organic Wastes and Pathogens

Most agricultural causes of high **biological oxygen demand (BOD)** in waters are a result of organic materials—manures, sewage wastes, food-processing wastes, and municipal wastewaters and garbage.

The only way to reduce the water's BOD is to reduce the organic materials in the water. Usually this requires ponding in aerobic conditions (aerate the water) to speed the organic matter decomposition. Filtration through soils to use the water to recharge groundwaters is common.

By July 1987 industries needed to comply with EPA standards for control of all pollutants by using the **best available technology (BAT)** to remove offending substances. (BATs are general terms for BMPs; in many instances they may be identical.) The guide used to regulate wastewater disposal is that the water must not exceed the allowable contamination limits for **drinking-water standards.** These standards designate allowable maximum levels of many metals, anions, toxic organic substances (such as PCBs and cyanide), pesticides, strong-tasting phenols, coliform bacteria, radioactivity, total dissolved solids, and even maximum water temperature and pH.

---

[3]Sam Brungardt, "Deeply Rooted Perennials Can Retrieve Nitrates from Soil Depths," *Minnesota Science* **47** (1996), No. 3; 4.

Tuscola, Illinois, shifted from an overhead sprinkler system (which was not useful during winter) for disposal of wastewater to an underground filtration-drainage system.[4] The 65-hectare system distributes water at low pressure into an underground network of pipes. The water filters through the soil to drainage lines placed 1.5 m deep, which move the water to a nearby creek. The overhead sprinkler was more effective in use, but it was only usable part of the year. The new system meets current water quality standards.

Sewage Sludge and Sewage Wastewaters

Sewage sludges (the solids settled out in sewage treatment plants) are much like farm manures (see Chapter 6). They can be applied to turf grasses on golf courses, in cemeteries, and around public buildings. Commercial nurseries are also potential users of sludge. Increasing quantities of sludge are expected to be used for the revegetation of soils drastically disturbed by surface mining and construction. Forests, pastures, and rangelands are also suitable sites for applying large amounts of sewage sludges and wastewaters.[5]

Although variable in chemical composition, sludges contain about 4.0 percent nitrogen (N), 2.0 percent phosphorus (P) (4.6% P_2O_5), and 0.4 percent K (0.5% K_2O) on a dry-weight basis. They also contain toxic and heavy metals, such as boron, cadmium, copper, mercury, nickel, lead, selenium, and zinc. There is real concern that toxic quantities of these metals will be detrimental to plants, animals, and people.

Pathogenic (disease-producing) organisms that are present in some sewage wastewaters and sludges include amoeba- and bacteria-dysentery, pinworms, and tapeworms. Sludges can also help spread cholera and poliomyelitis virus.

The techniques used to reduce to near zero the hazard of pathogens from soil-applied sludges are as follows:

1. Compost the sludge outside for at least 21 days. During composting, the heat of microbial decomposition normally reaches 55° C (131° F).
2. Store the semiliquid, anaerobically digested sludge for at least 60 days at 20° C (68° F), for 120 days at 4° C (39° F), or for some combination in between.
3. Treat the sludge, when moist, with lime [CaO or $Ca(OH)_2$] for at least 3 hours. This pH adjustment also helps to reduce plant uptake of heavy metals.
4. Pasteurize for 30 minutes at 70° C (158° F).[6]

Environmentally safe disposal of large tonnages of sewage and septage is of immediate concern. Cities near oceans have been dumping human and other solid wastes into the ocean, but this polluting practice was banned at the end of 1981. Sewage sludge can be burned, but the fuel costs to do so are now prohibitive and air is polluted thereby. (Sludge contains too much water to be burned without additional fuel.) Sludge has been buried in landfills along with other solid wastes; but again, costs of hauling are increasing and suitable soils nearby are scarce. The principal remaining alternative is to spread sludge on soils, either before or after composting. Before composting, sewage sludge has inherent hazards of spreading pathogens, polluting surface and underground waters, and being toxic to plants, animals, and people.

---

[4]Cooperative Extension Service, "New Wastewater Treatment Method," *Science of Food and Agriculture* **5** (1993), No. 2; 14.

[5]Alex Hershaft and J. Bruce Truett, *Long-Term Effects of Slow-Rate Land Application of Municipal Wastewaters,* EPA-600/57-81-152, Environmental Protection Agency, Washington, DC, 1981.

[6]Norman E. Kowal, *Health Effects of Land Application of Municipal Sludge,* EPA/660/1-85/015, Environmental Protection Agency, Washington, DC, 1985.

Treated sewage wastewater can be recycled back onto the land. Sewage wastewater can substitute for expensive irrigation water, and sewage solids and dissolved nutrients can replace part or all of the application of chemical fertilizers. If water is dispersed by sprinkler irrigation on forests and croplands and applied at a rate no faster than the soil's infiltration capacity, the use can be environmentally safe. One recycling program in Pennsylvania disposed of wastewater and produced these positive results:

- Field crop yields were increased. A 2.5-cm (1-in) depth of effluent per week increased grain yields 24 percent over land with 672–1120 kg/ha (600–1000 lb/a) of 10-10-10 commercial fertilizer added annually for 5 years.
- Growth of forest species, such as white pine and white spruce, was increased, but some species such as red pine were injured by heavier applications.
- Wastewaters replaced both chemical fertilizers and usual irrigation waters.
- Groundwaters were adequately recharged.

Animal Manures

Although animal manures are the oldest fertilizer used, the changes occurring when cattle, pigs, and poultry are raised in confined space and fed supplements and medication to increase appetites and control diseases, create large quantities of potential pollutants. We are learning more about manures, about their salt and metal contents, and about the tendency for them to provide too much mobile nitrate for leaching.

The environmentally safe management of manure from feedlots, to reduce pollution hazards, should be established with the following guidelines:

- All outdoor runoff and retention facilities should have the capacity to control any maximum 10-year, 24-hour storm water.
- All lagoons, ponds, and other animal-waste storage facilities should be watertight to eliminate runoff and deep seepage to surface water or groundwater.
- Animal wastes should be incorporated into the soil or, if surface-applied, be retained on sloping fields by terraces.
- The maximum safe application rate of animal wastes on fields should be established by research (Calculation 19-1).

Decomposition action (composting or decomposition in the soil) does not remove nonorganic materials such as lead, mercury, cadmium, arsenic, cyanide, strong acids, and other inorganics, nor do these prevent the potential formation of **mycotoxins**—substances that are toxic to plants or animals and form by fungal action of decomposing organic materials. In addition, *teratogenic* (deformity-causing), *carcinogenic* (cancer-causing), or other harmful organic substances may be produced naturally by the composting process.

Malodorous Gases

Bad odors are produced during early stages of decomposition of some organic materials, particularly in anaerobic conditions. These malodorous gases can often be dissipated by evaporation of the substances at the elevated temperatures while in the composting bed or with time. The most common offenders are hydrogen sulfide (H_2S, often called *rotten-egg gas*) and ammonia (NH_3), but other gases also are involved.

Removing odors by **biofiltration** moves the gases through a porous filter during composting. One such filter is made of wood chips, bark mulch, peat, lime, and sand (or

Calculation 19-1 Predicting Manure Addition Rates

The estimated amount of manure that can safely be added to soil without causing pollution is dependent upon (1) the nitrogen content of the manure, (2) the climate (warmer areas have more rapid manure decomposition), (3) the amount of residual manure left from previous additions, (4) storage and application methods (which affect nitrogen losses before manure is added to soils), and (5) the nitrogen requirement of the crop to be grown. Manure may best be used to supply only part of the total nitrogen requirement if the amount of nitrogen needed is large.

As an example, consider an operation in south-central Washington State that has nearby manures available for use. To calculate the weight of the available manure needed to add the amount of nitrogen suggested for the crop grown (fertilizer guide nitrogen = *FGN*), the following data could be used:

1. Nitrogen content (*NC*) in fresh manure, kg N/1000 kg manure (of 80% water):

Dairy cow	10 kg	Sheep feeder	17 kg
Beef feeder	9 kg	Layer hen	43 kg
Swine feeder	16 kg	Broiler chicken	52 kg

2. Fraction of nitrogen remaining (*NR*) from fresh manure after losses during handling and storage:

Anaerobic lagoon, oxidation ditch, liquid spreading	0.16%
Deep-pit storage, liquid spreading	0.34%
Open-stockpile storage, solid spreading	0.67%*
Fresh manure incorporated within 1–4 days, warm, dry soil	0.65%
Fresh manure incorporated within 1–4 days, warm, wet soil	0.85%

3. Fraction of initial manure nitrogen made plant available (*A*) each year after application:

Type of Manure	*Application Year*	*Years After Application*			
		1	*2*	*3*	*4*
Dairy, fresh	0.50	0.15	0.05	0.04	0.04
Beef, feedlot, piled	0.35	0.10	0.05	0.03	0.02
Dairy, liquid-manure tank	0.45	0.10	0.06	0.04	0.04
Poultry, fresh	0.75	0.05	0.05	0.05	0.03

4. Fraction of nitrogen left after denitrification losses (*D*):

Excessively drained soil	1.00
Well-drained soil	0.85
Poorly drained soil	0.70
Very poorly drained soil	0.60

Problem (a) How much stockpiled beef manure should be added per year to supply 100 kg of N per hectare (*FGN*) to corn on moderately well-drained soil?

(b) How much beef manure is needed each year after the fourth year in continuous corn cropping needing 100 kg N/ha yearly?

(c) How much fresh broiler poultry manure, rather than beef manure, would be needed each year after the fourth successive year of its use for 100 kg of N released per year? Assume wet soil.

Calculation 19-1, Cont'd Predicting Manure Addition Rates

Solution

1. Weight of manure to supply:

$$100 \text{ kg N/ha} = \frac{(FGN)}{(NC)(NR)(A)(D)}$$

$$= \frac{100}{(9)(0.67)(0.35)(0.85)}$$

$$= 55.7 \text{ Mg/ha } (24.8 \text{ t/a})$$

2. The fifth year would have 0.35 from the year of addition + 0.10 from manure added 2 years previous + 0.05 from third year previous + 0.03 from fourth year previous + 0.02 from fifth year previous = 0.55 for *A:*

$$= \frac{100}{(9)(0.67)(0.55)(0.85)} = 35.5 \text{ Mg/ha } (15.8 \text{ t/a})$$

3. Fresh poultry has 0.03 nitrogen fraction (*A*) for fifth year, so

$$A = 0.75 + 0.05 + 0.05 + 0.03 = 0.93:$$

$$\text{weight of poultry manure } = \frac{100}{(52)(0.85)(0.93)(0.85)} = 2.86 \text{ Mg/ha } (1.3 \text{ t/a})$$

This 2.86 Mg of broiler manure is considerably less than the 35.5 metric tons of beef manure with its lower N content and lower percentage availability.

*40% of original, but concentrated by drying and decomposition.

polystyrene spheres).[7] Each situation and the malodorous gases often react differently with each filter material. Some filters of equal amounts of pine bark and pine wood chips are used. The system is contained until it is desired to vent the system. The filter needs to absorb water readily to keep microbial populations active, yet have a high pore space for good air-flow rates.

19:5 Municipal Garbage, Composts, and Sanitary Landfills

Landfills receive most of our wastes, and a mixed lot it is. Because it contains such a variety of waste—some of it toxic—the major emphasis is to keep it from being leached with water that would move solubles to groundwaters and surface waters; thus the location and its properties are critical. Some EPA requirements for suitable landfills are summarized and briefly discussed in the following list.[8,9]

[7] Nora Goldstein, "Odor Control Experiences: Lessons From the Biofilter," *BioCycle* **37** (1996), No. 4; 70–75.
[8] Environmental Protection Agency, "Solid Waste Disposal Facilities," *Federal Register* **43** (Feb. 6, 1978), No. 25; 4952–4955.
[9] Environmental Protection Agency, "Landfill Disposal of Solid Waste: Proposed Guidelines," *Federal Register* **44** (Mar. 26, 1979), No. 59; 18138–18148.

FIGURE 19-4 This small landfill on Lakeland sand in South Carolina would seem to be on soil much too permeable and in danger of polluting the groundwater. Fortunately, just below the trench bottom is soil of only moderate permeability. The soil beneath the landfill is important if pollution is to be avoided. (*Source:* USDA—Soil Conservation Service.)

1. **Site selection and general information:** Avoid environmentally sensitive areas (for example, floodplains, permafrost areas, critical habitats of endangered species, and recharge zones that supply the sole source of culinary water), active earth faults, and **karst terrain** (limestone areas with caves and sinkholes). Soil should have low to moderate permeability and support vehicular traffic in bad weather.
2. **Soil, geology, and hydrology:** Determine the water balance for the site (rainfall, water movement through soil, the soil's permeability, and the location of the water table).
3. **Leachate control:** The bottom of the landfill should be at least 1.5 m (5 ft) above the seasonal high level of the groundwater table. Water on landfill areas should not have direct-flow connection to other standing or flowing surface waters. Determine the water-soil strata (hydrogeologic) conditions to predict possible contamination of groundwater. A slowdown barrier or a complete bottom-and-side sealing barrier may be needed (Figure 19-4).

   Lining materials can be clays, soil cements, crushed limestone,[10] and artificial materials such as asphaltic and plastic membranes. Synthetic membranes should be 20 mils or more thick.
4. **Gas control:** Control methane gas concentration in the atmosphere (produced by anaerobically decomposing wastes) to no more than 5% at the property boundary or 1.25% in buildings on site. Compacted clays, asphaltic and plastic liners, and cement help to retain gases within the landfill. The addition of perforated pipe in gravel-filled trenches with or without air pressure or suction can be used to remove unwanted gases before high concentrations build up within the landfill.

[10] Juan Artiola and Wallace H. Fuller, "Limestone Liner for Landfill Leachates Containing Beryllium, Cadmium, Iron, Nickel, and Zinc," *Soil Science* **129** (1980), 167–179.

5. **Runoff control:** Locate the landfill site so as to avoid accumulations of runoff water from adjacent lands. Surface grading and an impermeable soil surface can reduce leaching; ditches and dikes can be constructed to move and hold runoff water.

6. **Monitoring:** In conformance with laws passed in 1975, monitors continuously sample and analyze groundwaters from deep bore holes and check for gases and for pollution in groundwaters sampled from nearby wells.

Composting nontoxic organic wastes is a recent practice of some municipalities that helps to dispose of some garbage volume usefully. Leaves that are collected in large amounts without toxic wastes mixed with them are frequently used. Such compost is sometimes given away as a reward for hauling it. The problem in using composts rather than burying the materials is that composting costs more than land burial.

Graven County, North Carolina, started a pay-as-you-go program to increase recycling of municipal waste at the grass roots level—the home. Each household must attach a sticker costing $1.25 to each 33-gal bag of garbage.[11] A 90-gal trash cart can be used with three stickers attached. Curbside recycling was made available and use doubled in a short while. To encourage recycling, a household is sent only four stickers a month. Any additional stickers must be purchased at specified outlets, mainly grocery stores.

Landfills are one of the cheapest ways to hide wastes, but landfills can be timebombs in various ways. They will still be there after hundreds of years; even wastes that can decompose do so very slowly. Landfills hinder us from recycling organic materials and plastics.

Sorted municipal solid wastes (MSW) can be recycled to farmlands, home gardens, and parkland. Presently only 23 percent is recycled; most of the rest goes to landfills at costs of about $9 (New Mexico) to $83 (New Jersey) per metric ton. Additions of MSW to soils increased C, N, P, K, Ca, Mg, Mn, Fe, and Zn. Mixing MSW of a wide C:N ratio, such as 150:1, with animal manures of C:N ratio 10:1 helped retain nitrates and improved the general soil quality.

Paper makes up about 40 percent of solids that go into landfills. Recycling paper would be a big help to reduce needed space. Paper added to soil is a problem usually because of its high C:N ratio of about 150:1. A new approach mixes a higher N material (manure, food waste) with the shredded paper and pellets it (using hay pelleting equipment) into 1-cm-diameter pellets.[12] These can be incorporated into soil to increase organic matter and the porosity of clayey soils. Larger pellets (5 cm $\times$ 2 cm to 10 cm $\times$ 2 cm) can be used on top of the soil to help reduce erosion by wind.

Another system puts waste paper into water and beats it to fibrous pulp which can then be pressed into forms or blocks that are very much like the wood it came from.[13] Additives can make the end product waterproof, flame-retardant (for insulation), or even insect-proof. It can be sawed, sanded, glued, and nailed.

The methane gas (CH_4) evolved from landfills is a concern (Detail 19-2); however, the methane can be recovered, cleaned, and used.[14] Methane, a greenhouse gas, traps heat about 20 times more effectively than does carbon dioxide; fortunately, there is a lot less methane than CO_2 in the atmosphere. Landfills evolve CH_4 plus volatile organics that combine with

---

[11] Robert Bracken, "North Carolina County Institutes Sticker System," *Biocycle* **33** (1992), No. 2; 35–37.
[12] Don Comis, "Trash to Treasure," *Agricultural Research* **41** (1993), 18–21.
[13] Derick Schermerhorn, "Changing Paper Back into Wood," *Compressed Air* **101** (1996), No. 4; 10–15.
[14] Mary Nichols, "Landfill Gas Energy Recovery: Turning a Liability into an Asset," *Waste Age* **27** (1996), No. 8; 89, 90, 92, 94, 96.

Detail 19-2 Landfill Gases Are a Problem

Near Commerce City, Colorado, a 150-cm (5-ft)-diameter water conduit was constructed in 1977 to carry water to the eastern part of Denver. During inspection and cleaning to begin use of the conduit, a welder requested two workmen to install a fan at a distant manhole. At the manhole a worker lit a match and touched off an explosion that sent flames 12 m (39 ft) into the air. This burning created a vacuum that pulled more air (oxygen) into the conduit. Four explosions during 90 seconds rocked the area and knocked the welder about 10 m (33 ft). Four of the firemen who came to fight the fire were sent to the hospital for treatment of carbon monoxide inhalation; one workman was dead on arrival, and another died 3 weeks later.

The exploding gas was **methane** produced by anaerobic bacteria decomposing organic matter in a nearby landfill. The gas diffused laterally through the soil and entered the eastern end of the closed conduit. Tests of gas from the landfill indicated 31.8%–52.5% methane by volume (lower explosive limits are near 5% for methane); the eastern end of the conduit (near the landfill) contained 6.1%–14.4% methane. Large gas concentrations occurred in the gravelly soil at distances of 120 m (394 ft) from the landfill.

Source: J. W. Martyny et al., "Landfill-Associated Methane Gas a Threat to Public Safety," *Journal of Environmental Health* **41** (1979), No. 4; 194–197.

nitrogen oxides. Sunlight energy combined with these organics produces ground-level ozone, or *smog*.

Across the United States about 140 landfills convert landfill gases (predominately methane) to useful energy by (1) converting it to electricity through an internal combustion engine, (2) cleaning the gas to inject it into natural gas supplies, (3) direct use in boilers or industry or greenhouses, and (4) a few other emerging technologies.

It is actually this long-term (even hundreds of years) production of gases from landfills that prompts some people to suggest that incineration is more favorable. This is not because incineration produces less of the hazardous gases per year, but because incineration just releases the gases once; landfills continue to release gases for decades.

19:6 Managing Toxic Wastes

The toxic wastes critical in agriculture are pesticides, petroleum products, and natural poisons. The variety of substances and problems are enormous.

Reducing Pesticide Contamination

The obvious way to reduce pollution with pesticides is to reduce pesticide use. New regulations will reduce pesticide use. Pollution control laws greatly influence both the health and the livelihood of many people. Laws need to be made with reason, logic, and fairness and based on the most accurate data possible. Many proponents of stopping the use of pesticides may be the first to select unblemished fruits in the market, to buy sprays to control home mosquitoes, and to complain about increased food prices. The world existed without pesticides until the 1940s; obviously we could exist again without them. But all alternatives have a cost. The cost without pesticides would be higher food costs, more blemished products, less control of nuisance insects (mosquitoes, cockroaches, weevils, locusts, etc.), more people working on the farms, and lower crop yields for a hungry world (see Figure 19-5).

FIGURE 19-5 Without pesticides, insects like the Japanese beetle (top left), Mediterranean fruit fly (top right), and gypsy moth (larva and adult), which infest more than 300 species of trees, vegetables, flowers, and fruits, would have many opportunities to chew their way through our crops. (*Source:* Photos courtesy of USDA—*Agricultural Research* **27**, No. 9; and California Department of Food and Agriculture; photos by D. Zadig and M. Pendrak.)

Benefits from the banning of organic pesticides are harder to document unequivocally. Clearly, numerous accidental deaths and poisonings would be eliminated. Damage of a carcinogenic and teratogenic nature, attributed to some pesticides, would be reduced. People would be forced to adopt farming practices that are more self-sustaining and compatible with nature. Undoubtedly greater effort would be expended toward developing suitable cultural and biological methods of pest control, which would be helpful in self-sustaining agriculture.

Integrated Pest Management (IPM)

There are alternatives—or perhaps more correctly termed *partial substitutes*—to the use of chemical pesticides. The use of natural predators (**biological control**) is increasing. The cottony-cushion scale insect, which devastated citrus trees, is controlled by two Australian insects: a small parasite fly and a predatory ladybird beetle. Another pest, the sugarcane leafhopper, threatened to devastate the Hawaiian sugarcane industry. A total of five insects parasitizing the leafhopper eggs finally brought the problem under control. Other control techniques involve disease-causing bacteria that kill certain pest larvae and larval that feed on certain weeds, thus partially controlling the larvae. But biological controls are usually slow (may take many days or weeks for populations to build up) and incomplete (may have only 70%–90% control). Pesticides, in contrast, may kill 95 percent to 99 percent in a few hours or days. Unfortunately, not many of these biological-control systems are known and developed.

Biological control is slow to be developed and has a lag time when it is needed. When new pests, such as the Russian wheat aphid, suddenly infest an area they had not infested before, they cause severe damage. Pesticides suitable to control the aphid may not be developed for several years after their appearance.[15] Estimated crop damage from this aphid in 1987 in the ten Great Plains states was $53.8 million. Texas alone lost $100 million in 1988 and expected a loss of $140 million in 1989. No dependable biological control is known yet.

Numerous biocontrols are known.[16] A bacterium has been marketed since 1945. A fungus controls sweet potato whitefly. A new nematode product is an enemy of pink bollworms, corn earworms, and fall armyworms. A yeast from tomato peel outcompetes rot-causing fungi on citrus and some other fruits allowing longer storage without rot on wound sites.

The water hyacinth, given out as a souvenir at the 1884 New Orleans Cotton Exposition, soon got into waterways of the Southern states and clogged whole canals that even stopped small boat travel. The use of pesticides to control the hyacinth is expensive and polluting. As a complement to chemical control, weevils and a moth have been imported to help control the hyacinth (Figure 19-6).

Cultural control is the control of pests by management and physical activities. We may chop and bury old plant residues to kill insect eggs. We eliminate weeds in adjacent areas before they go to seed (to reduce weed seeds). We select crops that have fewer pest problems in our area. We rotate crops to remove the *host plant* for several years, causing the pest to die out or be reduced to low numbers. We bury residues by tillage; no-till and reduced-tillage practices have more pest problems. We are then faced with the question, Do we reduce soil erosion, use less fuel, and add more pesticides, or do we use the methods employing extensive tillage that require less pesticides? This is an example of more compromises that need good scientific data from which to make rational decisions.

Other ways to reduce pesticide use include (1) **breeding disease-resistant plant varieties,** (2) **male sterilization by irradiation,** and (3) use of **natural pesticides.** One example of natural chemicals is the use of **juvenile hormones,** which keep the larval stage from becoming an adult. **Sex attractants** (included as **pheromones**) have been used to attract insects to traps and to confuse males so that they cannot find mates, thus reducing viable eggs.

Control of grasshoppers has been a recurring problem for decades. With infestations, as in 1984 to 1987, more than 5.6 million hectares (13.8 million acres) were sprayed with pesticides. This is costly and may contaminate sensitive areas, such as wetlands. Recent field work has shown that, by focusing on spreading pesticides on the egg beds, adult hoppers were reduced.[17] This approach boosted the effectiveness and reduced overall quantity of the chemicals by 95 percent. A managed grazing program will reduce grasshopper numbers, also. One new chemical intended to replace malathion requires only about 0.37 percent of the concentration of malathion needed. A *Grasshopper IPM User Handbook* and model (on a computer disk) for grasshopper problems is available.[18]

Red dyes used by industry can be deadly to many insects.[19] The Mediterranean fruit fly (medfly) will eat FDA-approved red dye number 28, and on exposure to sunlight the fly dies. The dye replaces the pesticide malathion. Other related pests may react similarly. The medfly is attracted to traps and baited with lure chemicals; hundreds of traps are set in the fruit area.

---

[15] Vernon M. Stern and Steve B. Orloff, "Controlling Russian Wheat Aphid in California," *California Agriculture* **45** (1991), No. 1; 6–8.

[16] Sean Adams, "Biologicals Hit the Marketplace," *Agricultural Research* **44** (1996), No. 1; 16–17.

[17] Kathryn Barry Stelljes, "IPM Targets Grasshoppers," *Agricultural Research* **44** (1996), No. 1; 4–10.

[18] The Handbook is *USDA, APHIS Technical Bulletin No. 1809,* Stephen A. Knight, Unit 134, 4700 River Road, Riverdale, MD, 20737. For the model request *Hopper* version 4.0, or download from http://www.USDA.Montana.edu.

[19] Anonymous, "Red Dye, Updated Traps," *Agricultural Research* **44** (1996), No. 1; 20–22.

(a)

(b)

FIGURE 19-6 Canal cleaned of water hyacinth (see Figure 18-1, Chapter 18). The water hyacinth grows large and vigorous but the natural weevil and moth enemies shown in photo b help control the growth as shown in photo a. (Courtesy of USDA—Agricultural Research Service.)

A unique nonpesticide control for some crops is **vacuum cleaning** the plants.[20] Special vacuums to suck pests off strawberry plants and leaf lettuce have worked well. Vacuuming two or three times a week usually has controlled the major problem—lygus bugs—on strawberries and cut pesticide expenses by 73 percent. But vacuums cost more than $20,000 each, remove only 30 percent to 40 percent of adult lygus bugs per pass, and miss a lot of immature bugs that are still very damaging. Vacuums are being tried on grapes. Vacuuming will be useful for flying insects, but it is probably less valuable for nonflying types, for compact crops, such as head lettuce, and for very open crops, such as cotton, where the vacuum can't get close to much of the plant.

All of these possible ways to control pests work best if they are used in the correct *combination,* called **integrated pest management (IPM).** Smaller amounts of chemical pesticides

[20]Richard Steven Street, "Is Vacuum Pest Control for Real?" *Agricultural Age* **34** (1990), No. 2; 22–23, 26.

are needed with IPM, but IPM allows immediate control of infestations. Unfortunately, suitable biological and cultural controls are available for only a relatively few pests (Figure 19-7). *Chemical pesticides are likely to continue to be used in the next few decades.* We can reduce the amounts of the chemicals needed by IPM and by careful applications. As one scientist put it, IPM should mean *Integrated Pharm* (aceutical) *Management,* the careful, comprehensive, knowledgeable use of the best techniques for all practices in agriculture.

Petroleum Products and Other Organics

"Oil and water don't mix" is an old saying that emphasizes a major clean-up problem. If the problem substance doesn't mix with water, it is also nearly immune to attacks by microorganisms, which have water as their medium of life. In fact, petroleum and its products (gasoline, kerosene, heating oils, etc.) and other solvents, mostly immiscible in water, are deadly to most microorganisms and plants. What is possible on such sites?

One technique is to find microorganisms that are tolerant and do the job in situ, wherever the contamination of the soil exists. This use of microorganisms and related clean-up activities is called **bioremediation.** It is effective but slow. The emphasis is to find or genetically produce more effective organisms and to speed up their activities. This topic is discussed in more detail at the end of this chapter.

(a)

(b)

FIGURE 19-7 *Parlatoria oleae* (olive scale) infestation (a) can be parasitized by a small wasp to produce clean olives (b). The control is said to be among the most perfectly and closely documented for a biological control effort. (*Source: California Agriculture,* Division of Agricultural Science, University of California.)

One unique approach, unlikely to be used on large-scale field areas but perhaps suitable for localized cleanup (as a work area), is the use of high-temperature water.[21] Called **supercritical water oxidation (SCWO),** when water is heated above its critical point (where it is difficult to distinguish between liquid and vapor, at 374° C under 3200 lb/in^2 pressure), organic material will dissolve in the water. If O_2 is added it forms a combustion mixture that produces innocuous products—carbon dioxide and water and a few products from other goo-and-gunk materials left over in manufacturing. Destruction efficiencies of more than 99 percent have been achieved for almost all pollutants studied. The process has been effective on a large scale for solvents and fuel; PCB-contaminated oils; contaminated groundwater, soils, and sediments; process wastewater from organic chemicals, petroleum refining, pulp and paper, textiles, and pharmaceuticals; biological sludge; propellents and explosives; medical wastes; and organic components of radioactive wastes. From an industrial perspective the temperatures and pressures used are not unusual, but the process is expensive.

Remediation of groundwater using solar UV energy involves another oxidation approach.[22] Platinum-titanium oxide (Pt-TiO_2) catalysts in water and illuminated with reflected UV rays from sunlight produce hydroxyl radicals, which are strong oxidants. The process has been very effective on benzene, toluene, ethylbenzene, and xylenes in fuel-contaminated groundwaters.

Resolving the Heavy Metals Problem

Heavy metals (As, Cd, Cr, Co, Hg, Pb, Ni, U, and Zn) are elements; they cannot be further decomposed or oxidized. The only solutions to removing their hazardous condition are to do the following:

- Tie them up into insoluble substances.
- Remove them to another depository.
- Reclaim them, thus removing them.
- Bury them, thereby removing them from circulation in life cycles.

All of these techniques are either costly or somewhat ineffective.

A partial help for metal tailings is to reclaim some of the metals so that the contaminant is less. It is also beneficial in reclaiming more of the metal in tailings.[23] Figure 19-8 shows a gold-mining heap using bioxidation to break down pyrites to ease extraction of the gold with cyanide or other compounds. It is reported that at least 25 percent of world copper production uses bioprocessing to help liberate the metal from the ore.

Metal-scavenging plants (hyperaccumulators) is the term some use for plants that have abnormally high uptake of some of the problem metals.[24] For example, alpine pennycress plants accumulate up to 30,000 ppm Zn without yield reduction; most plants have Zn toxicity at about 500 ppm. If such plants work, the cost is much less than digging and hauling away the contaminated soil. The plant ash is often similar to commercial metal ore. This approach of plant extraction is feasible on soils at smelters and mining sites, landfills, nuclear waste dumps, farmlands, and other sites. Unfortunately, the cleansing process is slow, probably even decades needed to remove heavy metals to acceptable levels.

---

[21] Anonymous, "Pressure Cooking Toxic Wastes," *Compressed Air* **100** (1995), No. 4; 16–20.
[22] Anonymous, "Fun Fuels Groundwater Remediation," *Water, Environment & Technology* **7** (1995), No. 2; 15–16.
[23] Anonymous, "Bring in the Bugs," *Compressed Air Magazine* **101** (1996), No. 1; 18–24.
[24] Don Comis, "Metal-Scavenging Plants to Cleanse the Soil," *Agricultural Research* **43** (1995), No. 11; 4–9.

FIGURE 19-8 Bioprocessing is not quite the same as bioremediation, but it involves a similar concept that microorganisms might solubilize low-solubility contaminants. In this instance it is solubilizing pyrite to make gold extraction easier, recovering perhaps 25 percent more gold. (Courtesy of Newmont Metallurgical Services.)

Chromium (Cr) has several valences [Cr(VI) and Cr(III) are the most common]; *chromate* (CrO_4^{2-}) with Cr in 6+ valence, is most mobile and toxic. Attempts to reduce the Cr(VI) to the less toxic and less mobile Cr^{3+} form were accomplished by indigenous soil microorganisms.[25] Additions of organic amendments help reduce and bind the Cr(VI), immobilizing the chromium somewhat.

Some superfund sites have both toxic organics and heavy metals. Incineration can destroy the organics but the heavy metals remain in the ash, sometimes more soluble than prior to burning. This metal-rich ash can be enclosed in small nodules (0.8 cm–2 cm size) if the temperature feed preparation, fluxing agents, moisture, and time are controlled in the kiln to have enough molten material to glue the nodules together. The ideal state is between fine-grained ash and molten slag. Metals in these nodules are resistant to leaching.[26]

For heavy metals, avoiding contamination wherever practical is by far the best approach. A good solution for areas polluted with heavy metals is not available; it is difficult to visualize any simple solution. Where the metals occur in large land areas, control of pH (above 6.5 to 7.0) and growth of plants providing the least hazard as food (grains or fiber rather than vegetative foods—pasture, silage, green vegetables) should be grown (Figure 19-9).

Reducing Damages from Natural Toxins

The damages from some natural toxins (poisonous plants, toxic elements in plants) are difficult to eliminate. Usually eradicating the plants (a temporary solution) or fencing animals from the site are the most common techniques. Study of the halogeton-caused deaths of hun-

[25] F. R. Cifuentes, W. C. Lindemann, and L. L. Barton, "Chromium Sorption and Reduction in Soil with Implications to Bioremediation," *Soil Science* **161** (1996), No. 4; 233–238.

[26] John N. Lees, Marta K. Richards, Thomas F. McGowan, and Richard A. Carns, "Thermal Encapsulation of Metals in Superfund Soils," *Journal of the Air and Wastes Management Association* **45** (1996), 514–520.

FIGURE 19-9 Contamination of soils by some pollutants can only be eliminated by expensive soil excavation. An entire block in Midvale, Utah, is contaminated by lead and arsenic from a metal smelter that has long since ceased operation. This is a supersite cleanup that is costing enormous amounts of money. (Courtesy of Duane T. Gardiner, Texas A & M University–Kingsville.)

gry sheep led to the following general suggestions for dealing with certain toxic range plants:[27]

- Do not drive or unload hungry animals into areas low in good range plants but high in poisonous plants. Dumping trucked sheep in overgrazed unloading areas where halogeton is the common ground cover has caused heavy losses.
- Maintain free access to salt to limit grazing of salty poisonous plants (halogeton).
- Animals low in phosphorus tend to eat abnormally. Provide all needed minerals to the animals.
- Fence animals out of extremely toxic plant areas.
- Maintain good range feed and avoid putting new arrivals on bad areas of range until they become accustomed to low levels of the scattered toxic plants.

It is important, generally, to know what plants are poisonous, what they do, and to be able to recognize them. Then, it is easier to keep animals from the plant, the plants can be removed, or the area can be fenced off, whichever solution best fits the situation. In some situations (large range areas) there may not be an economical solution except to absorb the livestock losses.

19:7 Particulates and Dust

Controlling particulates and dust usually requires no-burn regulations, more efficient internal combustion engines, and wind-erosion-control practices. Burning of sugarcane fields prior to harvest to reduce trash, burning grain stubble to reduce residues and pests, and burning old range grass (in many areas of the world) to initiate new growth when rains are due are all practices that can be reduced or eliminated. Eliminating them will increase costs to the operator of the land.

[27] William C. Krueger and Lee A. Sharp, "Management Approaches to Reduce Livestock Losses from Poisonous Plants on Rangeland," *Journal of Range Management* **31** (1978), 347–351.

FIGURE 19-10 These dunes in North Africa are widespread, and they are promoted by arid years, overgrazing, and cutting vegetation for fuel. Containment followed by reclamation is slow, precarious, and expensive. (By permission of Food and Agriculture Organization, Rome.)

Wind-erosion controls are presently practiced. Maintaining a soil cover, moist soil, vegetation barriers to wind, and rough land aid in control (Figure 19-10). These conditions are discussed in Chapter 15.

19:8 Evolved Gases and Acidic Rain

Once oxides of sulfur and nitrogen get into the air, there is little that can be done to reverse the process of acidic rain. To eliminate evolution of gases is the practical solution. The use of low-S coal reduces the amount of sulfur oxides produced during burning. More efficient fuel burning produces less carbon monoxide and reactive hydrocarbons that produce smog. The only ways to reduce carbon dioxide, the most abundant greenhouse gas, are (1) to burn less fuel (coal, petroleum, wood, crops, garbage) and (2) to grow plants that will use the carbon dioxide. Other sources of gases include swamps, animals, composting, and flooded rice paddy (Detail 19-3). Extensive planting of forests is recommended worldwide.

19:9 Radioactive Materials

There is no known way to alter the rate or kind of decay of radioactive materials. Where radioactive fallout has occurred, there is no known way to eliminate the hazard, except by scraping and carrying off contaminated soil to which the cations have adsorbed. Radioactive elements in the soil are absorbed into plants, as are other elements, and become a part of whatever consumes the plants. Wildlife grazing on forages on radioactive tailing piles or on plants containing fallout elements and later eaten for food is one way that isolated radioactivity may affect people. The inability to mitigate contamination by radioactive substances except by the

Detail 19-3 Gases from Flooded Rice

When rice soils are flooded, O_2 becomes in short supply and aerobic bacteria are replaced by facultative anaerobes, which in a few days are replaced by obligate anaerobes.

In aerobic biological oxidation, free molecular O_2 is the ultimate electron acceptor. Facultative anaerobes use nitrate, manganic oxide, ferric oxide, carbonate, and other compounds with high oxidation levels as the electron acceptors. The most striking difference between aerobic and anaerobic decomposition lies in the nature of the end products. In normal, well-drained soils the main end products of organic-matter decomposition are CO_2, H_2O, nitrate, sulfate, and resistant humus; in submerged anaerobic soils the end products are CO_2, H_2O, H_2, CH_4, NH_3, H_2S, mercaptans, and partially humified residues. The successive microbial changes are accompanied by a lowering of the redox potential (from +600 to −300 mV) and a change in pH to near neutral. The pattern of changes and the accompanying formation of gases are related to soil properties, environmental factors, and management.

The range in the percentages of the gases in submerged rice soils is as follows: N_2 is 35%–98%; CH_4 is 4%–55%; CO_2 is 2%–10%; H_2 is 0%–9%; and O_2 is as low as about 1%–6%.

Source: Heinz-Ulrich Neue and Hans-Wilhelm Scharpenseel, "Gaseous Products of the Decomposition of Organic Matter in Submerged Soils," in *Organic Matter and Rice,* International Rice Research Institute, Los Baños, The Philippines, 1984.

passage of time makes many people strongly oppose development of nuclear energy systems and weapons testing.

The only known controls against damage from radioactive materials are as follows:

- Isolate the contaminated area from organisms.
- Shield organisms from the radiation (distance or heavy metal shields).
- Allow time for radiation to decline.
- Do not produce the contamination in the first place.

Encapsulation, forming insoluble materials, burning, and other chemical treatments do not change the radiation, only the substance's mobility in the system.

Radon gas is a radioactive daughter of radium from uranium decay. Some soils, and particularly uranium mine tailings, have excessive and dangerous radon production. Although radon gas (^{222}Rn) has a half-life of only 3.8 days, some confined areas (homes, other buildings) can accumulate dangerous levels. There is no control except to remove the soil high in the uranium or to move away. Venting the building can keep accumulation levels lower but in some localities that is not enough.

19:10 The Gordian Knot: Soluble Salts

Historically, an oracle proclaimed that whoever could untie the Gordian Knot, an unusual knot of bark into which the ends were slyly woven, would rule Asia. Alexander the Great is reported to have cut the knot with one stroke of his sword. The problem of soluble salts needs such a decisive strike. However, a realistic look at our inability to solve soluble salt problems is sobering (Figure 19-11).

Most dissolved inorganic chemicals in natural waters are soluble salts. They are found in all soil solutions. In high concentrations, salts are unwanted because they reduce or hinder

FIGURE 19-11 What can be done with the pollutant called *soluble salts?* Washing (leaching) salts may pollute groundwaters; not leaching them leaves a soil that will not grow plants—even plants that have some salt tolerance, such as the cotton in this photo. This soil near Bakersfield, California, is typical of what can happen without careful control when salty waters (wastewaters, sewage effluents, brackish waters), and even normal waters, are used. (*Source: Agricultural Research* **27** [1978], No. 11; 14.)

plant growth, speed corrosion of metals, make drinking water unpalatable, and interfere in many other uses of water. California has already established regulations that wastewaters must have total dissolved solids (TDS) reduced to 500 ppm before they can be discharged. In the near future more controls likely will be placed on irrigation runoff water disposal. Salt pollution from agricultural runoff water is largely **nonpoint pollution.** This means that the pollution does not always derive from one source, or point, but from a combination of sources. For example, lateral seepage flow and salt carried from the fields in irrigation wastewater is a widespread source difficult to pinpoint to a field or location.

Soluble Salts: Is There a Solution?

There is no simple, obvious solution to the soluble-salt dilemma. Removing salt by desalinization processes is too expensive, even for drinking water. People can reclaim the soil by washing out the salt, but the salt then goes into groundwaters or surface waters. The only uncontested disposal sites are the oceans, which are already salty, and a few salt basins that are near, and convenient, to a few areas. One such area is the originally dry Salton Sea of California, now used to collect drainage (Detail 19-2). However, even in such basins, increased salt often brings other problems, such as toxic metals, to wildlife. In the future, actions that will *not* be allowed include indiscriminant leaching of salts into groundwaters and the washing of salty irrigation wastewaters into surface waters. A number of states already have some regulations for this salt problem.

A permanent irrigation agriculture requires the sacrifice of some value elsewhere. If normal drainage runoff becomes more restricted, it will be difficult for regulators to identify and regulate the sources of contaminating salts found in surface waters. The more careful use of soil as a receptor of the salt (salt precipitation) will be necessary. Careful irrigation to avoid excess water application will allow precipitation of calcium and magnesium as carbonates, silicates, and sulfates during drying cycles. Many of these precipitated salts do not redissolve very easily, especially the carbonates and silicates. In some waters with low sodium but a high proportion of calcium and magnesium, as much as 60 percent to 80 percent of the soluble salts may be precipitated.

There is a catch to this seeming light at the end of the tunnel. The portion of salt that is *not* precipitated is high in sodium and chloride ions. The sodium causes soil dispersion and is highly corrosive to metals. The chloride is toxic to plants in high concentrations; thus periodically even these soils will need leaching and the exchangeable sodium will need to be reduced in the soil. Careful addition of water will, however, minimize salt moving into downstream waters and the total amount of salt in return-flow waters.

Individual states are to develop best management practices for on-farm improvements to control salt contents in waters leaving the farm. These studies have been in progress for several years.

Desalination: Plants or Mechanical Processes?

It seems inevitable that methods to reduce salt movement and remove salt from salty water will be used in some areas. Deep-rooted plants (alfalfa) have slowed surface-salt accumulations where saline seeps were forming. Ladak 65 alfalfa rooted 4.6 m (15 ft) and lowered seepage water about 3 m.[28] Numerous studies have been similar. However, these are *hold* positions and do not remove appreciable salts or change anything except the *progression* of seep development.

Waters may need to be ponded and desalted in order to dispose of some salts. Reverse osmosis is developed and seems the likely mechanism for mechanical desalting. Unfortunately the concentrated brine left over must still be disposed of somewhere. It may be necessary to evaporate brine to dryness if disposal to waters is not permitted. Although this approach is believed to be a leading possibility, the cost is high, about $0.88/m^3 ($1,090/acre-foot).[29]

19:11 Reducing Soil Sediments

Soil sediments are reduced when erosion is reduced (see Chapter 15). Federal regulations for water pollution control, recognition of the importance of soil as a nonrenewable resource, and the need to minimize other damages from soil erosion have increased restrictions on land management. Typical of these new land management restrictions is the North Carolina State Sedimentation Pollution Control Act of 1973 to control erosion from all land-distributing actions, particularly commercial development, including road construction. Some of its requirements follow:[30]

- Any land disturbance near a lake or natural watercourse must have a wide enough buffer zone that the 25% of the buffer zone nearest the activity collects all of the visible siltation. The law excludes constructions on, over, or under the water.

---

[28] M. R. Miller, P. L. Brown, J. J. Donovan, R. N. Bergatino, J. L. Sonderegger, and F. A. Schmidt, "Saline Seep Development and Control in the North American Great Plains—Hydrologic Aspects," *Agricultural Water Management* **4** (1981), 115–141.

[29] Edwin W. Lee, "Drainage Water Treatment and Disposal Options," Chapter 21, in Kenneth K. Tanji (ed.), *Agricultural Salinity Assessment and Management,* American Society of Civil Engineers, Publishers, 345 East 47th Street, New York, 10017, 1990, 450–468.

[30] Joseph A. Phillips and Jesse L. Hicks, "Sediment Control: The North Carolina Law," *Journal of Soil and Water Conservation* **31** (1976), 75–77.

- The angles for graded slopes are limited to those on which vegetative cover can be maintained; graded slopes must be planted within 30 working days after any phase of grading is completed.
- Any land area of more than 1 acre (0.4 hectare) that has been disturbed must be planted to ground cover within 30 working days on portions not to be further worked.

Minimum tillage is a management system that reduces soil erosion. It maintains residues on the soil surface and thus reduces the dispersing action of rainfall and decreases the speed of water runoff. However, the increased need for herbicides and other pesticides has caused some people to wonder if minimum tillage *is* reducing pollution. An increase in one pollutant (pesticides) is substituting for a decrease in a second pollutant (soil).

The 1985 Food Security Act (PL-99-198) has the following three conservation provisions that should help to reduce erosion:[31]

- **Conservation Reserve Program:** Persons seeding highly erodible land to grass for 10 years will receive a per-acre payment.
- **Swampbuster provision:** This act eliminates any federal incentives for all farm benefits to anyone converting wetlands to crop production without specified conservation treatments.
- **Sodbuster provision:** This act is similar to the swampbuster provision, but it refers to putting certain *fragile* (highly erosive) grasslands into crop production.

In 1976 the National Commission on Water Quality concluded that despite intense efforts to treat **point sources** of pollution (sources clearly identifiable, such as factory discharge lines), 92 percent of the total solids suspended in water would continue to come from uncontrolled sources. These uncontrolled sources would also furnish 79 percent of the total nitrogen, 53 percent of the total phosphorus, and 37 percent of the biological oxygen demand (BOD) material. This depressing estimate emphasizes the magnitude of uncontrollable pollution sources and the continual need to reduce water pollution, including that from suspended solids resulting from agricultural activity, natural erosion, or anthropic (human-caused) disturbance.

The Federal Water Pollution Control Act of 1972, amended in 1977 by the Clean Water Act and the Resource Conservation Act, was the first giant step toward national regulation of the pollution problem. Congress specified 1985 as the date by which the United States was to "restore and maintain chemical, physical and biological integrity to the nation's waters." There is general unanimity that more must be done to halt water pollution, but disparity of opinion still exists about the cost of the final phase of regulation versus the probable benefit (Figure 19-12).

Certainly the role of eroding cropland in causing pollution is important. Erosion control is a sensible approach in looking at conserved soil for the well-being of future generations. Soil erosion does come from many noncultivated crop areas, construction sites, and mine areas. The costs to control erosion and the social benefits of that control all need to be a part of the planning and legislative regulations.

---

[31] Clayton W. Ogg, "Erodible Land and State Water Quality Programs: A Linkage," *Journal of Soil and Water Conservation* **41** (1986), 371–373.

FIGURE 19-12 Without adequate regulation, some people build homes in places where they should not be built, as in this Layton, Utah, site. Then they clear the steep slopes nearby. With this lack of realism, pollution of drainage water is only one of the hazards here. (Courtesy of USDA—Soil Conservation Service.)

19:12 Bioremediation

Bioremediation (bio = organism, mediation from Latin *remedium* = correct or remove an evil) involves the ability of certain bacteria and fungi to degrade hazardous materials. The degradation is done to obtain metabolic energy by the decomposers. Its primary values are that it is natural and cheaper (cost savings of 30%–70%) than other remedies. But it can also be quite slow. Nevertheless, in 1996 there were more than 160 cleanup sites using bioremediation.

The breadth of applications using microorganisms is growing. Some of the common uses already familiar (composting, cleaning septic tanks, cleaning grease traps in restaurants and in sewage lagoons) now include predators on citrus root weevil, cotton bollworm, Colorado potato beetle, European corn borer, the release of copper from the ore on one-third of world copper production, a pretreatment for gold extraction, removal of malodors, removal of sulfur from petroleum, and degradation of PCBs, heavy oils, and creosote.[32]

Remediation costs are high.[33] Processes that flush the pollutant from the soil and then treat it (pump-and-treat) were about $10/ton in 1996. Treatment may be needed for many years. Low-temperature thermal desorption ran from $75/t to $125/t. High-temperature thermal desorption can be $300/t to $450/t. For radioactive sites and special problem areas, costs could be several thousand dollars per ton. It is not uncommon for the U.S. Government to spend $1 million to remediate 1 ha to a depth of 50 cm. With costs such as these, it is not surprising that considerable effort is made to find new and less costly techniques. Bioremediation will not solve the cost of cleaning up the potential superdamage site shown in Figure 19-13, if a major earthquake hits the San Andreas fault again soon.

Bioremediation is slow and the more atypical the material (non-water miscible, high concentrations, poison to microorganisms), the slower the process works. Enzymes are produced to do the bond breaking. Production and use of large amounts of specific enzymes might speed the processes generally.

---

[32] Lincoln Bates, "Bring in the Bugs," *Compressed Air Magazine* **101** (1996), No. 1; 18–24.

[33] Scott D. Cunningham, Todd A. Anderson, A. Paul Schwab, and F. C. Hsu, "Phytoremediation of Soils Contaminated with Organic Pollutants," *Advances in Agronomy* **56** (1996), 60–61.

FIGURE 19-13 Bioremediation on building on the San Andreas earthquake fault in California will have to involve people removing people, by zoning or other regulations. This is an unknown hazard (when is the next major quake?) so remediation is held off, because the hazard is uncertain. The fault runs vertically through the center of the photo. (Courtesy of U.S. Geological Survey, Department of the Interior.)

One author referred to microorganisms as the perfect workforce because they can clean up toxic waste, guard crops, mine precious metals, and leave behind only carbon dioxide, water, and salts. Is bioremediation that good? How is it done and what does it do?

Finding the Right Organisms

All organisms do not decompose all kinds of substances. The first requirement is to find one or more effective microorganism for each problem substance needing degrading. Usually two approaches have been used to find the best microorganism: (1) Search for high populations in soil already contaminated with the problem contaminant and (2) add the pollutant to a soil and then lock for an active decomposer. (Most microorganisms multiply numbers whenever a substance is added that they can degrade to obtain energy.) Each organism is then isolated, numbers increased, and the increase used as inoculum into a soil-pollutant mixture to see if the organism is the decomposer and is stable in the system.

White rot fungus (*Phanerochoete chrysosporium*) and other white rot fungi have been found to degrade a number of environmentally persistent compounds (DDT, 2,4,5-T, dioxin, PCP, TNT, and polycyclic aromatic hydrocarbons).[34] These are natural wood-rotting fungi and seem not to have competition problems from other organisms. The fungi do attack many persistent chemicals, but the bioremediation processes have been slow and degradation is incomplete. Because they are wood-rotting fungi, they may not be competitive in a contaminated soil without the addition of substantial quantities of growth substrate (wood chips, corncobs, straw). This can be costly and require additional manipulation.

[34] John A. Bumpus, "White Rot Fungi and Their Potential Use in Soil Bioremediation Processes," in Jean-Marc Bollog and G. Stotzky (eds.), *Soil Biochemistry* **8** (1993), 65–100.

Mechanisms of in situ Bioremediation

Bioremediation is favored and speeded by conditions that favor microbial growth. The first step is to clearly, and quite completely, identify the problem materials that must be degraded and to know which organisms will do that completely. The soil conditions are modified to favor growth of the organism or organisms. Usually additions of essential *nutrients* and *oxygen* (O_2) are minimums. Good soil aeration, adequate water, suitable pH, and suitable soil temperature may all need adjustment. Unless desired organisms exist in the soil already, inoculation (seeding) with the correct microorganisms may be necessary. Often the soil may need leaching after bioremediation, so general soil permeability is important.

Rate-limiting factors are numerous. It may be necessary to produce microbial cultures to add to the site. Severe conditions may produce *chemical shock, extremes of pH or temperature,* and the high pollutant concentrations may slow, kill, or inactivate microbial cells. Other limiting factors include limited mobility within the soil, competition by other soil microbes, and alternate carbon sources. It can be difficult to build and maintain a vigorous microbial population.

Enhancing Biodegradation

Correcting rate-limiting factors is simply improving the conditions for the organisms—fertility, pH, free oxygen (O_2), and water conditions. Isolated microbes that have shown ability to degrade pollutants can be added. Usually there is a lag period as the organisms accommodate to the new environment (pollution). The seeded inoculum might even die off. A cocktail (mixture of organisms, 10–20 strains) is sometimes added to a site with mixed pollutants, knowing that several different organisms will be necessary.

The capability and adaptability of an organism is often associated with its ability to mutate and produce new adapted enzymes. **Genetic engineering** is the changing of a microbe's ability to produce enzymes for new or multiple degradative activities. Gene manipulation on DNA makes it possible to exchange some enzymatic activity. *Multiple gene transfers* were done 20 years ago with *Pseudomonas* species that can degrade several hydrocarbons (octane, naphthalene, etc.). The search to find or synthesize a *superbug* continues.

Microorganisms may have some tough conditions to overcome. First, microbe predators can decimate a population in a hurry. Second, some partial degradation products may be very toxic and kill off the population. Third, different species may compete for the same energy sources or produce partial degradation products unfavorable to other decomposers. Fourth, microbes may develop biofilms (gels, slimes) that partially seal aeration and water flow channels. In closed systems (pipes, tanks) organisms may produce new tastes, turbidity, and odors that are unwanted.

Adsorption and Solubility of the Pollutant

Microorganisms work in water. Most degradation they accomplish must be in the soil water or aquifer water. Hydrophobic materials (petroleum products, PCBs, polyaromatic hydrocarbons) adsorb to soil and soil humus. Their solubility in water is low and access of enzymes to the pollutant is slow and limited. Pump-and-treat water (pumped out of the soil for treatment) carries little of such pollutants and could take tens to hundreds of years to clean up a contaminated soil. Wetting agents or other chemicals to increase solubility in water are essential to rapid bioremediation of such sites. Slow desorption of the pollutant into water is likely to be the rate-limiting factor in bioremediation of highly hydrophobic organic contaminants.

Engineering Enzymes[35]

Enzymes, the bond-breaking catalysts by which microorganisms accomplish bioremediation, might be effective by themselves, *without the microbe.* Free enzymes in soils are usually short-lived, but they can accomplish degradation. Since enzymes are not living things, toxins and other life-threatening conditions would generally not be problems. One could theoretically find a plant or microorganism that naturally degrades the toxic substance and then identify the active enzyme.

The technology for the modification of enzymes is in its infancy, but some results have been encouraging. Two nearly identical PCB-degrading enzymes showed dramatic differences in the range of PCBs they degraded. Altering a few amino-acids in the enzyme resulted in a strain that exhibits the best capabilities of both enzymes, able to attack a wide variety of PCBs.

The use of enzymes has limitations. Enzymes can be altered (and become inactive) by unfavorable pH, ionic strength (soluble salts), and unfavorable temperature. Enzymes are extremely specific, so the enzymes must match the pollutant's bonds to be broken or it will be ineffective. Each enzyme breaks a certain kind of bond, so numerous enzymes are required to degrade complex chemicals and mixtures. Obviously the use of enzymes in bioremediation will occur but it is slow to develop because there is so much to learn about them, how to change them, and how to produce and maintain them in a remediation site.

Phytoremediation[36]

Phytoremediation is remediation using green plants (Greek *phyto* = plant). More elaborately, phytoremediation is the use of green plants and associated microbiota, amendments, and agronomic activities to contain, remove, or render environmental contaminants harmless. The system has been used for a long time for treatment of polluted wastewaters.

One example of phytoremediation is testing for removal of uranium in water near the Chernobyl nuclear accident in Russia in 1986. The roots of sunflower (*Helianthus annuus* L.) accumulate heavy metals. Sunflowers lowered uranium contents in the water having 100 to 400 parts per *billion* by 95 percent the first 24 hours, which reduced uranium below the EPA groundwater standard of 20 ppb.[37] The sunflowers, grown on 1 m^2 Styrofoam® rafts, were floated in the pond for absorption of the metals. Cesium and strontium were also monitored. Costs were estimated at $2 to $6 per 1000 gal of water treated. However, there are many gaps in the information collected, such as the long-term survival of the plants and the reliability of the system. In fact, it is not known whether or not the metal uptake is incorporated into the plant or if it is bacterial uptake and the bacteria are just adhering to the plant.

The choice of plant and culture is different than common agronomic crop production. The plant is selected to provide active root growth and associated microbial activity. The roots and microbes produce the enzymes to degrade the pollutants. The agronomic practices involve several similarities and some differences to ordinary crop production.

- Select plants adapted to the climate and that readily absorb or degrade the contaminants.
- Prepare seedbeds and plant spacing (density) for maximum *root* density.

---

[35] Jeanne Trombly, "Engineering Enzymes for Better Bioremediation," *Environmental Science & Technology* **29** (1995), No. 12; 560A–563A.

[36] Scott D. Cunningham, et al., *ibid.*

[37] Catherine M. Cooney, "Sunflowers Remove Radionuclides from Water in Ongoing Phytoremediation Field Tests," *Environmental Science & Technology* **30** (1996), No. 5; 194A.

- Add nutrients, probably more (for microbes) than would be used for normal plant growth.
- Manage for optimum water, control the weeds, and cut or mow as needed to enhance *root activity.*

As with other techniques, phytoremediation has limitations. It is not applicable to all situations and has other drawbacks, some of which follow:

- The site must be accessible to heavy equipment for planting, etc.
- Phytoremediation can only be effective to 0.5 m–1.0 m deep in most cases. Root densities decrease rapidly at depths deeper than about 0.5 m.
- The soil must have good aeration to supply O_2 to roots and microbes.
- Phytotoxic contaminants at toxic levels cannot be treated; the plants die.
- A longer time (years) is usually needed than if using standard engineering methods. Cost versus time may be the issue.
- Contamination of the food chain must be considered.
- Soil erosion may occur during planting and crop establishment. Control measures are essential to inhibit pollution of waters with eroded soil-contaminant sediment.

The Future of Bioremediation

Bioremediation will continue. However, a survey of forty CEOs in bellwether firms in remediation and consulting indicates that the CEOs expect declining markets in the next few years and increased competition.[38] Less funding from Departments of Defense (DOD) and Energy (DOE) is one cause. This area of research is still funded in large part by government dollars.

The science itself has a number of challenges that take time to resolve. Permission to release genetically engineered microbes should require considerable scientific data on the organisms: What other roles might they have in a natural environment? Will they become dominant and harmful? An abbreviated list of conclusions and recommendations of one author on bioremediation for the future follows:[39]

- In situ biodegradation is still a developing technology and is a complex process.
- The organisms used must have a competitive edge over indigenous species in order to survive to accomplish their task. How is this to be accomplished?
- Development of, or selection for, organisms that use the toxic chemical *for energy* is essential. If they use other compounds for energy, they may be inefficient.
- It is essential to know what will happen to the organism and the environment after the target toxin is degraded. Are potentially pathogenic species being released?
- There is a need to isolate organisms (superbugs) that are effective in adverse conditions (cold, hot, salty, heavy metals, adhere to nonaqueous substrate).
- How will genetically engineered species survive, grow, and function in the natural environment?

---

[38] Jeff Johnson, "Tough Times Predicted for Remediation Industry," *Environmental Science & Technology* **29** (1995), No. 6; 253A.

[39] Eve Riser-Roberts, *Bioremediation of Petroleum Contaminated Sites,* C. K. Smoley, (publisher), 1992, 122–139.

- Careful attention is needed to potential carcinogenicity and mutagenicity of the intermediate products during biodegradation.
- The relationships of various organisms working at the same time and their interactions need to be known. This may be particularly important in hard-to-degrade (*refractory*) man-made nonnatural products.
- Enzyme remediation is new, unproven, and of unknown reliability. Much needs to be learned.

The microbial world is complex and seems to have a limitless metabolic potential. But, this seemingly limitless potential is also its major hindrance. The heterogeneous environment cannot be manipulated easily to ensure the survival and success of any one or few organisms. Thus, the persistence and activity of soil microbial changes, additions, or enhancements are difficult to predict. The history of the results when humans try to manipulate the environment is not replete with unqualified success stories. We know more now, but do we know enough?

19:13 Protecting Land from People

People tend to view the future with near-sighted and selfish eyes. We have exploited, even wasted, much of our natural resources of petroleum, timber, prime lands, wetlands, and erodible soils and polluted our waters. We are indeed hazardous to the land and water we wish to use. As an example of the concern some people have, the American Farmland Trust stated in 1985 that unless California's agricultural problems were addressed in the next 10 to 20 years, the state's farming industry would decline. The four major problems are as follows:

- Agricultural-land conversion to nonagricultural uses
- Soil erosion
- Increasing salinity of soil and water
- Diminishing water supplies and diversion to nonagricultural uses

More than 17,000 ha yearly are converted to urban uses, of which more than 80 percent were irrigated croplands.

Protection Mechanisms

How can farmland be protected from people? So far only limited controls are available. Many states have erosion-control legislation, which forces landowners to establish controls if the erosion exceeds a defined value (usually 4–10 Mg/ha [2–5 t/a]). Poor management practices that allow erosion are not tolerated. In addition, the EPA has strict regulations on pollution of water and air by excessive fertilizer and pesticide use and uncontrolled burning.

A number of states have varying regulations to restrict unlimited land conversion to nonagricultural uses. New York bought development rights to thousands of acres of farmland on Long Island. It was thought that the land as farmland had high economic and social values for (1) producing vegetables near a large market, (2) a way for city dwellers to quickly get into the country, and (3) to control somewhat the urban sprawl.

Prime, Unique, and Other Farmlands

Most land-control measures have attempted to pinpoint the saving of the best farmland for agricultural use. This best land is called **prime farmland** or **unique farmland** (see Detail 19-4).

Detail 19-4 Prime and Unique Farmlands

The criteria for the best farmlands are approximately those given by the Soil Conservation Service (SCS).

Prime Farmlands These lands have the best combination of physical and chemical characteristics for producing food, feed, forage, fiber, and oilseed crops, and they are available for these uses (not already covered with urban development, roads, etc.). The lands must meet *all of these criteria* (condensed and approximated by the authors):

1. Good temperature for most crops. Average summer temperatures above 15° C (59° F).
2. Sufficient natural rainfall or irrigation water 7 out of 10 years for the commonly grown crops
3. A soil depth of at least 1 m (40 in)
4. A pH in the top 1 m of between 4.5 and 8.4
5. No shallow water table to interfere with normal plant growth
6. The soil can be managed to keep within the top 1 m of soil (a) a low salt content (<4 dS/m) and (b) a low exchangeable-sodium percentage (<15%).
7. Soils are not flooded in the growing season more often than once in 2 years.
8. The soil is not highly erodible. The *K* (erodibility factor) times the percentage slope is less than 2.0 and, for wind erosion, *I* (erodibility) times climate factor does not exceed 60.
9. Permeability is not a limiting factor in using the soil. The permeability is >0.15 cm/h in the top 50 cm (20 in) of soil.
10. The soil is not rocky. It must have less than 10% of the surface with rocks larger than 7.6 cm (3 in) in diameter.

Unique Farmlands Unique farmlands are lands *other than prime farmlands* that are used to produce specific crops (such as citrus, tree nuts, olives, cranberries, and certain other fruits and vegetables). Some of the requirements for land to be classified as unique lands are as follows:

1. The land must be used for a specific high-value food or fiber crop (with unique soil and/or climatic requirements).
2. The land has adequate moisture available for the specific crop.
3. The land area combines the favorable factors of soil qualities, temperature, humidity, aboveground air drainage (frost control), steepness of slope, aspect (direction of slope), or other attributes (such as nearness to market) that favor the growth and/or distribution of the specific crop.

Lands Critical to the State Lands critical to the state do not qualify for either prime or unique lands. However, their extensiveness or locations may make them particularly valuable to the state. These soils are identified and their conservation is encouraged.

Source: Raymond W. Miller, with data from "Part 657—Prime and Unique Farmland," *Federal Register* **7CFR** (Jan. 1, 1986), Ch. VI, 657.1.

As an example, Ontario, Canada, proposed in 1988 new and stronger guidelines for the preservation of prime farmland; these new regulations replace its 1978 guidelines. The regulations are designed to minimize conflicting land uses within agricultural areas and preserve prime and unique agricultural land (Figure 19-14). Following are the methods by which these lands will be preserved:

FIGURE 19-14 Residential housing moving in on prime farmland, Tama silt loam, in Illinois. This is great soil for the home owner but a great loss, permanently, for agricultural production. Zoning controls are the only present solution. (Courtesy of USDA—Soil Conservation Service and Illinois Agricultural Experiment Station, Soil Survey of Whiteside County in Illinois.)

- Direct any new nonagricultural development to urban areas or onto marginal land.
- Require documented evidence to justify any nonagricultural development that is proposed to go onto prime farmland.
- Designate certain kinds of uses that are incompatible for any development on prime farmland. These could include many industrial uses, recreation facilities, conventional residential development, and mobile-home parks.

These examples and other laws already in use in many U.S. states emphasize the trend to protect agricultural lands. These lands are needed for food production for local people, to protect local economies, to maintain open space, and to increase exports to balance U.S. foreign-trade deficits.

As important as technology, politics, law, and ethics are to the pollution question, all such approaches are bound to have disappointing results, for they ignore the primary fact that pollution is primarily an economic problem, which must be understood in economic terms.

—Larry E. Ruff

Questions

1. (a) In which ways can soil act to remove pollutants from waters? (b) Are all pollutants filtered out by soil? Explain.
2. By what authorization does the EPA have some control over land use?

3. What can be done to reduce evolution of gases causing acidic rain?
4. Explain what BMPs are and discuss why they are used.
5. What approaches can be used to beneficially recycle manures and sewage sludge?
6. Give an illustration of some hypothetical BMP.
7. (a) What can be done to make landfills safer or less hazardous in the short term (50 years, for example)? (b) Which criteria are needed in selecting sites for new landfills?
8. Which are the approaches used to reduce pollution by nitrate and by phosphate?
9. (a) What are the alternatives to using pesticides? (b) What is required to make pesticide usage drop sharply in the next few decades?
10. (a) Discuss *biological pest control* and (b) *cultural pest control.*
11. What are the recommendations to minimize damages from poisonous plants in range areas?
12. What are possible solutions to the soluble salt problem globally? (This is not a question of reclaiming salty soils but of what do we do with the salt.)
13. What are the suggestions for reducing particulates and dust pollution?
14. What can be done to reduce sediment as a contaminant?
15. (a) What is *bioremediation?* (b) How effective is it? (c) What is its potential?
16. Is there a solution to the problem of methane evolution from landfills? Discuss.
17. How are the microorganisms that are effective in degradation of certain toxins located and identified for use?
18. In a polluted area with a mixture of toxins, can a single organism degrade the toxins or must there be two, three, or many different organisms? Discuss.
19. (a) What is possible to minimize the damages from radioactive pollution? (b) What is done to minimize the problem of radon?
20. How has minimum tillage and no-till reduced pollution by soil sediments?
21. How are some soil resources protected from people and their detrimental actions?
22. Define (a) *prime farmland* and (b) *unique farmland.*
23. Which activities are used in various localities or countries to protect prime farmlands and unique farmlands?

Soil Surveys, Interpretations, and Land-Use Planning

Our duty to the whole, including the unborn generations, bids us restrain an unprincipled present-day minority from wasting the heritage of these unborn generations.

—Theodore Roosevelt

20:1 Preview and Important Facts

PREVIEW

The first soil surveys were simple and brief. They were intended to answer practical agronomic questions of soil differences and limitations important in improving and expanding crop production. Was a new soil area suitable for crops that had never been planted there? How much additional fertilization did it need? What were the problems of water, salts, or acidity? Could other crops be grown more profitably? Soil surveys expanded in detail and concept with the increase in scientific knowledge and demand for more useful information.

The National Cooperative Soil Survey was organized in 1952 to coordinate and simplify soil-survey information. It is now part of the U.S. Department of Agriculture (USDA) within the Natural Resources Conservation Service (NRCS), originally the Soil Conservation Service (SCS). In most states the Land Grant University serves as the statewide cooperating agency. The U.S. Forest Service and the U.S. Bureau of Land Management (BLM) cooperate closely in surveying areas within their respective jurisdictions.

Surveys require several years to complete. A finished report contains details on the soils, their distribution in the area, the local crops, the most important uses for the soils, the climate, the soil classification, and other details. The information of the survey helps in engineering construction, locating sources of sand and gravel, forestry management, urban development, game management, recreation development, predicting erosion hazards, irrigation, urban housing, taxation, and land-use planning—in addition to the traditional agricultural evaluation for farms and ranches.

Although soils do not change very fast, agricultural practices and the information we collect about a soil do change. Some very productive agricultural areas have been surveyed several times in the past half-century. A comment attributed to an early soil surveyor indicates the continuing need to add information by some new surveys. R. W. Smith in 1928 in Illinois said,

> I hope that my answer to your question [when asked for an estimate of cost to complete the state soil survey, 1928] is clearly indicated . . . it is that the soil survey will never be completed because I cannot conceive of the time when knowledge of soils will be complete. Our expectation is that our successors will build on what has been done, as we are building on the work of our predecessors.

IMPORTANT FACTS TO KNOW

1. The general method used to prepare a soil survey
2. The measurements and descriptions included in soil surveys
3. The useful life of data included in soil surveys
4. The definition and equation for calculating soil potential index (SPI)
5. The definition and use of *corrective measures* (*CM*), *continuing limitations* (*CL*), and the *yield* or *performance standard* (*P*) in the soil potential index equation
6. The LESA approach to land-use planning
7. The definition and use of *benchmark* soils
8. The bases for rating soils as having *slight, moderate,* or *severe* limitations for each particular engineering use
9. The land capability class and subclass
10. The approaches available for governmental control of land use

20:2 Making a Soil Survey[1]

The soil survey of an area begins with numerous conferences, the collection of aerial photographs, and some initial field reviews to prepare a partial legend of properties and descriptions of some of the most extensive soils of the area. A **legend** is prepared during the initial field reviews from soil pits dug in the most extensive landforms, where large areas of similar soil are expected to occur. Then the detailed soil survey is ready to be done.

The surveyor needs to be a self-starter, adaptive, competent, and good at public relations. As surveyors work, they encounter rain, locked gates, unfriendly landowners, dogs, nuisance insects, and rocky soils. At times they work hard physically, and they constantly need to interpret correctly the soil profiles that they dig and probe, always looking for the boundary where different soils meet (or grade into) each other. Short travel is done by foot, but maps and materials usually require a car nearby. Helicopters have been used in many locations, including Vermont, Pennsylvania, New Mexico, and Alaska. Various kinds of remote sensing (satellite) maps have been used to assist in the mapping.

Working alone, a surveyor carries an aerial photograph of the area, digging tools, a hand level (the Abney level) for measuring slope percentages, a pH kit, Munsell color chip book, and 10 percent hydrochloric acid to identify the presence of lime. The surveyor carries the legend describing the profile characteristics of most of the extensive soils of the area. By frequent borings (with an auger or shovel) or pits (using shovels, or backhoes occasionally) and

---

[1]K. C. Thomas, "Computer Assisted Writing—Its Application in Soil Survey Manuscripts," *Agronomy Abstracts* (1985), 199.

observations and notations of soil color, horizon thicknesses, texture by feel, pH, soil structure, and other features, the surveyor will determine the soil mapping units and soil mapping unit boundaries. A **soil mapping unit** is an area of soil that is delineated from adjacent areas on a map. Those differences may be slope, erosion, or other features, as well as differences in the soil profile itself (Figure 20-1).

Continual modifications in the procedures and guidelines are made to improve the surveys. In Alaska power ice augers are used on permafrost layers. Backhoes are often used for pits. The boundaries of each mapping unit are drawn on the map and the area labeled with a mapping code symbol from a legend. The advent of global positioning systems (GPSs) has made locating sites on the Earth accurate, even where landmarks are not very close (see Chapter 2). Geophysical information systems (GISs) have made the accumulation, storage, and processing of data much easier and the preparation of many kinds of maps fast and accurate.

20:3 Soil Survey Reports

Soil surveys have traditionally been made on a county basis, but a survey may be a valley or part of a valley in mountainous areas. Typically, each survey has about the same contents, but these contents change periodically. Table 20-1 lists typical soil survey report contents and how the total report is distributed among the various items.

Contents of Soil Surveys

The present demands for more information for urban development, environmental control, and other uses have increased the kinds and quantity of detail recorded. About half of the volume of each report is **maps with delineation of the soils.** Much of the text is details of the **soil profiles,** including their classification. There are about 20 tables in current surveys, most of which are as follows:

- **Temperature and precipitation** and **freeze dates for fall and spring**
- **Acreage** and **proportionate extent** of the soils in the survey area

FIGURE 20-1 Basement homes and shallow bedrock don't mix well. The soil survey separates shallow-to-bedrock areas from loose, fractured-but-deep areas. This excavation for a basement in Muskingum County, Ohio, shows Frankstown shallow soil left of the measuring rod (poorly suited for basements) and the deeper, cherty Mertz soil to the right of the rod. Septic drainage fields may need to be doubled in these soils. (Courtesy of USDA—Natural Resources Conservation Service and the Ohio Department of Natural Resources, *Soil Survey of Muskingum County, Ohio.*)

Table 20-1 Contents of Soil Survey Reports in the United States and the Distribution of the Total Report Length Among the Various Contents

Soil Survey Contents	*Montana Report*[a]	*Indiana Report*[b]	*Florida Report*[c]	*Kansas Report*[d]
Index to map units, pages	1	1	1	1
Summary of tables, pages	1	1	2	2
Foreword, pages	1	1	1	1
Introduction, pages	5	4	10	3
General soil map units, pages	8	7	13	6
Detailed soil map units, pages	60	39	65	23
Prime farmland, pages	1	1	1	1
Use and management of the soils, pages	13	12	13	13
Soil properties, pages	4	4	7	4
Classification of soils, pages	19	14	40	13
Formation of the soils, pages	2	2	3	3
References, number	6	6	24	12
Glossary, number of items defined	159	177	163	110
Number of tables	14	19	21	18
Number of photographs	4	7	14	11
Drawings, diagrams	2	4	3	4
Total text, pages	193	149	257	119
Total double-page maps	254	64	119	72

Source: Raymond W. Miller, Utah State University.
[a]Roosevelt and Daniels Counties, Montana, 1985.
[b]Wayne County, Indiana, 1987.
[c]Alachua County, Florida, 1985.
[d]Jewell County, Kansas, 1984.

- The soils in the survey area that are **prime farmland**
- The **yield of the major crops** of the area (5 or 6 major crops)
- **Land-use-capability classes**
- **Woodland management** or **range** (concerns such as erosion, windthrow hazard, site index, production rate, trees to plant)
- **Windbreaks** and **environmental plantings** (Figure 20-2)
- **Recreation development** (camp areas, picnic areas, playgrounds, paths, trails)
- **Wildlife habitat** with numerous details of type of habitat
- **Building site development** (shallow excavation, dwellings with and without basements, small commercial buildings, local roads and streets, lawns, landscaping)
- **Sanitary facilities** (septic tank fields, sewage lagoons, trench and open landfills)
- **Construction materials** (roadfill, sand, gravel, topsoil)
- **Water management** (ponds, embankments, drainage, irrigation, grassed waterways)
- **Engineering Index Properties** (USDA texture, Unified, and AASHTO classification, 8–25 cm fragments, percentage passing sieve sizes, liquid limits, plasticity index)
- **Physical and chemical properties** (depths clay and bulk density, permeability, available water, soil reaction—pH, shrink–swell, erosion factors K and T, erodibility group, percentages clay and organic matter) (Figure 20-3)
- **Soil and water features** (flooding—frequency and duration, high-water table, subsidence, potential frost action, risk of corrosion of steel and concrete [high-medium-low]) (Figure 20-4)
- **Classification of soils** (taxonomy)
- **Interpretive groups** (land capability, prime farmland, range classification, woodland coordination symbol)

FIGURE 20-2 Windbreaks may be necessary to prevent the need to hold the soil and reduce the need to clear this soil from Morrow County Road in Oregon. Land speculators cleared the land and blowouts occur frequently on this Quincy loamy fine sand. (Courtesy of USDA—Soil Conservation Service; photo by Ed Weber.)

FIGURE 20-3 The erosion factors of a soil are important in anticipating problems and providing controls. This landscape in Minnesota has enough slope to be an erosion problem if not covered with permanent cover or by using other control practices. The windbreaks on the left indicate some wind problems if corrective control measures are not used. (Courtesy of USDA—Natural Resources Conservation Service, *Soil Survey of Faribault County, Minnesota.*)

Mapping Legends for Soils

Mapping units, delineated on the maps, are named for the most extensive soil series within each unit, but each unit is often a natural mixture of two to five soil mapping units. These mixtures are referred to as **soil associations.** The most likely landscapes to have extensive intermixing of series are small rolling hills or ridges-and-depressions landscapes. In these areas there will usually be one series on the ridges or hilltops (eroded, shallow), perhaps another series along the slope, and at least a third series in the bottoms (deep deposits or wetter than tops).

Each *detailed soil survey map* consists of many folded double-sheet maps that are prepared on an aerial photographic base on which mapping units that are **phases of series** are drawn. One scale used is 1:20,000 (3.16 in/mi, or 5 cm/km). Soil surveys are made at several

FIGURE 20-4 Flooding hazard should be known if you wish to build a home on the site—but knowing doesn't always change people's point of view. They say, "It won't happen to me." It did happen on this Nolin silt loam soil in Ohio, said to be occasionally flooded. The survey says, "This soil . . . is unsuited to building site development, camp areas, and septic tank absorption fields . . . flooding." Someone should have read the survey first. (Courtesy of USDA—NRSC and Ohio Department of Natural Resources, *Soil Survey of Jefferson County, Ohio.*)

detailed levels. With maps at 1:12,000 to 1:32,000, areas from 0.6 ha to 4 ha would be the smallest areas to delineate. The soil map has sufficient detail to be suitable for making most decisions on land use. Additional uses for urban planning, erosion control, and engineering are common.

Soil map units are identified within each boundary by a symbol consisting of numbers and letters keyed to a legend. A mapping unit (for example, HiE3) may consist of a geographic series name, such as Hickory (Hi); a textural phase of series, such as loam; a slope class, such as E (e.g., 18%–30%); and an erosion class, such as 3 (severely eroded). Each survey sets up its own legend for the meaning of the letters and numbers used.

In the above example, all soils within the delineated unit called *Hickory soil series* may not actually be Hickory soil series. Natural soil variations develop different series, but the areas may be too small and intermixed to separate them practically on the scale of map selected; thus, a map unit listed as the dominant soil in that delineated area may include more than just the soil named. The soils distributed in the area usually differ because of variations in soil-forming factors (Figure 20-5).

Value of Detailed Soil Survey Reports

Soil surveys are long-time inventories because soils are not easily changed rapidly by common soil management. It is true that carelessness may cause or permit rapid destruction of a few soils, such as shallow soils over bedrock that can be eroded away. Organic soil can be burned. Permanent toxins, such as heavy metals, may be dumped onto soils or soluble salts accumulated. However, *most soil properties recorded in soil surveys change slowly.* These properties include the following:

- Land relief
- Soil texture and coarse fragments
- General organic-matter contents

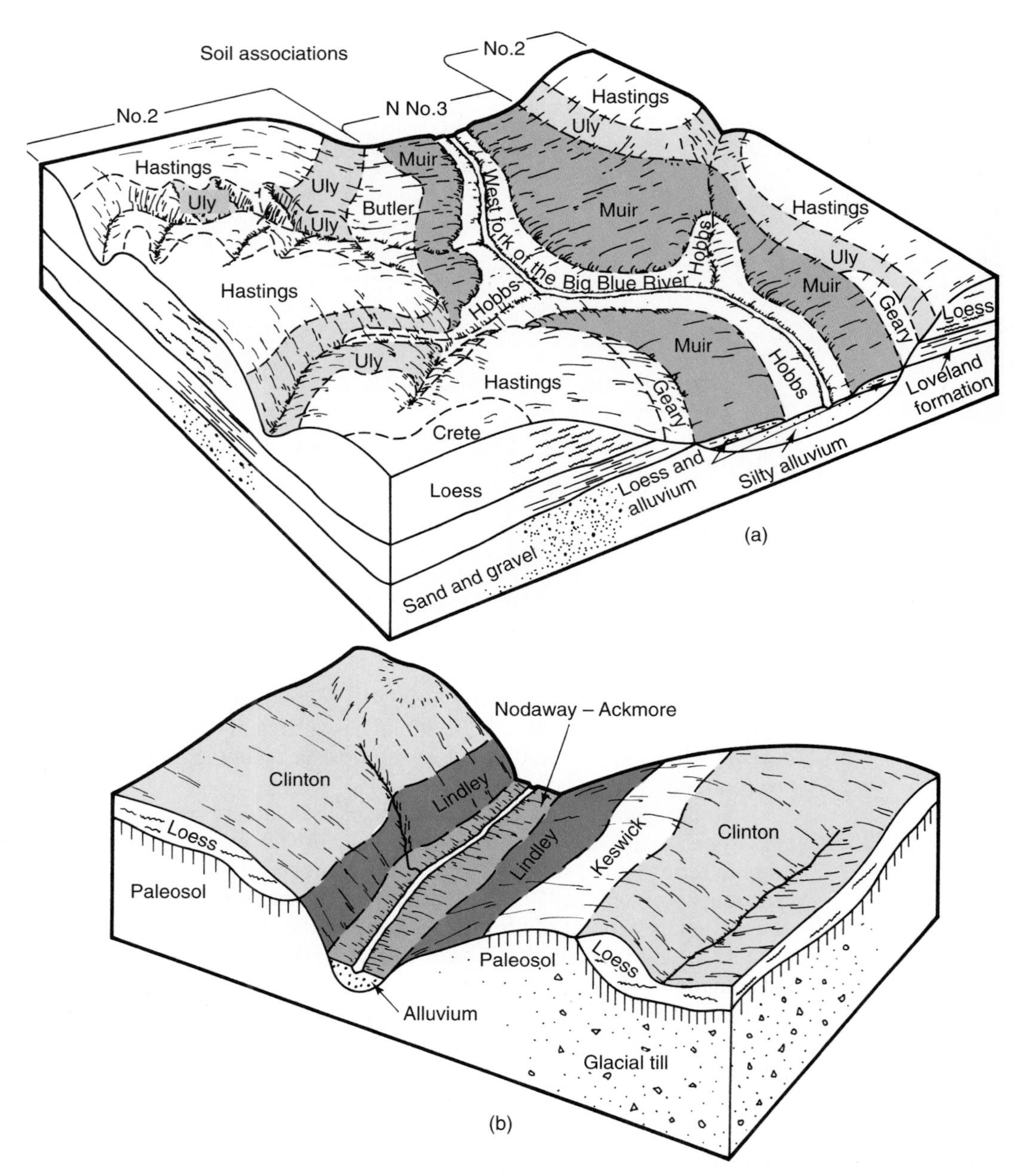

FIGURE 20-5 Soils differ from each other. In a given climatic area, these differences are greatly determined by (1) parent material and (2) extent of soil formation. In (a) the different soils follow boundaries between different parent materials (*Hastings* from loess, *Muir* from mixed loess and alluvium on a river terrace, and *Hobbs* on the bottom floodplain) and in the slopes of the material, thus the erosion occurring (*Uly* is on steeper loess slopes than is *Hastings*). In (b) even the portions of an ancient buried soil (paleosol), now exposed, will be more developed and more weathered (the *Keswick* series) than is the younger glacial till soil (*Lindley*) forming on substratum when the paleosol cover has been eroded off. (*Sources:* U.S. Soil Conservation Service, *Surveys of Fillmore County, Nebraska,* 1986; *Survey of Washington County, Iowa,* 1987.)

- Soil lime contents
- Geologic origin
- Natural fertility
- Soil depth
- The tendency to accumulate soluble salts
- Soil structure
- Soil engineering properties

Many other *unchanging* site factors tell much about the area. Some of these include the following:

- Climate—frost-free periods, average annual rainfall, average annual temperatures, windiness
- Natural vegetation
- Adaptable crops
- General productivity of the soils
- Land-use problems such as poor drainage or shallow depth to bedrock

Thus, soil surveys are a *permanent-to-slightly-changing record* of a very important world resource. Nevertheless, some areas have been resurveyed after only a few decades. This resurvey is usually done because many kinds of information were not gathered in a previous survey and are now needed. Certainly management practices, crops grown, and yields might be expected to change somewhat over several decades. More information is needed for environmental planning, urban development, roads, and multiple-use planning. Currently area-wide detailed information databases are being compiled and produced to make available the information to many more kinds of users.

Uses of Detailed Soil Survey Reports

Engineers and builders can use the soil and the corresponding interpretations to locate and design, scientifically and environmentally, suitable sites for roads and buildings. Soil and plant scientists, foresters, watershed scientists, range scientists, engineers, and horticulturists can establish field research plots on the most extensive of the suitable soils so that research results can be applied with better success to large areas of similar soils in other locations. Soil survey reports provide an enormous amount of detail to help researchers, county agents, and others to locate good or problem areas on which to work. The survey report provides details on drainage, parent materials, soil depths, some chemical data, and profile characteristics.

20:4 Benchmark Soils[2,3]

Benchmark soils are those few soils considered to be of *great importance* and of *extensive area,* or *occupying a key position in the USDA system of soil taxonomy.* There are about 1000 nationally designated benchmark soils. Detailed analyses and extensive characterization information are collected for these soils. The hope is that results of studies on these soils can be better extrapolated so that the results for those studies, when done on other soils, can be

[2]National Soil Survey Handbook 430-VI, USDA—Soil Conservation Service, Washington, DC, 1996, 630.
[3]James A. Silva, ed., "Soil-Based Agrotechnology Transfer," in *Benchmark Soils Project,* Department of Agronomy and Soil Science, Hawaii Institute of Tropical Agriculture and Human Resources, University of Hawaii, Honolulu, 1985, 269.

anticipated. These 1000 soils comprise less than 10 percent of the 15,000 series described in the United States. Five benchmark soils are shown in the color photos of soil orders in Chapter 7: the **Clarion** (IA), **Houston Black** (TX), **Mohave** (AZ), **Molokai** (HA), and **Valentine** (NE). A few other benchmark soils are the following:

Amarillo (TX)	Aridic Paleustalfs	**Minto** (AK)	Aeric Cryaquepts
Aravda (WY)	Ustollic Natrargids	**Palouse** (WA)	Pachic Ultic Haploxerolls
Beltsville (MD)	Typic Fragiudults		
Cecil (NC)	Typic Hapludults	**San Joaquin** (CA)	Abruptic Durixeralfs
Ft. Collins (CO)	Ustollic Haplargids	**Sharkey** (LA)	Vertic Haplaquepts
Houghton (MI)	Typic Medisaprists	**Tetonka** (SD)	Argiaquic Argialbolls
Mardin (NY)	Typic Fragiochrepts		

20:5 Prime and Unique Farmland and Farmlands Critical to the State

Soil surveys emphasize the importance of prime farmland with this statement in each soil survey:

> Prime farmland is one of several kinds of important farmland defined by the U.S. Department of Agriculture. It is of major importance in meeting the nation's short- and long-range needs for food and fiber. The acreage of high-quality farmland is limited, and the U.S. Department of Agriculture recognizes that government at local, state, and federal levels, as well as individuals, must encourage and facilitate the wise use of our nation's prime farmland.
>
> **Prime farmland soils,** as defined by the U.S. Department of Agriculture, are soils that are best suited to producing food, feed, forage, fiber, and oilseed crops. Such soils have properties that are favorable for the economic production of sustained high yields of crops. The soils need only to be treated and managed using acceptable farming methods. The moisture supply, of course, must be adequate, and the growing season has to be sufficiently long. Prime farmland soils produce the highest yields with minimal inputs of energy and economic resources, and farming these soils results in the least damage to the environment.
>
> Prime farmland soils may presently be in use as cropland, pasture, or woodland, or they may be in other uses. They either are used for producing food or fiber or available for these uses. Urban or built-up land and water areas cannot be considered prime farmland.
>
> Soils that have a high water table, are subject to flooding, or are droughty may qualify as prime farmland soils if the limitations or hazards are overcome by drainage, flood control, or irrigation.

Many states have low percentages of their land as prime farmland because of their cold areas with short growing seasons or their mountainous regions. Other states have large percentages of prime lands. As an example, Jewell County, Kansas, is 67 percent prime farmland, and Wayne County, Indiana, is nearly 74 percent prime farmland. These are not the highest percentages. Some counties in Iowa are more than 90 percent prime farmland.

Some criteria that soils must meet to qualify as *prime farmlands, unique farmlands,* and *lands critical to the state* are given in Chapter 19.

FIGURE 20-6 Land-use planning information will readily identify areas of unusual soils. This barren lava bed in Hawaiian *Aa rock* was covered only with a thin leaf mat prior to its very limited use for farming. However, in its location (warm, humid Hawaii), where land is scarce, and with crops that can adapt, this soil is better than it might be at some other site. These papaya trees do well in this soil. Macadamia nuts and coffee are also produced on these Typic Tropofolists (soils with thin leaf mats). (Courtesy of USDA—Soil Conservation Service; photo by R. P. Yonce.)

20:6 Land Evaluation for Land-Use Planning[4,5,6]

Land evaluation for land-use planning can be done for urban areas, rural areas, or both. The effects of land use on quality of the environment and environmental sustainability of agricultural production systems are major issues. The problems of concern include pollution with nitrate, phosphate, and biocides; erosion of land; declining soil fertility; low-input farming; exploitation of timber and range resources; and information for engineering uses. Land-use problems may have agronomic, economic, political, and social dimensions. There will be conflicting uses and multipurpose uses.

The USDA Natural Resources Conservation Service uses a system called **Land Evaluation and Site Assessment (LESA)** for purposes of *guiding the conversion of farmland to urban uses.* One major aim is to preserve the best agricultural land.

The land evaluation approach involves three procedures. First, land capability classifications make preliminary subdivisions and characterizations of the areas. Second, the current category of land use is documented. (Is the land prime farmland or of lower quality?) Third, each soil is given a rating according to capability of use and is sorted into about ten agricultural groups based on each soil's quality.

The site assessment portion of LESA considers the location of the farmland (Figure 20-6). How far is it from present urban areas and services? What are the zoning regulations now in effect? What are adjacent land uses? How large is the farmland? Do its physical properties match up with adopted plans of use (Figure 20-7)? Has the site any unique properties? When sites are given ratings, it is preferred that the sites with the lowest ratings be put to nonagricultural use.

The Canadian Land Evaluation Group developed a more mathematical program under the name *integral land evaluation.*[7] Its prototype model (LEM2) provides information on the flexibility of land used and the feasibility of land-use options. Other models for this and other uses are listed in **Appendix A.**

---

[4] C. A. Van Diepen, et al., "Land Evaluation," *Advances in Soil Science* **15** (1991), 145–203.
[5] R. W. Dunford, R. D. Roe, F. R. Steiner, W. R. Wagner, and L. E. Wright, "Implementing LESA in Whitman County, Washington," *Journal of Soil and Water Conservation* **38** (1983), 87–89.
[6] J. Bouma, "Using Soil Survey Data for Quantitative Land Evaluation," *Advances in Soil Science* **9** (1989), 177–213.
[7] B. Smit, M. Brklacich, J. Dumanski, K. B. MacDonald, and M. H. Miller, "Integral Land Evaluation and Its Application to Policy," *Canadian Journal of Soil Science* **64** (1984), 467–479.

FIGURE 20-7 Evaluating land allows potential users to foresee problems for each use. These "dream acres" near Canton, Ohio, have a problem of temporary ponding, for which homeowners and urban developers need to plan. (Courtesy of USDA—Soil Conservation Service; photo by Marion F. Bureau.)

Land Capability Classification

The USDA Natural Resources Conservation Service uses a uniform system of two levels of soil management (**land capability classes** and **land capability subclasses**) for all soil mapping units in the United States. A third level, **land capability units,** is also used in some surveys.

Soil management **land capability classes** are numbered from one (I) to eight (VIII). Classes I through IV can be used for cultivation; Classes V through VIII cannot be cultivated in their present state under normal management (Figure 20-8).

Class I land can be used continuously for intensive crop production with good farming practices.

Class II land has more limitations than Class I land for intensive crop production, due to such characteristics as moderately steep slopes (2%–5%).

Class III land has severe limitations and requires more special conservation practices than Class II land to keep it continuously productive. For example, the land may have shallow soil, slopes of about 6%–10%, or shallow water tables.

Class IV land has severe limitations for cropping use and needs a greater intensity of conservation practices for cultivated crops than Class III land. Most of the time this land should be in permanent crops, such as pastures.

Class V land is not likely to erode but has other limitations, such as boulders or wetness, which are impractical to correct, and thus the land cannot be cultivated. It should be used for pasture, range, woodland, or wildlife habitat.

Class VI land is suitable for the same uses as Class V land, but it has a greater need for good management to maintain production because of such limitations as steep slopes or shallow soils.

Class VII land has very severe limitations and requires extreme care to protect the soil, even with low intensity use for grazing, wildlife, or timber.

Class VIII land has such severe limitations (steep slopes, rock lands, swamps, delicate plant cover) that it can be wisely used *only* for wildlife, recreation, watersheds, or aesthetic appreciation.

FIGURE 20-8 Land capability classes used by the U.S. Department of Agriculture's Soil Conservation Service. (*Source:* USDA—Soil Conservation Service.)

Land capability subclasses are soil groups within the eight classes that explain the reasons for the limitations of intensive crop production. Subclasses are designated by lowercase letters that follow the Roman numeral of the soil class. The land capability subclasses recognized are the following:

e **Erosion hazard,** the main limitation (Figure 20-9)
w **Wetness**
s **Shallow, droughty, stony,** or **permafrost**
c **Climate,** too cold or too dry

Ratings for Soil Potential[8]

The USDA Natural Resource Conservation Service develops **soil potential ratings,** which are classes of soils that indicate the *relative* quality of a soil for a particular use compared

[8] U.S. Department of Agriculture, Soil Conservation Staff, *National Soils Handbook,* USDA—Soil Conservation Service, Washington, DC, 1983.

FIGURE 20-9 This spinach field in New Jersey has been mapped as Collington sandy loam, 2%–4% slopes, and classified in the subgroup Typic Hapludults and in the fine-loamy, mixed, mesic family. The capability classification is IIe-9, meaning soil Class II with Subclass e—erosion hazard. (*Source:* USDA—Soil Conservation Service.)

with other soils in a given area. The ratings are developed for planning purposes but are not intended as recommendations for soil use. These ratings supplement land capability classes, woodland suitability groups, range sites, soil limitation ratings, or other interpretations and technical guides.

Each **soil potential index (SPI)** is defined to equal *performance* or *yield* (*P*) (established locally) minus the sum of costs of *corrective measures* (*CM*) and costs from *continuing limitations* (*CL*).

$$SPI = P - (CM + CL)$$

Use the same units and time scale for all factors. The value of *P* can be chosen to be above the average soil of the area, yet below the *P* for the best soil of an area.

Costs of **corrective measures (CM)** can involve building terraces or drainage systems, prorated to an annual cost; increasing septic tank or drain fields; or controlling barriers to stop erosion (Figure 20-10).

Continuing limitations (CL) can include three types of costs:

- Lower yields, inconvenience, discomfort, probability of periodic failure, and limitations because of field size or location

FIGURE 20-10 This builder needed to know the soil potential index (SPI) for this land. These houses in El Dorado Estates in New Mexico are on salty and sodium-affected soil. Many remain vacant, probably partly because septic systems failed in the low permeability soil, and landscaping plants and gardens did poorly in the salty soil. (Courtesy of USDA—Soil Conservation Service; photo by D. S. Pease.)

- Regular maintenance costs, such as pumping, irrigation, removal of septic tank solids, and pollution control devices or repair
- Offsite damages from sediment of other pollution as part of corrective measures

Large amounts of data are needed to evaluate a soil for various potential uses. Two very simplified examples are given in Details 20-1 and 20-2.

California Storie Index[9,10]

A unique suitability soil-rating system, the **Storie Index,** was developed for use in southern California but is adaptable to many other arid or semiarid regions. The four soil characteristics used in the soil-plant rating scale are as follows:

Factor A: profile characteristics that influence the depth and quality of the root zone
Factor B: textural class of the surface soil as it relates to infiltration, permeability, water capacity, and ease of tillage
Factor C: slope as a soil-plant limitation related to irrigation potential
Factor X: other characteristics that limit plant growth, including poor drainage, excessive erosion, excessive salts, high exchangeable sodium, low inherent fertility, and high acidity

Factors A, B, C, and X are individually assigned a rating, with a maximum of 100 percent for each one. For example, a specific field mapping unit, according to the Storie Index, might be rated as follows:

---

[9] Roy H. Bowman, et al., *Soil Survey of the San Diego Area, California, Part II,* USDA—Soil Conservation Service, in cooperation with the University of California Agricultural Experiment Station, U.S. Marine Corps, 1973, 92.
[10] K. Koreleski, "Adaptations of the Storie Index for Land Evaluation in Poland," *Soil Surveys and Land Evaluation* **8** (1988), 23–29.

Detail 20-1 Soil Potential Rating for Dwellings without Basements

		Cost	Index*
P	Index of performance, set locally and arbitrarily for a given area—say, a 1200-sq-ft dwelling on a ¼-acre lot (40-yr value)	\$85,000	850 (21.2/yr)
CM	Cost of corrective measures, which could include		
	Drainage of footing and slab floor	\$600–800	7
	Excavation and grading of 8–15% slope	\$1000–1400	12
	Reinforced slab floor, moderate shrink–swell	\$1500–2000	17
	Areawide surface drainage needed, per lot	\$100–200	2
	Importing topsoil for lawn and garden, as needed	\$1000–1400	12
	Prorated for 40 years for value/year		50/40
CL	Costs of continuing limitations, which could include		
	Maintenance of off-lot drainage	\$25–50	0.5
	Repair of drainage system for footing and slab	\$50–100	1.0
	Replacement of eroded-topsoil loss	\$25–50	0.25
	Flooding insurance	\$40	0.4

$$SPI_{dwelling} = 21.2/yr - (1.25 + 2.15) = 17.8$$

*Index is 1% of dollar value (= dollars/100)

Factor A: profile is permeable and deep, rating of 100% (= excellent)
Factor B: texture of surface is a loam with high water capacity, rating of 100% (= excellent)
Factor C: slope is nearly level, rating of 100% (= excellent)
Factor X: total soluble salts is 10 dS m^{-1} (classified as a saline soil), rating of 10% (= poor) (see Figure 20-11)

The composite soil index rating is determined by multiplying the four separate ratings as decimal fractions: 1.0 × 1.0 × 1.0 × 0.1 = 0.1. Changed back to a percentage, the *0.1* becomes *10 percent* and, according to Table 20-2, has a Storie numerical index rating of *5, poor* for intensive agriculture.

Once the soils of a county have been mapped, laboratory analyses on collected samples are made to characterize the soils more scientifically. Using field observations, field notes, and laboratory data, each soil mapping unit is interpreted as to its relative suitability or degree of limitation for the anticipated major land uses. These interpretations are explained for cropland, pasture land, and rangeland; woodlands and windbreaks; engineering and environment; soil hydrology; and general soil fertility.

20:7 Interpretations for Various Soil Uses

There are many approaches to evaluating land for various uses. Some of them have been briefly discussed earlier. As we consider the specific kinds of uses that can be made of soils, we recognize that they are used for homes, roads, large and small buildings, sources of gravel,

Detail 20-2 Soil Potential Rating for Grain Corn Production

		Cost	*Index**
P	Index of yield, set locally at 130 bu/a, which is a bit above the average for area ($420)	130 bu	4.20
CM	Cost of corrective measures, which could include		
	Equipment costs to convert to no-till, prorated for 10 years' use to cost per year, per acre for 300 acres	$60,000	0.20
	Install center-pivot sprinkler, prorated for 10 years' use for 130 acres	$80,000	0.06
CL	Cost of continuing limitations, which could include		
	Yearly maintenance of equipment per acre	$30	0.3
	Lower yields than average, only 110 bushels per acre, a loss of 20 bushels = $65	$65	0.65
	Maintenance of runoff sedimentation control ponds required, per acre	$15	0.15

$$SPI_{corn} = 4.20 - (0.19 + 1.10) = 2.91$$

*Index is 1% of dollar value (= dollars/100). For corn yield, index could be bushels per acre for all factors or converted to value in dollars, as done here.

FIGURE 20-11 A saline soil with a Storie Index rating of 5 (poor for intensive agriculture) is suited only to the most salt-tolerant of crops, such as barley, sugar beets, canola, upland cotton, or safflower. *(Source:* USDA—Soil Conservation Service.)

septic drain fields, playgrounds, golf courses, ponds, and roads, to name just a few. The numerous tables in each soil survey report will list, in part, the following data:

- Areas and locations of all described soils
- Land capability classification
- Soil physical and chemical properties
- Crop and range productivity
- Nonagricultural-use potential
- Climatic data

Table 20-2 Composite Storie Index Rating Scale for Intensive Agriculture

Composite Soil Rating (Storie Index)	Soil Grade for Intensive Agriculture	
	Numerical	*Relative*
80–100%	1	Excellent
60–80%	2	Very good
40–60%	3	Good
20–40%	4	Moderately good
10–20%	5	Poor
Less than 10%	6	Very poor

- The soil's mechanical properties (engineering data) and suitability for building sites
- Construction materials
- Information for sanitation facilities
- Information for recreational development

For each kind of use, soil can be rated as having **slight, moderate,** or **severe** limitations because of its texture, depth, slope, position, and other factors. The soil factors that are important to consider will vary with the use intended. A soil may have many kinds of limitations. Physical problems include shallow to bedrock, low permeability, and slippage. Chemical problems include strong acidity, high corrosiveness, and high salts. Developing the soil may require high costs to use it (too rocky to dig easily, needs leveling, needs drainage, needs dikes to protect against flooding). In Tables 20-3 to 20-9 are shown a selection of eight tables of more than thirty used by soil conservationists to aid in the evaluation of soil areas for various engineering uses. A rating of *severe* indicates that the soil will require major soil reclamation, special design, high costs, or intensive maintenance to use for the item listed (Figure 20-12). *Very severe* indicates that the properties of the soil cause either or both great difficulty in use for that purpose or high expense (Figure 20-13).

As an illustration of how soils are rated, five contrasting soils, each from a different order and area within the United States, are classified in Table 20-10. Some general features about the five soils used will help in understanding that table.

- **Aridisol: Mohave series,** *Typic Haplargid.* Sandy clay loam, 0–5% slopes, deep, well-drained, neutral to alkaline, developed on alluvium of mixed origin. Mean annual precipitation is 380 mm (15 in); mean annual temperature is 14.4°C (57.9° F). Growing season ranges from 179 to 200 days, depending on elevation. It is a benchmark soil but not prime farmland. The name is from Mohave County, Arizona (Mohave Desert).
- **Mollisol: Marshall series,** *Typic Hapludoll.* Silty clay loam, 0–2% slopes, well-drained, deep, slightly acidic, and developed from upland loess. Mean annual precipitation is 400–800 mm (16–32 in). Average annual temperature is 10.6°C (51°F); the average growing season is 141 days. The name *Marshall* is from Marshall County, Iowa. The soil is a benchmark soil and is prime farmland.
- **Spodosol: Berryland series,** *Typic Haplaquod.* Loamy sand, 0–2% slopes, very deep, poorly drained, very strongly acidic, and developed on glacial outwash plains in Massachusetts. Mean annual precipitation is 1170 mm (46 in); mean annual temperature is 9.4°C (49°F). The average growing season is 155 days. It is not

Table 20-3 Evaluating the Use of Soils for **Septic Tank Absorption Fields** and **Sanitary Landfill Area**

Septic Tank Absorption Fields

	Limits			
Property	*Slight*	*Moderate*	*Severe*	*Restrictive Feature*
Surface texture	—	—	Ice	Permafrost
Total subsidence (in)	—	—	>24	Subsides
Flooding	None	Rare	Common	Flooding
Depth to bedrock (in)	>72	40–72	<40	Depth to rock
Depth to cemented pan (in)	>72	40–72	<40	Cemented pan
Depth to high water table (ft)	—	—	+	Ponding
	>6	4–6	<4	Wetness
Permeability (in/hr)				
(24–60 in)	2–6	0.6–2	<0.6	Percs slowly
(24–40 in)	—	—	>6.0	Poor filter
Slope (%)	<8	8–15	>15	Slope
Fraction >3 in[a] (wt %)	<25	25–50	>50	Large stones
Downslope movement	—	—	II/	Slippage
Formation of pits	—	—	III/	Pitting

Sanitary Landfill (Area)

	Limits			
Property	*Slight*	*Moderate*	*Severe*	*Restrictive Feature*
Surface texture	—	—	Ice	Permafrost
Flooding	None	Rare	Common	Flooding
Depth to bedrock[b] (in)	>60	40–60	<40	Depth to rock
Depth to cemented[b] pan (in)	>60	40–60	<40	Cemented pan
Permeability[b] (in/hr) (20–40 in)	—	—	>2.0	Seepage
Depth to high water table (ft)	—	—	+	Ponding
Apparent	>5	3.5–5	<3.5	Wetness
Perched	>3	1.5–3	<1.5	Wetness
Slope (%)	<8	8–15	<15	Slope
Downslope movement	—	—	Occurs	Slippage
Formation of pits	—	—	Occurs	Pitting
Differential settling	—	—	Occurs	Unstable fill

Source: National Soils Handbook, USDA—Soil Conservation Service, 1983, 603–661, 603–668.
[a]Weighted average to 40 in.
[b]Disregard (1) in all Aridisols except Salorthids and Aquic subgroups, (2) all Aridic subgroups, and (3) all Torric great groups of Entisols except Aquic subgroups.

considered as a benchmark soil and is not prime farmland because of poor drainage, low inherent fertility, and a strongly cemented high-iron **B** horizon.

- **Ultisol: Ruston series,** *Typic Paleudult.* Fine sandy loam, 2%–5% slopes, deep and well-drained, and acidic. It developed from sandy marine sediments in Mississippi. Mean annual precipitation is 1420 mm (55 in); mean annual temperature is 18°C (65°F). The mean growing season is 221 days. *Ruston* is from the town of Ruston, Louisiana. Ruston soil is prime farmland but is not a benchmark soil, and it exists in Mississippi, as well as Louisiana.

Table 20-4 Evaluating the Use of Soils for **Local Roads and Streets**

Property	Limits			Restrictive Feature
	Slight	Moderate	Severe	
Surface texture	—	—	Ice	Permafrost
Total subsidence (in)	—	—	>12	Subsides
Depth to bedrock (in)	>40	20–40	<20	Depth to rock
Depth to cemented pan (in)	>40	20–40	<20	Cemented pan
Shrink–swell*	Low	Moderate	High very high	Shrink–swell
Depth to high water table (ft)	>2.5	1.0–2.5	<1.0	Wetness
Slope percentage	<8	8–15	>15	Slope
Flooding	None	Rare	Common	Flooding
Potential frost action	Low	Moderate	High	Frost action
Fraction >3 in (wt %)	<25	25–50	>50	Large stones
Downslope movement	—	—	Occurs	Slippage
Formation of pits	—	—	Occurs	Pitting
Differential settling	—	—	Occurs	Unstable fill

Source: National Soils Handbook, USDA—Soil Conservation Service, 1983, 603–681.
*Thickest layer 10–40 in.

Table 20-5 Evaluating the Use of Soils for **Dwellings** (Homes) **with Basements**

Property	Limits			Restrictive Feature
	Slight	Moderate	Severe	
USDA texture	—	—	Ice	Permafrost
Total subsidence (in)	—	—	>12	Subsides
Flooding	None	—	Rare, common	Flooding
Depth to high water table (ft)	—	—	0.5	Ponding
	>6	2.5–6	<2.5	Wetness
Depth to bedrock (in)				Depth to rock
Hard	>60	40–60	<40	
Soft	>40	20–40	<20	
Depth to cemented pan (in)				Cemented pan
Thick	>60	40–60	<40	
Thin	>40	20–40	<20	
Slope (%)	<8	8–15	>15	Slope
Shrink–swell	Low	Moderate	High, very high	Shrink–swell
Unified (bottom level)	—	—	OL, OH, PT	Low strength
Fraction >3 in (wt %)	<25	25–50	>50	Large stones
Downslope movement	—	—	Occurs	Slippage
Formation of pits	—	—	Occurs	Pitting
Differential settling	—	—	Occurs	Unstable fill

Source: National Soils Handbook, USDA—Soil Conservation Service, 1983, 603–681.

Table 20-6 Evaluating the Use of Soils for **Lawns, Landscaping,** and **Golf Fairways**

Property	*Limits*			*Restrictive Feature*
	Slight	*Moderate*	*Severe*	
USDA texture	—	—	Ice	Permafrost
Flooding	None	Occasionally	Frequently	Flooding
Slope percentage	<8	8–15	>15	Slope
Depth to bedrock (in)	>40	20–40	<20	Depth to rock
Depth to cemented pan (in)	>40	20–40	<20	Cemented pan
Surface texture[a]	—	—	SiC, C, SC	Too clayey
Surface texture	—	—	Muck, peat	Excess humus
Surface texture	—	LCoS, S	CoS	Too sandy
Salinity, dS m^{-1}	<4	4–8	>8	Excess salt
Sodium adsorption ratio	—	—	>12	Excess sodium
Soil reaction (pH)	—	—	>3.6	Too acidic
Coarse fragments[b] (wt %)	<25	25–50	>50	Small stones
Fraction >3 in[b] (wt %)	<5	5–30	>30	Large stones
Depth to high water table (ft)	>2	1–2	<1	Wetness

Source: National Soils Handbook, USDA—Soil Conservation Service, 1983, 603–681.
[a]Thickest layer 0–40 in, Co = coarse, C = clay, L = loam(y), S = sand(y), Si = silt(y).
[b]Fraction all passes number 10 sieve.

Table 20-7 Evaluating the Use of Soils for **Topsoil**

Property	*Limits*			*Restrictive Feature*
	Slight	*Moderate*	*Severe*	
Surface texture	—	—	Ice	Permafrost
Depth to bedrock (in)	>40	20–40	<20	Depth to rock
Depth to cemented pan (in)	>40	20–40	<20	Cemented pan
Depth to bulk density >1.8 Mg/cm^3	>40	20–40	<20	Too compacted
Surface texture[a]	—	LCoS, LS, LFS, LvFS	CoS, S FS, vFS	Too sandy
Surface texture[a]	—	SCL, CL, SiCL	SiC, C, SC	Too clayey
Fraction >3 in[b] Surface (wt. %)	<5	5–30	>30	Large stones
Coarse fragments[b] Surface (wt. %)	<25	25–50	>50	Small stones
Salinity, dS m^{-1}	<4	4–8	>8	Excess salt
Layer thickness (in)	>40	20–40	<20	Twin layer
Depth to high water table (ft)	—	—	<1	Wetness
Sodium adsorption ratio (0–40 in)	—	—	>12	Excess sodium
Soil reaction (pH)	—	—	<3.6	Too acidic
Slope percentage	<8	8–15	>15	Slope

Source: National Soils Handbook, USDA—Soil Conservation Service, 1983, 603–689.
[a]Thickest layer 0–40 in; Co = coarse, L = loam(y), S = sand(y), Si = silt(y), C = clay, F = fine, vF = very fine.
[b]Sum >3 in, all passing number 10 sieve.

Table 20-8 Evaluating Soils for **Drainage**

Property	*Limits*	*Restrictive Feature*
Surface texture	Ice	Permafrost
Depth to high water table (ft)	>3	Deep to water
Permeability (in/hr)	<0.2	Percs slowly
Depth to bedrock (in) or cemented pan	<40	Depth to rock (cemented pan)
Flooding	Common	Flooding
Total subsidence	Any	Subsides
Fraction >3 in (wt %)	>25	Large stones
Potential frost action	High	Frost action
Slope percentage	>3	Slope
Texture, thickest layer 10–60 in*	CoS, S, FS, vFS, LCoS, LS, LFS, LvFS, SG, G	Cutbanks cave
Salinity (dS m^{-1})	>8	Excess salt
Sodium adsorption ratio	>12	Excess sodium
Soil reaction (pH)	<3.6	Too acidic
Downslope movement	Occurs	Slippage
Complex landscape	Occurs	Complex slopes
Availability of outlets	Difficult to find	Poor outlets

Source: National Soils Handbook, USDA—Soil Conservation Service, 1983, 603:p. 101.
*Co = coarse, S = sand(y), F = fine, vF = very fine, L = loam(y), G = gravel.

Table 20-9 Evaluating the Use of Soils for **Irrigation**

Property	*Limits*	*Restrictive Feature*
Surface texture	Ice	Permafrost
Slope percentage	>3	Slope
Fraction >3 in (wt. %)	>25	Large stones
Depth to high water table (ft)	<3	Wetness
Available water capacity (in/in)	<0.10	Droughty
Surface texture*	CoS, S, FS, vFS, LCoS, LS, LFS, LvFS	Fast intake
Surface texture*	SiC, C, SC	Slow intake
Wind erodibility group	1, 2, 3	Soil blowing
Permeability (in/hr)	<0.2	Percs slowly
Depth to bedrock (in) or cemented pan	<40	Depth to rock (cemented pan)
Bulk density (Mg/m^3)	>1.7	Rooting depth
Erosion factor (surface K)	>0.35	Erodes easily
Flooding	Common	Flooding
Sodium adsorption ratio	>12	Excess sodium
Salinity (dS m^{-1})	>4	Excess salt
Soil reaction (pH)	<3.6	Too acidic
Complex landscape	Exists	Complex slope
Formation of pits	Occurs	Pitting

Source: National Soils Handbook, USDA—Soil Conservation Service, 1983, 603:p. 103.
*Co = coarse, S = sand(y), F = fine, vF = very fine, L = loam(y), Si = silt(y), C = clay.

FIGURE 20-12 Definitions of *topsoil* differ from person to person. It is becoming more valuable and scarce to the extent that some people will stockpile it when it is available. This stockpile of topsoil has been mined from the Cattail Channel bottom land, in Illinois. Is it high quality? (Courtesy of USDA—Natural Resources Conservation Service and the Illinois Agricultural Experiment Station, *Soil Survey of Whiteside County, Illinois.*)

FIGURE 20-13 Consider what you need to know about this Berks County, Pennsylvania, landscape, if it were to have constructed on it (1) additional roads, (2) some houses with septic tanks, (3) a small park, and (4) a school and playground. At the least you would need good information on slope, internal and external drainage, depth to water table and hardpans, ability of soil to support small buildings and a roadbed, and whatever is needed to control erosion. A soil survey and land evaluation would provide that information. (Courtesy of USDA—Soil Conservation Service; photo by Ledbetter.)

- **Vertisol: Houston Black series,** *Udic Pellustert.* Clay, 0–1% slopes, deep, moderately well-drained, alkaline, and developed from marine chalk (impure calcite). Mean annual precipitation is 810 mm (32 in); mean annual temperature is 19°C (66°F). The average growing season is 202 days. The name *Houston Black* is from a county name in Texas and its dark black color. It is prime farmland and is a benchmark soil.

Table 20-10 Interpretive Uses of Five Soil Series in Five Soil Orders under Good Management

	Soil Order and Soil Series (State)				
Proposed Use	*Aridisols, Mohave (New Mexico)*	*Mollisols, Marshall (Iowa)*	*Spodosols, Berryland (Massachusetts)*	*Ultisols, Ruston (Mississippi)*	*Vertisols, Houston Black (Texas)*
Crop production potential					
Sorghum, corn (bu/a)	Not adapted	109 (corn)	Not adapted	65 (corn)	90 (sorghum)
Oats, wheat (bu/a)		62 (oats)		30 (wheat)	45 (wheat) 90 (oats)
Soybeans (bu/a)	Not adapted	41	Not adapted	25	Not adapted
Improved Pasture/ Hay (t/a)	Not adapted	7.6	Not adapted	12.0	8.0
Engineering uses					
Dwellings with basements	Fair (too clayey)	Fair (shrink–swell clay)	Very poor (wetness)	Good	Very poor (shrink–swell clay)
Local roads	Poor (low strength)	Poor (low strength)	Poor (wetness)	Fair (low strength)	Very poor (poor bearing)
Septic absorption fields	Severe (slow percolation)	Slight	Very poor (wetness)	Moderate (slow percolation)	Very poor (slow percolation)
Landfills	Fair	Fair	Very poor (wetness)	Moderate (too clayey)	Poor
Sewage lagoons/ponds	Moderate (slopes)	Moderate (seepage)	Poor	Moderate (slow perc)	Fair Good
Topsoil	Fair (too clayey)	Fair (slopes)	Poor (sandy, wet)	Fair (small stones)	Poor (too clayey)
Irrigation	Moderate (slopes)	Favorable	Poor	Fair (sloping)	Poor (slow percolation)
Playgrounds	Moderate (slopes)	Favorable	Poor (too wet)	Fair (slopes)	Moderate (too clayey)
Habitat potential for					
Openland wildlife	Very poor	Good	Poor	Good	Fair
Woodland wildlife	Not adapted	Good	Poor	Good	Poor (few trees)
Wetland wildlife	Not adapted	Very poor	Good	Very poor	Poor
Corrosion risk for					
Uncoated steel	High (too salty)	Moderate	High (acidic)	Moderate	High
Concrete	Low hazard	Moderate	High (acidic)	Moderate	Low hazard

Source: Ellis Knox, National Leader for Soil Research, USDA—Soil Conservation Service, Washington, DC.

In Table 20-10 notice particularly the following items:

- Differences exist in crop potential in different soils.
- The soils that are good croplands often are poor (low strength or high swell–shrink) as construction materials (see *local roads*).
- Poorly permeable soils may be good for water reservoirs (see *sewage lagoons/ponds*) but poor for septic absorption fields, irrigation, and playgrounds.

Any single soil will likely be good for some uses but poor for others. A permeable, well-drained soil may be good for crops but poor for landfills, lagoons, and ponds. A clayey soil without good, deep drainage will be good for lagoons and ponds but may be difficult to use

Table 20-11 Changes in the Percentage of the Population Living in Cities in the United States

Year	Population in Cities % of Total
1800	6
1860	20
1900	40
1960	60
1972	75
2000	80+

Source: Modified from M. J. Redding and B. T. Parry, "Land Use: A Vital Link to Environmental Quality," in V. Curtis, ed., *Land Use and the Environment,* Environmental Protection Agency, 1975, 3–39.

for crops, unstable for roads and small buildings, and poor topsoil. Soil evaluations (land assessments) not only indicate the best uses for each soil but also indicate the degree to which the good and bad characteristics of each soil will affect its use for a given purpose.

20:8 Controls in Land-Use Planning[11]

Land-use planning utilizes information, foresight, and a wise compromise of the possibilities to designate the best use of land for its manifold possibilities now and in the future. Difficulties arise because private and public rights can clash and there are not enough appropriate soils in the right locations to satisfy all demands. Unfortunately, *master land-use plans* usually are formulated to benefit urban populations. The concept of reserving prime land areas for agricultural, wildlife, or recreational use because they are particularly valuable for those purposes is not as common as it needs to be. Only recently have many of these kinds of land zoning decisions been made. A soil survey is one means of establishing a scientific basis for planning the most appropriate uses of every acre or hectare of land.

Conventional Controls on Land Use

Times change! The *Homestead Act* of 1862, the *Timber Culture Act* of 1873, and the *Desert Act* of 1877 gave free to the user 160, 160 and 640 acres if the user settled the land 5 years or planted trees on one-quarter of the land or irrigated one-eighth of it, respectively.[12] A U.S. county zoning and sanitation department had the following statement on their letterhead: "The land belongs to the people . . . a little of it to those living . . . but most of it to those yet to be borne. . . ."[13]

The gradual move of people from rural to urban living has increased the proportion of people who own no land or who own only home lots (Table 20-11). Today less than 2 percent of the U.S. population are owners of farmland (Figure 20-14).

[11] Gerald W. Olson and Arthur S. Lieberman, eds., *Proceedings of the International Symposium on Geographic Information Systems for Conservation and Development Planning,* Cornell University, Ithaca, NY, Apr. 4–6, 1984.

[12] M. J. Redding and B. T. Parry, "Land Use: A Vital Link to Environmental Quality," in V. Curtis, ed., *Land Use and Environment,* Environmental Protection Agency, 1975, 3–39.

[13] Virginia Curtis, ed., *Land Use and the Environment: An Anthology of Reading,* Environmental Protection Agency, U.S. Government Printing Office, 1975, 200 pp.

FIGURE 20-14 Housing moving onto valuable farmland in Dade County, Florida. The ground is rocky (see Figure 14-6), but it exists in such a good climate and location that it is *unique farmlands* and should be saved for the excellent crops of fruits and vegetables it can produce. (Courtesy of USDA—Natural Resources Conservation Service and Florida Agricultural Experiment Station, *Soil Survey of Dade County, Florida, 1996.*)

Land-use regulation began with the 1926 *Standard State Zoning Enabling Act.* This act allowed governmental agencies the right to zone areas for limited kinds of use as a protection to the general public. The U.S. Supreme Court said the following in a 1926 case:

> A regulatory zoning ordinance, which would be clearly valid as applied to the great cities, might be clearly invalid as applied to rural communities. In solving doubts, the maxim *"sic utere tuo ut alienum no laedas,"* [i.e., use your own property in such a manner as not to injure another], which lies at the foundation of so much of the common law of nuisances, ordinarily will furnish a fairly helpful clue.[14]

The Court went on to say that a nuisance may merely be a right thing in the wrong place—like a pig in the parlor instead of in the barnyard. However, the Supreme Court left no doubt that government agencies could establish some controls on the uses made of land. Some of the regulatory mechanisms follow.

Zoning laws passed by local, regional, or national government agencies are intended to guide the orderly development and use of land. Laws may soon be utilized to protect valuable, irreplaceable lands from other uses. Regulations about draining wetlands is one example. Hawaii, a popular tourist spot with limited land, has zoned the entire state by major land-use classes: Agricultural, urban, conservation, and residential, with soil survey information as the scientific base.

The prime objective of **differential taxation** is the preservation of land for specified uses, such as agricultural, timber, recreational, urban, and suburban. Taxes are assessed on current use values (using soil surveys as the scientific physical basis) rather than on the principle of *highest and best, potential future use,* or *politics* to set a tax value. Differential-taxation laws usually permit productive timber lands, as interpreted partly from soil survey reports, to be taxed primarily at the time of harvest rather than on an annual basis, as is the practice for most other land uses. Agricultural land is taxed on the basis of its use for farming and ranching and not at the inflationary rate of its potential as a subdivided residential or

[14]Patrick J. Rohan, *Zoning and Land Use Controls,* Volume 1, Matthew Bender, Publishers, New York, 1978, 1–12.

commercial area. These differential taxes may accumulate and need to be paid later when a change in use occurs.

Easements and **contracts** are used by some governmental bodies to implement land-use planning by mandatory action. A municipality, county, state, or national government may demand an easement or a contract to purchase (the *earnest money* valid for a period of perhaps 10 years) that specifies the future use of the land for public purposes, such as a park. An initial payment is made to the owner, and tax reductions are offered. Such easements and contracts can be broken by the public agency but not by the landowner.

Public purchase is the most direct method of implementing a land-use plan. Some federal and state departments have authority to purchase lands for national or state forests, parks, public hunting areas, flood control dams, and backwater areas and for other land-use purposes to promote the general welfare. The forced sale of private land to a public body may be voluntary or involuntary. The right of eminent domain prevails. **Eminent domain** is based on a 1926 law that states that a forced sale is legal because it *promotes the general welfare,* which action prevails over the private good. Condemnation proceedings result in a fair-market price being determined and compensation given to the private landowners for their property. This law has been used extensively for decades to acquire land for schools, highways, and other public projects.

Recent Trends in Controlling Land Use

Will there be enough first-class agricultural soils to sustain the growing world population? More than 400,000 hectares (1 million acres) of prime agricultural soils and three times this hectarage of all soils used for agriculture are being diverted to other uses in the United States each year. Concern over the magnitude of these losses has resulted in laws being passed by 48 of the 50 states to retain prime agricultural soils for agricultural use.

Nonurban land-use planning has had limited effect in reserving agricultural lands for agricultural use. Most plans *encouraged* compliance but *did not provide any enforcement capability.* The basis of zoning laws dating from 1926 required that regulations imposed on the use of private property must be for purposes of *promoting health, safety, morals,* or *the general welfare* of the community (people). With the passage of the Clean Air Acts and the Clean Water Acts in the 1970s, additional control of land was possible if that control was essential to hinder or reduce polluting of water or air. These laws and increasing public concern about vanishing agricultural lands have resulted in *preferential tax assessment, buying development rights, transferable development credits,* and *altering enforced planning* as methods to achieve the land use which legal authorities decide is for the common good.

Preferential tax assessments are made to fit the *current use* of the land instead of potential uses that usually are assessed at higher rates. The additional accumulated taxes it might have had if valued as its potential selling price are collected when the land use is changed (e.g., subdivided for a housing development). A lower tax scale in agricultural production allows the owner to continue to farm the land without the additional burden of high land taxes that might make an agricultural use unprofitable or that might make subdividing too attractive.[15,16]

Buying development rights is a mechanism by which governments control land use. The owner sells development rights to the government but retains ownership of the land in its present use and can sell it for that continued use. The private landowner is recompensed

[15] Richard Barrows and Douglas Yanggen, "The Wisconsin Farmland Preservation Program," *Journal of Soil and Water Conservation* **33** (1978), 209–212.

[16] K. R. Olson and G. W. Olson, "Use of Agronomic Data and Enterprise Budgets in Land Assessment Evaluations," *Journal of Soil and Water Conservation* **40** (1985), 455–458.

for lost potential profit, and the public retains the land in agricultural or other determined use. Suffolk County in Long Island, New York,[17] has begun long-range plans to buy development rights on 12,000 ha (30,000 a) at $13,000/ha ($5,260/a) of farmland to keep a rural area accessible to city residents. This technique is costly to governments but is continuing to be done.

Transferable development credits (TDCs) is a technique of allotting a specified amount of development per area. If a developer wanted a higher-than-normal density development, he or she could purchase development credits from other landowners who were not interested in developing their own properties but who could not later change the use once they sold development credits. This process allows more efficient urban planning and reimburses farmers for their lost development potential.[18]

Enforced planning grew out of a concern about present land use. Increasing numbers of states require local governments *to establish and enforce land-use plans.* Such plans include identifying the most valuable agricultural lands and housing-growth areas. The object is to plan for growth by saving those lands best suited for certain uses and reserve them for the *highest and best* purposes and to designate them as prime land.

Although these laws embody the many variations of the methods of preservation previously mentioned, the most common technique has been differential taxation.

20:9 A Decision Case Study

An introduction to the use of *decision cases* was given in Chapter 6. An additional decision case study for use follows:

D. L. Taack, H. Murray, and S. R. Simmons, "Minto-Brown Island Park: A Case Study of Farming the Urban-Agricultural Interface," *Journal of Natural Resources and Life Sciences Education* **23** (1994), 98–103.

The law is the last result of human wisdom acting upon human experience for the benefit of the public.

—Samuel Johnson

Questions

1. (a) Which equipment is used by soil surveyors? (b) Which profile features will a surveyor describe and measure in the survey?
2. What information is in a survey report? List at least seven items.
3. Which three categories constitute the bulk of each soil survey report?
4. How permanent is the information in a soil survey report? Discuss.
5. Describe the LESA approach to land-use planning.
6. (a) What is a *soil potential index* (SPI)? (b) Is there only one SPI for each soil? Explain.
7. Which information in a soil survey report might be useful to a building contractor or to area planners?

---

[17]John V. N. Klein, "Preserving Farmland on Long Island," *Environmental Comment,* No. 5, Jan. 1978; 11–13.
[18]Peter J. Pizor, "New Jersey's TDC Experience," *Environmental Comment,* Apr. 1978; 11.

8. What are some of the important soil properties that may result in *slight, moderate,* or *severe* ratings for (a) drainage of soils, (b) local roads and streets, and (c) lawns, landscaping, and golf fairways?
9. In general terms list several of the properties that soils must have to qualify as prime farmland. List information for (a) depth, (b) pH, (c) erosion hazard, (d) flooding permitted, (e) temperature, and (f) available water.
10. In general terms define *unique soils, land critical to the state,* and *benchmark soils.*
11. Define (a) *land capability class III land,* (b) *class V land,* (c) *subclass IIs land,* and (d) *subclass IVw land.*
12. What are three general requirements of a soil that are important in its use for (a) septic system drainage fields, (b) topsoil, and (c) small buildings? Explain the importance of each criterion (requirement) you list.
13. What might be a *severe* restriction in a soil that was considered for use for (a) topsoil, (b) lagoons or ponds, and (c) roads?
14. Controlling land use is of increasing interest and concern. Discuss the mechanisms of *eminent domain, preferential tax* assessments, and *buying development rights.*

Greenhouse Soils and Soilless Culture

He who knows what sweet and virtues are in the ground, and waters, the plants, the heavens, and how to come at these enchantments, is the rich and royal man.

—Ralph Waldo Emerson

21:1 Preview and Important Facts

PREVIEW

The business of growing ornamental plants (flowers, home landscaping shrubs and trees) for sale has increased rapidly, as more than three-fourths of the U.S. population are now urban dwellers. Large numbers of these people annually purchase small plants ready to be transplanted into their gardens, to be kept in their homes in pots for the plant's lifetime, or to keep the plant potted for landscaping. Plants may be grown to produce vegetables and flowers continually in the greenhouse. For those young plants taken home, flower and vegetable starts are available mostly from grocery stores and other outlet stores rather than from the grower directly; thus, a lot of moving and shipping of the young plants are done.

The soil or other media used to start plants must, for several weeks, provide adequate water and adequate aeration for seed germination, be pathogen-free, and be a suitable anchor for roots. Nutrients must be furnished by the media or in irrigation water. Soilless media are popular with growers because they are relatively lightweight, free from diseases, readily available, and more suitable in various ways than are mineral soils. Peat, vermiculite, perlite, bark, and rockwool are used most extensively.

A major difference between the management of plants in small containers and in a field is *drainage and aeration.* In small containers soil usually is inadequately drained and remains too wet for optimum, even good, growth. Many of the greenhouse mixes and soilless media are designed to improve aeration in the small container. That these materials are lightweight is a plus when transporting the young plants.

These special root media completely change the method of fertilization. The media also need periodic pasteurization to kill potentially damaging pathogens and other pests. These special needs make greenhouse culture an art and science all its own.

At the opposite extreme from culture in soils is the culture in water to which essential nutrients are added—hydroponics. The plant is held in the solution by a rigid or floating support so that its roots are in the solution but its tops are in the atmosphere. A coarse support medium, such as sand or perlite, can be used so that only films of nutrient solution are held for the roots. Hydroponics is the method planned for growing food in space stations.

IMPORTANT FACTS TO KNOW

1. Why soil in small containers may be too wet and have inadequate aeration
2. Materials used for mixing a root media, particularly *peat moss, vermiculite, perlite,* and *bark* and their more important properties for root media
3. The recommendations on using manures, compost materials, and sawdust in mixes
4. The frequent concern, and its basis, about media pH and how the pH is adjusted
5. Some approximate compositions of good soilless mixes
6. The reasons and methods to pasteurize root media
7. Three general rules in preparing a media and correctly watering it
8. Techniques for automatic watering of pots
9. How to assure adequate plant nutrition
10. Management of excess soluble salts
11. General actions of the different classes of chemical growth-regulating materials
12. Management of pathogens and pests
13. The techniques and reasons for using hydroponics

21:2 Pot Root Media[1]

Soils do not produce yields as high as soilless culture (Table 21-1). However, soilless culture is expensive in time and effort and cannot compete with soil for crop production on scales of tens of hectares.

Table 21-1 Comparisons of Yields Per Hectare for Various Crops Grown in Soil (in the Field) and in Soilless Culture

Crop	*Soil*	*Soilless*	*Increase in Soilless*
Soybeans	675 kg	1,700 kg	2.5 times more
Beans	11,000 kg	47,000 kg	4.2 times more
Wheat	675 kg	4,600 kg	6.8 times more
Rice	1,100 kg	5,600 kg	5.1 times more
Potatoes	18,000 kg	160,000 kg	8.9 times more
Tomatoes	11,000–22,000 kg	135,000–674,000 kg	6.1–30.6 times more
Cucumbers	7,900 kg	31,000 kg	3.9 times more

Source: Selected and modified data from Howard M. Resh, *Hydroponic Food Production,* Woodbridge Press Publishing Company, Santa Barbara, CA, 1995, 527 pp.

[1]Paul V. Nelson, *Greenhouse Operation and Management,* Prentice-Hall, Englewood Cliffs, NJ, 1991, 171–208.

Most plants grow well in good soils; why use anything else, particularly if it is more expensive? On the basis of volume, field soil is about half the cost of peat moss, about two-thirds the cost of manures and sawdust, and about one-fourth the cost of perlite. The problem is that drainage and aeration are poorer with most soils than with mixes or soilless media. Soil is also heavy.

The major disadvantages of various media are their higher costs and their lack of nutrients. These disadvantages are not as critical as are the need for good water retention and adequate aeration, which the media have. Thus, the root media mixes are widely utilized.

Important Properties of Root Media

Root media should have the following properties:

- Stable organic matter
- A relatively low C:N ratio
- Water retention but with adequate aeration
- Moderate-to-high cation exchange capacity
- A suitable pH

Some soil or root media mixes will not meet these requirements without adjustments.

Stable organic matter (low percentage decomposed during a few months) will keep the root medium from becoming finer during plant growth. Extensive decomposition will make the residue finer and decrease aeration. Straws and sawdust may be poor materials for small-pot root media for this reason; on greenhouse benches those materials in soil bulk may be less problematic.

Because organic materials of low nitrogen content will use nitrogen from the media for decomposition, a C:N ratio of wider than about 30:1 is undesirable if it decomposes very fast (see Chapter 6). The C:N ratios of many materials used are high—sawdust (800:1), bark (300:1), and peat moss (80:1). Of these materials sawdust will decompose moderately fast, but bark and peat moss decompose slowly.

For water to drain from the bottom of a pot, the water must be near *saturation* at the bottom or edge of the root media (see Chapter 4). Soils, unless they are sands, are likely to retain *too much water* and *too little air space.* This is one reason that coarse materials such as peat moss, bark, sands, perlite, and vermiculite are used with, or in place of, soils (Table 21-2). Sands and soils are useful to provide weight so that the pot can hold the plant up, but the other materials improve aeration. Notice in Table 21-2 that two critical properties—*percentage air* and *available water*—vary considerably among materials. The *percentage air* at container capacity (= wetted and drained) is low for soils and peat moss. The *available water* is low in perlite and polystyrene beads.

The root media tend to contain low amounts of nutrients; therefore, it is important that the media have a high cation exchange capacity (CEC) to retain calcium, magnesium, potassium, ammonium-N, and micronutrient metals. Most organic materials and clays have high CEC; sand, perlite, polystyrene, rockwool, and agricultural wastes (rice hulls, sugarcane, bagasse, etc.) have low CEC values.

The pH of root media can be critical. The pH should be about pH 5.5 to 6.5. Most sphagnum peat moss and pine bark are below pH 4. Some soils, also, are strongly acidic. In mixes with such material, lime must be added.

Table 21-2 Water and Air Volumes, Available Water, and Bulk Density of Various Root Media Ingredients and Mixes When in 17-cm-diameter Pots

Material/Mix	Water (%)		Air (%)		Available Water (%)	Bulk Density (Mg/m^3)
	cc*	−1500 kPa	cc*	−1500 kPa		
Soil (sandy loam)	40	6.4	6.9	40	34	1.7
Sand	35	4.4	5.3	36	31	1.7
Sphagnum peat moss	76	26	8.1	59	50	0.86
Vermiculite	53	29	19.5	44	24	0.74
Pine bark, $<$10 mm	59	30	20.4	49	29	0.81
Perlite	38	20	24.8	43	18	0.51
Polystyrene beads	11	1	24.9	34	10	0.12
1 soil:1 peat moss:1 sand	49	8	5.9	46	41	1.6
1 peat moss:1 vermiculite	70	24	16.6	63	46	0.85

Source: Modified from Paul V. Nelson, *Greenhouse Operation and Management,* Prentice-Hall, Englewood Cliffs, NJ, 1991, 171–208.
*cc = container capacity when wetted and drained.

Components of Root Media Mixes

Root media mixes can be made of many materials and proportions. Some contain soil, but the many that do not have soil are called **soilless media.** Some of the most used materials are soils, peat moss, bark, sawdust, manure, compost, vermiculite, sand, perlite, polystyrene foam, and rockwool (Table 21-3).

There are numerous kinds of **peat moss** available. These materials usually decompose very slowly and have high water-holding capacity (up to 60% of its volume). Sphagnum peat moss is the most acidic peat moss with a pH about 3.5 to 4.0. About 5 kg to 15 kg of finely ground lime (limestone) is needed per cubic meter of peat moss. Darker colored (more decomposed) peat mosses may be less acidic and may decompose more than does the sphagnum moss.

Bark (redwood, fir, pine, etc.) is used to replace the more expensive peat moss and is quite satisfactory. The bark needs to be composted to remove any allelopathic effects of the bark and to increase its CEC (can increase CEC from about 8 to 60 $cmol_c/kg$). If good composting conditions are maintained, about 4 to 8 weeks is sufficient composting time.

Sawdust, straws, and **manures** are used. Sawdust and straws must be composted for one to two months, to reduce their rapid use of soil nitrogen in decomposition (see Chapter 6). Sawdust may be acidic and need limestone added. Manures are not used much, mostly because of problems of ammonium toxicity after the medium is pasteurized. Manure has a high CEC and furnishes nutrients. Cow manures have been good, but poultry manures may be too high in ammonium and too fast to decompose. Horse manures are porous and not too high in nutrients; this low-nutrient manure has been used for decades as mushroom media.

Vermiculite is a mica-like silicate mineral that expands into an accordion-like structure when heated to high temperature. This expanded material has a high water-holding capacity *within the expanded particles* and yet good aeration *between the large particles.* It has a high CEC and provides some potassium, calcium, and magnesium.

Sand is used mostly to improve drainage and aeration. It has no appreciable CEC or value for nutrients. One should use fairly coarse materials.

Table 21-3 Properties of Materials Used for Root Media Mixes

	Source	Dry Wt.[a]	pH	Nutrients	CEC	Water Retention[b]	Aeration, Drainage	Comments
Inorganics								
Perlite	Heated volcanic silicate	L	6.5–7.5	None	None	Fair	Very good	Nasal irritant
Pumice	Volcanic silicate	L	Neutral	Slight K	Slight	Fair	Good	Limited availability
Sand	Soil mineral	H	Neutral	None	None	Poor	Good	Best is quartz sand
Styrofoam®	Polystyrene	L	Neutral	None	None	Poor	Very good	Pasteurization problem
Vermiculite	Heated micaceous mineral	L	6.5–7.5	Slight K, Mg	High	Fair	Good	Mix with perlite for long use
Organics								
Bark	Softwood tree waste	M	3.5–6.5	Slight, several	Low–med.	Good	Good	Age 30 days with N
Compost	Organic wastes	M	5.5–8.5	Low, many	High	Good	Good	Slow N release
Peat moss	Decayed vegetation	M	3.8–4.5	1% N	High	Good	Good	Use wetting agent
Sawdust	Softwood tree waste	M	3.5–6.8	Slight, several	Low–med.	Good	Fair	Age 30 days with N
Sphagnum peat moss	Fresh, dried	L	3.5	None	High	Very good	Fair–good	Contains natural fungicides

[a]L = under 128 kg/m^3, M = up to 224 kg/m^3, H = about 1600 kg/m^3.
[b]Fair = absorbs 3 to 4 times its weight in water, good = 5–10 times, very good = 10–20 times.

Perlite is a siliceous volcanic rock that, when heated to 980°C (about 1800°F), expands, forming white particles with many air-filled cells that are not open to outside water. Perlite is mostly sterile and inert and has a low pH. It is a lightweight substitute for sand.

Polystyrene foam, also known as Styrofoam® and other names, is a synthetic material with many air-filled pores. It is very lightweight and does not absorb water. Mostly it is a substitute for sand, providing good aeration. **Rockwool** has a similar inertness to polystyrene foam but holds a large amount of available water and it helps produce good aeration.

21:3 Root Media Mixes

There are hundreds of possible media mix combinations. However, in a practical sense they are somewhat limited if the criteria for the *important properties of a mix* are followed. It is essential to include some stable organic matter to boost CEC and water retention while maintaining adequate aeration. Add soil or fertilizer to help supply nutrients. Add inert material, such as vermiculite or perlite, for aeration and to keep the mix light in weight. Add lime as needed to adjust the pH. The following sections include examples of mixes that meet these requirements.

Table 21-4 General Composition of Some Popular Root Media Mixes

Media Components			Function
1 loam soil	1 peat moss	1 coarse sand	Pot and bench mix
1 vermiculite	1 peat moss		Germination mix
2 vermiculite	2 peat moss	1 perlite	Potted plant mix
2 vermiculite	1 peat moss, 1 pine bark	1 perlite	Potted plant mix
	1 peat moss, 3 pine bark	1 coarse sand	Potted plant mix
1 rockwool	1 peat moss		Potted plant mix

Source: Modified from Paul V. Nelson, *Greenhouse Operation and Management,* Prentice-Hall, Englewood Cliffs, NJ, 1991, 171–208.

Soil-Based Root Media[2]

Soils often contain good water-holding capacity and moderate CEC. Aeration when soils are used in pots is usually not adequate, so a mixture of one-third sand plus one-third peat moss (or coarse bark) and one-third soil is used. The clay content should *not be high.* If the mix has poor drainage or cracks excessively when drying, add another 10 percent to 20 percent of the soil volume of sand or calcined clay (clay fused by heating to high temperatures).

Soilless Root Media

Although there are many materials available to use, it is convenient to employ only 1 to 3 per mix. In selecting the materials to use, remember that the functions of root media are (1) to support the plant, (2) to maintain good aeration, (3) to retain nutrients, and (4) to retain adequate water. Use organic matter or vermiculite for CEC and coarse organic matter, sand, or perlite for aeration and drainage. Add organic materials or vermiculite (or rockwool) materials to hold water.

As a single material, peat moss is the best, assuming that pH adjustment is done. It has a high CEC, high water and nutrient retention, and can have adequate aeration. However, when it is dry, it can be difficult to wet and may need a wetting agent added to the peat or the water. This problem of wetting dried peat moss can be solved by adding coarse particles, such as sand or perlite, to the peat to provide large pores that allow water intake.

Numerous commercial mixes involve various proportions of peat moss, vermiculite, and sand or perlite. Two long-used combinations are the University of California (UC) mixes and the Cornell mixes. There are five UC mixes ranging from 100 percent peat moss to 100 percent fine sand, with intermediate combinations. Cornell University used only two mixes: Mix A of 50 percent peat moss and 50 percent vermiculite and Mix B of 50 percent peat moss and 50 percent perlite. Some additional popular mixes are shown in Table 21-4. Notice that there are three groupings of the materials. Soil, vermiculite, and rockwool in column 1, peat mosses and bark in column 2, and the bulking (coarse) materials sand and perlite in column 3. Also, a quantity of bark effectively eliminates the need for vermiculite (water retention).

Some growers are interested in using organic materials with more nutrient-supplying capabilities than those in peat moss or bark, both of which are very low. Composts and rotted leaves have been recommended in place of peat moss.[3] A standard mixture is made of 3 parts compost, 1 part vermiculite, and ½ part perlite. Sand could be added for good drainage.

---

[2]Paul V. Nelson, *ibid.*

[3]Heather McCargo, "Compost-Based Potting Soils," *Horticulture* ***LXXIV*** (1996), No. 3; 28–30.

A second mixture (a little more acidic for rhododendrons and other plants growing best in more acidic media) was 1 part rotted bark, 1 part rotted leaves, 1 to 2 parts coarse sand, and ½ part compost. There are many possibilities.

Chemical Amendments

The root medium will require lime and nutrients. Up to 6 kg/m^3 to 8 kg/m^3 of lime to medium may be needed. Phosphorus is added as triple superphosphate (45% P_2O_5) at a rate of about 1.3 kg/m^3. Other phosphate sources can be used at the same rate of P_2O_5. An addition of micronutrients is recommended. A commercial formulation is easiest to use. Some nitrogen as *slow-release* nitrogen could be mixed in also, but nitrogen is often applied in irrigation water.

Because soilless root media are mostly organic or relatively inert, they do not retain phosphorus well. Considerable leaching of phosphates can occur. There are attempts to reduce this leaching because growers soon may face state or federal regulation concerning the chemical content in their wastewaters. One approach has been to add aluminum sulfate or alumina with phosphate already adsorbed.[4, 5] The aluminum-phosphate bond is quite strong and will hold phosphorus from leaching out. It is still experimental and needs to have the concentrations determined that will supply adequate phosphate yet not provide toxic aluminum nor acidify the media too much.

21:4 Pasteurization of Media

The warm, moist (subtropical) conditions of a greenhouse encourage microbial growth, pathogens included. To control diseases most media must be new or be pasteurized at least yearly. The process helps to control weeds, insects, and nematodes. **Pasteurization** is heating to a temperature hot enough only to kill most pathogens. **Sterilization** is heating to hotter temperatures to kill all, or nearly all, organisms.

Steam pasteurization is most widely used. Normally steam temperature is about 100°C, a temperature that will kill most organisms, the beneficial ones as well as the detrimental ones. So, the steam is mixed with cooler air to lower the temperature. Because most pathogens are killed and most viruses denatured at about 60°C (140°F) for 30 minutes, it is common to heat the medium to a slightly higher temperature of 70°C (160°F) for 30 minutes to be sure. The root medium needs to be loosened to allow easy and complete entry and heating. The medium is covered and the air-steam heat (air mixing controls the temperature) is blown into the medium until it is at the desired temperature for 30 minutes.

Chemical pasteurization is cheaper to do unless the steam unit is already available. The most used chemicals are *methyl bromide* and *chloropicrin.*[6] These gases at normal temperature and pressure are *heavier than air,* so if the media is covered (plastic) the gases stay fairly well in the media. **Methyl bromide** treatment is at the rate of about 0.6 kg/m^3 for at least 24 hours (longer for temperature below about 15°C [60°F]). **Chloropicrin** is applied at the rate of about 100 to 150 cc/m^3 for 1 to 3 days, or longer if the temperature is below 21°C (70°F). After chemical treatment, the medium must be well aerated, usually for at least 10 days, to allow the chemicals to dissipate.

[4] Kimberly A. Williams and Paul V. Nelson, "Modifying a Soilless Root Medium With Aluminum Influences Phosphorus Retention and Chrysanthemum Growth," *HortScience* **31** (1996), No. 3; 381–384.

[5] Yuan-ling P. Lin, E. Jay Holcomb, and Jonathan P. Lynch, "Marigold Growth and Phosphorus Leaching in a Soilless Medium Amended with Phosphorus-Charged Alumina," *HortScience* **31** (1996), No. 1; 94–98.

[6] Paul V. Nelson, *ibid.*

Methyl bromide is injurious to humans and safety precautions are essential. It also damages plants, particularly carnations. It has been used in fields (covered after treatment with plastic sheeting) to fight weeds and many root infestations, particularly of root-rot nematodes. Its use was banned in California in 1996 and will be banned throughout the United States by 2001, under the Clean Air Act.[7] This change will be difficult for strawberry growers and fresh-market tomato growers.

Chloropicrin, also known as *tear gas,* is much safer for greenhouse operators because it is extremely irritating even at low concentrations; people will not breath it long without moving away. It is not damaging to carnations, so it is often used on media to be used for carnations. It cannot be used where plants are growing. Other materials being tested as substitutes for methyl bromide include Telone®, methyl iodide, and Dazomet®.

Pasteurization can cause some element toxicities. If root media contain organic matter relatively high in nitrogen (manures, highly decomposed [black] muck or peat, compost), pasteurization can result in rapid ammonium release, accumulating to toxic levels. The ammonification and nitrifying bacteria are mostly killed by sterilization. The ammonifying bacteria repopulate faster than the nitrifying bacteria, thus building up concentrations of ammonium. Perhaps, even the toxic *nitrite* accumulates before rapid nitrification levels are reestablished (see Chapters 6 and 9). Pasteurization can also cause an increase in soluble manganese to toxic levels.

21:5 Watering: A Critical Operation[8]

Watering incorrectly is the most frequent cause of reduced crop quality in greenhouse pot culture. This fact emphasizes the importance of a seemingly simple daily task. Too little water when needed slows growth, results in smaller leaves and plants, and produces a plant with a hardened appearance; yet, overwatering can produce large, succulent plants that easily wilt when set out in bright light or the sun, do not last well, and do not ship well. Excess water can waterlog root areas and cause root death.

Watering Guides

The guides (rules) for watering are intended to emphasize that a watering program needs to wet the entire root volume, allow good drainage, and not water again until the medium has had about half or more of its *available water* used. The following guides result in adequate water, adequate aeration, and control accumulation of soluble salts:

1. Prepare and use a root medium that drains well. Unless the medium drains immediately and well following wetting to allow good aeration, no pattern of watering can solve the aeration problem. Growth will be unsatisfactory.
2. Wet the entire root-medium volume each watering time. Some water should run out the bottom of each pot. This watering pattern is beneficial in at least two ways: (1) The leaching water will remove any accumulating soluble salts (excess fertilizer), and (2) it assures that all of the root media is wetted, not just an upper portion.
3. Irrigate just before drying stress occurs. Let the medium dry out but not to the point of the plant wilting, which will reduce growth. Commonly, about 50%–80% of the *available water* can be used (see Chapters 4 and 16). Some plants darken in color as they approach a condition when they need to be irrigated.

---

[7] Dan Stephens, "Surviving without Methyl Bromide," *American Vegetable Grower* **44** (1996), No. 1; 28–29.
[8] Paul V. Nelson, *ibid.*

Watering Systems

Watering is done both by hand and by automated systems. Watering by hand is time-expensive; often it is done too often or infrequently, and incorrect amounts may be applied. If water sprays are too strong, damage is done to the plants. Wet foliage encourages fungi pathogens and other problems.

Most automated systems use some modification of a drip system (see Chapter 16). The outlet holes vary greatly. Small spaghetti tubing may be led to individual pots (with weighted ends to keep them in each pot), and the flow at time of watering may be much faster than in true drip irrigation; it might even be a small stream. Absorbent mats may be wetted and pots placed on the mats to wet upward by capillarity. Even wick-like sleeves may hang from the supported pot to a nutrient solution 15 cm to 45 cm below the pot to wet the soil without saturating the media. Another example is the use of small pumps to meter fertilizer into a main water line to be distributed to all pots in a certain part of the system.

21:6 Fertilization

The most often limiting nutrients for plants growing in soils are, in decreasing order, *nitrogen, phosphorus,* and *potassium.* Although these three may still be critical, in root media there may be relatively low levels of calcium, magnesium, sulfur, and several micronutrients. In addition, the use of large proportions of peat moss and bark usually assures the need to add lime (dolomitic lime preferred, to ensure adding magnesium). If lime is not needed, gypsum can be used to supply calcium and sulfur, but it will not ensure adequate magnesium.

Needed Nutrients

Each root medium and the crop grown will have different nutrient needs, but a general guide is useful to know about how much has been found satisfactory in many situations. An example guide for applying nutrition into a root medium is given in Table 21-5. Most of the mi-

Table 21-5 General Guide to the Adding of Plant Nutrients to a Root Medium Prior to Use

	Rate Per Cubic Meter*	
Nutrient Source	*Soil-Based Media*	*Soilless Media*
For Calcium and Magnesium and to Adjust pH		
Dolomitic limestone	0–6 kg	6 kg
For Phosphorus		
Triple Superphosphate (0-45-0)	0.9 to 1.0 kg	1 to 1.5 kg
For Sulfur		
Gypsum (calcium sulfate)	0.9 to 1.0 kg	0.9 to 1.2 kg
For Micronutrients: Fe, Mn, Zn, Cu, B, Mo		
Use one of many commercial combinations	Follow directions on the label for each material. Add the same amount to soil-based or soilless media.	
For Nitrogen and Potassium (optional, can be added later)		
Calcium nitrate	0.6 kg	0.6 kg
Potassium nitrate	0.6 kg	0.6 kg

Source: Modified from Paul V. Nelson, *Greenhouse Operation and Management,* Prentice-Hall, Englewood Cliffs, NJ, 1991, 171–208.

*1 $kg/m^3 = 0.0623\ lb/ft^3 = 1.0\ oz/ft^3$

Table 21-6 Generalized Content of Essential Elements in Plants

Elements	*Parts per Million (ppm)*	*Concentration in Dry Tissue, in %*
Carbon	450,000	45
Oxygen	450,000	45
Hydrogen	60,000	6
Nitrogen	15,000	1.5
Potassium	10,000	1.0
Calcium	5,000	0.5
Magnesium	2,000	0.2
Phosphorus	2,000	0.2
Sulfur	1,000	0.1
Chlorine	100	0.01
Boron	20	0.002
Iron	100	0.01
Manganese	50	0.005
Zinc	20	0.002
Copper	6	0.0006
Molybdenum	0.1	0.00001

cronutrient mixes available are slow-release materials with micronutrients in fritted glass (frits), as low-solubility oxides, or impregnated into clay. Micronutrients can even be added in the irrigation water when plants are growing. Table 21-6 illustrates an approximate content of the elements in plants, demonstrating, by the varying amounts of each, the difficulty in assessing a good fertilizer solution that has the right amount of each nutrient available.

Usually nitrogen and potassium are added periodically during growth. Additions to make dilute concentrations of 90 ppm and 250 ppm, respectively, of nitrogen and potassium in the irrigation water each irrigation is one approach. Numerous fertilizer injectors are available (Figure 21-1).[9] Stronger doses could be used less frequently, such as weekly. If cheaper nitrogen sources than calcium nitrate are used, the grower should be sure to provide adequate calcium, either in lime or gypsum or as another soluble calcium carrier (see Chapter 11).

Slow-Release Fertilizers

Numerous slow-release fertilizers are available (see Chapter 11). They include plastic-encapsulated materials, urea-formaldehyde polymers, sulfur-coated materials, and chelated micronutrients (actually, the chelates are soluble). These can be put into the root medium and will release the nutrients slowly. By different thicknesses of coatings on granules, or other techniques, the materials can be formulated to release most of their nutrients by a certain time period. Release within the first 3 to 4 months is common, but some materials can take 6 to 15 months to release most of it.

Plastic-coated dry fertilizers are widely used. Water enters into the capsules, dissolving the fertilizer, and eventually the osmotic pressure inside cracks the plastic capsule and allows the dissolved fertilizer to move into the root zones. Names[10] of some of these materials are Osmocote®, Sierra®, and Nutricote®.

[9]Gerald Klingaman, "On-the-Fly 'Fertigation'," *American Nurseryman* **184** (1996), No. 2; 70–77.

[10]Use of trade names does not suggest endorsement of the product or lack of endorsement of other products not named. Trade names are given only for clarity and illustration.

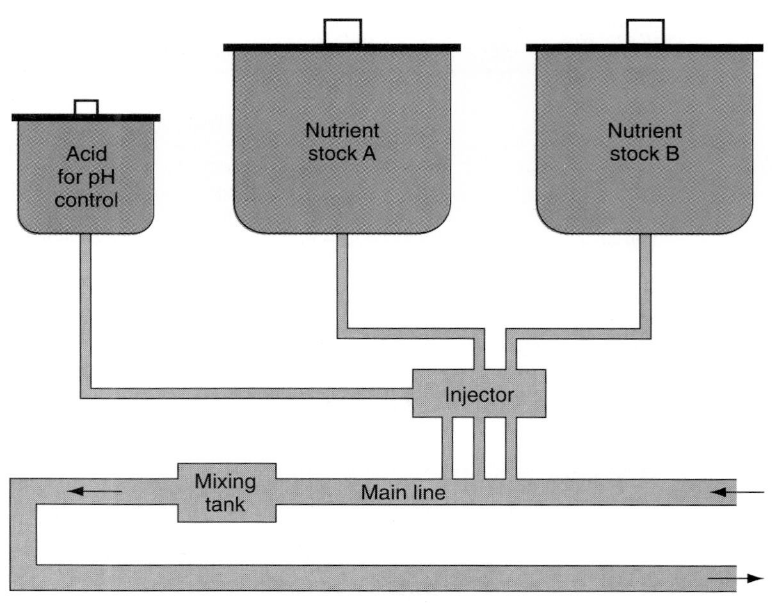

A basic nutrient injector system

FIGURE 21-1 A typical nutrient injection system. Nutrient stock A and stock B are to keep phosphorus and the ions that form precipitates with it (iron, calcium, others depending on concentration) separated. The nutrient feed lines could be divided into many spaghetti lines to individual pots or used in other ways. (*Source:* Modified and redrawn from Howard M. Resh, *Hydroponic Food Production,* Woodbridge Press Publishing Company, Santa Barbara, CA, 1995, 527 pp.)

Urea-formaldehyde materials supply only nitrogen, so are less used. Trade names include Ureaform® and Uramite®.

Sulfur-coated materials use a coating of molten sulfur plus a wax-like sealant. The thickness of the sulfur controls the rate of nutrient release. Microorganisms must oxidize (thus dissolving) the sulfur to release the nutrients inside. These materials can be used to supply any fertilizer mix but are used mostly to supply nitrogen, phosphorus, potassium, and sulfur.

21:7 Soil and Tissue Testing

A monitored fertilizer program that makes chemical nutrient tests of the media is basic to ensure that essential nutrients are not limiting plant growth and, conversely, are not in excess, creating an imbalanced or a toxic soil condition. Many problems in greenhouse production are related to soil fertility.

Sampling the media is done very much as described in Chapter 12. Sampling may be more frequently done in greenhouse media, even prior to each crop, or every 3 months or when a problem is evident. Sample to the depth of the pot or to 30 cm, whichever is reached first. Some peat and bark are not easily sampled with sampling tubes and sampling may need to be done with narrow spatulas or trowels. Wait at least 6 hours after watering or at least 5 days after applying dry fertilizer. Delay sampling 24 hours when fertilizer has been applied through a liquid injector system.

Soil Nutrient Levels

Table 21-7 shows recommended nutrient levels in soils to be used for floricultural crops. Levels considered to be deficient, optimum, or excessive (toxic) are shown for nitrate-nitrogen,

Table 21-7 General Guidelines for Greenhouse Soil Nutrient Levels and Their Interpretation*

Nutrient	Deficient	Optimum	Excessive
	(ppm in Media Extract)		
Nitrate-nitrogen	Below 39	100–279	Above 280
Phosphorus	Below 3	8–13	Above 20
Potassium	Below 59	150–249	Above 350
Calcium	Below 79	200–349	Above 500
Magnesium	Below 29	60–99	Above 150

Source: D. D. Warncke, "Testing Greenhouse Growing Media: Update and Research," Crop and Soil Sciences Department, Michigan State University, presented at the Seventh Soil-Plant Analysts' Workshop, Bridgeton, MO, 1979.
*Based on a saturated media extraction procedure.

Table 21-8 Soluble Salt Levels (Electrical Conductivity) in Soils for Greenhouse and Bedding Plant Production in Relation to Plant Growth

Conductivity Readings (dS/m)			
Saturated Media Extract	2:1 (Weight Basis) Water: Soil	5:1 (Volume Basis) Water: Solid	
All Soils	Mineral Soils	Organic Soils	Relation to Plant Growth
0–0.74	0–0.25	0–0.12	Very low salt levels; indicates very low nutrient status
0.75–1.99	0.25–0.50	0.12–0.35	Suitable range for seedlings and salt-sensitive plants
2.00–3.49	0.50–1.00	0.35–0.65	Desirable range for most established plants; upper range may reduce growth of some sensitive plants
3.50–4.99	1.00–1.50	0.65–0.90	Slightly higher than desirable; loss of vigor in upper range; suitable for high nutrient-requiring plants
5.00–5.99	1.50–2.00	0.90–1.10	Reduced growth and vigor; wilting and marginal leaf burn
6.00+	2.00+	1.10+	Severe salt symptoms—wilting; crop failure

Source: "Chemical Controls for Michigan Commercial Greenhouse and Bedding Plant Production," Extension Bulletin E-1275, Cooperative Extension Service, Michigan State University, East Lansing, 1978.

phosphorus, potassium, calcium, and magnesium. The nutrient values are expressed in parts per million (ppm or mg/kg) in the medium extract based on a saturated-medium extract procedure.

Soluble-Salt Problems

Any soluble fertilizer is also a soluble salt. When heavy fertilizer applications are used, their action as soluble salts must be considered. The soluble salt may be a combination of fertilizers and of other salts added in irrigation waters (see Chapters 9 and 16). The evaluation of the salt hazard in greenhouse soils is approximately as shown in Tables 21-8 and 21-9. Plants differ in their sensitivity to salt, so simple salt-tolerance tables such as these will always have some plants more tolerant and a few less tolerant than are listed. Unfortunately,

Table 21-9 Salt Tolerance of Selected Ornamental Shrubs, Trees, and Ground Cover

Common Name	*Botanical Name*	*Maximum Permissible EC_e in dS/m*
Very sensitive		
Star jasmine	*Trachelospermum jasminoides*	1–2
Pyrenees cotoneaster	*Cotoneaster congestus*	1–2
Oregon grape	*Mahonia aquifolium*	1–2
Sensitive		
Chinese holly, cv. Burford	*Ilex cornuta*	2–3
Rose, cv. Grenoble	*Rosa sp.*	2–3
Southern yew	*Podocarpus macrophyllus*	2–3
Tulip tree	*Liriodendron tulipifera*	2–3
Japanese pittosporum	*Pittosporum tobira*	3–4
Chinese hibiscus	*Hibiscus rosa-sinensis*	3–4
Crape myrtle	*Lagerstroemia indica*	3–4
Moderately sensitive		
Glossy privet	*Ligustrum lucidum*	4–6
Orchid tree	*Bauhinia purpurea*	4–6
Southern magnolia	*Magnolia grandiflora*	4–6
Japanese boxwood	*Buxus microphylla var. japonica*	4–6
Japanese black pine	*Pinus thunbergiana*	4–6
Moderately tolerant to tolerant		
Oleander	*Nerium oleander*	6–8
Spindle tree, cv. Grandiflora	*Euonymus japonica*	6–8
Rosemary	*Rosmarinus officinalis*	6–8
Sweet gum	*Liquidamabar styraciflua*	6–8
Ceniza	*Leucophyllum frutescens*	>8
Natal palm	*Carissa grandiflora*	>8
Bougainvillea	*Bougainvillea spectabilis*	>8
Very tolerant		
Iceplant	*Delosperma* and *Lampranthus*	>10

Selected plants from J. D. Rhoades, A. Kandiah, and A. M. Mashali, *The Use of Saline Waters for Crop Production,* FAO Irrigation and Drainage Paper 48, Food and Agriculture Organization of the United Nations, Rome, 1992, 32–33.

overwatering to ensure regular leaching of any accumulated salt also removes much of the soluble fertilizers from the root medium, thereby lowering available nutrients until more are added.

Even some other elements can be toxic to plants if their content in the water is too high. Boron is an element that is a problem in a few waters in arid regions. A list of plants with sensitivity to boron is in Table 21-10.

It is now possible to sample and analyze plants quickly, so that fertigation nutrient levels can be corrected frequently. The plant part to sample is discussed in Chapter 12 and some examples are given in Table 12-12; the *critical levels* of the nutrients in several flower crops are given in Table 12-14. Methods of treating the sampled plant (washing, drying) are also discussed in Chapter 12.

It is often a help to recognize visual deficiency symptoms to help in diagnosis of a nutrient deficiency problem. (An abbreviated outline of visual symptoms is given in Table 12-15.) Because many different plants have some unique coloring and odd patterns for certain deficiencies, it is helpful to look up deficiency symptoms for individual species.

Table 21-10 Boron Tolerances for Ornamentals (Scientific Names Omitted)

Common Name	Threshold, mg/L	Common Name	Threshold, mg/L
Very sensitive		**Moderately sensitive**	
Oregon grape	<0.5	Gladiolus	1.0–2.0
Photinia	<0.5	Marigold	1.0–2.0
Xylosma	<0.5	Poinsettia	1.0–2.0
Wax-leaf privet	<0.5	China aster	1.0–2.0
Spindle tree	<0.5	Gardenia	1.0–2.0
Japanese pittosporum	<0.5	Southern yew	1.0–2.0
Juniper	<0.5	Ceniza	1.0–2.0
Sensitive		**Moderately tolerant and tolerant**	
Zinnia	0.5–1.0	Bottlebrush	2.0–4.0
Pansy	0.5–1.0	California poppy	2.0–4.0
Violet	0.5–1.0	Oleander	2.0–4.0
Larkspur	0.5–1.0	Chinese hibiscus	2.0–4.0
Glossy abelia	0.5–1.0	Sweet pea	2.0–4.0
Rosemary	0.5–1.0	Carnation	2.0–4.0
Oriental arborvitae	0.5–1.0	Indian hawthorn	6.0–8.0
Geranium	0.5–1.0	Oxalis	6.0–8.0

Selected plants from J. D. Rhoades, A. Kandiah, and A. M. Mashali, *The Use of Saline Waters for Crop Production,* FAO Irrigation and Drainage Paper 48, Food and Agriculture Organization of the United Nations, Rome, 1992, 36.

21:8 Hydroponics: Culture in Solution

Hydroponics is a term used for the technique of supplying to plants nutrients dissolved in water. There may or may not be a solid support matrix in which roots may anchor themselves. Actually, the nutrients in soils and in many root media are in solution, too, before roots absorb them; thus, the major difference in hydroponic and soil or root media culture is that all nutrients in hydroponics are supplied as soluble chemicals in a water solution prepared and balanced for the plant. Using inert root media without soil, manure, or rotted organic material is essentially hydroponics with a solid matrix in which plant roots can grow.

The primary decisions that need to be made in hydroponics include (1) how best to keep nutrients in balanced and favorable concentrations, (2) how to physically support the plants, (3) how to keep the solution adequately aerated so roots have O_2, and (4) how to provide sufficient light for the plant density used. Temperature is important, too, but in a controlled environment it is assumed that temperature is controlled as desired.

The nutrient solution selected is undoubtedly important, but there is not agreement on a single solution (Table 21-11). Hoagland's solution has been a sort of classical *basic solution* for several decades. One author lists forty different solution compositions by various workers or for various stages of growth.[11]

Preferred solution compositions vary depending on the kind of plants grown and the stage of growth (maturity) of the plants. Many factors additionally can alter the solution composition chosen. One list of approximate amounts of each nutrient needed is given in Table 21-12 . The chemical formulas (carriers) used to supply these are only important in selecting adequately soluble materials. Table 21-13 shows the three solutions used for *seedling, vegetative growth,* and *seed fill* for wheat to be grown in space.[12]

[11]Howard M. Resh, *Hydroponic Food Production,* Woodbridge Press Publishing Company, Santa Barbara, CA, 1995, 80–81.
[12]Bruce Bugbee, Utah State University hydroponics Web page, "Current Nutrient Solution: Wheat—January 1996," http://www.usu.edu/~cpl/nutrwht.html.

Table 21-11 Comparison of Half-Strength Hoagland Solution with Utah Space Wheat Solutions Developed for Use in Space Flight

	Hoagland Solution	*Utah Wheat Solutions**		
		Starter Solution	*Vegetative Refill*	*Seed Fill Refill*
mM				
N	7.5	3	6	3
P	0.5	0.5	0.5	0.5
K	3	1.5	4.5	2.5
Ca	2	1	1	0.5
Mg	1	0.5	0.3	0.3
S	1	0.5	0.3	0.3
μM				
Fe	44.6	10	2.5	2.5
Fe-HEDTA	0	25	5	5
Mn	4.5	3	6	3
B	23	2	1	0.2
Zn	0.4	3	1	1
Cu	0.15	0.3	0.3	0.2
Mo	0.05	0.09	0.03	0.03
Cl	9	6	12	6
Si	0	100	100	0

Source: Bruce Bugbee, "Nutrient Management in Recirculating Hydroponic Culture," Internet, http://www.usu.edu/~cpl/hsapaper.html.

*The wheat growth chamber is initially filled with the *starter* solution. Nitric acid is used for pH control and about half of the nitrogen requirement is supplied in the pH control solution. Ammonium nitrate could be added to the pH control solution if wanted for higher levels of nitrogen. Ammonium reduces the uptake of other cations, so use it only if necessary.

Table 21-12 Approximate Amounts of the Nutrients Needed in Hydroponic Nutrient Solutions

Nutrient	*Amount*	*Nutrient*	*Amount*	*Nutrient*	*Amount*
Ca	200 ppm	N as NH_4^+	25 ppm	Cu	0.1 ppm
Mg	40 ppm	N as NO_3^-	165 ppm	Zn	0.1 ppm
Na	—	P as HPO_4^-	50 ppm	B as H_3BO_3	0.5 ppm
K	210 ppm	S as SO_4^{2-}	113 ppm	Mo as MoO_4^{2-}	0.05 ppm
		Fe	5 ppm	Cl	—

Using hydroponics in a limited-space, limited-weight, no-waste system, as in a space station, requires a careful nutrient balance tailored to the specific crop (Detail 21-1). The space program for wheat growth is attempting to develop hydroponics so that the solution can be balanced as needed. It will not be possible up there to throw out any unbalanced, used solution and start over, as is possible in a greenhouse. The solution must be changed by modifications that are reversible or otherwise allow reuse. Figures 21-2 and 21-3 illustrate some of the chambers and root growth obtained in these simple hydroponic growing chambers in inert solid media on Earth.

With computer-controlled injectors and more than one **stock solution** (concentrated nutrient solutions), the nutrients can be metered into the circulating solution as needed. The return solution can be monitored automatically for certain nutrients and the changes made by

Table 21-13 Complete Nutrient Solution for Three Stages of Growth in the Wheat Plant: Starter (Germination, Early Couple of Weeks), Pre-Anthesis (Major Size Growth Period), and Post-Anthesis (Seed Filling Stage). Notice the Salts Used as Nutrient Carriers.

Current Nutrient Solution: Wheat—April, 1996

		Starter		Pre-Anthesis		Post-Anthesis	
Salt	Stock Conc.	mL per 100 L	Final Conc.	mL per 100 L	Final Conc.	mL per 100 L	Final Conc.
$Ca(NO_3)_2$	1 M	100	1 mM	100	1 mM	50	0.5 mM
$K(NO_3)$	2 M	50	1 mM	200	4 mM	100	2 mM
KH_2PO_4	0.5 M	100	0.5 mM	100	0.5 mM	100	0.5 mM
$MgSO_4$	0.25 M	200	0.5 mM	200	0.5 mM	100	0.25 mM
K_2SiO_3	0.1 M	100	0.1 mM	100	0.1 mM	0	0 mM
$Fe(NO_3)_3$	50 mM	20	10 μM	5	2.5 μM	5	2.5 μM
Fe-HEDTA	100 mM	25	25 μM	5	5 μM	5	5 μM
$MnCl_2$	60 mM	5	3 μM	10	6 μM	5	3 μM
$ZnSO_4$	20 mM	20	4 μM	10	2 μM	10	2 μM
H_3BO_3	20 mM	10	2 μM	5	1 μM	2	0.2 μM
$CuSO_4$	20 mM	5	1 μM	5	1 μM	3	0.6 μM
Na_2MoO_4	0.6 mM	15	0.09 μM	5	0.03 μM	5	0.03 μM

Add HNO_3 or KOH as needed to control pH to about 5.6.

Source: Tim Grotenhuis, "Current USU Nutrient Solutions for Wheat," Internet, http://www.usu.edu/~cpl/nutrwht.html, April 26, 1996.

computer to keep nutrients somewhat balanced and at the near-planned concentrations. Stock solutions may be 50, 100, or 200 times the concentrations wanted in the dilute circulating solution. The stock solution concentration may be limited by the solubility of the chemicals used. The solid monocalcium phosphate and treble superphosphates are of low solubility. Because phosphate with iron and calcium form insoluble phosphates, phosphate is put into one stock solution and calcium and iron are put in a second stock solution (see Figure 21-1). Usually potassium, magnesium, and sulfur can be put into either or both stock solutions.

The differential rate of uptake of the various nutrients causes an imbalance in the solution. These variable uptakes are caused by different kinds of plants, stage of plant maturity, and environmental conditions, such as temperature, light intensity, and relative humidity. If the nutrient balance cannot be adequately monitored, the solution must periodically be discarded. The plant tissue can be analyzed to see if the tissue is within the adequate range (see Chapter 12). Each kind of plant has its own *critical* values. Some values for three vegetables are given in Table 21-14.

Plant Support

Plants can be supported by a solid support fixed above the water or attached to a solid base that will float on the water. They can also be grown in a porous, solid media (perlite, sand, etc.) through which nutrient solution percolates periodically. If the plant is supported so that roots grow in only solution, the solution must be aerated to provide adequate O_2 for the roots. This technique is often used in research work where roots will be analyzed and researchers need the roots to be free of soil or other materials.

Most hydroponics for commercial vegetables use a soilless medium, such as foam, gravel, sand, rockwool, sawdust, peat, perlite, pumice, peanut hulls, polyester netting, or

Detail 21-1 Wheat in Hydroponics in Space

For long stays in space and in space stations, it will be desirable to produce some of the needed food and at the same time recycle as many nutrients as possible. The recycling of wastes is still not part of this study. 'Apogee' wheat will likely be grown on the International Space Station scheduled for completion in 2002. Some important characteristics of Apogee are the following:

1. It is resistant to the severe leaf-tip chlorosis that is common in wheat under rapid growth, especially in continuous light. This problem often kills up to 30% of the leaf.
2. It has extremely rapid development rate, and heads emerge 23 days after seedling emergence.
3. It has better yields and equal bread-making qualities to competitive short-straw wheats.
4. It produces the equivalent of nearly 600 bu/a (37,700 kg/ha) on 45-cm-tall wheat.

The wheat in the study is grown in a new substrate to allow easy separation of roots from substrate for analysis and study. The medium used is a diatomaceous-earth material called **Isolite.** It is mined off the Japanese coast, mixed with 5% clay, and baked. This material is mostly silica, is relatively inert, and has not usually needed pasteurization for reuse. It is expensive ($1.22/L or $1.74/kg).

With nearly 24 hours a day of sunlight and increased CO_2 atmosphere, the plant yields will be abnormal compared to yields in fields on Earth. Although Apogee wheat seems to do well in the conditions in space, on Earth it is too short to combine (harvest) well or to compete with weeds.

Courtesy of Bruce Bugbee, Utah State University.

FIGURE 21-2 Apogee® wheat showing the root growth when grown in hydroponic nutrient solution in experiments in preparation for growing wheat in space, in the U.S. Space program. (Courtesy of Dr. Bruce Bugbee, Crop Physiology Laboratory, Utah State University.)

FIGURE 21-3 Small control chambers to test effects of carbon dioxide levels and nutrient solutions on the growth of wheat in anticipation of experiments in growing wheat in space in the near future. (Courtesy of Dr. Bruce Bugbee, Crop Physiology Laboratory, Utah State University.)

Table 21-14 Concentration Ranges for Nutrients in Healthy Tomatoes, Cucumbers, and Lettuce

Element	*Tomatoes*	*Cucumbers*	*Lettuce*
N, %	4.5–5.5	5.5–6.0	3.0–6.0
P, %	0.6–1.0	0.7–1.0	0.8–1.3
K, %	4.0–5.5	4.5–5.5	5.0–10.8
Ca, %	1.5–2.5	2.0–4.0	1.1–2.1
Mg, %	0.4–0.6	0.5–1.0	0.3–0.9
Fe, ppm	80–150	100–150	130–600
B, ppm	35–60	35–60	25–40
Mn, ppm	70–150	60–150	20–150
Zn, ppm	30–45	40–80	45–90
Cu, ppm	4–6	5–10	7–17
Mo, ppm	1–3	1–3	1–4
N/K ratio	0.9–1.2	1.0–1.5	—

Source: Modified and selected data from Howard M. Resh, *Hydroponic Food Production,* Woodbridge Press Publishing Company, Santa Barbara, CA, 1995, 108.

vermiculite. These materials have different costs and will retain different amounts of nutrient solution for the roots. Growers usually select a medium on the basis of *cost* and *solution retention.* A third concern with organic products as a medium is whether or not they have, or will produce, toxic materials. Sawdust may have soluble salts from being held as a log in saltwater. Other organic materials could have toxic products of decomposition.

Root Aeration and Darkness

Aeration in solution can be done by bubbling air through the nutrient solution. If the solution is circulated for reuse frequently or continuously, baffles at the ends of the beds over which the water tumbles will aerate the water. One or two solution changes per hour is required for about 30 m (about 100 ft) of bed kept 10 cm to 15 cm deep with nutrient solution. To keep out algae that would compete with the roots, the root system should be kept in darkness.

Nutrient Film Technique

Nutrient film technique (NFT) is the growing of plants with their roots in a plastic tube or trough through which a thin layer of water is continuously circulated. The water depth must be very thin to maintain good aeration without special techniques. The system is not air-tight; air-tight systems often build up ethylene that damages growth. The trough may be just a narrow length of black plastic pulled up around the seedlings (and stapled to make a semi-tube). The seedlings are started in peat pellets or other small root-holding blocks (Figures 21-4 and 21-5). Also, PVC tubing with holes for seedlings is used. With the plastics now available, the combinations of NFT are limited only by the imagination (Figure 21-6). For troughs PVC pipe, rain gutters, and plastic sheeting have been used. The important guides are the following:

- Start the seed in a small block to support the early root system (peat pellets, porous pots, rockwool, or other cohesive block). Then, move the seedling to the NFT trough for growth.
- Use a constant, shallow layer of flowing nutrient solution from which roots take water and nutrients.
- Have at least a 1:25 slope (one unit vertical to 25 units horizontal) to be sure solution flows and drains well. Use a trough that keeps most of the roots in the dark.
- A relatively flat trough bottom distributes the solution and roots better. Capillary matting works well on flat trough bottoms to hold more solution available for plants to be grown to maturity in the system.

Using Solid Media

As a support for plants growing to maturity in a bed, several substrates are used. Gravel was very popular in the past four decades. Sands are used, too. Any coarse material that drains well could be used, such as sawdust, bark, rockwool, synthetic foams, peat, vermiculite, perlite, and pumice. These become large-scale (bed) plantings in the same mixtures previously mentioned as soilless potting media for potted greenhouse plants. In beds with the solid, porous media, drain lines to move the solution well, and aid aeration as solution is drained away, are usually needed. The same, perhaps more severe, problems of pasteurization are encountered in the bed culture. Between crops bed media may need treatment for pathogen control.

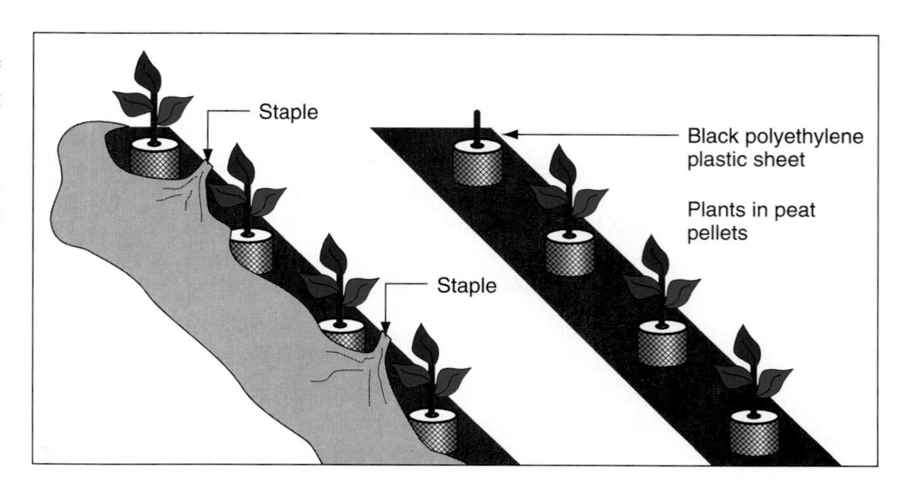

FIGURE 21-4 Nutrient film technique (NFT) using a simple strip of black polyethylene plastic, folding it up over plants started in peat pellets and stapling to keep it in place. The plastic unit is then placed on some inclined support and a thin layer of nutrient solution run along the bottom. (*Source:* Modified and redrawn from Howard M. Resh, *Hydroponic Food Production,* Woodbridge Press Publishing Company, Santa Barbara, CA, 1995, 527 pp.)

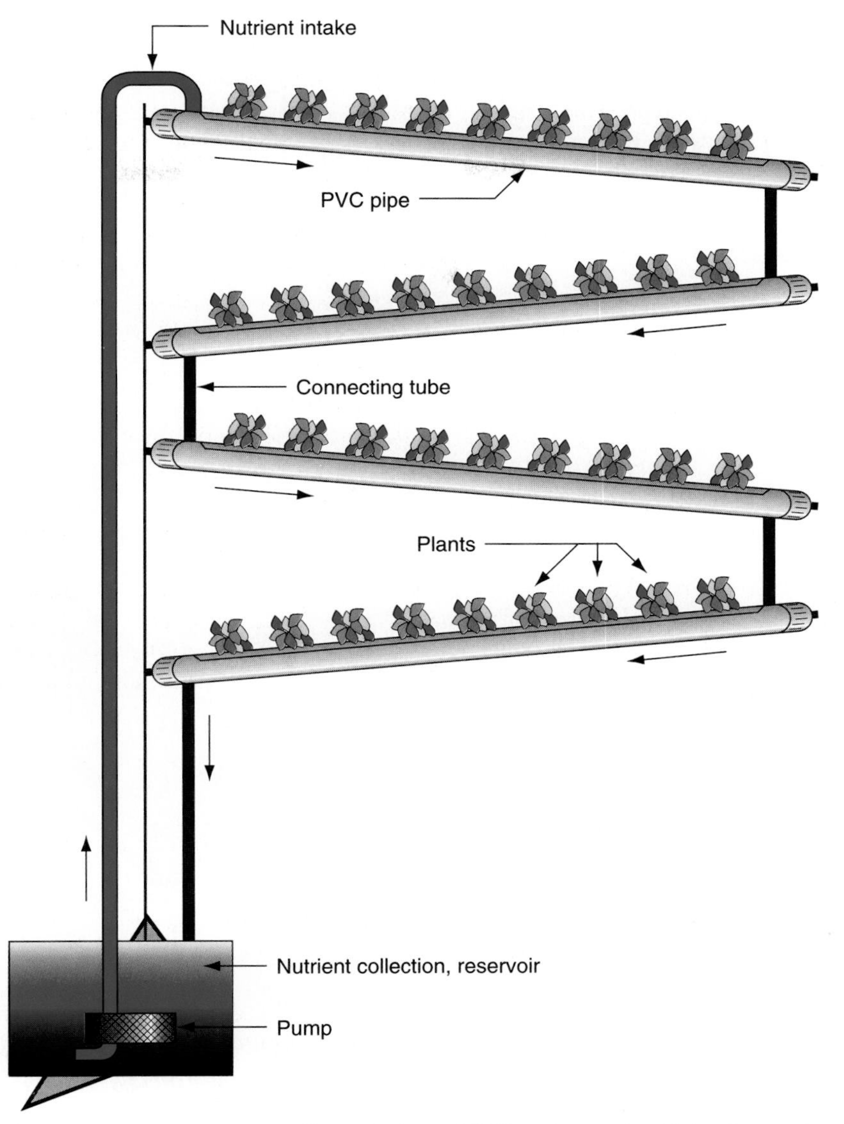

FIGURE 21-5 A cascade-style nutrient film technique (NFT) system using PVC pipe and a recycling system for the nutrient solution. The plant roots can be covered with black plastic to eliminate light in the root area. (*Source:* Modified and redrawn from Howard M. Resh, *Hydroponic Food Production,* Woodbridge Press Publishing Company, Santa Barbara, CA, 1995, 527 pp.)

FIGURE 21-6 A unique nutrient film technique (NFT) using a capillary matting sheet into which the nutrient solution can move and keep the young plant moist for germination and early, as well as later, growth. (*Source:* Modified and redrawn from Howard M. Resh, *Hydroponic Food Production,* Woodbridge Press Publishing Company, Santa Barbara, CA, 1995, 527 pp.)

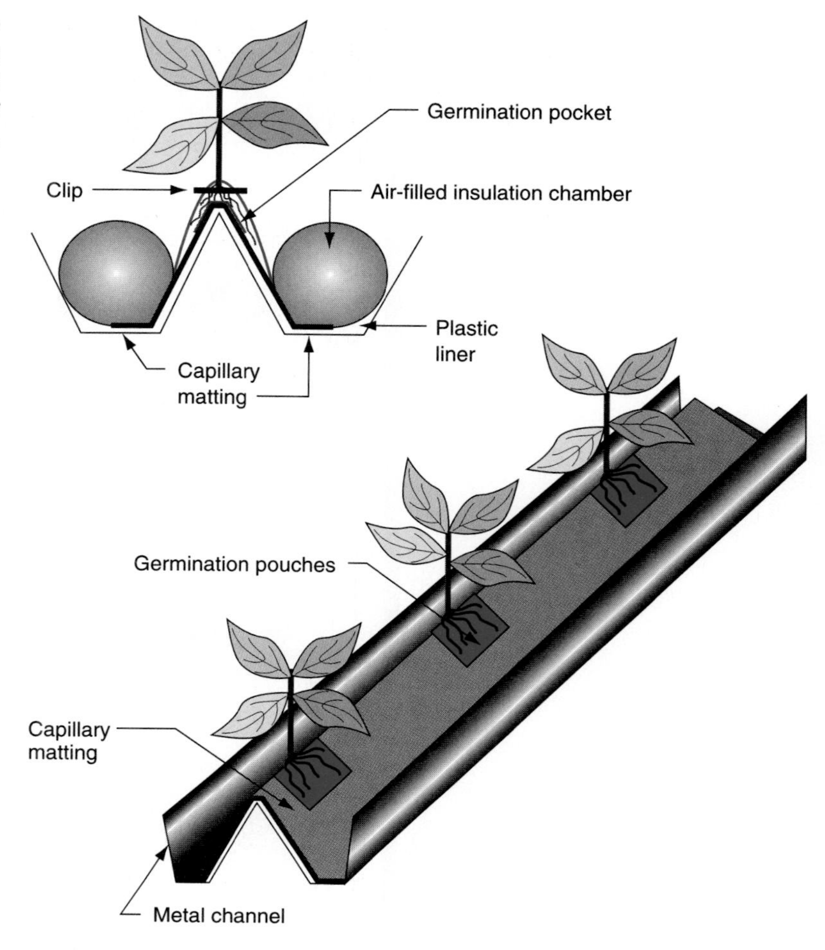

We know two things about the future: It cannot be known and it will be different from what now exists and from what we now expect.

—Peter Drucker

Questions

1. How do small pots affect root media aeration?
2. What is necessary to ensure that root media is adequately drained and aerated?
3. Describe the composition and characteristics of *peat moss, bark, perlite, vermiculite,* and *rockwool.*
4. List several root media and indicate why each material is used (which properties did it provide to the soil?).
5. What are the properties that a root medium should have?
6. What does each material used for root media provide that helps that media?
7. Describe a preferred watering schedule (frequency).
8. Which nutrients are added to the root media or later to the plant and how are they added?

9. Do most root media need lime? How much and why?
10. Why are root media pasteurized and how often must it be done?
11. How are soils or soilless root media pasteurized?
12. Describe the techniques of sampling greenhouse soils and plants.
13. What is nutrient film technique (NFT) and how is it done?
14. What is the general cultural system being used in preparation for growing wheat in space?

Appendix A: Models and Databases: Types and Sources

Models of the world seek to provide causal explanations for patterns and relations observed in data. . . . Data is a prod to our thinking.

—Kamran Parsaye and Mark Chignell

1. Water Flow, Irrigation, and Drainage

CIRF. Conceptual Irrigation Return Flow hydrosalinity model. **Sources:** K. K. Tanji, "A Conceptual Hydrosalinity Model for Predicting Salt Load in Irrigation Return Flow," *Managing Saline Water for Irrigation,* Texas Technical University, 1977, 419–470; and R. Aragues, K. K. Tanji, D. Quilez, and J. Faci, "Conceptual Irrigation Return Flow Hydrosalinity Model, Chapter 24, in Kenneth K. Tanji (ed.) *Agricultural Salinity Assessment and Management,* American Society of Civil Engineers, Publishers, 345 East 47th Street, New York, NY 10017, 1990, 504–529.

COMAX. An aid to irrigation timing and quantity of water. Cotton management expert (fertilization and irrigation). **Source:** Y. W. Jame and H. W. Cutforth, "Crop Growth Models for Decision Support Systems," *Canadian Journal of Plant Science* **76** (1996), 9–19.

CROPWAT. Program to calculate irrigation requirements and generate irrigation schedules, version 5.7, Oct. 1991. Linkage to CLIMWAT, a database. **Source:** Water Resources, Development, and Management Service/AGLW FAO, Via delle Terme di Carcella, 00100 Rome, Italy.

Note: Mention of a vendor or proprietary product does not constitute a guarantee or warranty of the vendor or product by the authors or publisher and does not imply its approval to the exclusion of vendors or products that may also be suitable but are not listed here. This list of models is not exhaustive; it is intended to be a starting point for those interested in gaining more information. The information here emphasizes the important role that models now have in general science and day-to-day operations in soils.

General Reference for Serious Modelers: John Hanks and J. T. Ritchie, co-editors, *Modeling Plant and Soil Systems,* No. 31 in the series **AGRONOMY,** American Society of Agronomy, Inc., publishers, 677 South Segoe Road, Madison, WI, 1991, 545 pp.

DRAINMOD. Developed to aid in the design and evaluation of shallow-water management systems. **Source:** R. W. Skaggs, "A Water Management Model for Shallow Water Table Soils," *Report No. 134,* Raleigh, NC, Water Resources Research Institute, University of North Carolina, 1978.

PET. A model to predict evaporation for row crops. **Source:** J. T. Ritchie, "A Model for Predicting Evaporation for a Row Crop with Incomplete Cover," *Water Resources Research* **8** (1972), No. 5; 1204–1213.

PREFLOW. To simulate macropore flow of water from the soil surface to the water table. **Source:** S. R. Workman and R. W. Skaggs, "Development and Application of a Preferential Flow Model PREFLOW," *Transactions of the ASAE* **34** (1991), No.5; 2053–2059.

SWATREN. A drainage simulation model for fluctuating water table. **Source:** S. R. Workman and R. W. Skaggs, "Comparison of Two Drainage Simulation Models Using Field Data," *Transactions of the ASAE* **32** (1989), 1933–1938.

2. Chemical Movement in Soils

CHEMISTRY. **Source:** R. J. Wagenet and J. L. Hutson, "Leaching Estimation and CHEMISTRY Model: A Process-Based Model of Water and Solute Movement, Transformations, Plant Uptake, and Chemical Reactions in the Unsaturated Zone," *Continuum* **2** (Version 2), Center for Environmental Research, Cornell University, Ithaca, NY.

CMLS. Simulating chemical movement in layered soils. **Sources:** D. L. Nofziger and A. C. Hornsby, "Chemical Movement in Layered Soil: Users Manual," *Florida Cooperative Extension Service Circular* **780,** Institute of Food and Agricultural Science, University of Florida, Gainesville, 1987; and John P. Wilson, William P. Inskeep, Jon M. Wraith, and Robert D. Snyder, "GIS-Based Solute Transport Modeling Applications: Scale Effects of Soil and Climate Data Input," *Journal of Environmental Quality* **25** (1996), 445–453.

GLEAMS. A one-dimensional deterministic conceptual root-zone model simulating flow and nitrate transport. **Source:** D. Knisel (ed.), *GLEAMS Manual* 2.10, version UGA-CPES-BAED, Publ. 5, University of Georgia, Coastal Plain Experimental Station, Tifton, GA.

LEACHM. Water and solute movement, transformations, and chemical reactions in unsaturated soil. **Sources:** R. J. Wagenet and J. L. Hutson, "LEACHM: Leaching Estimation and Chemistry Model," Centre for Environmental Research, Cornell University, Ithaca, NY; and R. J. Wagenet and J. L. Hutson, "Scale-Dependency of Solute Transport Modeling/GIS Applications," *Journal of Environmental Quality* **25** (1996), 499–510.

LEACHP. Similar to LEACHM but for pesticide movement. **Source:** R. J. Wagenet and J. L. Hutson, "Scale-Dependency of Solute Transport Modeling/GIS Applications," *Journal of Environmental Quality* **25** (1996), 499–510. Also, **LEACHA.**

NLEAP. Nitrate Leaching and Economic Analysis Package to help determine possible nitrate leaching problems. **Source:** Marvin J. Shaffer, USDA—ARS, Ft. Collins, CO; and James D. Kaap, USDA Natural Resources Conservation Service, Janesville, WI.

N-Show. Educational program displaying graphs of nitrogen in soil, using results generated by CERES—Maize. **Source:** Miguel L. Cabrera, "N-Show: An Educational Computer Program That Displays Dynamic Graphs of Nitrogen in Soil," *Journal of Natural Resources and Life Science Education* **23** (1994), No. 1; 43–45.

PRZM. A pesticide root-zone model to evaluate leaching threats. **Source:** R. F. Carsel, L. A. Mulkey, M. N. Lorber, and L. B. Baskin, "The Pesticide Root-Zone Model (PRZM): A Procedure for Evaluating Pesticide Leaching Threats to Groundwater," *Ecology Models* **30** (1985), 46–69.

RZWQM. Perhaps the most comprehensive in use for evaluating chemical movement in the vadose zone of agricultural soils. **Source:** Agricultural Research Services, "Root Zone Water Quality

Model (RZWQM) V. 1.0," Technical Documentation, Great Plains Systems Research Unit, USDA—ARS, Fort Collins, CO.

STELLA II. Teaching programs that allow graphic modeling of systems without an in-depth knowledge of computer languages. **Source:** H. A. Torbert, M. G. Huck, and R. G. Hoeft, "Simulation of Soil-Plant Nitrogen Interactions for Educational Purposes," *Journal of Natural Resources and Life Science Education* **23** (1994), No. 1; 35–42.

SUNDIAL. A simulation for nitrogen dynamics (turnover) in arable lands. Copies available from J. U. Smith (send disk), Dept. Soil Sci., Institute of Arable Crop Res., Rothamsted Exper. Stn., Harpenden, Herts, AL5 2JQ, United Kingdom. **Source:** J. U. Smith, N. J. Bradbury, and T. M. Addiscott, "SUNDIAL: A PC-Based System for Simulating Nitrogen Dynamics in Arable Land," *Agronomy Journal* **88** (1996), 38–43.

3. Physical Properties of Soils

Soil Compaction (no title given). User-friendly simulation of soil compaction caused by farm-machinery traffic. **Source:** C. Plouffe, S. Tessier, D. A. Angers, and L. Chi, "Computer Simulation of Soil Compaction by Farm Equipment," *Journal of Natural Resources and Life Science Education* **23** (1994), No. 1; 27–34.

4. Water Resources Inventory

SWAT. A soil and water assessment tool. **Source:** J. G. Arnold, P. M. Allen, and G. Bernhardt, "A Comprehensive Surface-Groundwater Flow Model," *Journal of Hydrology* **142** (1993), Nos. 1 and 4; 47–69.

SWRRB. A simulator for water resources in rural basins. **Source:** J. G. Arnold, J. R. Williams, A. D. Nicks, and N. B. Sammons, *SWRRB, A Basin Scale Simulation Model for Soil and Water Resources Management,* College Station, Texas A & M University Press, 1990.

5. Evaporation and Climatic Data

CLIGEN. Generates daily values for precipitation amount, duration, maximum intensity, time to peak intensity, temperatures, solar radiation, dew point temperature, wind speed and direction. **Source:** A. D. Nicks and L. J. Lane, "Weather Generator," in L. J. Lane and M. A. Nearing (eds.) *USDA—Water Erosion Prediction Project: Profile Model Documentation,* NSERL/Report No. 2, USDA—ARS—National Soil Erosion Research Lab., W. Lafayette, Indiana, Chapter 2, 1989.

WGEN. Weather generator. **Source:** C. W. Richardson and D. A. Wright, "WGEN: A Model for Generating Daily Weather Variables," USDA—ARS Rep. ARS–8, Washington, DC, 1984.

6. Water Suitability (Salts, Other)

Watsuit. Assessing the suitability of saline water for irrigation and crop production. **Source:** J. D. Rhoades, A. Kandiah, and A. M. Mashali, *The Use of Saline Waters for Crop Production,* FAO Irrigation and Drainage Paper 48, Food and Agriculture Organization of the United Nations, Rome, 1992, 52–59.

7. Erosion Prediction

CREAMS. Chemical movement, runoff, and erosion. **Source:** W. G. Knisel (ed.), "CREAMS: A Field-Scale Model for Chemicals, Runoff, and Erosion from Agricultural Management Systems," *Conservation Research Report No. 26,* Washington, DC: USDA—SEA.

EUROSEM. Erosion and sediment transport model. **Source:** R. P. C. Morgan, J. N. Quinton, and R. J. Rickson, "EUROSEM Documentation Manual," Unpublished Report, Silsoe College, Silsoe, Bedford, England, 1992, 34 pp.

GAMES. Evaluating the effects of agricultural management systems on erosion and sedimentation. **Source:** W. T. Dickinson and R. P. Rudra, "GAMES-User's Manual Version 3.01," School of Engineering, University of Guelph, Guelph, Ontario, Canada, Technical Report No. 126-86, 1990.

KINEROS. Erosion and sediment transport model. **Source:** D. A. Woolhizer, R. E. Smith, and D. C. Goodrich, "KINEROS, a Kinematic Runoff and Erosion Model: Documentation and User Manual," USDA—Agricultural Research Service, ARS-77 (1990), 130 pp.

KINEROS2. Erosion and sediment transport model. **Source:** R. E. Smith, D. C. Goodrich, D. A. Woolhizer, and C. A. Unkrich, "KINEROS: A KINematic Runoff and EROSion model, in V. P. Singh (ed.), *Computer Models of Watershed Hydrology,* Water Resources Publications, Highlands Ranch, CO, 1995.

RUSLE. **Sources:** Danielo Yoder and Joel Lown, "The Future of RUSLE: Inside the New Revised Universal Soil Loss Equation," *Journal of Soil & Water Conservation* **50** (1995), 484–489; and T. J. Toy and W. R. Osterkamp, "The Applicability of RUSLE to Geomorphic Studies," *Journal of Soil & Water Conservation* **50** (1995), 498–503; and K. G. Renard, G. R. Foster, D. C. Yoder, and D. K. McCool, "RUSLE Revisited: Status, Questions, Answers, and the Future," *Journal of Soil and Water Conservation* **49** (1994), 213–220.

WEPP. To predict water erosion. **Source:** J. M. Laflen, L. J. Lane, and G. R. Foster, "WEPP, A New Generation of Erosion Prediction Technology," *Journal of Soil and Water Conservation* **46** (1991), No. 1; 34–38.

WEPS. Submodels (**CROPS, EROSION,** others) that simulate important processes related to wind erosion. **Source:** L. J. Hagen, "A Wind Erosion Prediction System to Meet User Needs," *Journal of Soil & Water Conservation* **46** (1991), No. 2; 106–111.

8. Crop Growth Models

BEANGRO. A dry-bean growth model. **Source:** G. Hoogenboon, J. W. White, J. W. Jones, and K. J. Boote, "BEANGRO: a Process-Oriented Dry Bean Model with a Versatile User Interface," *Agronomy Journal* **86** (1994), 182–190.

DSSAT and **IBSNAT.** Organizations distributing a decision support system for agrotechnology transfer (to accommodate crop models), version 3, and the IBSNAT project to assemble and distribute a computerized decision support system to match crop biological needs to land characteristics. **Source:** G. Y. Tsuji, G. Uehara, and S. Balas (eds.), "DSSAT: A Decision Support System for Agrotechnology Transfer, Version 3, Vols. 1, 2, and 3." University of Honolulu, HI, 96822.

DSSAT and **IBSNAT** includes these models: **CERES Group**—A group of crop-growth programs (**Aroids, Cassava, Barley, Grain-Soybean, Grain-Peanut, Grain-Bean, Maize, Millet, Potato, Rice, Sorghum,** and **Wheat**). **Source:** International Benchmark Sites Network for Agrotechnology Transfer, 2500 Dole St., Krauss 22, Honolulu, HI 96822. About $500 each. Also, a **CROPGRO** group of models and **SUBSTOR** group of models.

EPIC. A generalized crop growth model. **Source:** J. R. Williams, C. A. Jones, J. R. Kiniry, and D. A. Spanel, "The EPIC Crop Growth Model," *Transactions of ASAE* **32** (1989), 497–511.

EXNUT. A management-growth program for peanut producers for irrigation and projected yields. **Source:** Tommy Bennett, Jim Davidson, and Harold Wilson (county agent) at USDA—ARS National Peanut Research Laboratory, Dawson, GA.

GOSSYM. Simulates the growth of the cotton plant. When linked with a weather generator it assesses crop productivity and risk before harvest. **Source:** D. N. Baker, J. R. Lambert, and J. M.

McKinion, "GOSSYM: A Simulator of Cotton Crop Growth and Yield," South Carolina Agricultural Experiment Station, Clemson, SC, *Technical Bulletin* 1089, 1983.

PLANTGRO. To predict dry-matter and grain production of field crops. **Source:** A. Retta and R. J. Hanks, *Utah Agricultural Experiment Station Research Report 48,* (1989). Contact Department of Plants, Soils, and Biometeorology, Utah State University, Logan, UT 84322-4820.

ROSESIM. A model for the growth of "Royalty" roses following hard-pinching. **Source:** D. A. Hopper, P. A. Hammer, and J. R. Wilson, "A Simulation Model of *Rosa hybrida* Growth Response to Constant Irradiance and Day and Night Temperatures," *Journal of American Society of Horticultural Science* **119** (1994), 903–914.

SUCROS, MACROS, WOFOST, LINTUL, and INTERCOM. Groups of crop-growth models. **Source:** P. Goldsworthy and F. Penning de Vries, "Opportunities, Use, and Transfer of Systems Research Methods in Agriculture to Developing Countries," in *Proceedings of the International Workshop on Systems Research Methods in Developing Countries,* 22–24 November 1993, ISNAR, The Hague, Kluwer Academic, The Netherlands.

9. Databases of Various Kinds

CAMPS. Soil Survey Division Staff, "State Soil Survey Database: New York, Rensselaer County Computer Aided Management Planning System (CAMPS)," USDA Soil Conservation Service, Syracuse, NY.

CLIMWAT. A climatic database of 3261 stations in 144 countries (not including the United States) worldwide, five groups by geography, on five 1.44 Mb discs. **Source:** Water Resources, Development, and Management Service/AGLW FAO, Via delle Terme di Carcello, 0010 Rome, Italy.

GRASS. Geographical Resources Analysis Support System, associated with GIS to provide a program to obtain physical parameters of complex soils and topography. **Sources:** U.S. Army, "GRASS-GIS Software and Reference Manual," U.S. Army Corps of Engineers, Construction Engineering Research Laboratory, Champaign, IL; and M. R. Savabi, D. C. Flanagan, B. Hebel, and B. A. Engel, "Application of WEPP and GIS-GRASS to a Small Watershed in Indiana," *Journal of Soil and Water Conservation* **50** (1995), No. 5; 477–483.

NATSGO. **National Soil Geographic Data Base.** Used for national, regional, and state resource appraisal, planning, and monitoring. From state general soil maps from 1982 National Resources Inventory. **Source:** William U. Reybold and Gale W. TeSelle, "Soil Geographic Data Bases," *Journal of Soil and Water Conservation,* (Jan.–Feb. 1989), 28–29.

NDCDB. National Digital Cartographic Data Base. Contains information on boundaries, of the U.S. Geological survey system (PLSS), topography, and use and land cover. **Source:** R. Jannace and C. Ogrosky, "Cartographic Programs and Transportation, Hydrography, Public Land of the U.S. Geological Survey," *The American Cartographer* **14** (1987), July, 197.

SSURGO. **Soil Survey Geographic Data Base.** USDA—NRCS (county) Soil Survey Geographic Database. **Source:** John P. Wilson, William P. Inskeep, Jon M. Wraith, and Robert D. Snyder, "GIS-Based Solute Transport Modeling Applications: Scale Effects of Soil and Climate Data Input," *Journal of Environmental Quality* (1996), 445–453.

STATSGO. **State Soil Geographic Data Base.** Geographical data from county soil survey maps (soil distribution and properties). **Sources:** W. U. Reybold and G. W. TeSelle, "Geographic Data Bases," *Journal of Soil and Water Conservation* **44** (1989), 28–29; and John P. Wilson, William P. Inskeep, Jon M. Wraith, and Robert D. Snyder, "GIS-Based Solute Transport Modeling Applications: Scale Effects of Soil and Climate Data Input," *Journal of Environmental Quality* **25**(1996), 445–453.

TIGER. New digital database of every street and road in the United States and its possessions. Has all railroads and significant hydrographic features. **Source:** Bureau of the Census and U.S. Geological Survey for 1990 census.

10. Weed Control

GWM. A decision support system for evaluating soil-applied and postemergence weed management options in row crops. **Source:** L. J. Wiles, USDA—ARS, Fort Collins, CO. **Reference:** L. J. Wiles, R. P. King, E. E. Schweizer, D. W. Lybecker, and S. M. Swinton, "GWM: General Weed Management Model," *Agricultural Systems* **50** (1996), 355–376.

HERB®. Aid for management of multiple weed species in soybean, North Carolina State. **Source:** G. G. Wilkerson, S. A. Modena, and H. D. Coble, "HERB: Decision Model for Postemergence Weed Control in Soybean," *Agronomy Journal* **83** (1991), 413–417.

WEEDCAM. Evaluation of management strategies for weeds in irrigated corn. **Source:** E. E. Schweizer, D. W. Lybecker, L. J. Wiles, and P. Westra, "Bioeconomic Models in Crop Production," Chapter 15 in *International Crop Science* **I** (1993), 103–107.

11. Environmental Contamination, Project Evaluation and Risk Evaluation

ALES. **Source:** D. Rossiter, "A Microcomputer Program to Assist in Land Evaluation," 113–116 in J. Bouma and A. K. Bregt (eds.), *Land Qualities in Space and Time,* Proceedings of the ISSS Symposium, Wageningen, Pudoc, Wageningen, The Netherlands, 1989.

CRIES. Comprehensive Resource Inventory and Evaluation System developed at Michigan State University for application in developing countries; land evaluation and alternative use. **Source:** G. Schultink, "The CRIES Resource Information System: Computer-Aided Resource Evaluation for Development Planning and Policy Analysis," *Soil Surveys and Land Evaluation* **7** (1987), 47–62.

FORVAL. FORest VALuation is for cash-flow analysis of natural resources investments. **Source:** Thomas J. Straka and Steven H. Bullard, "FORVAL—A Computer Software Package for Forestry and Natural Resources Project Valuation," *Journal of Natural Resources and Life Science Education* **23** (1994), No. 1; 51–55.

LUPLAN. A model developed by CSIRO in Australia and related to the LESA method but LUPLAN is more comprehensive. **Source:** J. R. Ive, J. R. Davis, and K. D. Cocks, "LUPLAN: A Computer Package to Support Inventory, Evaluation, and Allocation of Land Resources," *Soil Surveys and Land Evaluation* **5** (1985), 77–87.

Manure Application Planning (no title). **Source:** M. A. Schmitt, R. A. Levins, and D. W. Richardson, "A Comparison of Traditional Worksheet and Linear Programming Methods for Teaching Manure Application Planning," *Journal of Natural Resources and Life Science Education* **23** (1994), No. 1; 23–26.

SOILRISK. An integrated carcinogenic risk model for low levels of organic contaminants in soil. **Source:** Paula A. Labieniec, David A. Dzombak, and Robert L. Siegrist, "Soil Risk: Risk Assessment Model for Organic Contaminants in Soil," *Journal of Environmental Engineering* **122** (1996), No. 5; 388–398.

12. Soil Taxonomy Keys

Keys to Soil Taxonomy. The computer software and users' manual (1987) are available from the Department of Soil Science, North Carolina State University, P.O. Box 7619, Raleigh, NC 27695-7619. **Source:** *Agrotechnology Transfer* **2** (1986), 16.

SOILdisc. Produced by the University of Florida, uses audio, video, slides, and quiz-type games to learn about the 11 soil orders. $35. [Contact Ami Neiberger, University of Florida, Inst. of Food and Agric. Sciences, Bldg. 116 Mowry Road, P.O. Box 110810, Gainesville, FL 32611-0810; ph (904)392-2411, fax (904)392-7902.]

13. CD-ROM Materials

HYDRICsoils. Produced by the University of Florida, is for teaching about wetland communities. $75. [Contact Ami Neiberger, University of Florida, Inst. of Food and Agric. Sciences, Bldg. 116 Mowry Road, P.O. Box 110810, Gainesville, FL 32611-0810; ph (904)392-2411, fax (904)392-7902.]

SOILdisc. Produced by the University of Florida, uses audio, video, slides, and quiz-type games to learn about the 11 soil orders. $35. [Contact Ami Neiberger, University of Florida, Inst. of Food and Agric. Sciences, Bldg. 116 Mowry Road, P.O. Box 110810, Gainesville, FL 32611-0810; ph (904)392-2411, fax (904)392-7902.]

Appendix B

Plate 1 The rapid use of the rich natural resources in the United States helped fuel growth of one of the most affluent societies the world has known. These large tree cross-sections in a small lumber museum in western Oregon are evidence of old, mature forests cut during the past 100 years and which are seldom duplicated in size in forests today because some trees, such as Douglas fir, take as long as 800 years to reach such sizes. (Courtesy of Raymond W. Miller, Utah State University.)

Plate 2 Terraced, steep slopes in the Himalayan foothills in northern India. The Himalayan Mountains form an arc about 2400 km (1500 mi) from the Indus River in northern Pakistan across northern India, southern Tibet, Nepal, Sikkim, and Bhutan covering an area of 594,000 m^2 (229,500 mi^2). The foothills, with elevations up to 3000 m (9850 ft), are extensively terraced for many miles and indicate the importance of the land to these people, who have small flat areas. They obviously recognize the need to control erosion and to manage the water from runoff. (Courtesy of Raymond W. Miller, Utah State University.)

Plate 3 Where cheap natural resources are not available or people are too numerous, the standard of living is greatly reduced from that found in Europe and the United States. This view of a street scene in India, with its large number of bicycles, three-wheeled scooter taxis, motor scooters, and animal-drawn carts, is evidence of a limited-income population. (Courtesy of Raymond W. Miller, Utah State University.)

Plate 4 Planting corn in Mauritania, West Africa. The country has few natural resources and a somewhat hostile climate that forces farmers to use cheap and low-input techniques. The farmer pokes a hole in the rain-wetted soil, and a helper drops in one or two kernels of corn. (Courtesy of Keith R. Allred, Utah State University.)

Plate 5 Pegmatite rock. Slowly cooling magma allows the mineral components to grow slowly and form large crystals of its minerals. The dark, layered mineral is biotite; the glassy portion is quartz, and the chalky, white portion is feldspar. This, if cooled rapidly, would be granite. (Courtesy of Raymond W. Miller, Utah State University.)

Plate 6 Prismatic structure in soil from near the surface to 40 cm to 50 cm deep. Montana road cut. (Courtesy of Alvin R. Southard, Utah State University.)

Plate 7 Placic horizon in an Andisol south of Quito, Ecuador. The placic horizon is a thin (orange in photo) layer of iron and manganese hydrous oxides concentrated by solution, movement, and precipitation. (Courtesy of Stanley W. Buol, North Carolina State University.)

Plate 8 Mottling, probably early stages of plinthite, in the deeper layers in an Oxisol in Australia. The gray is the gley material caused by long periods of anaerobic conditions (wetness). The orange and yellow mottles are from various oxidized, hydrated iron hydrous oxides. (Courtesy of Alvin R. Southard, Utah State University.)

Plate 9 The unique and colorful canyonlands of the Western mountains are caused by the red-to-yellow color from cementing iron oxides and the variable hardness (degree of cementation) of the sandstones. Water rapidly erodes the steep slopes of softer (less well cemented) sandstones, forming deeply sculptured landscapes. (Courtesy of Raymond W. Miller, Utah State University.)

Plate 10 Typic Natrustalf in Queensland, Australia. The *natr* formative element indicates a high-sodium argillic horizon; such soils usually have columnar structure with thin, leached (whitish) caps. (Courtesy of Alvin R. Southard, Utah State University.)

Plate 11 A typical view of Monument Valley, Utah. The sandstones of varying cementation hardness allow erosion to cut and form the spectacular views as the different layers erode. Very limited vegetation is available for grazing animals. (Courtesy of Alvin R. Southard, Utah State University.)

Plate 12 Hardpans of many types occur in soils. These large masses of Croy soil hardpan, from digging an irrigation canal in northern Utah, are cemented mostly by calcium carbonate and some silica (petrocalcic grading toward a duripan). (Courtesy of Raymond W. Miller, Utah State University.)

Plate 13 A thick **E** horizon over a natric horizon (high exchangeable sodium) in Montana. The **E** horizon is caused by the leaching of dark clay from the upper layer because of easy clay dispersion. The **E** horizon is less clayey than the **B;** usually such horizons may be in precipitation areas less than 600 mm (25 in) per year. (Courtesy of Alvin R. Southard, Utah State University.)

Plate 14 A roadcut west of Denver, Colorado, exposing the Morrison and Dakota formation. The horizontal lines are roads and terraces to stabilize the steep slope from erosion and falling rock. The formations are the colored bands of soil in the near-vertical direction. (Courtesy of Alvin R. Southard, Utah State University.)

Plate 15 Oxisolic Spodosol from Australia (lateritic podzol). The nomenclature and genesis is debated. Some believe that sandy material alluvium on top is partly responsible for the unusual profile features. (Courtesy of Alvin R. Southard, Utah State University.)

Plate 16 Molokai soil series from Oahu, Hawaii, developed mostly from volcanic materials but with some remnant rocks in it. Many Oxisol and Ultisol soils have bright red colors because of the increase in residual iron oxides as these soils have weathered excessively. (Courtesy of Alvin R. Southard, Utah State University.)

Plate 17 Flooding in Davis Canyon, northern Utah, caused a rock slide that nearly covered this cabin. A lack of protective land cover and selection of a building site are probably both at fault here. (Courtesy of State of Utah and Bureau of Land Management.)

Plate 18 At 3300 m to 3600 m (11,000 ft–12,000 ft) elevation, this Altiplano area in Bolivia is continually cold and is a harsh environment for the growth of plants. Overgrazing by sheep and other animals keeps the sparsely vegetated landscape quite barren. Houses are of soil, because no timber is available, except when imported at high costs. (Courtesy of Raymond W. Miller, Utah State University.)

Plate 19 Vertisols, with more than 30 percent swelling clays, crack deeply and widely when dry. This Vertisol in India, near harvest time, is dry and cracked. The cracking damages (tears apart) plant root systems, yet, when the clayey soil is wet it has poor aeration and acts as a shallow soil. (Courtesy of Raymond W. Miller, Utah State University.)

Plate 20 Prismatic structure in Montana. The shallow **A** horizon above the prismatic structure is not well structured and the prisms are permeable mostly through the cracks between prisms. Most prisms are four- to six-sided in cross-section. (Courtesy of Alvin R. Southard, Utah State University.)

Plate 21 Unweathered layer of basalt over volcanic cinders deposited earlier than the molten basalt flow, in Achada Grande, Cape Verde Islands, off the coast of western Africa. (Courtesy of Alvin R. Southard, Utah State University.)

Plate 22 This root system is in a Haplustox, near Brasilia, Brazil. Deep root proliferation occurs only after calcium from added lime moves down into the low-**CEC** soil. This root system was possible only after about 10 years of adding lime and fertilizer to this soil. (Courtesy of Stanley W. Buol, North Carolina State University.)

Plate 23 Ironstone on the western coast of Africa in The Gambia. The soil above this hardened plinthite has eroded, leaving the rust-colored and rock-like ironstone exposed. (Courtesy of Raymond W. Miller, Utah State University.)

Plate 24 Manure, a valuable source of organic matter and nutrients for crops, is often used for other purposes in India. This animal manure is a cheap source of fuel for cooking, and much of it is made into thin cakes, dried, and sold for fuel. Without much money to buy fertilizers, the loss of this manure is unfortunate. (Courtesy of Raymond W. Miller, Utah State University.)

Plate 25 Well-weathered soil in Guatemala has a clay layer high in kaolinite and iron oxide clays that do not shrink or swell much when wet and dried. When formed into tile, brick, and ceramics and fired at high temperature, the material hardens but does not crack. (Courtesy of Raymond W. Miller, Utah State University.)

Plate 26 Cutting 2-year-old sugarcane in Venezuela. The cane at harvest time is set afire to burn the dry leaves and stem coatings, which helps to reduce the amount of trash to be hauled and helps keep the sugar recovery in the factory higher. Burning also kills and drives away snakes (some poisonous) that use the cane as a habitat during the 2 years of growth. The cutting is dirty and hard work, but the sugar is a valuable crop. (Courtesy of Raymond W. Miller, Utah State University.)

Plate 27 A lunch break for workers cleaning a ganat (tunnels into water-bearing strata) in Iran. (Courtesy of Bruce Anderson, Utah State University.)

Plate 28 The long, dry periods in central Africa, often up to 8 consecutive dry months, challenge many plants. The mango and cashew trees are two of these plants that have the ability to root deeply and extensively enough to survive, although those in the photo are irrigated. In dry soils the cashew is a small shrub. The single cashew nut, each growing on the end of a young pear-shaped fruit, has a covering of a blistering oil. With such a small harvest from a lot of work, it is amazing that cashews are not even more expensive than they are. They are related to poison ivy and poison sumac but are the only cultivated one of eight known species. (Courtesy of Raymond W. Miller, Utah State University.)

Plate 29 In Western Sao Paulo State in Brazil, this Argiudoll formed on calcareous sandstone grades into the reddish Hapludox formed in reworked sediments from mixed sources. (Courtesy of Stanley W. Buol, North Carolina State University.)

Plate 30 Ironstone cap over wavy, folded sediments in New South Wales, Australia, near Armi Lake. (Courtesy of Alvin R. Southard, Utah State University.)

Plate 31 A hardened titanium oxide crust on a highly weathered soil in Lanai, Hawaii, along the Munro trail, an access path into the nonpopulated Dole Pineapple Company plantation center. Only a few kinds of plants (such as eucalyptus trees) will growth on such crusted soils. (Courtesy of Alvin R. Southard, Utah State University.)

Plate 32 A Spodic Quartzipsamment near Perth, western Australia. The upper, white layer is a well-leached **E** horizon, partly because of its sandy texture and high rainfall. Vegetation supplies soluble organics in rainfall to help move iron to deeper horizons. The dark brownish-yellow **Bs** horizon lies beneath the **E,** and a lighter, yellow layer of the parent sandy material (quartz and sandy) is beneath that layer. (Courtesy of Alvin R. Southard, Utah State University.)

Plate 33 Top Iron-deficiency symptoms on peach trees viewed at a distance in northern Utah. (Courtesy of R. L. Smith, Utah State University.)

Plate 33 Bottom Iron-deficiency symptoms include interveinal chlorosis of younger leaves. The close-up view indicates three stages of severity of the deficiency. The peach trees in the distant view have most of the leaves chlorotic. (Courtesy of R. L. Smith, Utah State University.)

Plate 34 Iron deficiency in cherry. Only this single tree was deficient in a complete orchard. No analyses produced an answer. Since it is the only tree and is completely chlorotic, the only suggested cause is that this one tree has a genetic difference that makes it susceptible to being iron deficient. (Courtesy of David W. James, Utah State University.)

Plate 35 Top Potassium deficiency in clover. The necrotic edges (leaf-tip scorch) on older leaves is the most typical deficiency. The white spots or tissues throughout the leaf are typical of many legumes. In cotton the spots may be reddish and even be on younger leaves. Bronzing of thick, leathery leaves—with no necrosis—that snap when bent is found in cotton in the San Joaquin Valley of California on *younger, not older, leaves.* (Courtesy of Raymond W. Miller, Utah State University.)

Plate 35 Center Potassium deficiency in alfalfa in Utah. Edge scorch on older leaves is the more typical reported deficiency symptom. The white necrotic tissue throughout the leaves is common in many alfalfa and clover varieties. (Courtesy of David W. James, Utah State University.)

Plate 35 Bottom Potassium deficiency in potato in the Columbia basin, Washington, where the deficiency is found only in calcareous soil. The bronzing is not typical of what are considered the classic deficiency symptoms of necrotic tissue (edge scorch) along the edges of older leaves. (Courtesy of David W. James, Utah State University.)

Plate 36 Phosphorus deficiency in corn in Iowa. The purple-red coloring is typical of this deficiency and is usually on the older leaves. (Courtesy of R. L. Smith, Utah State University.)

Plate 37 *Crinkle leaf* of cotton is caused by toxic levels of manganese. Toxic levels of both manganese and aluminum usually occur only in strongly acidic soils. (Courtesy of Potash and Phosphate Institute, PPI.)

Plate 38 Zinc deficiency in corn. One extension soils specialist said this is one of only two areas of Utah in which zinc deficiency had been definitely identified. Both areas were on cut soil (during land leveling) where calcareous subsoil became the exposed soil to farm. (Courtesy of David W. James, Utah State University.)

Plate 39 Manganese deficiency in cotton. (Courtesy of Potash and Phosphate Institute, PPI.)

Plate 40 Boron toxicity in potato. Notice the leaf curl and necrotic edges of leaves. Often the deficiency is most severe on developing buds and new leaves. (Courtesy of David W. James, Utah State University.)

Plate 41 The dark green strips are the fertilized strips of lawn of a poorly done, uneven spread of fertilizer. Usually nitrogen is the element most needed on lawns for greening them up. (Courtesy of Keith R. Allred, Utah State University.)

Plate 42 Equipped with a computer-controlled system, this Ag-Chem Terra Gator unit applies potassium in solution at varying rates, according to a previously generated map of soil potassium levels measured with soil tests. (Courtesy of USDA—ARS Information Staff.)

Plate 43 A nitrogen distribution map near Bow Island, Alberta, for an irrigated soft-wheat field based on soil sampling (0 cm–60 cm). Comparison maps for salinity (derived from a GPS-based electromagnetic induction (EM) method) and for yield (yield monitor) were also prepared. Areas of high electrical conductivity corresponded to areas of high N and low wheat yields. High N may be a partial cause of the high salt and poor yields where salinity is high. (Courtesy of Colin McKenzie and *Better Crops with Plant Food* [**80** (1996, No. 3), 34–36].)

Plate 44 Farming the Palouse loess in southeastern Washington; crops are wheat, barley, peas, and lentils. Tillage erosion has lowered fields compared to the soil at the E–W fence row. Slopes up to 45 percent are farmed. Much of the landscape is shaped by erosion from winter snow melt and rainfall runoff. (Courtesy of Bruce E. Frazier, Washington State University.)

Plate 45 Levels of organic matter percentage and Soil Bray-P mapped using more than 2000 soil samples (on a 40 ft $\times$ 80 ft grid) in a quarter-section (0.635 km^2). The purpose was to evaluate aerial photography as a tool to help direct soil sampling. The organic matter varies from 1 percent to 5 percent and Bray-P ranges from 1 ppm to more than 350 ppm. Variable-rate treatments could certainly be considered for fields with such differences. (Courtesy of Tracy M. Blackmer, James S. Schepers, and *Better Crops with Plant Food* [**80** (1996, No. 3), 18, 19, 23].)

Plate 46 An illustration of topography and spatial distribution of corn yield within a 1.6-ha portion (4 a) of the field fertilized with 151 kg/ha of N (135 lb/a). The yield differences are expected in such a topography, but a low-yield area will not necessarily be caused by low N nor respond to more N. All that is known is what the yield is at a certain place. The need after yield differences are known is to determine the problems and how to correct them. (Courtesy of R. G. Kachanoski, I. P. O'Halloran, D. Aspinall, P. Von Bertoldi, and *Better Crops with Plant Food* [**80** (1996, No. 3), 20–23].)

Plate 47 This space satellite photo over Texas, taken with infrared film, shows the extent of center-pivot systems. Each red circle is about 52 ha (130 a, or 0.25 mi^2). A pivot this size will have about 14 moving towers. (Courtesy of Brantwood Publications, cover of *Irrigation Journal* [July–August 1974].)

Plate 48 Left The photo is of a uniform sudangrass field planted before imposing a long-time experiment with differentially treated plots on the area. The *normalized difference vegetation index* (NDVI) used shows the best (good) health plants in white (see color bar scale) grading to the poorest, black. The photo illustrates the barren control plot, black (upper left center), growth problem areas along the upper left edge (grays), some plot border effects adjacent to field roads, with purple around green centers (upper right) and problem spots in gray and orange (upper center right). This before-treatment photo of a supposedly quite uniform sudangrass field shows the value of crop imaging to find locations of different (problem or good) plant growth areas. (Courtesy of R. Ford Denison, Robert O. Miller, Dennis Bryant, Akbar Abshahi, and William E. Wildman, "Image Processing Extracts More Information from Color Infrared Aerial Photos," *California Agriculture* **50** (1996), No. 3, 9–13. Aerial photo by William E. Wildman and computer-enhanced by R. Ford Denison.)

Plate 48 Right The upper right one-quarter of the area of *Plate 50 Left* with treatments imposed on 1-acre plots: The photo shows some *normalized difference vegetation index* (NDVI) of unfertilized wheat (U), fertilized wheat (F), and legume cover crop (CC) plots. The color bar code in *Plate 50 Left* photo applies to *Plate 50 Right,* also. Notice the plot edge effects around the CC plots, particularly. Plots U-1 and U-2 were managed identically but the growth is different (see color). Plots U-1 and U-2 also have growth differences in the original planting of sudangrass (*Plate 50 Left*). It is apparent from these views that aerial photos can locate plant growth differences. (Courtesy of R. Ford Denison, Robert O. Miller, Dennis Bryant, Akbar Abshahi, and William E. Wildman, "Image Processing Extracts More Information from Color Infrared Aerial Photos," *California Agriculture* **50** (1996), No. 3, 9–13. Aerial photo by William E. Wildman and computer-enhanced by R. Ford Denison.)

Plate 49 Gravel pieces on the tops of pedestals in Hawaii. The nonsoil cover (gravel) protects the soil from raindrop splash erosion and allows the remnant pedestals to form as surrounding soil is eroded away. The soil has a natric horizon (high exchangeable sodium), which is readily dispersed with the force of raindrops. Naturally scattered rocks and gravel do help reduce further erosion. (Courtesy of Alvin R. Southard, Utah State University.)

Plate 50 Serious gulley erosion along Shinglehouse Creek in New South Wales, Australia. The *hard-setting surface* protects the soil until water breaks through at some area, and then erosion is rapid and severe. (Courtesy of Alvin R. Southard, Utah State University.)

Plate 51 Serious gulley erosion on Natrustalf soils in Queensland, Australia, near Chinchilla. The *hard-setting surface* protects the soil until water breaks through at some area and then erosion is rapid and severe. The high sodium-saturated **B** is very susceptible to dispersion and water erosion. (Courtesy of Alvin R. Southard, Utah State University.)

Plate 52 The area along the Li River near Guilin, where a Chinese commune production brigade was located. The tall, narrow peaks are limestone pedestals, perhaps from karst formation. (Courtesy of Alvin R. Southard, Utah State University.)

Plate 53 A badlands-type erosion of a Haplargid in New Mexico. For scale notice the man in the center of the background. Low rainfall results in poor vegetative cover and, thus, poor protection from the forces of water erosion. (Courtesy of Alvin R. Southard, Utah State University.)

Plate 54 Machu Picchu, one of the last Incan settlements and about 80 km (50 mi) northwest of Cuzco, Peru, was a remote and somewhat self-sustaining terraced mountain top. The city covers about 13 km^2 (5 mi^2) and was built with stone, which was laid by unique engineering skills. It was first discovered by the American explorer Hiram Bingham in 1911. It is not mentioned in the writings of Spanish conquerors, probably because of its remote location. View A shows the extent of terracing. View B shows the remote nature of the settlement and the steepness of the area. Even the tall peak to the right, up a very narrow steep ridge, had terraces built on it, on which to produce food. (Courtesy of Raymond W. Miller, Utah State University.)

Plate 55 Salt accumulation (the white covering) in Nevada, referred to by some persons as *summer snow.* Greasewood, a salt-tolerant shrub, barely survives the high salt content. (Courtesy of Alvin R. Southard, Utah State University.)

Plate 56 Terraced land on steep slopes, such as these 14 percent to 25 percent slopes in Missouri, require costly construction. The steep backslopes are planted to permanent grass cover to protect against erosion by water. The farmed areas are terraced to have a gentle slope with low potential for erosion. (Courtesy of USDA—NRSC and the Missouri Agricultural Experiment Station, *Soil Survey of Atchison County, Missouri.*)

Plate 57 Crop residues left on the soil are an aid in protecting the soil from raindrop splash on surface runoff. The residue also slows soil warming and planting operations. Control of erosion is more critical than the other problems in the long run for most soils. (Courtesy of Yetter Manufacturing Company, Colchester, IL.)

Plate 58 Left A 36-v 7-hp electric tractor pulling a 7000-lb refrigeration trailer up a steep slope. This converted Yanmar tractor can haul such a trailer for 8 hours, or rototill for 3 to 4 hours, on one charge of its batteries. The tractor originally had a 16-hp diesel engine. The electric motors have more torque than fuel motors, are quiet, are almost pollution free, and the batteries (shown exposed) can be placed to add stability and traction. (Courtesy of Stephen Heckeroth, owner, Homestead Enterprises, Albion, CA 95410.)

Plate 58 Right Solar cells on this south-facing barn roof provide enough electricity to charge the batteries of an electric tractor or to power a single-family home, an apartment and on sunny summer days will charge an electric car as well. (Courtesy of Stephen Heckeroth, owner, Homestead Enterprises, Albion, CA 95410.)

Plate 59 Center-pivot sprinklers can each cover about 0.65 km^2 (0.25 mi^2). From the air the center pivots make an interesting pattern of circles and half-circles of various colors of vegetation (wheat, alfalfa, potatoes, melons) and stages of maturity along the Columbia River near Hermiston, Oregon. A central computer-controlled system regulates irrigation of more than 100 center pivots. (Courtesy of USDA—ARS Information Staff.)

Plate 60 In the Andes mountains of Venezuela, many of the local people in the mountains live off the little food they can grow on the steep slopes near their houses. They locate their homes within walking distance of roadways to have access to produce trucks and buses. They clear and farm the slopes of 20 percent to 40 percent, the only land around them, without terracing. Such cultivation encourages enormous erosion of slopes. (Courtesy of Raymond W. Miller, Utah State University.)

Plate 61 Deep peat fields in Vancouver, Canada. The water table during the growing season may still be no deeper than 1 m. After the cultivated soil has been disturbed and dries, a truck-like vacuum sweeper will traverse the fields, picking up shredded, loose peat, to sell as a potting or other horticultural growth medium. (Courtesy of Raymond W. Miller, Utah State University.)

Plate 62 Homesites and canals were created by dredging and filling on Marco Island to establish housing developments near the water. This development was done in 1973, prior to strong pressures against such development. Such areas are also subject to damages from serious storms. (Courtesy of EPA-DOCUMERICA—Flip Schulke, photographer.)

Plate 63 Solar Aquatics™ wastewater treatment facility in Beaverbank Villa, Nova Scotia, Canada. The system claims to duplicate, under controlled conditions, the purifying processes of fresh-water streams, meadows, and wetlands. Greenhouses enhance the growth of bacteria, algae, plants, and aquatic animals. Water flows through a series of clear-sided tanks, engineered streams, and constructed marshes. It has been used for treatment of sewage, septage, landfill leachate, boat waste, and ice cream processing waste. (Courtesy of Ecological Engineering Associates, Marion, MA.)

Plate 64 Newly finished well in The Gambia, Africa. Household, and sometimes home garden, water is obtained from shallow wells. The few wells, because cost is high for low-income families, are important to everyone. The construction of a new well, such as this one under the direction of a Peace Corps volunteer, is an important event in the lives of villagers. (Courtesy of Raymond W. Miller, Utah State University.)

Plate 65 The Kennecott open-pit copper mine, southwest of Salt Lake City, Utah. It is one of the largest open-pit mines in the world, and the enormous amounts of tailings have been piled and settled from water over extensive areas. Heavy metals in acidic leaching waters from the tailings are concerns that have been watched and studied for decades. (Courtesy of Raymond W. Miller, Utah State University.)

Plate 66 A dairy farming landscape on the Capac-Wixom soil association in Michigan. These soils are nearly level, somewhat poorly drained, loamy and sandy soils on lake plains, till plains, and moraines. The soils have an **Ap,** often a thin **E,** and a thick **B** and **Bt.** The major problems are poor drainage and, sometimes, wind erosion. The soils are Aqualfs and Aquods. (Courtesy of USDA—Natural Resources Conservation Service, *Soil Survey of Mason County, Michigan;* photo by Roger Howell, NRCS retired.)

Plate 67 A thin brown volcanic ash layer (colored from weathering that produces iron hydrous oxides) over white calcium carbonate in Granite County, Montana. The profile is not well developed and might be listed as **A, 2C** without a **B** horizon in most locations. (Courtesy of Alvin R. Southard, Utah State University.)

Plate 68 Large lysimeters are used to study exact water gains and losses in a volume of soil. In Pawnee, Colorado, a 20,430-kg (45,040-lb) undisturbed soil core was prepared for rangeland studies. The undisturbed soil core in its lysimeter is excavated and ready to be lifted onto a truck for transport to its research location. (Courtesy of International Biological Programs Grassland Biome Study, Colorado State University; photo by Alan Brooks.)

Plate 69 Ice chunks from winter flooding of a city park in New Brunswick, New Jersey. Parks are a good use of such areas that would have much more damage if homes or other structures were built on these flood plains along the river. (Courtesy of Raymond W. Miller, Utah State University.)

Plate 70 A homestead in Nebraska, depicting the windbreaks and nonfarmland areas nearby. The area is of Kuma-Duroc-Keith soils. (Courtesy of USDA_Natural Resources Conservation Service, Nebraska, from Soil Survey of Keith County, Nebraska.)

Plate 71 Horticulturalists use both soil and soilless media for flowers. Plots such as these shown are the most economical for large quantities of some flowers. The value of extensive land areas is very evident when attempts to produce large amounts of plant materials are attempted. (Courtesy of C. Earl Israelson, Utah State University.)

Appendix C

Units of Measurement

The American Society of Agronomy (Plant and Soil Sciences) uses the **International System of Units (SI).** This system is a decimal metric system. Some of the other conventions follow. First, to write units in the denominator, use a **−1** [or **−2** or **−3** for square or cubed], such as pounds per gallon [$\mathbf{lb\ gal^{-1}}$] or meters per second per second [$\mathbf{m\ s^{-2}}$]. Second, use a space between units for clarity; "dots" should be used to separate units only if confusion may occur without their use [micrograms per gram would be $\mathbf{mg\ g^{-1}}$ or $\mathbf{mg \cdot g^{-1}}$]. The use of the "fraction" or solidus (/) as used previously is still acceptable, i.e. $\frac{g}{cm^2}$ or $\mathbf{g/cm^2}$. Generally, abbreviated units are not followed by periods. Some prefer to use periods with English units, such as ft., lb., and some others. Periods will not be used in this text.

A. The Seven Basic Units of the SI System

Measurement	*Basic Unit*	*Symbol*
amount of substance	mole	mol
electrical current	ampere	A
length	meter	m
luminous intensity	candela	cd
mass	gram (kilogram)*	g (kg)
temperature	kelvin†	K
time	second	s

*The kilogram is actually the basic reference unit of mass (weight), even though *gram* is the name used with prefixes. One gram equals 0.001 kg.

†The Kelvin temperature scale has the same size of degree as the Celsius scale, but 0°K is equal to 0°C − 273.2°. Therefore, Celsius degrees = °K + 273.2°.

B. *Common Units in Use with SI; Some Preferred Uses*

Quantity	*Unit*	*Symbol*	*Definition*
area	hectare	ha	1 ha = 10^4 m^2
concentration	mole per cubic meter	mol m^{-3}	mol [m × m × m]$^{-1}$
OR	mole charge per liter	mol_c L^{-1}	[at. wt./ion charge]L^{-1}
density	kilograms per cubic meter	kg m^{-3}	kg [m × m × m]$^{-1}$
electrical	siemens per meter	S m^{-1}	1 S m^{-1} = 10 mmho cm^{-1}
conductivity	decisiemens/meter	dS m^{-1}	1 dS m^{-1} = 1 mmho cm^{-1}
force	newton	N	1 m kg s^{-1}
mass	metric tonne	Mg	1 Mg = 10^3 kg
pressure	pascal	Pa	1 kg m^{-1} s^{-1}
	megapascal	MPa	1 MPa = 10^6 Pa = 10^3 kPa = 10 bar
time	second	s	1 s = 1/60 min
	minute	min	1 min = 60 s
	hour	h	1 h = 60 min
	day	d	1 d = 24 h
	year	yr	1 yr = 365 d
volume	cubic meter	m^3	m × m × m
	liter	L‡	1 L = 1 dm^3 = 10^{-3} m^3
	milliliter	mL	1 mL = 0.001 L
water potential	joules per kg	J kg^{-1}	1 J kg^{-1} = 1 kPa
yield	megagrams per hectare	Mg ha^{-1}	(Mg is a metric tonne)

C. *Prefixes: To indicate fractions or multiples of the basic or derived units, prefixes are used according to the following list:*

Prefix	*Multiple*	*Abbreviation*	*Multiplication Factor*
exa	10^{18}	E	1 000 000 000 000 000 000
peta	10^{15}	P	1 000 000 000 000 000
tera	10^{12}	T	1 000 000 000 000
giga	10^9	G	1 000 000 000
mega	10^6	M	1 000 000
kilo	10^3	k	1 000
hecto	10^2	h	100
deca	10	da	10
deci	10^{-1}	d	0.1
centi	10^{-2}	c	0.01
milli	10^{-3}	m	0.001
micro	10^{-6}	μ	0.000 001
nano	10^{-9}	n	0.000 000 001
pico	10^{-12}	p	0.000 000 000 001
femto	10^{-15}	f	0.000 000 000 000 001
atto	10^{-18}	a	0.000 000 000 000 000 001

‡The liter is now defined as the volume of 1 kilogram of pure water at its maximum density (which is about 4°C) and is equal to 1.000 028 dm^3, although it is often defined in general books as 1 dm^3.

D. Example Relationships for Some Unfamiliar Units

1 siemen equals 1 mho. The boundary for saline vs. nonsaline soils (4 mmhos/cm) is equivalent to 4 dS m^{-1} (= 4 dS/m).
1 pascal equals 9.87×10^{-6} atmosphere = 10^{-5} bar.

Factors for Converting Units

A Units		*B Units*
(Multiply this)	*(By this)*	*(To obtain this)*
(To obtain this)	*(By this)*	*(Divide this)*
AREA		
acre, a	0.404 685	hectare, ha ($10^4 m^2$)
acre, a	4046.85	square meter, m^2
acre, a	0.004 05	square km, km^2
square feet	0.092 9	square meter
square mile	2.54	square kilometer
LENGTH		
Angstrom unit, Å	0.1	nanometer, nm (10^{-9} m)
foot, ft	0.304 8	meter, m
inch, in	25.40	millimeter, mm (10^{-3} m)
micron, μ	1.00	micrometer, um (10^{-6} m)
mile	1.609 34	kilometer, km (10^3 m)
yard, yd	0.914	meter, m
MASS		
ounce, oz	28.349 5	gram, g (10^{-3} kg)
pound, lb	453.592	gram, g (10^{-3} kg)
pound, lb	0.453 5	kilogram, kg
ounce (avdp), oz	28.4	gram, g
quintal (metric), q	100	kilogram, kg
tonne, metric ton, Mg	1000	kilogram, kg
ton (2000 lb), t	907.185	kilogram, kg
ton (2000 lb), t	0.907 2	megagram, Mg (tonne)
VOLUME		
bushel, bu (U.S.)	35.238 1	liter, L (10^{-3} m^3)
cubic feet, ft^3	0.028 317	cubic meter, m^3
cubic feet, ft^3	28.316 5	liter, L (10^{-3} m^3)
cubic inch, in^3	0.000 016 4	cubic meter, m^3
gallon, gal	3.785 31	liter, L (10^{-3} m^3)
ounce, fluid oz	0.029 573	liter, L (10^{-3} m^3)
PRESSURE		
atmosphere, atm	14.70	lb in^{-2}, psi
bar	14.50	lb in^{-2}, psi

Note: A few older terms, which have special convenience, continue to be used by some workers in preference to new **SI** terms. One of these is the use of *milliequivalents* (meq) instead of *moles* in cation exchange discussions. The most likely problem will be the confusion of *t* used in past editions with *acre* to mean U.S. conventional short tons (2000 pounds) per acre. It should refer to metric tonnes (1000 kg) when used with hectares (as 15 t/ha). Using megagrams per hectare (Mg/ha), rather than t/ha, will reduce confusion.

bar	0.986 9	atmosphere, atm
bar	100.0	kilopascal, kPa (10^3 Pa)
kg (weight) cm^{-2}	14.22	lb in^{-2}, psi
pounds per square foot, lb ft^{-2}	47.88	pascals, Pa
YIELD AND RATE		
bushel per acre, 48 lb	53.75	kilogram per hectare, kg ha^{-1}
bushel per acre, 56 lb	62.71	kilogram per hectare, kg ha^{-1}
bushel per acre, 60 lb	67.19	kilogram per hectare, kg ha^{-1}
gallon per acre, gal a^{-1}	9.35	liter per hectare, L ha^{-1}
ounces per cubic foot, oz ft^{-3}	0.997	kilograms per cubic meter, kg m^{-3}
gram per cubic centimeter	1.00	megagram per cubic meter, Mg m^{-3}
hectare-30 cm of soil	4×10^6 kg	kg per hectare-30 cm of soil
pound per acre, lb a^{-1}	1.121	kilogram per hectare, kg ha^{-1}
ton (2000 lb) per acre, t a^{-1}	2.242	metric tons (1000 kg) per hectare, Mg ha^{-1}
miles per hour, mile	0.447	meters per second, m s^{-1}
CONCENTRATION		
milliequivalents per 100 g soil	10.0	millimole per kilogram, $mmol_c$ kg^{-1} soil
milliequivalents per 100 g soil	1.0	centimole per kilogram, $cmol_c$ kg^{-1} soil
milligram per kilogram, mg kg^{-1}	1.0	parts per million, ppm
TEMPERATURE		
Celsius temperature, °C	1.80	+32 = Fahrenheit temperature, °F
Fahrenheit temperature, °F − 32°	0.555 5	Celsius temperature, °C
Kelvin temperature, °K − 273.15	1.00	Celsius temperature, °C
PRESSURE		
atmosphere	0.101	megapascal, MPa (10^6 Pa)
bar	0.10	megapascal, MPa (10^6 Pa)
pound per square inch, lb in^2	6.90×10^3	pascal, Pa
pound per cubic foot, lb ft^{-3}	16.02	kilogram per cubic meter, kg m^{-3}
WATER MEASUREMENT		
acre-inch	102.8	cubic meter, m^3
cubic feet per second, ft^3 s^{-1}	101.9	cubic meter, per hour, m^3 h^{-1}
quart, qt	0.946	liter, l (10^{-3} m^3)
acre inches, acre-in	1.03×10^{-2}	hectare-meters, ha-m
acre feet, acre-ft	12.33	hectare-centimeters, ha-cm
OTHER CONVERSIONS		
British thermal unit, Btu	1.05×10^3	joule, J
bushel	32	lb of oats
bushel	45	lb of rice
bushel	56	lb of corn, sorghum
bushel	60	lb of wheat, soybeans, potatoes
calorie, cal	4.19	joule, J
cubic feet	0.028	cubic meters
cubic feet	28.3	liters
cubic inches	1.64×10^{-5}	cubic meters
dyne	1×10^{-5}	newton, N
erg	1×10^{-7}	joule, J
foot-pound	1.356	joule, J

PERIODIC TABLE OF THE ELEMENTS

PERIODS	IA	IIA	IIIB	IVB	VB	VIB	VIIB	VIII			IB	IIB	IIIA	IVA	VA	VIA	VIIA	VIIIA
1	1.008 H 1																	4.003 He 2
2	6.941 Li 3	9.012 Be 4											10.81 B 5	12.011 C 6	14.007 N 7	16.999 O 8	18.998 F 9	20.179 Ne 10
3	22.990 Na 11	24.305 Mg 12											26.982 Al 13	28.086 Si 14	30.9738 P 15	32.06 S 16	35.453 Cl 17	39.948 Ar 18
4	39.102 K 19	40.08 Ca 20	44.956 Sc 21	47.90 Ti 22	50.941 V 23	51.996 Cr 24	54.938 Mn 25	55.847 Fe 26	58.933 Co 27	58.71 Ni 28	63.546 Cu 29	65.37 Zn 30	69.72 Ga 31	72.59 Ge 32	74.922 As 33	78.96 Se 34	79.904 Br 35	83.80 Kr 36
5	85.468 Rb 37	87.62 Sr 38	88.906 Y 39	91.22 Zr 40	92.9064 Nb 41	95.94 Mo 42	98.906 Tc 43	101.07 Ru 44	102.906 Rh 45	106.4 Pd 46	107.868 Ag 47	112.40 Cd 48	114.82 In 49	118.69 Sn 50	121.75 Sb 51	127.60 Te 52	126.904 I 53	131.30 Xe 54
6	132.906 Cs 55	137.34 Ba 56	138.906 La 57	178.49 Hf 72	180.948 Ta 73	183.85 W 74	186.2 Re 75	190.2 Os 76	192.22 Ir 77	195.09 Pt 78	196.967 Au 79	200.59 Hg 80	204.37 Tl 81	207.2 Pb 82	208.981 Bi 83	(209) Po 84	(210) At 85	(222) Rn 86
7	(223) Fr 87	226.025 Ra 88	(227) **Ac 89	(261) [Rf] 104	(260) [Ha] 105	(263) [ ] 106												

TRANSITION ELEMENTS

* Lanthanides	140.12 Ce 58	140.908 Pr 59	144.24 Nd 60	(145) Pm 61	150.4 Sm 62	151.96 Eu 63	157.25 Gd 64	158.925 Tb 65	162.50 Dy 66	164.930 Ho 67	167.26 Er 68	168.934 Tm 69	173.04 Yb 70	174.97 Lu 71
** Actinides	232.038 Th 90	231.031 Pa 91	238.029 U 92	237.048 Np 93	(244) Pu 94	(243) Am 95	(247) Cm 96	(247) Bk 97	(251) Cf 98	(254) Es 99	(253) Fm 100	(256) Md 101	(253) No 102	(257) Lr 103

Numbers below the symbol of the element indicate the atomic numbers. Atomic masses, above the symbol of the element, are based on the assigned relative atomic mass of ^{12}C = exactly 12; ( ) indicates the mass number of the isotope with the longest half-life. [ ] indicates not officially approved or named.

Plant nutrients | Heavy metals | Plant nutrients and heavy metal | Beneficial plant nutrients | Radioactive | Other toxic elements

Glossary*

A

A horizon. *See* Soil horizon.

AASHTO engineering classification system. Originally a Public Roads system modified and adopted by the American Association of State Highway and Transportation Officials.

AB horizon. *See* Soil horizon.

Abiotic. The nonliving factors in the biosphere: rainfall, minerals, temperature, wind, and others.

Absorption, active. Movement of ions and water into the plant root as a result of metabolic processes by the root, frequently *against* an activity gradient. *See also* Adsorption.

Absorption, passive. Movement of water into roots resulting from pulling forces on the water column in the plant as water is lost to the atmosphere through the leaves by transpiration (a kind of *wick* action).

AC horizon. *See* Soil horizon.

Acceptable daily intake (ADI). The amount of material, including a safety factor, that is determined safe to ingest.

Acidic cations. Hydrogen ions or cations in water that undergo hydrolysis to form an acidic solution, as do Al^{3+} and Fe^{3+}.

Acid rain. Rainfall or melting snow with pH <5.7.

Acidic soil. Soil with a pH lower than 7.0.

Acidic sulfate soils. Soils strongly acidic (pH 3.5) or potentially so and with relatively large amounts of sulfides (or jarosite) when anaerobic or sulfates when aerobic.

**Sources:*

(1) *Glossary of Soil Science Terms,* Soil Science Society of America, 1979, 1987. Selections used with permission.

(2) *Resource Conservation Glossary,* Soil Conservation Society of America, 1976. The complete glossary is available from 7515 Northeast Ankeny Road, Ankeny, IA 50021. These selections have been used with permission.

(3) *Soil Series of the United States, Puerto Rico, and the Virgin Islands: Their Taxonomic Classification.* USDA—Soil Conservation Service, 1972.

(4) Soil Survey Staff, Soil Conservation Service, *Soil Taxonomy: A Basic System of Soil Classification for Making and Interpreting Soil Surveys,* Agriculture Handbook 436, USDA, Washington, DC, 1975.

(5) Robert L. Bates and Julia A. Jackson, eds., *Glossary of Geology,* 2nd ed., 1980. American Geologic Institute, Falls Church, VA.

(6) Soil Survey Staff, "Soil Families and Their Included Series," and "Classification of Soil Series," USDA—Soil Conservation Service, Washington, DC, 1988.

Acidity, active. The acidity in the soil solution; the free hydrogen measured as pH.

Acidity, residual. Exchangeable acidity (H^+ and aluminum forms) that can be neutralized by alkaline materials (lime).

Acre. An area of 43,560 ft^2, or 0.405 ha.

Actinomycetes. A nontaxonomic term applied to a group of filamentous bacteria with characteristics intermediate between simple bacteria and the true fungi.

Activation energy. The energy required before the reaction of concern will proceed. Often, heating will supply the needed energy. Enzymes or other catalysts lower the activation energy of a reaction.

Acute toxicity. Pronounced and immediate toxicity. Symptoms are exhibited by the organism in a short period of time. *See* chronic toxicity.

Adhesion. Molecular attraction that holds the surfaces of two dissimilar substances in contact, such as water and rock particles. *See also* Cohesion; H bond.

Adsorption. The bonding, usually temporary, of ions or compounds to the surfaces of a solid, such as a calcium ion held on the surface of a clay crystal.

Aeration, soil. The process by which air in the soil is replaced by air from the atmosphere. Poorly aerated soils usually contain a much higher percentage of carbon dioxide and a correspondingly lower percentage of oxygen than the atmosphere above the soil.

Aerobic. 1. Having molecular oxygen as a part of the environment. 2. Growing only in the presence of molecular oxygen, as aerobic organisms. 3. Occurring only in the presence of molecular oxygen (said of certain chemical or biochemical processes, such as aerobic decomposition).

Aflatoxins. Potent carcinogens in nature that are antibodies produced by fungal growth on many grains and nuts.

Aggregation, soil. The cementing or binding together of several soil particles into a secondary unit, aggregate, or granule.

Aglime. *See* Lime.

Agronomy. A specialization of agriculture concerned with field crop production and soil management. The scientific management of land.

Albic horizon. A strongly leached, light-colored mineral horizon, designated by the letter *E. See* Chapter 2.

Alfisols. *See* Soil classification: Order.

Algal bloom. Rapid increase in algae growth because of increased nutrients, particularly phosphorus. Water is turbid and colored green or reddish by the algae.

Alkali. (obsolete) *See* Saline-sodic soil; Sodic soil.

Alkaline soil. Any soil having a pH greater than 7.0. A *basic* soil.

Allelopathy. The action of some substance produced by or in one plant species that reduces the growth of another plant species. Mostly these are toxic organic materials produced either by the plant or during decomposition of plant residues.

Allophane. Amorphous (noncrystalline) clay-sized aluminosilicate.

Alluvial fan. Fan-shaped alluvium deposited at the mouth of ravines or canyons as the slope lessens and water-carried debris settles out of the slowed water.

Alluvium. Eroded soil sediments deposited from flowing water.

Alumina. The complex of aluminum and oxygen (and hydroxyl) that averages to Al_2O_3 or some combination of OH for O.

Aluminosilicate. Minerals dominantly of aluminum (Al), silicon (Si), and oxygen (O).

Amendment, soil. Any substance (gypsum, lime, manures, sewage sludge, sawdust, composts, etc.) added to the soil to improve plant growth.

Amensalism. Production by one organism of a substance that is inhibitory to other organisms. Similar in use to the term *allelopathy.*

Ammonia volatilization. Loss of nitrogen to the atmosphere as gaseous ammonia (NH_3), increasing with pH above 8–8.5.

Ammonification. The biochemical process whereby ammoniacal nitrogen is released from nitrogen-containing organic compounds.

Ammonium fixation. Adsorption of ammonium ions (NH_4^+) by the soil mineral fraction into interlayer positions that cannot be replaced by a neutral potassium salt solution (e.g., 1 *N* KCl).

Amorphous materials. Lacking adequate crystal orientation to diffract x-rays. Without regular crystalline form in molecular sizes.

Anaerobic. 1. The absence of molecular oxygen. 2. Growing in the absence of molecular oxygen (such as anaerobic bacteria). 3. Occurring in the absence of molecular oxygen (as a biochemical process).

Anaerobic respiration. Metabolic transfer of electrons to ions other than oxygen, predominantly to nitrate, manganese oxide, ferric ion, sulfate, and carbon forms.

Andisol. Soils with weakly developed horizons; formed in volcanic ejecta parent materials.

Animalia. One of the five kingdoms categorizing all living organisms. *See* Chapter 7.

Anion. Negatively charged ion; ion that during electrolysis is attracted to the anode (positively charged electrode).

Anion exchange capacity. The sum total of exchangeable anions that a soil can adsorb.

Antagonism. An effect of substance A to reduce the detrimental effects of substance B below the effect substance B would have if it were alone in the system.

Anthosols. In archaeology, ancient soils or horizons greatly influenced by people's behavior, such as yearly manuring, dumping garbage, or exhausting nutrients.

Anthropogenic, anthropic. Changes in soils caused by people, such as plowing, fertilizing, and construction.

Aquic conditions. A soil-water regime, mostly too wet (reducing conditions, waterlogged) for parts of the year. *See* Chapter 7.

Aquifer. A geologic formation that transmits water underground, usually sands, gravel, and fractured, porous, cavernous, and vesicular rock.

Arable land. Land suitable for the production of cultivated crops.

Argillan. *See* Clay film.

Argillic horizon. A diagnostic horizon of clay accumulation often designated as *Bt. See* Chapter 2.

Arid. A term applied to regions or climates that lack sufficient moisture for crop production without irrigation. The upper annual limit for cool regions is 250 mm (10 in) or less and for tropical regions as much as 380–510 mm (15–20 in). *See also* Semiarid.

Aridic. A soil water regime with long dry periods. *See* Chapter 7.

Aridisols. *See* Soil classification: Order.

Artesian well. A dug well that flows without pumping because the well top is below the water table in the aquifer.

Aspect. The direction that a slope faces.

Atterberg limits. Water content of manipulated soil at different consistency. *See* Liquid limit, Plastic limit.

Attribute. 1. A characteristic of a geographic feature described by numbers or characters, typically stored in tabular format and linked to the feature by a user-assigned identifier. 2. A numeric, text, or image data field in a relational database table that describes a spatial feature such as a point, line, node, area, or cell.

Autotrophic. Capable of utilizing carbon dioxide and/or carbonates as a sole or major source of carbon and of obtaining energy for carbon reduction and biosynthetic processes from radiant energy (photoautotroph) or oxidation of inorganic substances (chemoautotroph).

Available nutrient. That portion of any nutrient in the soil that can be absorbed readily by growing plants. Same meaning as *plant-available.*

Available water. The portion of water in a soil that can be readily absorbed by plant roots. Considered by most workers to be that water held in the soil with a water potential of −33 kPa to approximately −1500 kPa. Same as plant-available. *See also* Field (water) capacity; Permanent wilting point; Soil water potential.

Available water capacity. The capacity of soils to hold water available for use by most plants, usually defined as water between −33 kPa and −1500 kPal. In a 2 m profile, or a more shallow limiting layer, the values are as following:

Very low	0–3 in	0–7.5 cm
Low	3–6 in	7.5–15 cm
Moderate	6–9 in	15–23 cm
High	9–12 in	23–30 cm
Very high	More than 12 in	More than 30 cm

B

B horizon. *See* Soil horizon.

BA horizon. *See* Soil horizon.

Bacteroid. Clump of bacteria cells, particularly the swollen, vacuolated mass of bacteria in *nodules* of legumes.

Badland. A highly eroded area with little vegetation, often with narrow ravines and sharp ridges. Common in arid regions with deep alluvium deposits.

Bajada. Coalescing adjacent alluvial fans; a bajada is almost a continuous slope of alluvium along the side of a mountain.

Banding. Applying fertilizer or other amendment into the soil (7–15 cm, or 2.7–6 in, deep) in a thin narrow strip (band), as beside or beneath a planted row of seeds or plants.

Bar. A unit of pressure equal to 1 million dynes/cm^2 or 100 kilopascals.

Base saturation percentage. (basic cation saturation) The degree to which the adsorption complex of a soil is saturated with basic cations (cations other than hydrogen and aluminum), usually expressed in percentage.

BAT. *See* Best available technology.

BC horizon. *See* Soil horizon.

Bedrock. The solid rock underlying soils.

BE horizon. *See* Soil horizon.

B/E horizon. *See* Soil horizon.

Bedder. A sweep that looks somewhat like a small moldboard plow with the curved sides on both sides. It is used to build ridges in ridge-furrow cultivation.

Bentonite. A highly plastic clay consisting of the minerals montmorillonite and beidellite (smectites), which swell extensively when wet. Also called *volcanic clay, soap clay,* and *amargosite.*

Best available technology (BAT). The technological practices determined as best to reduce pollution or other undesirable effects.

Best management practices (BMP). The management practices determined to permit the least pollution or other undesirable effect.

Bioaccumulate (biomagnify). This term indicates an accumulation of a substance in increased concentrations over time by an organism. Biomagnify usually means increasing the concentration upward in the food chain.

Biodegradable. Decomposed by natural biochemical processes.

Biological magnification. (biomagnification) The accumulation of increasing amounts of a substance in organisms as one progresses up the food chain (e.g., low in plankton, higher in small fish, large fish, and highest in fish-eating birds).

Biological oxygen demand (BOD). The amount of dissolved oxygen in water that will be consumed as the organic matter present is decomposed. High BOD means low water quality and probably the development of anaerobic waters. It usually results when waters have received organic wastes. *See also* Chemical oxygen demand.

Biological pest management. Control of pests by organism enemies (insects, microorganisms).

Biomagnify. Increasing the concentration of a substance upward in the food chain.

Biomass. Total amount of living organisms and their residues in a volume or mass of the environment.

Bioremediation. Reduction or elimination of a pollutant by the use of organisms. *See* Phytoremediation; Remediation.

Biosphere. The environment in which living organisms live: the Earth's crust, vegetation, and atmosphere near the Earth.

Blocky soil structure. Aggregates with block-like (equidimensional) shapes.

BMP. *See* Best management practices.

Brackish. Slightly salty. Water with a content of 1.5–3% salts; seawater has more than 3% salts.

Breccia. Rock of coarse fragments cemented together; sometimes it is a cracked rock recemented.

Broadcast. Fertilizer is uniformly spread on the soil surface. It may or may not be incorporated into the soil.

Buffering. 1. The capacity of the soil solids and liquids to resist appreciable change in pH of the soil solution. 2. The ability to maintain the approximate concentration desired of any ion in the soil solution.

Bulk density, soil. The mass (weight) of dry soil per unit bulk volume.

Buried soil. A soil covered by transported soil material to a depth of at least 50 cm.

C

C horizon. *See* Soil horizon.

Cablegation. A surface-flow furrow irrigation system that uses a long cable connected to a sliding plug to control opening and closing of irrigation pipe gates.

Calcareous soil. Soil with sufficient carbonates (mostly calcium) to effervesce visibly with cold 0.1 *N* hydrochloric acid.

Calcic horizon. A diagnostic mineral horizon of carbonate accumulation. Indicated by the letter *k.*

Caliche. A pedologic layer near the surface, cemented by secondary carbonates of calcium or magnesium precipitated from the soil solution.

Cambic horizon. A weakly developed diagnostic subsoil horizon. Indicated by the letter *w.*

Cancer potency factor. A negligible risk, or potentially causing no more than *one additional cancer per million people.*

Capillary water. Plant-available water held in the capillary or small pores of a soil, usually with attraction forces exceeding the pull of a 60-cm (2-ft) column of water.

Carbon:nitrogen ratio. The ratio of the weight of organic carbon to the weight of total nitrogen in the soil or in organic material.

Carbonation. A chemical weathering process combining the action of acidity and formation of soluble bicarbonates from carbon dioxide dissolved in water to form carbonic acid.

Carboxyl group. The acid group of soil organic matter (—COOH), which is the most active group in soil humus in adsorbing cations.

Carcinogen. A substance that causes cancer.

Cat clays. Strongly acidic clays resulting from aeration of anaerobic soils containing many sulfide minerals. The sulfides are oxidized to sulfuric acid in aerated soil. In the Netherlands called *katteklei.*

Catena. A sequence of different soils formed under similar soil-forming factors except for the effect of their relief positions which altered erosion and drainage of each different soil.

Cation exchange. Replacement by a cation in solution for an adsorbed cation on negatively charged sites of a solid.

Cation exchange capacity (CEC). The sum total of exchangeable cations that a soil can adsorb, expressed in $centimoles_c$ per kg of soil or colloid.

CB horizon. *See* Soil horizon.

Celsius degree. A degree (°C) such that the difference between the freezing and boiling temperatures is divided into 100 Celsius degrees. Celsius temperature times 1.80 plus 32 equals degrees Fahrenheit (°F).

Cemented (indurated). Having a hard, brittle consistency, even when wet, because the particles are cemented together. *See* Hardpan; Indurated.

Chelation. The formation of strong bonds between metals and organic compounds. Some chelates are insoluble, such as in soil humus.

Chemical oxygen demand (COD). A measure of the oxygen-consuming capacity of inorganic and organic matter present in water or wastewater. *See also* Biological oxygen demand.

Chemigation. Application of pesticides and fertilizers through mixing them into irrigation water, usually in sprinkler or drip systems. *See also* Fertigation.

Chisel. A narrow shank, usually with a narrow sweep (with no wings or only small wings) fastened to its tip. The chisel is pulled through the soil to rip it.

Chisel plow. Large chisels, usually about 0.6–1 m (2–3 ft) apart, pulled through the soil to rip it to 15–30 cm (6–12 in) deep.

Chiseling. Tillage with an implement having one or more soil-penetrating points that shatter or loosen hard, compacted layers, often to a depth below normal plow depth.

Chlorinated hydrocarbon. Any of a number of organic pesticides with chloride groups on carbon atoms. Nearly all chlorinated hydrocarbons have long half-lives in the environment, are of low water solubility, and are nonselective in their action. The best known is DDT (**d**ichloro**d**iphenyl**t**richloroethane). Others include chlordane, aldrin, heptachlor, and dieldrin.

Chlorite. A nonexpanding clay mineral having a silica tetrahedral, an alumina octahedral, a silica tetrahedral, and a magnesium hydroxide (brucite) octahedral layer; has a 2:2 or 2:1:1 crystal structure.

Chlorosis. A loss of normal green color of the plant.

Chroma. The relative purity, strength, or saturation of a color.

Chromatography. The movement of substances in gas or liquid through a stationary matrix, usually solid, at different rates because of differential attraction of the substances to the stationary matrix.

Chronic toxicity. Low-level toxicity. Distinct effects or symptoms may be very slow to develop or not be evident at all; they may appear months or years after the initial exposure. *See* Acute toxicity.

Chronosequence. A sequence of soils related to each other in physical location but having different times of development.

Classification, soil. *See* Soil classification.

Clay. 1. A mineral soil separate consisting of particles less than 0.002 mm in equivalent diameter. 2. A soil textural class. 3. A fine-grained soil that has a high plasticity index in relation to the liquid limits (engineering). 4. A specific mineral structure.

Clay films. Coatings of clay on the surfaces of soil peds and mineral grains and in soil pores. Also called *clay skins, cutans, argillans,* or *tonhautchen.*

Claypan. A dense, compact layer in the subsoil having a much higher clay content than the overlying material. It usually impedes the movement of water and air and the growth of plant roots. Compare with Hardpan.

Cleavage of minerals. The smoothness of regular breaking surfaces of minerals; *perfect* is very smooth; *conchoidal* is convex and not smooth.

Clod. A compact, coherent mass of soil ranging in size from 5 to 10 mm (0.2–0.4 in) to as much as 200–250 mm (8–10 in); produced artificially by digging or tillage. *See also* Ped.

Coarse fragments in soils. Includes rock fragments between 2 mm and 25 cm (0.08–9.8 in) in diameter.

Coarse texture. Texture is sands, loamy sands, and sandy loams.

Cobblestone. Rounded rock and mineral particles of 8–25 cm (3–10 in) diameter.

COD (chemical oxygen demand). The oxygen-consuming capacity of all the substances in water. *See* Biological oxygen demand.

Cohesion. Force holding a solid or liquid together because of attraction between like molecules.

Colloid. A substance that, when suspended in water, diffuses not at all or very slowly through a semipermeable membrane; a substance in a state of fine subdivision with particles from 1 micrometer to 1 nanometer.

Colluvium. A deposit of rock fragments and soil material accumulated at the base of steep slopes as a result of gravitational action.

Color. *See* Munsell color notation system.

Columnar structure. Vertically oriented, round-topped structural prisms.

Complex, soil. A map unit of two or more kinds of soil in such an intricate pattern or so small in area that it is not practical to map them separately at the scale of mapping.

Compost. Organic residues or a mixture of organic residues and soil that have been piled and allowed to undergo biological decomposition.

Concretion. A local concentration of a chemical compound, such as calcium carbonate or iron oxide, in the form of an aggregate, a pellet, or nodule.

Conductivity, hydraulic. *See* Hydraulic conductivity.

Conservation. The protection, improvement, and use of natural resources according to principles that will ensure their highest economic, social, and psychological benefits to people in perpetuity.

Conservation tillage. The U.S. Natural Resource Conservation Service defines *conservation tillage* as any tillage system that leaves at least 30% of the surface covered by plant residues for control of erosion *by water;* for controlling erosion *by wind,* it means leaving at least 1120 kg/ha (1000 lb/a) of small-grain-straw-equivalent during the critical wind erosion period. The amount of residue needed depends on the kind of residue and whether it is standing or flat.

Consistence. 1. The resistance of a material to deformation or rupture. 2. The degree of cohesion or adhesion of the soil mass. Used for describing consistency of soil materials at various soil moisture contents and degrees of cementation.

Consumptive use. The total quantity of water transpired by vegetation plus that evaporated from soil plus that in the plant material. About 0.5% to 1.0% larger than evapotranspiration.

Contamination. Any substance added or accumulating in air, water, or soil that makes the air, water, or soil less desirable for people's use. Same as *pollution.*

Contour. 1. An *imaginary* line on the surface of the Earth connecting points of the same elevation. 2. A *true* line drawn on a map connecting points of the same elevation.

Contour stripcropping. Layout of crops in comparatively narrow strips in which the farming operations are performed approximately on the contour. Usually, strips of grass, close-growing crops, or fallow are alternated with those in cultivated crops.

Control section. A defined depth of the soil profile used for a specific determination (e.g., water control section, texture control section, or temperature control section).

Conventional tillage. By the most general definition, conventional tillage is a combination of tillage operations traditionally used for a given region and crop. A typical sequence might be moldboard plowing, following by two disking operations, then harrowing and seeding. By another definition conventional tillage is any tillage that leaves less than 15% of soil covered with crop residues at the time of planting.

Coulter. A sharpened disk, with a straight or fluted edge, often running in front of a chisel or small sweep to cut the surface vegetation and to open a slit in the soil.

Cover crop. A dense-cover crop grown when other crops are not grown (winter, other) to protect soil and to be plowed under for improving the soil.

Creep. (soils) Slow mass movement of soil and soil material down relatively steep slopes, primarily under the influence of gravity but facilitated by saturation with water and by alternate freezing and thawing.

Cross-slot drill. This planter opens a slot (furrow), drops the seed, and then covers the slot and seed with soil to retain soil moisture and to firm the soil over the seed.

Crust. A surface layer on soils, ranging in thickness from a few millimeters to perhaps as much as 25 mm (1 in), that is much more compact, hard, and brittle when dry than the material immediately beneath it.

Cryic. An annual soil temperature regime (0–8°C) in quite cold environments.

Cultivator. An implement fitted with several small sweeps that are pulled 5–10 cm (2–4 in) deep between crop rows to cut and kill weeds. Wings (sides of sweeps) are typically 10–25 cm (4–10 in) long.

Cultural control. (pest management) Physical methods of controlling pests (e.g., cleaning, screens, plowing, removal of trash).

Cyanobacteria. New name for N_2-fixing microbes; previously called *blue-green algae.*

D

Darcy's law. A law describing the saturated flow of water through a porous media.

Database. A logical collection of interrelated information, managed and stored as a unit, usually on some form of mass-storage system such as magnetic tape or computer disk (e.g., a GIS database includes data about the spatial location and shape of geographic features recorded as points, lines, areas, grid cells, or pixels as well as their attributes).

Deep banding. Fertilizer is placed in a strip at a depth of 14–18 cm (or deeper); usually the strips are 30–70 cm apart, placed at planting or in preplant with regard to rows.

Degraded soil. Soil that is less productive than previously, usually because of damages from erosion, loss of humus, loss of fertility, or accumulated salts or other pollutants.

Delta. An alluvial deposit formed where a stream or river drops its sediment load on entering a body of more quiet water.

Denitrification. The biochemical reduction of nitrate or nitrite to gaseous nitrogen, either as molecular nitrogen or as an oxide of nitrogen.

Desert pavement. The layer of gravel or stones left on the land surface in desert regions after the removal of the fine material by wind and water erosion.

Desertification. The decline in productivity and vegetative cover of arid and semiarid soils caused by natural or person-made (anthropogenic) stresses.

DGPS. *See* Global Positioning System.

Diagnostic horizons. Horizons with combinations of specific substances that are indicative of certain kinds of soil development.

Diatomaceous earth. Geologic fine mineral material of siliceous skeletons of algae (diatoms).

Diffuse double layer. A theoretical zone of the soil solution near the soil particles' negatively charged surface, only a few ionic diameters thick, where cations are strongly adsorbed.

Diffusion. The slow movement of an ion in water mostly by its own kinetic motion.

Dinitrogen fixation. Conversion of gaseous dinitrogen (N_2) in the air to organic nitrogenous substances by certain bacteria, algae, and actinomycetes.

Disk. A combination of concave-shaped circular blades. Disks within a set are parallel to each other. Some disks have multiple sets (or gangs) of disks. The common arrangement is a front set and a rear set arranged opposite each other and slightly offset from being perpendicular to the direction of movement through the field. The *disk plow* used for primary tillage usually has only one large set of disks.

Disk harrow. A combination set of closely spaced disks to break the large clods to smaller units. Sometimes a vertical *tine harrow* is attached.

Disk plow. A combination of large concave-shaped disks capable of turning soil to depths of 15–40 cm (6–16 in).

Dispersion. The breaking down of soil aggregates into individual particles. Generally, the more easily dispersed the soil, the more erodible it is. Favored by high exchangeable sodium.

Drainage class. The description of the ease with which a soil drains off excess water by percolation. Terms such as *well drained, imperfectly drained,* or *very poorly drained* are used.

Drumlin. Long elongated convex hills cut by glaciers and left covered by glacial till.

Dry weight. (soils) The equilibrium weight of the solid soil particles after the water has been vaporized by heating to 105°C (221°F).

Dryland farming. Rainfed farming in arid and semiarid regions without the use of irrigation.

Duff. The partially-decomposed organic-residue mat lying on the forest floor.

Duripan. *See* Diagnostic horizons.

E

E horizon. A strongly leached layer, often referred to as an *albic horizon. See also* Soil horizons.

Earth flow. The process of soil moving downslope because of the pull of gravity, usually lubricated by water.

EB horizon. *See* Soil horizons.

EC (electrical conductivity). Measured in Siemens/meter. EC_e = value from a saturated soil paste extract.

Ecology. The totality of relationships among organisms and their ambient (surrounding) environment.

Edaphology. The science that deals with the influence of soils on living things, particularly plants, including human use of land for plant growth.

Efflorescence. Accumulation of dried soluble salts on the soil surface, left as the water carrying the salts has evaporated.

Effluent. 1. The discharge or outflow of water from ground or subsurface storage. 2. The fluids discharged from domestic, industrial, and/or municipal waste collection systems or treatment facilities.

Eh. The electrical potential between a half-cell oxidation or reduction reaction and the H electrode.

Eluviation. The downward removal of soil material in suspension (or in solution) from a layer of soil.

Entisols. *See* Soil classification: Order.

Enzyme. A protein mass with surface chemical groups arranged so that atoms around a certain kind of bond can attach to the enzyme, allowing the enzyme to act as a catalyst to reduce the activation energy for a chemical reaction (such as splitting of a bond).

Eolian (aeolian) soil material. Soil material accumulated through wind action.

Epipedon. A term used with diagnostic horizons meaning that the horizon formed while at or near the soil surface.

Erosion. The wearing away of the land surface by water, wind, ice, or other geological agents, including such processes as gravitational creep.
Accelerated erosion. Erosion much more rapid than normal, natural, or geologic, primarily as a result of the influence of the activities of humans or other animals or natural catastrophes, such as fires and earthquakes.
Geological erosion. The normal erosion caused by geological processes acting over long geologic periods. Synonym: *natural erosion.*
Gully erosion. Erosion whereby water accumulates in narrow channels and over short periods removes the soil from this narrow area (gully) to depths ranging from 30 cm to 30 m (1–100 ft) or more.
Rill erosion. An erosion process in which numerous small channels several centimeters deep are formed. *See also* Rill.
Saltation. Bouncing or jumping action of particles falling from wind or water or impacted by other falling particles in wind or water flow.
Sheet erosion. The removal of a fairly uniform layer of soil from the land surface.
Splash erosion. The spattering of small soil particles caused by the impact of raindrops on wet soils.
Surface creep. Rolling or slow movement of particles too large to be carried in wind or flowing water (also gravity *flow*).

Erosion pavement. *See* Desert pavement.

Esker. Narrow ridge of sands and gravels deposited from flowing glacial waters through ice tunnels under the glacier.

Essential element. (plant nutrition) A chemical element required for the normal growth and reproduction of plants, people, or other animals.

Eucaryote. (you-car′-ry-ot) Organism cell type that has the genetic material enclosed inside a nucleus.

Eutrophic (eutrophication). Nutrient enrichment of water. It is usually most affected by phosphorus; less so by nitrogen. Increased algal growth eventually loads the water with dead algae, which, during microbial decomposition, results in consumption of the water's dissolved oxygen, causing water life to die.

Evaporites. Residue of gypsum plus all salts more soluble than gypsum, which solidify as water evaporates.

Evapotranspiration (ET). Water transpired by vegetation plus that evaporated from the soil. Approximate synonym: consumptive use.

Exchangeable cation percentage (ECP). The percentage of the cation adsorption complex of a soil occupied by a particular cation. It is expressed as follows:

$$\text{ECP} = \left(\frac{\text{exchangeable cation}}{\text{cation exchange capacity}}\right)(100)$$

Exchangeable ion. Any ion held through electrical attraction to a charged surface; can be displaced by other ions from surrounding solution.

Exchangeable sodium percentage (ESP). The percentage of the cation exchange capacity occupied by sodium.

Extrusive rocks. Volcanic igneous rocks formed from molten magma that cooled rapidly (has glassy nature of small crystals) by exposure above the earth's crust.

Exudation. The excretion or natural elimination of substances from the plant.

F

Fahrenheit degree. A degree such that the temperature between freezing of water and boiling of water is divided into 180 degrees. Water freezes at 32°F and boils at 212°F. Degrees Fahrenheit minus 32 times 0.555 5 equals °C.

Fallow. Land left without a crop for a period of several months to a year, with weeds or volunteer plants killed. Fallow is used in marginal rainfall areas (25–50 cm, or 10–20 in) to accumulate a few extra centimeters of water for the next crop. To kill weeds V-sweeps are used. They have wings 45–60 cm (18–24 in) long, but some up to 100 cm (39 in) are used.

Family, soil taxonomy. *See* Soil classification: Family.

Feldspars. Minerals made up of silicates of aluminum, with potassium, calcium, or sodium ions; very low solubility.

Ferrihydrite. An amorphous or poorly crystalline iron hydrous oxide mineral.

Fertigation. A term coined for application of fertilizers in irrigation waters, usually through sprinkler systems. *See also* Chemigation.

Fertilizer. Any material, except lime, added to soil to supply one or more essential elements.

Fertilizer analysis. The actual composition of a fertilizer as determined in a chemical laboratory by standard methods.

Fertilizer grade. The guaranteed minimum analysis in whole numbers, in percent, of the nitrogen, phosphorus, and potassium in a fertilizer material. For example, a fertilizer with a grade of 20-10-5 is guaranteed to contain 20% *total* nitrogen (N), 10% *available* phosphoric acid (P_2O_5), and 5% *water-soluble* potash (K_2O).

Fertilizer, starter. A small application of fertilizer applied with or near the seed for accelerating early growth of the crop.

Fibric material. Organic material, only slightly decomposed.

Field (water) capacity. The amount of water remaining in a soil after the soil layer has been saturated and the free (drainable) water has been allowed to drain away (a day or two). Estimated at −33 kPa water potential.

Fine texture. A general term to indicate a soil with large portions of clay and silt.

Fixation. The processes by which chemical elements are converted from a soluble or exchangeable form to a much less soluble or to a nonexchangeable form.

Flocculate. To aggregate, or clump together, small soil particles into small clumps or granules that usually settle out of suspension quickly.

Floodplain. Nearly level land situated on either side of a channel that is subject to overflow flooding, unless protected artificially.

Flux. The time rate of flow of a quantity (substance) across a given area cross-section.

Foliar diagnosis. An estimate of a plant's nutrient deficiencies or sufficiencies by analytical measurement of selected parts of the plants.

Foliar spray. The application of liquid fertilizer to the foliage of plants, just enough to wet the leaves.

Folic. A thin organic soil of leaves over solid rock.

Food chain (food web). The transfer of energy and material through a number of organisms, one being food for the next (small fish eaten by larger fish eaten by a bird, then eaten by a hunter). The *food web* is the complex intermeshing of many food chains.

Formative element. A syllable indicating a soil property that is added to a soil order name's root to make a suborder name (or to a suborder name to make a great group name).

Fossil soils. Buried paleosols; unchanged since deep burial. *See* Paleosols, relict soils.

Fragipan. A hard, dense, brittle-when-dry soil pan.

Fragmental. Stones, cobbles, gravel, and very coarse sand with too little fine sand, silt, and clay to fill soil interstices larger than 1 mm.

Friable. Soils that crumble easily.

Frigid. A cool soil temperature regime (0–8°C) with warm summers. *See* Chapter 7.

Fritted trace elements. Sintered silicates (glass fragments) having guaranteed analyses of micronutrients with slow release characteristics.

Frost heaving. The lifting of the surface soil by growing ice crystals in the underlying soil. Such action often pushes plants out of the ground.

Fulvic acid. An indefinite term for the mixture of organic substances remaining soluble after a soil extract, using dilute alkali, has been acidified.

Fungi. One of five kingdoms into which all organisms are fitted.

G

Ganat. (also *Qanat*) Gently sloping underground tunnels into shallow aquifers to obtain gravity flow streams.

Gated pipe. Portable pipe with small gates (slide openings) installed along one side for uniformly distributing irrigation water to corrugations or furrows.

Genesis, soil. The mode of origin of the soil, especially to the processes or soil-forming factors responsible for the formation of the solum (true soil) from the unconsolidated parent material.

Genetic engineering (organisms). The process of altering the genes of organisms to improve their action for people's purposes.

Genetic (soil) horizon. Horizons formed by natural processes (weathering, translocations, transformations, and additions) in place.

Geographical Information System (GIS). An organized collection of computer hardware, software, geographic data, and personnel designed to efficiently capture, store, update, manipulate, analyze, and display all forms of geographically referenced information.

Geological erosion. *See* Erosion.

Geophagy. The eating of soil by animals and people. *See also* Pica.

Gibbsite, $Al(OH)_3$. Aluminum trioxide mineral, often the residue in extensively weathered soils in hot wet climates.

Gilgai. The microrelief of clayey (montmorillonitic) soils produced by expansion and contraction with wetting and drying. A succession of microbasins and microknolls. Diagnostic for Vertisols.

GIS. *See* Geographical Information Systems.

Glacial outwash. Gravel, sand, and silt deposited by glacial meltwater; it is commonly stratified.

Glacial till. Unstratified, nonsorted materials deposited from melting glaciers and consisting of clays, silts, sands, gravels, and boulders. Also called Glacial drift.

Gley. Some layer of mineral soil developed under conditions of poor drainage (poor aeration), resulting in reduction of iron and other elements and in gray colors and mottles (blobs of variously colored soils).

Global Positioning System (GPS). A satellite-based device that records *x-y-z* coordinates and other data. GPS devices can be taken into the field to record the location of data while driving, flying, or hiking. Ground locations are calculated by signals from satellites orbiting the Earth. A **differential GPS (DGPS)** is interacting two or more receiving units together which increases the accuracy of site location.

Glossic horizon. A degrading argillic, kandic, or natric horizon, having 15%–85% albic-like materials in it. At least 5 cm thick.

Goethite. A yellowish-brown iron oxide mineral, a common cause of yellowish-brown soil color.

Graded. (coarse fragments). Sorted by many sizes. *Poorly graded* have few sizes and do not pack well. *Well graded* have many sizes and compact well.

Granule. A natural soil ped or aggregate.

Grassed waterway. A natural or constructed waterway, typically broad and shallow, seeded to grass as protection against erosion. Usually conducts surface water away from cropland.

Gravel. Rounded or angular fragments of rock 2 mm to 7.6 cm (up to 3 in) in diameter. An individual piece is a pebble.

Gravitational potential. The amount of work an infinitesimal amount of pure free water can do at the site of the soil solution as a result of the force of gravity.

Great soil group. A category of the U.S. Soil Taxonomy system. Soils are placed according to soil moisture, temperature, base saturation status, and expression of horizons.

Greenhouse effect. Entrapment of heat, thus warming the Earth, by an increase in amounts of gases (carbon dioxide, methane) and water vapor in the atmosphere. If this is occurring, serious consequences are likely.

Green-manure crop. Any crop grown for the purpose of being turned under while green or soon after maturity for soil improvement.

Groundwater. Subsurface water in the zone of saturation.

Gully. Large eroded channels. *See also* Erosion.

Gypsic horizon. A diagnostic horizon having an accumulation of gypsum.

Gypsum requirement. The quantity of gypsum or its equivalent required to reduce the exchangeable-sodium percentage of a soil to an acceptable level.

H

H bond. (hydrogen bond) The bond of a hydrogen atom already strongly attached to one electronegative atom (such as oxygen, nitrogen, sulfur), less strongly to a second electronegative ion. H bonds are common between water molecules, between water and mineral oxygens, and between organic-substance hydrogens, and mineral oxygens.

Half-life. The time for half of a substance to be destroyed or inactivated or, for radioactive substances, to lose half of its radiation.

Halophyte. Any organism that grows in salty environments and readily absorbs salts.

Hardpan. A hardened soil layer in the lower A or a deeper horizon caused by cementation of soil particles with organic matter, silica, sesquioxides, or calcium carbonate.

Harrow. A set of vertical tines that, when pulled over the soil, break the larger clods into smaller pieces. Often the harrow is the final equipment pulled over soil to prepare the seedbed.

Heavy metals. Those metals of high atomic weight having densities greater than 5.0 Mg/m^3. Many are toxic when accumulated into animal bodies; some heavy metals are cadmium, chromium, copper, beryllium, lead, manganese, mercury, nickel, and zinc, among many others.

Heavy soil. (obsolete scientifically) An inexact term still widely used to indicate clayey soils (hard to till).

Hectare. (ha) A metric unit equalling 10,000 square meters or 2.471 acres.

Hematite. A red iron oxide (Fe_2O_3, rust) that contributes red coloring to soils.

Hemic material. Organic residues in organic soils (Histosols) having an intermediate stage of decomposition. *See* Histosols in Chapter 7.

Heterotroph. An organism capable of deriving energy for life processes from the oxidation of organic compounds.

Histic horizon. A thin diagnostic epipedon of organic material formed by periods of excess wetness (anerobic, waterlogged).

Histosols. *See* Organic soils; Soil classification: Order.

Horizon. *See* Soil horizon.

Hue. One of the three variables of color, the rainbow color of light reflected from each soil.

Humic acid. A mixture of dark-colored organic materials of indefinite composition extracted from soil with dilute alkali and precipitated upon acidification.

Humin. Soil organic matter insoluble in dilute alkali solution.

Humus. The fraction of the soil organic matter remaining, usually amorphous and dark colored, after the major portion of added residues have decomposed.

Hydraulic conductivity. (K) The proportionality factor in Darcy's law, indicating the soil's ability to transmit flowing water.

Hydric soil. A soil with periods of wetness that exhibits evidence of that wetness (mottles, gleying, redox conditions at times).

Hydrogen bond. *See* H bond.

Hydrologic cycle. The circuit of water movement from the atmosphere to the Earth and return to the atmosphere through various stages or processes, as precipitation, runoff, percolation, storage, evaporation, and transpiration.

Hydrologic soil groups. Soils are grouped according to their runoff-producing characteristics, closely associated with bare soil to permit infiltration. Slope and vegetation are not considered. All soils fit four groups. **Group A:** high infiltration rate when wet with low runoff potential (deep, sandy, or gravelly). Other extreme is **Group D:** very slow infiltration and high runoff potential (clayey, claypan, permanent high water table, shallow to rock, etc.).

Hydrolysis. That chemical reaction involving double displacement in which hydrogen of water combines with the anion of the mineral and hydroxyl of water combines with the cation of the mineral to form an acid and a base.

Hydroponics. Growing of plants in nutrient solution without a solid medium.

Hydrous mica. (illite) *See* illite.

Hydroxyl. Oxygen with one hydrogen forming OH^-, the anion of bases.

Hyperthermic. Continuously hot temperature regime (15–22°C). *See* Chapter 7.

Hypha (pl. hyphae). Thread-like filament of some fungi that penetrate the substrate being decomposed.

I

Igneous rock. Formed by solidification from a molten or partially molten state. Synonym: primary rock. Example: granite.

Illite. A hydrous aluminosilicate clay mineral with structurally mixed mica and smectite or vermiculite, similar to montmorillonite but containing potassium between the crystal layers. Also referred to as *hydrous mica* or *mica.*

Illuvial horizon. A soil layer or horizon in which material carried from an overlying layer has been precipitated from solution or deposited from suspension. The layer of accumulation. Contrast to eluviation.

Immobilization. The transfer of an element from the soluble inorganic into the organic form of microbial or plant tissues.

Imogolite. Slightly crystalline allophane.

Inceptisols. *See* Soil classification: Order.

Indurated. (soil) Soil material cemented into a hard mass that will not soften on wetting.

Infiltration. Entry of water downward into the soil surface.

Injector-point fertilization. *See* Point injector.

Inoculation. The process of introducing cultures of microorganisms into soils or culture media, such as by adding *Rhizobia* bacteria coated on legume seed.

Intake rate. Average rate of water entering the soil under irrigation. Usually beginning rate is variable (faster at the beginning than after some time). Intake rate, based in inches per hour, is as follows:

Very low	Less than 0.2 in (<0.5 cm)
Low	0.2–0.4 in (0.5–1.0 cm)
Moderately low	0.4–0.75 in (1.0–1.9 cm)
Moderate	0.75–1.25 in (1.9–3.1 cm)
Moderately high	1.25–1.75 in (3.1–4.38 cm)
High	1.75–2.5 in (4.38–6.25 cm)
Very high	More than 2.5 in (>6.25 cm)

Integrated pest management (IPM). The use of many different techniques in combination to control pests, such as the combined uses of resistant plant varieties, natural predators of the pest, specific chemical pesticides, good preventive measures, and good management practices, such as crop rotations.

Internal soil drainage. The downward movement of water through the soil profile. *See also* Percolation, soil water.

Intrusive rocks. Rocks formed from molten magma that cooled slowly (has large crystals) because it cooled beneath the earth's crust. Also called *plutonic igneous rocks.*

Ions. Atoms or groups of atoms that are electrically charged as a result of the loss of electrons (cations) or the gain of electrons (anions).

IPM. *See* Integrated pest management.

Iron-pan. An indurated soil horizon in which iron oxides are the principal cementing agent along with aluminum oxides. If from plinthite it is called *ironstone.*

Ironstone. *See* Iron-pan.

Irrigation application efficiency. Percentage of irrigation water applied to an area that is stored in the soil for crop use.

Irrigation methods. The manner in which water is artificially applied to an area. *See* Chapter 16.

Isomorphous substitution. The replacement of one atom by another of similar size in a crystal lattice during crystal growth without changing the crystal structure.

Isotopes. Atoms of the same element that have different atomic mass because of differences in neutrons, e.g., carbon-12, carbon-13, and carbon-14.

J

Joule. The force of 1 newton applied over a distance of 1 meter.

K

Kandic horizon. A diagnostic argillic horizon having mostly low activity (1:1 crystal) clays, such as kaolinite.

Kaolinite. Hydrous aluminosilicate clay mineral of the 1:1 crystal structure group—that is, consisting of one silicon tetrahedral sheet and one aluminum oxide-hydroxide octahedral sheet.

Karst relief. Weathered limestone areas and likely to have sinkholes and tunnels.

L

Labile pool. The total sum of a nutrient that readily solubilizes or exchanges to become available to plants during a season.

Lacustrine deposit. Sediments deposited in fresh (nonsaline) lake water and later exposed either by lowering of the water level or by the elevation of the land.

Land capability class. One of eight classes of land in the land capability classification of the USDA—Soil Conservation Service, distinguished according to the risk of land damage or the difficulty of land use.

Land capability subclass. The four kinds of limitations recognized at the subclass level are risks of erosion, designated by the symbol *e;* wetness, drainage, or over-flow *w;* other root-zone limitations *s,* and climatic limitations *c.*

Land capability unit. A group of soils that are nearly alike in suitability for plant growth and that respond similarly to the same kinds of soil management.

Land evaluation and site assessment (LESA). An evaluation of land for guiding the conversion of farmland to urban areas and to protect prime farmland from other uses.

Landform. A discernible natural landscape, such as a floodplain, stream terrace, plateau, or alluvial fan.

Latent heat. The amount of heat involved per unit of mass to undergo a phase change, such as heat of melting.

Layer. 1. (*GIS*) A logical set of thematic data described and stored in a map library. Layers organize a map library by subject matter and extend over the entire geographic area defined by the spatial index of the map library. 2. (*Mineralogy*) A combination of sheets in minerals in a 1:1, 2:1, or 2:2 combination.

Leaching. The downward removal of materials in solution from the soil.

Leaching requirement. The extra fraction of the amount of water needed to wet the soil that must be added to keep soil salinity below a predetermined tolerance concentration.

Legume. A member of the legume or pulse family *Leguminosae:* peas, beans, peanuts, clovers, alfalfas, sweetclovers, vetches, lespedezas, and kudzu. Most are nitrogen-fixing plants.

Legume inoculation. *See* Inoculation.

LESA. *See* Land evaluation and site assessment.

Lichen. A symbiotic relationship of a fungus and an alga whereby the fungus supplies water and dissolved nutrients and the alga photosynthesizes carbohydrates and fixes nitrogen. Lichens colonize bare minerals, rocks, and large trees, a first step in rock weathering and soil formation.

Ligand. An organic molecule that can bond to metals through two or more bonds. The ligand-metal is called a *chelate.*

Lignin. Cell wall material that helps cement cells together; very resistant to decomposition and constitutes much of the residual humus in soils.

Light soil. (obsolete scientifically) Indicates sandy textures (easy to till).

Lime. 1. *Chemistry:* calcium oxide (CaO); 2. *Agriculture:* a variety of acid-neutralizing materials; most are the oxide, hydroxide, or carbonate of calcium, or of calcium and magnesium (ground limestone, marl, oyster shells [carbonates], wood ashes).

Lime (calcium) requirement. The amount of agricultural limestone required to raise the pH of the surface soil to about pH 6–6.5.

Limestone. A sedimentary rock composed of more than 50% calcium carbonate ($CaCO_3$).

Liquid lime. Lime materials that have been pulverized and added to soil as a liquid suspension.

Liquid limit. In engineering, the water percentage between a soil's defined liquid and plastic states (consistence).

Lithic contact. A boundary between soil and continuous, coherent underlying material that has a hardness of 3 or more (Mohs' scale).

Lithosequence. Related soils that differ from each other primarily because of differences in parent material from which each formed.

Loam. Soil material that is 7% to 27% clay, 28%–50% silt, and less than 52% sand.

Loess. Material transported and deposited by wind; predominantly silt sized.

Longwave radiation. Radiation of infrared and radio wavelengths emitted from the earth's surface.

Low strength. The soil is not strong enough to support loads (weight).

Lysimeter. Container of soil to measure the water movement, gains, or losses through that block of soil, usually undisturbed or in situ.

M

Macronutrient. A chemical element needed in amounts usually >1 part per 500 in the plant for plant growth. Examples: C, N, O, K, Ca, Mg, S, H.

Mapping unit. *See* Soil mapping unit.

Marl. An earthy, unconsolidated deposit formed in freshwater lakes, consisting chiefly of calcium carbonate mixed with clay or other impurities.

Marsh. Periodically or constantly wet or ponded with hydric soil and wet-loving vegetation such as cattails, rushes, sedges, and willows.

Mass flow. Movement of nutrients with the overall flow of water to plant roots.

Mass water content. The water content expressed as the weight of water in a soil divided by the oven-dry weight of soil.

Mass water percentage. The mass water content times 100.

Massive. Lack of soil structure in coherent materials; structureless but holds together.

Matric potential. The amount of work an infinitesimal quantity of water in the soil can do as it moves from the soil to a pool of free water of the same composition and at the same location. This work is less than zero, or *negative* work, thus reported in negative values. Matric potential nearly equals water potential in nonsalty soils.

Mechanical analysis. Older term for particle-size analysis.

Melanic horizon. Deep, black surface horizon, over 10% organic carbon, formed in volcanic material (Andisols).

Mesic. A soil temperature regime of intermediate range (8–15°C).

Metal oxides. *See* Sesquioxides.

Metamorphic rock. Igneous or sedimentary rock that has changed because of high temperature, high pressure, and the chemical environment while deep in the crust of the earth. Examples: marble, slate, gneiss.

Methane, CH_4. An odorless, colorless gas produced by anaerobic decomposition of organic materials. Produced in landfills and swamps. Contributes to global warming.

Methemoglobinemia. "Blue baby disease," caused by high nitrate intake in very young mammals, which when nitrate is reduced to nitrite can reduce the blood's ability to carry oxygen.

Micronutrient. A chemical element necessary in only small amounts (usually less than several parts per million in the plant) for the growth of plants. Examples: boron, copper, iron, and zinc.

Mineral soil. Soil that is mainly mineral material and low in organic material.

Mineralization. The conversion of an element from an organic combination to an inorganic form as a result of microbial decomposition.

Minimum tillage. Nonspecific term for the minimum soil manipulation necessary to produce a crop or meet a particular objective.

Mixed. (family mineralogy) No single mineral makes up more than 40% of that soil.

Model. 1. An abstraction of reality. A model is structured as a set of rules and procedures to derive new information that can be analyzed to aid in problem solving and planning. 2. Data representation of reality.

Mohs' scale of hardness. Relative hardness of minerals ranging from a rating of 1 for the softest (talc) to 10 for the hardest (diamond).

Moisture suction. *See* Matric potential; Soil water potential.

Moisture tension (or pressure). (obsolete) *See* Soil water potential.

Moldboard plow. An implement for deep (15–40 cm, or 6–16 in), primary tillage. One side of the implement has a curved surface to turn the plowed strip almost upside down. This inverts the plowed layer and buries most of the surface plant residues.

Mole drain. Drains formed by pulling a bullet-shaped cylinder through the soil at a depth of 30–91 cm (1–3 ft).

Mollic horizon. A diagnostic epipedon of dark color, of moderate pH, and quite deep.

Mollisols. *See* Soil classification: Order.

Monera. (moan-ee′-ra) One of five kingdoms for organisms and having procaryotic cells. Includes bacteria, actinomycetes, and cyanobacteria.

Montmorillonite. A hydrous aluminosilicate clay mineral with 2:1 expanding crystal structure—that is, with two silicon tetrahedral sheets enclosing an aluminum octahedral sheet. Considerable expansion may be caused by water.

Mor. Forest surface humus in which the **Oa** horizon has little mixing into mineral soil beneath it.

Moraine. An accumulation of glacial drift formed chiefly by the direct deposition from glacial ice. Examples are ground, lateral, recessional, and terminal moraines.

Morphology. The physical nature of the soil as exhibited by horizon differences and such physical properties as texture, porosity, and color.

Mottles. Irregular soil mass spots of various colors. A common cause of mottling is impeded drainage.

Muck. Organic soil whose organic material is too decomposed to be recognizable.

Mulch. A natural or artificial layer of crop residues, leaves, sand, plastic, or paper on the soil surface.

Mulch-till. A conservation tillage system in which the entire soil surface is tilled before planting, using such implements as the chisel plow, rod weeder, disk, or field cultivator. At least 30% of the soil surface is covered with residue.

Mull. Forest surface humus with or without **Oe** and without an **Oa.** Organic matter is intimately mixed with mineral soil so the transition to an A is gradual.

Munsell color notation system. A system that specifies the relative degrees of the three simple variables of color: hue, value, and chroma. For example: 10YR 6/4 is a color with hue 10YR, value 6, and chroma 4.

Mutation. A random but inheritable gene change, which can occur naturally but may be caused by increased exposure to radioactive radiation and/or certain chemicals. Usually, a mutation is a physical or health defect.

Mycelia. A cottony mass of individual hypha (filaments or elongated threads) typical of growth structures of many fungi.

Mycorrhiza. The association, usually symbiotic, of specific fungi with the roots of specific higher plants.

N

Natric horizon. An argillic horizon with >15% exchangeable sodium.

Necrosis. Death of plant tissue, often with darkened color, i.e., brown dead edges of leaves.

Nematodes. Small (0.5–1.5 mm long) nonsegmented worms. Some are parasitic on plant roots; others parasitize insects.

Newton. (N) Unit of force required to accelerate a mass of 1 kg 1 meter per second per second. It equals 100,000 dynes.

Nitrate depression period. A week to several weeks following incorporation of low-N residues when decomposers use soil nitrogen and plants may not have sufficient for their needs.

Nitrate reduction. Reduction process converting nitrate to ammonium for use by plants and microorganisms.

Nitrate toxicity of forage. Forage containing more than 6000 ppm of nitrate (NO_3^-) may be toxic to cattle. Causes of high nitrate include drought, cool weather, cloudy weather, and heavier-than-recommended applications of manures, sludges, and nitrogen fertilizers.

Nitrification. The biological oxidation of ammonium salts to nitrites and the further oxidation of nitrites to nitrates.

Nitrogen cycle. Interrelations of biological and chemical changes which nitrogen undergoes as it cycles through the air-organisms-soil-water system.

Nitrogen fixation. *See* Dinitrogen fixation.

Nitrogenase. The enzyme involved in biological dinitrogen fixation.

No-till. (also called *zero-till* or *slot-plant*) A farm management scheme using no tillage except to insert seed, and sometimes fertilizer, in a slit. Often a leading coulter is followed by a chisel. The seed, and sometimes the fertilizer, is dropped through a small tube immediately behind the chisel.

Nodule. A growth developed on the roots of plants in response to the stimulus of root nodule bacteria or actinomycetes.

Nonpoint (pollution) source. Coming from a general area (as in sheet erosion) but unable to identify to a ditch, field, or home lot.

Nutrient, plant. *See* Essential element.

O

O horizon. Surface organic layers. *See also* Soil horizon.

Ochric horizon. A diagnostic epipedon of light color, low humus, or shallow depth.

Order, soil. *See* Soil classification.

Organic phosphorus. Phosphorus present as a part of organic compounds, such as glycerophosphoric acid, inositol phosphoric acid, and cytidylic acid.

Organic soils. Histosols 1. Saturated with water for prolonged periods unless artificially drained and having at least 12% or 18% organic carbon by weight, depending on the mineral fraction and the kind of organic materials. 2. Never saturated with water for more than a few days and having 20% or more organic carbon by weight.

Fibrists. The least decomposed of all the organic soils; high amounts of fiber are well preserved and readily identifiable as to botanical origin; a bulk density of less than 0.1 Mg/m^3 (6.2 lb/ft^3).

Folists. Freely drained Histosols that consist primarily of leaf litter that rests on rock or on gravel, stones, and boulders whose voids are filled with organic matter; contain 20% or more organic carbon.

Hemists. Histosols that are intermediate in degree of decomposition between the less decomposed Fibrists and the more decomposed Saprists; saturated with water for 6 months or more in a year.

Saprists. The most highly decomposed of the Histosols; a bulk density of more than 0.2 Mg/m^3 (12.5 lb/ft^3); saturated for 6 months or more in a year.

Osmotic potential. (solute potential) The amount of work an infinitesimal quantity of water will do in moving from a pool of free water the same composition as the soil water to a pool of pure water at the same location. The effect of dissolved substances. Usually very small.

Outwash. Stratified glacial drift, often high in sands and gravels, that is deposited by meltwaters near glaciers.

Oven-dry soil. Soil dried at about 105–110°C (221–230°F).

Oxic horizon. A diagnostic horizon common to Oxisols. It is thick, low in weatherable minerals, and contains low CEC clays.

Oxidation. 1. Combination with oxygen. 2. Removal of electrons from an atom, ion, or molecule during a reaction.

Oxisols. *See* Soil classification: Order.

Ozone shield. A concentration of ozone gas (O_3) in the upper atmosphere, which reacts with and reduces much of the harmful ultraviolet radiation from the sun that reaches the Earth. The shield is believed to be reduced or destroyed by certain chemicals (e.g., chlorofluorocarbons) in polluted air.

P

Paleosols. Soils formed under (ancient) climates different from those that now exist. *See* Relict soils and Fossil soils.

Pan. A layer in soils that is strongly compacted, indurated, or very high in clay content.

Pan, pressure or induced. A subsurface soil layer having a high bulk density and a lower total porosity than the soil directly above or below it as a result of pressure. Frequently referred to as *plow pan, plowsole, tillage pan,* or *traffic pan.*

PANs. (Air pollutants) A group of chemicals (**p**eroxy**a**cetyl-**n**itrates) of various origins present in photochemical smog. They are extremely toxic to plants and irritating to eyes and membranes of the nose and throat.

Paralithic contact. Soil contact at a hard layer, which can be dug by a spade, though with difficulty. *See also* Lithic contact.

Paraplow. A series of chisels on a frame. Each chisel leg goes straight down and then bends to the side at a 45° angle. As the plow is pulled, soil flows over each bent wing (leg) and falls back, causing the soil to shatter but not to be mixed or inverted.

Parent material. (soils) The unconsolidated, chemically weathered mineral or organic matter from which the **A** and **B** horizons (solum) of soils may have developed by pedogenic processes.

Particle density. The mass per unit volume of the soil particles. *See also* Bulk density, soil.

Particle size analysis. Laboratory procedure to determine the amounts of particle-size groups; originally called *mechanical analysis.*

Particle-size classes for family groupings. *See* Chapter 7.

Pascal. (Pa) A unit of pressure equal to 1 newton per square meter.

Peat. Undecomposed or only slightly decomposed organic matter accumulated under conditions of excessive moisture.

Peat moss. Dried peat from various plants, slow to decompose, used for potting mixtures.

Ped. A unit of soil structure; an aggregate, such as prism, block, or granule, formed by natural processes.

Pediment. A uniformly sloped (graded) rock surface cut by erosion, such as a mountain slope, and with only a shallow cover of alluvium.

Pedogenesis. Caused by the natural processes of soil development. Synonyms: *soil genesis, soil development,* and *soil formation.*

Pedology. The study of soil as a geologic entity.

Pedon. The smallest volume that can be called *a soil.* It has three dimensions. It extends downward to the depth of plant roots or to the lower limit of the genetic soil horizons. Its lateral cross-section is roughly hexagonal and ranges from 1 to 10 m^2 in size, depending on the variability in the horizons.

Peneplain. A near plain; the alluvium-covered low areas that have almost formed a plain near the level of water drainage.

Perc test. A rate-of-percolation test for site evaluation for drainage, waste disposal, septic drain fields, and so on.

Percolation, soil water. The downward movement of water through soil, especially the downward flow of water in saturated or nearly saturated soil.

Pergelic. A soil temperature regime (0°C) averaging subzero temperature (permafrost).

Perlite. A siliceous volcanic rock that, when heated to 980°C, expands and forms white, air-filled, lightweight particles.

Permafrost. A permanently frozen layer.

Permanent wilting point. The largest water content in soil at which plants will wilt and not recover when placed in a humid chamber. It is estimated at about –1.5 MPa matric potential.

Permeability. The amount of water that moves downward through the saturated soil. Described in inches per hour, terms are:

Very slow	Less than 0.06 in (<0.15 cm)
Slow	0.06 to 0.2 in (0.15–0.5 cm)
Moderately slow	0.2 to 0.6 in (0.5–1.5 cm)
Moderate	0.6 to 2.0 in (1.5–5.0 cm)
Moderately rapid	2.0 to 6.0 in (5.0–15 cm)
Rapid	6.0 to 20 in (15–50 cm)
Very rapid	More than 20 in (>50 cm)

Permeability, soil. The quality of a soil layer that enables water or air to move through it.

Pesticide. Any chemical designed to kill pests (weeds, insects, mites, nematodes, fungi, rodents, algae, bacteria). The term now includes many natural chemicals (juvenile hormones or attractants) to trap or confuse insects, hindering them from finding mates (as pheromones).

Petrocalcic horizon. A diagnostic horizon with carbonate accumulation and hard cementation.

Petrogypsic horizon. A diagnostic horizon with gypsum accumulation and hard cementation.

pH, soil. A numerical measure of the acidity or hydrogen ion activity of a soil. Exactly, the negative logarithm of the hydrogen-ion activity of a soil. *See also* Reaction, soil.

pH-dependent charge. The portion of the cation or anion exchange capacity that varies with pH.

Phase, soil. A subdivision of a soil taxon, usually a soil series or other unit of classification, based on characteristics that affect the use and management of the soil; phases include degree of slope, degree of erosion, content of stones, and texture of the surface.

Phosphorus or potassium fixation. *See* Fixation.

Phytoremediation. Removal of part or all of contaminants in water and soils by absorption into plants.

Pica. An unnatural appetite of animals and people that results in eating soil, bark, bones, and/or hair.

Pixel. A contraction of the words *picture element.* The smallest unit of information in an image or map.

Plane of atoms. A flat (planar) array of one atomic thickness. Example: in soil mineralogy, a plane of basal oxygen atoms within a tetrahedral sheet.

Plant-available nutrient. *See* Available nutrient.

Plantae. One of five kingdoms into which all organisms are fitted. Includes such plants as algae, mosses, grasses, and trees.

Plastic limit. (Engineering) The water content at which the soil rolled to a 1/3-cm wire begins to crumble.

Plasticity index. (Engineering) The water content percentage between the liquid limit and the plastic limit.

Platy structure. Soil aggregates developed along the horizontal direction; flaky.

Playa. A shallow basin on a plain where water gathers and is evaporated.

Plinthite. A nonindurated mixture of iron and aluminum oxides, clay, quartz, and other diluents that commonly occurs as red soil mottles, usually arranged in platy, polygonal, or reticulate patterns. Plinthite changes irreversibly to ironstone hardpans or irregular ironstone aggregates on exposure to repeated cycles of wetting and drying.

Plowpan. *See* Pan, pressure or induced.

Plutonic rocks. *See* Intrusive rocks.

Point injector. A narrow cylindrical tube (injector) penetrates the soil and allows fertilizer to be inserted at the depth of penetration. Usually liquid fertilizer is used. The injectors may be on wheels (rolling-point injectors) with frequency and spacing set by the operator.

Point (pollution) source. Identifiable source of pollution, such as a smokestack or discharge from a pipe or channel.

Polder. A low-lying land area normally under water but reclaimed from the water and protected from resubmersion by dikes, as in the Netherlands.

Pollution. *See* Contamination.

Pollution (pollutant). *See* Contamination.

Polypedon. Two or more contiguous pedons, all of which are within the defined limits of a single soil series.

Ponding. Standing water on soils in closed depressions.

Poorly graded. (Engineering) *See* Graded.

Porosity. In engineering and soils, the ratio of the volume of voids to the total volume of the soil. In soils, usually it is given in percentage.

Pore space. Total space not occupied by soil particles in a bulk volume of soil.

Potassium fixation. *See* Fixation.

Pressure potential. The amount of work an infinitesimal amount of soil water can do in moving from a pool of pure water under the pressures common to that soil position to a pool of pure water at the same location and at normal atmospheric pressure.

Primary mineral. A mineral that has not been chemically altered since it crystallized from molten magma.

Primary nutrients. The nutrients N, P, and K which are most often deficient in growing plants unless plants are fertilized.

Prime agricultural lands. Soils capable of the highest production levels with the least hazard of erosion or damage.

Prion. Virallike proteins without a protective coat. Carries genetic material.

Prismatic structure. A soil structure type with a long vertical axis that is prism shaped.

Procaryote. (pro-car′-ry-ot) Organism cell type without nuclear materials enclosed in a nuclear membrane (bacteria, cyanobacteria, actinomycetes).

Profile, soil. A vertical section of the soil through all its horizons and extending into the parent material.

Protista. One of five kingdoms into which all organisms are fitted. Protista includes mostly one-celled organisms with eucaryotic cells: amoeba, sporozoans, slime molds, ciliates, and flagellates.

Puddling. The act of destroying natural soil structure by intensive tillage when soil is saturated with water.

Pyroclastics. Volcanic materials explosively or aerially ejected from a volcanic vent.

Pyrophyllite. An aluminosilicate mineral of 2:1 layer structure and with no isomorphous substitution.

Q

Qanat. *See* Ganat.

Quick test, soil. Simple, routine analysis on soils, usually to measure pH, soluble salts, and nutritional status.

R

R horizon. Solid-rock horizon, usually found below parent material, but may be at great depth.

Radiation. The process of emitting, from atoms, energy as waves and particles through space.

Radioactive substance. Substances of, or containing, unstable chemical isotopes of elements and that give off radiation (alpha, beta, gamma rays, etc.) at some specific half-life rate. These are destructive to biological tissues and can cause cancer or mutations in some instances.

Reaction, soil. The degree of acidity or alkalinity (basicity) of a soil, usually expressed as a pH value. Descriptive terms commonly associated with certain ranges in pH are extremely acid, less than 4.5; very strongly acid, 4.5–5.0; strongly acid, 5.1–5.5; medium acid, 5.6–6.0; slightly acid, 6.1–6.5; neutral, 6.6–7.3; mildly alkaline, 7.4–7.8; moderately alkaline, 7.9–8.4; strongly alkaline, 8.5–9.0; and very strongly alkaline, more than pH 9.0.

Redox. 1. A term for the overall reactions in which one substance is oxidized while another is reduced by the electron transfers. 2. The electron density of the media. Redox is measured in units of millivolts.

Reduced tillage. By the most general definition, *reduced tillage* is any combination of tillage operations that do less soil manipulation than conventional tillage. Reducing the number of operations may be accomplished by eliminating one or more operations such as a plowing or disking operation. Various terms are used for different degrees of reduced tillage: *minimum tillage, conservation tillage,* and *no-till.*

Reduction. Atoms or ions that gain electrons. Often associated with very wet, waterlogged soil conditions.

Reference dose (RFD). Level of daily exposure that, over a 70-year human life, is believed to have no negative effect.

Regolith. The layer or mantle of loose, noncohesive or cohesive rock material that nearly everywhere forms the surface of the land and rests on bedrock.

Relict soils. Exposed paleosols; changed from time of deep burial by present soil-forming factors.

Relief. The difference between the high and low areas of a landscape. Approximate synonyms: *topography, Earth surface contour, elevation differences.*

Remediation. Eliminating a contaminant, partially or wholly, to improve the degraded soil.

Residual material. Unconsolidated and partly weathered mineral materials derived from rock in place.

Resistance, developed. Organisms, through natural selection, develop populations that are no longer killed by a particular pesticide. These may be genetic changes to a few organisms. These changed organisms breed new populations which are resistant to the pesticide.

Revised Universal Soil Loss Equation (RUSLE). An update of the universal soil loss equation (USLE).

Rhizobia. Collective name for bacteria of the genus *Rhizobium,* which are capable of symbiotic nitrogen fixation with legume plant roots.

Rhizosphere. The zone of soil immediately adjacent to plant roots in which microbial numbers and kinds may be much different than in the bulk soil in general.

Ridge-furrow. The formation of alternate ridges and furrows, often with ridge tops 15–25 cm (6–10 in) above the furrow bottom. This system is needed for furrow irrigation of row crops. It is also useful for a larger or more aerated root zone (1) in shallow soils, (2) in poorly drained soils, and (3) in cold soils where ridges warm faster and are drier than nonridged soil. Ridges are built and rebuilt by a special cultivation sweep (bedder).

Ridge-till. (also called *ridge-plant* or *till-plant*) The soil is left undisturbed from harvest until planting except for nutrient injection. Crops are planted and grown on ridges formed by sweeps or other implements.

Rill. A small, eroded ditch, usually only a few inches deep and hence no great obstacle to tillage operations.

Risk-benefit analysis. An evaluation of pesticides or any other practice when considering their registration (or use) and changes in residues (or effects).

Rockwool. Mineral strands woven into a wool-like mat, very light-weight, water-absorbent, and inert to most chemicals.

Rod weeding. Pulling a sweep or blade through the subsoil at a depth of 6–10 cm (2.5–4 in) to cut and kill weeds. Sometimes called *blading.*

Roller harrow. A large set of toothed disks close together (10 cm, or 4 in, apart) on a wide-diameter drum. As the drum disk rolls over the soil, it pulverizes and smooths the seedbed.

Root nodule. A swelling formed on the roots of leguminous plants, caused by the symbiotic nitrogen-fixing bacteria *Rhizobium.*

Root of soil order. Syllable of the order name prior to the vowel connecting to *-sol.* Examples: *od* from Spodosol, *ert* from Vertisol, and *ept* from Inceptisol.

Rotary plow. Blades that rotate into soil to mix it (rototill). Often only *narrow strips* of soil are rotary tilled in front of a seeder (*rotary strip tillage*).

S

Salic horizon. A mineral horizon with enough accumulated soluble salts to affect plant growth.

Salination. The process whereby soluble salts accumulate in soil. It replaces the obsolete term *salinization.*

Saline seep. Accumulation of salt by local movement of salts into subsoil at a higher location, movement downslope at a shallow depth, and accumulation into the surface soil in downslope areas as salt moves up and water is evaporated.

Saline-sodic soil. 1. A soil containing an exchangeable-sodium percentage greater than 15 (or SAR greater than 13) and a conductivity of the saturation extract greater than 4 dS m^{-1} (25°C). Growth of most crop plants is reduced.

Saline soil. A nonsodic soil containing sufficient soluble salts to impair its productivity; conductivity of the saturation paste extract is greater than 4 dS m^{-1} (25°C).

Salinization. (obsolete) *See* Salination.

Salt. Any mineral composed of the cation of a base and the anion of an acid. Soluble salts are mostly of sodium, potassium, and magnesium with the anions sulfate, chloride, carbonates, and hydroxides.

Salt balance. The quantities of dissolved salts removed by drainage water minus the quantities of dissolved salts carried to that area in irrigation water.

Salt-affected soil. Soil that has been adversely modified for the growth of most crop plants by the presence of soluble salts, exchangeable sodium, or both.

Saltation. Synonym: *vaultation. See* Erosion.

Sapric materials. *See* Organic soils.

SAR. *See* Sodium adsorption ratio.

Saturated flow. Movement of water through soil by gravity flow, as in irrigation or during a rainstorm.

Saturation paste extract. The extract from a saturated soil paste. Field capacity times 2 approximates saturation percentage.

Savannah. Mostly grasslands with individual and clumps of trees.

Secondary mineral. A mineral formed by the precipitation of the products of weathered primary minerals.

Secondary nutrients. The nutrients Ca, Mg, and S used in large amounts by plants but less often deficient than the primary nutrients N, P, and K.

Sedimentary rock. Rock made up of particles deposited from suspension in water. Some wind-deposited sand is consolidated into sandstone.

Seedbed. Prepared soil, usually finely pulverized, to promote growth of planted seeds.

Seed inoculation. *See* Inoculation.

Self-mulching soil. A soil with a high shrink–swell potential in the surface layer so that portions of its surface yearly fall deeply into cracks (self-mixing).

Semiarid. Regions or climates where moisture is normally greater than under arid conditions but still definitely limits the growth of most crops. The upper limit of average annual precipitation in the cool semiarid regions is as low as 38 cm (15 in), whereas in tropical regions it is as high as 114–127 cm (45–50 in).

Separate. *See* Soil separates.
Septage. The anaerobic residues from septic tanks.
Septic tank. Sewage disposal tank for homes. The tank (often cement) allows decomposition and settling of solids and drainage of liquid off into the soil through perforated drainage lines.
Sequestrene. *See* Chelation; Ligand.
Sequum. A soil profile sequence consisting of an illuvial horizon (as a **B**) and the overlying eluvial horizon (as an **E**).
Series, soil. *See* Soil classification.
Sesquioxides. (metal oxides) A term for minerals containing 1.5 atoms of oxygen per atom of the metal, particularly Al_2O_3 and Fe_2O_3. Often TiO_2 is included, although it does not strictly fit the meaning of *sesqui* (= 1.5 times).
Sewage sludge. *See* Sludge.
Sheet erosion. *See* Erosion.
Sheet of atoms. (mineralogy) A flat array of more than one atomic thickness and composed of one or more levels of linked coordination polyhedra. A sheet is thicker than a plane and thinner than a layer. Examples: tetrahedral sheet, octahedral sheet.
Shelterbelt. A row, or usually several rows, of trees and tall shrubs to protect against wind.
Shifting cultivation. Farming, mostly in tropics and subtropics, in which trees and brush are cleared and burned. After a few years cultivation, weeds and low fertility require the area be left to grow back for a dozen or more years while other plots are cleared and used every few years. Also called slash-and-burn.
Shortwave radiation. High energy radiation (ultraviolet, visible, and near-infrared). Sun's light. It readily passes through air.
Shovels. A term often used for narrow-winged sweeps.
Shrinkage limit. (Engineering) The water content below which no further change in soil volume occurs.
Shrink–swell potential. Susceptibility to volume change due to loss or gain in moisture content.
Side dressing. Side dressing can involve insertion of the fertilizer into the soil beside the growing crop.
Sideraphore. A nonporphyrin metabolite, secreted by certain microorganisms, that forms highly stable bonds with iron, thereby mobilizing iron in soil. The two major types are catecholate and hydroxamate.
Siemen. A unit of electrical inductance; in SI metric units, electrical conductance is measured in siemens per meter ($S\ m^{-1}$). Decisiemens per meter ($dS\ m^{-1}$) is equivalent to millimhos per centimeter (mmhos/cm).
Silica. Amorphous or slightly crystalline mineral masses with the approximate formula SiO_2.
Siliceous. *See* Soil mineralogy classes for family groupings.
Silt. 1. A soil separate consisting of particles between 0.05 and 0.002 mm in equivalent diameter. 2. A soil textural class.
Single grain. (obsolete when called *structure*) A term for a lack of structure in which individual particles do not cohere together. Example: coarse sands.
Sink. 1. Sunken depression in the land surface. 2. A material or reaction that acts as an infinite reservoir or removal mechanism. Example: The ocean is a large reservoir (a sink) to absorb carbon dioxide.
Site index. A designation of the quality of a forest site based on the height of the dominant stand at an arbitrarily chosen age.
Skeletal. A textural term used in soil families to indicate that the soil has greater than 35% coarse fragments (larger than sand).
Slick spots. Small areas in a field that appear wet longer and are slick due to a high content of clay with high exchangeable sodium (low permeability).
Slickensides. Polished and grooved clayey surfaces produced by one soil mass sliding past another. Common in Vertisols. *See also* Clay films.
Slip. The downslope movement of a soil mass under wet or saturated conditions; a microlandslide that produces microrelief in soils.
Slope. The inclination of the land surface from the horizontal. A slope of 15% is a vertical drop of 15 meters in 100 meters of horizontal distance.
Slot mulcher. Cuts a slot into the soil about 8 cm (3 in) wide and 25 cm (10 in) deep along the contour; slots are 4–6 m (13–20 ft) apart. The mulcher fills the slot with grain straw mulch left in the fields. Flowing water is readily absorbed into these slots.
Sludge. A general term for solid wastes, usually collected by sedimentation from water. Common sludges are sewage sludge, food-processing sludges, boiler sludge, electroplating sludge, and sugar-processing sludge.
Smectite. A group of minerals of 2:1 layer silicates having high cation exchange capacities and variable interlayer spacing. Typified by montmorillonite.
Sodic soil. A nonsaline soil that contains an exchangeable sodium percentage of 15 or more or a saturation extract SAR of 13 or more.
Sodium adsorption ratio (SAR). A value representing the relative hazard of irrigation water because of a

Table G-1 Comparison of the Present U.S. Soil Classification System Adopted in 1965 with the Approximate Equivalents in Use Before 1965

Soil Order (Adopted in 1965)	*Approximate Equivalents (In Use before 1965)*
Alfisols	Gray Brown Podzolic, Gray Wooded soils, Noncalcic Brown soils, Degraded Chernozem, and associated Planosols and some Half-Bog soils
Andisols	Volcanic materials formed into weakly developed soils, soils with *and-* and *vitr-* prefixes, and selected soils from most categories
Aridisols	Desert, Reddish Desert, Sierozem, Solonchak, some Brown and Reddish Brown soils, and associated Solonetz soils
Entisols	Azonal soils and some Low-Humic Gley soils
Histosols	Bog soils
Inceptisols	Ando, Sols Bruns Acides, some Brown Forest, Low-Humic Gley, and Humic Gley soils
Mollisols	Chestnut, Chernozem, Brunizem (Prairie), Rendzina, some Brown, Brown Forest, and associated Solonetz and Humic Gley soils
Oxisols	Laterite soils, Latosols
Spodosols	Podzols, Brown Podzolic soils, and Ground-Water Podzols
Ultisols	Red-Yellow Podzolic soils, Reddish Brown Lateritic soils of the United States, and associated Planosols and Half-Bog soils
Vertisols	Grumusols, Rendzinas

Source: Soil Survey Staff, *Soil Taxonomy: A Basic System of Soil Classification for Making and Interpreting Soil Surveys,* Agriculture Handbook 436, USDA—Soil Conservation Service, Washington, DC, 1975, 433–435. Andisols inserted by authors of this textbook in 1988.

high sodium content relative to its calcium plus magnesium content.

$$SAR = \frac{Na^{+}}{\sqrt{(Ca^{2+} + Mg^{2+})/2}}$$

The ions are in $millimoles_c$ per liter.

Soil. 1. The unconsolidated mineral and organic material on the immediate surface of the Earth that serves as a natural medium for the growth of land plants. 2. The unconsolidated mineral matter on the surface of the Earth that has been subjected to and influenced by genetic and environmental factors of parent material, climate, macro- and microorganisms, and topography, all acting over a period of time and producing a product—soil—that differs from the material from which it is derived in many physical, chemical, biological, and morphological properties and characteristics.

Soil amendment. *See* Amendment, soil.

Soil association. A mapping unit in which two or more defined taxonomic units occurring together are combined because the scale of the map or the purpose for which it is being made does not require delineation of the individual soils.

Soil classification. The systematic arrangement of soils into classes in one or more categories or levels of classification for a specific objective. The relationship between the orders of the present system and approximate equivalents of the 1938 system used in the United States are shown in Table G-1.

Order. The category at the highest level of generalization in the soil classification system. The properties selected to distinguish the orders are reflections of the degree of horizon development and the kinds of horizons present.

Suborder. This category narrows the ranges in soil moisture and temperature regimes, kinds of horizons, and composition, according to which is most important. Moisture, temperature, or soil properties associated with them are used to define suborders of Alfisols, Mollisols, Oxisols, Ultisols, and Vertisols. Kinds of horizons are used for Aridisols, composition for Histosols and Spodosols, and combinations for Entisols and Inceptisols.

Great group. The classes in this category contain soils that have the same kinds of horizons in the same sequence and have similar moisture and temperature regimes. Exceptions to the horizon sequences are made for horizons near the surface that may get mixed or lost by erosion if plowed. Formative elements used to formulate great group names are given in Chapter 7.

Subgroup. The great groups are subdivided into subgroups that show the central properties of the great group, intergrade subgroups that show properties of

Table G-2 Designation and Description of Master Soil Horizons and Layers (See Also Table G-3) and Horizons by Name

Horizon Designation		
Current	*Old*	*Description*
O	**O**	Organic horizons of minerals soils. Horizon (i) formed or forming in the upper part of mineral soils above the mineral part, (ii) dominated by fresh or partly decomposed organic materials, and (iii) containing more than 30% organic matter if the mineral fraction is more than 50% clay, or more than 20% organic matter if the mineral fraction has no clay. Intermediate clay content requires proportional organic-matter content.
Oi	**O1**	Organic horizons in which essentially the original form of most vegetative matter is visible to the naked eye. The **Oi** corresponds to the L (litter) and **Oe** or **Oa** to fermentation layers in forest soils designations and to the horizon formerly called **Aoo** or **O1.**
Oa or **Oe**	**O2**	Organic horizons in which the original form of most plant or animal matter cannot be recognized with the naked eye. The **Oa** corresponds to the H (humification) and **Oe** to F (fermentation) layers in forest soils designations and to the horizon formerly called **Ao** or **O2.**
A	**A**	Mineral horizons consisting of (i) horizons of organic-matter accumulation formed or forming at or adjacent to the surface, (ii) horizons that have lost clay, iron, or aluminum with resultant concentration of quartz or other resistant minerals of sand or silt size, or (iii) horizons dominated by (i) or (ii) but transitional to an underlying **B** or **C.**
A	**A1**	Mineral horizons, formed or forming at or adjacent to the surface, in which the feature emphasized is an accumulation of humified organic matter intimately associated with the mineral fraction.
E	**A2**	Mineral horizons in which the feature emphasized is loss of clay, iron, or aluminum, with resultant concentration of quartz or other resistant minerals in sand and silt sizes.
AB or **EB**	**A3**	A transitional horizon between **A** or **E** and **B,** dominated by properties characteristic of an overlying **A** or **E** but having some subordinate properties of an underlying **B.**
AB	**AB**	A horizon transitional between **A** and **B,** having an upper part dominated by properties of **A** and a lower part dominated by properties of **B;** the two parts cannot be conveniently separated into **A** and **B.**
E/B	**A & B**	Horizons that would qualify for **E** except for included parts constituting less than 50% of the volume that would qualify as **B.**
AC	**AC**	A horizon transitional between **A** and **C,** having subordinate properties of both **A** and **C** but not dominated by properties characteristic of either **A** or **C.**
B	**B**	Horizons in which the dominant feature or features are one or more of the following: (i) an illuvial concentration of silicate clay, iron, aluminum, or humus, alone or in combination, (ii) a residual concentration of sesquioxides or silicate clays, alone or mixed, that has formed by means other than solution and removal of carbonates or more soluble salts, (iii) coatings of sesquioxides adequate to give conspicuously darker, stronger, or redder colors than overlying and underlying horizons in the same sequum but without apparent illuviation of iron and not genetically related to **B** horizons that meet requirements of (i) or (ii) in the same sequum, or (iv) an alteration of material from its original condition in sequums lacking conditions defined in (i), (ii), and (iii) that obliterates original rock structure; that forms silicate clays, liberates oxides, or both; and that forms granular, blocky, or prismatic structure if textures are such that volume changes accompany changes in moisture. A sequum is an **E** horizon and its related **B** horizon.
BA or **BE**	**B1**	A transitional horizon between **B** and **A** or between **B** and **E** in which the horizon is dominated by properties of an underlying **B** but has some subordinate properties of an overlying **A** or **E.**
B/E	**B & A**	Any horizon qualifying as **B** in more than 50% of its volume, including parts that qualify as **E.**
B or **Bw**	**B2**	That part of the **B** horizon where the properties on which the **B** is based are clearly expressed characteristics, indicating that the horizon is related to an adjacent overlying **A** or an adjacent underlying **C** or **R.**

Continued.

Table G-2 Designation and Description of Master Soil Horizons and Layers (See Also Table G-3) and Horizons by Name *Cont'd*

Horizon Designation		
Current	*Old*	*Description*
BC or **CB**	**B3**	A transitional horizon between **B** and **C** in which the properties diagnostic of an overlying **B** are clearly expressed but are associated with clearly expressed properties characteristic of **C.**
C	**C**	A mineral horizon or layer, excluding bedrock, that is relatively little affected by pedogenic processes, and lacking properties diagnostic of **A** or **B** but including materials modified by (i) weathering outside the zone of major biological activity, (ii) reversible cementation, development of brittleness, development of high bulk density, and other properties characteristic of fragipans, (iii) gleying, (iv) accumulation of calcium or magnesium carbonate or more soluble salts, (v) cementation by accumulations, as calcium or magnesium carbonate or more soluble salts, or (vi) cementation by alkali-soluble siliceous material or by iron and silica.
R	**R**	Underlying consolidated bedrock, such as granite, sandstone, or limestone. If presumed to be like the parent rock from which the adjacent overlying layer or horizon was formed, the **R** is used alone. If presumed to be unlike the overlying material, the **R** is preceded by an arabic numeral denoting lithologic discontinuity, such as **2R.**

Sources: Modified from (1) Soil Survey Staff, *Soil Taxonomy: A Basic System of Soil Classification for Making and Interpreting Soil Surveys,* Agriculture Handbook 436, USDA—Soil Conservation Service, Washington, DC, 1975, 461–462; (2) Soil Survey Service directive 430-V-SSM, May 1981; (3) USDA—Soil Conservation Service draft of *Soil Survey Manual,* Chapter 4, 1981, 4-39 to 4-51.

more than one great group, and other subgroups for soils with atypical properties that are not characteristic of any great group.

Family. Families are defined largely on the basis of physical and mineralogical properties of importance to plant growth.

Series. The soil series is a group of soils having horizons similar in differentiating characteristics and arrangement in the soil profile, except for texture of the surface, slope, gravel, stones, and erosion.

Soil complex. Much like a soil association, but the different soils occur in a regular pattern and as smaller, discrete units.

Soil conservation. *See* Conservation.

Soil creep. *See* Creep.

Soil development. 1. The breakdown of rocks to unconsolidated materials called *soil.* 2. The changes in the soil profile brought about by natural processes of leaching, translocation of colloids, accumulation of organic materials, and continued mineral and rock weathering.

Soil erosion. *See* Erosion.

Soil-formation factors. The variables (parent material, climate, organisms, topography, and time) active in and responsible for the formation of soil.

Soil genesis. Formation of the soil with special reference to the processes or soil-forming factors responsible for the development of the solum or true soil from the unconsolidated parent material. Synonyms: *pedogenesis, soil formation.*

Soil health. *See* Soil quality.

Soil horizon. A layer of soil or soil material approximately parallel to the land surface and differing from adjacent genetically related layers in physical, chemical, and biological properties or characteristics, such as color, structure, texture, consistency, amount of organic matter, and degree of acidity or alkalinity. Tables G-2 and G-3 list the designation and description of the master soil letter horizons and subscript horizons.

Soil individual. *See* Pedon; Polypedon.

Soil interpretations. Predictions of soil behavior under specific uses or management and based on inferences from the soils' characteristics as reported in a soil survey report.

Soil loss tolerance. (*T*) In conservation farming, the maximum average annual soil loss in tons per acre per year that should be permitted on a given soil. In general the rate of soil formation should equal or exceed soil erosion loss.

Soil mapping unit. A kind of soil, a combination of kinds of soil, or miscellaneous land type or types that can be shown at the scale of mapping for the defined purposes and objectives of the survey. The soil mapping legend lists all mapping units for the survey of an area.

Table G-3 Suffixes to Master Letter Horizons to Indicate Subordinate Distinctions within Master Horizons. Current and Old Symbols

Current	*Old*	*Description of Symbol*	*Example*
a	—	Highly decomposed organic matter	**Oa**
b	b	Buried genetic soil horizon. Must be by at least 50 cm (20 in) of sediment	**Ab**
c	cn	Accumulation of concretions or other hard nodules enriched in sesquioxides	**Bc**
d	—	Physical root restriction	**Bd**
e	—	Intermediately decomposed organic matter	**Oe**
f	f	Frozen soil; mostly for permanently frozen (dry permafrost)	**Af**
fm	—	Ice-cemented permafrost	**A2fm**
g	g	Strong gleying. Intense reducing (anaerobic) conditions due to stagnant water. Base colors approaching neutral (grays), with or without mottles (brighter orange and yellow colors of oxidized iron)	**Cg**
h	h	Illuvial humus, usually coating on sands or as pellets. Mostly used with **B** horizons	**Bh**
i	—	Slightly decomposed organic matter (litter)	**Oi**
j	—	Jarosite mineral	**Bj**
jj	—	Cryoturbated horizon	**Ajj**
k	ca	Accumulation of carbonates, usually calcium and magnesium salts	**Bk**
m	m	Strong cementation by various materials	**Cm**
n	sa	Accumulation of sodium salts (not just any salts)	**Bn**
o	—	Residual accumulation of sesquioxides	**Bo**
p	p	Plowing or other disturbance of soil	**Ap**
q	si	Accumulation of silica, usually cementation	**Bq**
r	r	Weathered or soft bedrock	**Cr**
s	ir	Illuvial accumulation of sesquioxides and organic matter	**Bs**
ss	—	Presence of slickensides	**Bss**
t	t	Accumulation of clay, usually large amounts	**Bt**
v	—	Plinthite present	**Bv**
w	—	Color or structural **B** (weakly developed)	**Bw**
Wfm	—	More than 75% water	**Wfm**
x	x	Fragipan character; the horizon has firmness, is brittle, and has a high density	**Bx**
y	cs	Accumulation of gypsum	**Cy**
z	sa	Accumulation of salts more soluble than gypsum	**Bz**

Sources: Modified from (1) Soil Survey Staff, *Soil Taxonomy: A Basic System of Soil Classification for Making and Interpreting Soil Surveys,* Agriculture Handbook 436, USDA—Soil Conservation Service, Washington, DC, 1975, 461–462; (2) Soil Survey directive 430-V-SSM, May 1981; (3) USDA—Soil Conservation Service draft of *Soil Survey Manual,* Chapter 4, 1981, 4-39 to 4-51.

Soil mineralogy classes for family groupings. The family category includes mineralogy classes for specific control sections that are similar to those used for particle-size classes for family groupings.

Soil moisture regimes. Moisture regimes indicate the time periods during which the soil "control section" (loam soil between about 10 and 30 cm [4–12 in] deep) is either moist or dry (wilting percentage). In addition to details below, *see also* aquic, udic, ustic, xeric, aridic, and torric moisture regimes. To simplify this tabulation the data given are not precise in all particulars. All temperatures are at the 50-cm (20-in) depth in soils; *parts* refers to parts of the control section (CS).

Aquic conditions. Must have reducing conditions for at least long enough to exhibit one or more of (1) redoximorphic features (wetness-caused mottles), (2) redox concentrations of Fe and Mn, (3) redox depletions of Fe and Mn (leaves low-chroma wetness mottles), and (4) reduced matrix that changes color when exposed to oxygen (air).

Aridic regime. A dry and arid soil environment. Dry in all parts more than half the time above 5°C. Never moist in any part as long as 90 consecutive days when above 8°C.

Udic regime. Wet but aerated soil; moist year around. Not dry in any part as long as 90 cumulative days. If colder than 22°C, the control section is not dry in all parts 45 consecutive days in the 4 months after the summer solstice (June 22) in 6 of 10 years.

Ustic regime. Between aridic and udic; moist summers. Above mean annual temperature of 22°C,

CS is dry in some parts 90 or more days. Moist in some parts for more than 180 cumulative days *or* continuously moist in some parts for 90 consecutive days. Below MAT of 22°C, soil is dry in some parts 90 or more accumulative days. Not dry in all parts more than half the time that it is above 5°C. Not dry in all parts 45 consecutive days in the 4 months after the summer solstice (June 22) in 6 of 10 years.

Xeric regime. Mediterranean climate: moist winters, dry summers. Control zone is dry 45 or more consecutive days in the 4 months after the summer solstice (June 22) in 6 of 10 years. Control zone is moist 45 or more consecutive days in the 4 months after the winter solstice (December 22) in 6 of 10 years. Moist in some parts more than half the time (accumulative) when soil is above 5° C or during 90 consecutive days when soil is above 8° C. Mean annual temperature is below 22° C.

Soil moisture suction. *See* Soil water potential.

Soil monolith. A soil profile removed from the soil and mounted for display, often 10–20 cm (4–8 in) wide, 5–10 cm (2–4 in) thick, profile depth (1.5 m), and stabilized with transparent resin (vinyl).

Soil morphology. The constitution of the soil, including the texture, structure, consistency, color, and other physical, chemical, and biological properties of the various soil horizons that make up the soil profile.

Soil order. The most inclusive category of the U.S. soil classification system.

Soil phase. *See* Phase, soil.

Soil potential index (SPI). Numerical ratings of soils for each of various uses, using the equation:

$$SPI = P - (CM + CL)$$

where P = performance or yield, CM is costs of corrective measures, and CL is costs of continuing limitations.

Soil profile. *See* Profile, soil.

Soil quality. The capability of the soil to supply the needs for the planned use, usually to grow plants. Quality is often called *health* and is related to the planned use.

Soil salinity. The quantity of soluble salts in soils; salts hinder plant growth.

Soil science. That science dealing with soils as a natural resource on the surface of the Earth, including soil formation, classification and mapping, and the physical, chemical, biological, and fertility properties of soils per se; and these properties in relation to their management for the growth of plants and to cleanse the environment.

Soil separates. Mineral particles, less than 2.0 mm in equivalent diameter, ranging between specified size limits.

Soil series. *See* Soil classification.

Soil structure. The combination or arrangement of primary soil particles into secondary particles, units, or peds.

Soil subgroup. *See* Soil classification; Chapter 7.

Soil suborder. *See* Soil classification; Chapter 7.

Soil survey. The systematic examination, description, classification, and mapping of soils of an area.

Soil test. A chemical, physical, or microbiological operation that estimates a property of the soil.

Soil texture. The relative proportions of the various soil size separates.

Soil variant. A kind of soil that differs enough from recognized series to justify a new series name but is so limited in area that creation of a new series is not justified at the given time.

Soil water potential. The amount of work (usually given in kilopascals) an infinitesimal quantity of soil water can do in moving from the soil to a pool of pure, free water at the same location and at normal atmospheric pressure. It is mostly matric potential.

Solifluction. The slow downhill flowage or creep of saturated soil and other loose materials.

Solum. (plural: sola) The **O, A, E,** and **B** horizons of a soil profile, above the parent material.

Solute potential. *See* Osmotic potential.

Sombric horizon. An illuvial humus subsurface horizon, darker than the horizon above it but without characteristics of a spodic.

Spatial analysis. The process of extracting or creating new information about a set of geographic features.

Spatial variability. The changes that occur in a property with distance along a direction.

Specific gravity. The relative weight of a given volume of any kind of matter (volume occupied by solid phase, pore space excluded) compared with an equal volume of distilled water at a specified temperature. The average specific gravity for soil is about 2.65 Mg/m^3.

SPI. *See* Soil potential index.

Split application. The fertilizer is added in two or more portions at different times during the season.

Spodic horizon. A diagnostic **B** horizon accumulating amorphous alumina and/or humus colloids, with or without iron, designated **Bs.**

Spodosols. *See* Soil classification: Order.

Spoilbank. A pile of soil, subsoil, rock, or other material excavated from a drainage ditch, pond, road cut, or surface mining.

Starter (pop-up). Fertilizer is uniformly spread on the soil surface. It may or may not be incorporated into the soil.

Stokes's law. Equation to calculate settling rates: fall of sands, silts, and clay when they are suspended in water; used to determine soil texture.

Stones. Rock fragments 25–60 cm (10–24 in) in diameter if rounded or 38 to 60 cm (15–24 in) in length if flat.

Strip cropping. The practice of growing crops for controlling erosion that requires different types of tillage, such as row and sod, in alternate strips along contours or at right angles to the prevailing direction of erosive winds.

Strip tillage. Tillage of only strips of soil, perhaps 15–25 cm (6–10 in) wide, with nontilled strips between. The tilled strip is prepared into a loose seedbed, which allows incorporation of fertilizer and seed and leaves protective untilled strips between planted areas.

Structural charge. The negative charge on a clay mineral caused by isomorphous substitution within the layer.

Structure, soil. *See* Soil structure.

Stubble mulch. Stubble mulching is one of the early erosion control practices in which only part of the crop residue is incorporated somewhat into the soil. Most of the residue was left anchored in the soil but exposed at the soil surface. Frequently, this was accomplished by pulling wide sweeps through the soil.

Subgroup, soil. *See* Soil classification: Subgroup.

Suborder. *See* Soil classification.

Subsidence. A lowering of the land elevation caused by solution and collapse of underlying soluble deposits, reduction of fluid pressures within an aquifer or petroleum reservoir, or decomposition of organic soils.

Subsoiler bedder. When a bedder is built onto a subsoil chisel, it is called a *subsoiler bedder.*

Subsoiling. The tillage of subsurface soil without inversion, to break up dense soil layers that restrict water movement and root penetration.

Substrate. 1. An underlying layer. 2. Substance or nutrient on which an organism grows. 3. Substances acted upon by enzymes and changed to other compounds in the reaction.

Substratum. Any layer lying beneath the solum, which is the A and B horizons.

Suction. *See* Soil water potential.

Sulfuric horizon. A diagnostic horizon with both a pH <3.5 and yellow jarosite (sulfate) mottles.

Summer fallow. *See* Fallow.

Surface soil. The **A, E, AB,** and **EB** horizons and all their subdivisions.

Sweep. In tillage, an implement with a point and flat or curved wings going back from the point at 30–60° angles. *See* Chapter 14.

Sweep plow. A sweep, up to 2 m (6.5 ft) wide, fitted on its underside with four or more shanks about 15 cm (6 in) long. The shanks, inclining toward the center, break up the soil to a depth of 20–25 cm (8–10 in).

Sweeps. Any chisel tip that has *wings,* curved surfaces to throw or disrupt the soil, or flat surfaces simply to cut through the soil to kill weeds with minimum soil disturbance. Small sweeps are used for cultivators; large sweeps are used in stubble-mulch or rod-weeding operations on fallow fields.

Symbiosis. The living together in intimate association of two dissimilar organisms, the cohabitation being mutually beneficial, such as *Rhizobia* legume bacteria with the host leguminous plant.

Synergism. The action of two or more substances, organs, or organisms to achieve an effect which is greater than the sum of the two effects when acting separately.

T

Talus. Fragments of rock and other soil material accumulated by gravity at the foot of cliffs or steep slopes.

Taxon. In the context of soil survey, a class at any categorical level in the U.S. system of soil taxonomy.

Tensiometer. Instrument used for measuring the water potential (suction or negative pressure) of soil water.

Tension, water. *See* Water tension.

Teratogenic. The substance causes tissue deformations, when eaten by the mother at certain stages during gestation, usually in early stages of the development of the fetus.

Terminal moraine. A belt of thick glacial drift that generally marks the termination of important glacial advances.

Terrace. 1. A natural level plain bordering a river, lake, or sea. 2. A raised, level strip of earth usually constructed on or nearly on a contour designed to make the land suitable for tillage and to prevent accelerated erosion.

Texture. *See* Soil texture.

Thermic. A soil temperature regime averaging between 15° and 22°C.

Thermophilic organisms. Organisms that grow readily at temperatures of about 113°F (45°C).

Threshold level. The maximum level of a substance or condition that can be tolerated without ill effects. Often these values are difficult to determine unequivocally.

Tile drain. Lines of concrete or ceramic (clay) pipe placed in the subsoil to collect and drain water from the soil to an outlet.

Till. 1. Unstratified glacial drift deposited directly by the melting ice and consisting of clay, sand, gravel, and boulders intermingled in any proportion. 2. To plow and prepare for seeding; to seed or cultivate the soil.

Tillage. Breaking or manipulating soil.

Tillage, conventional. Tillage in the 1960s to 1980s, which usually involved preparing a pulverized seedbed and burial or burning of plant residues—*clean fields.* Often this involved plowing, disking once or twice, harrowing, and even smoothing.

Tillage, primary. The major breaking of the soil, usually with a plow or with plow disks.

Tillage, rotary. Loosening and mixing soil with small forward rotating blades (rototiller).

Tillage, secondary. Any operations following primary tillage (disking, harrowing, rototilling) to prepare a finer seedbed.

Tillage pan. *See* Pan, pressure or induced.

Tilth. The physical condition of soil as related to its ease of tillage, fitness as a seedbed, and impedance to seedling emergence and root penetration. Tilth is a result of tillage.

Topdressing. An application of fertilizer to a soil after the crop stand has been established.

Toposequence. A sequence of soils that differ primarily because of differences in topography as a soil-forming factor.

Topsoil. The upper part of the soil, which is the most favorable material for plant growth. Ordinarily, it is rich in organic matter.

Torric. A water-deficient soil water regime defined like aridic but used in a different category level of the U.S. soil taxonomy than is aridic.

Tortuosity. The winding, twisting, and intermeshing of soil pores.

Trace elements. *See* Micronutrient.

Transpiration. The process by which water vapor is released from plants to the atmosphere primarily through the leaf stomata.

Tundra. A treeless plain with mosses, shrubs, and grasses found in arctic areas.

U

Udic. *See* Soil moisture regimes.

Ultisols. *See* Soil classification: Order.

Umbric horizon. A dark-colored, high-organic-matter, diagnostic epipedon similar to mollic but more acidic.

Unique agricultural lands. Soils uniquely suited to the high production of certain desired crops. The uniqueness may be aided by a lack of frost, a needed wetness or acidity, a peculiar climate, or a particular location relative to markets.

Universal Soil Loss Equation (USLE). An equation used to predict soil erosion due to water.

Unified engineering soil classification system. The Airfield Classification adopted by the Corps of Engineers and the Bureau of Reclamation and by A.S.T.M.

Unsaturated flow. The movement of water in a soil that is not filled to capacity with water. Water moves because of water-potential differences toward areas of lower water potentials (drier soil).

Ustic. *See* Soil moisture regimes.

V

Vadose zone. The unsaturated soil between the soil surface and the underlying water table. The rooting depth is often called the *root zone.*

Value, color. The relative lightness or darkness of color.

Vesicular arbuscular mycorrhiza. Common endomycorrhiza with small *arbuscles* within root cells and storage *vesicles* between root cells.

Vermiculite. 1. A clay similar to hydrous mica and having 2:1 layers of 2 tetrahedral sheets to 1 octahedral sheet. Vermiculite has the layers held together by hydrated cations and has less swelling than montmorillonite. 2. A mica-like silicate mineral that expands into an accordion-like structure when heated to high temperature; it has a high water-holding capacity *within the expanded particles* and yet good aeration *between the large particles.*

Vertical mulching. A subsoiling operation in which a vertical band of mulching material is placed into a vertical slit in the soil immediately behind a special chisel.

Vertisols. *See* Soil classification: Order.

Virgin soil. A soil essentially undisturbed from its natural environment; never cultivated.

Viroid. A virallike RNA protein without a protective coating.

Virus. Any of numerous submicroscopic pathogens consisting of a single nucleic acid and one or two protein coats, capable of being replicated only inside a living cell. *See also* Prion; Viroid.

Void ratio. (Engineering) The ratio of the volume of voids to the volume of solids.

Volcanic ash. Fine particles of rock blown into the air from a volcano and settled on land, often in layers several meters thick.

Volumetric water content. The ratio of the volume of water in a soil to the total bulk volume of the soil, in decimal form.

Volumetric water percentage. Volume water ratio multiplied by 100.

W

Water content. *See* Mass water content and volumetric water content.

Water potential. *See* Soil water potential.

Water retention curve. A graph showing the soil-water content versus applied tension, suction, or water potential. Also called *water release characteristic curve.*

Water rights. The legal rights to the use of water. They consist of riparian rights and those acquired by appropriation and prescription. Riparian rights are those by virtue of ownership of the banks on the stream, lake, or ocean. Appropriated rights are those acquired by an individual to the exclusive use of water, based strictly on priority of appropriation and application of the water to beneficial use. Prescribed rights are those to which legal title is acquired by long possession.

Water suction. *See* Water tension.

Water table. The upper surface of groundwater; that level below which the soil is saturated with water.

Water table, perched. The surface of a local zone of saturation held above the main body of groundwater by an impermeable layer, usually clay or rock, and separated from the main body of groundwater by an unsaturated zone.

Water tension. The equivalent negative pressure in the soil water. It is equal to pressure needed to bring the water into hydraulic equilibrium.

Water use efficiency. Crop production per unit of water used, irrespective of water source, expressed in units of water depth per unit weight of crop; for example, 25 cm (10 in) of water used per ton of alfalfa hay produced.

Waterlogged. Saturated with water, usually developing anaerobic conditions. Most or all pore space filled with water; natural or induced by humans.

Weathering. The group of processes (such as chemical action of air and rainwater and the biological action of plants and animals) whereby rocks and minerals change in character, disintegrate, decompose, and synthesize new compounds and clay minerals.

Well graded. *See* Graded.

Wetlands. Soil areas that have evidence of saturated conditions part of the year (ponded water, hydric soil, wet-area plants, such as cattails).

Windbreak. *See* Shelterbelt.

Wilting point. *See* Permanent wilting point.

Wind strip cropping. The production of crops in relatively narrow strips placed perpendicular to the direction of prevailing erosive winds.

X

Xeric. *See* Soil moisture regimes.

Xenobiotic. These are compounds foreign to natural ecosystems, such as many plastics and some chlorinated hydrocarbons and that do not degrade easily. Many are toxic materials.

Xerophytes. Plants that tolerate and grow in very dry soils.

Z

Zero tillage. *See* No-till.

Index

B

C

D

E

F

G

H

I

J

K

L

M

N

O

P

Q

R

S

T

U

V

W

X

Y

Z

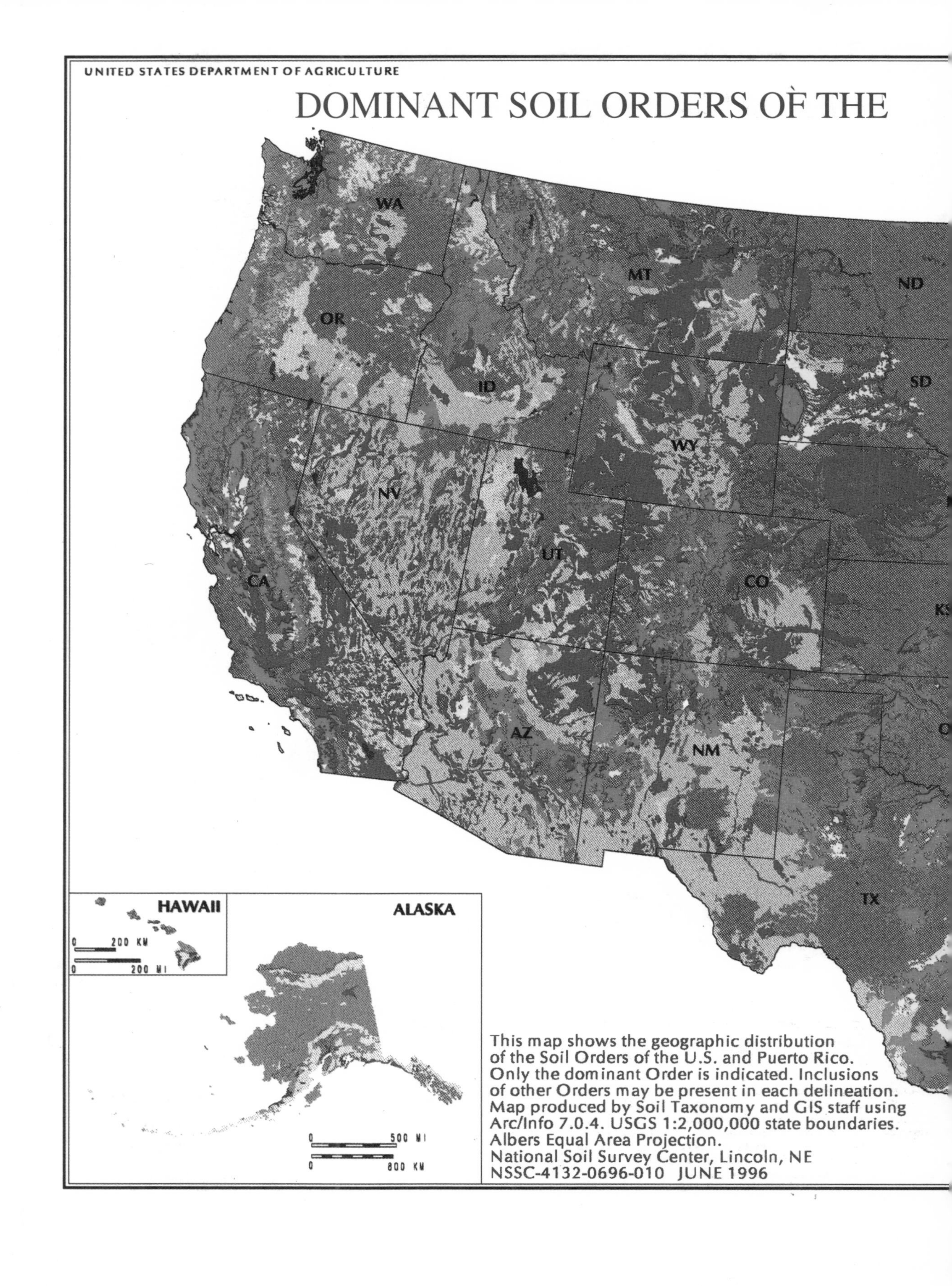
UNITED STATES DEPARTMENT OF AGRICULTURE
DOMINANT SOIL ORDERS OF THE
WA
MT
ND
OR
ID
SD
WY
NV
UT
CA
CO
KS
AZ
NM
TX
HAWAII
ALASKA
0 200 KM
0 200 MI
0 500 MI
0 800 KM
This map shows the geographic distribution of the Soil Orders of the U.S. and Puerto Rico. Only the dominant Order is indicated. Inclusions of other Orders may be present in each delineation. Map produced by Soil Taxonomy and GIS staff using Arc/Info 7.0.4. USGS 1:2,000,000 state boundaries. Albers Equal Area Projection.
National Soil Survey Center, Lincoln, NE
NSSC-4132-0696-010 JUNE 1996